Michael Mulholland

Applied Process Control

Related Titles

Michael Mulholland

Applied Process Control

Efficient Problem Solving

2016

Print ISBN: 978-3-527-34118-4

Michael Mulholland

Applied Process Control Set

2016

Print ISBN: 978-3-527-34116-0

Svrcek, W.Y., Mahoney, D.P., Young, B.R.

A Real-Time Approach to Process Control

3rd Edition

2014

Print ISBN: 978-1-119-99388-9

Rangaiah, G.P., Bonilla-Petriciolet, A. (eds.)

Multi-Objective Optimization in Chemical Engineering - Developments and Applications

2013

Print ISBN: 978-1-118-34166-7

Kruger, U.U.

Statistical Monitoring of Complex Multivatiate Processes - With Applications in Industrial Process Control

2012

Print ISBN: 978-0-470-02819-3

Rangaiah, G.P.

Plantwide Control - Recent Developments and Applications

2011

Print ISBN: 978-0-470-98014-9

Buzzi-Ferraris, G., Manenti, F.

Fundamentals and Linear Algebra for the Chemical Engineer

Solving Numerical Problems

2010

Print ISBN: 978-3-527-32552-8

Buzzi-Ferraris, G., Manenti, F.

Interpolation and Regression Models for the Chemical Engineer

Solving Numerical Problems

2010

Print ISBN: 978-3-527-32652-5

Buzzi-Ferraris, G., Manenti, F.

Nonlinear Systems and Optimization for the Chemical Engineer

Solving Numerical Problems

2014

Print ISBN: 978-3-527-33274-8

Buzzi-Ferraris, G., Manenti, F.

Differential and Differential-Algebraic Systems for the Chemical Engineer

Solving Numerical Problems

2015

Print ISBN: 978-3-527-33275-5

Michael Mulholland

Applied Process Control

Essential Methods

WILEY-VCH
Verlag GmbH & Co. KGaA

Author

Professor Michael Mulholland
University of KwaZulu-Natal
Chemical Engineering
4041 Durban
South Africa

Library of Congress Card No.: applied for

British Library Cataloguing-in-Publication Data
A catalogue record for this book is available from the British Library.

Bibliographic information published by the Deutsche Nationalbibliothek
The Deutsche Nationalbibliothek lists this publication in the Deutsche Nationalbibliografie; detailed bibliographic data are available on the Internet at <http://dnb.d-nb.de>.

Print ISBN: 978-3-527-34119-1
ePDF ISBN: 978-3-527-80161-9
ePub ISBN: 978-3-527-80167-1
Mobi ISBN: 978-3-527-80166-4
oBook ISBN: 978-3-527-80162-6

Cover Design Formgeber, Mannheim, Germany
Typesetting Thomson Digital, Noida, India
Printing and Binding CPI books GmbH, Germany

Printed on acid-free paper

Essential Methods Contents

Preface

Material in this book is sequenced for the process engineer who needs 'some' background in process control (Chapters 1–5) through to the process engineer who wishes to *specialise* in advanced process control (Chapters 1–9). The theory needed to properly understand and implement the methods is presented as succinctly as possible, with extensive recourse to linear algebra, allowing multi-input, multi-output problems to be interpreted as simply as single-input, single-output problems.

Before moving on to the more advanced algorithms, an essential practical background is laid out on plant instrumentation and control schemes (Chapters 2, 4 and 5). Chapter 3 builds modelling abilities from the simplest time-loop algorithm through to discrete methods, transfer functions, automata and fuzzy logic. By the end of Chapter 5, the engineer has the means to design simple controllers on the basis of his or her models, and to use more detailed models to test these controllers. Moreover, ability has been developed in the use of the multi-element control schemes of 'advanced process control'.

Chapter 6 focuses on observation. Whereas Chapter 3 reveals the tenuous chain of preparation of plant signals, Chapter 6 aims to make sense of them. Important issues on the plant are signal conditioning, data reconciliation, identification of model parameters and estimation of unmeasured variables.

Chapter 7 addresses more advanced control algorithms, drawing on a wide range of successful modern methods. To a large extent, continuous and discrete versions of an algorithm are presented in parallel, usually in multi-input, multi-output formats – which simply devolve to the single-input, single-output case if required. State–space, input–output, fuzzy, evolutionary, artificial neural network and hybrid methods are presented. There is a strong emphasis on model predictive control methods which have had major industrial benefits.

A review of the classical methods of stability analysis is delayed until Chapter 8. This has been kept brief, in line with reduced application in the processing industries. One recognises that stability criteria, such as pole locations, do underlie some of the design techniques of Chapter 7. Certainly, frequency domain concepts are part of the language of control theory, and essential for advanced investigation. But with the slower responses and inaccurate models of processing plants, controllers are not predesigned to 'push the limits' and tend to be tuned up experimentally online.

A review of a range of optimisation techniques and concepts is given in Chapter 9. Although not a deep analysis, this imparts a basic working knowledge, enabling the development of simple applications, which can then later be built upon. Topics covered include *linear*, *integer*, *mixed*, and *nonlinear* programming, search techniques, global optimisation, simulated annealing, genetic algorithms and multi-objective optimisation. These methods, and *dynamic programming*, underlie the predictive control and optimal scheduling topics in Chapter 7, and are also important as static optimisers in such applications as supply chain, product blending/distribution and plant economic optimisers.

This book tries to make the methods practically useful to the reader as quickly as possible. However, there is no shortcut to reliable results, without a basic knowledge of the theory. For example, one cannot make proper use of a Kalman filter, without understanding its mechanism. Complex multi-input, multi-output applications will require a good theoretical understanding in order to trace a performance problem back to a poorly calibrated input measurement. Hence, an adequate theoretical background is provided.

A few distinctions need to be clarified:

1) Modelling is a particular strength of the process engineer, and is a basis of all of the algorithms – especially model predictive control. The reader needs to distinguish *state-based* models versus *input–output* models. The state-based models can predict forward in time knowing only the initial state and future inputs. Some algorithms rely on this. In contrast, *input–output* models will need additional information about past inputs and outputs, in order to predict future outputs. To use state-based algorithms on these, a state observer algorithm (e.g. Kalman filter) will be required to estimate the states.
2) The forward shift operator $z = e^{Ts}$ is used to relate discrete versions of systems to their transfer function forms $G(s)$ in the s (Laplace/frequency) domain. In a lot of what follows, this theoretical connection is not significant, and the data sampling shift parameter q could be used, but sometimes it is not in this text.
3) The text consistently uses bold characters to signify matrices [$\boldsymbol{A}$], vectors [$\boldsymbol{x}$] and matrix transfer functions [$\boldsymbol{G}(s)$, $\boldsymbol{G}(z)$]. Non-bold characters are used for scalars.

A number of examples are presented in this book in order to clarify the methods. In addition, the separate accompanying book *Applied Process Control: Efficient Problem Solving* presents 226 solved problems, using the methods of this text. These often make use of MATLAB® code which is arranged in obvious time loops, allowing easy translation to the real-time environment. There will, however, be the challenge to provide additional routines such as matrix inversion.

A simple interactive simulator program has been made available at https://sourceforge.net/projects/rtc-simulator/. It includes 20 different applications for such aspects as PID and DMC controller tuning, advanced level control, Smith prediction, Kalman filtering and control strategies for a furnace, a boiler and a hybrid system. No support is available for the simulator.

Although I have personally used a variety of methods on industrial and research applications, in writing this book I have been fascinated to discover the brilliant ideas of many other workers in the field. To all of those people who get excited about process control, I wish you an optimal trajectory.

University of KwaZulu-Natal — Michael Mulholland
March, 2016

Acknowledgements

Many of the problems in this book are dealt with using the MATLAB® program, which is distributed by the MathWorks, Inc. They may be contacted at

The MathWorks, Inc.
3 Apple Hill Drive
Natick, MA 01760–2098, USA
Tel: 508-647-7000
Fax: 508-647-7001
E-mail: info@mathworks.com
Web: mathworks.com
How to buy: http://www.mathworks.com/store

A few problems are dealt with in the GAMS® optimisation environment, distributed by

GAMS Development Corporation
1217 Potomac Street, NW
Washington, DC 20007, USA
General Information and Sales: (+1) 202 342-0180
Fax: (+1) 202 342-0181
Contact: sales@gams.com

Some problems make use of the LPSOLVE mixed integer linear programming software which is hosted on the SourceForge Web site at

http://sourceforge.net/projects/lpsolve/

Abbreviations

A/D	analogue to digital
AC/FO	air to close/fail open
ANN	artificial neural network
AO/FC	air to open/fail closed
APC	advanced process control
ARIMAX	autoregressive integrated moving average exogenous
ARMAX	autoregressive moving average exogenous
ARX	autoregressive exogenous
BB	branch and bound
BFW	boiler feedwater
BIDP	backward iterative dynamic programming
CEM	cause and effect matrix
CRT	cathode ray tube
CV	controlled variable
CVP	control variable parameterisation
CW	cooling water
D/A	digital to analogue
DAE	differential and algebraic equations
DCS	distributed control system
DMC	dynamic matrix control
DP	differential pressure
DV	disturbance variable
E{ . . . }	expectation of . . .
EKF	extended Kalman filter
ES	evolutionary strategy
FFT	fast Fourier transform
FIDP	forward iterative dynamic programming
FIMC	fuzzy internal model controller
FIR	finite impulse response
FRM	fuzzy relational model
FRMBC	fuzzy relational model-based control
FSQP	feasible sequential quadratic programming

Fuzzy	fuzzy logic
GA	genetic algorithm
GAMS	General Algebraic Modelling System®
GM	gain margin
GPC	generalised predictive control
HP	high pressure (port)
HS	high select
I	identity matrix
I/O	input–output
I/P	current to pressure (pneumatic) converter
IAE	integral of absolute error
IDP	iterative dynamic programming
IMC	internal model control
INA	inverse Nyquist array
IO	input–output
IP	integer programming
ISE	integral of squared error
KO	knockout (separation drum)
LAN	local area network
LBT	lower block triangular
LCD	liquid crystal display
LDMC	linear dynamic matrix control
LP	linear programming
LP	low pressure (port)
LPG	liquefied petroleum gas
LPSOLVE	MILP program (http://sourceforge.net/projects/lpsolve/)
LQR	linear quadratic regulator
LS	low select
LS	least squares
MATLAB	MATLAB® program, distributed by the MathWorks, Inc.
MEK	methyl ethyl ketone
MIDO	mixed integer dynamic optimisation
MILP	mixed integer linear programming
MIMO	multi-input, multi-output
MINLP	mixed integer nonlinear programming
MIP	mixed integer programming
MIQP	mixed integer quadratic programming
MLD	mixed logical dynamical
MM	molecular mass
MPC	model predictive control
MRI	Morari resiliency index
MTBF	mean time between failures
MV	manipulated variable
NC	normally closed
NLP	nonlinear programming
NO	normally open

NSGA-II	fast non-dominated sorting genetic algorithm II
OA	outer approximation
ODE	ordinary differential equation
OHTC	overall heat transfer coefficient
P	proportional
P/I	pressure (pneumatic) to current converter
PCA	principal components analysis
PDE	partial differential equation
PI	proportional integral
PID	proportional integral derivative
PLC	programmable logic controller
PM	phase margin
PV	process variable
QDMC	quadratic dynamic matrix control
RAID	redundant array of independent discs
RGA	relative gain array (Bristol array)
RLS	recursive least squares
RTD	resistance temperature detector
RTO	real-time optimisation
SCADA	supervisory control and data acquisition
SG	specific gravity
SISO	single-input single-output
SP	setpoint
SQP	sequential quadratic programming
VPC	valve position control
WABT	weighted average bed temperature
WAN	wide area network (e.g. using telecommunication, radio)
WG	water-gauge
ZOH	zero-order hold

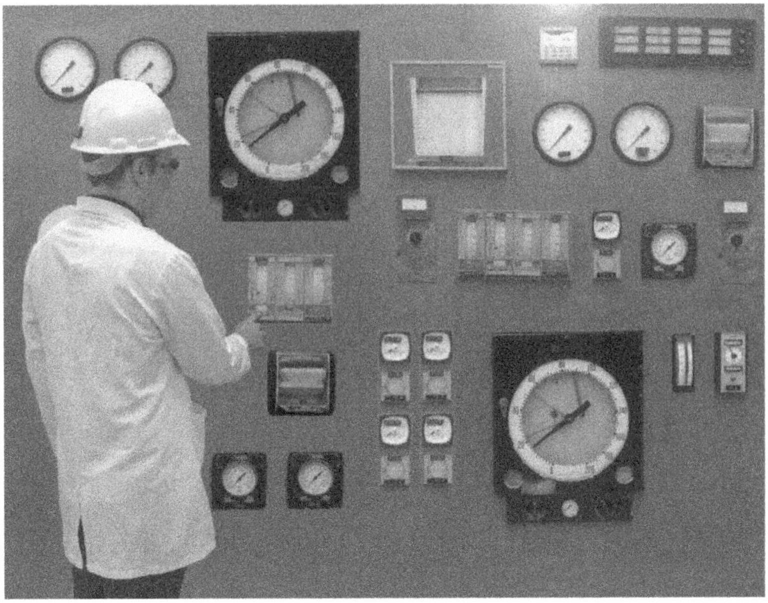

Old analogue control panel

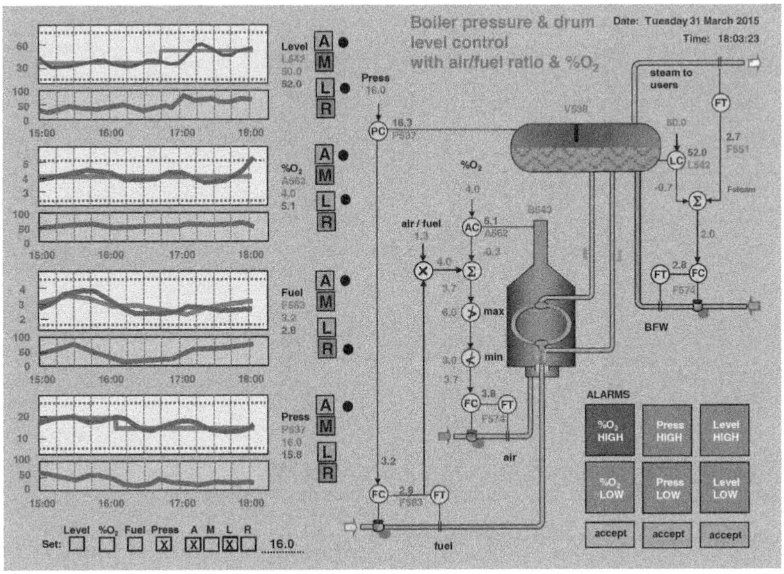

Modern digital control display

1
Introduction

1.1
The Idea of Control

Plant and animal life relies on numerous control mechanisms. Here one thinks of an 'open-loop' *causal* system in which cause (input) generates effect (output). Then the 'control' part operates around this in order to modify the effect (Figure 1.1).

All autonomous entities need 'feedback' paths like this. One might shut a door to reduce the noise level, eat to reduce hunger or turn the car steering wheel to keep in a lane. The control decision can be discrete, such as when the geyser thermostat switches on or the reserve bank adjusts the prime lending rate in order to control inflation, or it can be continuous, such as the variation of one's iris with available light. The 'system' part can have various behaviours which make the decision difficult. Thus, when the shower it too hot, one cautiously increases the cold water, knowing that there is a ('dead-time') delay in the pipe. The doctor will adjust blood pressure medication in small steps as he/she waits to observe the body response. One also recognises that there are possibly multiple 'causes' at play. It may only be possible to observe a limited range of the effects, and usually one is only manipulating a small selection of the causes. Humans are not adept at coordinating multiple inputs.

What is common in these cases of feedback control is that one does not know where the inputs should be set exactly. The control problem amounts to 'given desired levels of the outputs, at what levels should the inputs be set in order to best achieve this?'. So the 'decision' is a task of *inversion*, in just the same way as one might want to find an x such that a function $f(x) = 0$. A simple control law for this mathematical task was provided by Newton (Equation 1.1).

$$x_{n+1} = x_n - \frac{f(x_n)}{f'(x_n)} \tag{1.1}$$

This is closely related to the 'dead-beat' controller (Section 7.1.1), in which one attempts to hit the target on every time step.

It is intuitive that one can expect difficulty with automatic feedback adjustments. The decision to adjust the inputs will affect the outputs which in turn will affect the next decision. If one is over-reacting on each step, the output would be driven past its desired level by successively larger amounts. So adjusting the shower water in too big steps would successively cause scalding and freezing by greater amounts. The possibility of such endless growth makes considerations of *closed-loop stability* important in the study of process control.

Applied Process Control: Essential Methods, First Edition. Michael Mulholland.

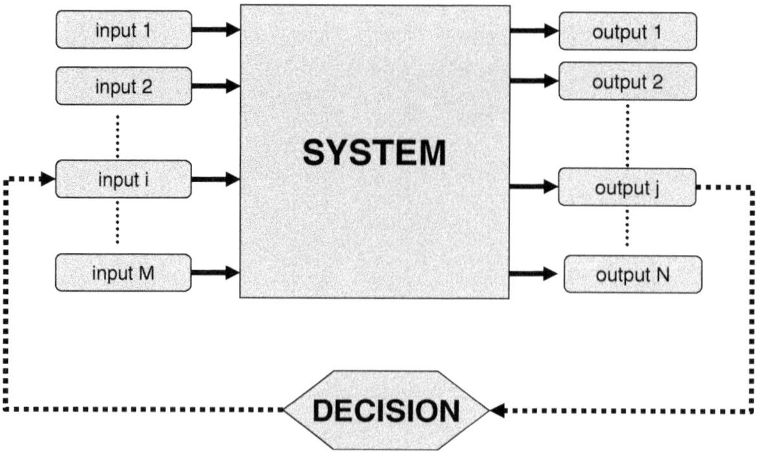

Figure 1.1 Feedback control mechanism.

A situation often arises where one tries to diminish the impact of disturbed inputs to a system by manipulating other inputs (Figure 1.2). This is 'feedforward' control. It is important to note that the outputs are not involved in this decision at all. Of course, one needs to be able to observe the relevant inputs first. Thus, the fuel to a boiler may be increasing in order to maintain pressure. Though the flue gas oxygen content may be unmeasured, a feedforward controller can increase the combustion air flow in proportion to the fuel flow in order to maintain a margin of oxygen excess. One realises that feedforward control will always require some kind of model, for example the air/fuel stoichiometric ratio. Models are never perfect, so it is likely that the relevant output may not be quite where it was planned. Often this error can be tolerated. In other cases, a feedback loop may be superimposed to provide the correction (Figure 1.3). There is nevertheless a benefit in using feedforward to eliminate most of the upset.

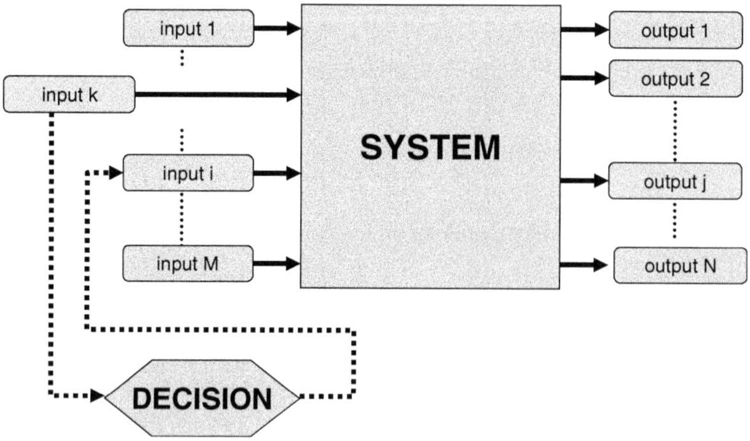

Figure 1.2 Feedforward control mechanism.

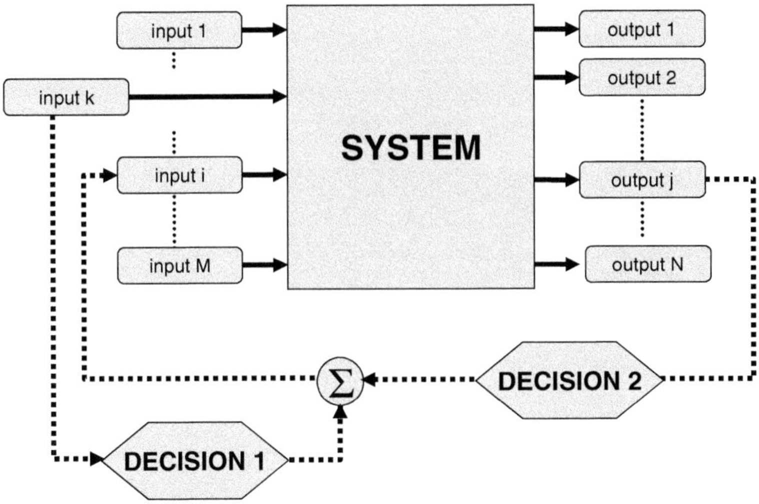

Figure 1.3 Combined feedforward and feedback.

The control engineer must learn to recognise 'information flows' in a system, that is from 'cause' to 'effect'. Sometimes these are not intuitive and have little to do with the physical arrangement. For example, the flow into a tank and the flow out of a tank could equally be used to affect the level in a tank.

1.2 Importance of Control in Chemical Processing

This book will focus on modelling, estimation, control and optimisation in the processing industries. There are unique challenges here to do with the inaccuracy of models and undefined disturbances. In addition, the widespread use of computers to handle process instrumentation in recent decades has spurred the concept of 'advanced process control' (APC), which has become a specialised process engineering domain. The objective is to take advantage of the plant-wide view of outputs, and access to inputs, of these computers, in order to enhance regulation and optimisation. In this way, industries have been able to work safely with narrower specifications and less loss (Figure 1.4). With the increasing globalisation of markets, industries which do not seek such efficiency improvements will soon find themselves uncompetitive and out of business.

In the processing industries, the automatic control aspects are viewed to constitute a pyramid of three main layers in which each layer achieves its objectives by supervising the layer below. Generally, this means that the control loop setpoints (SPs) are passed downwards (Figure 1.5).

Usually the base layer becomes the responsibility of instrumentation technicians, but more advanced inputs are required from control engineers in the upper layers. Of course, the overall control scheme, including the base layer, must be specified by engineers in the design phase. At that stage, additional specifications may be made, such as increased vessel hold-ups to facilitate 'advanced level control' (Section 4.11). Indeed, there is a growing trend to integrate the equipment design and control design at an early stage (Sakizlis, Perkins and Pistikopoulos, 2004). Increasing

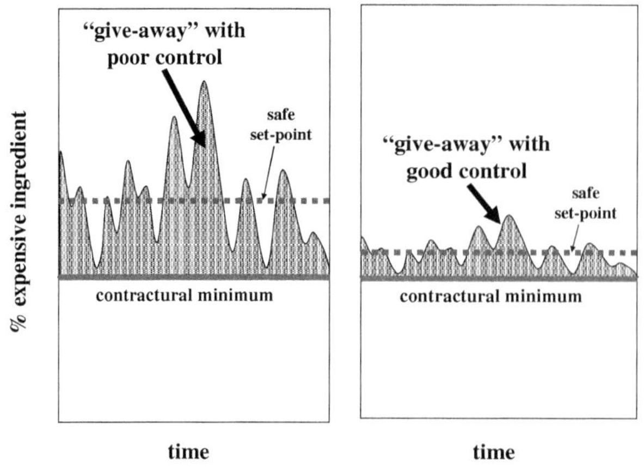

Figure 1.4 Reduction of an expensive ingredient through better control.

integration of processes through 'pinch' analysis often renders the internal regulation highly interactive, requiring special control approaches. Another lesson that has been learned is that the advanced control algorithms cannot simply be installed and left to operate without ongoing knowledgeable oversight.

The advanced algorithms focus on criteria such as throughput, product specifications and economics, not necessarily smooth process operation, and thus they can be unpopular with operating personnel. All too often such unwelcome behaviour can cause operators to switch off these algorithms. Thus, education is important, as well as investigation of downtime incidences and constant reviewing of performance. On one level, one aims to make a control scheme as simple and transparent as possible, to facilitate understanding. However, some algorithms are unavoidably complex,

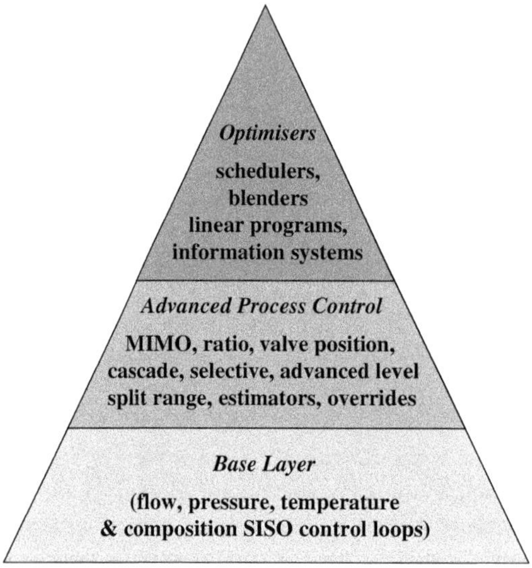

Figure 1.5 Main conceptual layers in a processing plant control scheme.

referring to a number of measurements as the basis of their output decisions. A specially trained control engineer is required to diagnose poor performance that might arise from a poorly calibrated measurement. Industries which recognise the need for ongoing care, and provide the necessary resources, have successfully increased the fractional online time of their optimisers and advanced controllers (Karodia, Naidoo and Appanah, 1999).

1.3 Organisation of This Book

With the increasing use of computers and digital communications in the processing industries, the employers' requirements of a process engineer are rapidly evolving. An engineer working day to day on troubleshooting and debottlenecking on a plant must already have a high proficiency in the use of computers for data extraction and analysis. He/she must comprehend a complex plant control scheme in order to identify the relationships between variables arising from such a scheme. Where a process engineer is specifically employed in the area of 'advanced process control', there will be even more of an expectation that good computer skills will be brought to bear on processing issues. The general brief given to such an engineer by the employer is likely to be: 'Do whatever you can with this computer know-how to safely maximise the profitability of this process!'. Well, that is a very open-ended request, extending far beyond the traditional 'process control' skills of drawing Nyquist plots to assure closed-loop stability.

At the outset, the strength of the process engineer in this environment is his/her understanding of the physics and chemistry at play, leading on to his/her ability to mathematically model the process. The model can be used effectively as the mathematical basis of the control scheme design, or simply to test proposed strategies developed in other ways. Regardless, it is the model that gives this engineer a clear insight into the process behaviour, and it provides an interpolative bridge between what is usually too few plant measurements.

Beyond the model, and now thinking of an alternative 'black box' approach, what will be important is to get an appreciation of a range of *ad hoc* methods of identification, control and optimisation which have proved useful industrially. Some of these defy mathematical treatment, and will not easily lead on to proofs of stability, which has been a major preoccupation of the field. Rather, the main purpose of mathematical treatments will be to promote understanding, and to allow one to move on quickly to useful algorithms for online implementation.

In order to establish the context of the considered algorithms, a view will initially be developed of the instrumentation and computer hardware required for their implementation. In connection with this, it will be important to understand conventions for representation of instrumentation and control on plant piping and instrumentation diagrams.

The present-day field of process control is built conceptually on remnants and artefacts of the past. For example, a computer representation of a loop controller has the equivalent switches and adjustments of the preceding panel-mounted analogue device. An engineer might claim that a control loop that oscillates a lot and won't settle down 'has a poorly damped closed-loop pole location'. One would be extremely hard put to find an application of such frequency response methods in the processing industries, yet this is the language that is naturally used. Why is this so? Well, the reason is that one's mind picture of the phenomena is built mostly on classical control theory. Though most of what is presently done in industry is based on the time domain, one ignores classical theory such as 'frequency response' at one's peril, because it is part of the language, and in many instances it is the route forward to deeper analysis and research. In some cases, for example the use of Laplace domain transfer functions, classical approaches give a much clearer view of relationships. The classical methods will thus also be used in parallel where appropriate.

Working on from simple controllers, more advanced algorithms for estimation and control will be considered, finally viewing the application of optimisation algorithms. Along the way, skills will be developed in the overall instrumentation and control of a process, effectively what is necessary to specify the key plant document, namely the piping and instrumentation diagram. Methods of quantifying and describing control performance and stability will be presented, largely connecting to the classical theory.

1.4 Semantics

Some concepts and related vocabulary in process control need to be clarified initially to avoid confusion:

System: all or part of a process which can be viewed in isolation (provides output values in response to input values).

Dynamic: the mathematical description involves a derivative with respect to time, or a time delay.

Static or *algebraic*: input values immediately determine output values.

Lumped: no spatial derivatives are involved in the mathematical description.

Distributed: variations also occur in space (e.g. position within a reactor bed), requiring spatial derivatives in the mathematical description.

Order: number of time derivatives of different variables involved in the mathematical description (each higher derivative also contributes 1 to the count).

States: a selection of variables describing a system such that if their initial values are known, and all future inputs are known, all future values of the states can be predicted. Effectively these are the variables in the set of first-order derivatives describing the system, so the *order* is equal to the number of *states*.

Open loop: information generated in the output does not influence the input.

Closed loop: information generated in the output is used to influence the input.

Stable: a system is stable if its outputs are bounded (non-infinite) for all bounded inputs.

Unstable: at least one bounded input excitation can cause an unbounded output – usually manifested as exponentially increasing oscillation or magnitude. Usually this type of behaviour is restricted to a *limit cycle* or final magnitude because of the physical limits of the equipment – unless failure occurs before this point.

Step response: output variation resulting from a step in one of the inputs.

Frequency response: output characteristics when the input is a steady oscillation (varies with frequency).

Tuning: choice of free parameters for controllers, estimators or optimisers, to obtain desired performance.

Controlled variable (CV): one of the outputs for which tracking of a *setpoint* is required.

Manipulated variable (MV): one of the inputs which is available to be varied by a controller.

Disturbance variable (DV): one of the inputs which is not available for manipulation.

Dead time: this is a time delay (usually caused by a plug flow transport lag).

Inverse response: the initial direction of the response (up/down) differs from the final position.

References

Karodia, M.E., Naidoo, S.G. and Appanah, R. (1999) Closed-loop optimization increases refinery margins in South Africa. *World Refining*, **9** (5), 62–64.

Sakizlis, V., Perkins, J.D. and Pistikopoulos, N. (2004) Recent advances in optimization-based simultaneous process and control design. *Computers & Chemical Engineering*, **28** (10), 2069–2086.

2
Instrumentation

Plant instrumentation constitutes 6–30% of the cost of all purchased equipment for the plant, and thus up to 5% of total project costs for new plants (Peters and Timmerhaus, 1980). A perusal of the instrumentation trade magazines shows what a fiercely competitive market this is. Suppliers are developing a never-ending range of measurement and actuation devices based on new technologies and materials. In order to understand the context of the control schemes and algorithms addressed in this book, one needs a basic appreciation of the devices available.

2.1
Piping and Instrumentation Diagram Notation

The piping and instrumentation diagram is the most important reference document in both the construction and functioning of a processing plant (Davey and Neels, 1991). Not only is it the means of understanding the operation, but its indexing system also allows location of documents pertaining to individual plant items (such as reactors and pumps), the pipework and the measurement and control equipment and philosophy (Figure 2.1).

Conventions for showing instrumentation vary with company and national standards (e.g. BS 1646 and DIN 28004). Some examples are presented in Figure 2.2. Up until the 1980s, analogue instrumentation data handling was still common, requiring some specific conventions such as in Figure 2.3.

The development of instrumentation that could perform the required operations (such as PID control or square-root extraction) in a purely analogue context became a very sophisticated art, leading to ingenious fluidic (pneumatic) or electronic circuitry. Apart from the high purchase costs, there was a heavy burden of maintenance, so the advent of data-handling systems based on digital computers was warmly welcomed. On economic grounds, most analogue systems were quickly dumped in favour of modern DCS (distributed computer systems), SCADA (supervisory control and data acquisition) and PLCs (programmable logic controllers) (Figure 2.4). Such systems are highly configurable. No longer would the recording of a signal require a dedicated panel-mounted strip or circular chart (frontispiece). All data could be continuously recorded in a historical database, and accessed at will. Alarm limits could be defined for every signal. The control panel along which sociable operators used to stroll, tapping recorders to ensure free pen movement, disappeared, and was replaced by one or more CRT or LCD displays. The art of being an operator changed, and quite a few of the 'old school' moved on. In the digital systems, the functional distinctions 'I' and 'R' in Figure 2.3 have been lost and are omitted from tags.

Applied Process Control: Essential Methods, First Edition. Michael Mulholland.

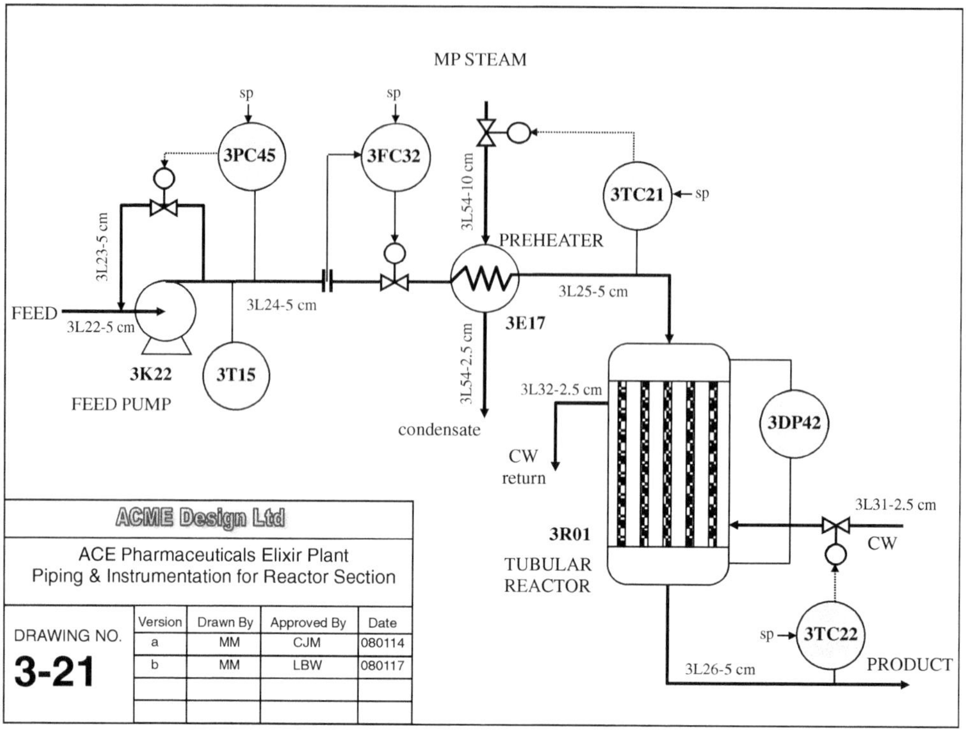

Figure 2.1 Example of a piping and instrumentation diagram.

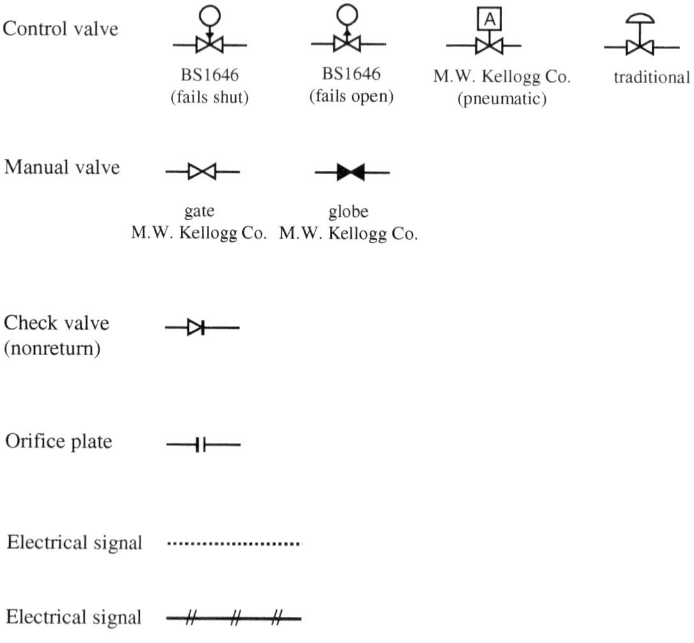

Figure 2.2 Examples of instrumentation symbols.

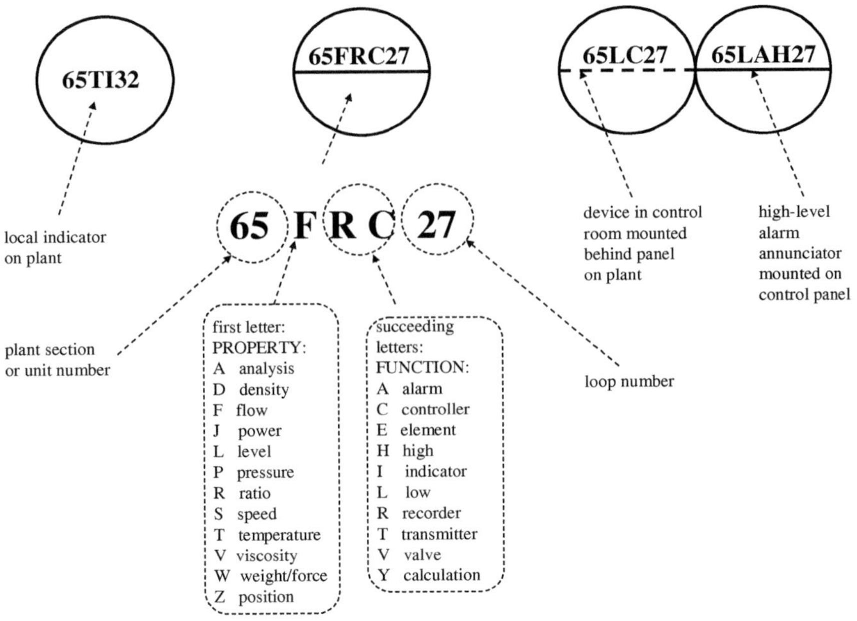

Figure 2.3 Instrumentation tag bubble notation.

The growth of the digital systems out of the old analogue systems constrained manufacturers (probably needlessly) into a sort of item-per-item replacement. Clearly, this would reduce the shock for existing personnel. The square-root extraction, ratio, control, alarm and trip calculations stayed in functional blocks which could be interconnected just like the old analogue signals. The computer 'faceplates' of PID controllers had the recognisable adjustments from the old control room panel. However, beyond the direct replacement of the old functionality, industries realised that they were now in a completely new situation where every measurement and every control actuator could be accessed simultaneously, and there was almost no constraint on the complexity of any calculation. This opened up a whole new world for control engineers to take the concept of 'advanced process control' a lot further.

2.2 Plant Signal Ranges and Conversions

One legacy of the analogue era is the handling of measurement and actuator signals in conventional ranges such as

4–20 mA;

3–15 psig;

20–100 kPag.

With new fieldbus devices, instruments can communicate with the computer control system digitally, avoiding the need to convert current or pneumatic signals into the accepted ranges.

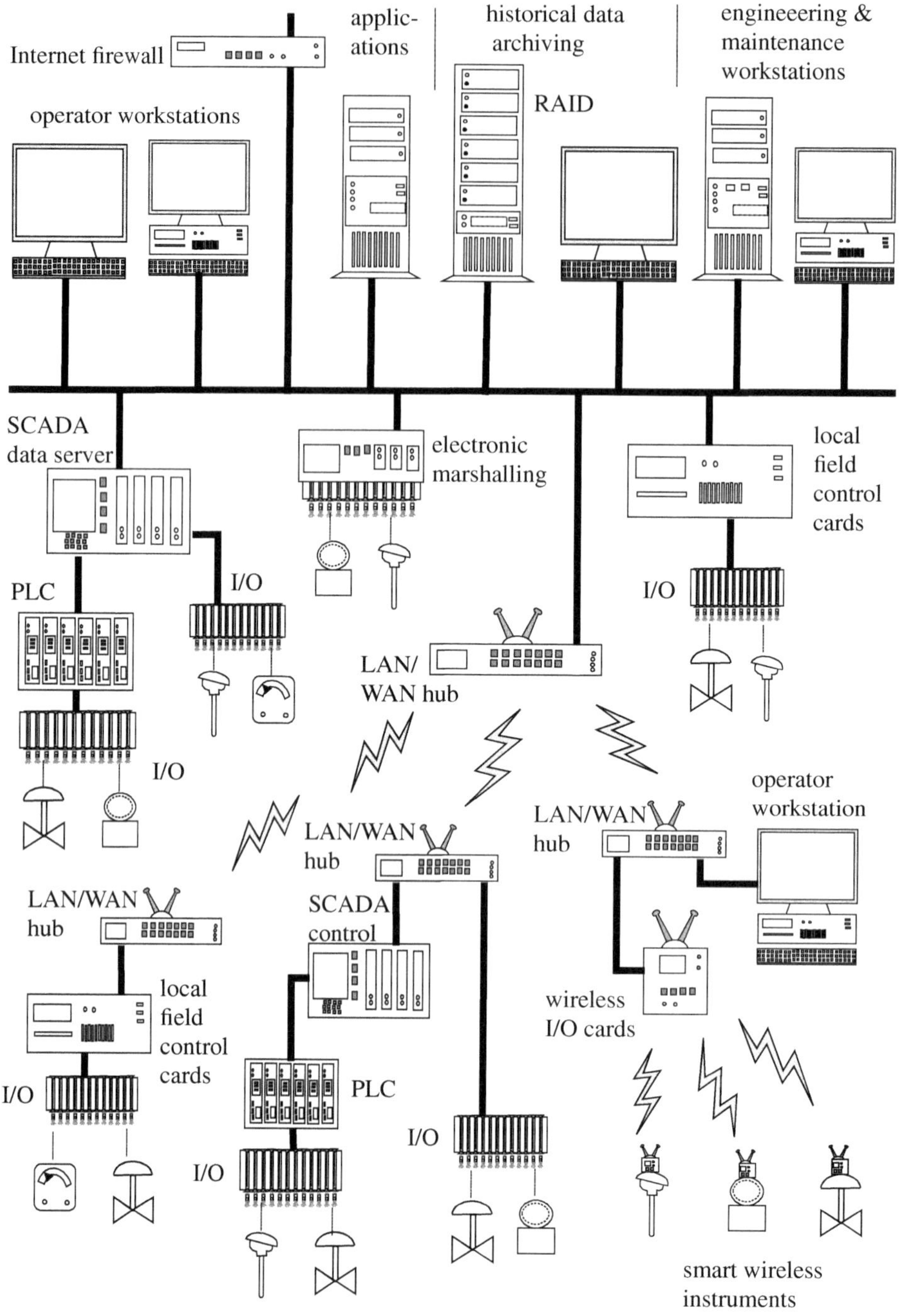

Figure 2.4 Distributed control system structure illustrating a range of features.

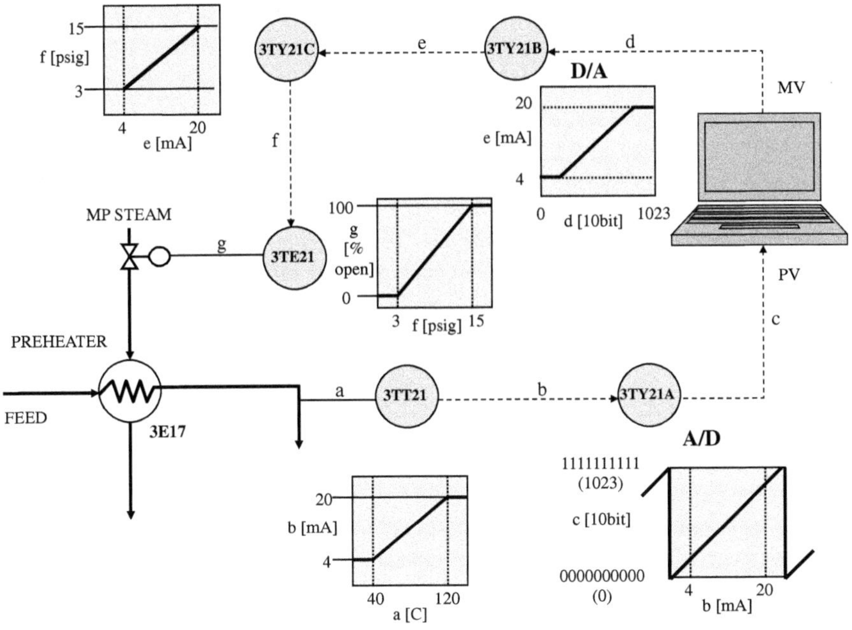

Figure 2.5 Typical signal conversion sequence.

Nevertheless, much instrumentation is still being designed around the analogue signal range concept, and one has to be aware that any device will anyway have lower and upper saturation limits. On large plants, these analogue signals are handled as close to the associated plant equipment as possible in signal substations. Signal conditioning (such as smoothing) and A/D or D/A (analogue-to-digital or digital-to-analogue) conversion are handled there, so that communication with distributed and centralised computer facilities thereafter occurs on digital buses (e.g. coaxial, fibre-optic or 'wireless' transmissions).

A typical signal conversion sequence is given in Figure 2.5. Note that the information flow, starting from the temperature sensor (a), passes through a sequence of calibrations, in this case each involving an intercept and a slope. Each of these is subject to drift and error, so it is only a foolhardy engineer who simply accepts the engineering values presented in the computer display. A nonlinearity arises from the saturation conditions determined by the conventional signal ranges and the A/D and D/A conversion windows. For example, a 20 mA signal (b) might represent an original temperature measurement (a) anywhere above 120 °C. Regarding the 20% offset of the minimum transmission signal values from zero (e.g. 4 mA, 3 psig), this was originally to aid signal loop continuity checking, since the most likely fault would be an open circuit.

For a pneumatically actuated control valve (air-to-open) as suggested here, the intercept and slope (*zero* and *span*) are determined by two nuts on the valve stem which position and tension the return spring. It is conceivable that wear or incorrect set-up might result in a 3 psig signal not quite closing the valve, say leaving it 4% open. There is no feedback of this position, so the operator would be unaware that live steam is still in the system, for example. Often the nuts might be set to close the valve at, say, 4 psig, to ensure a really tight shut-off when a 0% (3 psig) open position is requested. These are but some of the pitfalls to be expected, so the engineer would do well to maintain a healthy scepticism regarding the validity of data on the system and should cross-check wherever possible.

2.3
A Special Note on Differential Pressure Cells

The differential pressure (DP) cell is the ubiquitous workhorse of industrial instrumentation, playing a part in the measurement of flow, pressure and level (Figure 2.6). It is a transmitter in the sense that it receives a pressure signal, converts it linearly into the desired signal range (e.g. 4–20 mA, 20–100 kPag) and retransmits it. In cases where it transmits a current signal, it is sometimes referred to as a P/I transducer (pressure/pneumatic-to-current). Flow measurement techniques relying on the creation of a pressure difference (e.g. orifice plates) usually attempt to work with as small a Δp as possible to minimise pumping power. So in this case the DP cell might receive a signal of, say, 0–20 ″WG, and retransmit in the 3–15 psig range. Here one can think of the device as a high-gain amplifier. Conversely, the pressure inside an ammonia synthesis loop (e.g. 150 barg) might be compared with atmospheric pressure, and the Δp retransmitted in the 3–15 psig range.

DP cells will be installed close to the point of measurement, so there could still be a need to use a completely pneumatic device (as above) for intrinsic safety in the presence of flammable gases. Commercial electrical DP cells are usually provided in explosion-proof housing, so these are likely to be acceptable in hazardous areas as well.

A DP cell might have a very sensitive diaphragm to measure flow or furnace draught (e.g. 20 ″WG), yet all DP cells are supplied in extremely robust steel housing, to allow for the measurement to occur at a high pressure (e.g. flow in an ammonia synthesis loop). In this case, special precautions need to be taken to ensure that the diaphragm is not inadvertently exposed to a high Δp, for example if one of the impulse lines is disconnected from the plant piping.

Figure 2.7 highlights some important considerations of DP cell installation. When it is being attempted to measure small Δp values, as in the use of an orifice plate for flow measurement, a

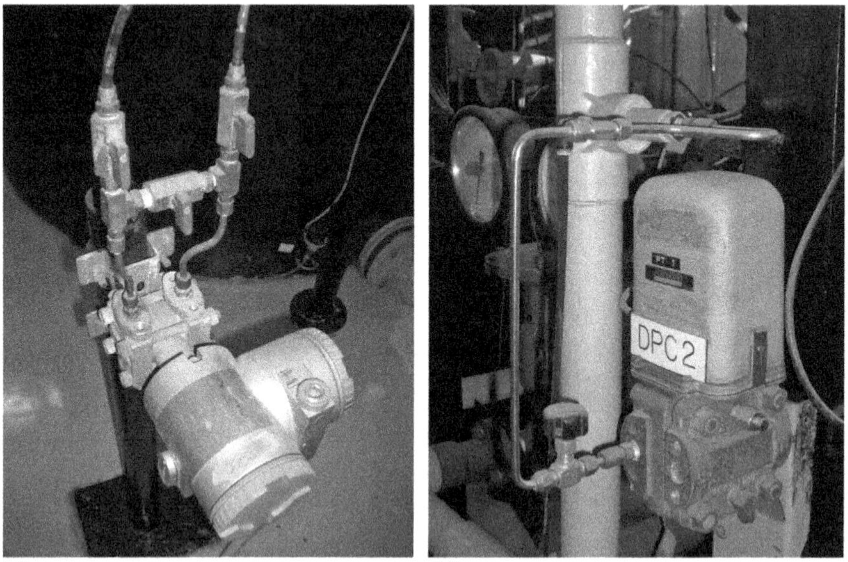

(a) (b)

Figure 2.6 Installed DP cells: (a) electronic and (b) pneumatic (measuring pressure).

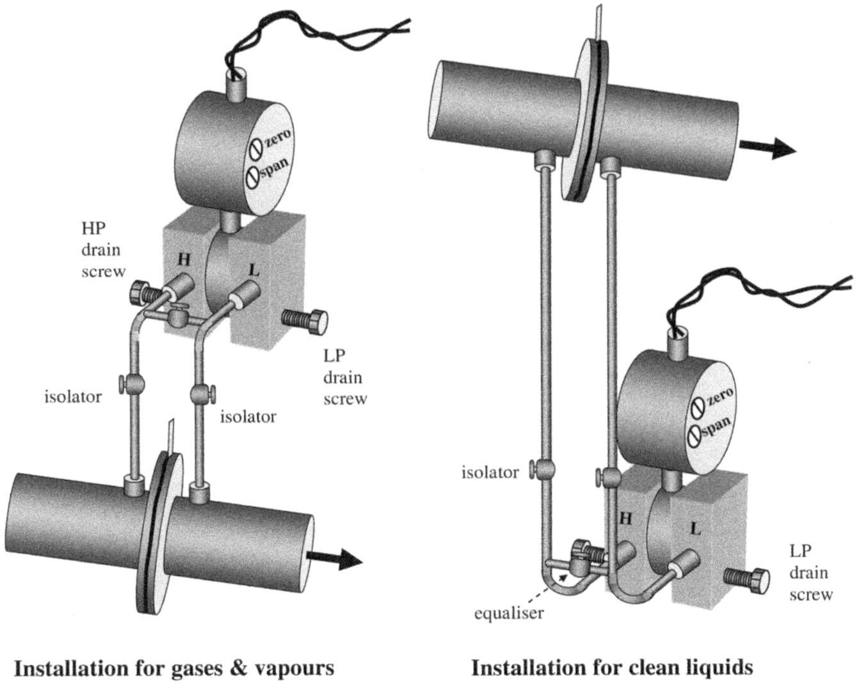

Figure 2.7 DP cell installation for flow measurement.

situation where the impulse lines themselves can exert significant and unknown pressure through vapour locks (in the case of liquids) or liquid slugs (in the case of gases and vapours) must be avoided. After all, the cell is seeking to measure the equivalent of only a few inches of water. Even a horizontal connection, if it undulates, becomes subject to the *sum* of all slugs trapped in the troughs. Thus, for gases and vapours the lines must be arranged to allow clear drainage of any condensate back into the duct, whilst for liquids one seeks a continuity of liquid all the way back into the liquid flow duct. Assuming the liquid duct is above atmospheric pressure, at start-up one loosens the drain screws on the two cell chambers, to allow the liquid to run freely through each chamber and displace any air present. Such an operation of course must take due regard of any excessive Δp that might arise across the diaphragm.

An equalising valve is normally installed between the HP and LP ports of the cell. This is the means of setting a zero Δp for the purpose of zeroing the cell. With the equalising valve open, the zeroing adjustment is turned until the cell transmits at the threshold of its signal range, namely 4 mA, 3 psig or 20 kPag. Similarly, the span adjustment can be used to effectively set the top end of the scale, for example by matching correctly to a known applied Δp.

Electronic DP cells may use potentiometric, capacitance, piezoelectric, strain gauge, silicon resonance or differential transformer techniques for interpretation of the diaphragm deflection. In the common 4–20 mA type, the device acts as a variable resistance in the current loop, and actually draws its own power from the same loop.

The principle of a pneumatic DP cell is illustrated in Figure 2.8. The diaphragm deflection varies the distance of a flapper from a nozzle, causing a variable back-pressure. This signal cannot be used directly, as additional movement of air out of the nozzle cavity will vary the calibration. Instead, a

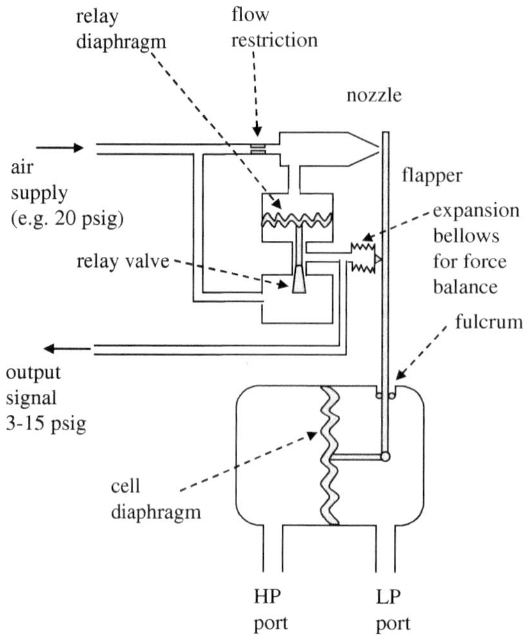

Figure 2.8 Principle of a pneumatic DP cell.

relay is used to create a balancing pressure by supplying air to an expansion bellows acting on the same flapper. The latter source is able to supply a greater flow of air at a pressure proportional to the nozzle pressure.

2.4 Measurement Instrumentation

An initial division of measurement instruments is into the categories 'local' and 'remote'. A local device needs to be read at its point of installation, and has no means of signal transmission. Typical candidates are thermodials and Bourdon-type pressure gauges. Occasionally, manometers might be used to indicate differential pressures. Up until the 1980s, local measurements were prolific, and important, with operators patrolling the whole plant at hourly intervals, jotting down the readings on a clipboard. Increasingly now the operational picture is built up entirely from electronic information, whether it is updated on graphical mimic diagrams of the plant, or archived for future analysis. It seems that the additional investment in signal transduction, marshalling, conversion and capture is worthwhile in comparison with manual reading, transcription and data entry.

In the processing industries, there are few measurements requiring the high speeds of response often needed in electrical or mechanical systems. Nevertheless, it is important to bear in mind the impact of the response time constant of an instrument considered for each application. For example, a flue gas O_2 measurement might output a smoothed version of the actual composition. A brief O_2 deficiency might not be seen, but could be enough to start combustion when the uncombusted vapours meet O_2 elsewhere in the ducting.

2.4.1 Flow Measurement

2.4.1.1 Flow Measurement Devices Employing Differential Pressure

Bernoulli's equation for incompressible fluids gives

$$\frac{\rho v^2}{2} + \rho g h + p = \text{constant} \tag{2.1}$$

which expresses the conservation of energy in the flow direction (ρ: density; v: velocity; h: height; g: gravity; p: pressure). For small pressure changes, this is also an adequate representation of gas flows. So between two points in a flow system the changes must balance as follows:

$$\frac{\rho}{2}\Delta\left[v^2\right] + \rho g \Delta h + \Delta p = 0 \tag{2.2}$$

For systems in which some of the energy is used up in overcoming frictional resistance,

$$\frac{\rho}{2}\Delta\left[v^2\right] + \rho g \Delta h + \Delta p = \Delta p_{\mathrm{f}} \tag{2.3}$$

Measurements under turbulent conditions show that Δp_{f} is almost proportional to the square of the total flow rate, whether it be flow through ducts, restrictions, plant equipment, particulate beds or free objects moving through fluids. For example, the Darcy equation for pressure loss along a duct is

$$\Delta p_{\mathrm{f}} = \left(\frac{4f' L}{D}\right)\rho\frac{v^2}{2} \tag{2.4}$$

where L is the duct length and D is the equivalent diameter. There is a minor secondary dependence on the velocity v since the friction factor f' is generally correlated using the Reynolds number $N_{\mathrm{Re}} = \rho v D/\mu$. However, as an approximation, frictional pressure losses are usually thought of in terms of a number of kinetic 'velocity heads' (one head $= v^2/2g$). For example, the disorderly expansion of a flow with sudden enlargement of a duct incurs a loss of approximately one velocity head, that is $\Delta p_{\mathrm{f}} = \rho g(v^2/2g)$.

Venturi Meter

As noted above, the flow rate in a duct, or stream velocity passing an object, can be estimated from the pressure drop occurring along a duct section or across the object. Usually these features will cause a net loss of pressure due to friction. However, the venturi flow meter is an attempt to obtain a measurable pressure difference, yet minimise the overall frictional loss. This becomes important in low-pressure systems such as vacuum distillation pipework, furnace draught or flue gas ducting, where the friction loss might compete with the available delivery pressure, alter the density or affect the vapour–liquid equilibrium significantly. It is common to see tapered venturi duct sections in place on the suctions of forced-draught furnace fans, for the purpose of combustion air flow measurement.

Assuming that the velocity profiles at positions 1 and 2 in Figure 2.9 are uniform, and that the flow is frictionless, a general equation for the total mass flow of incompressible and compressible fluids through a venturi tube is

$$w = \rho_2 A_2 v_2 = C_{\mathrm{d}}\rho_2 A_2 \sqrt{\frac{-2\int_1^2 (1/\rho)\mathrm{d}p}{1 - \left(\dfrac{\rho_2 A_2}{\rho_1 A_1}\right)^2}} \tag{2.5}$$

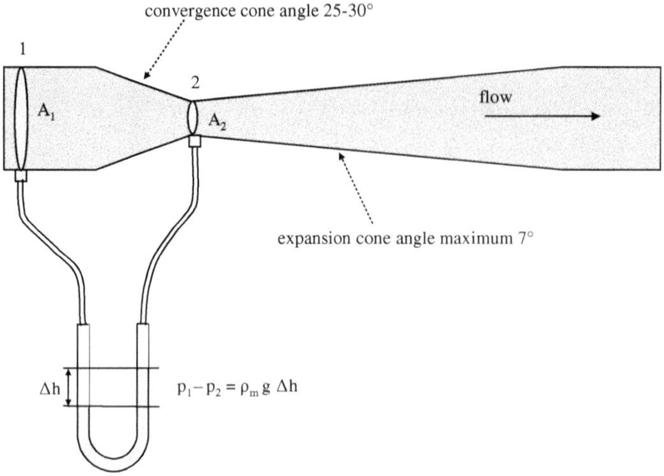

Figure 2.9 Venturi flow meter.

For an ideal gas under adiabatic conditions, this gives

$$w = C_d \rho_2 A_2 \sqrt{\frac{2\left(\frac{p_1}{\rho_1}\right)\left(\frac{\gamma}{\gamma-1}\right)\left[1-\left(\frac{p_2}{p_1}\right)^{(\gamma-1)/\gamma}\right]}{1-\left(\frac{A_2}{A_1}\right)^2\left(\frac{p_2}{p_1}\right)^{2/\gamma}}} \tag{2.6}$$

where $\gamma = c_p/c_v$ and the pressures are absolute. For p_2/p_1 close to 1, this reduces to

$$w = C_d A_2 \sqrt{\frac{2\rho_1(p_1 - p_2)}{1-\left(\frac{A_2}{A_1}\right)^2}} \tag{2.7}$$

which is also valid for an incompressible fluid. For well-designed venturi meters, the discharge coefficient C_d is found to be around 0.98 for both compressible and incompressible fluids.

Orifice Plate

The orifice plate is the most common flow measurement device employed in the processing industries. There are variations in the installation (e.g. flange tappings versus corner tappings versus upstream/downstream tappings, bevelled edge versus square edge, upstream/downstream free run, etc.), some complying with strict standards.

The principle is the same as for the venturi meter, except that here the precise area of the liquid jet represented by the flow streamlines at the *vena contracta* in Figure 2.10 is not known. Using instead the area of the orifice itself, one obtains for an incompressible fluid

$$w = C_d A_O \sqrt{\frac{2\rho(p_1 - p_2)}{1-\left(\frac{A_O}{A_1}\right)^2}} \tag{2.8}$$

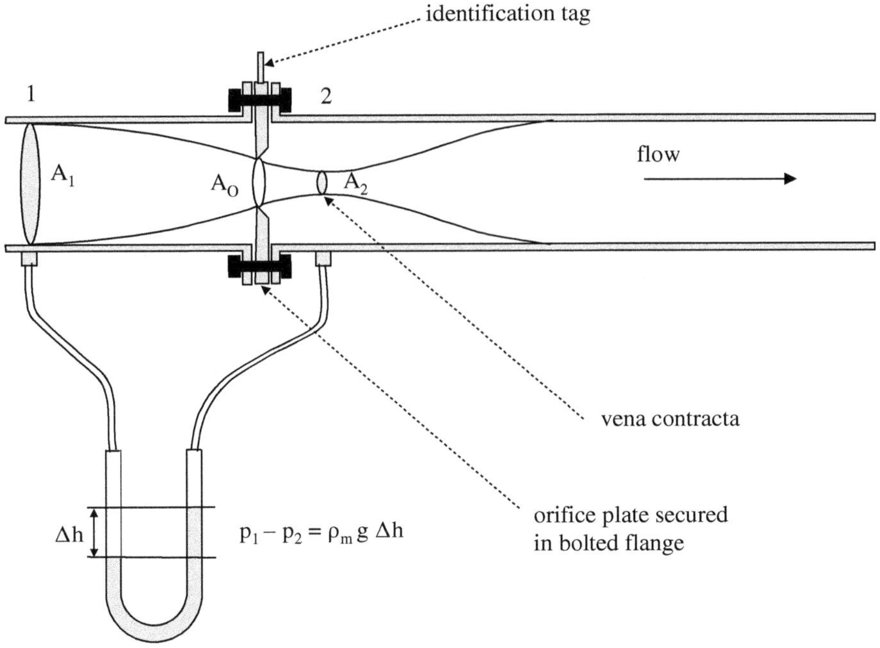

Figure 2.10 Orifice plate flow meter.

but now the discharge coefficient is around 0.61 (for high Reynolds numbers), the difference from its value in Equation 2.7 largely compensating for the deviation between A_2 and A_O.

Equation 2.6 suggests that adjustment of Equation 2.8 for compressible flow will depend primarily on γ and the ratios A_O/A_1 and p_2/p_1. Indeed, data are available correlating a compressibility factor Y in terms of γ, d_O/d_1 and $\Delta p/p_1$ $(= 1 - p_2/p_1)$.

For a particular fixed installation, users generally rely on the square-root behaviour

$$w = k\sqrt{\rho_1(p_1 - p_2)} \tag{2.9}$$

where it is seen that

$$k = YC_dA_OA_1\sqrt{\frac{2}{A_1^2 - A_O^2}} \tag{2.10}$$

where the compressibility factor $Y \rightarrow 1$ for an incompressible fluid.

Calibration of Orifice Plate, Venturi and Similar Flow Meters

In practice, k in Equation 2.9 might be estimated using measured plant displacements. Note that the total volumetric flow rate is obtained as

$$F = k\sqrt{\frac{(p_1 - p_2)}{\rho}} \tag{2.11}$$

with the assumption that $\rho_1 \approx \rho_2$, and it is sensible to calibrate *in situ* to determine the approximately constant k. Indeed, most installations have no compensation for variations in density ρ, with

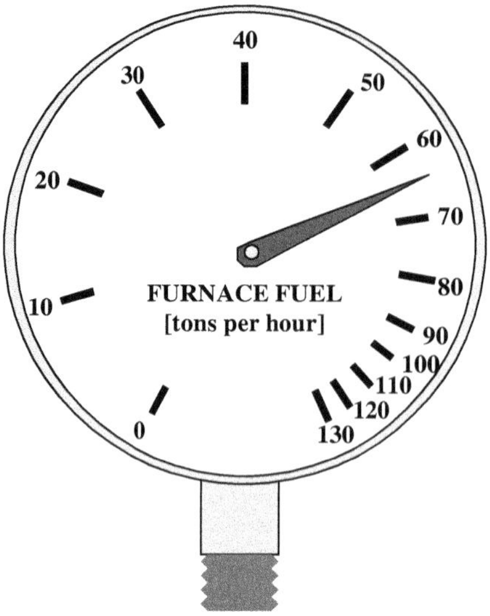

Figure 2.11 Square-root extracting flow gauge.

a fixed multiplying constant representing $k\sqrt{\rho}$ determined at plant *design conditions* applied to $\sqrt{\Delta p}$ for the form in Equation 2.9.

$$w = K\sqrt{\Delta p} \tag{2.12}$$

Prior to the widespread use of digital computers, the signal for Δp would often arrive at a display panel to be shown on the dial of a simple pressure gauge as in Figure 2.11. The graduations on the dial were simply arranged to extract the square root of the signal, and apply the multiplier K.

Engineers need to be very wary of this type of mass flow or volumetric flow indication based on a fixed K value. The density of streams on a plant will of course vary with composition, as well as operating pressure and temperature in the case of gases and vapours. The indicated flows must be corrected relative to the calibration condition as in Equations 2.13–2.14 for mass w and volumetric F flows respectively.

$$w_{\text{ACTUAL}} = w_{\text{INDICATED}}\sqrt{\frac{\rho_{\text{ACTUAL}}}{\rho_{\text{CALIBRATION}}}} \tag{2.13}$$

$$F_{\text{ACTUAL}} = F_{\text{INDICATED}}\sqrt{\frac{\rho_{\text{CALIBRATION}}}{\rho_{\text{ACTUAL}}}} \tag{2.14}$$

It follows for ideal gases that

$$w_{\text{ACTUAL}} = w_{\text{INDICATED}}\sqrt{\frac{M_{\text{ACTUAL}}}{M_{\text{CALIBRATION}}}\cdot\frac{P_{\text{ACTUAL}}}{P_{\text{CALIBRATION}}}\cdot\frac{T_{\text{CALIBRATION}}}{T_{\text{ACTUAL}}}} \tag{2.15}$$

$$F_{\text{ACTUAL}} = F_{\text{INDICATED}}\sqrt{\frac{M_{\text{CALIBRATION}}}{M_{\text{ACTUAL}}}\cdot\frac{P_{\text{CALIBRATION}}}{P_{\text{ACTUAL}}}\cdot\frac{T_{\text{ACTUAL}}}{T_{\text{CALIBRATION}}}} \tag{2.16}$$

where M is molecular mass, P is absolute pressure and T is absolute temperature.

A properly calibrated and installed orifice flow meter working at its design conditions can be expected to be accurate to within 2–4% of full span, whilst the accuracy of a venturi meter is around 1% of full span.

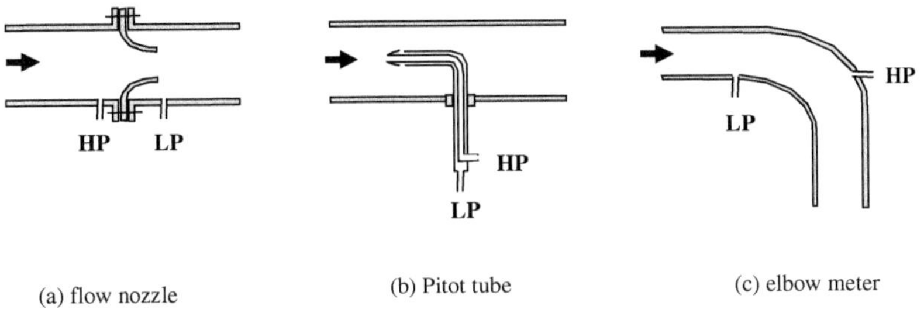

Figure 2.12 Some other flow measurement devices based on differential pressure.

Other Flow Measurement Devices Employing Differential Pressure

Figure 2.12 shows three additional flow measurement devices relying on differential pressure. The flow nozzle (a) can be expected to comply with the general equations for venturi or orifice plate meters, and has an accuracy of 2% of span. The Pitot tube (b) has a dynamic port on the tip at which the local pressure will rise by $\rho v^2/2$ according to Bernoulli's equation (Equation 2.3), as a result of the velocity being reduced to zero. The static ports just behind the tip provide the offset which allows the local v to be resolved. The device can be simplified using a separate tapping on the pipe wall for the static measurement. The elbow meter (c) is based on the same principle as the Pitot tube, and achieves an accuracy of 5–10% of full span.

Flow Control Loop Signals

It is useful at this stage to review a detailed piping and instrumentation diagram of a flow control loop, in order to identify the various elements and interconnections arising in the configuration. In Figure 2.13, where only analogue signals are shown, a measurement device requiring square-root extraction is considered, as above. Depending on the audience for an instrument and control

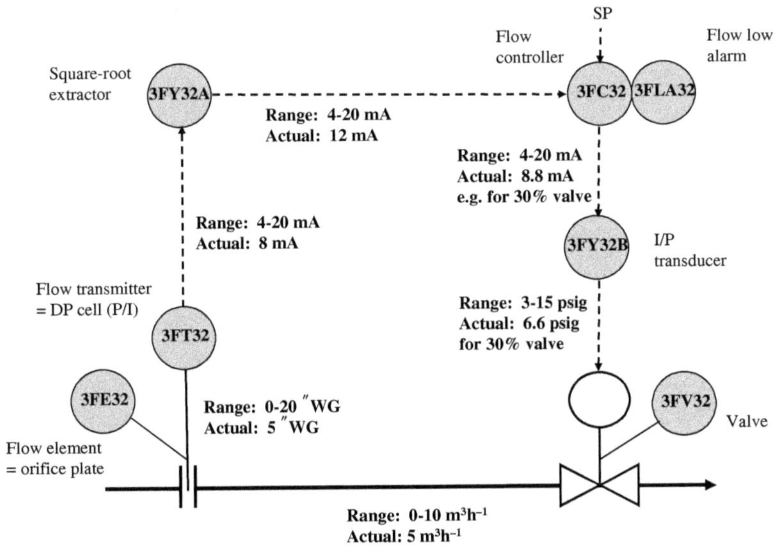

Figure 2.13 Analogue signals around a flow control loop.

scheme representation of this nature, one includes more or less detail. For example, Figure 2.13 might be shown with just a single '3FC32' bubble, and it would generally be understood that the various conversions and the square-root extraction are occurring in this simplified loop. In modern control systems, there are many optional settings for such instrumentation (such as ranges and controller parameters), and this information is usually held in a database, amounting to a number of pages for one loop.

It is interesting to follow the sequence of conversions around the control loop in Figure 2.13 for an operating flow of 5 $m^3 h^{-1}$:

$$\text{From orifice plate}: \quad \left(\frac{5-0}{10-0}\right)^2 (20-0) = 5''\text{WG}$$

$$\text{From DP cell}: \quad \left(\frac{5-0}{20-0}\right)^2 (20-4)+4 = 8\ \text{mA}$$

$$\text{From square-root extractor}: \quad \left(\frac{8-4}{20-4}\right)^{1/2} (20-4)+4 = 12\ \text{mA}$$

$$\text{From controller (e.g. for 30\% valve)}: \quad 0.3 \times (20-4)+4 = 8.8\ \text{mA}$$

$$\text{From I/P converter (for 30\% valve)}: \quad \left(\frac{8.8-4}{20-4}\right)(15-3)+3 = 6.6\ \text{psig}$$

2.4.1.2 Other Flow Measurement Devices

As with most industrial instrumentation, there is a plethora of devices based on alternative technologies for flow measurement. The diversity is driven along by manufacturers' attempts to develop new niches based on patents. Indeed, some are well suited to particular environments such as dirty fluids and high-accuracy measurement. Some of these devices are shown in Figure 2.14.

Gas mass flow (thermal) devices (g) work on the principle of adding heat to the stream, and detecting its temperature change. In constant-current I mode, the flow rate is inferred from ΔT, and in constant ΔT mode from I. *Coriolis* meters (h) provide both a flow rate and density. The flow passes through a U-shaped tube which is subjected to a lateral vibration by electromagnets. The flow momentum resists the lateral movement as it enters, and increases that of the other arm of the U as it leaves – causing a phase lag between the two arms which is related to the mass velocity. In turn, the vibration amplitude depends on the stream density.

In Table 2.1, Swearingen (1999, 2001) compares the attributes of selected flow measurement devices. Here 'variable area' refers to rotameter-like devices in which an object is displaced (e.g. against gravity or a spring) in order to increase the flow area.

2.4.2 Level Measurement

2.4.2.1 Level Measurement by Differential Pressure

Differential pressure is often used to determine level. Three scenarios are shown in Figure 2.15.

Measurements of liquid level for vessels open to atmosphere are obtained from $\Delta p = \rho g h$, with atmospheric pressure acting on both the liquid surface and the LP port of the DP cell. For enclosed vessels, it is necessary to locate a second sensing point above the liquid surface, to eliminate the effect of any absolute pressure fluctuations. This will involve an extra vertical section of impulse

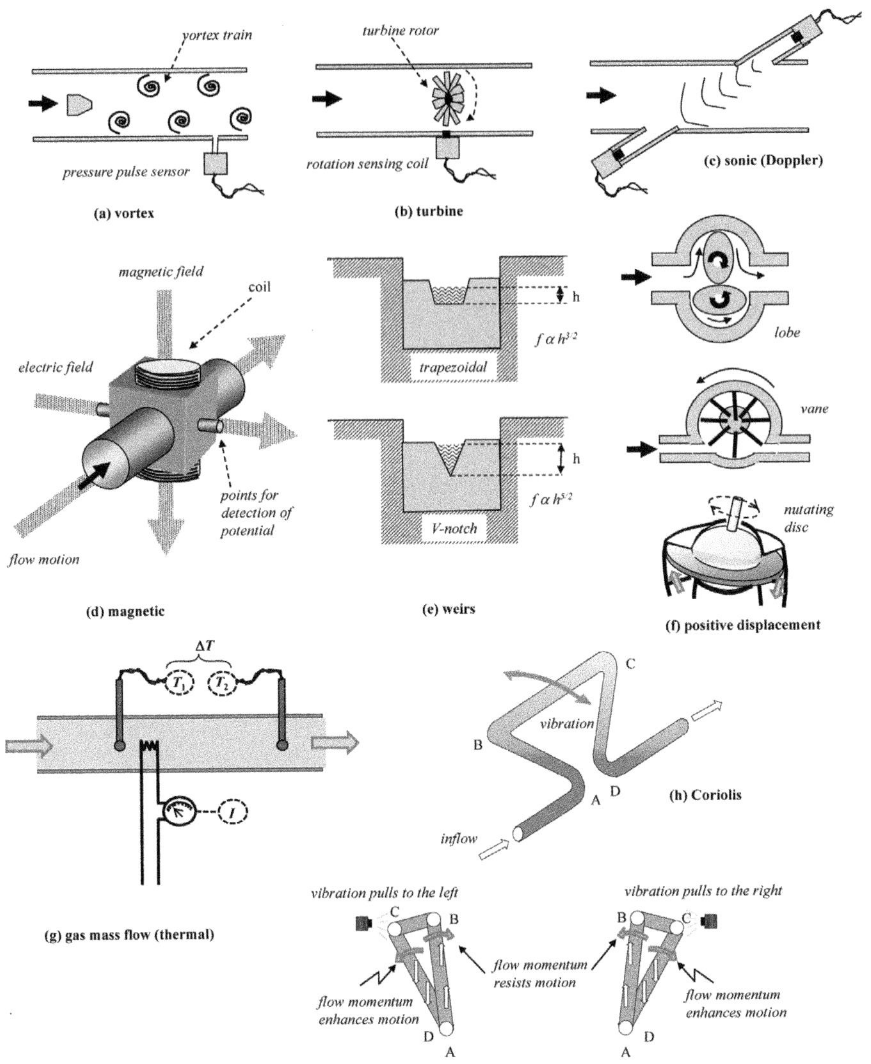

Figure 2.14 Flow measurement devices based on various principles.

tubing, which introduces the problem of possible condensate collection, for example if the tubing temperature falls below that of the vessel. A serious situation can arise where the actual liquid level is much higher than that indicated. In practice, this problem is dealt with by reversing the DP cell port allocation, and ensuring that the contents of the high leg are known. Typically, the high leg is preloaded with the same liquid as in the vessel. A different, less volatile liquid might be used to avoid re-evaporation as process conditions change, for example glycerine is used in wax crystallisation from MEK.

It is clear that levels obtained from $h = \Delta p/\rho g$ will be dependent on the average density of the liquid. Where this varies much, or froth forms or bubbles are present (as in the boiler 'swell' effect),

Table 2.1 Comparison of selected flow metering devices (Swearingen, 1999, 2001).

Attribute	Variable area	Coriolis	Gas mass flow	DP	Turbine	Oval gear	Doppler	Vortex	Magnetic
Clean gases	Yes	Yes	Yes	Yes	Yes	—	Yes[a)]	Yes	No
Clean liquids	Yes	Yes	—	Yes	Yes	Yes	Yes	Yes	Yes
Viscous liquids	Yes[a)]	Yes	—	No	Yes[a)]	Yes[b)]	Yes	Yes	Yes
Corrosive liquids	Yes	Yes	—	No	Yes	Yes	Yes	Yes	Yes
Accuracy (+/−)	2–4% S	0.05–0.15% F	1.5% S	2–3% S	0.25–1% F	0.1–0.5% F	2% F	0.75–1.5% F	0.5–1% F
Repeatability	0.25% S	0.05–0.10% F	0.5% S	1% S	0.1% F	0.1% F	0.5% F	0.2% F	0.2% F
Maximum P (psig)	>200	>900	>500	100	>5000	>4000	Noncontact	300–400	600–800
Maximum T (°F)	>250	>250	>150	122	>300	>175	Noncontact	400–500	250–300
Pressure drop	Medium	Low	Low	Medium	Medium	Medium	Negligible	15–20 psi	Negligible
Turndown ratio	10 : 1	100 : 1	50 : 1	20 : 1	10 : 1	25 : 1	50 : 1	20 : 1	20 : 1
Average cost ($)	200–600	2500–5000	600–1000	500–800	600–1000	600–1200	2000–5000	800–2000	2000–3000

% S : % of full scale; % F : % of measured flow.
a) With special calibration.
b) >10 centistokes.

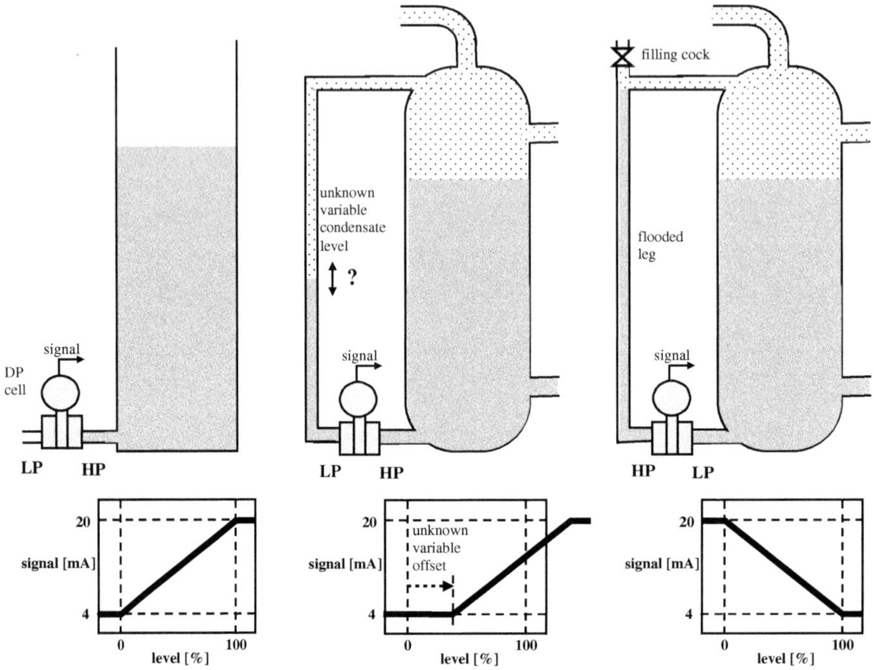

Figure 2.15 Level measurement by DP cell.

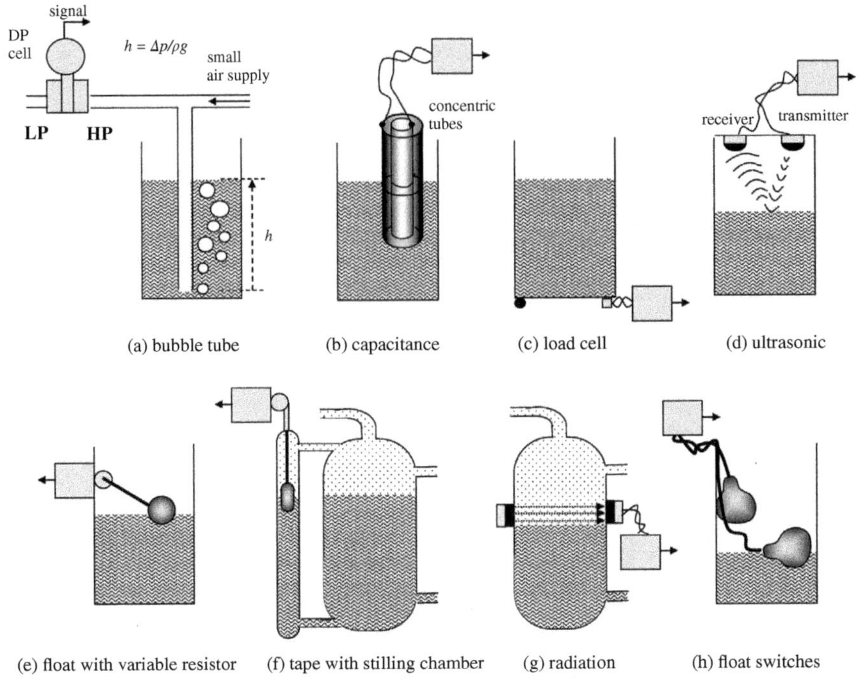

Figure 2.16 Various level measurement techniques.

procedures need to make allowance for the possible deviation, or an alternative level measurement technique must be sought.

2.4.2.2 Other Level Measurement Techniques

Figure 2.16 illustrates a few alternative level measurement methods. The bubble tube (a) has an advantage in the case of corrosive liquids or those containing solids or other material that might block impulse lines. Air is bled through the tube to keep the interface at its tip, and this flow must not be so great as to cause an additional frictional pressure drop in the line. Ultrasonic sensors (d) are used for solids hoppers or material with unknown or variable density, for example ice cream. The ball float moving a circular variable resistance (e) is very common, for example in motor vehicle fuel tanks. Radiation (g) and float switches (h) only provide on/off signals as the level crosses the mounting height.

2.4.3 Pressure Measurement

Figure 2.17 illustrates various pressure measurement devices. As noted in Section 2.3, for low-pressure measurements, collection of liquid in impulse lines must be avoided or compensated for. In all cases shown, the pressure in the duct is being compared with atmospheric pressure, so it is important to qualify the result as a 'gauge' pressure, for example 120 kPag or 3.7 barg. A few devices, for example the ionisation gauge for very low absolute pressures, give an absolute pressure. (In this case, a heating element causes ionisation of molecules present, giving a current flow between electrodes which will depend on the number of molecules per unit volume.)

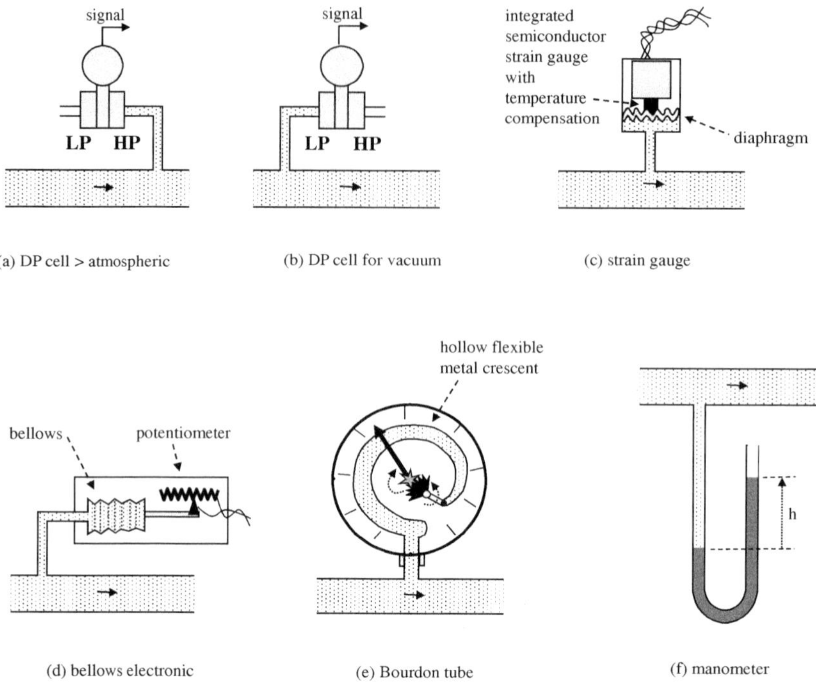

Figure 2.17 Various pressure measurement techniques.

Differential pressure cells (a), or a variation thereof with only one port, are frequently used for pressure measurement. The strain gauge pressure measurement (c) will commonly use an integrated semiconductor incorporating the diaphragm, strain gauge and temperature compensation. The Bourdon tube (e), invented by the French engineer Eugene Bourdon, is the basis of the very common local pressure gauge. The hollow, thin-walled metal crescent flexes as the internal pressure varies. The resultant rotatory motion is captured by a gear system which moves the needle around the dial. Another example of a local indicator is the manometer (f), which may also be used in vacuum applications.

2.4.4 Temperature Measurement

Common devices used for temperature measurement are thermocouples and resistance temperature detectors (RTDs). These are usually encapsulated within a stainless steel rod which can be mounted with or without sheath protection as in Figure 2.18, depending on the application.

2.4.4.1 Thermocouple Temperature Measurement

In Figure 2.19a, two wires consisting of different metals are connected at their ends. If the two ends of, say, metal A are at different temperatures, the atomic vibration and electron motion will differ, causing the more energetic hot electrons to tend to drift towards the cooler end. This creates a small potential (emf) across the wire. The same will occur in the metal B wire, but because the metal is different, the characteristic potential differs. As a result, when the two wires are looped as shown, a small current

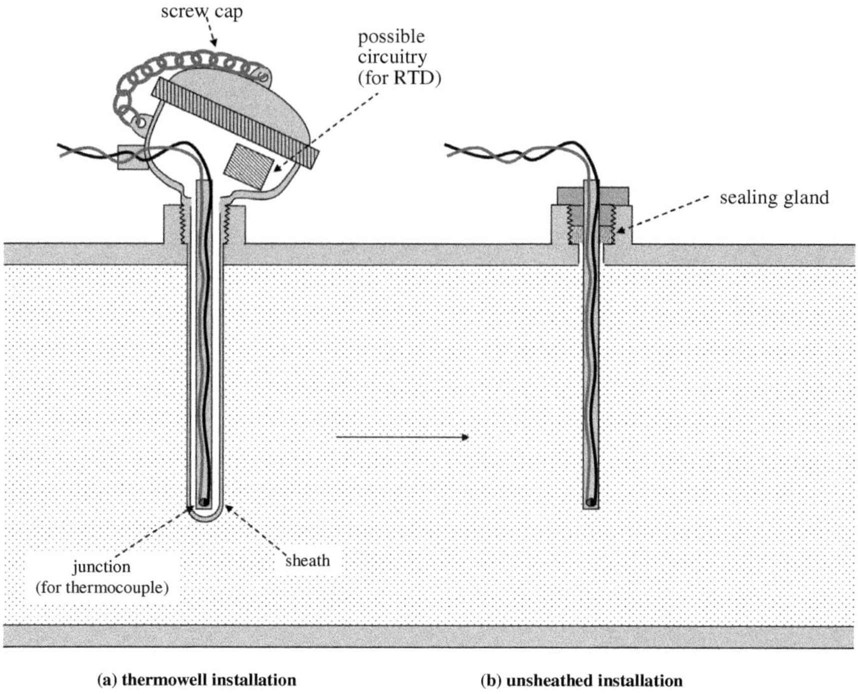

Figure 2.18 Sheathed and unsheathed temperature probes.

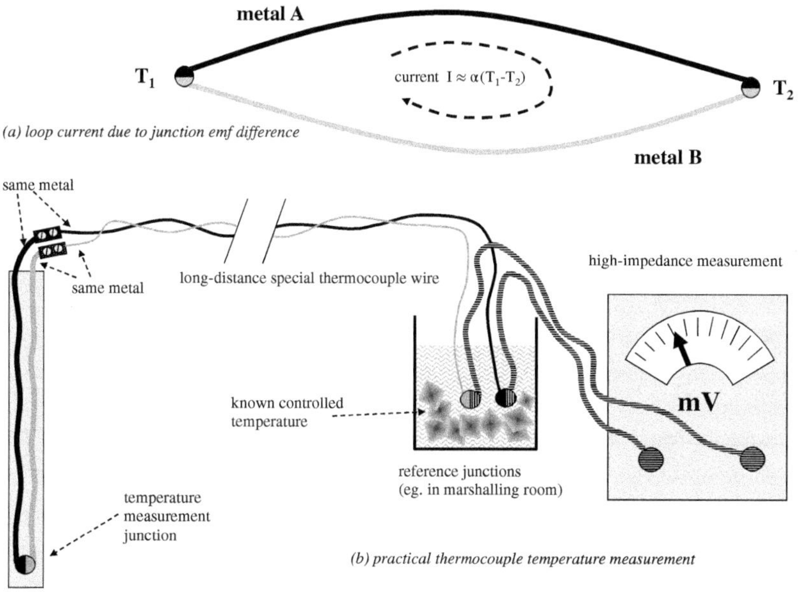

Figure 2.19 Principle of thermocouple temperature measurement.

Table 2.2 Common thermocouple metal combinations.

Type	Metal A	Metal B	Normal range
J	Iron	Constantan®	−190 to 760 °C
T	Copper	Constantan®	−200 to 371 °C
K	Chromel®	Alumel®	−190 to 1260 °C
E	Chromel®	Constantan®	−100 to 1260 °C
S	90% platinum + 10% rhodium	Platinum	0–1482 °C
R	87% platinum + 13% rhodium	Platinum	0–1482 °C

will flow around the loop to balance the net potential. The development of the net potential is known as the Seebeck effect. The emf is almost proportional to the temperature difference. To measure it, the loop is opened and a high-impedance voltage measurement is used (millivoltmeter), to ensure that no further emf changes occur with current flow through the wire resistance. This immediately introduces a problem, because attaching the voltage measurement will create two new dissimilar metal junctions, which will contribute emf depending on their local temperatures. The solution is to make this the reference temperature junction as in Figure 2.19b. A further consideration is that the controlled environment required for the reference temperature will be some distance from the actual plant measurement. Special marked thermocouple cabling of the same metals is used to carry the signal to the reference point, thus avoiding any intermediate dissimilar junctions.

The common industrial metal alloy combinations used for thermocouples are given in Table 2.2. Type J and E thermocouples give the greatest emf change per unit change in temperature, and are thus suited to more sensitive measurements. Typically, type J varies at 0.05 mV $°C^{-1}$ whilst type R varies at 0.006 mV $°C^{-1}$. The variation of emf with the ΔT between the junctions is almost linear for all thermocouples, but not quite. The relationships of ΔT to emf for each type are held in thermocouple tables, which are generally preloaded on plant computer systems, and which are 'looked up' with automatic interpolation as the mV signals come in from the plant. Of course, the system needs to know the reference temperature in order to offset the result.

2.4.4.2 Metal Resistance Temperature Measurement

As the temperature of a metal increases, the vibration of atoms increasingly impedes the movement of electrons through the material, resulting in a nearly linear increase of resistance with temperature. This is the principle employed in a class of resistance temperature detectors. Platinum and sometimes nickel wire are used for this type of temperature measurement. The fractional change of resistance is around 0.004 $°C^{-1}$ for platinum and 0.005 $°C^{-1}$ for nickel, so a resistance bridge with amplifier is used to obtain a usable signal. The resistance is mounted within a protecting tube or sheath, and a small current is used to minimise self-heating under conditions of low dissipation. The device can be set up in a similar way to the thermocouple probe in Figure 2.18, with the additional circuitry for conversion of the resistance to a standard 4–20 mA in a current loop mounted in the cap. For smaller temperature ranges, a linear or quadratic fit is adequate for interpretation of the temperature.

2.4.4.3 Temperature Measurements Using Other Principles

In Figure 2.20, some examples of alternative temperature measurements are given. The thermodial (a), based on a bimetallic spiral in which strips of two dissimilar metals are fastened together, is the

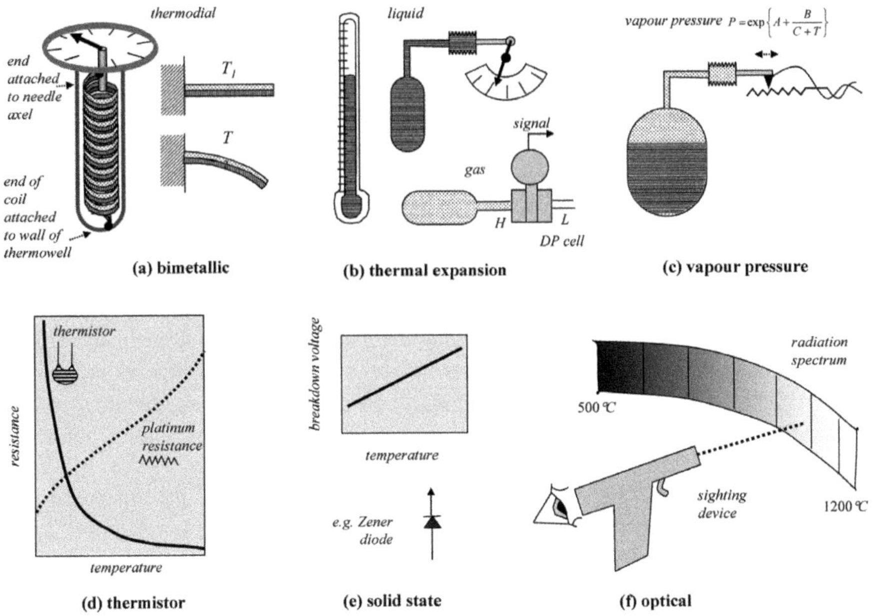

Figure 2.20 Temperature measurement using various principles.

very common local measurement seen on plants. As the temperature rises, the two metals expand differently, imparting a turning moment in the spiral. One end of the spiral is fastened to the thermowell wall, whilst the free end rotates the needle on the dial. Liquid temperature indicators, such as mercury or alcohol thermometers, are well known, and can be arranged to impart mechanical movement for indicator dials, or for electrical signals, as in (b). The pressure exerted by a fixed volume of gas can similarly be interpreted as a temperature, for example by the ideal gas law, whereas that exerted by a vapour in equilibrium with its liquid could be used to infer its temperature from a known relationship such as the Antoine equation (c).

The thermistor (d) is made from semiconductor material which has a low electrical conductivity compared with metals, because electrons are more tightly bound to the molecules. As the temperature increases, the molecules vibrate more, imparting energy to the electrons and thus increasing their ability to escape and move through the material. Thus, unlike metals, the apparent resistance decreases as temperature increases. A table or graph is required to infer the temperature from the resistance. Other solid-state devices such as transistors or Zener diodes (e) can be doped so as to give significantly varying voltage as temperature changes. Finally, one notes that any body radiates at a frequency dependent on its temperature, so that optical techniques such as infrared sensors and optical pyrometers can be used to infer temperature. For example, a pyrometer 'gun' might be used to sight on individual furnace tubes to determine hot spots.

2.4.5 Composition Measurement

Even in the laboratory, the measurement of composition is based on indirect techniques such as flame ionisation, light transmission or polarisation. On a chemical plant, one usually seeks a simpler indirect measure on the composition which is more robust and does not require the controlled environment and sophisticated calibration inherent in accurate laboratory measurements. Thus, gas density might be used to infer the average molecule size in a hydrocarbon vapour, or liquid density used to determine dissolved sugar (brix), alcohol or HCl. Other indirect techniques might use pH or conductivity. In distillation, where mixtures are binary, or at least have characteristic distributions, a temperature measurement might be used, possibly with a correction for pressure (e.g. Figure 4.23). Other 'composition'-type measurements required include Kappa value, turbidity, viscosity, knock, octane, flue O_2, and heating value. Manufacturers have developed a range of ingenious devices to provide online indications of most of these properties. Only occasionally is an online gas or liquid chromatograph installed. Such an installation tends to be very expensive, and of course can only provide an intermittent measurement update following the injection and purge cycles. A remarkable achievement of the old analogue period was Foxboro's intrinsically safe, plant-mounted process gas chromatograph, powered and programmed pneumatically, with a steam-heated oven, and with eluted gases identified by pressure-drop variations across an orifice (Annino et al., 1976).

Figure 2.21 shows the H^+ sensing electrode and reference electrode for pH measurement. These are usually housed within one probe, which also has a sensor for temperature compensation. Note that the nonlinear relationship of $pH = -\log_{10}([H^+])$ to the molar concentration of H^+ ions [gmol L^{-1}] is a recognised nonlinear control problem.

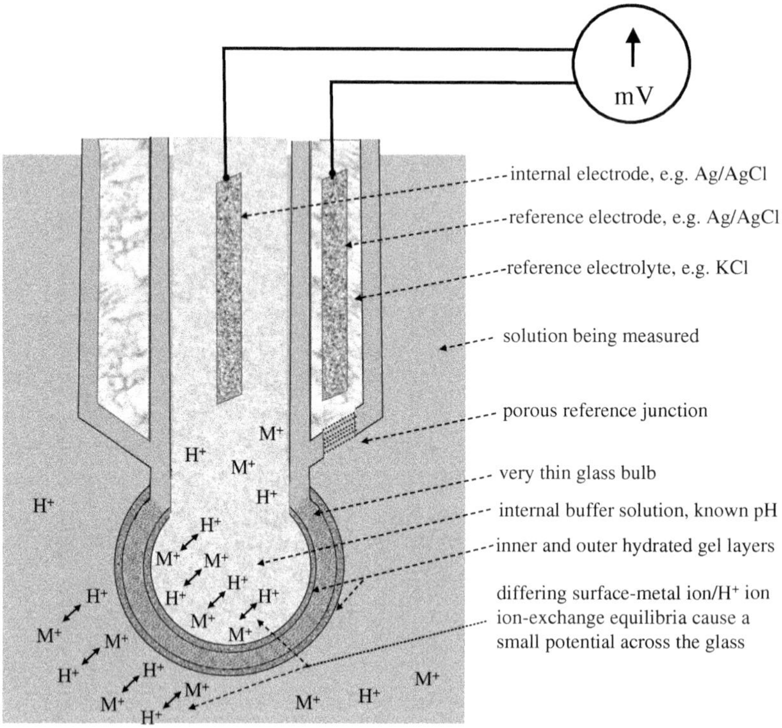

Figure 2.21 Composite pH probe with both electrodes in one housing.

2.5 Current-to-Pneumatic Transducer

Current-to-pneumatic (I/P) converters are very common industrially. Though some control actuators are electrically driven, most still rely on transmitted instrument air for their motive power. Apart from anything else, air signals are intrinsically safe in the presence of flammable materials. At some stage in the transmission of an output signal from the controller, a conversion must be made to a pneumatic range (3–15 psig or 20–100 kPag). Typically, this conversion is from current to pneumatic, which is illustrated in Figure 2.22.

The similarity to Figure 2.8 for the pneumatic DP cell is obvious. Again, a relay is used to equate a high-flow output signal to the more sensitive nozzle pressure. Typically, one does not wish to upset the pressure balance around the nozzle whilst a control-valve diaphragm is being inflated.

2.6 Final Control Elements (Actuators)

Most of the final control elements in the processing industries are valves used to regulate the motion of fluids. The focus in this section is on remotely operated valves, but at the outset one notes that there will be many manually operated valves on a plant for less frequent use, as in the common 'double-block and bypass' arrangement in Figure 2.23. Control valves sometimes need maintenance, so that they are often located in accessible spots near floor level, to allow disconnection and removal. A means must be provided to prevent residual process fluids from escaping through the vacancy – that is the purpose of manual block valves. Quite often, with communication to an operator at the location, the process can continue to function temporarily with occasional manual adjustments of the bypass valve. A quite undesirable practice is sometimes seen where debottlenecking of a process

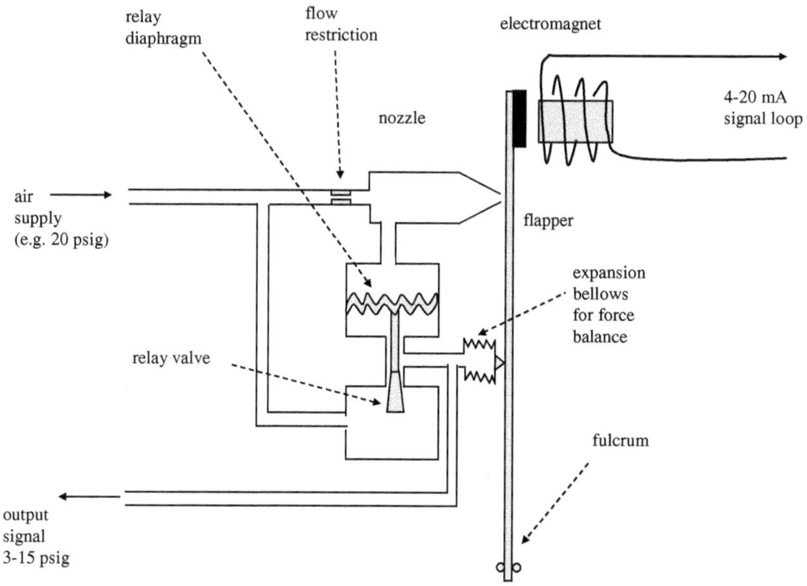

Figure 2.22 Current-to-pneumatic converter.

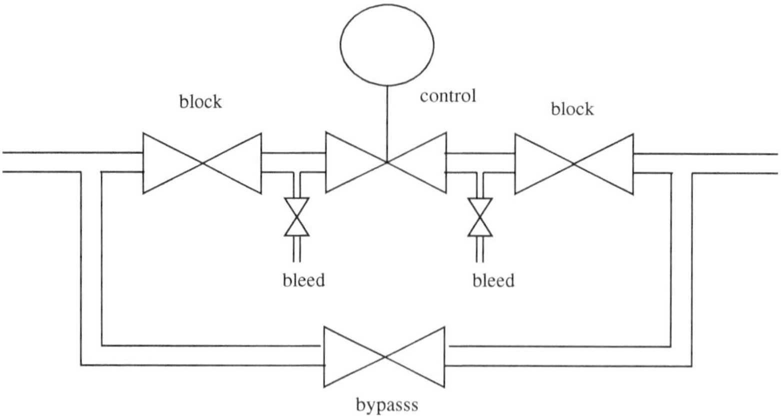

Figure 2.23 Double block and bypass.

has led to bypass valves being left partially open in normal operation, just to increase throughput. In addition to the three manual valves involved, two manual bleed cocks will sometimes be provided in liquid systems not only to empty the closed-off sections in a controlled way, but also to avoid trapping of liquid which on thermal expansion might rupture the pipework.

2.6.1 Valves

2.6.1.1 Pneumatically Operated Globe Control Valve

The globe control valve (Figure 2.24) is the most common device used for precise flow regulation, and the simple and safe source of motive power for this valve is usually the instrument air system.

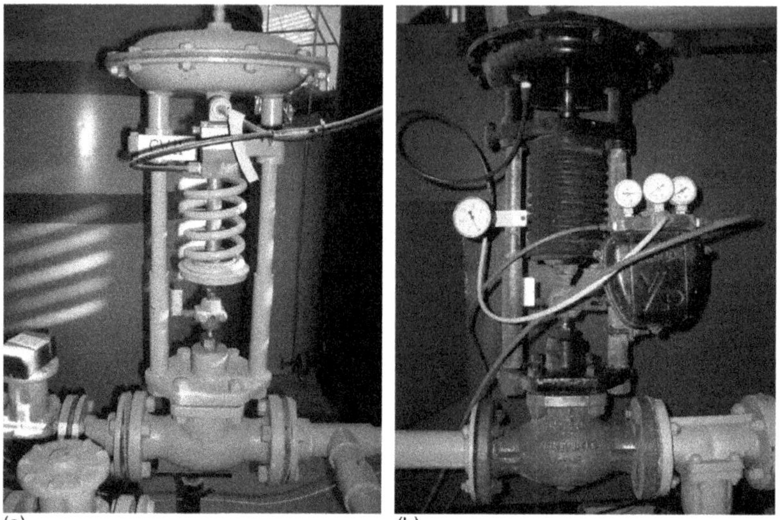

(a) (b)

Figure 2.24 Air-to-open globe control valves without (a) and with (b) a positioner.

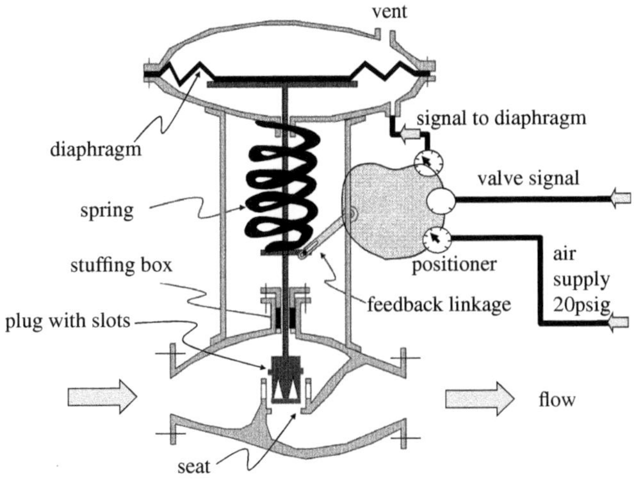

Figure 2.25 Air-to-open globe control valve with positioner.

Figure 2.25 illustrates the workings of an air-to-open globe control valve with positioner. Unlike domestic taps, note that the flow normally approaches from the stem side of the valve plug. An adjustable stuffing box with a gland prevents leakage to the outside.

Most control valves can be switched between *air-to-open* and *air-to-close* service by simple mechanical adjustments as indicated in Figure 2.26. Bearing in mind that the most likely disaster

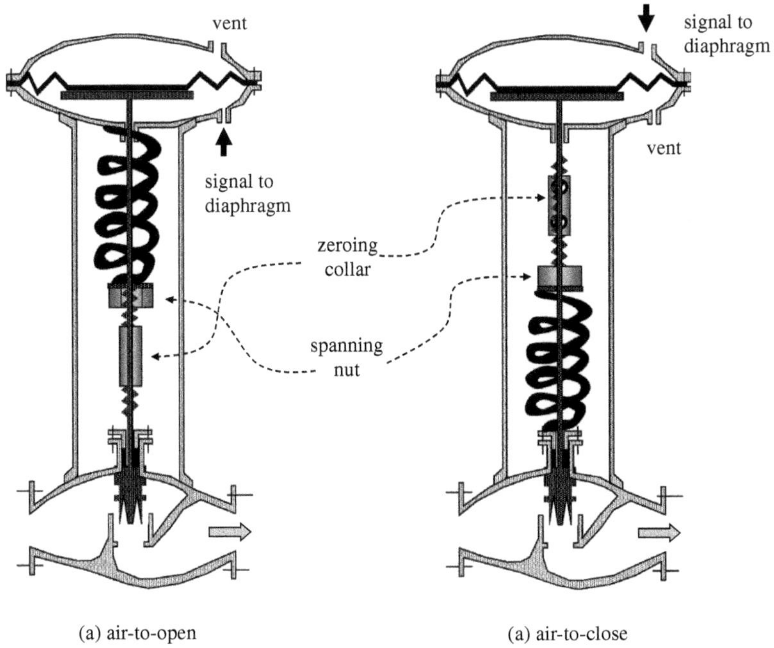

Figure 2.26 Valve air-to-open and air-to-close arrangements with adjustments.

would be loss of instrument air pressure, or breakage of the pneumatic signal line, the *air-to-open* (*AO*) is taken as *fail-closed* (*FC*) and the *air-to-close* (*AC*) is taken as *fail-open* (*FO*). In control scheme design, it is important to make this specification to ensure as orderly a shutdown as possible in the event of disaster, for example furnace fuel supplies will typically be fail closed. For motor-operated valves, or electrical solenoid valves, the fail situation would correspond to loss of the electrical power or signal.

In Figure 2.26, a typical range adjustment provision is shown. A collar allows variation of the stem length, whilst one or two nuts may vary the position and tension on the return spring. It is important to check these adjustments, so that the actual motion of the stem between 0% open and 100% open corresponds as well as possible to the instrument signal range, say 3 and 15 psig (or 15 and 3 psig for an *air-to-close* valve). A point of caution is that valve signals are occasionally reported as '% closed', so it is wise to qualify the units of measurement as either '% open' or '% closed' when valve position is being discussed.

In Figure 2.27, the arrangement inside a globe control valve with cage and plug is illustrated. The arriving fluid passes through the holes in the cage, around its full circumference. The plug passes up and down within the cage, with the various holes cut into the plug's cylindrical surface overlapping in a predetermined way with the holes in the cage. The fluid passes through this overlap into the hollow interior of the plug, and thus moves downward and out of the valve. It is the way in which the overlap area varies with stem movement that determines the *characteristic* of the valve, which will be discussed shortly. Whereas the perforated part of the plug passes through a

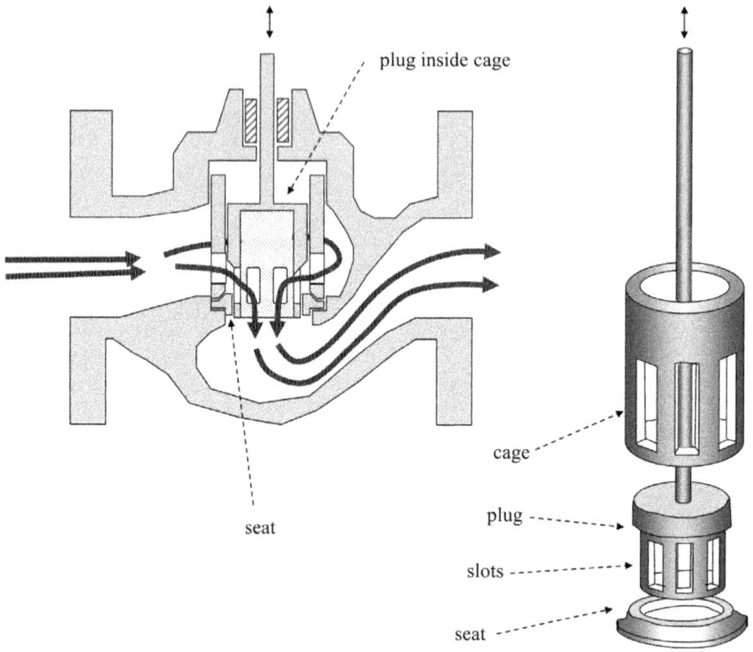

Figure 2.27 Globe valve showing flow through cage and plug.

hole in the seat, a protruding edge higher up on the plug provides the means for a tight seal against the seat when the valve is closed.

2.6.1.2 Valve Characteristics

Any valve, whether manual, automatic, globe or butterfly, has a characteristic. This is a plot of the fraction of the maximum flow achieved whilst the valve 'position' is varied from 0 to 100% open, *with a constant pressure difference across the valve.* The position will be based on either linear stem movement or rotation.

For most process control applications, the linear characteristic would appear to be more suitable, whilst a quick-opening valve might be useful in pressure relief. This characteristic determined under constant valve pressure drop, ΔP_V, is called the *inherent characteristic.* In practice, the actual flow achieved is also dependent on other frictional pressure losses in the remainder of the line flow route including equipment, ΔP_L. These will increase as the flow rate increases, and if the overall driving pressure difference $\Delta P_T = \Delta P_V + \Delta P_L$ is fixed, it must mean that the ΔP_V available across the valve decreases. The effect is to distort the operating characteristic as in Figure 2.28, which is called the *installed characteristic* (Figure 2.29).

When the valve is shut, the maximum possible pressure is exerted across the valve, and this reduces as it opens. For lines with large resistances, such as a long hosepipe, this is quite noticeable, as the water first rushes out, and then drops to a lower rate as the available ΔP_T becomes distributed elsewhere. Depending on the resistance of the remaining equipment in the line, and the ΔP_T available, notice that it should be possible to choose an equal-percentage valve (e.g. $\alpha = 30$ in Figure 2.28) which restores the $f(x)$ behaviour to approximate linearity. Such a valve would of course be designed to reveal flow area at a greater rate *per % opening* as it opens further, for example with plug slots broadening towards the bottom.

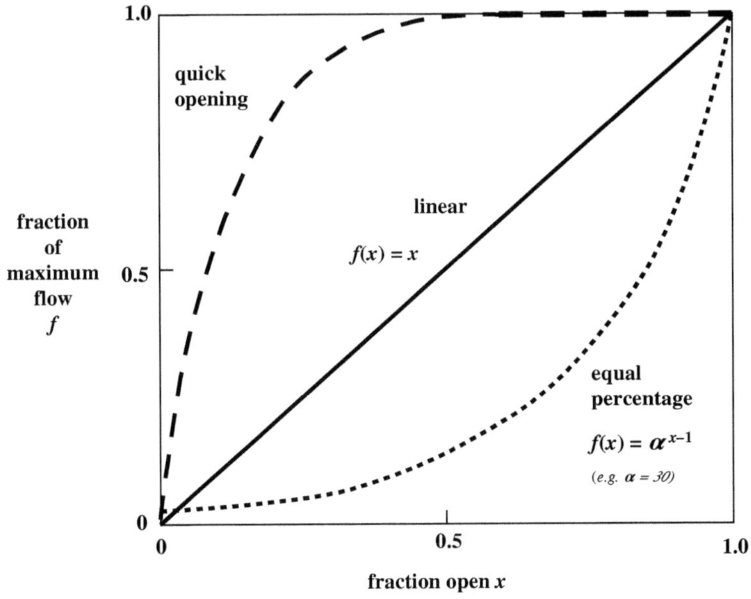

Figure 2.28 Valve inherent characteristics.

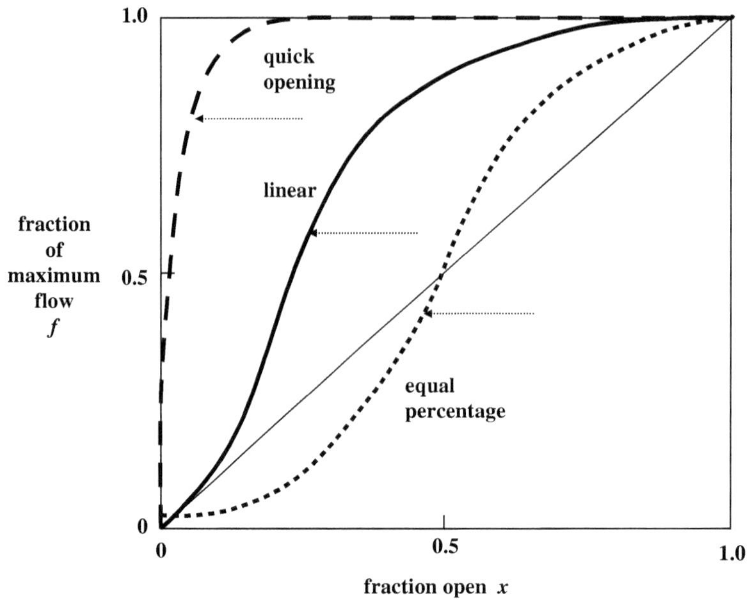

Figure 2.29 Valve installed characteristics.

2.6.1.3 Valve C_V and K_V

Incompressible Fluids

For incompressible fluids, the volumetric and mass flows through a valve are determined by Equations 2.17 and 2.18, which follow from the orifice equation (Equation 2.11), with the characteristic scaling $f(x)$ for valve opening.

$$F = kf(x)\sqrt{\frac{\Delta p_V}{\rho}} = C_V f(x)\sqrt{\frac{\Delta p_V}{SG}} \quad \text{(volumetric flow)} \tag{2.17}$$

$$w = kf(x)\sqrt{\rho \Delta p_V} \quad \text{(mass flow)} \tag{2.18}$$

The coefficient C_V is usually quoted as a numerical value without units, but its units are implied by the following definition:

> C_V is the number of US gallons per minute of water at 60 °F that flow through a fully open valve with a pressure drop of 1 psi across the valve.

This follows because water has a specific gravity of 1 at 60 °F. The effective units are

$$C_V \text{ (US gallon min}^{-1}\text{ psi}^{-1/2}\text{)}$$

and for the more general form involving ρ, any units may be used based on

$$\begin{aligned} k &= \rho_{W,60\,°F}^{1/2} C_V \\ &= (8.33\,\text{lb US gallon}^{-1})^{1/2} \times C_V \end{aligned} \tag{2.19}$$

There is an equivalent 'flow factor' K_V based on SI units, either in $m^3 h^{-1}$ or $l \, min^{-1}$:

> K_V is the number of cubic metres per hour of water between 5 and 40 °C that flow through a fully open valve with a pressure drop of 1 bar across the valve.

So

$$\begin{aligned} K_V &= 0.003785 \, m^3 \text{ US gallon}^{-1} \times 60 \text{ min h}^{-1} \times (14.503 \text{ psi bar}^{-1})^{1/2} \times C_V \\ &= 0.865 C_V \end{aligned} \quad (2.20)$$

Manufacturers give their valve sizes in terms of C_V of K_V.

Compressible Fluids

From Equations 2.9 and 2.10 for an orifice of area A_O in a duct of area A_1, one has

$$w = Y C_d A_O A_1 \left(\frac{2}{A_1^2 - A_O^2} \right)^{1/2} \sqrt{\rho_1 (p_1 - p_2)} \quad (2.21)$$

The passage of fluid through the valve restriction behaves in the same way, except that only a fraction $f(x)$ of the fully open area A_O is available. Manufacturers (e.g. Fisher Controls International LLC, 2005; Smith and Corripio, 1997) have developed correlations and charts for two corrections, namely a 'piping geometry factor' F_P and the compressibility factor ('expansion factor') Y. These are applied in the form

$$w = Y F_P C_V f(x) \sqrt{\rho_1 (p_1 - p_2)} \quad (2.22)$$

The expansion factor Y depends on $\gamma = c_p/c_v$, p_2/p_1 and $(p_2/p_1)_{CRIT}$, the critical absolute pressure ratio above which flow becomes independent of p_2 (0.53 for air).

2.6.1.4 Specification of Valves for Installed Performance

The dependence of valve performance on other line resistances has been mentioned in Section 2.6.1.2. The flow resistance imposed by a valve has to be 'significant' compared with the line resistance, for the whole of the range 0–100% open, otherwise there may be no flow variation as the valve position changes.

In the example shown in Figure 2.30, one notes that a fixed total pressure difference is available to drive the process liquid through line section AB:

$$\Delta p_T = (p_1 - p_2) + \rho g h \quad (2.23)$$

This will be absorbed in the line pressure loss Δp_L, dependent on the geometry of the piping and process equipment, and the flow rate, as well as the valve pressure drop, Δp_V, dependent on the valve C_V, inherent characteristic $f(x)$ and flow rate (Figure 2.31).

$$\Delta p_T = \Delta p_L + \Delta p_V \quad (2.24)$$

The discussion in Section 2.4.1.1 led to the approximation of frictional pressure losses as a number of kinetic 'velocity heads', $\rho v^2/2$, which we can re-express as some multiple of density times the square of the total volumetric flow rate.

$$\Delta p_L = k_L \rho F^2 \quad (2.25)$$

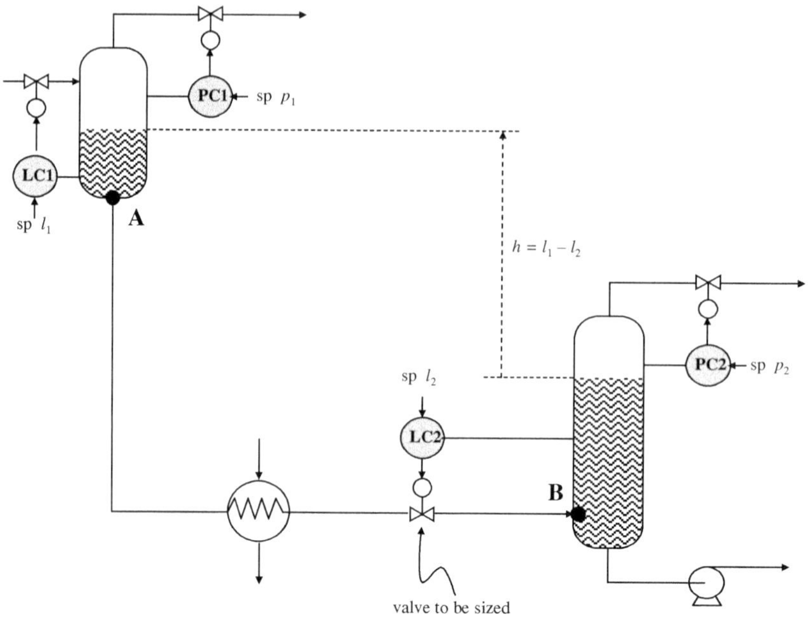

Figure 2.30 Valve to be specified for LC2 on line section AB.

Though adequate, this is not strictly true, as seen in the Darcy equation (Equation 2.4), where the friction factor f would have a secondary dependence on F. For liquid passing through a control valve, Equation 2.17 gives

$$\Delta p_{\mathrm{V}} = \frac{k_{\mathrm{V}} \rho F^2}{f^2(x)} \tag{2.26}$$

where $k_{\mathrm{V}} = 1/k^2$ for the k in Equation 2.17, or 0.12 US gallon $\mathrm{lb}^{-1}/C_{\mathrm{V}}^2$ according to Equation 2.19. Then Equation 2.24 can be rewritten as

$$\Delta p_{\mathrm{T}} = \rho F^2 \left(k_{\mathrm{L}} + \frac{k_{\mathrm{V}}}{f^2(x)} \right) \tag{2.27}$$

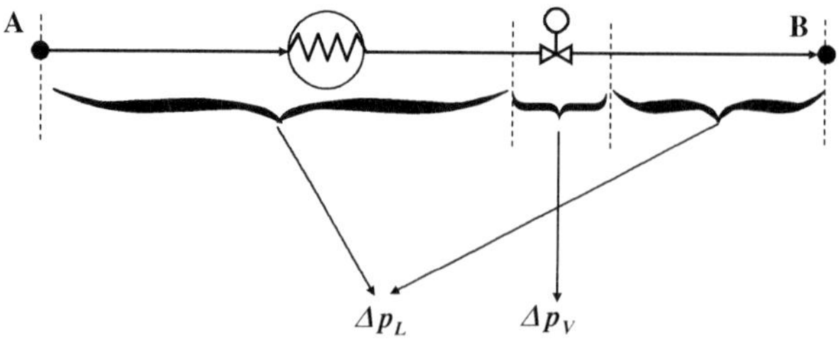

Figure 2.31 Pressure loss contributions for line and valve.

Table 2.3 Examples of valve sizing rules of thumb.

Δp_V to be 20–50% of $(\Delta p_V + \Delta p_L)$ for $x = 1$
Δp_V to be 25% of $(\Delta p_V + \Delta p_L)$ or 10 psi, whichever is greater, for $x = 1$
Δp_V to be 50% of $(\Delta p_V + \Delta p_L)$ for $x = 0.5$

The situation $k_L \gg k_V$ could arise where too large a valve C_V has been specified, bearing in mind that $k_V \propto 1/C_V^2$. Clearly there would only be a significant change of flow as the valve starts to open (low x giving low $f(x)$). On the other hand, too large a k_V (small C_V) would impede flow unnecessarily, giving high energy costs. Various 'rules of thumb' have been proposed to preserve a balance between these two resistances, as in Table 2.3.

2.6.1.5 Control Valve Hysteresis

Generally, when a control system commands a valve to go to a particular position, there is no feedback evidence of what position is actually achieved, that is the valve operates in *open loop*. Because of their mechanical linkages and friction, valves are relatively unreliable in this respect. Note at the outset that only the signal sent to the valve is known. In practice, if one varies this signal from 0 to 100% open, and then back to 0% open, the true valve stem position follows as in Figure 2.32.

The hysteresis behaviour of control valves will arise to some extent from play in the linkages, but one expects that it is largely a result of 'slip-stick' friction in the valve stem movement through a tight stuffing box. In Figure 2.32, when the upstroke starts, the signal to the valve increases to 20% open before any stem movement occurs. Effectively, the upward force applied to the stem has

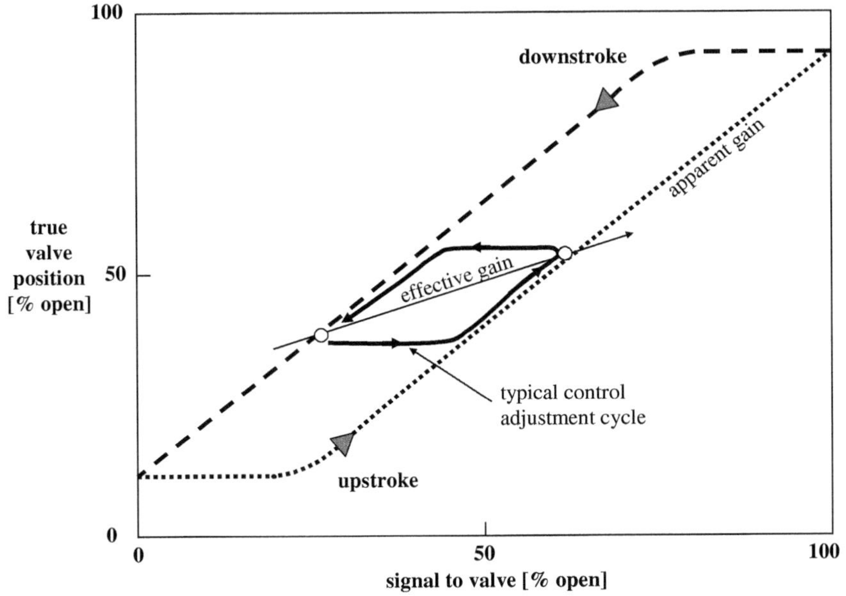

Figure 2.32 Control valve hysteresis.

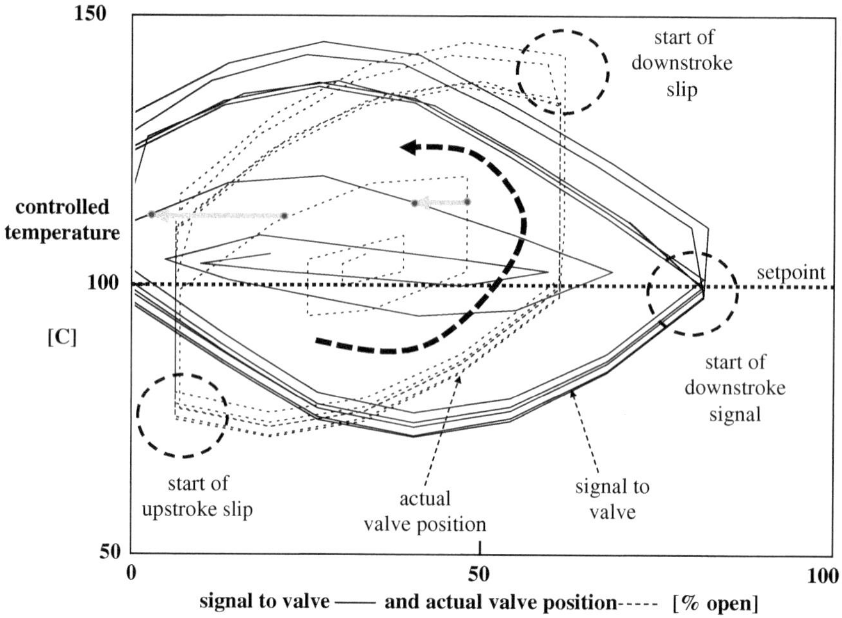

Figure 2.33 Typical phase-plane plot revealing valve hysteresis.

increased to 20% of range, before the static friction is overcome and the stem 'gives'. This excess force has to stay in place for the entire upstroke travel, allowing the stem to move with each incremental signal increase thereafter. As might be expected, the true valve position might never reach 100% open. When the valve signal is reversed, to go downwards from 100% open, the excess force must first be removed, and a similar excess force applied in the reverse direction before the 'stick' friction is overcome, and the stem begins to 'slip' in the reverse direction.

The misleading aspect of this phenomenon is that one usually obtains system 'gains', for example flow per % open, using big valve adjustments. The actual gain experienced by a controller will depend on the size of its adjustment cycles, and these will usually be smaller as in the Figure 2.32 example, giving much smaller effective gains, and often no valve movement at all! The phenomenon is variable and nonlinear, and bound to cause problems in control loops which rely on reasonably linear behaviour. de Vaal, Eggberry and Jones (2006) used phase-plane plots to detect hysteresis in plant control (Figure 2.33).

Industrially, the problem of slip-stick friction in valves is reduced using a valve positioner, as in Figures 2.24 and 2.25. Here the signal transmitted to the valve becomes a position setpoint, and a mechanical controller mounted on the valve compares this value with the actual stem position, obtained using a mechanical link. The local control loop has access to its own air supply, and it applies more or less of this air to the valve diaphragm to bring the feedback position to the setpoint position.

2.6.1.6 Various Flow Control Devices

A great variety of flow control devices exists, suited to different applications. Some are discrete in the sense that they only use two positions, some are designed for accurate regulation and some are

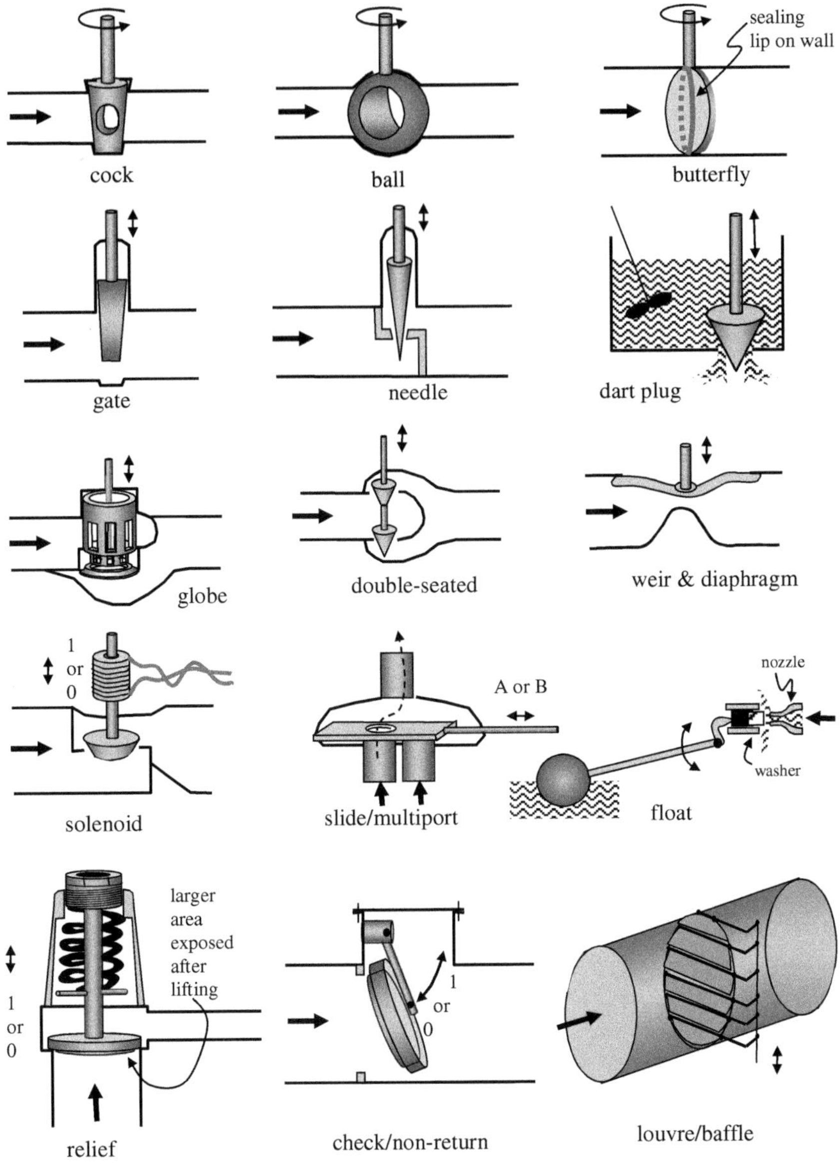

Figure 2.34 Various flow control devices.

designed to handle fluids containing solid particles. In Figure 2.34, the weir and diaphragm valve can be used for particulate-bearing flows, because it is able to flush itself clear. Likewise, the dart plug valve is typically used to control the flow of slurries through flotation cells. Gate or ball valves can be arranged to leave the entire pipe flow area clear, and are thus suited to special cleaning requirements such as the passage of a 'pig'. Louvres and baffles would be suited to low-pressure systems such as furnace ducting.

2.6.2
Some Other Types of Control Actuators

Having considered the major process control requirement of actuators for flow regulation, it is worthwhile before proceeding to mention a few other actuator devices that could play a part in control schemes. Fans, pumps, compressors, electrical heaters, solenoid-driven actuators and electrical shut-off valves normally only have two states – on and off. There are, however, a few electrically driven devices offering scope for intermediate settings. These include thyristor power supplies, variable-speed drives and stepper motors. Hydraulic actuators are not that common in the processing industries. Though mostly discrete in nature, hydraulic devices also exist for intermediate positioning.

2.7
Controllers

The centrifugal governor, developed by James Watt and Matthew Boulton in 1788, was one of the earliest control devices (Figure 2.35). The same principle is still in use for setting the speed of steam-powered turbine drives.

Until the 1980s, panel-mounted analogue controllers similar to Figure 2.36 were common. Sometimes these were integrated within a circular or strip chart recorder. This is the SISO controller which could service a single control loop. With the conversion to computer-based systems, the same concepts and terminology have been retained. The same functions are carried out by programs which run on PLCs or distributed computer cards, and these can continue to function independently if part of the system is damaged. When the operator views a controller on the control

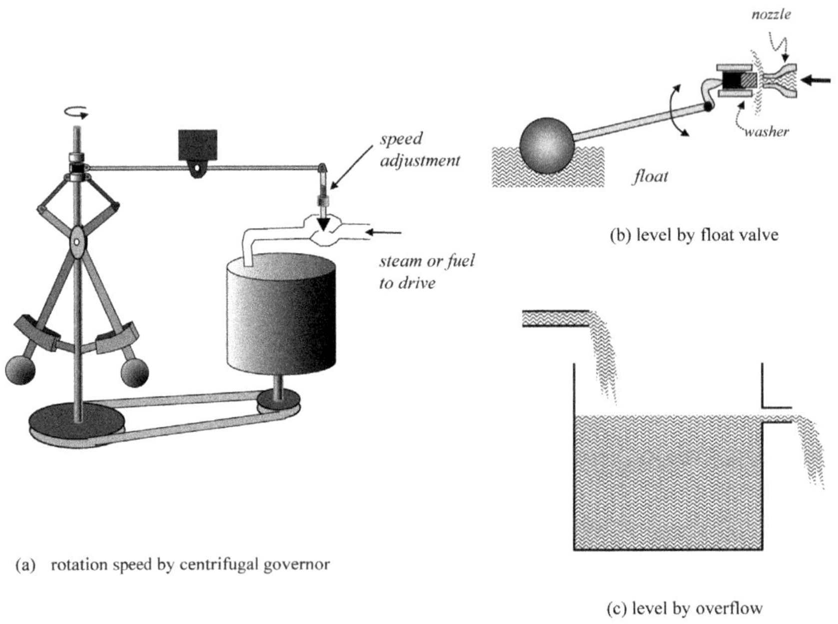

Figure 2.35 Early control devices.

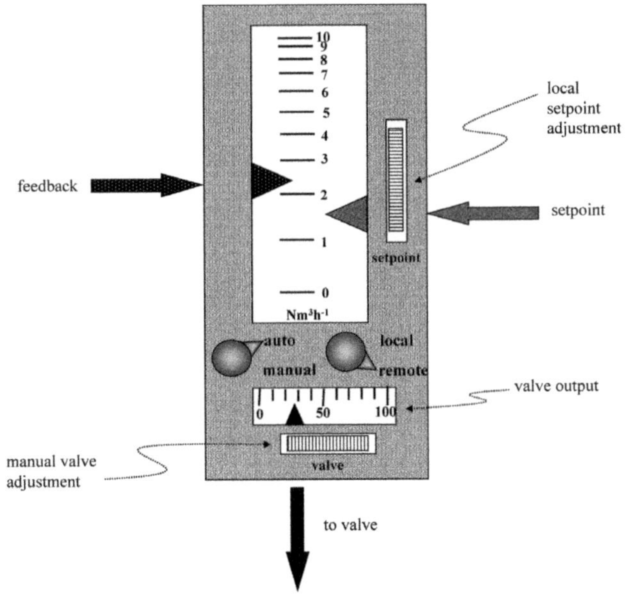

Figure 2.36 Typical panel-mounted analogue controller.

room displays of a DCS or SCADA system, it is represented in analogous format. On the conversion of plants to the much cheaper computer-based systems, the control loops were replaced on a one-for-one basis. The many tasks requiring this type of control are referred to as the 'base layer' control system, and it will be seen later that the modern *advanced* control algorithms facilitated by computers tend to communicate with the plant through this layer.

The loop controller has a switch to select either AUTO or MANUAL operation. In the MANUAL mode, it is possible to adjust the control action output directly, using the 'manual loading station' at the bottom of the controller in Figure 2.36. In AUTO mode, the device makes use of its internal computation to set a suitable control action output which will bring the feedback signal to the setpoint value.

The other switch on the controller allows selection of LOCAL or REMOTE. In LOCAL mode, the setpoint can be adjusted directly by the operator. In REMOTE mode, this is no longer possible, because the setpoint is manipulated from elsewhere. This could be in a cascade format from a similar SISO controller, or possibly from an 'advanced' control algorithm based on several PVs and manipulating several MVs.

The original analogue controllers were used in one of the three control modes: proportional (P), proportional–integral (PI) and proportional–integral–derivative (PID). Rather more features are possible today in the computer equivalent controller. For the analogue devices, one accessed the back of the cabinet to adjust the proportional gain, integral time constant or derivative time constant. These can of course be set remotely in the computer devices.

For computer-based controllers, one usually works in engineering units (e.g. % open, °C, tph, barg), a change from the fractions of standard instrument ranges of the past. The important thing to note with the SISO controller is that it has three different signals. Two of these signals refer to the same variable and will have the same units of measurement, namely the controlled variable feedback and its setpoint. The third signal leg that must attach to the controller bubble on a P&I diagram is the control action, which refers to a different variable and is nearly always in different units. In this sense, the controller can be viewed as a 'translator', for example 'At what [% open] must this valve be set to bring the temperature to 30 [°C]?' So the algorithm inside the device must iteratively solve an implicit problem, and translate from the 'language' of the input to the 'language' of the output.

2.8 Relays, Trips and Interlocks

The on/off signals emanating from a control system cannot in general power plant equipment. These are passed to relays designed to close or open circuits carrying the required power. Such a relay typically uses an electromagnet to open and/or close contacts. Such relays were also found useful for 'latching' circuits as in the electrical motor start (NO - normally open) and stop (NC - normally closed) button circuit shown in Figure 2.37. The functioning of this latching circuit, involving several digital 0/1 states, can be described in Boolean algebra as a series of interdependent events X and non-events $\overline{X}$:

G *GO* pushbutton pressed and released

S *STOP* pushbutton pressed and released

C coil energised closing both relay switches

M motor receives power

$$C = G \cdot \overline{S} + C \cdot \overline{S} = (G + C) \cdot \overline{S} \tag{2.28}$$

$$M = C \tag{2.29}$$

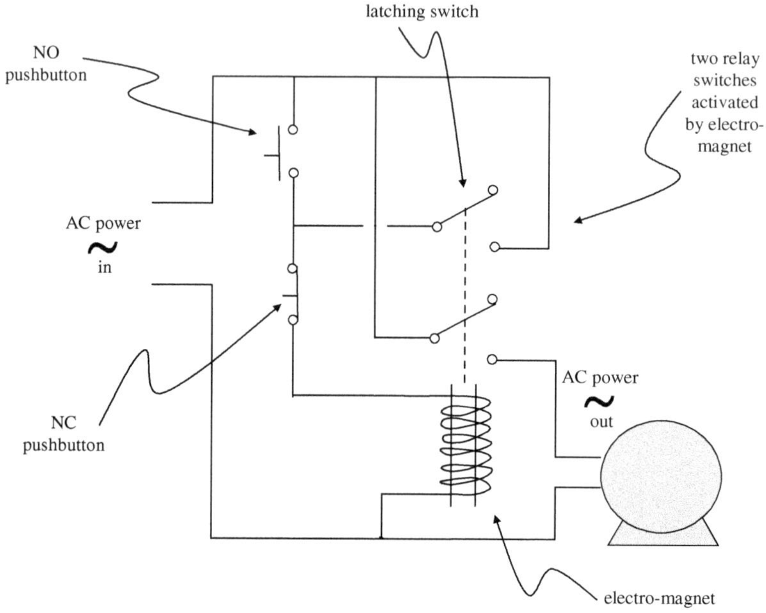

Figure 2.37 Latching circuit for motor power.

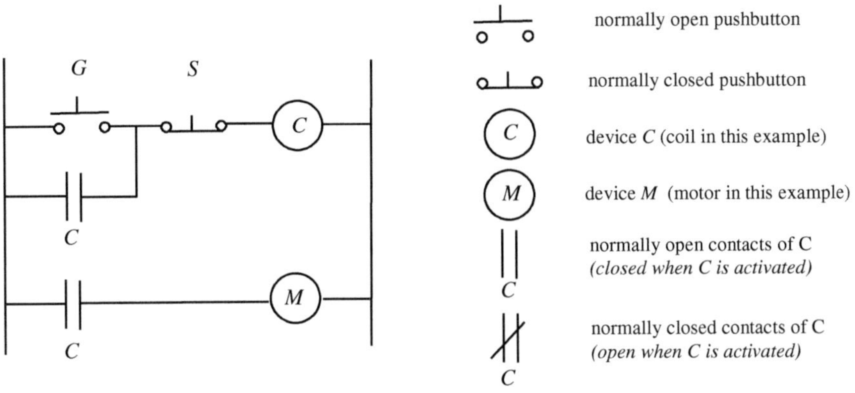

Figure 2.38 Ladder diagram for discrete logic of circuit in Figure 2.37.

Although the emphasis of this text is on the continuous algorithms which smoothly vary plant parameters, one has to be aware of the network of discrete overrides, interlocks and trips which become superimposed on such algorithms in practice. These generally aim to avoid dangerous or destructive situations from developing; for example, *if the tank level exceeds its high limit, switch off the pump*. In the past, these protective systems made use of electromagnetic relays as in Figure 2.37. Nowadays, the discrete decision system is programmed directly on the plant computers. Although Boolean or 'expert system' forms may be used, *ladder diagrams* based on the original alternating current relay systems are still widely used to represent the sequence of events and the hardware involved. The equivalent of the circuit in Figure 2.37 then becomes the simple ladder in Figure 2.38. For larger systems, many more rungs can be added to the ladder, forming a clear record of the interrelationship of events. Additional symbols exist for normally open and normally closed limit switches for pressure, temperature, level and position, as well as for timing delays. One way used to 'trip' an air-to-open control valve into its shut position is to vent the air signal at the valve by energising a solenoid valve.

2.9 Instrument Reliability

An important step in the design of new plants, or plant modifications, is the *operability study* as described by Lawley (1974). This aims to detect problems that could arise from *unusual* plant operation, for example higher flows or temperatures than expected. These may be resolved by plant equipment changes, but very often lead to the addition of further instrumentation, as well as control and interlocking systems for monitoring and protection. It is not surprising then that when the *hazard analysis* is performed, the degree of risk becomes highly dependent on instrument reliability. It is useful then to consider some typical failure rates for standard instrumentation (Tables 2.4 and 2.5). Such data might more realistically be estimated simply from experience on the processing plant of interest. Moreover, one notes that most 'controllers' are nowadays computer-based, giving failure rates much lower than 0.29 pa on typical high-integrity systems. The intention now is to use these standard *component* failure rates to estimate the overall probability of failure of a system constructed from these components.

Table 2.4 Typical failure rates for processing plant instruments from Lees (2001).

Instrument	Failure (faults per year)
Control valve	0.60
Pressure measurement	1.41
Flow measurement (fluids)	1.14
Level measurement (liquids)	1.70
Thermocouple temperature measurement	0.52
Controller (analogue)	0.29
Pressure switch	0.34
pH meter	5.88
Gas–liquid chromatograph	30.6
Impulse lines	0.77

Table 2.5 Causes of analogue control loop failure from Lees (2001).

Element in loop	Faults (%)
Sensing/sampling	21
Transmitter	20
Transmission	10
Receiver (e.g. indicators and recorders)	18
Controller (analogue)	7
Control valve	7
Others	17

Figure 2.39 represents a fired boiler control scheme in which two protective trip systems activate as excessive pressure levels are reached. When the high limit P_A is reached, fuel to the furnace is cut off, and when the next high limit P_B ($P_B > P_A$) is reached, steam is vented directly from the drum. What one is concerned about in a hazard analysis is the possibility of the pressure rise occurring and going completely unchecked, based on a few possible failures, which might be defined as follows:

V1	fuel valve (open)
FT1	fuel flow measurement (too low)
FC1	analogue fuel flow controller (valve output too high)
PC2	analogue pressure controller (output sp to FC1 too high)
PT2	pressure measurement (too low)
PAH2A	first pressure trip (does not activate)
PAH2B	second pressure trip (does not activate)
Fhigh	fuel flow higher than setpoint
Fsphigh	fuel flow setpoint too high
Phigh	occurrence of excessive pressure in drum
Pexplode	final high-pressure failure

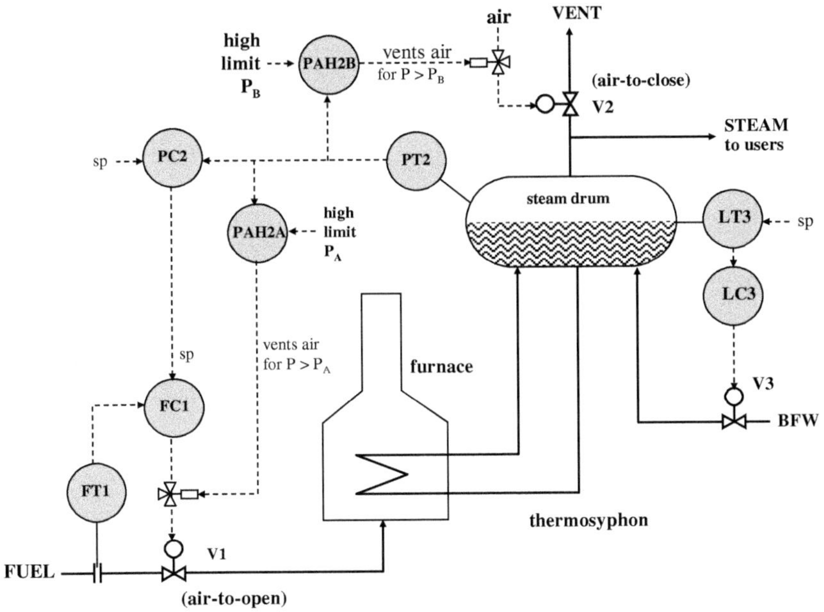

Figure 2.39 Boiler pressure control with two trip levels.

The various events can be related by Boolean logic as follows:

$$\text{Fhigh} = \text{V1} + \text{FT1} + \text{FC1} \tag{2.30}$$

$$\text{Fsphigh} = \text{PT2} + \text{PC2} \tag{2.31}$$

$$\text{Phigh} = \text{Fhigh} + \text{Fsphigh} = \text{V1} + \text{FT1} + \text{FC1} + \text{PT2} + \text{PC} \tag{2.32}$$

$$\text{Pexplode} = \text{Phigh} \cdot \text{PAH2A} \cdot \text{PAH2B} \tag{2.33}$$

$$\text{Pexplode} = (\text{V1} + \text{FT1} + \text{FC1} + \text{PT2} + \text{PC2}) \cdot \text{PAH2A} \cdot \text{PAH2B} \tag{2.34}$$

A fault tree diagram represents this sequence as in Figure 2.40. A few useful results when evaluating fault trees include

$$\overline{1} = 0 \quad \overline{0} = 1 \quad \text{A} \cdot \overline{A} = 0 \quad \text{A} + \overline{A} = 1 \tag{2.35–2.38}$$

$$\text{A} + \text{A} = \text{A} \quad \text{A} \cdot \text{A} = \text{A} \quad 1 + \text{A} = 1 \quad 1 \cdot \text{A} = \text{A} \tag{2.39–2.42}$$

$$\text{A} \cdot (\text{B} + \text{C}) = \text{A} \cdot \text{B} + \text{A} \cdot \text{C} \tag{2.43}$$

$$(\text{A} + \text{B}) \cdot (\text{A} + \text{C}) = \text{A} + (\text{B} \cdot \text{C}) \tag{2.44}$$

$$\text{A} + \overline{A} \cdot \text{B} = \text{A} + \text{B} \tag{2.45}$$

De Morgan's theorem:

$$\overline{\text{A} + \text{B}} = \overline{A} \cdot \overline{B} \tag{2.46}$$

$$\overline{\text{AB}} = \overline{A} + \overline{B} \tag{2.47}$$

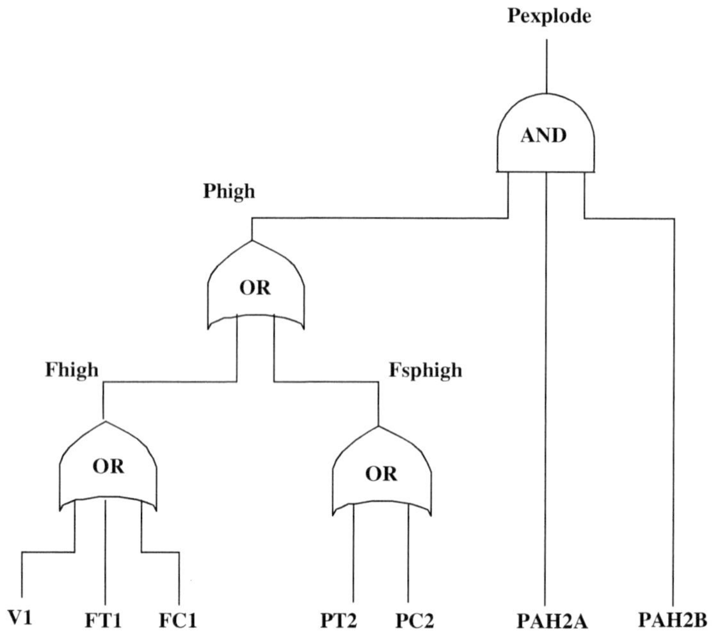

Figure 2.40 Fault tree for boiler pressure control with two trip levels.

Quite often evaluation of the Boolean expression for the main event reveals a simple underlying cause, for example a fire which both overheats a vessel and destroys shutdown trip cabling. It turns out that the same expressions can be used to determine the *probability* of an event occurring. Here one has to think of 'A' as the probability that this state will exist at any instant in time.

The available data such as in Table 2.4 are unlikely to have such detail as the *direction* of failure (shown in parentheses above). However, if one uses the frequency as stated, one will certainly arrive at a probability representing the worst-case scenario. Take for example the equivalent probability for V1. According to Table 2.4, a control valve can be expected to fail 0.6 times per year. One needs to think of an average time for detection and rectification of such a situation – say 6 h. So in 1 year, a single valve will spend 3.6 h in its failure state, representing a time fraction of 0.0004. One notes that the probability of this event *not* occurring $\overline{V1} = 0.9996$, and that the probability that *both* V1 and V2 will be in a failed state simultaneously is $V1 \cdot V2 = 16 \times 10^{-8}$, or that *either* will be in a failed state is $V1 + V2 = 0.0008$, and so on. Arbitrarily taking a detection and rectification time of 6 h for all of the events in Table 2.4, the Boolean expression (Equation 2.34) determined above gives the frequency:

$$\begin{aligned}\text{Pexplode} &= (V1 + FT1 + FC1 + PT2 + PC2) \cdot PAH2A \cdot PAH2B \\ &= \left(\frac{(0.6 + 1.14 + 0.29 + 1.41 + 0.29) \times 6}{365 \times 24}\right) \times \left(\frac{0.34 \times 6}{365 \times 24}\right) \times \left(\frac{0.34 \times 6}{365 \times 24}\right) \\ &= 1.4 \times 10^{-10} \text{ times per year}\end{aligned} \quad (2.48)$$

Of course, one is unlikely to derive a single final equation like Equation 2.48 for large systems. The computational algorithm used will more likely be built on a sequence of intermediate evaluations as in Equations 2.30–2.33.

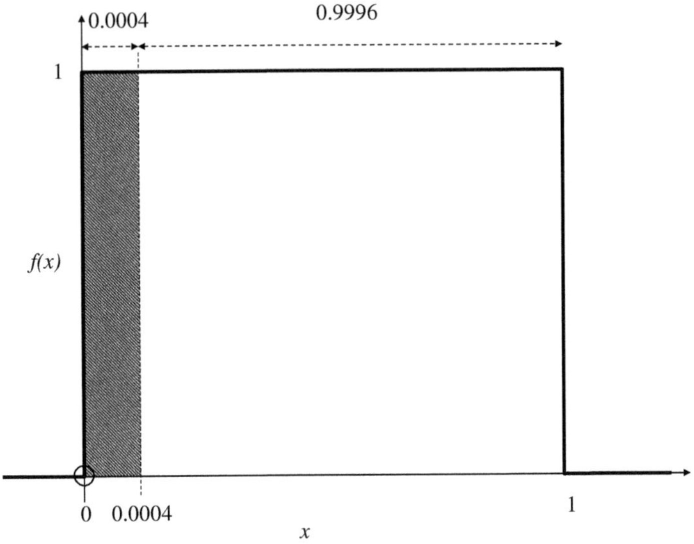

Figure 2.41 'Top-hat' probability density function.

An equivalent way of doing this calculation is to simulate the underlying phenomena using a Monte Carlo technique. Consider the vector of component status Boolean variables

$$\{\text{V1}, \text{FT1}, \text{FC1}, \text{PT2}, \text{PC2}, \text{PAH2A}, \text{PAH2B}\}$$

to be randomly drawn from a 'top-hat' probability density function as in Figure 2.41 according to the rule that a value of 1 results for the sample x lying in the range 0–*probability of state*, and 0 otherwise. For example, if the sample drawn for V1 lies between 0 and 0.0004, then the result is 1, and 0 otherwise. Clearly the frequency of a 'TRUE' result will match each variable's individual probability. Each fresh 'realisation' of the Boolean vector is inserted into Equations 2.30–2.33 to get the state of Pexplode (0 or 1) for this realisation. As more realisations are used, the frequency of a TRUE result (i.e. 1) for Pexplode must approach 1.4×10^{-10} times per year.

This idea has been taken further by Hauptmanns and Yllera (1983) to describe a particular situation where either the failure of a component item is not detected or rectification cannot be effected. It has been noted that the distribution of times of failure of such items follows an exponentially decreasing function as in Figure 2.42. This can be understood in the sense that 100 equal items all starting out at time zero will have an ever-decreasing failure time density as items are eliminated. For human beings, the mean time to failure is around 75 years!

Samples of failure time are drawn from a distribution like Figure 2.42 for each item i. It is noted that this distribution is easily simulated by drawing a sample x from a unit top-hat distribution as in Figure 2.41, and subjecting the result to the conversion

$$t = -T_i \ln(x) \tag{2.49}$$

For the valve V1 in Figure 2.39, Table 2.4 gives a failure frequency of 0.6 pa, suggesting a mean time to failure $T_{\text{V1}} = 1/0.6 = 1.7$ years.

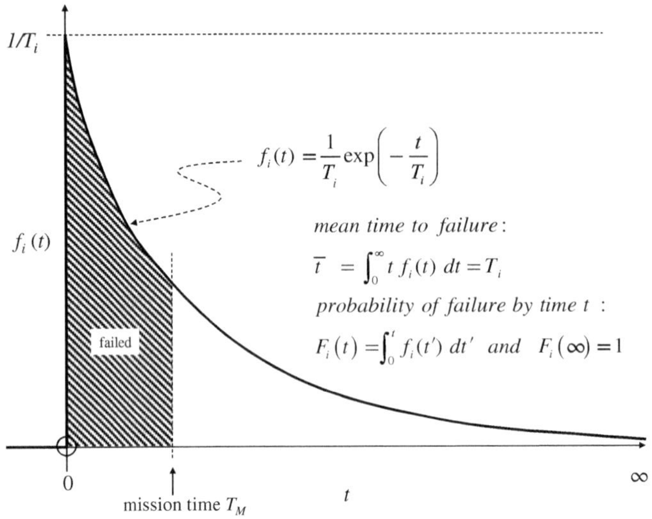

Figure 2.42 Unrecoverable failure time distribution for an item type '*i*'.

If the failure time t thus obtained is less than some defined mission time T_M, then the failure of that component is considered to have occurred, and the corresponding Boolean value is set to 1, else it is set to 0. In this way, a single realisation of the component status Boolean vector is evaluated as before, and used to evaluate the overall failure status of interest. Repeated sample realisations of the component vector will give further values of the overall failure status (0 or 1), from which the probability of failure by a chosen time T_M may be evaluated. Repeating this at a series of mission times T_M allows the *cumulative* probability $F(T_M)$ of the failure of interest within the mission time T_M to be evaluated. As shown in Figure 2.43, this can be differentiated to get the probability density function $f(T_M)$ of the overall failure of interest with time.

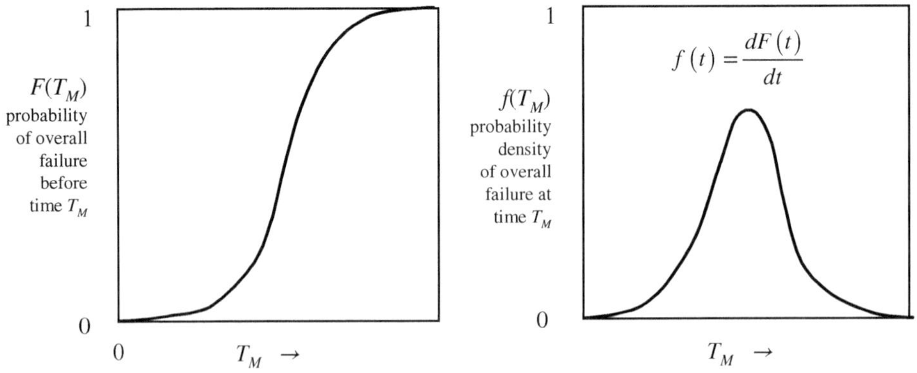

Figure 2.43 Cumulative probability and probability density of overall failure with mission time.

References

Annino, R., Curren, J., Kalinowiski, R., Karas, E., Lindquist, R. and Prescott R. (1976) Totally pneumatic gas chromatographic process stream analyzer. *Journal of Chromatography*, **126**, 301–314.

Davey, W.L.E. and Neels, J. (1991) The engineering of complex projects. 6th National Meeting of the South African Institution of Chemical Engineers, Durban, August 7–9, 1991.

de Vaal, P.L., Eggberry, I. and Jones, M. (2006) Methods to detect causes of control loop oscillations. SACEC 2006: Engineering Africa in the 21st Century, South African Institution of Chemical Engineers, p. 12.

Fisher Controls International LLC (2005) *Control Valve Handbook*, 4th edn, Fisher/Emerson Process Management.

Hauptmanns, U. and Yllera, J. (1983) Fault-tree evaluation by Monte-Carlo simulation. *Chemical Engineer*, **90**, 91–103.

Lawley, G. (1974) Operability studies and hazard analysis. *Chemical Engineering Progress*, **70**, 45–55.

Lees, F.P. (2001) *Loss Prevention in the Process Industries*, 2nd edn (revised), vol. **1**, Butterworth-Heinemann.

Peters, M.S. and Timmerhaus, K.D. (1980) *Plant Design and Economics for Chemical Engineers*, McGraw-Hill.

Smith, C.A. and Corripio, A.B. (1997) *Principles and Practice of Automatic Process Control*, 2nd edn, John Wiley & Sons, Inc., p. 204.

Swearingen, C. (1999) Selecting the right flowmeter – part 1. *Chemical Engineering Magazine*, July 1999.

Swearingen, C. (2001) Selecting the right flowmeter – part 2. *Chemical Engineering Magazine*, January 2001.

3
Modelling

The most important skill that the process engineer brings to bear on the field of process control and optimisation is his/her ability to describe the dynamics and relationships of process variables. The model (Figure 3.1) may serve several purposes:

- insight and understanding;
- basis for controller/optimiser design;
- offline testing of controller/optimiser;
- basis of filter for online estimation of process variables.

In the distant past, models sometimes ran on analogue computers – using capacitors and resistors to convert signals. However, what is being thought of here is an algorithm which will run on a digital computer. Variations to bear in mind include

- theoretical versus regressed (black box);
- continuous versus discrete *equations*;
- logical versus analogue;
- online versus offline;
- linear versus nonlinear;
- lumped versus distributed;
- continuous versus discrete versus mixed *inputs and outputs*;
- single versus multiple *behaviour regimes* (*modes*);
- numerical versus analytical *solution*;
- multi-input, multi-output (MIMO) versus single-input, single-output (SISO);
- differential versus algebraic;
- open loop versus closed loop;
- state-space versus input–output;
- deterministic versus stochastic;
- approximate versus accurate;
- stable versus unstable;
- transfer function form versus equation form.

In this chapter, the focus will be on the modelling of the process itself. At the outset, an important distinction should be noted between *input–output* model forms and *recursive* (or *autoregressive*) model forms (Figure 3.2). The former typically arise from observation of the process as a 'black box', whereas the latter are usually based on physical principles and involve the states of the system. Additionally, there is the idea of the process input $\boldsymbol{u}(t)$ being *exogenous*, meaning that it is being

Applied Process Control: Essential Methods, First Edition. Michael Mulholland.

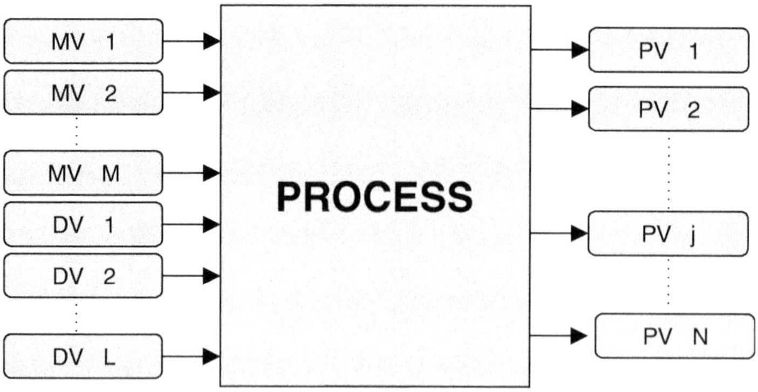

Figure 3.1 General open-loop process model.

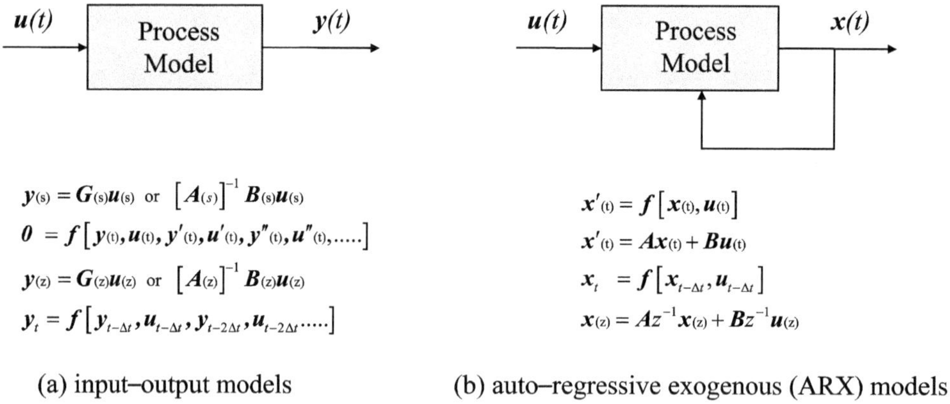

Figure 3.2 Basic model forms.

imposed from outside of the process. Without an exogenous input, a process cannot be controlled, so this form could only be used to describe variation from an initial state, for example decay or equilibrium processes varying towards final asymptotic states.

Most of the effort will be directed at developing the model of the process itself, that is the *open-loop* model without the additional effects of feedback control, optimisation or identification. When it is desired to use the model equations as the basis of a control/optimisation/identification design, one normally makes simplifying assumptions (e.g. linearisation). However, the resultant algorithms need to be tested on as accurate a model as possible. Such an accurate model may only be representable in a series of program code steps including decision points, saturation tests, clipping of negative flows and so on. The discussion that follows will attempt to be as general as possible within the above variations.

3.1 General Modelling Strategy

Control engineers are particularly interested in the *dynamics* of processes, that is outputs (PVs) that change over a period of time once a change occurs in a process input (MV or DV). In some situations, processes can have continuous variations (limit cycles, chaotic behaviour or instability) even if all input variations have ceased! It is the slowly responding processes (e.g. temperature of a large catalytic bed) that are particularly problematic, because it is difficult to predict exactly where they will end up. Fast processes effectively obey algebraic equations, so problems such as overshoot are insignificant. For example, one fills car tyres at the garage using a very quick feedback from the pressure gauge.

Variables that respond over a period of time store important information that is required to predict the ongoing changes in a system. If a flow is introduced to fill a tank, one needs to know the initial level in order to predict the future level variation. This type of variable is called a *state* of the system:

> The *state variables* of a system comprise a sufficient selection of variables such that if their initial values are known, and all future input values are known, it is possible to predict all future values of these selected variables.

This idea is instinctive to engineers – the states are apparently those variables which have to be integrated to solve for the response. However, the set of states may not be unique, and may include discrete variables, such as the status of a bursting disc resulting from a past state value. The tank level and flow system in Figure 3.3 has some of these features.

The dynamic modelling of a system like this is best tackled in several steps:

1) *Determine which variables are constant and which could vary in time.*
   In this example, the only time-varying quantities are f_1, f_2, f_3, h and b. There is no indication that the flow coefficients k_1, k_2, k_3 (see below), H and H_b are likely to change.
2) *Determine which of the time variables are independent inputs (MV or DV, possibly discrete), and which of the rest, if any, should be chosen as states.*

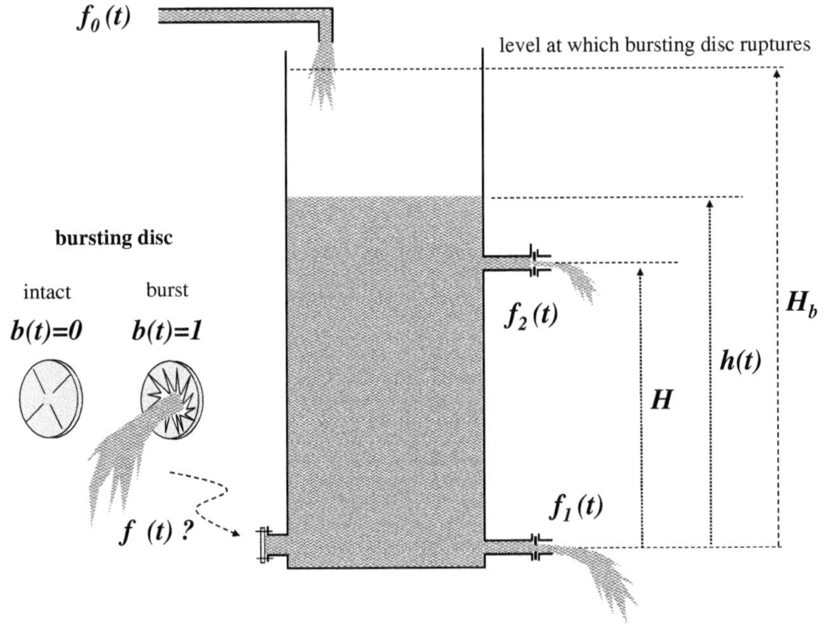

Figure 3.3 Tank with two restricted outflows and a bursting disc.

Here f_0 is an independent input. h is an obvious state, because its starting value is required to predict its future values. On reflection, that would not be enough, as the starting value of b would be required as well. In fact, the starting values of both h and b are required to predict all future values of both h and b, provided the input $f_0(t)$ is known for $t > 0$. So a sufficient set of states is $\{h, b\}$. In this problem, an alternative selection of states may be made, namely $\{f_1, b\}$. This is made possible by the monotonic algebraic relationship between f_1 and h. The particular choice of states in a given problem will depend on the focus. If the focus is on the implications of the varying exit flow f_1, this might well be chosen as a state instead of h.

3) *For each continuous state variable, use a balance of the form 'Accumulation = In − Out' to obtain its time derivative.*

The dynamic balance will normally involve mass or energy, or possibly momentum. Often the equations will involve other states, meaning that the system is 'coupled' or 'interactive'. Typical items involved in balances are listed in Table 3.1.

In the present example, the volume balance gives

$$A\frac{dh(t)}{dt} = f_0(t) - f_1(t) - f_2(t) - f_3(t) \tag{3.1}$$

since the liquid surface area in the vessel A is constant.

Table 3.1 Typical quantities involved in dynamic balances.

Extensive property	Accumulation rate	Inflows	Outflows	Units of balance
Mass	dW/dt	Streams in Reaction generation Desorption, permeation, diffusion, evaporation and so on	Streams out Reaction consumption Absorption, permeation, diffusion, evaporation and so on	$kg\ s^{-1}$
Moles A (one species in a flow)	dm_A/dt	Streams in Reaction generation Desorption, permeation, diffusion, evaporation, dissolution and so on	Streams out Reaction consumption Absorption, permeation, diffusion, evaporation, crystallisation, filtration, precipitation and so on	$kg\ mol\ A\ s^{-1}$
Volume (liquids)	dV/dt	Streams in Permeation, condensation and so on	Streams out Permeation, evaporation and so on	$m^3\ s^{-1}$
Energy	$d\{\rho V c_P T\}/dt$	Streams in (enthalpy) Exothermic reaction heat Transfer in (convection, conduction, radiation) Mechanical work Condensation, freezing Heat of solution (exo)	Streams out (enthalpy) Endothermic reaction heat Transfer out (convection, conduction, radiation) Evaporation, melting Heat of solution (endo)	kW
Momentum	$W\ d^2y/dt^2$	Applied forces Shear	Friction Potential	$kg\ m\ s^{-1}$

4) *For each discrete state variable, determine the logic governing its value.*

$$b(t) = \begin{cases} 1 & \text{for} \quad h(t) > H_b \\ b(t)\,\text{unchanged} & \text{for} \quad h(t) \le H_b \end{cases} \tag{3.2}$$

5) *The remaining time variables (ancillary variables), which are neither states, nor MVs nor DVs (but which could be discrete), must be related to each other and the states, MVs and DVs using algebraic and logical expressions.*

   Following the discussion in Section 2.4.1.1, for a fixed liquid density one can take

$$f_1(t) = k_1\sqrt{h(t)} \tag{3.3}$$

$$f_2(t) = \begin{cases} k_2\sqrt{h(t) - H} & \text{for} \quad h(t) > H \\ 0 & \text{for} \quad h(t) \le H \end{cases} \tag{3.4}$$

$$f_3(t) = \begin{cases} 0 & \text{for} \quad b(t) = 0 \\ k_3\sqrt{h(t)} & \text{for} \quad b(t) = 1 \end{cases} \tag{3.5}$$

6) *A stepwise solution can then be set up for the period 0 to t_f using a simple Euler integration.*

Several aspects of the above procedure (steps 1–6) should be noted. Real problems will always involve logical tests, whether they be for empty or overflowing tanks, limits of valve ranges or signal saturation. Since the solution is typically performed as a series of computer statements, there is no point in attempting to eliminate variables, for example by substituting Equations 3.3–3.5 into Equation 3.1. In fact, one would lose useful information by doing this. Another point is that the algorithmic approach in Figure 3.4 easily adapts to real-time implementations by synchronising the timing loop. More sophisticated integration schemes can be substituted once the basic algorithm works, but modern computer power does not warrant a lot of effort on this aspect.

In the processing industries, there are many problems that are well described by a set of DAEs (differential and algebraic equations). Typically, the lumped differential part describes accumulations in vessels, and the algebraic part describes stream interconnections. The algebraic equations may be implicit (i.e. the dependent variable appears on both sides of the equation), and in any case, the differential equations become very unwieldy if substitutions are attempted to get a set of differential equations alone. Thus, many workers have developed software for solution (integration or optimisation) of a system described by DAEs. The above *integration* solution for the simple tank problem might appear not to warrant anything more sophisticated. It seems in this example that most complications could be dealt with just by decreasing the step size Δt. But in general the algebraic equations could be implicit, and there could be a large set of coupled DEs, possibly with problems of stiffness (fast and slow responses together). Moreover, in an *optimisation* mode, one might, for example, seek the best $f_0(t)$ variation to bring h to its setpoint (SP), so the equations have to be solved more or less backwards.

However, the preceding discussion of DAE solutions applies to systems which have no logical equations. It has been noted above that real systems will in general require description by what one might call DALEs (differential, algebraic and logical equations). Certainly that was the case in the tank example above. The effect of the logical equations is to create discontinuities in the functions describing the behaviour. A few workers such as Mao and Petzold (2002) have developed *integration* solutions for DALE systems. However, the *optimisation* problem is difficult because of the branching caused by the logical expressions. Typically, a MINLP (mixed integer nonlinear programming) solution is required in a commercial package such as GAMS®.

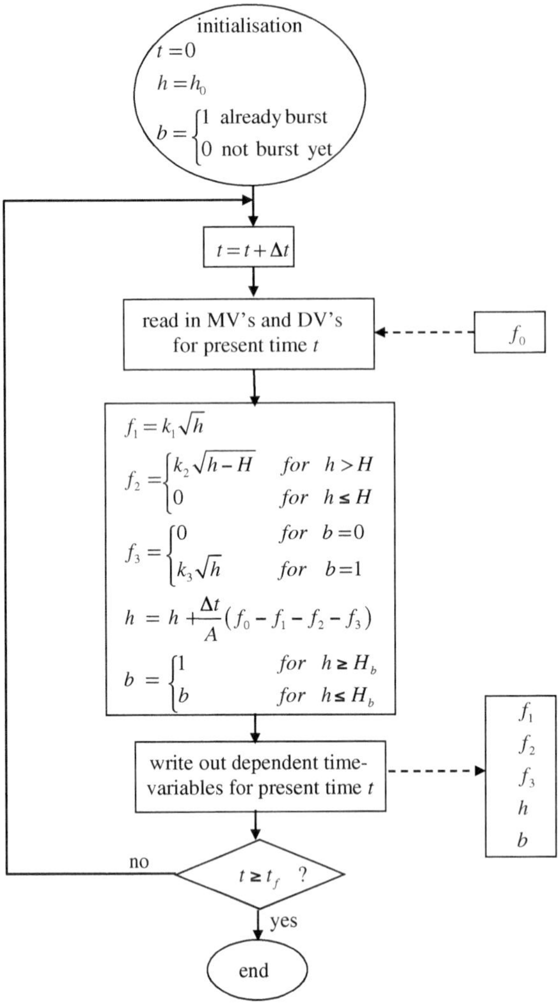

Figure 3.4 Stepwise algorithm for open-loop system in Figure 3.3 using Euler integration (note: '=' implies assignment).

Define vectors $\boldsymbol{x}$ to contain all of the continuous states, $\boldsymbol{w}$ the discrete states, $\boldsymbol{y}$ the ancillary variables and $\boldsymbol{u}$ the input MVs and DVs. It is noted that $\boldsymbol{y}$ and $\boldsymbol{u}$ may contain both continuous and discrete variables. The *integration* problem amounts to solving

$$\boldsymbol{x}(0) = \boldsymbol{x}_0 \tag{3.6}$$

$$\boldsymbol{w}(0) = \boldsymbol{w}_0 \tag{3.7}$$

$$\dot{\boldsymbol{x}} = \boldsymbol{f}(\boldsymbol{x}, \boldsymbol{w}, \boldsymbol{y}, \boldsymbol{u}) \tag{3.8}$$

$$\boldsymbol{0} = \boldsymbol{g}(\boldsymbol{x}, \boldsymbol{w}, \boldsymbol{y}, \boldsymbol{u}) \tag{3.9}$$

for a given $\boldsymbol{u}(t)$, $0 < t < t_f$, where $\boldsymbol{f}$ and $\boldsymbol{g}$ are vectors of functions. On the other hand, a typical constrained *optimisation* problem might involve for

$$\boldsymbol{x}(0) = \boldsymbol{x}_0 \tag{3.10}$$

$$\boldsymbol{w}(0) = \boldsymbol{w}_0 \tag{3.11}$$

$$\dot{\boldsymbol{x}} = \boldsymbol{f}(\boldsymbol{x}, \boldsymbol{w}, \boldsymbol{y}, \boldsymbol{u}) \tag{3.12}$$

$$\boldsymbol{0} = \boldsymbol{g}(\boldsymbol{x}, \boldsymbol{w}, \boldsymbol{y}, \boldsymbol{u}) \tag{3.13}$$

find $\boldsymbol{u}(t)$, $0 < t < t_f$, such that $\boldsymbol{0} < \boldsymbol{h}(\boldsymbol{x}, \boldsymbol{w}, \boldsymbol{y}, \boldsymbol{u})$ and $\phi(\boldsymbol{x}, \boldsymbol{w}, \boldsymbol{y}, \boldsymbol{u})$ is minimised.

Here the vector of functions $\boldsymbol{h}$ represents the constraints, whilst the scalar function ϕ is the objective function for the optimisation.

3.2 Modelling of Distributed Systems

There are many instances of distributed systems in the process industries. This is where conditions vary with both time and position, requiring the system to be described using partial differential equations (PDEs). Examples include reactors which are not mixed, packed absorption, extraction and distillation towers, fixed bed leaching and filtration, and heat exchangers. Usually one is interested in conditions at the exit of such equipment, but quite often there is interest also in values at intermediate positions. Regardless, the only way to model the behaviour is by solution of the PDE. This is usually done by discretisation in the spatial dimensions, such as x in the axial flow reactor in Figure 3.5. So instead of modelling just one value of C_A, now one has to model n values just to get

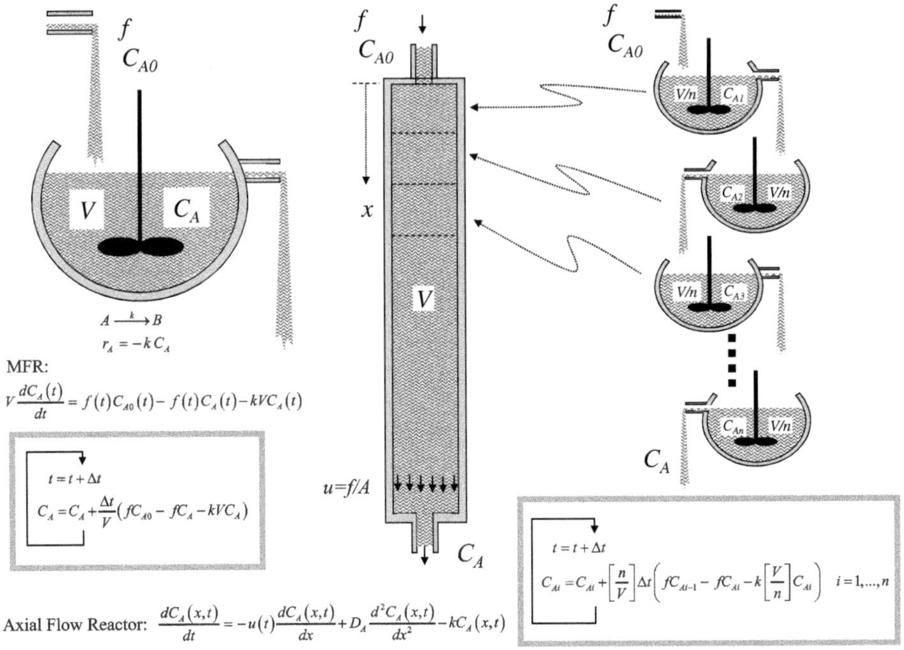

Figure 3.5 Lumped and distributed systems: mixed flow and axial flow reactors.

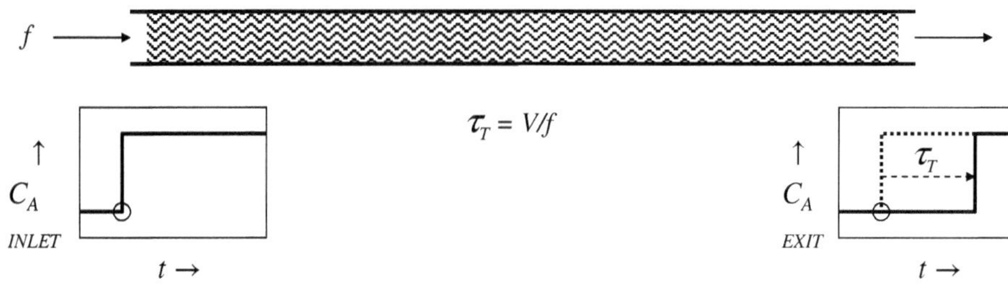

Figure 3.6 Transport lag (dead time).

the one or two results required. This confirms the idea that the state C_A has become *distributed.* The approximate solution based on discretisation effectively creates n lumped states C_{A1}, C_{A2}, . . . , C_{An} and these must be solved simultaneously using the resultant n ordinary differential equations (ODEs). Mathematicians have developed various schemes for these solutions (ADI, Crank–Nicholson, tridiagonal), but in Figure 3.5, a simple sequential Euler integration is again shown which ignores changes in neighbouring elements *during* each time step.

One notes that the one-dimensional discretisation procedure shown divides up the volume into n completely mixed compartments, that is the 'tanks-in-series' model. The greater the value of n, the more closely the plug flow is approached. Actually, n can be set to simulate a degree of axial dispersion according to $n = Lu/D_A$ approaching plug flow for large n (>50) and approaching mixed flow for small n ($n = 1$ being ideal mixed flow). In this expression, L is the length of the flow path, u is the superficial velocity and D_A is the (axial) eddy diffusivity in the flow direction.

In the processing industries, dead time, also known as 'transport lag', is a common phenomenon related to distributed systems (Figure 3.6). This is typically caused by flow through long

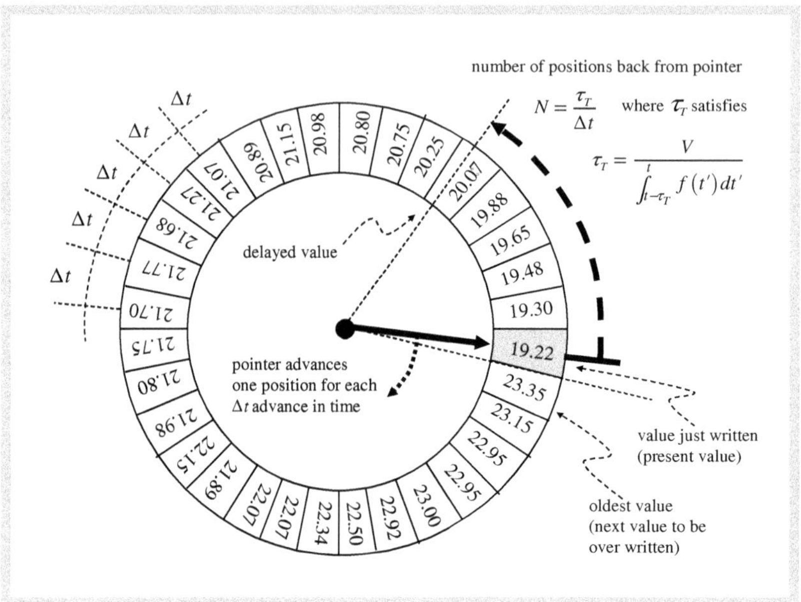

Figure 3.7 Use of a cyclical file and moving pointer to simulate a transport lag.

pipelines, or large volumes that are unmixed. Another source of dead time is travel on conveyor belts. To model dead time dynamically, one could follow the same procedure as for the axial flow reactor in Figure 3.5, without any reaction of course. A large number of compartments, and thus states, would be required to avoid serious blunting of the shapes of signals passing through. A typical computer algorithm for achieving a pure delay is given in Figure 3.7. A cyclical ('wrap-around' or 'stack') file is achieved by the pointer jumping back to the start, once it reaches the end. When the delayed value is found by moving backwards from the pointer, interpolation could be used to improve the 'looked-up' value. The file must be long enough to handle the longest expected delay, or at least the oldest value should be returned for an unusually long delay (e.g. zero flow).

3.3
Modelling Example for a Lumped System: Chlorination Reservoirs

Consider the pair of drinking water conditioning reservoirs in Figure 3.8. The treated water enters the first reservoir at flow f_0 and with chlorine concentration C_0. An interconnecting pipe between the two reservoirs transfers water either way ('tidal flow'), depending on which level is lower, which is determined by the rates f_1 and f_2 at which water is drawn from each compartment, as well as the feed rate f_0. Assuming that each reservoir is well mixed, the varying levels and flows will cause a varying residence time, and thus a varying residual chlorine content at each exit, since the dissolved chlorine is gradually lost.

The problem posed is to develop an algorithm for prediction of the behaviour of this system over a period of time, $t = 0$ to $t = t_f$.

loss of Cl_2 : $r = -kC$

$$f_{12} = \begin{cases} K(h_1-h_2)^{1/2} & \text{for } h_1 > h_2 \\ -K(h_2-h_1)^{1/2} & \text{for } h_1 <= h_2 \end{cases}$$

$$C_{12} = \begin{cases} C_1 & \text{for } h_1 > h_2 \\ C_2 & \text{for } h_1 <= h_2 \end{cases}$$

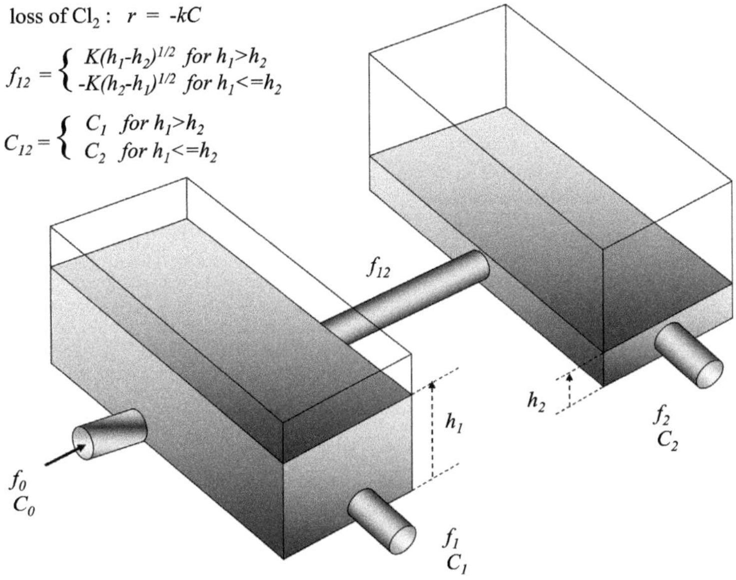

Figure 3.8 Interconnected chlorine conditioning reservoirs for drinking water.

Solution:

Variables:

$h_1(t)$, $h_2(t)$, $C_1(t)$, $C_2(t)$	continuous states
$f_0(t)$, $C_0(t)$, $f_1(t)$, $f_2(t)$	continuous MVs and DVs
$f_{12}(t)$, $C_{12}(t)$	ancillary continuous variables
k	first order rate constant for chlorine decay
K	constant coefficient for pipe flow
A_1, A_2	constant water surface area in each reservoir

Volume balances:

$$\frac{dV_1}{dt} = A_1\frac{dh_1}{dt} = f_0 - f_1 - f_{12} \tag{3.14}$$

$$\frac{dV_2}{dt} = A_2\frac{dh_2}{dt} = f_{12} - f_2 \tag{3.15}$$

Chlorine balances:

$$\frac{dV_1C_1}{dt} = A_1\frac{dh_1C_1}{dt} \tag{3.16}$$

$$\frac{dV_1C_1}{dt} = A_1C_1\frac{dh_1}{dt} + A_1h_1\frac{dC_1}{dt} = f_0C_0 - f_1C_1 - f_{12}C_{12} - A_1h_1kC_1 \tag{3.17}$$

so

$$A_1h_1\frac{dC_1}{dt} = f_0C_0 - f_1C_1 - f_{12}C_{12} - (f_0 - f_1 - f_{12})C_1 - A_1h_1kC_1 \tag{3.18}$$

$$\frac{dC_1}{dt} = \frac{f_0}{A_1h_1}(C_0 - C_1) + \frac{f_{12}}{A_1h_1}(C_1 - C_{12}) - kC_1 \tag{3.19}$$

$$\frac{dV_2C_2}{dt} = A_2\frac{dh_2C_2}{dt} \tag{3.20}$$

$$\frac{dV_2C_2}{dt} = A_2C_2\frac{dh_2}{dt} + A_2h_2\frac{dC_2}{dt} = f_{12}C_{12} - f_2C_2 - A_2h_2kC_2 \tag{3.21}$$

so

$$\frac{dC_2}{dt} = \frac{f_{12}}{A_2h_2}(C_{12} - C_2) - kC_2 \tag{3.22}$$

Algebraic equations:

$$f_{12} = \begin{cases} K\sqrt{h_1 - h_2} & \text{for} \quad h_1 > h_2 \\ -K\sqrt{h_2 - h_1} & \text{for} \quad h_1 \le h_2 \end{cases} \tag{3.23}$$

$$C_{12} = \begin{cases} C_1 & \text{for} \quad h_1 > h_2 \\ C_2 & \text{for} \quad h_1 \le h_2 \end{cases} \tag{3.24}$$

The algorithm is given in Figure 3.9.

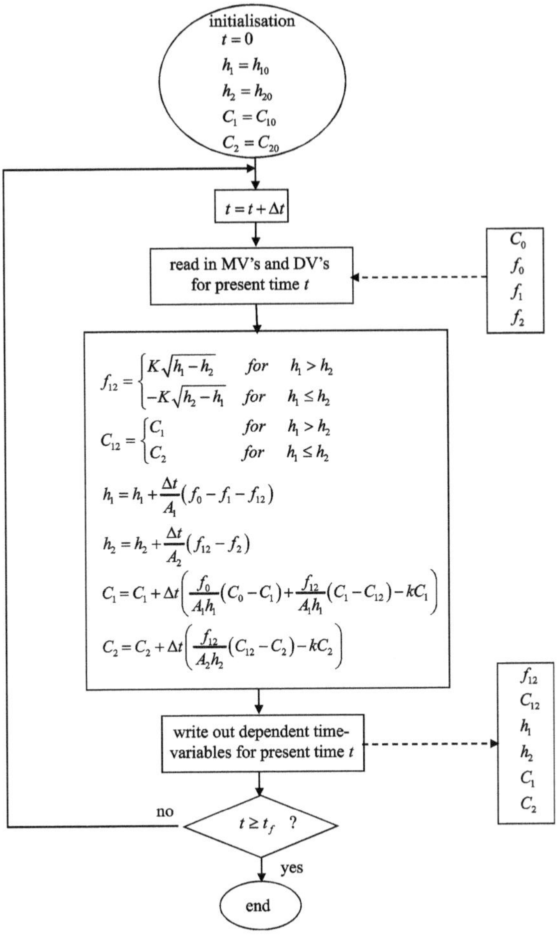

Figure 3.9 Stepwise algorithm for open-loop reservoir system in Figure 3.8 using Euler integration (note: '=' implies assignment).

3.4
Modelling Example for a Distributed System: Reactor Cooler

Figure 3.10 shows the combined reactor and cooler used in the BASF process for formaldehyde production from methanol. The reaction gases pass through a 2 cm thick catalytic bed lying on a perforated crucible. As the reaction product gases enter the tubes of the cooler, they are around 650–700 °C, and must be cooled rapidly to avoid the formation of by-products. Boiler feed water is fed through an equal-percentage valve at the bottom of the shell side of the cooler. As the water moves up, it becomes steam at some point, and the steam is allowed to proceed to users through an equal-percentage valve connected to the top of the shell.

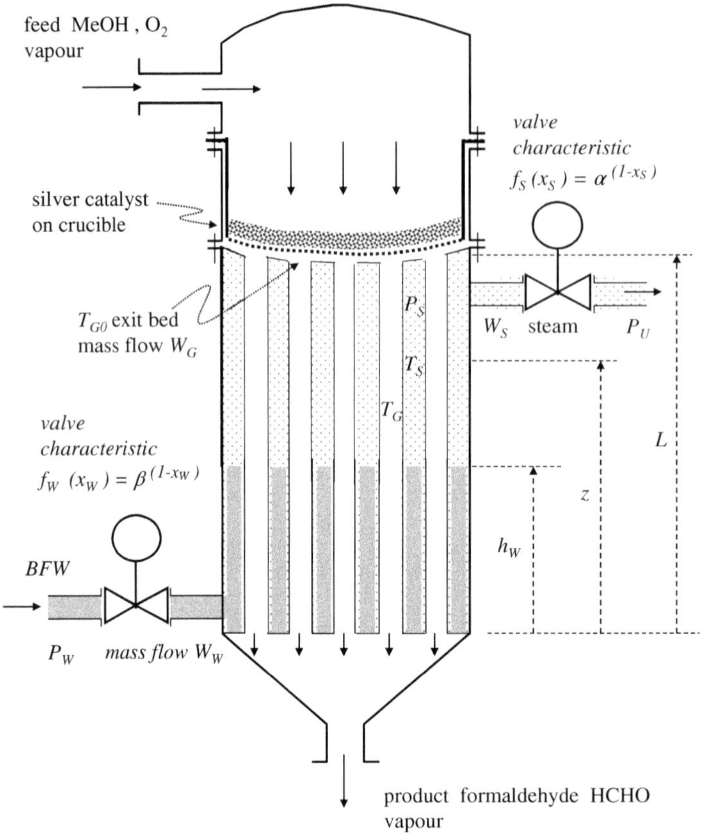

Figure 3.10 Reactor crucible and tubular cooler for BASF process: formaldehyde by dehydrogenation of methanol over silver catalyst.

Since both the reaction gas side and the water/steam side are distributed, the cooler will be represented by a series of elements (1, . . . , *N*) as in Figure 3.11, which interconnect the two sides by virtue of a heat transfer surface.

The art of creating a model is (a) to record as many equations as possible which interrelate the variables and (b) to recognise reasonable approximations which simplify the model as far as possible.

To simplify the solution on the water/steam side, the following assumptions will be made:

1) Water enters at its boiling point, which is determined by the steam pressure.
2) Sensible heat transfer to the water is negligible – all heat added creates steam.
3) Steam bubbles rising in the water occupy little volume.
4) The water volume is well mixed.
5) The steam volume is well mixed.
6) The variation of c_P with temperature is ignored.
7) Mass flows W_G (gas moving down) and $W_S = W_W$ (steam/water moving up) are taken constant with height.

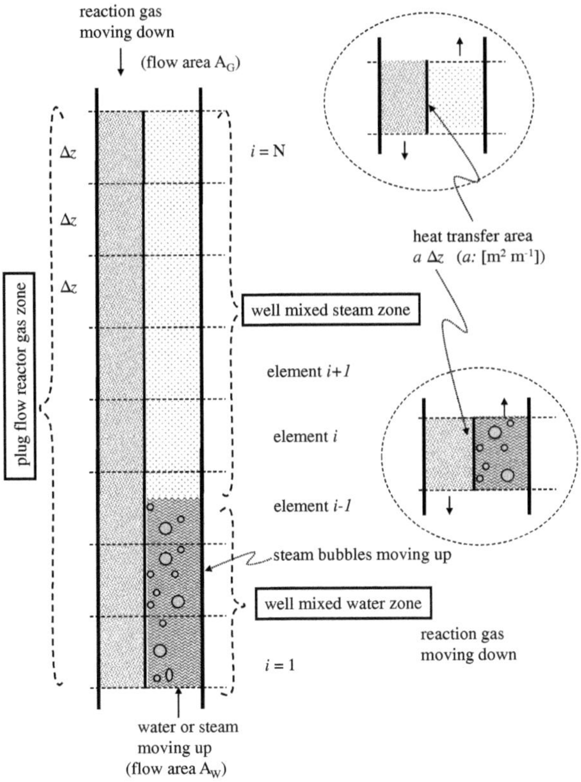

Figure 3.11 Conversion of distributed system into multiple lumped systems by discretisation of spaces for reaction gas, water and steam for reactor cooler in Figure 3.10.

Solution:

Variables:

$x_W(t)$, $x_S(t)$, $T_{G0}(t)$, $W_G(t)$	continuous MVs and DVs
P_W, P_U	pressures of BFW supply and steam users assumed constant
α, β, k_W, k_S	valve characteristic constants and flow coefficients
A_G, A_W, a	constant flow areas and heat transfer area per unit height
U_W, U_S	constant overall heat transfer coefficient from gas to water and gas to steam
ρ_W, c_{PW}, c_{PS}, c_{PG}, λ	constant properties of fluids (including latent heat)
A, B, C	constant Antoine coefficients for water
$P_S(t)$, $T_S(t)$, $h_W(t)$	single continuous state variable: pressure on water/steam side, temperature of steam, height of water

For each element 'i'

$T_{Gi}(t)$	continuous state variables: temperatures of reaction gas
$q_i(t)$	ancillary time-dependent variable: heat transferred from reaction gas to water/steam in element i

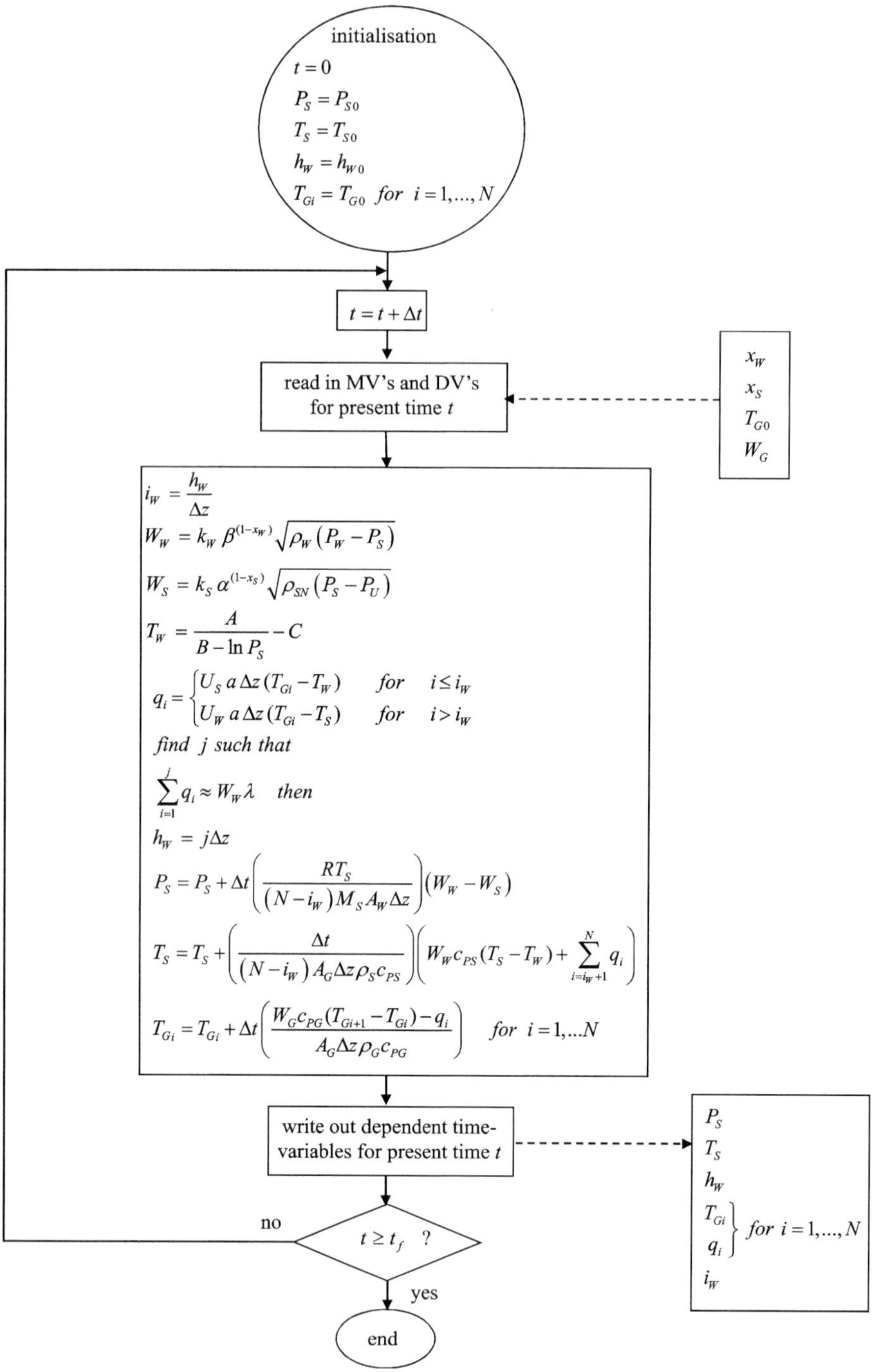

Figure 3.12 Stepwise algorithm for open-loop distributed reactor cooler system in Figure 3.10 using Euler integration (note: '=' implies assignment).

Water level:

$$i_W = \frac{h_W}{\Delta z} \tag{3.25}$$

Energy balances:

$$A_G \Delta z\, \rho_G c_{PG} \frac{dT_{Gi}}{dt} = W_G c_{PG}(T_{Gi+1} - T_{Gi}) - q_i \quad \text{for} \quad i = 1, \ldots, N \quad (\text{ignores } \rho_G \text{ changes}) \tag{3.26}$$

$$(N - i_W) A_G \Delta z\, \rho_S c_{PS} \frac{dT_S}{dt} = W_W c_{PS}(T_S - T_W) + \sum_{i=i_W+1}^{N} q_i \quad (\text{ignores } \rho_S \text{ changes}) \tag{3.27}$$

$$W_W \lambda = \sum_{i=1}^{i_W} q_i \tag{3.28}$$

Steam balance (with molecular mass $M_S = 18$):

$$\frac{d}{dt}\left(\frac{(N - i_W) M_S P_S A_W \Delta z}{R T_S}\right) = W_W - W_S \quad (\text{approximating vapour as ideal gas}) \tag{3.29}$$

$$\frac{dP_S}{dt} = \left(\frac{R T_S}{(N - i_W) M_S A_W \Delta z}\right)(W_W - W_S) \quad \text{treating } \frac{dT_S}{dt} \text{ and } \frac{di_W}{dt} \text{ as slow in comparison} \tag{3.30}$$

Algebraic equations:

$$W_S = k_S \alpha^{(1-x_S)} \sqrt{\rho_{SN}(P_S - P_U)} \quad (\text{approximating steam flow as incompressible}) \tag{3.31}$$

$$T_W = \frac{A}{B - \ln P_S} - C \tag{3.32}$$

$$q_i = \begin{cases} U_S a \Delta z (T_{Gi} - T_W) & \text{for} \quad i \le i_W \\ U_W a \Delta z (T_{Gi} - T_S) & \text{for} \quad i > i_W \end{cases} \tag{3.33}$$

$$W_W = k_W \beta^{(1-x_W)} \sqrt{\rho_W (P_W - P_S)} \quad (\text{ignoring static head}) \tag{3.34}$$

for 'T_{GN+1}' use T_{G0}.

The algorithm is given in Figure 3.12.

3.5 Ordinary Differential Equations and System Order

The modelling problems considered so far have been somewhat 'open-ended', requiring rather *ad hoc* approaches. It was intended merely to obtain as close a representation of the physical phenomena as possible, bearing in mind that the algorithmic approach (sequential computer instructions) gave a lot of freedom to deal with state-dependent behaviour, discontinuities, logical/discrete issues, saturation and nonlinearity. The models developed were based on physical principles, giving access to meaningful parameters (e.g. heat transfer coefficients) which could be adjusted to get a good match to real plant behaviour. It is a good idea to develop skills in this type of algorithmic modelling, because it allows one to simulate real process behaviour more closely.

Moving on from the strictly algorithmic approach, it needs to be recognised that the useful theoretical ideas that are going to be developed later in this text for control, identification and

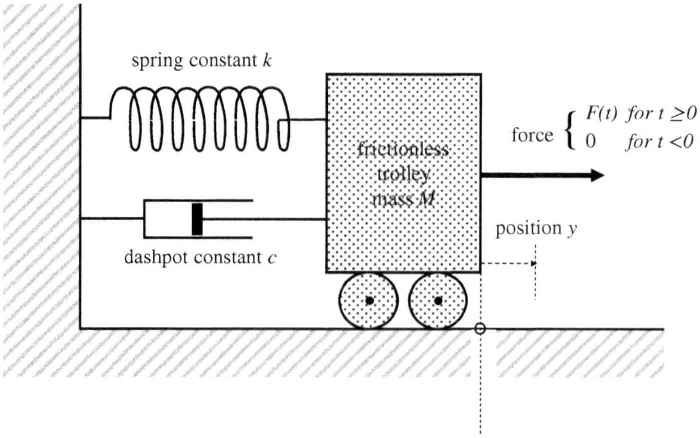

Figure 3.13 Force applied to trolley with spring and dashpot resisting.

optimisation will usually rely on more restricted types of models – typically those that can be expressed directly as a set of first-order ODEs. (In fact, a lot of useful ideas are based on the specific case of a system of *linear* ODEs.)

The *order* of a system is the number of equations using a first derivative (d/dt) that one needs to represent its dynamics. In other words, following on from Section 3.1, the order is determined by the number of states. In the lumped chlorination reservoir problem of Section 3.3 it was 4, and in the distributed reactor cooler problem of Section 3.4 it was $N+2$. (So the process of discretising the spatial dimension of a system described by PDEs leads to extra states, and an increase in order by the same amount.) In the processing industries, virtually all of the individual differential equations found in mass and energy balances for lumped systems will arise as first derivatives. There are a very few situations where this type of theoretical modelling of physical phenomena leads initially to a second-order differential equation. Usually this is where there is inertia and momentum involved, for example compressor shaft rotation, pipeline flow or mercury in a manometer. To illustrate this point, consider the well-known mechanical mass, spring and dashpot example in Figure 3.13

A force balance leads to the equation

$$M\frac{\mathrm{d}^2y}{\mathrm{d}t^2} + c\frac{\mathrm{d}y}{\mathrm{d}t} + ky = F \tag{3.35}$$

that is

$$\tau^2\frac{\mathrm{d}^2y}{\mathrm{d}t^2} + 2\zeta\tau\frac{\mathrm{d}y}{\mathrm{d}t} + y = KF \tag{3.36}$$

which is the standard form of a *second-order system*, where

$$\text{time constant } \tau = \sqrt{\frac{M}{k}} \tag{3.37}$$

$$\text{damping factor } \zeta = \frac{c}{2\sqrt{kM}} \tag{3.38}$$

$$\text{gain } K = \frac{1}{k} \tag{3.39}$$

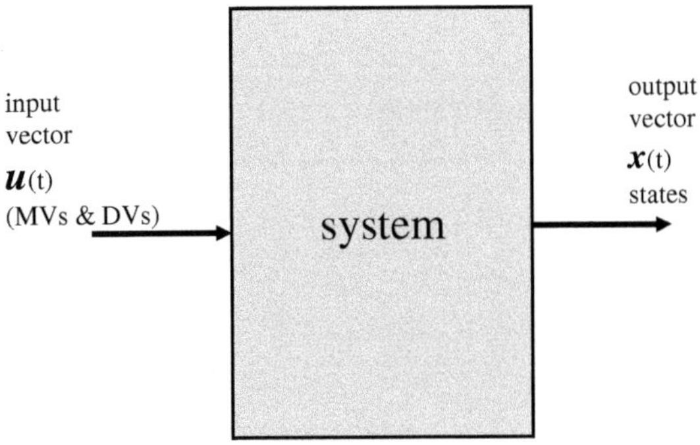

Figure 3.14 General state-space system.

This system has two states, namely $x_1 = \mathrm{d}y/\mathrm{d}t$, velocity, and $x_2 = y$, position. Initial values are required for both of these in order to solve for the continued variation of the system with $F(t)$. Equation 3.36 can then be written as the system of first-order ODEs

$$\dot{\boldsymbol{x}} = \boldsymbol{f}(\boldsymbol{x}, F) \tag{3.40}$$

where

$$f_1 = -\frac{2\zeta}{\tau}x_1 - \frac{1}{\tau^2}x_2 + \frac{K}{\tau^2}F \tag{3.41}$$

$$f_2 = x_1 \tag{3.42}$$

and since this happens to be linear it can be expressed as

$$\frac{\mathrm{d}}{\mathrm{d}t}\begin{pmatrix} x_1 \\ x_2 \end{pmatrix} = \underbrace{\begin{bmatrix} -\dfrac{2\zeta}{\tau} & -\dfrac{1}{\tau^2} \\ 1 & 0 \end{bmatrix}}_{\boldsymbol{A}} \begin{pmatrix} x_1 \\ x_2 \end{pmatrix} + \underbrace{\begin{bmatrix} \dfrac{K}{\tau^2} \\ 0 \end{bmatrix}}_{\boldsymbol{B}} F \tag{3.43}$$

which is clearly a second-order system. Obviously, if independent differential equations arise in the modelling, these can be solved separately. However, the higher order systems one is contemplating here are those that have interdependent differential equations, that is they share the state variables $x_1, x_2, \ldots$

More generally, the 'state-space' representation of a continuous system (Figure 3.14) is

$$\dot{\boldsymbol{x}} = \boldsymbol{f}(\boldsymbol{x}, \boldsymbol{u}) \tag{3.44}$$

and if it happens to be linear one can use the common form

$$\dot{\boldsymbol{x}} = \boldsymbol{A}\boldsymbol{x} + \boldsymbol{B}\boldsymbol{u} \tag{3.45}$$

Referring to Section 3.1, one notes that Equation 3.44 is a special case of Equations 3.6–3.9, with no ancillary algebraic equations and variables shown. A form like Equation 3.44 could of course still be obtained from Equations 3.6–3.9 where ancillary algebraic equations exist, provided all of the ancillary variables could be eliminated from the expression by substitution.

3.6 Linearity

In process control one spends a lot of time thinking about linearity, because most of the robust and powerful methods assume linear process behaviour. One needs to be able to find linear versions of process models and to deal with the problems of mismatch to the actual process.

Table 3.2 Principles of linearity.

Principle	Implication	Responses
Superposition	If $\boldsymbol{u}_1(t) \to \boldsymbol{x}_1(t)$ and $\boldsymbol{u}_2(t) \to \boldsymbol{x}_2(t)$, then $[\boldsymbol{u}_1(t) + \boldsymbol{u}_2(t)] \to [\boldsymbol{x}_1(t) + \boldsymbol{x}_2(t)]$	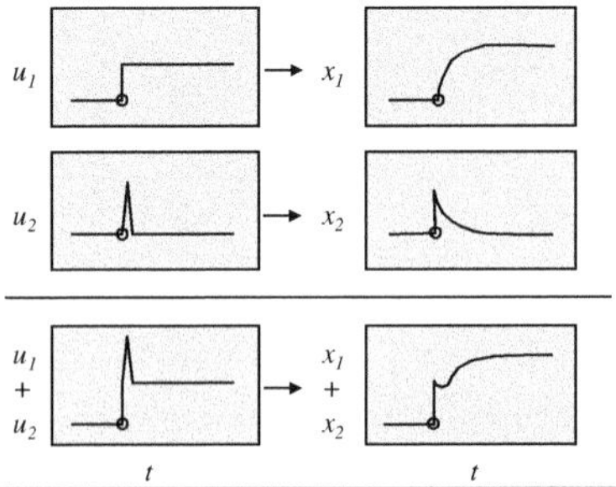
Homogeneity	If $\boldsymbol{u}_1(t) \to \boldsymbol{x}_1(t)$, then $a \times u_1(t) \to a \times x_1(t)$	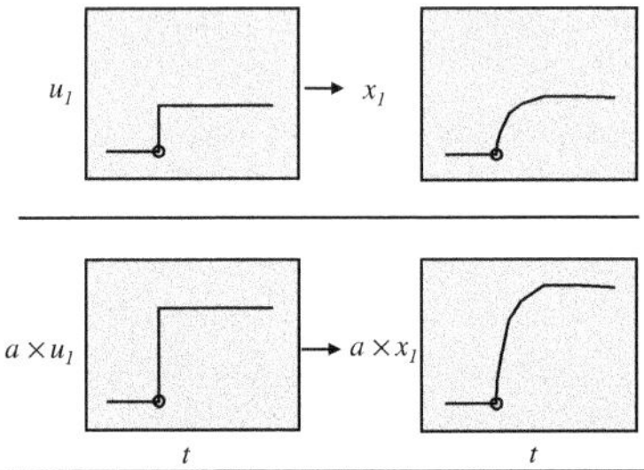
Stationarity	If $\boldsymbol{u}_1(t) = A\sin(\omega t)$, then eventually $x_1(t)$ will be sinusoidal with the same frequency ω	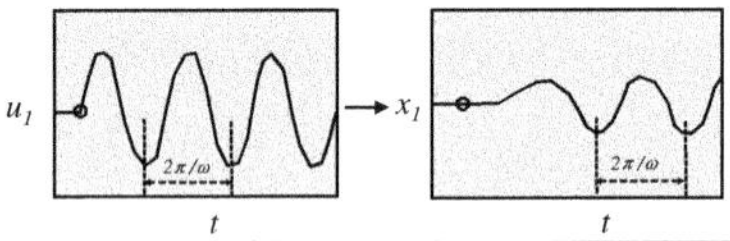

As in Section 3.1, let vector $\boldsymbol{x}$ contain the continuous states, $\boldsymbol{y}$ the ancillary variables and $\boldsymbol{u}$ the input MVs and DVs. With the restriction that discrete states cannot be considered, nor discrete variables in $\boldsymbol{y}$ and $\boldsymbol{u}$, the system of Equations 3.6–3.9 becomes

$$\dot{\boldsymbol{x}} = \boldsymbol{f}(\boldsymbol{x}, \boldsymbol{y}, \boldsymbol{u}) \tag{3.46}$$

$$\boldsymbol{0} = \boldsymbol{g}(\boldsymbol{x}, \boldsymbol{y}, \boldsymbol{u}) \tag{3.47}$$

where $\boldsymbol{f}$ and $\boldsymbol{g}$ are vectors of continuous functions. Considerations of linearity will focus on the input–output relationship $\boldsymbol{u}(t) \to \boldsymbol{x}(t)$ (Table 3.2).

Two examples will serve to illustrate the test for linearity by superposition.

Example 3.1

(Figure 3.15)

Volume balance:

$$A\frac{\mathrm{d}h}{\mathrm{d}t} = F_0 - F_1 \tag{3.48}$$

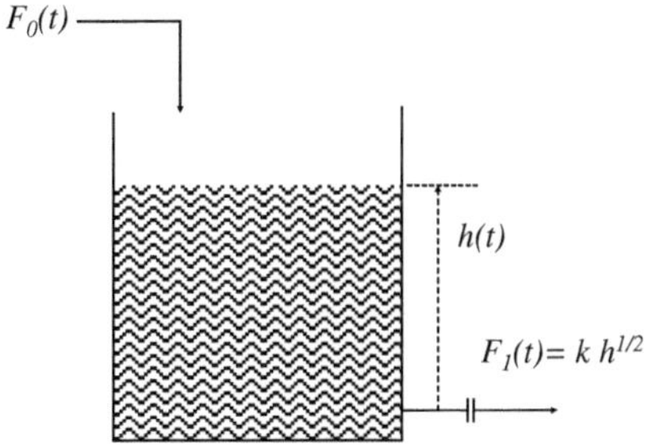

Figure 3.15 Tank with restriction orifice at exit.

Algebraic equation:

$$F_1 = kh^{1/2} \tag{3.49}$$

Substituting,

$$A\frac{dh}{dt} = F_0 - kh^{1/2} \tag{3.50}$$

$$\text{For input } F_{01}(t): \quad A\frac{dh_1}{dt} = F_{01} - kh_1{}^{1/2} \tag{3.51}$$

$$\text{For input } F_{02}(t): \quad A\frac{dh_2}{dt} = F_{02} - kh_2{}^{1/2} \tag{3.52}$$

Summing for $[F_{01}(t) + F_{02}(t)]$:

$$A\frac{d[h_1 + h_2]}{dt} = [F_{01} + F_{02}] - k\left[h_2{}^{1/2} - h_2{}^{1/2}\right] \tag{3.53}$$

Since $k\left[h_2{}^{1/2} - h_2{}^{1/2}\right] \neq k[h_2 - h_2]^{1/2}$, which should be the form of the last term (for linearity), it is concluded that this system is nonlinear.

Instead of doing the substitution, alternatively view Equations 3.48 and 3.49 in the form of Equations 3.6–3.7:

$$\dot{h} = f(h, F_1, F_0) = \left(\frac{1}{A}\right)F_0 - \left(\frac{1}{A}\right)F_1 \tag{3.54}$$

$$0 = g(h, F_1) = F_1 - kh^{1/2} \tag{3.55}$$

Now the differential equation is linear in the time functions $h(t)$, $F_0(t)$ and $F_1(t)$ but the nonlinearity is in the algebraic equation, because, by homogeneity,

$$g(h, F_1) = F_1 - kh^{1/2} \tag{3.56}$$

$$\begin{aligned} g(ah, aF_1) &= aF_1 - k(ah)^{1/2} \\ &\neq a\left[F_1 - kh^{1/2}\right] \end{aligned} \tag{3.57}$$

Example 3.2

(Figure 3.16)

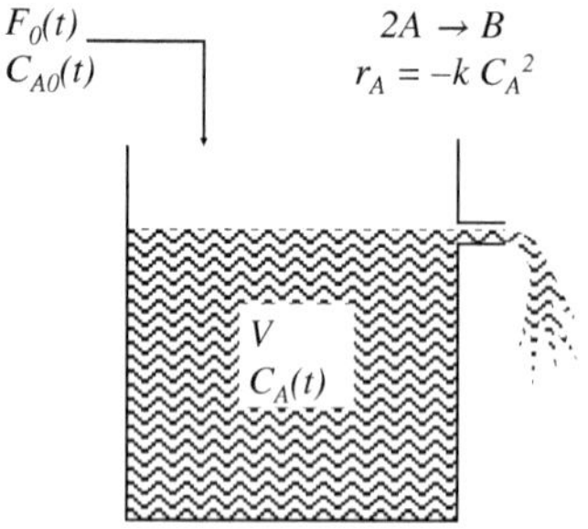

Figure 3.16 Fixed volume MFR with second-order reaction.

Variables:

$F_0(t)$, $C_{A0}(t)$	input MVs or DVs
$C_A(t)$	state

Molar balance on species A:

$$V\frac{dC_A}{dt} = F_0C_{A0} - F_0C_A - kVC_A^2 \tag{3.58}$$

$$\text{For input } F_{01}(t): \quad V\frac{dC_{A1}}{dt} = F_{01}C_{A0} - F_{01}C_{A1} - kVC_{A1}^2 \tag{3.59}$$

$$\text{For input } F_{02}(t): \quad V\frac{dC_{A2}}{dt} = F_{02}C_{A0} - F_{02}C_{A2} - kVC_{A2}^2 \tag{3.60}$$

Summing for $[F_{01}(t) + F_{02}(t)]$:

$$V\frac{d[C_{A1} + C_{A2}]}{dt} = [F_{01} + F_{02}]C_{A0} - [F_{01}C_{A1} + F_{02}C_{A2}] - kV\left[C_{A1}^2 + C_{A2}^2\right] \tag{3.61}$$

Since $[F_{01}C_{A1} + F_{02}C_{A2}] \neq [F_{01} + F_{02}][C_{A1} + C_{A2}]$ or alternatively since $\left[C_{A1}^2 + C_{A2}^2\right] \neq [C_{A1} + C_{A2}]^2$, the system is nonlinear.

Note: According to the principle of superposition, it would appear that an equation like

$$2\frac{dx}{dt} = -3x + u - 2 \tag{3.62}$$

is nonlinear. However, it is noted that it is easily linearised by substituting either a new input variable $w = u - 2$ or a new state variable $y = x + 2/3$.

3.7 Linearisation of the Equations Describing a System

Again, the discussion here will focus on continuous systems which do not involve discrete states or inputs, that is as in Equations 3.46–3.47:

$$\dot{\boldsymbol{x}} = \boldsymbol{f}(\boldsymbol{x}, \boldsymbol{y}, \boldsymbol{u}) \tag{3.63}$$

$$\boldsymbol{0} = \boldsymbol{g}(\boldsymbol{x}, \boldsymbol{y}, \boldsymbol{u}) \tag{3.64}$$

Define a Jacobian matrix symbolically $(\partial \boldsymbol{f}/\partial \boldsymbol{x})$ as

$$\frac{\partial \boldsymbol{f}}{\partial \boldsymbol{x}} = \begin{bmatrix} \frac{\partial f_1}{\partial x_1} & \frac{\partial f_1}{\partial x_2} & \cdots & \frac{\partial f_1}{\partial x_N} \\ \frac{\partial f_2}{\partial x_1} & \frac{\partial f_2}{\partial x_2} & \cdots & \frac{\partial f_2}{\partial x_N} \\ \vdots & \vdots & \ddots & \vdots \\ \frac{\partial f_N}{\partial x_1} & \frac{\partial f_N}{\partial x_2} & \cdots & \frac{\partial f_N}{\partial x_N} \end{bmatrix} \tag{3.65}$$

and similarly for $\partial \boldsymbol{f}/\partial \boldsymbol{y}$, $\partial \boldsymbol{f}/\partial \boldsymbol{u}$, $\partial \boldsymbol{g}/\partial \boldsymbol{x}$, $\partial \boldsymbol{g}/\partial \boldsymbol{y}$ and $\partial \boldsymbol{g}/\partial \boldsymbol{u}$, but it is noted, however, that the latter Jacobian matrices will not in general be $N \times N$. Choosing a point $(\boldsymbol{x}_0, \boldsymbol{y}_0, \boldsymbol{u}_0)$, where $\boldsymbol{f}=\boldsymbol{f}_0$ and $\boldsymbol{g}=\boldsymbol{g}_0$, about which to perform the linearisation, a Taylor series expansion to the second term yields

$$\dot{\boldsymbol{x}} \approx \boldsymbol{f}_0 + \left[\frac{\partial \boldsymbol{f}}{\partial \boldsymbol{x}}\right]_0 (\boldsymbol{x} - \boldsymbol{x}_0) + \left[\frac{\partial \boldsymbol{f}}{\partial \boldsymbol{y}}\right]_0 (\boldsymbol{y} - \boldsymbol{y}_0) + \left[\frac{\partial \boldsymbol{f}}{\partial \boldsymbol{u}}\right]_0 (\boldsymbol{u} - \boldsymbol{u}_0) \tag{3.66}$$

$$\boldsymbol{0} \approx \boldsymbol{g}_0 + \left[\frac{\partial \boldsymbol{g}}{\partial \boldsymbol{x}}\right]_0 (\boldsymbol{x} - \boldsymbol{x}_0) + \left[\frac{\partial \boldsymbol{g}}{\partial \boldsymbol{y}}\right]_0 (\boldsymbol{y} - \boldsymbol{y}_0) + \left[\frac{\partial \boldsymbol{g}}{\partial \boldsymbol{u}}\right]_0 (\boldsymbol{u} - \boldsymbol{u}_0) \tag{3.67}$$

Using *deviation ('perturbation') variables* $\boldsymbol{x}' = (\boldsymbol{x} - \boldsymbol{x}_0)$, $\boldsymbol{y}' = (\boldsymbol{y} - \boldsymbol{y}_0)$ and $\boldsymbol{u}' = (\boldsymbol{u} - \boldsymbol{u}_0)$, and choosing the point $(\boldsymbol{x}_0, \boldsymbol{y}_0, \boldsymbol{u}_0)$ such that it satisfies $\boldsymbol{g}(\boldsymbol{x}_0, \boldsymbol{y}_0, \boldsymbol{u}_0) = \boldsymbol{0}$ and causes the system to lie at steady state, that is $\boldsymbol{f}(\boldsymbol{x}_0, \boldsymbol{y}_0, \boldsymbol{u}_0) = \boldsymbol{0}$,

$$\dot{\boldsymbol{x}}' = \left[\frac{\partial \boldsymbol{f}}{\partial \boldsymbol{x}}\right]_0 \boldsymbol{x}' + \left[\frac{\partial \boldsymbol{f}}{\partial \boldsymbol{y}}\right]_0 \boldsymbol{y}' + \left[\frac{\partial \boldsymbol{f}}{\partial \boldsymbol{u}}\right]_0 \boldsymbol{u}' \tag{3.68}$$

$$\boldsymbol{0} = \left[\frac{\partial \boldsymbol{g}}{\partial \boldsymbol{x}}\right]_0 \boldsymbol{x}' + \left[\frac{\partial \boldsymbol{g}}{\partial \boldsymbol{y}}\right]_0 \boldsymbol{y}' + \left[\frac{\partial \boldsymbol{g}}{\partial \boldsymbol{u}}\right]_0 \boldsymbol{u}' \tag{3.69}$$

To resolve the ancillary variables, $[\partial \boldsymbol{g}/\partial \boldsymbol{y}]_0$ has to be square and nonsingular, so

$$\boldsymbol{y}' = -\left[\frac{\partial \boldsymbol{g}}{\partial \boldsymbol{y}}\right]_0^{-1} \left\{ \left[\frac{\partial \boldsymbol{g}}{\partial \boldsymbol{x}}\right]_0 \boldsymbol{x}' + \left[\frac{\partial \boldsymbol{g}}{\partial \boldsymbol{u}}\right]_0 \boldsymbol{u}' \right\} \tag{3.70}$$

$$\boldsymbol{y}' = \boldsymbol{C}\boldsymbol{x}' + \boldsymbol{D}\boldsymbol{u}' \tag{3.71}$$

which on substitution in Equations 3.68–3.69 yields the linear equation

$$\dot{\boldsymbol{x}}' = \boldsymbol{A}\boldsymbol{x}' + \boldsymbol{B}\boldsymbol{u}' \tag{3.72}$$

where

$$\boldsymbol{A} = \left[\frac{\partial \boldsymbol{f}}{\partial \boldsymbol{x}}\right]_0 - \left[\frac{\partial \boldsymbol{f}}{\partial \boldsymbol{y}}\right]_0 \left[\frac{\partial \boldsymbol{g}}{\partial \boldsymbol{y}}\right]_0^{-1} \left[\frac{\partial \boldsymbol{g}}{\partial \boldsymbol{x}}\right]_0 \tag{3.73}$$

$$\boldsymbol{B} = \left[\frac{\partial \boldsymbol{f}}{\partial \boldsymbol{u}}\right]_0 - \left[\frac{\partial \boldsymbol{f}}{\partial \boldsymbol{y}}\right]_0 \left[\frac{\partial \boldsymbol{g}}{\partial \boldsymbol{y}}\right]_0^{-1} \left[\frac{\partial \boldsymbol{g}}{\partial \boldsymbol{u}}\right]_0 \tag{3.74}$$

In many situations, the nonlinear or implicit form of $\boldsymbol{g}$ does not permit easy substitution (prior to linearisation). So it is worthwhile remembering that separate linearisation of Equations 3.68–3.69 as above leads to the same result.

Example 3.3

(Figure 3.17)

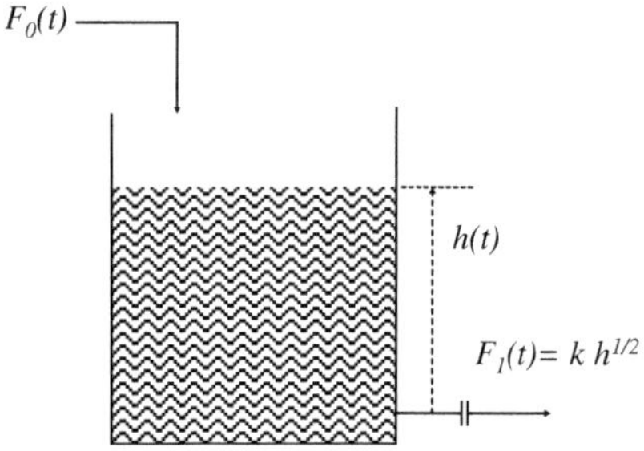

Figure 3.17 Tank with restriction orifice at exit.

$$\frac{dh}{dt} = f(h, F_1, F_0) = \frac{1}{A}(F_0 - F_1) \tag{3.75}$$

$$0 = g(h, F_1, F_0) = F_1 - kh^{1/2} \tag{3.76}$$

Linearise *f*:

$$\frac{dh'}{dt} = \left.\frac{\partial f}{\partial h}\right|_0 h' + \left.\frac{\partial f}{\partial F_1}\right|_0 F'_1 + \left.\frac{\partial f}{\partial F_0}\right|_0 F'_0 \tag{3.77}$$

$$\frac{dh'}{dt} = 0 + \left\{-\frac{1}{A}\right\}F'_1 + \left\{\frac{1}{A}\right\}F'_0 \tag{3.78}$$

Linearise *g*:

$$0 = \left.\frac{\partial g}{\partial h}\right|_0 h' + \left.\frac{\partial g}{\partial F_1}\right|_0 F'_1 + \left.\frac{\partial g}{\partial F_0}\right|_0 F'_0 \tag{3.79}$$

$$0 = \left\{-\frac{k}{2\sqrt{h_0}}\right\}h' + 1\,F'_1 + 0 \tag{3.80}$$

So

$$F_1' = \left\{\frac{k}{2\sqrt{h_0}}\right\} h' \tag{3.81}$$

and thus

$$\frac{dh'}{dt} = \left\{-\frac{k}{2A\sqrt{h_0}}\right\} h' + \left\{\frac{1}{A}\right\} F_0' \tag{3.82}$$

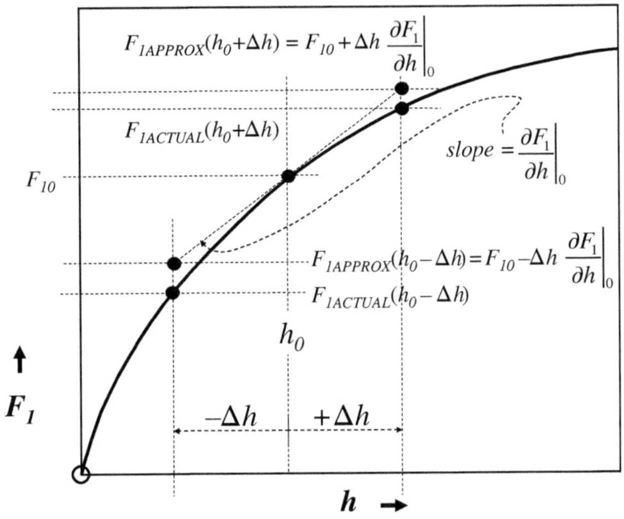

Figure 3.18 Linearisation of orifice flow characteristic for tank flow example (Example 3.3).

Example 3.3 follows the general procedure without a direct substitution of the ancillary variable F_1 in the original differential equation, which, as mentioned, is often problematic. Furthermore, in the process of assuming a steady-state point $(h, F_1, F_0)_0$ which satisfies $g = 0$, it is important to identify the implications:

$$F_{10} = kh_0^{1/2} \text{ for } g = 0 \tag{3.83}$$

$$F_{00} = F_{10} \text{ for steady state} \tag{3.84}$$

So F_0' represents deviations from the steady-state inflow F_{00}. Moreover, if it required to establish the actual absolute level in the tank for a particular h' value, it must be added back to its offset, namely $h = h' + h_0$.

In this example, the linearisation entails an approximation for F_1 as represented in Figure 3.18.

3.8 Simple Linearisation 'Δ' Concept

At the risk of repeating what has already been recommended in Section 3.7, it is worth suggesting an equivalent 'delta' procedure for linearisation of systems of DAEs. Taking the restricted case of the continuous system

$$\dot{\boldsymbol{x}} = \boldsymbol{f}(\boldsymbol{x}, \boldsymbol{y}, \boldsymbol{u}) \tag{3.85}$$

$$\boldsymbol{0} = \boldsymbol{g}(\boldsymbol{x}, \boldsymbol{y}, \boldsymbol{u}) \tag{3.86}$$

(which has no discrete variables or 'logical' mode changes), one recognises that

$$\Delta(f_i) = \Delta(\text{term 1 of } f_i) + \Delta(\text{term 2 of } f_i) + \Delta(\text{term 3 of } f_i) + \cdots \tag{3.87}$$

$$\Delta(g_j) = \Delta(\text{term 1 of } g_j) + \Delta(\text{term 2 of } g_j) + \Delta(\text{term 3 of } g_j) + \cdots \tag{3.88}$$

where 'Δ' represents the linear partial derivative chain with respect to all of the time variables present in any term. Bearing in mind the assumption of *linearisation about the steady-state operating point,* one simply passes the Δ operator through all of the available differential and algebraic equations. The equivalent treatment of Example 3.3 is then as in Example 3.4. Again, it must be remembered that implicitly the resultant new *deviation* (or *perturbation*) *variables* $(\Delta x_k, \Delta y_l, \Delta u_m)$ are deviations from a particular set of values (x_{k0}, y_{l0}, u_{m0}) which cause $f_{i0} = 0$ and $g_{j0} = 0$.

Example 3.4

Given

$$A\frac{dh}{dt} = F_0 - F_1 \tag{3.89}$$

$$F_1 = kh^{1/2} \tag{3.90}$$

then

$$\Delta\left\{A\frac{dh}{dt} = F_0 - F_1\right\} \tag{3.91}$$

$$\Rightarrow \Delta\left\{A\frac{dh}{dt}\right\} = \Delta\{F_0\} - \Delta\{F_1\} \tag{3.92}$$

$$\Rightarrow A\frac{d\Delta h}{dt} = \Delta F_0 - \Delta F_1 \tag{3.93}$$

because operator Δ is linear, that is

$$A\frac{dh'}{dt} = F_0' - F_1' \tag{3.94}$$

and

$$\Delta\left\{F_1 = kh^{1/2}\right\} \tag{3.95}$$

$$\Rightarrow \Delta\{F_1\} = \Delta\left\{kh^{1/2}\right\} \tag{3.96}$$

$$\Delta\{F_1\} = \frac{\partial}{\partial h}\left\{kh^{1/2}\right\}\Delta h + \cancel{\frac{\partial}{\partial F_1}\left\{kh^{1/2}\right\}}\Delta F_1 + \cancel{\frac{\partial}{\partial F_0}\left\{kh^{1/2}\right\}}\Delta F_0 \tag{3.97}$$

$$\Delta\{F_1\} = \left\{\frac{k}{2\sqrt{h}}\right\}\Delta h \tag{3.98}$$

so

$$F_1' = \left\{\frac{k}{2\sqrt{h}}\right\}_0 h' \tag{3.99}$$

3.9 Solutions for a System Response Using Simpler Equations

In the lumped and distributed system examples of Sections 3.3 and 3.4, stepwise algorithmic approaches were used to obtain the output response to time variations of the input MVs and DVs. Special logical tests were required in the integration cycle $t \to t + \Delta t$ to handle such occurrences as state-dependent changes in behaviour. In dynamic systems, these solutions are clearly integrations of the defining equations. In many cases, it is satisfactory to consider operation in a restricted range where variables can be treated as unbounded (no saturation) and no logical branches need to be handled. Most control systems are based on models where such assumptions have been made, usually with an additional assumption of linearity. It is worthwhile to consider several forms of mathematical solution of such systems, because (a) the resultant formulae are often useful, and (b) some ideas arising in these solutions form part of the conceptual basis and language of control theory. So, at the outset, consider a linear system described by the general form of Equations 3.70–3.72:

$$\dot{\boldsymbol{x}} = \boldsymbol{A}\boldsymbol{x} + \boldsymbol{B}\boldsymbol{u} \tag{3.100}$$

$$\boldsymbol{y} = \boldsymbol{C}\boldsymbol{x} + \boldsymbol{D}\boldsymbol{u} \tag{3.101}$$

Here the 'prime' has been dropped from the time-variable vectors $\boldsymbol{x}(t)$, $\boldsymbol{u}(t)$ and $\boldsymbol{y}(t)$ for convenience, as is common practice, but one must obviously remain conscious that the values are deviations from the steady-state condition. In certain systems, the matrices $\boldsymbol{A}$, $\boldsymbol{B}$, $\boldsymbol{C}$ and $\boldsymbol{D}$ can be time dependent, but that case will not be considered here. It is noted that $\boldsymbol{A}$ is an $N \times N$ matrix, where N is the order of the system, that is the number of states needed to describe it. The matrix $\boldsymbol{B}$ is $N \times M$, where M is the number of inputs to be considered (MVs and DVs). Any number P of ancillary 'output' variables $\boldsymbol{y}$ may be involved, with $\boldsymbol{C}$ and $\boldsymbol{D}$ being $P \times N$ and $P \times M$, respectively. The latter concept is often useful when the only measurable observation or feedback is based indirectly on the states, for example 'weighted average bed temperature (WABT). Often it is not possible to observe all of the states $\boldsymbol{x}$, in which case one is considering an *input–output system* $\boldsymbol{u}(t) \to \boldsymbol{y}(t)$, where $\boldsymbol{y}$ is a (linear) combination of some selection of the states.

3.9.1 Mathematical Solutions for a System Response in the *t*-Domain

One system readily lending itself to time-domain solution is the SISO case of Equation 3.100:

$$\dot{x} = ax + bu \tag{3.102}$$

where a and b are merely scalar constants. In the case of the tank flow example (Example 3.3), it was seen that

$$x = h \tag{3.103}$$

$$u = F_0 \tag{3.104}$$

$$a = \left\{ -\frac{k}{2A\sqrt{h}} \right\}_0 \tag{3.105}$$

$$b = \left\{ \frac{1}{A} \right\} \tag{3.106}$$

Integration of Equation 3.102 is possible for certain forms of $u(t)$ by separation of variables. Noting that

$$\frac{\mathrm{d}}{\mathrm{d}t}(x\,\mathrm{e}^{-at}) = \mathrm{e}^{-at}\left(\frac{\mathrm{d}x}{\mathrm{d}t} - ax\right) \tag{3.107}$$

it follows that

$$\frac{\mathrm{d}}{\mathrm{d}t}(x\,\mathrm{e}^{-at}) = \mathrm{e}^{-at}\,bu \tag{3.108}$$

Integrating from 0 to t

$$\left[x\,\mathrm{e}^{-at}\right]_0^t = \int_0^t \mathrm{e}^{-at'}\,bu(t')\mathrm{d}t' \tag{3.109}$$

so

$$x(t) = \mathrm{e}^{at}\,x(0) + b\int_0^t \mathrm{e}^{a(t-t')}\,u(t')\mathrm{d}t' \tag{3.110}$$

Two specific cases are considered in Table 3.3.

Table 3.3 Time-domain response solutions for a first-order linear system.

Input	Solution	Output response plot
$u(t) = 0$ for $t > 0$	$x(t) = \mathrm{e}^{at}\,x(0)$ where a is negative for stable systems (else unlimited growth)	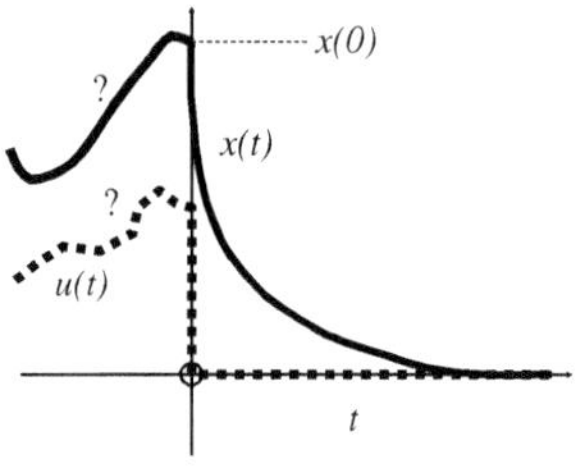
$u(t) = \alpha$ (const) for $t > 0$	$x(t) = \mathrm{e}^{at}\,x(0) + b\alpha\int_0^t \mathrm{e}^{a(t-t')}\,\mathrm{d}t'$ $= \mathrm{e}^{at}\,x(0) + b\alpha\left[-\frac{1}{a}\mathrm{e}^{a(t-t')}\right]_0^t$ $= \mathrm{e}^{at}\,x(0) + \frac{b\alpha}{a}\left[\mathrm{e}^{at} - 1\right]$ (a is negative for stable systems)	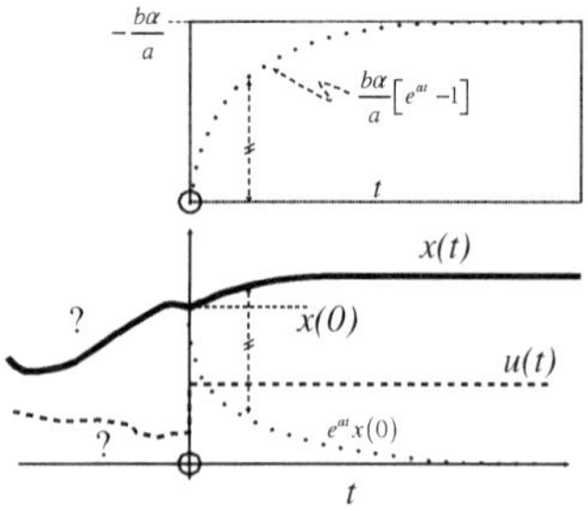

Moving on to a second-order system with constant coefficients $\boldsymbol{A}$ and $\boldsymbol{B}$, the time-domain solution becomes more difficult, and is developed as the sum of a *complementary solution* ($\boldsymbol{u}(t) = \boldsymbol{0}$) and *particular solution* ($\boldsymbol{u}(t) \neq \boldsymbol{0}$). For such larger linear models, it will be found easier to obtain output responses using Laplace s-domain methods in the next section.

3.9.2 Mathematical Solutions for a System Response in the *s*-Domain

Laplace transform methods using the parameter 's' are seldom used in the processing industries, yet they are very important on a conceptual level. A lot of the useful theory refers to aspects of these methods, whether the context involves 'pole locations', 'frequency response', 'stability margins', 'transfer functions' or 'integrators'. It will be found that a good background in these methods enables one to build up a mental picture of key aspects of process dynamics and control.

3.9.2.1 Review of Some Laplace Transform Results

The Laplace transform of a function of time $x(t)$ is defined as

$$X(s) = L\{x(t)\} = \int_0^{\infty} e^{-st}\, x(t)\mathrm{d}t \tag{3.111}$$

Only behaviour at times $t \geq 0$ is considered, so it is implicit in the approach that all time functions are zero up until $t = 0$.

Example 3.5

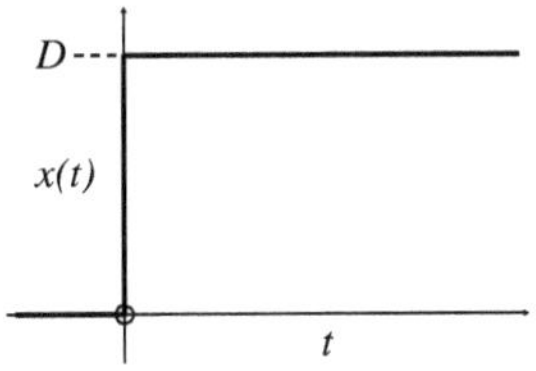

$$\text{Step function size } D: \quad x(t) = \begin{cases} 0 & \text{for} \quad t < 0 \\ D & \text{for} \quad t \geq 0 \end{cases} \tag{3.112}$$

$$X(s) = L\{x(t)\} = \int_0^{\infty} e^{-st}\, D\,\mathrm{d}t \tag{3.113}$$

$$X(s) = D\left[-\frac{e^{-st}}{s}\right]_0^{\infty} \tag{3.114}$$

$$X(s) = \frac{D}{s} \tag{3.115}$$

Apart from the fact that s-domain versions of various functions may be found in tables, one notes that the operator $L\{\cdot\}$ is linear, so that if $f = \text{term } 1 + \text{term } 2 + \cdots$, then $F(s) = L\{f\} = L\{\text{term } 1\} + L\{\text{term } 2\} + \cdots$. Furthermore, $L\{af(t)\} = aF(s)$.

Now consider the transform of the time derivative:

$$L\left\{\frac{\mathrm{d}x(t)}{\mathrm{d}t}\right\} = \int_0^{\infty} \mathrm{e}^{-st}\,\frac{\mathrm{d}x(t)}{\mathrm{d}t}\,\mathrm{d}t \tag{3.116}$$

Integrating by parts

$$L\left\{\frac{\mathrm{d}x(t)}{\mathrm{d}t}\right\} = \left[\mathrm{e}^{-st}x(t) - \int \left(-s\,\mathrm{e}^{-st}\right)x(t)\mathrm{d}t\right]_0^{\infty} \tag{3.117}$$

$$L\left\{\frac{\mathrm{d}x(t)}{\mathrm{d}t}\right\} = -x(0) + sX(s) \tag{3.118}$$

A similar treatment for the second derivative yields

$$L\left\{\frac{\mathrm{d}^2x(t)}{\mathrm{d}t^2}\right\} = -\left.\frac{\mathrm{d}x(t)}{\mathrm{d}t}\right|_{t=0} - sx(0) + s^2X(s) \tag{3.119}$$

In general, provided $x(0) = 0$ and $\left(\mathrm{d}^k x(t)/\mathrm{d}t^k\right)\big|_{t=0} = 0$ for $k = 1, \ldots, n-1$, then

$$L\left\{\frac{\mathrm{d}^n x(t)}{\mathrm{d}t^n}\right\} = s^n X(s) \tag{3.120}$$

For integration, note that

$$L\left\{\int x(t)\mathrm{d}t\right\} = \frac{1}{s}\left[\int x(t)\mathrm{d}t\right]_{t=0} + \frac{1}{s}X(s) \tag{3.121}$$

These results will shortly prove useful for the conversion of ODEs with constant coefficients into transfer functions. However, transport (dead-time) lag (Figure 3.19) cannot be described using ODEs, and warrants a special treatment.

$$y(t) = x(t - \tau_{\mathrm{T}}) \tag{3.122}$$

$$L\{y(t)\} = L\{x(t - \tau_{\mathrm{T}})\} \tag{3.123}$$

$$L\{y(t)\} = \int_0^{\infty} \mathrm{e}^{-st}\,x(t - \tau_{\mathrm{T}})\mathrm{d}t \tag{3.124}$$

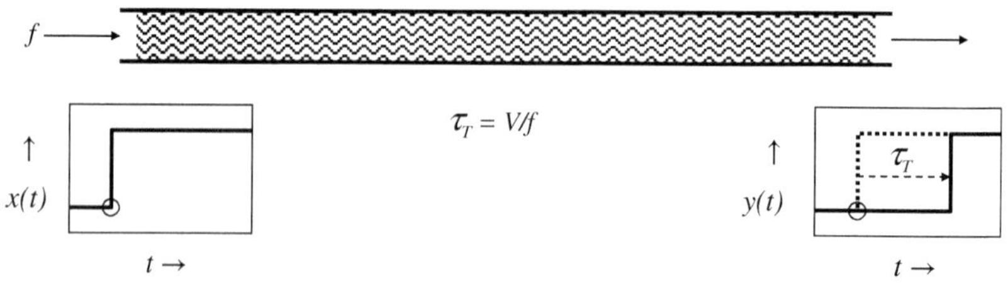

Figure 3.19 Transport lag (dead-time lag).

Substituting u for $t - \tau_T$,

$$Y(s) = \int_{-\tau_T}^{\infty} e^{-s(u+\tau_T)}\, x(u)\mathrm{d}u \tag{3.125}$$

$$Y(s) = e^{-s\tau_T} \int_{0}^{\infty} e^{-su}\, x(u)\mathrm{d}u \quad \text{since } x(t) = 0 \text{ for } t < 0 \tag{3.126}$$

$$Y(s) = e^{-s\tau_T}\, X(s) \tag{3.127}$$

So the transfer function of a dead-time lag τ_T is

$$G(s) = \frac{Y(s)}{X(s)} = e^{-s\tau_T} \tag{3.128}$$

Several useful Laplace transform results are summed up in Table 3.5.

Consider an arbitrarily high-order SISO linear system with constant coefficients similar to the trolley with spring and dashpot in Section 3.5.

$$a_n \frac{\mathrm{d}^n x}{\mathrm{d}t^n} + a_{n-1} \frac{\mathrm{d}^{n-1} x}{\mathrm{d}t^{n-1}} + \cdots + a_1 \frac{\mathrm{d}x}{\mathrm{d}t} + a_0 x = b_m \frac{\mathrm{d}^m u}{\mathrm{d}t^m} + b_{m-1} \frac{\mathrm{d}^{m-1} u}{\mathrm{d}t^{m-1}} + \cdots + b_1 \frac{\mathrm{d}u}{\mathrm{d}t} + b_0 u \tag{3.129}$$

Consider the particular circumstance of

$$\left.\frac{\mathrm{d}^{n-1}x}{\mathrm{d}t^{n-1}}\right|_{t=0} = 0, \quad \left.\frac{\mathrm{d}^{n-2}x}{\mathrm{d}t^{n-2}}\right|_{t=0} = 0, \quad \ldots, \quad \left.\frac{\mathrm{d}x}{\mathrm{d}t}\right|_{t=0} = 0, \quad x(0) = 0 \tag{3.130}$$

$$\left.\frac{\mathrm{d}^{m-1}u}{\mathrm{d}t^{m-1}}\right|_{t=0} = 0, \quad \left.\frac{\mathrm{d}^{m-2}u}{\mathrm{d}t^{m-2}}\right|_{t=0} = 0, \quad \ldots, \quad \left.\frac{\mathrm{d}u}{\mathrm{d}t}\right|_{t=0} = 0, \quad u(0) = 0 \tag{3.131}$$

This requires that the system starts at $t=0$ with both input u and output x at zero, and at a 'complete' *steady state* where the indicated time derivatives are all zero. Then Equation 3.120 gives

$$\begin{aligned} &a_n s^n X(s) + a_{n-1} s^{n-1} X(s) + \cdots + a_1 s X(s) + a_0 X(s) \\ &= b_m s^m U(s) + b_{m-1} s^{m-1} U(s) + \cdots + b_1 s U(s) + b_0 U(s) \end{aligned} \tag{3.132}$$

Thus, in the s-domain, a transfer function $G(s)$ can be used as a multiplier to represent Equation 3.129.

$$G(s) = \frac{X(s)}{U(s)} = \frac{b_m s^m + b_{m-1} s^{m-1} + \cdots + b_1 s + b_0}{a_n s^n + a_{n-1} s^{n-1} + \cdots + a_1 s + a_0} \tag{3.133}$$

$$X(s) = G(s)U(s) \tag{3.134}$$

For a system to be *physically realisable,* it is necessary that $n \geq m$. Indeed, in the trolley, spring and dashpot example of Section 3.5, $n=2$ and $m=0$. Few physical systems would be modelled with the derivative terms on the right-hand side of Equation 3.129. Mathematically, one might propose a system like

$$a_0 x = b_1 \frac{\mathrm{d}u}{\mathrm{d}t} + b_0 u \tag{3.135}$$

but one would be asking for an impossible response, for example if u were a step function.

The standard input function transforms in Table 3.4 suggest that, in general, the input $U(s)$ would be expressed as a ratio of two polynomials in s. Since the transfer function $G(s)$ in

Table 3.4 Selected Laplace transforms.

x(t)	**Plot**	**X(s)**
$\delta(t)$	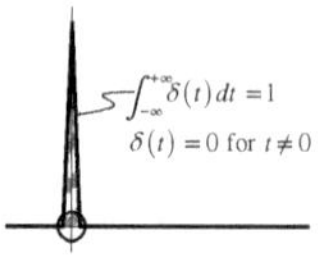	1
1	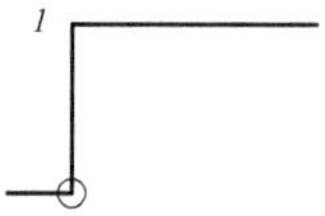	$\frac{1}{s}$
t	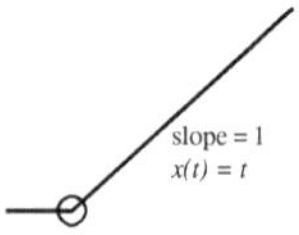	$\frac{1}{s^2}$
e^{-at}	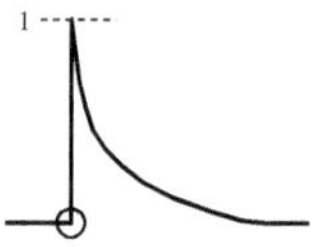	$\frac{1}{s+a}$
$\sin(\omega t)$		$\frac{\omega}{s^2+\omega^2}$
$\cos(\omega t)$		$\frac{s}{s^2+\omega^2}$

Equation 3.133 is similarly a ratio of two polynomials in *s*, one expects that usually the output *X*(*s*) will arise as a ratio of two polynomials in *s*. These will be more complex than the simple transforms in Table 3.4, so they must be broken down into more elemental pieces using a *partial fraction expansion*.

Say

$$X(s) = \frac{d_p s^p + d_{p-1}s^{p-1} + \cdots + d_1 s + d_0}{c_k s^k + c_{k-1}s^{k-1} + \cdots + c_1 s + c_0} \tag{3.136}$$

Letting

$$N(s) = \left(d_p s^p + d_{p-1}s^{p-1} + \cdots + d_1 s + d_0\right)/c_k \tag{3.137}$$

$$X(s) = \frac{N(s)}{\left(s-q_1\right)\left(s-q_2\right)\cdots\left(s-q_k\right)} \tag{3.138}$$

where q_i, $i = 1, \ldots, k$, are the roots obtained by setting the denominator to zero. If these roots are all distinct, one can write

$$\frac{N(s)}{\left(s-q_1\right)\left(s-q_2\right)\cdots\left(s-q_k\right)} = \frac{A}{\left(s-q_1\right)} + \frac{B}{\left(s-q_2\right)} + \cdots + \frac{Z}{\left(s-q_k\right)} \tag{3.139}$$

Otherwise, if a root is repeated – say q_1 occurs three times – write

$$\frac{N(s)}{\left(s-q_1\right)^3\left(s-q_2\right)\cdots\left(s-q_k\right)}=\frac{A_1}{\left(s-q_1\right)}+\frac{A_2}{\left(s-q_1\right)^2}+\frac{A_3}{\left(s-q_1\right)^3}+\frac{B}{\left(s-q_2\right)}+\cdots+\frac{Z}{\left(s-q_k\right)} \tag{3.140}$$

In the case of distinct roots, one multiplies by each denominator factor in turn, and simultaneously sets s to the root value, for example

$$\left.\begin{matrix}\times\left(s-q_1\right)\\ s\to q_1\end{matrix}\right\}:\quad \frac{N\left(q_1\right)}{1\left(q_1-q_2\right)\cdots\left(q_1-q_k\right)}=A+0+\cdots+0 \tag{3.141}$$

For the repeated roots, first multiply by the highest power denominator

$$\begin{aligned}R(s)&=\frac{N(s)}{1\left(s-q_2\right)\cdots\left(s-q_k\right)}\\&=A_1\left(s-q_1\right)^2+A_2\left(s-q_1\right)+A_3+\left[\frac{B}{\left(s-q_2\right)}+\cdots+\frac{Z}{\left(s-q_k\right)}\right]\left(s-q_1\right)^3\end{aligned} \tag{3.142}$$

Now obtain

$$R\left(q_1\right)=A_3 \tag{3.143}$$

$$\left.\frac{\mathrm{d}R(s)}{\mathrm{d}s}\right|_{s=q_1}=A_2 \tag{3.144}$$

$$\left.\frac{1}{2}\frac{\mathrm{d}^2R(s)}{\mathrm{d}s^2}\right|_{s=q_1}=A_1 \tag{3.145}$$

In general, s is a complex number, and complex roots q_i certainly can arise in the above procedure. These are associated with oscillation in the response. Such roots will occur in complex conjugate pairs, and the associated coefficients must then also be expected to occur in complex conjugate pairs so that the complex variable $j=\sqrt{-1}$ does not remain in an expression like Equation 3.139 if a real value of s is substituted.

Say

$$\frac{A}{\left(s-q_1\right)}+\frac{B}{\left(s-q_2\right)}=\frac{A}{(s-a-jb)}+\frac{B}{(s-a+jb)}\quad\left(q_1=a+jb,\ q_2=a-jb\right) \tag{3.146}$$

Then it is required that $A=c+jd$ and $B=c-jd$ so that

$$\frac{c+jd}{(s-a-jb)}+\frac{c-jd}{(s-a+jb)}=\frac{2c(s-a)-2bd}{(s-a)^2+b^2} \tag{3.147}$$

which is real for real s.

Going a little further, one notes that Tables 3.4 and 3.5 allow the following inversion ($L^{-1}\{\cdot\}$) of these first two terms of the expansion to the time domain:

$$L^{-1}\left\{\frac{2c(s-a)-2bd}{(s-a)^2+b^2}\right\}=\mathrm{e}^{at}\,L^{-1}\left\{\frac{2cs-2bd}{s^2+b^2}\right\} \tag{3.148}$$

$$L^{-1}\left\{\frac{2c(s-a)-2bd}{(s-a)^2+b^2}\right\}=2\mathrm{e}^{at}[c\cos(bt)-d\sin(bt)] \tag{3.149}$$

Table 3.5 Selected Laplace transform results.

First derivative	$L\left\{\frac{\mathrm{d}x(t)}{\mathrm{d}t}\right\} = -x(0) + sX(s)$
Second derivative	$L\left\{\frac{\mathrm{d}^2x(t)}{\mathrm{d}t^2}\right\} = -\left.\frac{\mathrm{d}x(t)}{\mathrm{d}t}\right\|_{t=0} - sx(0) + s^2X(s)$
Integral	$L\left\{\int x(t)\mathrm{d}t\right\} = \frac{1}{s}\left[\int x(t)\mathrm{d}t\right]_{t=0} + \frac{1}{s}X(s)$
Transport lag	$G(s) = \mathrm{e}^{-s\tau_T}$
s associated with a	$L\{\mathrm{e}^{-at}x(t)\} = X(s+a)$
Complex conjugate partial fractions	$L\left\{\frac{2t^{n-1}}{(n-1)!}\mathrm{e}^{-k_1t}[a\cos(k_2t) + b\sin(k_2t)]\right\} = \left[\frac{a+jb}{(s+k_1+jk_2)^n} + \frac{a-jb}{(s+k_1-jk_2)^n}\right]$
Final value theorem	$\lim_{t\to\infty} x(t) = \lim_{s\to 0} sX(s)$

Note that it is implicit in all of these developments that s, a and b have units of inverse time. In the angular sense this is understood as radians per unit time.

The above discussion is based on the premise that the function to be inverted will occur as a ratio of two polynomials in s. One notable exception to this occurs with the transport lag $G(s) = \mathrm{e}^{-\tau s}$ in Table 3.5. If this is simply a multiplier of the expression, it can be used subsequently to time shift the result. If it is embedded, special procedures such as the *Padé* approximation will be required (Section 8.2.1.1).

3.9.2.2 Use of Laplace Transforms to Find the System Response

Now consider the use of Laplace transforms in solution of the modelling problem represented by the *linear state equation* (Equation 3.100).

$$\dot{\boldsymbol{x}} = \boldsymbol{A}\boldsymbol{x} + \boldsymbol{B}\boldsymbol{u} \tag{3.150}$$

If the elements of the matrices $\boldsymbol{A}$ and $\boldsymbol{B}$ are constant, transformation using the result (Equation 3.118) yields

$$\begin{aligned} s\boldsymbol{X}(s) &= \boldsymbol{x}(0) + \boldsymbol{A}\boldsymbol{X}(s) + \boldsymbol{B}\boldsymbol{U}(s) \\ [s\boldsymbol{I} - \boldsymbol{A}]\boldsymbol{X}(s) &= \boldsymbol{x}(0) + \boldsymbol{B}\boldsymbol{U}(s) \end{aligned} \tag{3.151}$$

$$\begin{aligned} \boldsymbol{X}(s) &= [s\boldsymbol{I} - \boldsymbol{A}]^{-1}\boldsymbol{x}(0) + [s\boldsymbol{I} - \boldsymbol{A}]^{-1}\boldsymbol{B}\boldsymbol{U}(s) \\ \boldsymbol{x}(t) &= L^{-1}\{[s\boldsymbol{I} - \boldsymbol{A}]^{-1}\}\boldsymbol{x}(0) + L^{-1}\{[s\boldsymbol{I} - \boldsymbol{A}]^{-1}\boldsymbol{B}\boldsymbol{U}(s)\} \end{aligned} \tag{3.152}$$

The matrix of time functions resulting from the first inversion $L^{-1}\{[s\boldsymbol{I} - \boldsymbol{A}]^{-1}\}$ is known as the *state transition matrix* (or 'matrix exponential'– Section 3.9.3.2) and the result of its multiplication with the numerical vector of initial values $\boldsymbol{x}(0)$ will be the *complementary solution*. One can get an idea of the structure of the *state transition matrix* by examining a 2×2 system:

$$\begin{aligned} \boldsymbol{A} &= \begin{bmatrix} a_{11} & a_{12} \\ a_{21} & a_{22} \end{bmatrix} \\ [s\boldsymbol{I} - \boldsymbol{A}] &= \begin{bmatrix} s - a_{11} & -a_{12} \\ -a_{21} & s - a_{22} \end{bmatrix} \end{aligned} \tag{3.153}$$

Recall that

$$M^{-1} = \frac{\text{adj}\{M\}}{\det\{M\}} \tag{3.154}$$

where

$$\text{adj}\{M\} = [\text{cofactor}\{M\}]^{\text{T}} \tag{3.155}$$

where

$$\text{cofactor}\{M\} = \begin{bmatrix} +M_{11} & -M_{12} & \cdots & (-1)^{1+N}M_{1N} \\ -M_{21} & +M_{22} & \cdots & (-1)^{2+N}M_{2N} \\ \vdots & \vdots & \ddots & \vdots \\ (-1)^{N+1}M_{N1} & (-1)^{N+2}M_{N2} & \cdots & +M_{NN} \end{bmatrix} \tag{3.156}$$

and the M_{ij} are minors, that is the determinant of what is left after eliminating row i and column j.

$$M_{ij} = \det\begin{bmatrix} m_{11} & \cdots & m_{1,j-1} & m_{1j+1} & \cdots & m_{1N} \\ \vdots & \ddots & \vdots & \vdots & \ddots & \vdots \\ m_{i-1,1} & \cdots & m_{i-1,j-1} & m_{i-1,j+1} & \cdots & m_{i-1,N} \\ m_{i+1,1} & \cdots & m_{i+1,j-1} & m_{i+1,j+11} & \cdots & m_{i+1,N} \\ \vdots & \ddots & \vdots & \vdots & \ddots & \vdots \\ m_{N1} & \cdots & m_{N,j-1} & m_{N,j+1} & \cdots & m_{NN} \end{bmatrix} \tag{3.157}$$

where the m_{ij} are the original elements of M. Applying this to $[sI-A]$ obtain

$$\begin{aligned} [sI-A]^{-1} &= \frac{1}{(s-a_{11})(s-a_{22})-a_{21}a_{12}}\begin{bmatrix} s-a_{22} & a_{12} \\ a_{21} & s-a_{11} \end{bmatrix} \\ &= \begin{bmatrix} \dfrac{s-a_{22}}{(s-a_{11})(s-a_{22})-a_{21}a_{12}} & \dfrac{a_{12}}{(s-a_{11})(s-a_{22})-a_{21}a_{12}} \\ \dfrac{a_{21}}{(s-a_{11})(s-a_{22})-a_{21}a_{12}} & \dfrac{s-a_{11}}{(s-a_{11})(s-a_{22})-a_{21}a_{12}} \end{bmatrix} \end{aligned} \tag{3.158}$$

So the *complementary solution* $x_C(t)$ (for $u(t)=0$), given $x(0) = \begin{pmatrix} x_{10} \\ x_{20} \end{pmatrix}$, requires evaluation of

$$x_C(t) = L^{-1}\left\{\begin{pmatrix} \dfrac{(s-a_{22})x_{10}}{(s-a_{11})(s-a_{22})-a_{21}a_{12}} + \dfrac{a_{12}x_{20}}{(s-a_{11})(s-a_{22})-a_{21}a_{12}} \\ \dfrac{a_{21}x_{10}}{(s-a_{11})(s-a_{22})-a_{21}a_{12}} + \dfrac{(s-a_{11})x_{20}}{(s-a_{11})(s-a_{22})-a_{21}a_{12}} \end{pmatrix}\right\} \tag{3.159}$$

Each term arises as a ratio of two polynomials in s. It has been noted in reference to Table 3.4 that the terms in the forcing vector $U(s)$ in Equation 3.151 will likewise be ratios of polynomials in s. Thus, the *particular solution* $x_P(t)$ (for $u(t) \neq 0$ but $x(0)=0$) will be similar to Equation 3.159 with larger polynomials. The final solution $x(t)=x_C(t)+x_P(t)$ requires inversion of these expressions using the *partial fraction expansion* methods presented in Section 3.9.2.1.

The idea of a *transfer function* for linear systems with constant coefficients was built up in Section 3.9.2.1 based on a SISO system. Now one sees that it easily extends to MIMO systems for

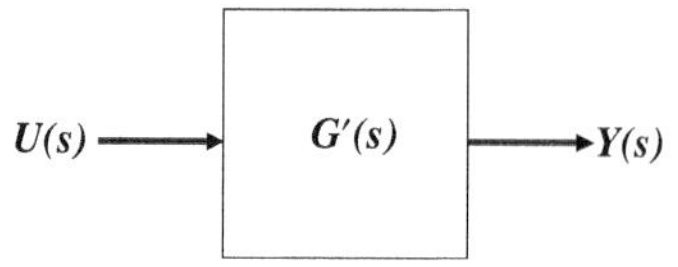

Figure 3.20 Representation of a general SISO or MIMO *input–output* transfer function.

the case $\boldsymbol{x}(0)=\boldsymbol{0}$ (Figure 3.20). Then the state Equation 3.151 becomes the *input–output* form

$$\boldsymbol{X}(s)=[s\boldsymbol{I}-\boldsymbol{A}]^{-1}\boldsymbol{B}\boldsymbol{U}(s) \tag{3.160}$$

that is

$$\boldsymbol{X}(s)=\boldsymbol{G}(s)\boldsymbol{U}(s) \quad \text{where } \boldsymbol{G}(s)=[s\boldsymbol{I}-\boldsymbol{A}]^{-1}\boldsymbol{B} \tag{3.161}$$

Following Equation 3.154, note that a single scalar polynomial

$$D(s)=\det[s\boldsymbol{I}-\boldsymbol{A}] \tag{3.162}$$

will apply as a denominator throughout this transfer function, and that the adjoint matrix of $[s\boldsymbol{I}-\boldsymbol{A}]$, multiplied by $\boldsymbol{B}$, will yield the set of numerator polynomials $\boldsymbol{N}(s)$, expressed here as a constant coefficient matrix for each power of s, that is

$$\boldsymbol{X}(s)=\frac{1}{(s^n+d_{n-1}s^{n-1}+\cdots+d_1s+d_0)}\left[\boldsymbol{N}_{n-1}s^{n-1}+\boldsymbol{N}_{n-2}s^{n-2}+\cdots+\boldsymbol{N}_1s+\boldsymbol{N}_0\right]\boldsymbol{U}(s) \tag{3.163}$$

that is

$$\boldsymbol{X}(s)=D^{-1}(s)\boldsymbol{N}(s)\boldsymbol{U}(s) \tag{3.164}$$

where

$$D(s)=(s-\lambda_1)(s-\lambda_2)\cdots(s-\lambda_n) \tag{3.165}$$

and the λ_i are clearly the eigenvalues of $\boldsymbol{A}$.

Since in the above development $\boldsymbol{x}$ contains all of the states, the requirement that $\boldsymbol{x}(0)=\boldsymbol{0}$ implies complete steady state at $t=0$.

More general linear *input–output* forms may not involve all of the states (Figure 3.20). These are stated directly as

$$\boldsymbol{Y}(s)=\boldsymbol{G}'(s)\boldsymbol{U}(s) \tag{3.166}$$

From Equation 3.71 for linear systems, $\boldsymbol{Y}(s)$ can effectively arise as some combination of $\boldsymbol{X}(s)$ and $\boldsymbol{U}(s)$

$$\boldsymbol{Y}(s)=\boldsymbol{C}\boldsymbol{X}(s)+\boldsymbol{D}\boldsymbol{U}(s) \tag{3.167}$$

$$\boldsymbol{Y}(s)=\left[D^{-1}(s)\boldsymbol{C}\boldsymbol{N}(s)+\boldsymbol{D}\right]\boldsymbol{U}(s) \tag{3.168}$$

$$\boldsymbol{Y}(s)=\boldsymbol{G}'(s)\boldsymbol{U}(s) \tag{3.169}$$

Though input–output forms are usually not derived like this, one clearly expects from Equation 3.168 that the transfer function $\boldsymbol{G}'(s)$ for this state-based system will be a similar matrix of polynomial ratios, with the denominator factors for this state-based system arising similarly from the same root values of det$[sI - A] = 0$ (i.e. the *characteristic equation* of the state open-loop system). However, this is not generally the case for input-output systems, where arbitrary polynomial ratios can occur in G', requiring more d_i and N_i terms (see Section 7.8.1)

The following examples illustrate the use of Laplace transforms to obtain the output response functions of several systems, for some standard input excitation functions. The response to an oscillating input is dealt with later in Chapter 8, for example Example 8.4.

Example 3.6

First-order system with step input.

Example 3.3 considered the tank flow system repeated in Figure 3.21 and provided a linearisation for operation near level h_0, with h the deviation from h_0:

$$\frac{dh}{dt} = \left\{-\frac{k}{2A\sqrt{h_0}}\right\}h + \left\{\frac{1}{A}\right\}F_0 \tag{3.170}$$

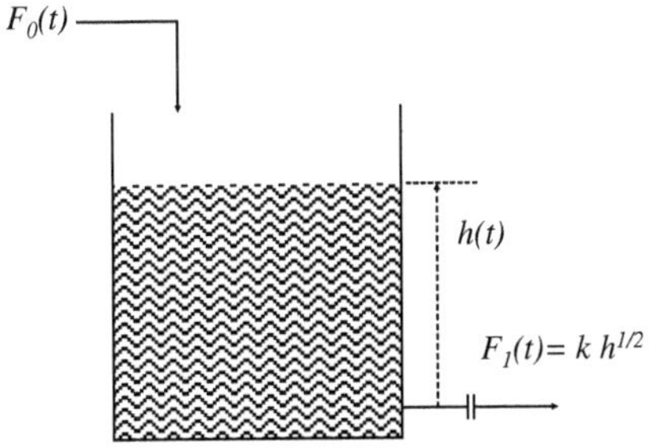

Figure 3.21 Tank with restriction orifice at exit.

Rearranging,

$$\left\{\frac{2A\sqrt{h_0}}{k}\right\}\frac{dh}{dt} + h = \left\{\frac{2\sqrt{h_0}}{k}\right\}F_0 \tag{3.171}$$

This is the standard form of a *first-order system*

$$\tau\frac{dh}{dt} + h = KF_0 \tag{3.172}$$

where

$$\text{time constant } \tau = \left\{\frac{2A\sqrt{h_0}}{k}\right\} \tag{3.173}$$

and

$$\text{gain } K = \left\{\frac{2\sqrt{h_0}}{k}\right\} \tag{3.174}$$

Under the condition that the system is initially steady at $h = 0$ (i.e. at the linearisation point $[F_0, h_0]$), Equation 3.172 can be transformed to

$$\tau s H(s) + H(s) = K F_0(s) \tag{3.175}$$

that is

$$\boxed{G(s) = \frac{H(s)}{F_0(s)} = \frac{K}{\tau s + 1}} \tag{3.176}$$

which is the standard form of a *first-order system* transfer function. Now consider a step of size D in the inlet flow F_0 at time $t = 0$:

$$F_0(s) = \frac{D}{s} \tag{3.177}$$

$$H(s) = G(s)F_0(s) = \frac{KD}{s(\tau s + 1)} \tag{3.178}$$

Partial fraction expansion:

$$\frac{KD}{s(\tau s + 1)} = \frac{A}{s} + \frac{B}{(\tau s + 1)} \tag{3.179}$$

$$\left.\begin{matrix} \times s \\ s \to 0 \end{matrix}\right\} \quad \frac{KD}{1(\tau 0 + 1)} = A \tag{3.180}$$

$$\left.\begin{matrix} \times(\tau s + 1) \\ s \to -\dfrac{1}{\tau} \end{matrix}\right\} \quad \frac{KD}{-1/\tau} = B \tag{3.181}$$

So

$$h(t) = L^{-1}\left\{\frac{KD}{s} - \frac{KD}{\left(s + \dfrac{1}{\tau}\right)}\right\} \tag{3.182}$$

$$h(t) = KD\left(1 - e^{-t/\tau}\right) \tag{3.183}$$

as a perturbation variable. This result is plotted in Figure 3.22.

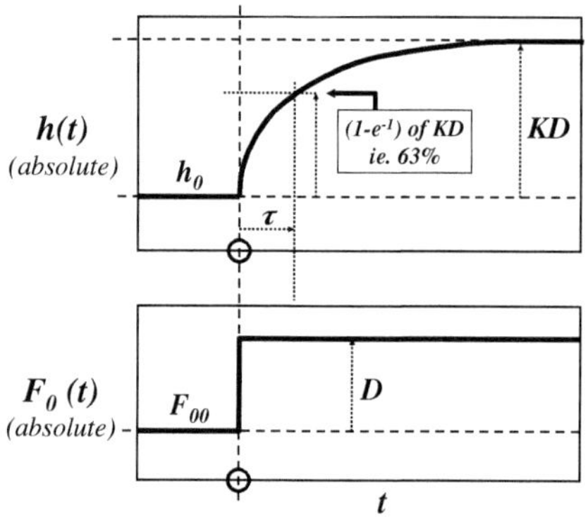

Figure 3.22 Tank level response to a step in the inflow.

The absolute level asymptotically reaches $h_0 + KD$ at large times. This gives meaning to the word 'gain' for K, which is seen to be the multiplying factor applied to the input change to achieve the steady-state output change. It is noted that 63% of the final level change is reached at $t = \tau$. Clearly, a large time constant implies a slow response. This is to be expected from Equation 3.173, where it is seen that τ is proportional to the liquid surface area A and inversely proportional to the orifice flow coefficient.

Example 3.7

First-order system with impulse input.

The equivalent of an impulse in the inflow $F_0(t)$, to the tank system considered in Example 3.6, is the sudden dumping of a bucket of liquid into the tank at $t = 0$ (Figure 3.23).

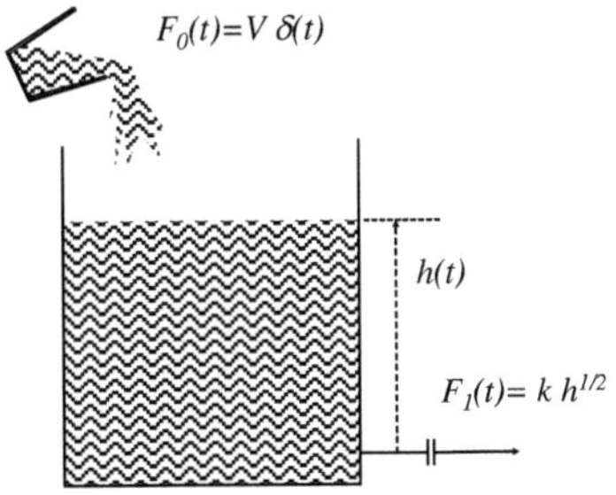

Figure 3.23 Impulse input to tank with restriction orifice at exit.

Following the same procedure as in Example 3.6, one obtains

$$F_0(s) = V \tag{3.184}$$

$$H(s) = G(s)F_0(s) = \frac{KV}{(\tau s + 1)} \tag{3.185}$$

This inverts directly as

$$h(t) = KV\,e^{-t/\tau} \tag{3.186}$$

as a perturbation variable. This result is plotted in Figure 3.24. In this case it is clear that a steady base level h_0 would need to be sustained by a steady underlying through-flow F_{00}.

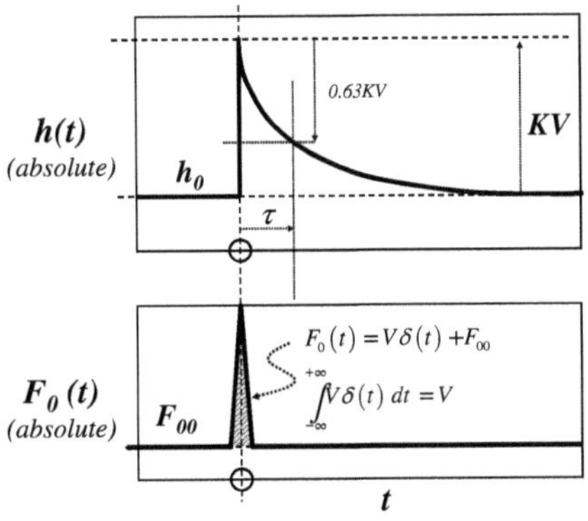

Figure 3.24 Tank level response to an impulse in the inflow.

Example 3.8

Second-order system with step input: (a) non-interactive; (b) interactive.

Consider the two cases of interconnected tanks in Figure 3.25. In case (a), the conditions in tank 2 have no effect on the operation of tank 1. Conversely, in case (b), tank 2 is said to 'load' tank 1 because of the effect of tank 2 level on the flow from tank 1.

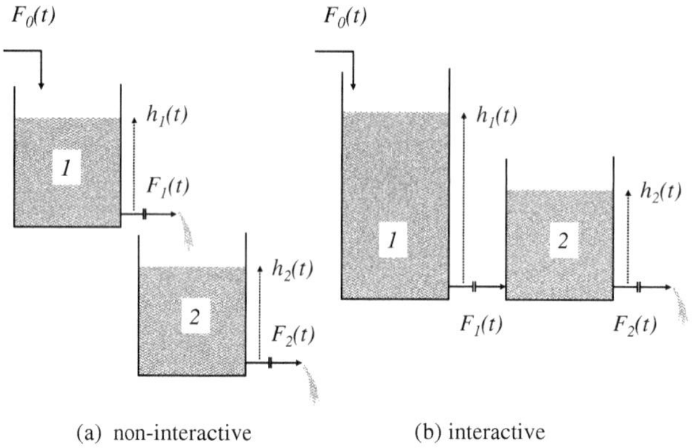

Figure 3.25 Two interconnected tanks with restriction orifices.

Volume balances:

$$A_1\frac{dh_1}{dt} = F_0 - F_1 \tag{3.187}$$

$$A_2\frac{dh_2}{dt} = F_1 - F_2 \tag{3.188}$$

Linearise the algebraic equation for F_2:

$$F_2 = k_2\sqrt{h_2} \tag{3.189}$$

$$\Delta F_2 = \Delta\left\{k_2\sqrt{h_2}\right\} = \left\{\frac{k_2}{2\sqrt{h_2}}\right\}_0 \Delta h_2 \tag{3.190}$$

and let

$$\alpha_2 = \left\{\frac{k_2}{2\sqrt{h_2}}\right\}_0 \tag{3.191}$$

Linearise the algebraic equation for F_1:

case (a):

$$F_1 = k_1\sqrt{h_1} \tag{3.192}$$

$$\Delta F_1 = \Delta\left\{k_1\sqrt{h_1}\right\} = \left\{\frac{k_1}{2\sqrt{h_1}}\right\}_0 \Delta h_1 \tag{3.193}$$

and let

$$\alpha_1 = \left\{\frac{k_1}{2\sqrt{h_1}}\right\}_0 \tag{3.194}$$

case (b):

$$F_1 = k_1\sqrt{h_1 - h_2} \tag{3.195}$$

$$\Delta F_1 = \Delta\left\{k_1\sqrt{h_1 - h_2}\right\} = \left\{\frac{k_1}{2\sqrt{h_1 - h_2}}\right\}_0 (\Delta h_1 - \Delta h_2) \tag{3.196}$$

and let

$$\alpha_{12} = \left\{\frac{k_1}{2\sqrt{h_1 - h_2}}\right\}_0 \tag{3.197}$$

Then, reverting to h_1, h_2, F_0, F_1 and F_2 as deviation variables:

$$\text{case (a)}: \quad \frac{\mathrm{d}}{\mathrm{d}t}\begin{pmatrix} h_1 \\ h_2 \end{pmatrix} = \begin{bmatrix} -\frac{\alpha_1}{A_1} & 0 \\ \frac{\alpha_1}{A_2} & -\frac{\alpha_2}{A_2} \end{bmatrix}\begin{pmatrix} h_1 \\ h_2 \end{pmatrix} + \begin{bmatrix} \frac{1}{A_1} \\ 0 \end{bmatrix} F_0 \tag{3.198}$$

$$\text{case (b)}: \quad \frac{\mathrm{d}}{\mathrm{d}t}\begin{pmatrix} h_1 \\ h_2 \end{pmatrix} = \begin{bmatrix} -\frac{\alpha_{12}}{A_1} & \frac{\alpha_{12}}{A_1} \\ \frac{\alpha_{12}}{A_2} & -\frac{(\alpha_{12} + \alpha_2)}{A_2} \end{bmatrix}\begin{pmatrix} h_1 \\ h_2 \end{pmatrix} + \begin{bmatrix} \frac{1}{A_1} \\ 0 \end{bmatrix} F_0 \tag{3.199}$$

Because of the unique algebraic relationships between flows and levels in this system,

case (a):

$$F_1 = \alpha_1 h_1 \tag{3.200}$$

and

$$F_2 = \alpha_2 h_2 \tag{3.201}$$

case (b):

$$F_1 = \alpha_{12}(h_1 - h_2) \tag{3.202}$$

and

$$F_2 = \alpha_2 h_2 \tag{3.203}$$

it is interesting to use the flows as the states instead of the levels:

$$\text{case (a)}: \quad \frac{\mathrm{d}}{\mathrm{d}t}\begin{pmatrix} F_1 \\ F_2 \end{pmatrix} = \begin{bmatrix} -\frac{\alpha_1}{A_1} & 0 \\ \frac{\alpha_2}{A_2} & -\frac{\alpha_2}{A_2} \end{bmatrix}\begin{pmatrix} F_1 \\ F_2 \end{pmatrix} + \begin{bmatrix} \frac{\alpha_1}{A_1} \\ 0 \end{bmatrix} F_0 \tag{3.204}$$

$$\text{case (b)}: \frac{\mathrm{d}}{\mathrm{d}t}\begin{pmatrix} F_1 \\ F_2 \end{pmatrix} = \begin{bmatrix} -\dfrac{\alpha_{12}}{A_1} - \dfrac{\alpha_{12}}{A_2} & \dfrac{\alpha_{12}}{A_2} \\ \dfrac{\alpha_2}{A_2} & -\dfrac{\alpha_2}{A_2} \end{bmatrix}\begin{pmatrix} F_1 \\ F_2 \end{pmatrix} + \begin{bmatrix} \dfrac{\alpha_{12}}{A_1} \\ 0 \end{bmatrix} F_0 \tag{3.205}$$

The lower triangular format in case (a) reflects the fact that in the *non-interacting* case, each tank can be solved for in turn. In terms of SISO transfer functions,

$$G_i(s) = \frac{F_i(s)}{F_{i-1}(s)} = \frac{K_i}{\tau_i s + 1} \quad \text{with} \quad \begin{cases} K_i = 1 \\ \tau_i = \dfrac{A_i}{\alpha_i} = \left\{\dfrac{2A_i\sqrt{h_i}}{k_i}\right\}_0 \end{cases} \tag{3.206}$$

and

$$F_N(s) = \left[\prod_{i=1}^{N} G_i(s)\right] F_0(s) \tag{3.207}$$

In Figure 3.26, the increasing lag is noted in the flow response as the order increases. Only the first-order response has a nonzero slope at $t = 0^+$. Applying Equation 3.160 to Equation 3.204 (or directly from Equation 3.206), one obtains

$$\text{case (a)}: \quad F_2(s) = \left(\frac{1}{s^2 + \left(\dfrac{A_1}{\alpha_1} + \dfrac{A_2}{\alpha_2}\right)s + 1}\right) F_0(s) \quad \text{(non-interacting)} \tag{3.208}$$

and applying Equation 3.160 to Equation 3.205 one obtains

$$\text{case (b)}: \quad F_2(s) = \left(\frac{1}{s^2 + \left(\dfrac{A_1}{\alpha_{12}} + \dfrac{A_2}{\alpha_2} + \dfrac{A_1}{\alpha_2}\right)s + 1}\right) F_0(s) \quad \text{(interacting)} \tag{3.209}$$

where one notes that α_{12} plays the same role as α_1. Thus, the effect of the 'loading' of the first tank by the second is to add the term A_1/α_2 to the coefficient of s in the denominator, which will be seen to increase the value of the damping factor ζ (Equation 3.36). As expected, this system will then have a more sluggish response than the case of the non-interacting tanks.

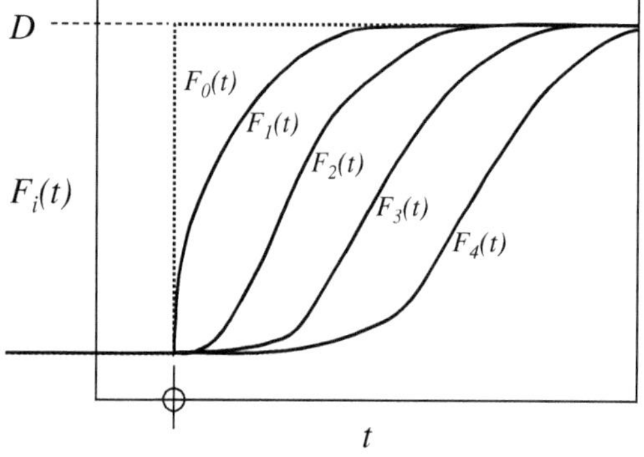

Figure 3.26 Step input flow response through non-interacting tanks in series.

Example 3.9

Analysis of a general SISO second-order system.

The consideration of the trolley with spring and dashpot in Section 3.5 led to the general second-order system in Equation 3.36:

$$\tau^2 \frac{d^2y}{dt^2} + 2\zeta\tau\frac{dy}{dt} + y = KF \tag{3.210}$$

Under the particular circumstance of $y(0) = 0$ and $(dy/dt)|_0 = 0$, this equation transforms to

$$\tau^2 s^2 Y(s) + 2\zeta\tau s Y(s) + Y(s) = KF(s) \tag{3.211}$$

that is

$$G(s) = \frac{Y(s)}{F(s)} = \frac{K}{\tau^2 s^2 + 2\zeta\tau s + 1} \tag{3.212}$$

which is the standard form of the transfer function for a *second-order system*.

In Section 3.5, this was then represented in the state-space form (Equation 3.43)

$$\frac{d}{dt}\begin{pmatrix} x_1 \\ x_2 \end{pmatrix} = \begin{bmatrix} -\dfrac{2\zeta}{\tau} & -\dfrac{1}{\tau^2} \\ 1 & 0 \end{bmatrix}\begin{pmatrix} x_1 \\ x_2 \end{pmatrix} + \begin{bmatrix} \dfrac{K}{\tau^2} \\ 0 \end{bmatrix} F \tag{3.213}$$

In the case of the trolley, the first state x_1 is recognised as the velocity, and the second x_2 as the position. From Equation 3.160,

$$\boldsymbol{X}(s) = [s\boldsymbol{I} - \boldsymbol{A}]^{-1}\boldsymbol{BU}(s) \tag{3.214}$$

$$\boldsymbol{X}(s) = \begin{bmatrix} s + \dfrac{2\zeta}{\tau} & \dfrac{1}{\tau^2} \\ -1 & s \end{bmatrix}^{-1} \begin{bmatrix} \dfrac{K}{\tau^2} \\ 0 \end{bmatrix} F(s) \tag{3.215}$$

$$\boldsymbol{X}(s) = \frac{K}{\tau^2 s + 2\zeta\tau s + 1}\begin{bmatrix} s \\ 1 \end{bmatrix} F(s) \tag{3.216}$$

One notes that regardless of the form of the input $F(t)$, the denominator in Equation 3.216 will contribute two terms in any partial fraction expansion of the terms of $\boldsymbol{X}(s)$, for example

$$X_1(s) = \frac{A}{\left(s + \dfrac{\zeta + \sqrt{\zeta^2 - 1}}{\tau}\right)} + \frac{B}{\left(s + \dfrac{\zeta - \sqrt{\zeta^2 - 1}}{\tau}\right)} + \cdots \tag{3.217}$$

The denominator factors are determined using the roots of

$$\tau^2 s^2 + 2\zeta\tau s + 1 = 0 \tag{3.218}$$

which is known as the *characteristic equation* of the system. More generally, for any linear state system in the form of Equation 3.214, the characteristic equation determining the factors (contributed by the system itself) in the partial fraction expansion of any response is

$$\det[s\boldsymbol{I} - \boldsymbol{A}] = 0 \tag{3.219}$$

which obviously yields the *eigenvalues* of $\boldsymbol{A}$. The importance of these eigenvalues lies in their strong impact on the types of response which will occur. If the quantities

$$\lambda = -\frac{\zeta \pm \sqrt{\zeta^2 - 1}}{\tau} \tag{3.220}$$

are real, Table 3.4 shows that these terms of the expansion will invert as

$$x_1(t) = A\,\mathrm{e}^{\lambda_1 t} + B\,\mathrm{e}^{\lambda_2 t} + \cdots \tag{3.221}$$

which will decay to zero if the λ_i are negative, and grow unboundedly if they are positive.

There is a possibility that the eigenvalues are complex, in which case they will be complex conjugates of each other. It is seen that this will occur for damping factors $\zeta < 1$, when the factors can be written more conveniently as

$$X_1(s) = \frac{A}{\left[\left(s + \frac{\zeta}{\tau}\right) + j\frac{\sqrt{1-\zeta^2}}{\tau}\right]} + \frac{B}{\left(\left(s + \frac{\zeta}{\tau}\right) - j\frac{\sqrt{1-\zeta^2}}{\tau}\right)} + \cdots \tag{3.222}$$

Since A and B will be complex conjugates, these two terms can be combined to produce a real function of s (no j's present) as a numerator polynomial in s over a common denominator polynomial in s.

$$X_1(s) = \frac{a\left(s + \frac{\zeta}{\tau}\right) + \left(b - a\frac{\zeta}{\tau}\right)}{\left(s + \frac{\zeta}{\tau}\right)^2 + \left(\frac{\sqrt{1-\zeta^2}}{\tau}\right)^2} + \cdots \tag{3.223}$$

So, from Table 3.5, since 's' always appears as $(s + \zeta/\tau)$,

$$\begin{aligned} &x_1(t) = \text{const}_1 \times \mathrm{e}^{-\zeta t/\tau}\cos(\omega t) + \text{const}_2 \times \mathrm{e}^{-\zeta t/\tau}\sin(\omega t) \\ &\text{where}\quad \text{const}_1 = a, \quad \text{const}_2 = \left(b - a\frac{\zeta}{\tau}\right)\frac{\tau}{\sqrt{1-\zeta^2}} \quad \text{and} \quad \omega = \frac{\sqrt{1-\zeta^2}}{\tau} \end{aligned} \tag{3.224}$$

As will be seen later (Example 8.4), the sum of cos and sin can be represented by a single sinusoidal oscillation with a phase shift. It is important to note that this oscillation will be multiplied by exp $(-\zeta t/\tau)$. It is now clear why ζ is referred to as the 'damping factor'. The oscillations become slower with maximum decay rate as ζ increases to 1. For $\zeta \geq 1$, there are no oscillations. As ζ decreases from 1 to 0, the oscillations become faster and the decay rate reduces, until a permanent steady oscillation is obtained at $\zeta = 0$ with frequency $1/\tau$.

Figure 3.27 shows some responses of a second-order system to a step input for various ζ values. In the case of the trolley in Figure 3.13, it is initially at rest, and then a step occurs in the force that is applied. If a load is placed in or on a motor vehicle, a similar effect is achieved, with the springs and

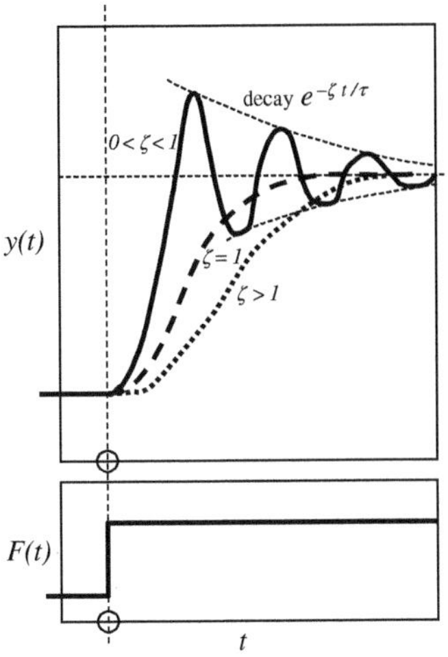

Figure 3.27 Response of a second-order system to a step input.

shock absorber playing the same parts. If the shock absorbers are worn, an oscillation can persist for some time. Table 3.6 summarises the damping categories. As ζ decreases from 1 to 0, the *size* of the negative real part of the root of the characteristic decreases (slower decay) whilst the imaginary part increases (faster oscillation). This is a useful way of describing the effect of roots arising from the *characteristic equation* (Equation 3.219) for any system, not just second-order systems. They are characterised by their locations in the complex plane.

Table 3.6 Damping factor ranges.

Category	ζ	Effect
Overdamped	$\zeta>1$	Slow, sluggish response, no oscillation, like tanks in series
Critically damped	$\zeta=1$	No oscillation, equal roots, like equal tanks in series
Underdamped	$0<\zeta<1$	Oscillation with decay
Undamped	$\zeta=0$	Steady oscillation – no decay
Unstable	$\zeta<0$	Exponential growth of oscillating output

3.9.2.3 Open-Loop Stability in the *s*-Domain

The idea of stability becomes very important when controllers are left in charge of a process, and this will be considered in detail later (Chapter 8). The problem is that closing a control loop around a process invites a number of problems, for example overreaction to a disturbance. For the meantime, just the open loop will be considered. Most processes are quite well behaved on their own (in open loop); for example, in all of the tank flow examples in Section 3.9.2.2, if the inflow is stepped up, the levels will rise until the exit flows balance the new inflows, where a new equilibrium is found. However, a few naturally unstable systems exist in the processing industries – notably cooled or heated reactors (exothermic or endothermic). Reaction rates are strongly dependent on temperature according to the Arrhenius relationship for rate constants:

$$k = k_0\, e^{-E/RT} \tag{3.225}$$

so if the cooling is reduced on an exothermic reactor, the temperature increases, giving higher reaction rates, and thus even more heat is generated, so temperature increases rapidly and a runaway reaction occurs (provided the reagent supply is sufficient). Conversely, an endothermic reaction will die if heating is reduced.

The formal definition of stability is as follows:

> A system is stable if its outputs remain bounded for all bounded inputs.

The important thing is that if just *one* finite input function can be found that causes unbounded growth of the output, then the system must be declared *unstable.* An example of this would be a bridge or chimney that might only resonate in its vortex street at a particular wind speed.

The open-loop MIMO systems considered in Section 3.9.2.2 were of the forms

$$\text{state}: \quad \boldsymbol{X}(s) = \boldsymbol{G}(s)\boldsymbol{U}(s) \tag{3.226}$$

$$\text{input–output}: \quad \boldsymbol{Y}(s) = \boldsymbol{G}'(s)\boldsymbol{U}(s) \tag{3.227}$$

For these systems derived from the state equation it was noted that both transfer functions involve denominator factors based on roots of the characteristic equation of the state system, namely $\det[s\boldsymbol{I} - \boldsymbol{A}] = 0$, that is the eigenvalues λ_i of $\boldsymbol{A}$. The system itself thus contributes corresponding partial fractions to any response, in the form of factors $\exp\{\mathrm{Re}(\lambda_i) \times t\}$. It is thus clear that the eigenvalues of $\boldsymbol{A}$ must all have negative real parts for open-loop stability (Figure 3.28). More generally, for *input–output* systems, the denominator factors in the elements of $\boldsymbol{G}'(s)$ in Figure 3.2a can all differ, giving more factors $(s - a_i)$ requiring $\mathrm{Re}(a_i) < 0$ for open-loop stability.

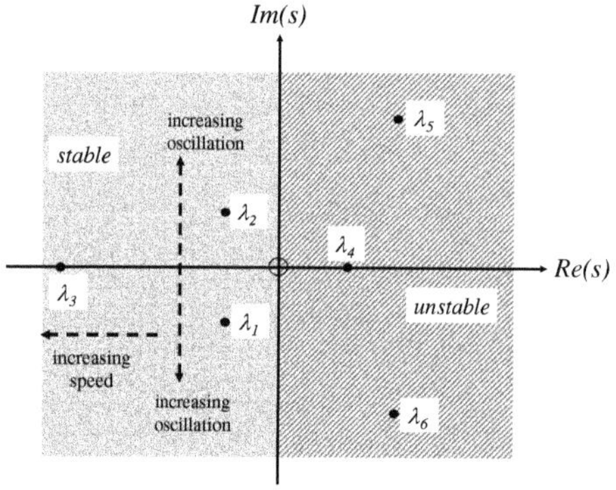

Figure 3.28 *s*-domain characteristic equation roots (stable: λ_1, λ_2, λ_3; unstable: λ_4, λ_5, λ_6).

Example 3.10

Stability of a second-order system.

$$\frac{d}{dt}\begin{pmatrix} x_1 \\ x_2 \end{pmatrix} = \begin{bmatrix} -2 & 1 \\ 9 & -2 \end{bmatrix}\begin{pmatrix} x_1 \\ x_2 \end{pmatrix} + \begin{bmatrix} 0 \\ 3 \end{bmatrix} u_1 \tag{3.228}$$

$$\det[sI - A] = \det\begin{bmatrix} s+2 & -1 \\ -9 & s+2 \end{bmatrix} \tag{3.229}$$

$$\det[sI - A] = s^2 + 4s - 5 \tag{3.230}$$

So the roots of the characteristic equation are

$$\lambda = \frac{-4 \pm \sqrt{16+20}}{2} = +1, -5 \tag{3.231}$$

of which the positive real root indicates instability.

3.9.3 Mathematical Solutions for System Response in the *z*-Domain

The continuous mathematical descriptions of a process considered in Section 3.9.1 (t-domain) and Section 3.9.2 (s-domain) are useful starting points for development of the control, identification and optimisation ideas of importance in the processing industries. In practice, of course, only an analogue computer or controller could deal with a process on this basis, and some degree of *discretisation* is necessary in modern monitoring and control systems. The sequence of events in these systems is much like the general algorithms presented in Section 3.1 (Figure 3.29). There are a number of timing and synchronisation issues to be considered in a *real-time program* like this. The main loop will execute at reasonably small intervals Δt, but it would be wasteful to execute all of the tasks on every pass. Usually sampling intervals ('scan cycles') are set individually depending on the speed of variation of individual variables. Variables with long responses and their associated controllers will execute infrequently, whilst the continuous and logical variables in safety trip systems will be scanned in and out on a rapid cycle (or trigger 'interrupts').

There are some mathematical implications for *sampled data systems* like this. Chief of these is that the computer settings going back to the plant move in a series of steps. This effect would be minimal for fast sampling, and the continuous theories would apply quite well. However, there are frequently situations where data are only updated on large intervals (e.g. gas–liquid chromatograph measurements) or a control algorithm must work on a large interval (e.g. dynamic matrix control, Section 7.8.2). Moreover, since new data are only being reconsidered at discrete times, there is no point in repeating calculations between these times. *Sampled data systems'* theory and representations based on the *z-transform* allow one to properly describe behaviour from one sampling instant to the next, and to derive and analyse useful recursive formulae for real-time implementations.

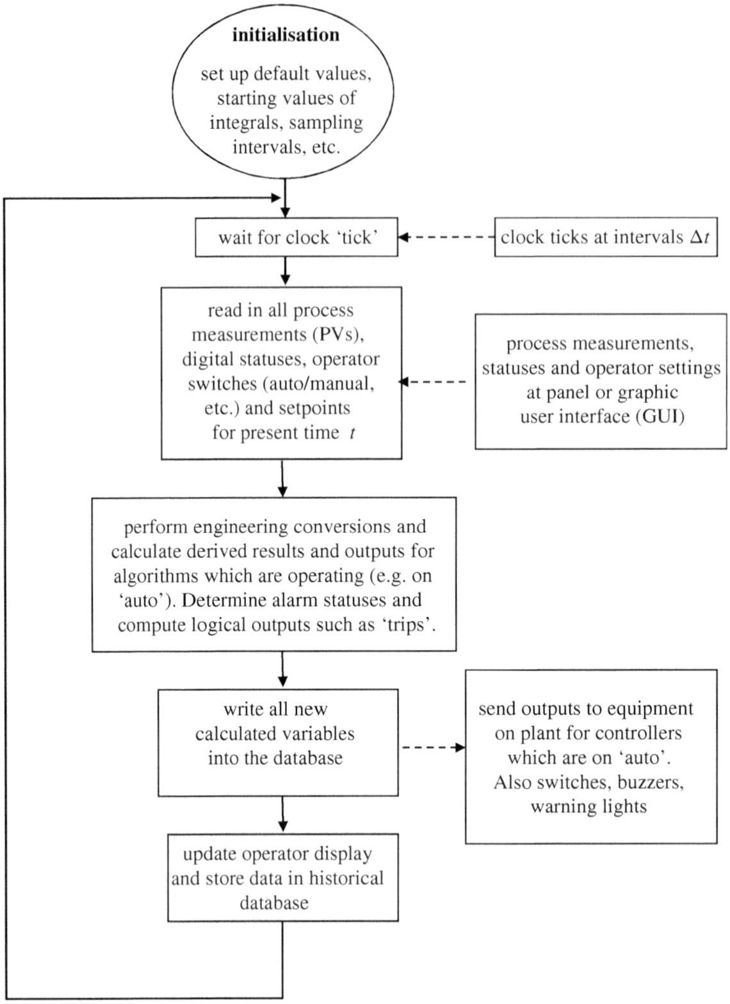

Figure 3.29 Typical timed loop of tasks on a plant computer.

3.9.3.1 Review of Some *z*-Transform Results

The initial requirement is to develop a formal way of describing a time series of sampled values. This is more or less just a vector of numbers to which a new value is added at each sampling instant. It is conceptually useful to represent these numbers as a series of delayed Dirac impulses of size *numerically equal* to the original signal values at the time instants. In Figure 3.30, some license has been taken to represent these impulse sizes by the heights of equal-base triangles. So the sampled signal then becomes the *impulse-modulated* function

$$x_*(t) = \sum_{i=0}^{\infty} x(iT)\delta(t - iT) \tag{3.232}$$

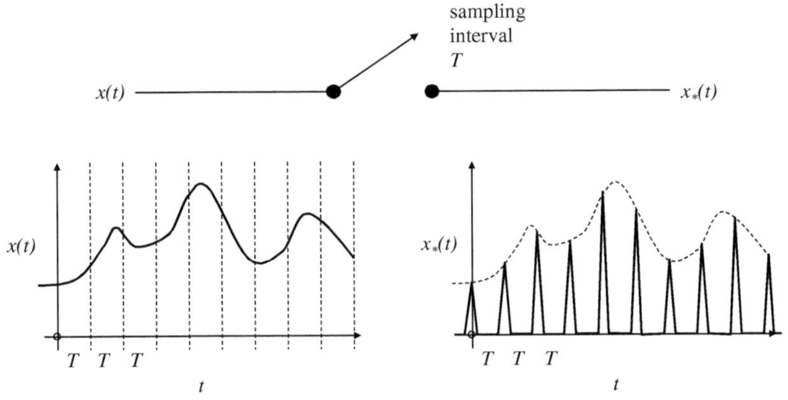

Figure 3.30 Representation of sampled values as a series of proportional impulses.

To transform this to the s-domain, one makes use of the transport (or dead-time) lag e^{-Ts} (Table 3.5) to delay the impulses successively. It is also noted in Table 3.4 that the $\delta(t)$ transforms to 1.

$$X_*(s) = L\{x_*(t)\} = \sum_{i=0}^{\infty} x(iT)e^{-iTs} \tag{3.233}$$

The idea of *z-transforms* arises from the substitution

$$z = e^{Ts} \tag{3.234}$$

which is just a single forward time shift in the s-domain. Then the notation $x(z)$ is used for the transform $X_*(s)$ of the impulse-modulated signal, so that

$$x(z) = \sum_{i=0}^{\infty} x(iT)z^{-i} \tag{3.235}$$

A simple example would be a unit step at $t=0$ (Figure 3.31).

Here

$$x(z) = \sum_{i=0}^{\infty} z^{-i} = \frac{z}{z-1} \tag{3.236}$$

Some additional useful z-transforms and their corresponding Laplace transforms are included in Table 3.7. Some care must be taken in the use of these functions. The z functions are merely

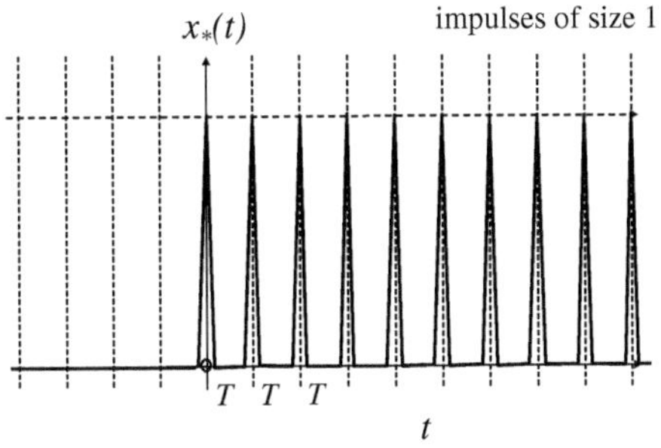

Figure 3.31 Impulse modulation of a unit step at $t=0$.

Table 3.7 Selected z-transforms and corresponding Laplace transforms.

x(t)	**x(z)**	**X(s)**	**Plot**
$\delta(t)$	1	1	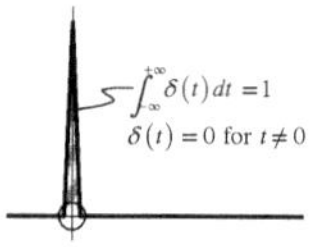
1	$\frac{z}{z-1}$	$\frac{1}{s}$	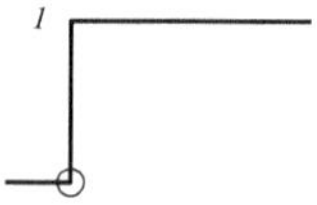
t	$\frac{Tz}{(z-1)^2}$	$\frac{1}{s^2}$	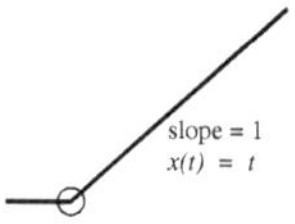
e^{-at}	$\frac{z}{z-e^{-aT}}$	$\frac{1}{s+a}$	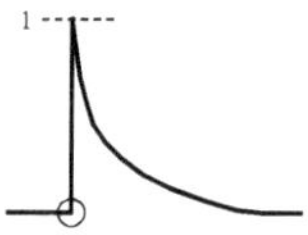
$\sin(\omega t)$	$\frac{z\sin(\omega T)}{z^2-2z\cos(\omega T)+1}$	$\frac{\omega}{s^2+\omega^2}$	
$\cos(\omega t)$	$\frac{z[z-\cos(\omega T)]}{z^2-2z\cos(\omega T)+1}$	$\frac{s}{s^2+\omega^2}$	
Delay nT	z^{-n}	e^{-nTs}	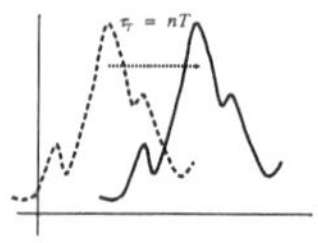
Zero-order hold of $F(s)$	$\frac{z-1}{z}Z\left\{\frac{F(s)}{s}\right\}$	$\frac{1-e^{-sT}}{s}F(s)$	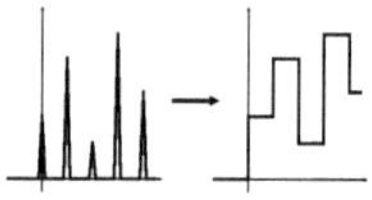

s-domain functions in disguise (where z replaces e^{Ts}). In particular, the $x(z)$ functions in Table 3.7 are the Laplace transforms of the *impulse-modulated* signals, and only 'represent' the smooth functions such as t and e^{-at} at discrete points in time. The transformation operator $Z\{\cdot\}$ implies that the argument is to be replaced by the corresponding z function from this table.

To use the z notation to solve for the way continuous signals move through a system, one needs to 'hold' values between the impulses. The most common way of doing this is by means of the 'zero-order hold', which keeps the signal constant at the last value, rather than trying to interpolate or extrapolate it in some other way according to higher order holds.

Consider the general signal of Equation 3.235:

$$x(z) = \sum_{i=0}^{\infty} x(iT)z^{-i} \tag{3.237}$$

that is

$$X_*(s) = \sum_{i=0}^{\infty} x(iT)e^{-iTs} \tag{3.238}$$

The *zero-order hold* transfer function is

$$G_h(s) = \frac{1 - e^{-Ts}}{s} \tag{3.239}$$

Then

$$X_h(s) = G_h(s)X_*(s) = \frac{1 - e^{-Ts}}{s}\sum_{i=0}^{\infty} x(iT)e^{-iTs} \tag{3.240}$$

$$X_h(s) = \sum_{i=0}^{\infty} x(iT)\left[\frac{e^{-iTs}}{s} - \frac{e^{-(i+1)Ts}}{s}\right] \tag{3.241}$$

The term in square brackets is seen to be a unit step of +1 delayed until $t = iT$ with an equivalent unit step at $t = (i + 1)T$ subtracted from it after an interval of T. The resultant square pulse is also scaled by its own factor $x(iT)$. All of these functions are added together as in Figure 3.32.

The procedure to convert a given impulse-modulated form to its step function, and feed it to a system $G(s)$, then involves the arrangement in Figure 3.33. Everything between the two sampling switches must be included if a z-domain transfer function $G(z)$ is required to convert $u(z)$ to $x(z)$. Individual transfer functions between the switches $G_1(s)$, $G_2(s)$, $G_3(s)$, . . . cannot be individually transformed to the corresponding $G_1(z)$, $G_2(z)$, $G_3(z)$, . . . because the latter are only phased with individual impulses, and do not recognise variations between these values.

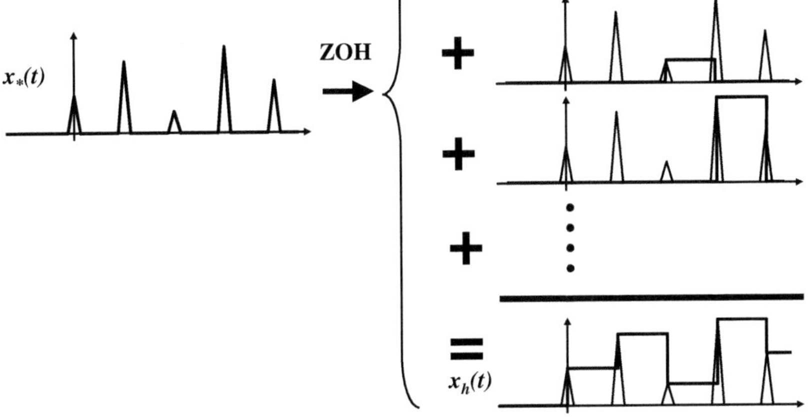

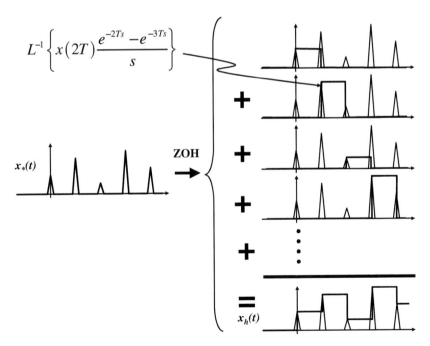

Figure 3.32 Operation of a zero-order hold on an impulse-modulated function.

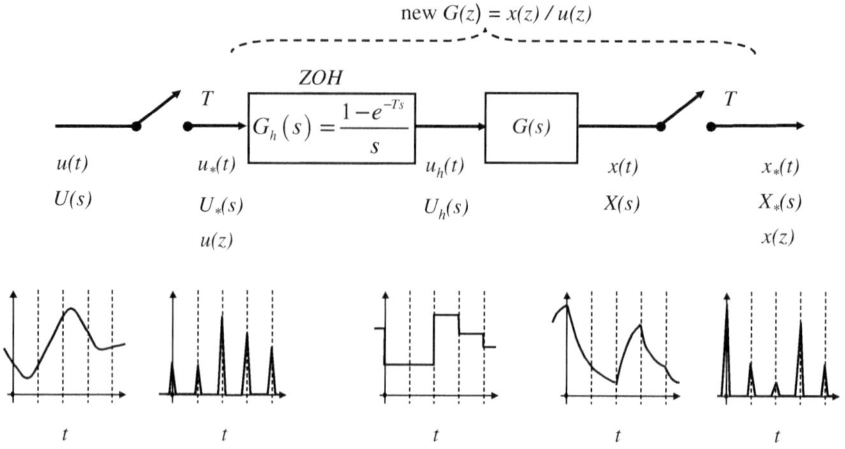

Figure 3.33 The transformation to create a $G(z)$ must include everything between sampling switches.

In general, the original signal is altered in the process of *sampling and holding*. Following a zero-order hold, it will vary in a series of steps at interval T. The ramp function t will become a staircase. One exception is of course the step function, so it is interesting to repeat Example 3.6 using the z notation.

Example 3.11

First-order system with step input using *z*-transforms.

Repeating Example 3.6, in the s-domain this was

$$G(s) = \frac{H(s)}{F_0(s)} = \frac{K}{\tau s + 1} \tag{3.242}$$

So

$$G(z) = Z\left\{\frac{1 - e^{-Ts}}{s}\left(\frac{K}{\tau s + 1}\right)\right\} \tag{3.243}$$

where 'Z' is the z-transform operator. Thus,

$$G(z) = \frac{z-1}{z} Z\left\{\frac{K}{s(\tau s + 1)}\right\} \tag{3.244}$$

Expanding the s function so that simpler transforms can be used,

$$\frac{K}{s(\tau s + 1)} = \frac{A}{s} + \frac{B}{(\tau s + 1)} \tag{3.245}$$

$$\left.\begin{matrix} \times s \\ s \to 0 \end{matrix}\right\} \quad K = A \tag{3.246}$$

$$\left.\begin{matrix} \times(\tau s + 1) \\ s \to -\dfrac{1}{\tau} \end{matrix}\right\} \quad -K\tau = B \tag{3.247}$$

So

$$G(z) = \frac{z-1}{z} Z\left\{ \frac{K}{s} - \frac{K}{s + \dfrac{1}{\tau}} \right\} \tag{3.248}$$

$$G(z) = K\left(\frac{z-1}{z}\right)\left(\frac{z}{z-1} - \frac{z}{z - e^{-T/\tau}}\right) \tag{3.249}$$

$$G(z) = K\left(1 - \frac{z-1}{z - e^{-T/\tau}}\right) \tag{3.250}$$

Now it is noted that the transform of the input is

$$u(z) = Z\{F_0(s)\} = Z\left\{\frac{D}{s}\right\} = \frac{Dz}{z-1} \quad \text{(step of size } D \text{ at } t = 0) \tag{3.251}$$

so

$$x(z) = G(z)u(z) \tag{3.252}$$

$$x(z) = K\left(1 - \frac{z-1}{z - e^{-T/\tau}}\right)\frac{Dz}{z-1} \tag{3.253}$$

$$x(z) = KD\left(\frac{z}{z-1} - \frac{z}{z - e^{-T/\tau}}\right) \tag{3.254}$$

Inverting directly to the t-domain

$$x(t) = KD\left(1 - e^{-t/\tau}\right) \tag{3.255}$$

but it is understood in this inverse z-transform that the only information available is that

$$x(iT) = KD\left(1 - e^{-iT/\tau}\right) \quad \text{for } i = 0, 1, 2, \ldots \tag{3.256}$$

It is a mistake to assume that the intermediate values of x can be obtained by substituting intermediate values of t, and in general these must be obtained by *dual-rate sampling.* In this example, there is no access to $x(t)$ in Figure 3.33 because the discrete transfer function $G(z)$ stretches across it to the next sampling switch. Nevertheless, it is noted in this instance that there is agreement with the complete response in Example 3.6, because the zero-order hold recreates the original step for the single step input.

The response $x(z)$ of a system will generally occur in the form of a ratio of two polynomials as in Equation 3.254. It is worth noting an alternative to seeking inverses in tables of z-transforms, namely *long division*. For example,

$$\frac{z+1}{z^2-2z+1} \rightarrow \begin{array}{r} z^{-1}+3z^{-2}+5z^{-3}+\cdots \\ z^2-2z+1\,\overline{)\,z+1\qquad\qquad\qquad} \\ \underline{z-2+z^{-1}} \\ 3-z^{-1} \\ \underline{3-6z^{-1}+3z^{-2}} \\ 5z^{-1}-3z^{-2} \\ \vdots \end{array} \tag{3.257}$$

is a response that will have a value of 1 at the end of the first interval T, 3 at the second, 5 at the third and so on.

The idea of an *impulse response* arises from feeding a unit impulse input $\delta(t)$ to a system. In the z-domain, it is seen that the output values at the sampling instants arise as the coefficients of the original transfer function. For example, consider a unit impulse fed to an integrator, giving a finite impulse response which is a series of 1's:

$$u(z) = 1 \tag{3.258}$$

$$G(z) = Z\left\{\frac{1}{s}\right\} = \frac{z}{1-z} \tag{3.259}$$

$$x(z) = G(z)u(z) = \frac{z}{z-1}1 \tag{3.260}$$

$$x(z) = 1 + 1z^{-1} + 1z^{-2} + 1z^{-3} + 1z^{-4} + \cdots \tag{3.261}$$

In the case of integrating (or unstable/undamped) systems, one expects an infinite series of nonzero coefficients. However, for non-integrating, stable and damped systems, the coefficients arising from an impulse input become insignificant after a finite number of steps, and one describes this as a *finite impulse response* (FIR). An example would be the first-order system

$$G(z) = \frac{z}{z - \mathrm{e}^{-aT}} \tag{3.262}$$

where stability and damping require $a > 0$ and thus $0 < \mathrm{e}^{-aT} < 1$.

In Example 3.11, the focus was on the handling of actual signals, subjecting impulses to a zero-order hold to create continuous step functions, and so on. In practice, the concept is often used as a sort of shorthand to represent a sequence of data values. A *forward shift operator* q, or *backward shift operator* q^{-1}, is used to shift a time sequence of values in the same way as z or z^{-1}, without invoking the theoretical basis of z-transforms.

3.9.3.2 Use of z-Transforms to Find the System Response

The use of z-transforms here will focus on the time-shifting properties of z or z^{-1}. One notes that

$$z^{-1}x(z) \tag{3.263}$$

returns the value of x at a time T before the present. If a transport lag from inlet u to outlet x is exactly $3T$, then one could write

$$x(z) = z^{-3}u(z) \tag{3.264}$$

Consider a linear system which is expressed as a set of first-order differential equations with constant coefficients:

$$\dot{\boldsymbol{x}} = \boldsymbol{A}\boldsymbol{x} + \boldsymbol{B}\boldsymbol{u} \tag{3.265}$$

As in Equation 3.152, the Laplace transform yields

$$\boldsymbol{X}(s) = [s\boldsymbol{I} - \boldsymbol{A}]^{-1}\boldsymbol{x}(0) + [s\boldsymbol{I} - \boldsymbol{A}]^{-1}\boldsymbol{B}\boldsymbol{U}(s) \tag{3.266}$$

The inversion will depend on the time functions in the input vector. A special case will be considered here where the inputs $\boldsymbol{u}(t)$ are held constant at their starting values $\boldsymbol{u}(0)$, so

$$\boldsymbol{U}(s) = \frac{1}{s}\boldsymbol{u}(0) \tag{3.267}$$

Recall that the s-domain functions represent zero values up until time zero, so that the input being considered here is a step from zero in each input variable, at time zero. Then

$$\boldsymbol{X}(s) = [s\boldsymbol{I} - \boldsymbol{A}]^{-1}\boldsymbol{x}(0) + [s\boldsymbol{I} - \boldsymbol{A}]^{-1}\frac{1}{s}\boldsymbol{B}\boldsymbol{u}(0) \tag{3.268}$$

$$\boldsymbol{X}(s) = [s\boldsymbol{I} - \boldsymbol{A}]^{-1}\boldsymbol{x}(0) + [s\boldsymbol{I} - \boldsymbol{A}]^{-1}[s\boldsymbol{I}]^{-1}\boldsymbol{B}\boldsymbol{u}(0) \tag{3.269}$$

It is useful to expand the inverse matrices of the second term using partial fractions, that is

$$[s\boldsymbol{I} - \boldsymbol{A}]^{-1}[s\boldsymbol{I}]^{-1} = [s\boldsymbol{I} - \boldsymbol{A}]^{-1}\boldsymbol{\alpha} + \boldsymbol{\beta}[s\boldsymbol{I}]^{-1} \tag{3.270}$$

where $\boldsymbol{\alpha}$ and $\boldsymbol{\beta}$ are constant matrices.

$$[s\boldsymbol{I} - \boldsymbol{A}] \times \rightarrow \{[s\boldsymbol{I} - \boldsymbol{A}]^{-1}[s\boldsymbol{I}]^{-1} = [s\boldsymbol{I} - \boldsymbol{A}]^{-1}\boldsymbol{\alpha} + \boldsymbol{\beta}[s\boldsymbol{I}]^{-1}\} \leftarrow \times [s\boldsymbol{I}] \tag{3.271}$$

$$\left.\begin{array}{c} \times[s\boldsymbol{I} - \boldsymbol{A}] \text{ from the left and} \\ s\boldsymbol{I} \rightarrow \boldsymbol{A} \end{array}\right\} \quad \boldsymbol{A}^{-1} = \boldsymbol{\alpha} \tag{3.272}$$

$$\left.\begin{array}{c} \times[s\boldsymbol{I}] \text{ from the right and} \\ s\boldsymbol{I} \rightarrow \boldsymbol{0} \end{array}\right\} \quad -\boldsymbol{A}^{-1} = \boldsymbol{\beta} \tag{3.273}$$

so

$$[s\boldsymbol{I} - \boldsymbol{A}]^{-1}[s\boldsymbol{I}]^{-1} = [s\boldsymbol{I} - \boldsymbol{A}]^{-1}\boldsymbol{A}^{-1} - \boldsymbol{A}^{-1}[s\boldsymbol{I}]^{-1} \tag{3.274}$$

Since $[s\boldsymbol{I}]^{-1}$ is diagonal, and $\boldsymbol{A}$ is square,

$$[s\boldsymbol{I} - \boldsymbol{A}]^{-1}[s\boldsymbol{I}]^{-1} = \{[s\boldsymbol{I} - \boldsymbol{A}]^{-1} - [s\boldsymbol{I}]^{-1}\}\boldsymbol{A}^{-1} \tag{3.275}$$

Thus,

$$\boldsymbol{X}(s) = [s\boldsymbol{I} - \boldsymbol{A}]^{-1}\boldsymbol{x}(0) + \{[s\boldsymbol{I} - \boldsymbol{A}]^{-1} - [s\boldsymbol{I}]^{-1}\}\boldsymbol{A}^{-1}\boldsymbol{B}\boldsymbol{u}(0) \tag{3.276}$$

Taking the inverse Laplace transform

$$\boldsymbol{x}(t) = L^{-1}\{[s\boldsymbol{I} - \boldsymbol{A}]^{-1}\}\boldsymbol{x}(0) + (L^{-1}\{[s\boldsymbol{I} - \boldsymbol{A}]^{-1}\} - \boldsymbol{I})\boldsymbol{A}^{-1}\boldsymbol{B}\boldsymbol{u}(0) \tag{3.277}$$

This result involves the *state transition matrix* $L^{-1}\{[s\boldsymbol{I} - \boldsymbol{A}]^{-1}\}$, which is known as the *matrix exponential*, represented symbolically as

$$\boldsymbol{e}^{\boldsymbol{A}t} = L^{-1}\{[s\boldsymbol{I} - \boldsymbol{A}]^{-1}\} \tag{3.278}$$

It can be evaluated directly in mathematical programs such as MATLAB®, and has similar properties to a normal scalar exponential, for example

$$e^{A0} = I \tag{3.279}$$

$$e^{At}\, e^{-At} = I \tag{3.280}$$

$$\begin{aligned} &e^{At}\, e^{At} = e^{2At} \\ &\vdots \end{aligned} \tag{3.281}$$

There is an obvious resemblance to the SISO case of Equation 3.110 and Table 3.3. The frame of reference so far has been $0 < t < \infty$, but the integration can be used in the same way for successive intervals T as follows:

$$\begin{aligned} &x((i+1)T) = e^{AT}\, x(iT) + \left[e^{AT} - I\right]A^{-1}Bu(iT) \\ &\quad \text{where } e^{At} = L^{-1}\{[sI - A]^{-1}\} \end{aligned} \tag{3.282}$$

The form of this equation has a strong resemblance to the continuous system case (Equation 3.100). Here one defines similar matrices for the discrete system

$$A_* = e^{AT} \tag{3.283}$$

$$B_* = \left[e^{AT} - I\right]A^{-1}B \tag{3.284}$$

so that

$$x((i+1)T) = A_*x(iT) + B_*u(iT) \tag{3.285}$$

In terms of the time-shift operator z, this is

$$zx(z) = A_*x(z) + B_*u(z) \tag{3.286}$$

so

$$[zI - A_*]x(z) = B_*u(z) \tag{3.287}$$

$$x(z) = [zI - A_*]^{-1}B_*u(z) \tag{3.288}$$

The transfer function matrix for this discrete linear state system is thus determined by

$$x(z) = G(z)u(z) \tag{3.289}$$

with

$$G(z) = [zI - A_*]^{-1}B_* \tag{3.290}$$

This transfer function is taking a series of input values at time intervals T and providing the output values at corresponding intervals T. It replaces the integration step of the continuous system in Equation 3.100 over each of these intervals, under the specific condition that the input values vary in a series of steps, remaining fixed at the initial value for each interval. It is quite clear that this formulation replaces the combination of *zero-order hold* and continuous system $G(s)$ in Figure 3.33, and provides a general MIMO approach to the SISO example (Example 3.11). The successive multiplication of several transfer functions on this basis would imply that the intermediate values are sampled and held between each system. In practice, this is very much how intermediate values from various calculations are only updated periodically in a processing plant computer database.

Following the same treatment as in the s-domain (Section 3.9.2.2)

$$G(z) = \frac{1}{\det[zI - A_*]} \operatorname{adj}[zI - A_*] B_* \tag{3.291}$$

Again, the eigenvalues λ_i of A_* are the roots of the *characteristic equation*

$$D(z) = \det[zI - A_*] \tag{3.292}$$

$$D(z) = z^n + d_{n-1}z^{n-1} + \cdots + d_1 z + d_0 \tag{3.293}$$

$$D(z) = (z - \lambda_1)(z - \lambda_2)\cdots(z - \lambda_n) = 0 \tag{3.294}$$

and will contribute factors $(z - \lambda_1)$, $(z - \lambda_2)$, . . . in the denominators of the partial fraction expansion of any response. Table 3.7 shows the importance of these denominators in determining the response. Each term in the adjoint matrix shown will also be a polynomial in z, so Equation 3.289 can be expressed as

$$\left(z^n + d_{n-1}z^{n-1} + \cdots + d_1 z + d_0\right) x(z) = \left[N_{n-1}z^{n-1} + N_{n-2}z^{n-2} + \cdots + N_1 z + N_0\right] u(z) \tag{3.295}$$

where the matrices N_i group the coefficients of the relevant powers of z. This relationship is analogous to the s-domain expression (Equation 3.163). The equation may be divided by z successively until the highest power is z^0. In this form, it provides an alternative recursive predictor for x based on past values of x and u. The transfer function is often expressed using a matrix of polynomial ratios. For this state system it is seen that the matrix $G(z)$ has a common denominator for all elements (but these may all differ for a general input-output system).

$$D(z)x(z) = N(z)u(z) \tag{3.296}$$

$$G(z) = D^{-1}(z)N(z) \tag{3.297}$$

$$x(z) = G(z)u(z) \tag{3.298}$$

Example 3.12

Second-order continuous system using z-transforms.

Develop discrete equations for the system

$$\frac{d}{dt}\begin{pmatrix} x_1 \\ x_2 \end{pmatrix} = \begin{bmatrix} -2 & 1 \\ 2 & -3 \end{bmatrix}\begin{pmatrix} x_1 \\ x_2 \end{pmatrix} + \begin{bmatrix} 0 & 2 \\ 1 & 1 \end{bmatrix}\begin{pmatrix} u_1 \\ u_2 \end{pmatrix} \tag{3.299}$$

using a time interval of 0.5.

$$[sI - A] = \begin{bmatrix} s+2 & -1 \\ -2 & s+3 \end{bmatrix} \tag{3.300}$$

$$[sI - A]^{-1} = \frac{1}{(s+2)(s+3) - 2}\begin{bmatrix} s+3 & 1 \\ 2 & s+2 \end{bmatrix} \tag{3.301}$$

Consider

$$(s+2)(s+3)-2=s^2+5s+4=0 \tag{3.302}$$

Roots are

$$s=\frac{-5\pm\sqrt{25-16}}{2}=-1,-4 \tag{3.303}$$

Partial fraction expansion:

$$\frac{1}{(s+1)(s+4)}=\frac{a}{(s+1)}+\frac{b}{(s+4)} \tag{3.304}$$

$$\left.\begin{matrix}\times(s+1)\\ s\to -1\end{matrix}\right\}\quad \frac{1}{3}=a \tag{3.305}$$

$$\left.\begin{matrix}\times(s+4)\\ s\to -4\end{matrix}\right\}\quad \frac{1}{-3}=b \tag{3.306}$$

So

$$[sI-\boldsymbol{A}]^{-1}=\frac{1}{3}\begin{bmatrix}\dfrac{s+3}{(s+1)}-\dfrac{s+3}{(s+4)} & \dfrac{1}{(s+1)}-\dfrac{1}{(s+4)}\\ \dfrac{2}{(s+1)}-\dfrac{2}{(s+4)} & \dfrac{s+2}{(s+1)}-\dfrac{s+2}{(s+4)}\end{bmatrix} \tag{3.307}$$

$$[sI-\boldsymbol{A}]^{-1}=\frac{1}{3}\begin{bmatrix}1+\dfrac{2}{(s+1)}-1+\dfrac{1}{(s+4)} & \dfrac{1}{(s+1)}-\dfrac{1}{(s+4)}\\ \dfrac{2}{(s+1)}-\dfrac{2}{(s+4)} & 1+\dfrac{1}{(s+1)}-1+\dfrac{2}{(s+4)}\end{bmatrix} \tag{3.308}$$

and

$$L^{-1}\left\{[sI-\boldsymbol{A}]^{-1}\right\}=\frac{1}{3}\begin{bmatrix}2e^{-t}+e^{-4t} & e^{-t}-e^{-4t}\\ 2e^{-t}-2e^{-4t} & e^{-t}+2e^{-4t}\end{bmatrix} \tag{3.309}$$

so with $t\to T=0.5$

$$\boldsymbol{e}^{AT}=\frac{1}{3}\begin{bmatrix}2e^{-0.5}+e^{-2} & e^{-0.5}-e^{-2}\\ 2e^{-0.5}-2e^{-2} & e^{-0.5}+2e^{-2}\end{bmatrix}=\begin{bmatrix}0.449 & 0.157\\ 0.314 & 0.292\end{bmatrix} \tag{3.310}$$

Now use Equations 3.283 and 3.284:

$$\boldsymbol{A}_*=\boldsymbol{e}^{AT}=\begin{bmatrix}0.449 & 0.157\\ 0.314 & 0.292\end{bmatrix} \tag{3.311}$$

$$\boldsymbol{B}_*=\left[\boldsymbol{e}^{AT}-I\right]\boldsymbol{A}^{-1}\boldsymbol{B}=\left\{\begin{bmatrix}0.449 & 0.157\\ 0.314 & 0.292\end{bmatrix}-\begin{bmatrix}1 & 0\\ 0 & 1\end{bmatrix}\right\}\begin{bmatrix}-2 & 1\\ 2 & -3\end{bmatrix}^{-1}\begin{bmatrix}0 & 2\\ 1 & 1\end{bmatrix} \tag{3.312}$$

$$B_* = \begin{bmatrix} -0.551 & 0.157 \\ 0.314 & -0.708 \end{bmatrix} \begin{bmatrix} \frac{-3}{4} & \frac{-1}{4} \\ \frac{-2}{4} & \frac{-2}{4} \end{bmatrix} \begin{bmatrix} 0 & 2 \\ 1 & 1 \end{bmatrix} = \begin{bmatrix} 0.059 & 0.729 \\ 0.276 & 0.513 \end{bmatrix} \tag{3.313}$$

So the recursive integration formula (Equation 3.285) is

$$\boldsymbol{x}_{i+1} = \boldsymbol{A}_*\boldsymbol{x}_i + \boldsymbol{B}_*\boldsymbol{u}_i = \begin{bmatrix} 0.449 & 0.157 \\ 0.314 & 0.292 \end{bmatrix} \boldsymbol{x}_i + \begin{bmatrix} 0.059 & 0.729 \\ 0.276 & 0.513 \end{bmatrix} \boldsymbol{u}_i \tag{3.314}$$

and one recalls that it can only be used in cases where $\boldsymbol{u}$ is piecewise constant in each interval $T = 0.5$. Going on to find the equivalent z-domain transfer function, Equation 3.290 gives

$$\boldsymbol{G}(z) = [z\boldsymbol{I} - \boldsymbol{A}_*]^{-1}\boldsymbol{B}_* \tag{3.315}$$

$$\boldsymbol{G}(z) = \begin{bmatrix} z - 0.449 & -0.157 \\ -0.314 & z - 0.292 \end{bmatrix}^{-1} \begin{bmatrix} 0.059 & 0.729 \\ 0.276 & 0.513 \end{bmatrix} \tag{3.316}$$

$$\boldsymbol{G}(z) = \frac{1}{(z - 0.449)(z - 0.292) - 0.0493} \begin{bmatrix} z - 0.292 & 0.157 \\ 0.314 & z - 0.449 \end{bmatrix} \begin{bmatrix} 0.059 & 0.729 \\ 0.276 & 0.513 \end{bmatrix} \tag{3.317}$$

$$\boldsymbol{G}(z) = \frac{1}{z^2 - 0.741z + 0.0818} \begin{bmatrix} z - 0.292 & 0.157 \\ 0.314 & z - 0.449 \end{bmatrix} \begin{bmatrix} 0.059 & 0.729 \\ 0.276 & 0.513 \end{bmatrix} \tag{3.318}$$

$$\boldsymbol{G}(z) = \frac{1}{z^2 - 0.741z + 0.0818} \begin{bmatrix} 0.059z + 0.0261 & 0.729z - 0.1323 \\ 0.276z - 0.1054 & 0.513z - 0.0014 \end{bmatrix} \tag{3.319}$$

Dividing numerator and denominator by z, another way of expressing $\boldsymbol{x}(z) = \boldsymbol{G}(z)\boldsymbol{u}(z)$ is

$$\begin{pmatrix} x_1 \\ x_2 \end{pmatrix}_{i+1} = 0.741 \begin{pmatrix} x_1 \\ x_2 \end{pmatrix}_i - 0.0818 \begin{pmatrix} x_1 \\ x_2 \end{pmatrix}_{i-1} + \begin{bmatrix} 0.059 & 0.729 \\ 0.276 & 0.513 \end{bmatrix} \begin{pmatrix} u_1 \\ u_2 \end{pmatrix}_i + \begin{bmatrix} 0.0261 & -0.1323 \\ -0.1054 & -0.0014 \end{bmatrix} \begin{pmatrix} u_1 \\ u_2 \end{pmatrix}_{i-1} \tag{3.320}$$

but this merely repeats Equation 3.314 over two time steps.

3.9.3.3 Evaluation of the Matrix Exponential Terms

Equations 3.283 and 3.284 give the coefficient matrices of the discrete system

$$\boldsymbol{A}_* = \boldsymbol{e}^{\boldsymbol{A}T} \tag{3.321}$$

$$\boldsymbol{B}_* = \left[\boldsymbol{e}^{\boldsymbol{A}T} - \boldsymbol{I}\right]\boldsymbol{A}^{-1}\boldsymbol{B} \tag{3.322}$$

The matrix exponential $\boldsymbol{e}^{\boldsymbol{A}T}$ is provided directly in environments such as MATLAB® ('expm'), but it is worth noting that its Taylor expansion is

$$\boldsymbol{A}_* = \boldsymbol{e}^{\boldsymbol{A}T} \approx \boldsymbol{I} + \frac{\boldsymbol{A}T}{1!} + \frac{(\boldsymbol{A}T)^2}{2!} + \frac{(\boldsymbol{A}T)^3}{3!} + \cdots \tag{3.323}$$

since $\boldsymbol{e}^0 = \boldsymbol{I}$. It follows then that

$$\boldsymbol{B}_* = T\left[\boldsymbol{I} + \frac{(\boldsymbol{A}T)}{2!} + \frac{(\boldsymbol{A}T)^2}{3!} + \frac{(\boldsymbol{A}T)^3}{4!} + \cdots\right]\boldsymbol{B} \tag{3.324}$$

This latter result for $\boldsymbol{B}_*$ is particularly useful when the matrix $\boldsymbol{A}$ is singular.

3.9.3.4 Shortcut Methods to Obtain Discrete Difference Equations

The procedure used in Section 3.9.3.2 to obtain an exact discrete equivalent of a continuous system with a piecewise-constant input has been laborious, though one notes that the matrix exponential is readily available in mathematical programs. Useful transfer functions are easily found in the s-domain, so several methods have been devised to convert these directly to approximate discrete equivalent equations. Noting that s represents a derivative, the following substitutions are used:

- *Forward difference (implicit Euler)*:

$$s \approx \frac{z-1}{T} \tag{3.325}$$

- *Backward difference (explicit Euler)*:

$$s \approx \frac{1-z^{-1}}{T} \tag{3.326}$$

- *Tustin (bilinear or trapezoidal)*:

$$s \approx \left(\frac{2}{T}\right)\frac{z-1}{z+1} \tag{3.327}$$

  Inverting the Tustin approximation,

$$z = e^{Ts} \approx \frac{1+\left(\frac{T}{2}\right)s}{1-\left(\frac{T}{2}\right)s} \tag{3.328}$$

  which is the first-order Padé approximation, sometimes used in the s-domain to deal with transport lags in the quest for polynomial-ratio forms. The inverse form of these approximations (like Equation 3.328: $z = fn(s)$) is sometimes used for *frequency analysis* of discrete systems, by substitution of $j\omega$ for s in the resulting equation, as discussed in Section 8.6.1.

Example 3.13

Integration of a linear system using the Tustin approximation.

Equation 3.265 is

$$\dot{\boldsymbol{x}} = \boldsymbol{A}\boldsymbol{x} + \boldsymbol{B}\boldsymbol{u} \tag{3.329}$$

so for $\boldsymbol{x}(0) = \boldsymbol{0}$

$$s\boldsymbol{X}(s) = \boldsymbol{A}\boldsymbol{X}(s) + \boldsymbol{B}\boldsymbol{U}(s) \tag{3.330}$$

Using the Tustin approximation

$$\left(\frac{2}{T}\right)\frac{z-1}{z+1}x(z) \approx Ax(z) + Bu(z) \tag{3.331}$$

$$z\left[I - \frac{T}{2}A\right]x(z) \approx \left[I + \frac{T}{2}A\right]x(z) + \left(\frac{T}{2}\right)B\{zu(z) + u(z)\} \tag{3.332}$$

$$\times\left[I - \frac{T}{2}A\right]^{-1} : \quad zx(z) \approx \left[I - \frac{T}{2}A\right]^{-1}\left[I + \frac{T}{2}A\right]x(z) + \left(\frac{T}{2}\right)\left[I - \frac{T}{2}A\right]^{-1}B\{zu(z) + u(z)\} \tag{3.333}$$

It is noted that the equivalent first-order Padé approximation for the matrix exponential is

$$e^{AT} \approx \left[I - \frac{T}{2}A\right]^{-1}\left[I + \frac{T}{2}A\right] \tag{3.334}$$

which on substitution in Equation 3.282 yields

$$zx(z) \approx \left[I - \frac{T}{2}A\right]^{-1}\left[I + \frac{T}{2}A\right]x(z) + \left[I - \frac{T}{2}A\right]^{-1}TBu(z) \tag{3.335}$$

It is noted that Equation 3.333 only differs from Equation 3.335 by the averaging of the present and past input vectors.

3.9.3.5 Open-Loop Stability in the *z*-Domain

Equations 3.289–3.295 in the previous section examined the impact of the discrete system transfer function $\boldsymbol{G}(z)$ itself on the output $\boldsymbol{x}(z)$, regardless of the particular input $\boldsymbol{u}(z)$. The *full-state* representation is

$$\boldsymbol{x}(z) = \boldsymbol{G}(z)\boldsymbol{u}(z) \tag{3.336}$$

and

$$\boldsymbol{G}(z) = \frac{1}{\det[z\boldsymbol{I} - \boldsymbol{A}_*]}\operatorname{adj}[z\boldsymbol{I} - \boldsymbol{A}_*]\boldsymbol{B}_* \tag{3.337}$$

So the characteristic equation for this open loop is

$$\det[z\boldsymbol{I} - \boldsymbol{A}_*] = z^n + a_{n-1}z^{n-1} + \cdots + a_1 z + a_0 \tag{3.338}$$

$$\det[z\boldsymbol{I} - \boldsymbol{A}_*] = (z - \lambda_1)(z - \lambda_2)\cdots(z - \lambda_n) = 0 \tag{3.339}$$

It is noted that the factors $(z-\lambda_1)$, $(z-\lambda_2)$, ... will occur as denominators in the partial fraction expansion of any output $\boldsymbol{x}(z)$.

In Table 3.7, one needs to set $\lambda = \mathrm{e}^{-\alpha T}$, and it is seen that

$$Z^{-1}\left\{\frac{z}{z - \mathrm{e}^{-\alpha T}}\right\} = \mathrm{e}^{-\alpha t} = L^{-1}\left\{\frac{1}{s+\alpha}\right\} \tag{3.340}$$

The possibility does exist that λ is complex:

$$\lambda = a + jb = |a + jb|\{\cos(\theta) + j\sin(\theta)\} \quad \text{where } \theta = \angle(a + jb) \tag{3.341}$$

$$\lambda = \left(a^2 + b^2\right)^{1/2} e^{j\theta} = e^{(1/2)\ln\left(a^2+b^2\right)+j\theta} \tag{3.342}$$

that is

$$\alpha = -\frac{\ln\left(a^2 + b^2\right)}{2T} - j\frac{\theta}{T} \tag{3.343}$$

Thus, in the time domain, terms of the following form will occur in the system response:

$$e^{+\ln\left(a^2+b^2\right)(t/2T)}\left\{\cos\left(\frac{\theta t}{T}\right) + j\sin\left(\frac{\theta t}{T}\right)\right\} \tag{3.344}$$

The presence of a complex conjugate root λ will cause the imaginary values to disappear in the characteristic equation. Whether or not there is oscillation, there will be unbounded growth in the output if $\ln(a^2+b^2)$ is positive (making α negative), meaning that *all roots of the characteristic equation of a discrete system must lie within the unit circle for the system to be stable* (Figure 3.34). This result is not surprising when it is recalled that $z = e^{sT}$, since it has been established in Section 3.9.2.3 that the real part of the solutions of the s-domain characteristic equation must all be negative for stability, that is to ensure that no bounded input can cause an unbounded output.

As in the case of continuous systems (Section 3.9.2.3), discrete systems more generally can be represented in the *input–output* form

$$\boldsymbol{y}(z) = \boldsymbol{G}'(z)\boldsymbol{u}(z) \tag{3.345}$$

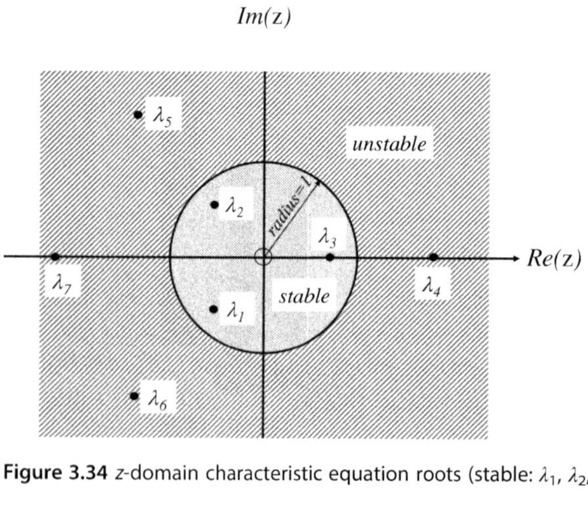

Figure 3.34 z-domain characteristic equation roots (stable: λ_1, λ_2, λ_3; unstable: λ_4, λ_5, λ_6, λ_7).

and again one expects the terms in matrix $G'(z)$ to be ratios of polynomials in z, but here the denominators may all differ, giving more factors (see Section 7.8.1). For the same reason as above, the factors of these denominators, $(z-a_i)$, require $|a_i|<1$ for stability. The input–output form (Equation 3.345) is often represented as

$$y(q) = G'(q)u(q) \tag{3.346}$$

where the *backward shift operator* q^{-1} replaces z^{-1}. This reflects models based directly on data sequences, rather than implying any theoretical relationship to continuous systems.

3.9.4 Numerical Solution for System Response

From Equations 3.6–3.9, the general form of the system model based on physical principles is

$$x(0) = x_0 \tag{3.347}$$

$$w(0) = w_0 \tag{3.348}$$

$$\dot{x} = f(x, w, y, u) \tag{3.349}$$

$$0 = g(x, w, y, u) \tag{3.350}$$

for a given $u(t)$, $0 < t < t_f$, where x is the vector of continuous states, w the discrete states, y the continuous and discrete ancillary variables, and u the continuous and discrete manipulated and disturbance variables. The vectors of functions f and g may be nonlinear, and may have logical conditions within them which change the behaviour depending on the other variables.

The examples in Sections 3.1, 3.2 and 3.3 considered systems which had some of these complications. An algorithmic 'freehand' form of modelling was suggested that employed a simple Euler integration. This type of solution can be 'improved' (faster, more accurate, less computation) if some restrictions are imposed. Some approaches are

- linear and logical (e.g. Bemporad and Morari, 1999);
- differential, algebraic and logical (e.g. Mao and Petzold, 2002);
- differential and algebraic (e.g. MATLAB®, Ascher and Petzold, 1998);
- stepwise re-linearisation and linear solution (e.g. Becerra, Roberts and Griffiths, 2001).

After developing a model, the process engineer's first task is to check how well the model represents the process, so that at least some basic tools are needed to integrate it. However, it is unwise to complicate the solution too early, as it is easy to lose sight of the basics and risk algebraic errors. So this discussion will restrict itself to some simple ideas concerning the integration of the system of first-order ODEs

$$x(0) = x_0 \tag{3.351}$$

$$\dot{x} = f(x, u) \tag{3.352}$$

for a given $u(t)$, $0 < t < t_f$, which excludes discrete and ancillary variables. The solution will be based on past and present values of x and u, and can be expressed after time discretisation of Equation 3.352 in the general form

$$0 = h(x_{i+1}, x_i, x_{i-1}, \ldots, x_{i-n}, u_i, u_{i-1}, \ldots, u_{i-m}) \tag{3.353}$$

for given x_0 and $u_0, u_1, \ldots, u_i$.

If it possible to separate out $\boldsymbol{x}_{i+1}$ onto the left-hand side of Equation 3.353, it is said to be *explicit*, otherwise it is *implicit*.

3.9.4.1 Numerical Solution Using Explicit Forms

An explicit Euler integration formula for Equation 3.352, using a time step of T, is obtained as

$$\frac{\boldsymbol{x}_{i+1} - \boldsymbol{x}_i}{T} = \boldsymbol{f}(\boldsymbol{x}_i, \boldsymbol{u}_i)$$

$$\boldsymbol{x}_{i+1} = \boldsymbol{x}_i + T\boldsymbol{f}(\boldsymbol{x}_i, \boldsymbol{u}_i) \quad \text{(one-step predictor)} \tag{3.354}$$

Example 3.14

Explicit Euler integration of a linear system.

Equation 3.265 is

$$\dot{\boldsymbol{x}} = \boldsymbol{A}\boldsymbol{x} + \boldsymbol{B}\boldsymbol{u} \tag{3.355}$$

so

$$\begin{aligned} \boldsymbol{x}_{i+1} &\approx \boldsymbol{x}_i + T(\boldsymbol{A}\boldsymbol{x}_i + \boldsymbol{B}\boldsymbol{u}_i) \\ &= [\boldsymbol{I} + T\boldsymbol{A}]\boldsymbol{x}_i + T\boldsymbol{B}\boldsymbol{u}_i \\ z\boldsymbol{x}(z) &= [\boldsymbol{I} + T\boldsymbol{A}]\boldsymbol{x}(z) + T\boldsymbol{B}\boldsymbol{u}(z) \end{aligned} \tag{3.356}$$

so that the transfer function is obvious in

$$\boldsymbol{x}(z) = [z\boldsymbol{I} - \boldsymbol{I} - T\boldsymbol{A}]^{-1} T\boldsymbol{B}\boldsymbol{u}(z) \tag{3.357}$$

Example 3.12 considered the system

$$\frac{\mathrm{d}}{\mathrm{d}t}\begin{pmatrix} x_1 \\ x_2 \end{pmatrix} = \begin{bmatrix} -2 & 1 \\ 2 & -3 \end{bmatrix}\begin{pmatrix} x_1 \\ x_2 \end{pmatrix} + \begin{bmatrix} 0 & 2 \\ 1 & 1 \end{bmatrix}\begin{pmatrix} u_1 \\ u_2 \end{pmatrix} \tag{3.358}$$

with an integration interval of $T = 0.5$. Substitution of $\boldsymbol{A}$ and $\boldsymbol{B}$ in Equation 3.356 yields

$$\boldsymbol{x}_{i+1} = \begin{bmatrix} 0 & 0.5 \\ 1 & -0.5 \end{bmatrix}\boldsymbol{x}_i + \begin{bmatrix} 0 & 1 \\ 0.5 & 0.5 \end{bmatrix}\boldsymbol{u}_i \tag{3.359}$$

whereas the exact solution from Equation 3.314 is

$$\boldsymbol{x}_{i+1} = \begin{bmatrix} 0.449 & 0.157 \\ 0.314 & 0.292 \end{bmatrix}\boldsymbol{x}_i + \begin{bmatrix} 0.059 & 0.729 \\ 0.276 & 0.513 \end{bmatrix}\boldsymbol{u}_i \tag{3.360}$$

The explicit Euler integration method is seen to use a single gradient vector evaluated at the start (or left) of the interval. The effects of this bias can be reduced by means of a smaller integration step T. However, it is worth noting another well-known *explicit* technique that seeks to eliminate this left-hand bias, the *fourth-order Runge–Kutta* method. In this technique, the estimates of the gradient are successively updated as new estimates are obtained of the change in $\boldsymbol{x}$ across the interval (Equations 3.361–3.364). It is seen that the final estimate (Equation 3.365) is based on a gradient that is more heavily weighted towards the centre of the interval.

$$\Delta \boldsymbol{x}_{\mathrm{a}} = T\boldsymbol{f}(\boldsymbol{x}_i, \boldsymbol{u}_i) \tag{3.361}$$

$$\Delta \boldsymbol{x}_{\mathrm{b}} = T\boldsymbol{f}\left(\boldsymbol{x}_i + \frac{1}{2}\Delta \boldsymbol{x}_{\mathrm{a}}, \frac{\boldsymbol{u}_i + \boldsymbol{u}_{i+1}}{2}\right) \tag{3.362}$$

$$\Delta \boldsymbol{x}_{\mathrm{c}} = T\boldsymbol{f}\left(\boldsymbol{x}_i + \frac{1}{2}\Delta \boldsymbol{x}_{\mathrm{b}}, \frac{\boldsymbol{u}_i + \boldsymbol{u}_{i+1}}{2}\right) \tag{3.363}$$

$$\Delta \boldsymbol{x}_{\mathrm{d}} = T\boldsymbol{f}(\boldsymbol{x}_i + \Delta \boldsymbol{x}_{\mathrm{c}}, \boldsymbol{u}_{i+1}) \tag{3.364}$$

$$\boldsymbol{x}_{i+1} = \boldsymbol{x}_i + \frac{1}{6}\{\Delta \boldsymbol{x}_{\mathrm{a}} + 2\Delta \boldsymbol{x}_{\mathrm{b}} + 2\Delta \boldsymbol{x}_{\mathrm{c}} + \Delta \boldsymbol{x}_{\mathrm{d}}\} \tag{3.365}$$

A general observation regarding explicit methods is that the exclusion of $\boldsymbol{x}_{i+1}$ from the formulation leaves them prone to overshoot resulting in instability. MIMO systems of ODEs often have widely varying time constants ('modes') in the equations, so that if a single time step is used, it might have to be very small to deal with this *stiffness*.

3.9.4.2 Numerical Solution Using Implicit Forms

An *implicit* form of stepwise integration would appear like Equation 3.353:

$$\boldsymbol{0} = \boldsymbol{h}(\boldsymbol{x}_{i+1}, \boldsymbol{x}_i, \boldsymbol{x}_{i-1}, \ldots, \boldsymbol{x}_{i-n}, \boldsymbol{u}_i, \boldsymbol{u}_{i-1}, \ldots, \boldsymbol{u}_{i-m}) \tag{3.366}$$

The equation might be written like this, but occasionally some manipulation allows extraction of $\boldsymbol{x}_{i+1}$. However, the general case will require an iterative solution for $\boldsymbol{x}_{i+1}$, for example using the Newton–Raphson method:

$$\{\boldsymbol{x}_{i+1}\}_{n+1} = \{\boldsymbol{x}_{i+1}\}_n - \left[\frac{\partial \boldsymbol{h}}{\partial \boldsymbol{x}_{i+1}}\right]_n^{-1} \{\boldsymbol{h}\}_n \tag{3.367}$$

where $\left[\partial \boldsymbol{h}/\partial \boldsymbol{x}_{i+1}\right]_n$ is a Jacobian matrix obtained by differentiating each function in the $\boldsymbol{h}$ vector by each element in the $\boldsymbol{x}_{i+1}$ vector, and evaluating the result at the $\{\boldsymbol{x}_{i+1}\}_n$ condition. Similarly, $\{\boldsymbol{h}\}_n$ represents the vector $\boldsymbol{h}$ evaluated with the estimate $\{\boldsymbol{x}_{i+1}\}_n$ (and of course the earlier solutions $\boldsymbol{x}_i$, $\boldsymbol{x}_{i-1}, \ldots$ and $\boldsymbol{u}_i, \ldots, \boldsymbol{u}_{i-m}$).

An implicit Euler integration for Equation 3.352, using a time step of T, is

$$\frac{\boldsymbol{x}_{i+1} - \boldsymbol{x}_i}{T} = \boldsymbol{f}(\boldsymbol{x}_{i+1}, \boldsymbol{u}_{i+1}) \tag{3.368}$$

However, it is noted that the gradient being used is now biased to the *right* of the time interval. Since one is already committed to an implicit solution, why not attempt an 'average' value of the gradient using average values of the variables? Thus, a possibility is

$$\frac{\boldsymbol{x}_{i+1} - \boldsymbol{x}_i}{T} = \boldsymbol{f}\left(\frac{\boldsymbol{x}_{i+1} + \boldsymbol{x}_i}{2}, \frac{\boldsymbol{u}_{i+1} + \boldsymbol{u}_i}{2}\right) \tag{3.369}$$

Example 3.15

Implicit integration of a linear system.

Taking an 'average' gradient according to Equation 3.369:

$$\mathbf{x}_{i+1} = \mathbf{x}_i + T\left(\mathbf{A}\frac{\mathbf{x}_{i+1} + \mathbf{x}_i}{2} + \mathbf{B}\frac{\mathbf{u}_{i+1} + \mathbf{u}_i}{2}\right) \tag{3.370}$$

$$\left[\mathbf{I} - \frac{1}{2}T\mathbf{A}\right]\mathbf{x}_{i+1} = \left[\mathbf{I} + \frac{1}{2}T\mathbf{A}\right]\mathbf{x}_i + \frac{1}{2}T\mathbf{B}(\mathbf{u}_{i+1} + \mathbf{u}_i) \tag{3.371}$$

$$\mathbf{x}_{i+1} = \left[\mathbf{I} - \frac{1}{2}T\mathbf{A}\right]^{-1}\left[\mathbf{I} + \frac{1}{2}T\mathbf{A}\right]\mathbf{x}_i + \frac{1}{2}T\left[\mathbf{I} - \frac{1}{2}T\mathbf{A}\right]^{-1}\mathbf{B}(\mathbf{u}_{i+1} + \mathbf{u}_i) \tag{3.372}$$

So the original implicit form has been made explicit in this case. Example 3.12 considered the system

$$\frac{\mathrm{d}}{\mathrm{d}t}\begin{pmatrix} x_1 \\ x_2 \end{pmatrix} = \begin{bmatrix} -2 & 1 \\ 2 & -3 \end{bmatrix}\begin{pmatrix} x_1 \\ x_2 \end{pmatrix} + \begin{bmatrix} 0 & 2 \\ 1 & 1 \end{bmatrix}\begin{pmatrix} u_1 \\ u_2 \end{pmatrix} \tag{3.373}$$

with an integration interval of $T = 0.5$. Substitution of $\mathbf{A}$ and $\mathbf{B}$ in Equation 3.372 yields

$$\mathbf{x}_{i+1} = \begin{bmatrix} 0.4 & 0.2 \\ 0.4 & 0.2 \end{bmatrix}\mathbf{x}_i + \frac{1}{2}\begin{bmatrix} 0.05 & 0.75 \\ 0.3 & 0.5 \end{bmatrix}(\mathbf{u}_{i+1} + \mathbf{u}_i) \tag{3.374}$$

whereas the explicit Euler integration (Equation 3.359) gave

$$\mathbf{x}_{i+1} = \begin{bmatrix} 0 & 0.5 \\ 1 & -0.5 \end{bmatrix}\mathbf{x}_i + \begin{bmatrix} 0 & 1 \\ 0.5 & 0.5 \end{bmatrix}\mathbf{u}_i \tag{3.375}$$

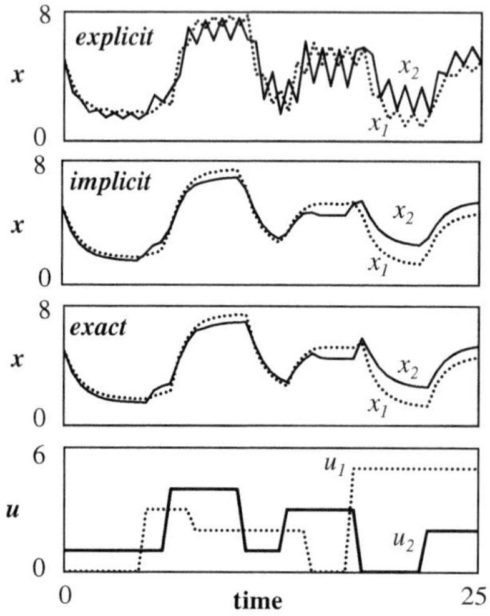

Figure 3.35 Comparison of explicit (Equation 3.359) and implicit (Equation 3.374) solutions with exact integration (Equation 3.314).

and the exact solution from Equation 3.314 is

$$\mathbf{x}_{i+1} = \begin{bmatrix} 0.449 & 0.157 \\ 0.314 & 0.292 \end{bmatrix} \mathbf{x}_i + \begin{bmatrix} 0.059 & 0.729 \\ 0.276 & 0.513 \end{bmatrix} \mathbf{u}_i \tag{3.376}$$

The explicit (Equation 3.359) and implicit (Equation 3.374) numerical solutions are compared with the exact stepwise integration (Equation 3.314) in Figure 3.35. Significant overshoot is obvious in the explicit solution for this step size $T = 0.5$.

3.9.5 Black Box Modelling

The open-loop modelling of the process discussed above has focused on mathematical descriptions of the physical phenomena occurring in the system. This type of modelling has the advantage that correction and adjustment of the model to match the process is based on meaningful parameters. In addition, the model operation can be extrapolated over new ranges with some confidence. Setting up such a mathematical description can, however, require a lot of skilled manpower, and there are cases where 'black box' models, based mainly on input and output observations, have proved quite adequate for process optimisation and control. A brief review of some of the popular black box methods follows. The view taken here is that a large amount of historical process data is available (e.g. step test measurements), and can be used offline to develop the required models. A similar problem, in which this type of model is identified in real time, will be discussed in Section 6.5.

3.9.5.1 Step Response Models

As will be seen in Section 7.8.2, models based on measured process step responses have become very important as they form the basis of common controllers such as the *dynamic matrix control* algorithm. However, the *model* part of this *model predictive control* technique needs to be recognised as a useful open-loop modelling method in its own right.

Consider the two-input, two-output system in Figure 3.36 as being representative of MIMO systems in general. Because the observable outputs are not necessarily all of the *states*, or indeed the states at all, these will be represented by the vector $\boldsymbol{y}$ instead of $\boldsymbol{x}$. With the system at steady state, each input is stepped in turn to obtain a matrix of step responses. For example, consider the effect of u_1 on y_1. The step of u_1 from u_{1SS} to $u_{1SS} + \Delta u_{1(0)}$ at $t = 0$ produces y_1 values at *subsequent* intervals T of $y_{1SS} + \Delta y_{1(1)}$, $y_{1SS} + \Delta y_{1(2)}$, . . . Normalising these with respect to the input step,

$$b_{11(1)} = \frac{\Delta y_{1(1)}}{\Delta u_{1(0)}}, \quad b_{11(2)} = \frac{\Delta y_{1(2)}}{\Delta u_{1(0)}}, \quad \ldots \tag{3.377}$$

one obtains the unit step response function

$$b_{11}(q) = b_{11(1)}q^{-1} + b_{11(2)}q^{-2} + \cdots + b_{11(N-1)}q^{-(N-1)} + b_{11(N)}q^{-N}\left(\frac{q}{q-1}\right) \tag{3.378}$$

Here q^{-1} is being used to represent a backward shift of one time interval, rather than z^{-1}, as is conventional when none of the theoretical z-transform properties are intended. So far, a non-integrating

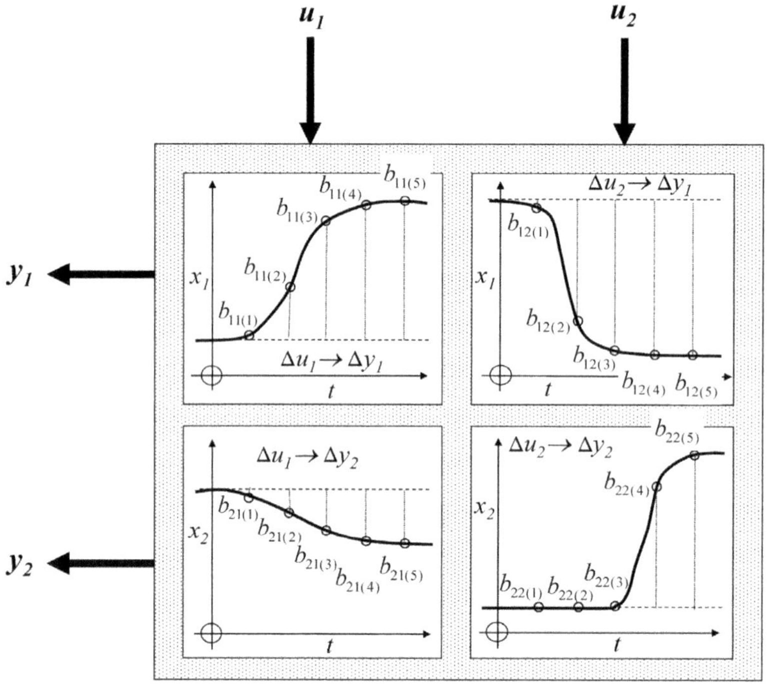

Figure 3.36 Step response measurement matrix for a MIMO system.

system has been assumed, and the interval T and number of points N in each response have been chosen to both give good definition to the variations, and ensure that the final point is close to the new equilibrium of the system. In Equation 3.378, the final point has been extended indefinitely with a delayed unit step (see Table 3.7).

The important assumption one makes in step response modelling is that the system is *linear*. So if a system is initially at steady state $y_{1(0)}$, a positive or negative step of any size $\Delta u_{1(0)}$ at $t = 0$ must produce the following x_1 output:

$$y_1(q) = b_{11}(q)\Delta u_{1(0)} + y_{1(0)}\left(\frac{q}{q-1}\right) \tag{3.379}$$

Notice that the product $b_{11}(q)\Delta u_{1(0)}$ only gives the change in y_1 from its initial value, so a step function is used to add back the offset at each future point in time. If there are now subsequent 'moves' of u_1 at the intervals T, that is

$$\Delta u_1(q) = \Delta u_{1(0)} + \Delta u_{1(1)}q^{-1} + \Delta u_{1(2)}q^{-2} + \Delta u_{1(3)}q^{-3} + \cdots \tag{3.380}$$

then the appropriate responses are just delayed in time before being summed:

$$\begin{aligned} y_1(q) &= b_{11}(q)\Delta u_1(q) + y_{1(0)}\left(\frac{q}{q-1}\right) \\ &= \left\{ b_{11(1)}q^{-1} + b_{11(2)}q^{-2} + \cdots + b_{11(5)}q^{-5}\left(\frac{q}{q-1}\right)\right\}\{\Delta u_{1(0)} + \Delta u_{1(1)}q^{-1} + \Delta u_{1(2)}q^{-2} + \cdots\} + y_{1(0)}\left(\frac{q}{q-1}\right) \end{aligned} \tag{3.381}$$

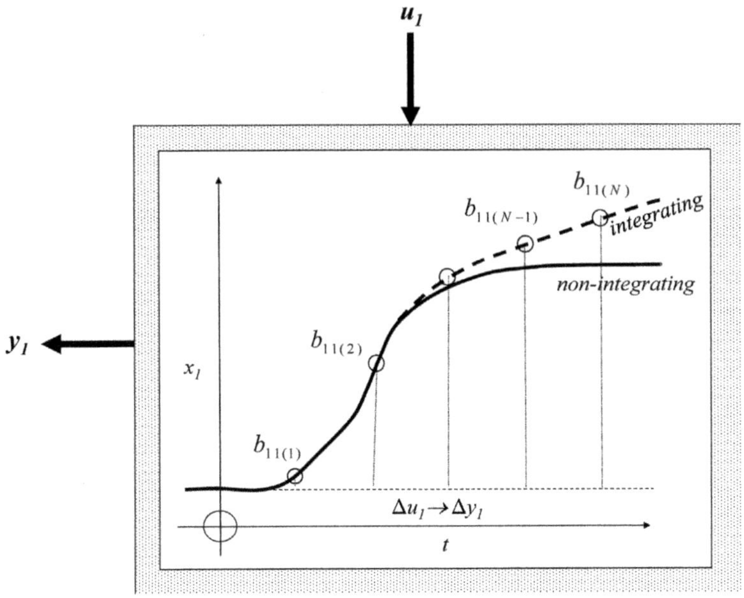

Figure 3.37 Integrating step response.

Now include the effects of moves in u_2, and treat y_2 similarly to obtain

$$\begin{pmatrix} y_1(q) \\ y_2(q) \end{pmatrix} = \begin{bmatrix} b_{11}(q) & b_{12}(q) \\ b_{21}(q) & b_{22}(q) \end{bmatrix} \begin{pmatrix} \Delta u_1(q) \\ \Delta u_2(q) \end{pmatrix} + \left\{ \frac{q}{q-1} \right\} \begin{pmatrix} y_1 \\ y_2 \end{pmatrix}_{(0)} \tag{3.382}$$

that is

$$\boldsymbol{y}(q) = \boldsymbol{b}(q)\Delta\boldsymbol{u}(q) + \left\{ \frac{q}{q-1} \right\} \boldsymbol{y}_{(0)} \tag{3.383}$$

Another way of viewing this is

$$\begin{aligned} &\boldsymbol{y}_{(0)} + \boldsymbol{y}_{(1)}q^{-1} + \boldsymbol{y}_{(2)}q^{-2} + \boldsymbol{y}_{(3)}q^{-3} + \cdots \\ &= \left\{ \boldsymbol{b}_{(1)}q^{-1} + \boldsymbol{b}_{(2)}q^{-2} + \cdots + \boldsymbol{b}_{(N-1)}q^{-(N-1)} + \boldsymbol{b}_{(N)}q^{-N}\left(\frac{q}{q-1}\right) \right\} \\ &\quad \left\{ \Delta\boldsymbol{u}_{(0)} + \Delta\boldsymbol{u}_{(1)}q^{-1} + \Delta\boldsymbol{u}_{(2)}q^{-2} + \cdots \right\} + \boldsymbol{y}_{(0)}\left(\frac{q}{q-1}\right) \end{aligned} \tag{3.384}$$

where

$$\boldsymbol{b}_{(1)} = \begin{bmatrix} b_{11(1)} & b_{12(1)} \\ b_{21(1)} & b_{22(1)} \end{bmatrix}, \quad \boldsymbol{b}_{(2)} = \begin{bmatrix} b_{11(2)} & b_{12(2)} \\ b_{21(2)} & b_{22(2)} \end{bmatrix}, \quad \ldots \tag{3.385}$$

Matching up the time-shift coefficients

$$\begin{aligned}
\boldsymbol{y}_{(1)} &= \boldsymbol{y}_{(0)} + \boldsymbol{b}_{(1)}\Delta\boldsymbol{u}_{(0)}\\
\boldsymbol{y}_{(2)} &= \boldsymbol{y}_{(0)} + \boldsymbol{b}_{(2)}\Delta\boldsymbol{u}_{(0)} + \boldsymbol{b}_{(1)}\Delta\boldsymbol{u}_{(1)}\\
\boldsymbol{y}_{(3)} &= \boldsymbol{y}_{(0)} + \boldsymbol{b}_{(3)}\Delta\boldsymbol{u}_{(0)} + \boldsymbol{b}_{(2)}\Delta\boldsymbol{u}_{(1)} + \boldsymbol{b}_{(1)}\Delta\boldsymbol{u}_{(2)}\\
\boldsymbol{y}_{(4)} &= \boldsymbol{y}_{(0)} + \boldsymbol{b}_{(4)}\Delta\boldsymbol{u}_{(0)} + \boldsymbol{b}_{(3)}\Delta\boldsymbol{u}_{(1)} + \boldsymbol{b}_{(2)}\Delta\boldsymbol{u}_{(2)} + \boldsymbol{b}_{(1)}\Delta\boldsymbol{u}_{(3)}\\
\boldsymbol{y}_{(5)} &= \boldsymbol{y}_{(0)} + \boldsymbol{b}_{(5)}\Delta\boldsymbol{u}_{(0)} + \boldsymbol{b}_{(4)}\Delta\boldsymbol{u}_{(1)} + \boldsymbol{b}_{(3)}\Delta\boldsymbol{u}_{(2)} + \boldsymbol{b}_{(2)}\Delta\boldsymbol{u}_{(3)} + \boldsymbol{b}_{(1)}\Delta\boldsymbol{u}_{(4)}\\
\boldsymbol{y}_{(6)} &= \boldsymbol{y}_{(0)} + \boldsymbol{b}_{(5)}\Delta\boldsymbol{u}_{(0)} + \boldsymbol{b}_{(5)}\Delta\boldsymbol{u}_{(1)} + \boldsymbol{b}_{(4)}\Delta\boldsymbol{u}_{(2)} + \boldsymbol{b}_{(3)}\Delta\boldsymbol{u}_{(3)} + \boldsymbol{b}_{(2)}\Delta\boldsymbol{u}_{(4)} + \boldsymbol{b}_{(1)}\Delta\boldsymbol{u}_{(5)}\\
\boldsymbol{y}_{(7)} &= \boldsymbol{y}_{(0)} + \boldsymbol{b}_{(5)}\Delta\boldsymbol{u}_{(0)} + \boldsymbol{b}_{(5)}\Delta\boldsymbol{u}_{(1)} + \boldsymbol{b}_{(5)}\Delta\boldsymbol{u}_{(2)} + \boldsymbol{b}_{(5)}\Delta\boldsymbol{u}_{(3)} + \boldsymbol{b}_{(3)}\Delta\boldsymbol{u}_{(4)} + \boldsymbol{b}_{(2)}\Delta\boldsymbol{u}_{(5)} + \boldsymbol{b}_{(1)}\Delta\boldsymbol{u}_{(6)}\\
&\vdots
\end{aligned} \tag{3.386}$$

This is conveniently represented using a *matrix of matrices* and several *vectors of vectors*:

$$\overline{\boldsymbol{y}} = \overline{\boldsymbol{y}}_0 + \overline{\boldsymbol{B}}\,\overline{\Delta\boldsymbol{u}} \tag{3.387}$$

with

$$\overline{\boldsymbol{y}} = \begin{pmatrix} \boldsymbol{y}_{(1)}\\ \boldsymbol{y}_{(2)}\\ \boldsymbol{y}_{(3)}\\ \boldsymbol{y}_{(4)}\\ \boldsymbol{y}_{(5)}\\ \boldsymbol{y}_{(6)}\\ \boldsymbol{y}_{(7)}\\ \vdots \end{pmatrix}, \quad \overline{\boldsymbol{y}}_0 = \begin{pmatrix} \boldsymbol{y}_{(0)}\\ \boldsymbol{y}_{(0)}\\ \boldsymbol{y}_{(0)}\\ \boldsymbol{y}_{(0)}\\ \boldsymbol{y}_{(0)}\\ \boldsymbol{y}_{(0)}\\ \boldsymbol{y}_{(0)}\\ \vdots \end{pmatrix}, \quad \overline{\Delta\boldsymbol{u}} = \begin{pmatrix} \Delta\boldsymbol{u}_{(0)}\\ \Delta\boldsymbol{u}_{(1)}\\ \Delta\boldsymbol{u}_{(2)}\\ \Delta\boldsymbol{u}_{(3)}\\ \Delta\boldsymbol{u}_{(4)}\\ \Delta\boldsymbol{u}_{(5)}\\ \Delta\boldsymbol{u}_{(6)}\\ \vdots \end{pmatrix} \quad \text{and}$$

$$\overline{\boldsymbol{B}} = \begin{bmatrix}
\boldsymbol{b}_{(1)} & \boldsymbol{0} & \boldsymbol{0} & \boldsymbol{0} & \boldsymbol{0} & \boldsymbol{0} & \boldsymbol{0} & \cdots\\
\boldsymbol{b}_{(2)} & \boldsymbol{b}_{(1)} & \boldsymbol{0} & \boldsymbol{0} & \boldsymbol{0} & \boldsymbol{0} & \boldsymbol{0} & \cdots\\
\vdots & \boldsymbol{b}_{(2)} & \boldsymbol{b}_{(1)} & \boldsymbol{0} & \boldsymbol{0} & \boldsymbol{0} & \boldsymbol{0} & \cdots\\
\boldsymbol{b}_{(N-1)} & \vdots & \boldsymbol{b}_{(2)} & \boldsymbol{b}_{(1)} & \boldsymbol{0} & \boldsymbol{0} & \boldsymbol{0} & \cdots\\
\boldsymbol{b}_{(N)} & \boldsymbol{b}_{(N-1)} & \vdots & \boldsymbol{b}_{(2)} & \boldsymbol{b}_{(1)} & \boldsymbol{0} & \boldsymbol{0} & \cdots\\
\boldsymbol{b}_{(N)} & \boldsymbol{b}_{(N)} & \boldsymbol{b}_{(N-1)} & \vdots & \boldsymbol{b}_{(2)} & \boldsymbol{b}_{(1)} & \boldsymbol{0} & \cdots\\
\boldsymbol{b}_{(N)} & \boldsymbol{b}_{(N)} & \boldsymbol{b}_{(N)} & \boldsymbol{b}_{(N-1)} & \vdots & \boldsymbol{b}_{(2)} & \boldsymbol{b}_{(1)} & \cdots\\
\vdots & \vdots & \vdots & \vdots & \vdots & \vdots & \vdots & \ddots
\end{bmatrix} \tag{3.388}$$

Note that the equations developed to this point rely on the system being at steady state at $\boldsymbol{y}_{(0)}$ at $t = 0$. The matrix $\overline{\boldsymbol{B}}$ is referred to as the *dynamic matrix*, and will later form the basis of *dynamic matrix control* (Section 7.8.2). It is obvious that if there had been input moves $\Delta\boldsymbol{u}$ prior to the starting time $t = 0$, these would have an effect extending past $t = 0$, and would have to be included.

Up to this point only non-integrating systems have been considered, that is systems which reach a steady state within the N times represented in the dynamic matrix (with $N=5$ in the preceding 2×2 example). Now consider a simple strategy for dealing with integrating systems. A possible *coding* of the final gradient of a response might be in terms of the last two points ($N-1$ and N) of the step response (Figure 3.37). So the integrating gradient of y_1 for a unit step input $\Delta u_{1(0)}$ is $(b_{11(N)} - b_{11(N-1)})/T$.

Then an appropriate delayed ramp function is included in $\boldsymbol{b}(z)$ in Equation 3.384 to obtain

$$\begin{aligned} &\boldsymbol{y}_{(0)} + \boldsymbol{y}_{(1)}q^{-1} + \boldsymbol{y}_{(2)}q^{-2} + \boldsymbol{y}_{(3)}q^{-3} + \cdots \\ &= \left\{ \begin{aligned} \boldsymbol{b}_{(1)}q^{-1} + \boldsymbol{b}_{(2)}q^{-2} + \cdots \quad &+ \quad \boldsymbol{b}_{(N-1)}q^{-N+1} + \boldsymbol{b}_{(N)}q^{-N}\left(\frac{q}{q-1}\right) \\ &+ \quad \left(\boldsymbol{b}_{(N)} - \boldsymbol{b}_{(N-1)}\right)q^{-N}\left(\frac{Tq}{(q-1)^2}\right) \end{aligned} \right\} \\ &\quad \left\{\Delta\boldsymbol{u}_{(0)} + \Delta\boldsymbol{u}_{(1)}q^{-1} + \Delta\boldsymbol{u}_{(2)}q^{-2} + \cdots\right\} + \boldsymbol{y}_{(0)}\left(\frac{q}{q-1}\right) \end{aligned} \tag{3.389}$$

The equivalent dynamic matrix for Equation 3.387 then becomes

$$\overline{\boldsymbol{B}} = \begin{bmatrix} \boldsymbol{b}_{(1)} & \boldsymbol{0} & \boldsymbol{0} & \boldsymbol{0} & \boldsymbol{0} & \boldsymbol{0} & \boldsymbol{0} & \cdots \\ \boldsymbol{b}_{(2)} & \boldsymbol{b}_{(1)} & \boldsymbol{0} & \boldsymbol{0} & \boldsymbol{0} & \boldsymbol{0} & \boldsymbol{0} & \cdots \\ \vdots & \boldsymbol{b}_{(2)} & \boldsymbol{b}_{(1)} & \boldsymbol{0} & \boldsymbol{0} & \boldsymbol{0} & \boldsymbol{0} & \cdots \\ \boldsymbol{b}_{(N-1)} & \vdots & \boldsymbol{b}_{(2)} & \boldsymbol{b}_{(1)} & \boldsymbol{0} & \boldsymbol{0} & \boldsymbol{0} & \cdots \\ \boldsymbol{b}_{(N)} & \boldsymbol{b}_{(N-1)} & \vdots & \boldsymbol{b}_{(2)} & \boldsymbol{b}_{(1)} & \boldsymbol{0} & \boldsymbol{0} & \cdots \\ 2\boldsymbol{b}_{(N)} - \boldsymbol{b}_{(N-1)} & \boldsymbol{b}_{(N)} & \boldsymbol{b}_{(N-1)} & \vdots & \boldsymbol{b}_{(2)} & \boldsymbol{b}_{(1)} & \boldsymbol{0} & \cdots \\ 3\boldsymbol{b}_{(N)} - 2\boldsymbol{b}_{(N-1)} & 2\boldsymbol{b}_{(N)} - \boldsymbol{b}_{(N-1)} & \boldsymbol{b}_{(N)} & \boldsymbol{b}_{(N-1)} & \vdots & \boldsymbol{b}_{(2)} & \boldsymbol{b}_{(1)} & \cdots \\ \vdots & \vdots & \vdots & \vdots & \vdots & \vdots & \vdots & \ddots \end{bmatrix} \tag{3.390}$$

Equation 3.387, taken together with the possibility of Equations 3.389 and 3.390 for integrating systems, constitutes the important results of the step response modelling approach. One notes in Equation 3.388 that the output vector values correspond to the *end* of each interval, whilst the input vector *moves* are at the *start* of each interval. The response at the end of an interval T is independent of the input move at that same time, owing to the finite response time required, so the output vector starts at one interval later in time.

The step response modelling approach easily handles transport lag (dead time), since an arbitrary sequence of response values can be specified. In practice, several step response measurements should be done on a plant, to ensure that the features used in the model do not include random and temporary disturbances. Some degree of averaging and smoothing is necessary, and in some installations, a standard response such as *first-order plus dead time* may be fitted to the measurements for use in a controller. Many industries represent their dynamic matrix as an array of s-domain transfer functions, that is those functions that would convert each input step into the observed output responses.

As mentioned, a limitation of the method is that it assumes linearity. Model validity will thus be improved if the step responses are determined close to the normal operating point. One method used to handle severe process nonlinearity is to superimpose a separate nonlinear model, for example an artificial neural network (Example 3.17), just for the *residual* nonlinearity.

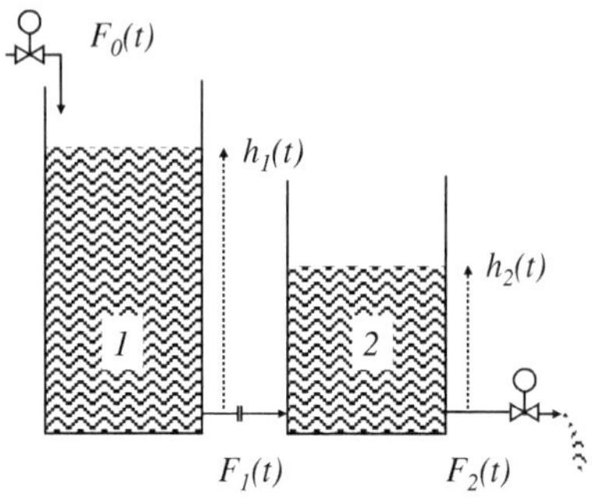

Figure 3.38 Step response modelling cannot find the equilibrium with both valves shut.

Another limitation to bear in mind is that the method does not explicitly recognise the relationships between variables in the same way as a mathematical model based on physical principles. For example, consider the two-tank flow system in Figure 3.38.

The separate step responses to each valve do not carry the information that if both valves shut, the levels must equilibrate. The nonlinearity of the level/flow relationship will cause the step response model to find two different tank levels at steady state, with both valves shut (assuming the step response measurements were not made at this state). Even if this relationship were linear, one would have to ensure that the initial output $\boldsymbol{y}_{(0)}$ used to start the model represented an equilibrium to avoid this situation.

3.9.5.2 Regressed Dynamic Models

Historical plant measurements stored at intervals T constitute an input–output data set of the form

$$\overline{\boldsymbol{u}} = \begin{pmatrix} \boldsymbol{u}_{(0)} \\ \boldsymbol{u}_{(1)} \\ \boldsymbol{u}_{(2)} \\ \boldsymbol{u}_{(3)} \\ \boldsymbol{u}_{(4)} \\ \boldsymbol{u}_{(5)} \\ \vdots \\ \boldsymbol{u}_{(L)} \end{pmatrix} \tag{3.391}$$

$$\overline{\boldsymbol{y}} = \begin{pmatrix} \boldsymbol{y}_{(0)} \\ \boldsymbol{y}_{(1)} \\ \boldsymbol{y}_{(2)} \\ \boldsymbol{y}_{(3)} \\ \boldsymbol{y}_{(4)} \\ \boldsymbol{y}_{(5)} \\ \vdots \\ \boldsymbol{y}_{(L)} \end{pmatrix} \tag{3.392}$$

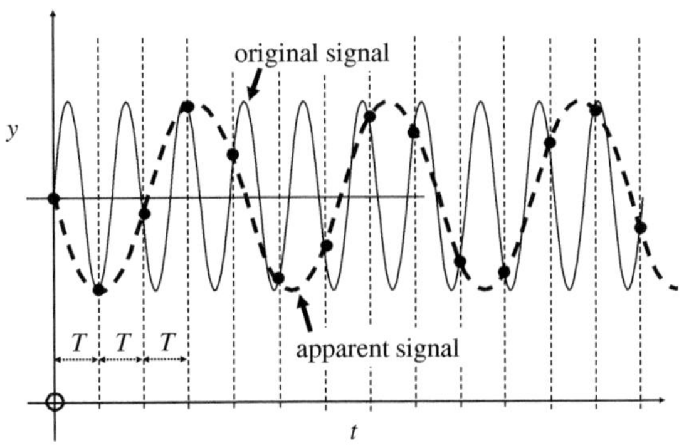

Figure 3.39 Aliasing due to too large a sampling interval.

where $\boldsymbol{u}_{(i)}$ is a vector of the input variables (MVs and DVs) at time iT and $\boldsymbol{y}_{(i)}$ is a vector of the output variables (PVs or CVs) at time iT. Proper identification of the dynamic behaviour of this system will only be possible if the sampling interval T is not larger than about one-tenth of the shortest time constant in the system, that is one requires about 10 data points to define the shortest transient. A problem arising from too large a sampling interval is that of *aliasing* (Figure 3.39), where a higher frequency signal manifests at a lower frequency. Usually one is not concerned with frequency signals, but the same effect may cause misinterpretation of any signal.

A *form* of model must initially be selected, requiring identification of the defining variables and the system order, plus an appropriate interval T for a discrete model.

$$\boldsymbol{0} = \boldsymbol{f}_{(i)}(\boldsymbol{p}) = \boldsymbol{f}\left(\boldsymbol{y}_{(i)}, \boldsymbol{y}_{(i-1)}, \ldots, \boldsymbol{y}_{(i-N)}, \boldsymbol{u}_{(i)}, \boldsymbol{u}_{(i-1)}, \ldots, \boldsymbol{u}_{(i-M)}, \boldsymbol{p}\right) \tag{3.393}$$

Here $\boldsymbol{p}$ is a set of constant parameters used in the model. The determination of the model then consists in finding $\boldsymbol{p}$ which minimise some performance criterion, for example a least square deviation:

$$\text{find } p \text{ such that } \phi(\boldsymbol{p}) = \sum_{i=N}^{L} \left\{\boldsymbol{f}_{(i)}(\boldsymbol{p})\right\}^{\mathrm{T}} \left\{\boldsymbol{f}_{(i)}(\boldsymbol{p})\right\} \text{ is minimised} \tag{3.394}$$

Example 3.16

Fitting of a linear discrete state-space model to measurement data.

In this case, one is using the autoregressive form, so the measured 'outputs' $\boldsymbol{y}$ are actually the states $\boldsymbol{x}$:

$$\begin{pmatrix} x_1 \\ x_2 \end{pmatrix}_{(i)} = \begin{bmatrix} a_{11} & a_{12} \\ a_{21} & a_{22} \end{bmatrix} \begin{pmatrix} x_1 \\ x_2 \end{pmatrix}_{(i-1)} + \begin{bmatrix} b_{11} & b_{12} \\ b_{21} & b_{22} \end{bmatrix} \begin{pmatrix} u_1 \\ u_2 \end{pmatrix}_{(i-1)} \tag{3.395}$$

Define

$$\boldsymbol{p} = \begin{pmatrix} a_{11} \\ a_{12} \\ a_{21} \\ a_{22} \\ b_{11} \\ b_{12} \\ b_{21} \\ b_{22} \end{pmatrix} \tag{3.396}$$

and measurement matrix and vector

$$\boldsymbol{M}_{(i)} = \begin{bmatrix} x_{1(i-1)} & x_{2(i-1)} & 0 & 0 & u_{1(i-1)} & u_{2(i-1)} & 0 & 0 \\ 0 & 0 & x_{1(i-1)} & x_{2(i-1)} & 0 & 0 & u_{1(i-1)} & u_{2(i-1)} \end{bmatrix} \tag{3.397}$$

$$\boldsymbol{y}_{(i)} = \begin{pmatrix} x_{1(i)} \\ x_{2(i)} \end{pmatrix} \tag{3.398}$$

so

$$\boldsymbol{f}_{(i)} = \boldsymbol{y}_{(i)} - \boldsymbol{M}_{(i)}\boldsymbol{p} \tag{3.399}$$

$$\phi = \sum_{i=N}^{L} \boldsymbol{f}_{(i)}^{\mathrm{T}} \boldsymbol{f}_{(i)} = \sum_{i=N}^{L} \left(\boldsymbol{y}_{(i)} - \boldsymbol{M}_{(i)}\boldsymbol{p}\right)^{\mathrm{T}} \left(\boldsymbol{y}_{(i)} - \boldsymbol{M}_{(i)}\boldsymbol{p}\right) \tag{3.400}$$

$$\phi = \sum_{i=N}^{L} \left(\boldsymbol{y}_{(i)}^{\mathrm{T}}\boldsymbol{y}_{(i)} - \boldsymbol{p}^{\mathrm{T}}\boldsymbol{M}_{(i)}^{\mathrm{T}}\boldsymbol{y}_{(i)} - \boldsymbol{y}_{(i)}^{\mathrm{T}}\boldsymbol{M}_{(i)}\boldsymbol{p} + \boldsymbol{p}^{\mathrm{T}}\boldsymbol{M}_{(i)}^{\mathrm{T}}\boldsymbol{M}_{(i)}\boldsymbol{p}\right) \tag{3.401}$$

Since $\boldsymbol{p}^{\mathrm{T}}\boldsymbol{M}_{(i)}^{\mathrm{T}}\boldsymbol{y}_{(i)}$ and $\boldsymbol{y}_{(i)}^{\mathrm{T}}\boldsymbol{M}_{(i)}\boldsymbol{p}$ are scalar and transposes of each other, they are equal, so

$$\phi = \sum_{i=N}^{L} \left(\boldsymbol{y}_{(i)}^{\mathrm{T}}\boldsymbol{y}_{(i)} - 2\boldsymbol{p}^{\mathrm{T}}\boldsymbol{M}_{(i)}^{\mathrm{T}}\boldsymbol{y}_{(i)} + \boldsymbol{p}^{\mathrm{T}}\boldsymbol{M}_{(i)}^{\mathrm{T}}\boldsymbol{M}_{(i)}\boldsymbol{p}\right) \tag{3.402}$$

Differentiate ϕ with respect to $\boldsymbol{p}$ and set to the zero vector to minimise ϕ

$$\frac{\partial \phi}{\partial \boldsymbol{p}} = \begin{pmatrix} \dfrac{\partial \phi}{\partial p_1} \\ \dfrac{\partial \phi}{\partial p_2} \\ \vdots \\ \dfrac{\partial \phi}{\partial p_8} \end{pmatrix} = \boldsymbol{0} = \sum_{i=N}^{L} \left(\boldsymbol{0} - 2\boldsymbol{M}_{(i)}^{\mathrm{T}}\boldsymbol{y}_{(i)} + 2\boldsymbol{M}_{(i)}^{\mathrm{T}}\boldsymbol{M}_{(i)}\boldsymbol{p}\right) \tag{3.403}$$

whence

$$\boldsymbol{p} = \left\{\sum_{i=N}^{L} \boldsymbol{M}_{(i)}^{\mathrm{T}}\boldsymbol{M}_{(i)}\right\}^{-1} \left\{\sum_{i=N}^{L} \boldsymbol{M}_{(i)}^{\mathrm{T}}\boldsymbol{y}_{(i)}\right\} \tag{3.404}$$

are the desired model coefficients.

Example 3.17

Using an artificial neural net (ANN) to create a nonlinear model.

Many applications and variations of the artificial neural network have made it a popular choice for the modelling of observation data. This is a data-fitting (or classification) procedure which attempts to mimic the parallel processing of the human brain. A single neuron in the brain can collect thousands of electrical signals through its *dendrites*. The output of the neuron through the *axon* is interpreted by *synapses*, which in turn stimulate dendrites to other neurons. In Figure 3.40, the *sigmoidal curve* is shown which is used to interpret the output in most ANNs.

The discussion here will focus on the *back-propagation* network. This has an input layer, output layer, and one or more hidden layers (Figure 3.41). It is arranged only in *feedforward*, that is signals are not allowed to loop backwards in the structure. The nonlinear transfer function in each neuron of the hidden layer and output layer allows the network to fit nonlinear behaviour. The network is *trained* by being presented with many sets of both the inputs and the outputs. In this process, the internal weights w applied to the interconnections are systematically adjusted to minimise the overall output error. Each iteration of this process involves a *forward step* to establish the present output error, followed by a *backward step* in which the weights used at the output nodes are first adjusted, and then the error is *back-propagated* to the hidden layer, to alter the weights there too. The method is described in Section 7.11.1. Only enough neurons should be added to the hidden layer to support justifiable complexity in the behaviour.

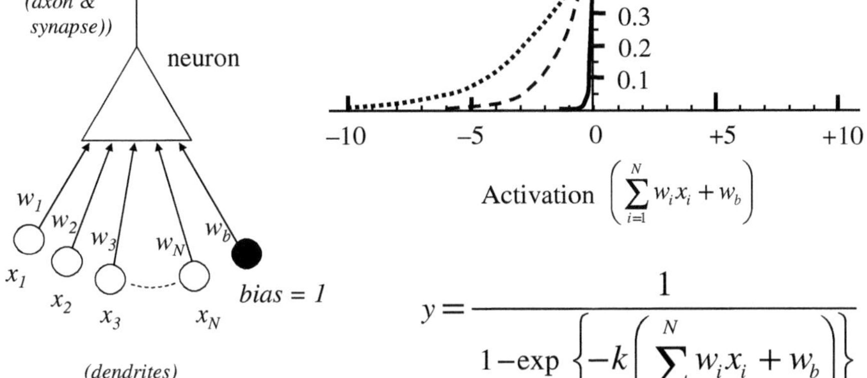

$$\left(\sum_{i=1}^{N} w_i x_i + w_b\right)$$

$$y = \frac{1}{1 - \exp\left\{-k\left(\sum_{i=1}^{N} w_i x_i + w_b\right)\right\}}$$

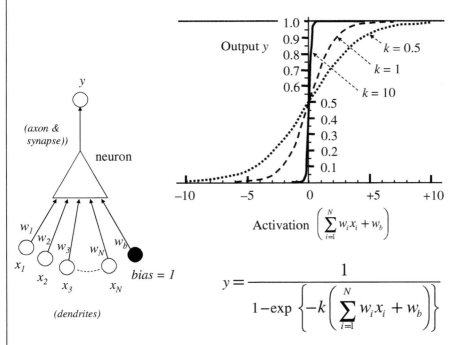

Figure 3.40 Interpretation of the weighted sum of inputs by an ANN neuron with sigmoidal function.

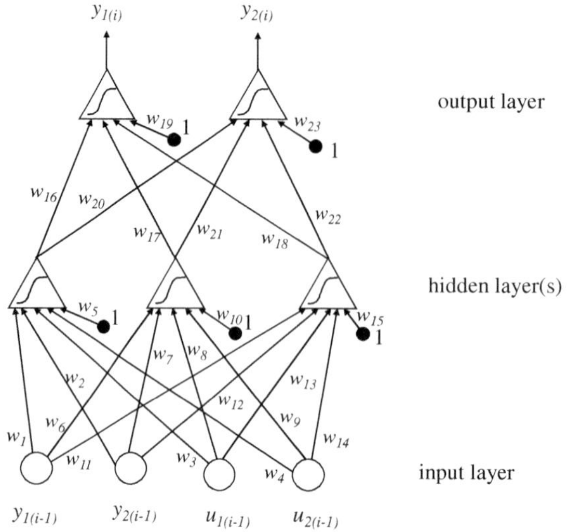

Figure 3.41 Artificial neural network for simulation of a two-input, two-output dynamic system.

Consider the problem of identifying an arbitrary discrete dynamic system

$$\boldsymbol{0} = \boldsymbol{f}\left(\boldsymbol{y}_{(i)}, \boldsymbol{y}_{(i-1)}, \ldots, \boldsymbol{y}_{(i-N)}, \boldsymbol{u}_{(i)}, \boldsymbol{u}_{(i-1)}, \ldots, \boldsymbol{u}_{(i-M)}\right) \tag{3.405}$$

For this example, it has been decided that the system can be described using two inputs and two outputs, that only the first time shift need be included and that the present inputs are not required:

$$\begin{pmatrix} 0 \\ 0 \end{pmatrix} = \boldsymbol{f}\left(y_{1(i)}, y_{2(i)}, y_{1(i-1)}, y_{2(i-1)}, u_{1(i-1)}, u_{2(i-1)}\right) \tag{3.406}$$

Thus, the data set $\left(y_{1(i)}, y_{2(i)}, y_{1(i-1)}, y_{2(i-1)}, u_{1(i-1)}, u_{2(i-1)}\right)$ at many times iT will be used to train the ANN. It is obvious that $y_{1(i)}$ and $y_{2(i)}$ should be chosen as the output values, in a cause-to-effect sense. If it is assumed that three neurons are required in the hidden layer, the structure in Figure 3.41 results. It is interesting to note that 23 weights w must be found, compared with the 8 parameters which were regressed in the equivalent linear state-space example (Example 3.16). Neurons should be minimised in ANN modelling, to avoid 'overfitting', that is assignation of more complex behaviour than warranted, just to fit measured data more tightly.

3.9.6 Modelling with Automata, Petri Nets and their Hybrids

Sometimes systems are too large and complex to be represented as a single monolithic set of equations and conditions, and it helps conceptually to divide them up into clear-cut entities which interact with each other according to well-defined rules. One approach is to use *automata*, or a particular form of these, the *Petri net*. These techniques originally grew around

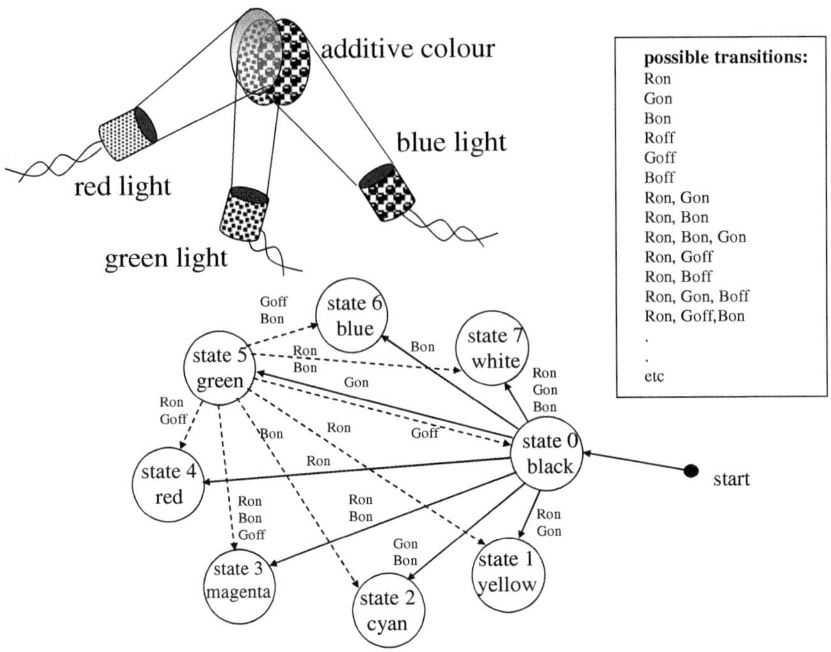

Figure 3.42 State and transition relationships for additive colours.

the idea of systems with a finite number of discrete states, but the methods were subsequently hybridised by the inclusion of continuous components. Along the way, a lot of useful theory and software has been developed, so if one is prepared to constrain one's approach to a problem to these established formalisms, one can take advantage of this background. The introductory discussion here will focus on how these approaches can be used to model concurrent systems which are linked by events.

An *automaton* is an entity for which a set of *states* and *transitions* are defined as in Figure 3.42. The *initial state* must be known in order to determine the outcome of a series of transitions. Conditions which must be satisfied before a transition can occur are called *guards.* Reconsider the tank problem of Figure 3.3, repeated in Figure 3.43, in terms of its possible states:

0: Level below H and disc *intact.*

1: Level above H and disc *intact.*

2: Level above H and disc *burst.*

3: Level below H and disc *burst.*

A *hybrid* automaton representing this system could be expressed as in Figure 3.44, where both the discrete states and continuous variable are handled simultaneously. Available treatments for such systems are somewhat restricted, for example *timed automata with linear equations.*

Petri nets similarly focus on states and transitions, using *tokens* to ensure that the conditions for a transition are met, including concurrency with other transitions. Since the original

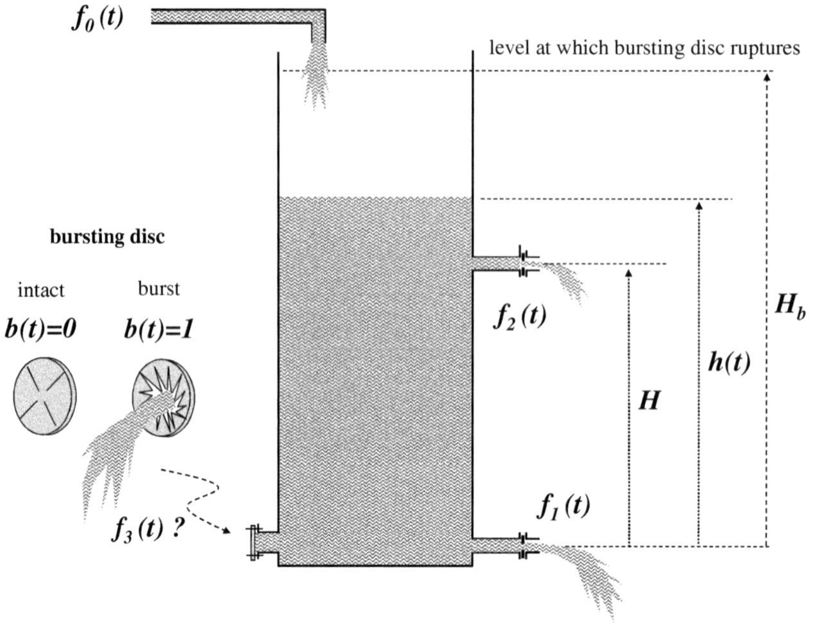

Figure 3.43 Tank with two restricted outflows and a bursting disc for automaton representation.

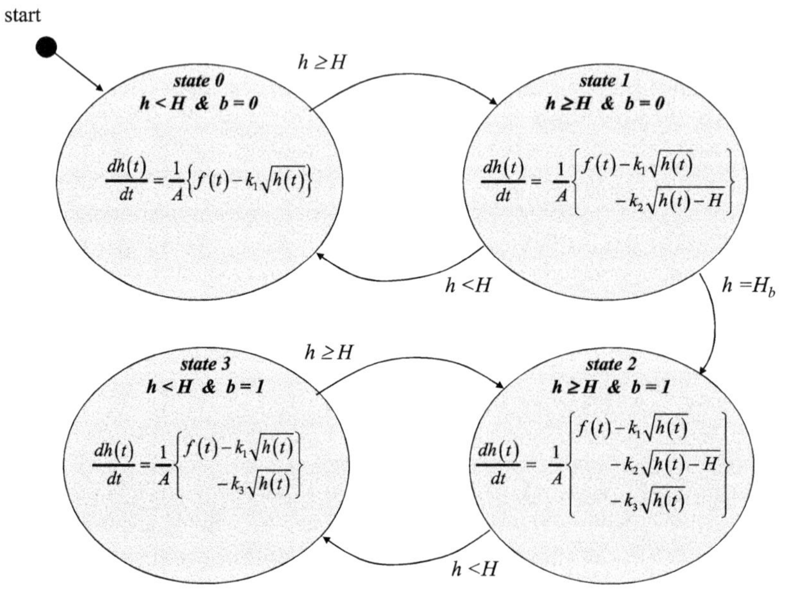

Figure 3.44 Hybrid automaton for tank in Figure 3.42, showing *guards* on transitions.

work by Carl Petri in 1962 (Petri, 1962), many permutations of this approach have been developed, including *stochastic, timed, coloured, fluid* and *hybrid* Petri nets. Initially, consider just the basic definitions:

A Petri net is a bipartite directed graph consisting of two types of nodes: *places* and *transitions.*
Each *place* represents a certain condition in the system.
Each *transition* represents an event which could change the condition of the system.
Input *arcs* connect places to transitions, and output *arcs* connect transitions to further places.
Tokens*Tokens* are dots (or integers) associated with places. The presence of a token means that the condition has been satisfied.
A transition *fires* when all of its input places have at least one token, and in so doing it removes one token (or more if specified) from each input place, and puts one token into each output place.
The *marking* of the system is the distribution of tokens in its places.
A marking is *reachable* from another marking if there exists a sequence of transition firings capable of taking the system from the original marking to the new marking.
An arrow-headed arc arriving at a transition enables the transition. If it has a small empty circle instead of an arrowhead, it *inhibits* the transition (assuming a token is available).

The original Petri net framework as defined above is well suited to the calculational representation of event-based systems, such as the motor start/stop system of Figure 2.36, represented in Figure 3.45. In this example, the initial *marking* has the motor off and the start button (NO) pressed, so the next state will be 'motor on' – the new marking will be a single token in the 'motor on' place.

Some extensions of these basic ideas include *timing* and *delays* which can be assigned to transitions (represented by a wider bar with a numeric indication of the delay period). Fixed *firing rates* can be used for *continuous variables, guards* can be placed on arcs and *hybrid* systems can include both discrete and continuous variable states. The drive towards *hybrid* representations has been fuelled by the usefulness of Petri nets in graphically representing biochemical reaction networks, for example for drug metabolism and interaction. These descriptions needed to include concentrations and reaction mass balances based on differential equations.

Figure 3.46 represents three levels of detail which might be used to represent a tank-filling operation using Petri nets. The version (a) only has the discrete states, and no information concerning the timing or intermediate values of the tank contents. Version (b) is a timed Petri net which makes use of a delay transition (wider black bar), which in an event-driven system will allow 'time stamping' of the occurrence of the represented events. However, this form relies on constant flow rates, and complete knowledge of the states, for example it could not start with the tank partially filled.

The *hybrid* version (c) shows the typical arrangement where the discrete network and continuous network function in parallel, with limited interconnection. *Continuous places* are shown as a double circle, and *continuous transitions* are shown as a double bar (empty narrow box). Associated with a continuous transition is a *firing speed* at which it automatically and continuously fires. The 'tank being filled' transition in Figure 3.46 will be firing at a rate *f*, which is the volumetric rate delivered by the stock pump. So when it is enabled, a volume $f\Delta t$ will be delivered from the stock to the tank over an interval Δt. Two loops cause this continuous transition to interact with the logical states. Firstly, the pump must be switched on, and secondly the receiving tank must not yet be full. On each firing, the 'tank being filled' transition must be

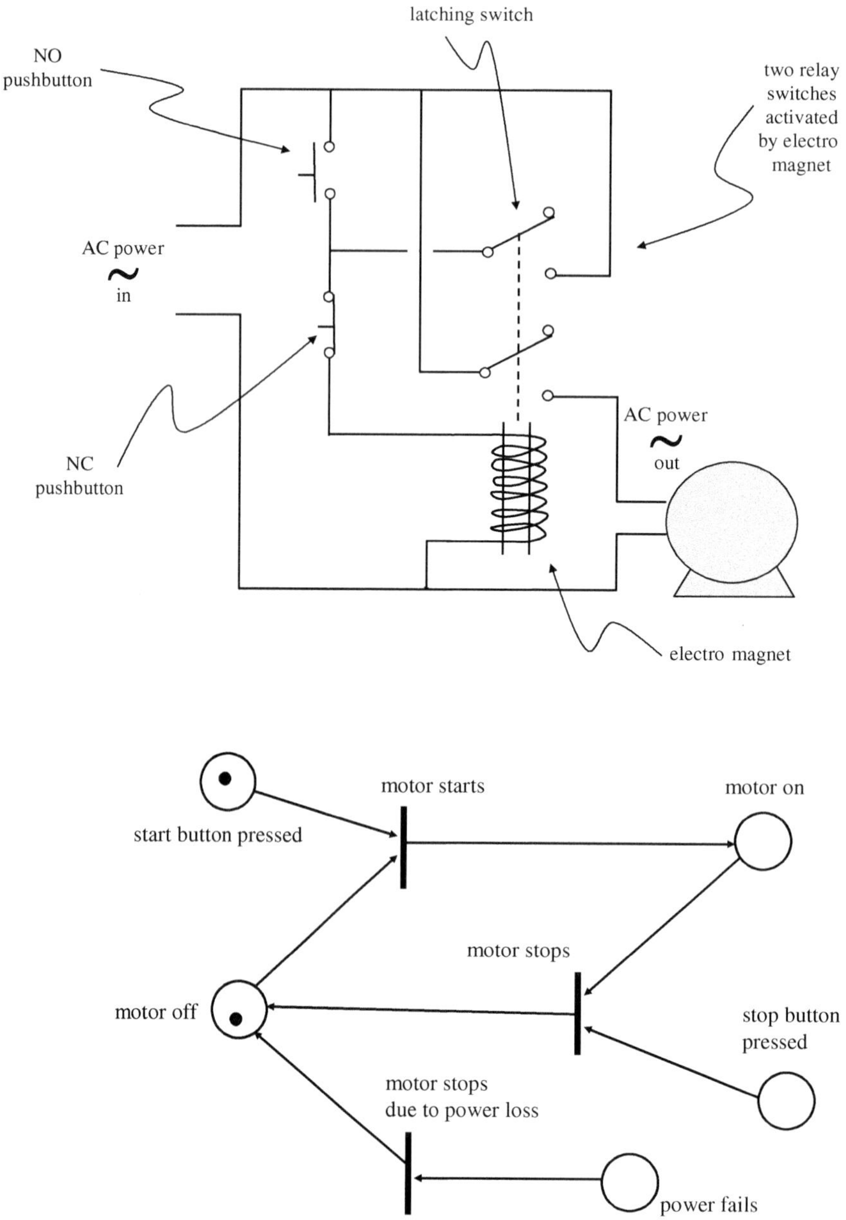

Figure 3.45 Petri net for motor power latching circuit.

able to take a token from each of the two logical places, but it returns them immediately. As can be seen, the continuous 'tank contents' place is able to supply a token to the 'tank filled' transition through the $V = V_{max}$ guard, that is once the level V reaches V_{max}. So at that point the 'tank not full' token would be lost and filling would cease.

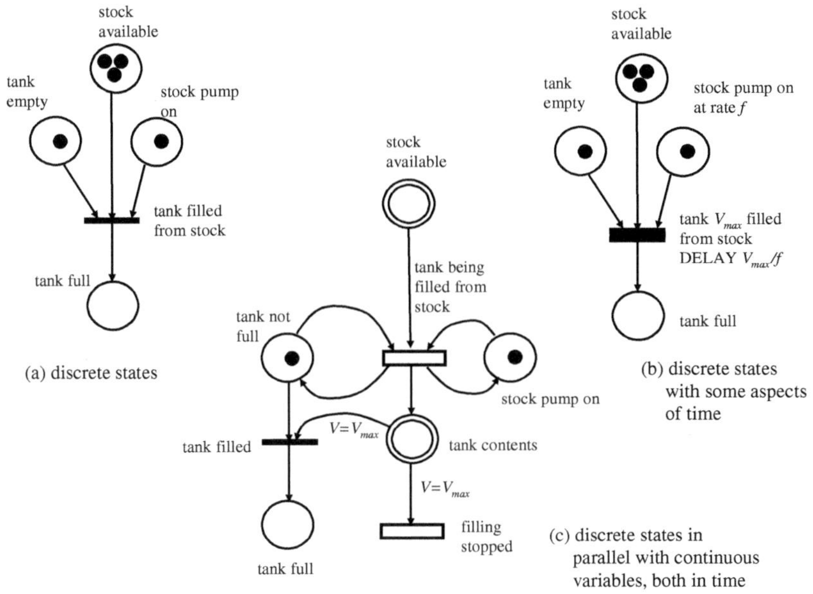

Figure 3.46 Petri nets with three levels of detail for a tank-filling operation.

Figure 3.47 is a *hybrid* Petri net representation of the tank with two restricted outflows and a bursting disc in Figure 3.43. As in version (c) of Figure 3.46, the hybrid arrangement may seem cumbersome, and indeed the ability to treat it efficiently in the Petri net context is limited. An alternative is simply to use a discrete Petri net to supervise the execution of an associated set of equations – usually

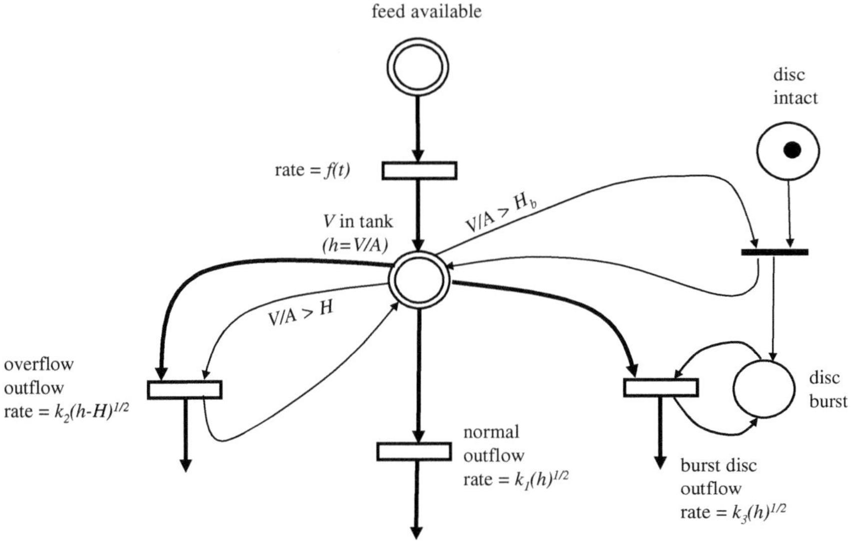

Figure 3.47 *Hybrid* Petri net for tank with two restricted outflows and a bursting disc in Figure 3.43.

in differential and algebraic form (Champagnat *et al.*, 1998). Much like the automaton in Figure 3.44 then, the place of the token in the discrete Petri net represents the particular set of equations which must be integrated as time moves on. The Petri net itself still provides the means to synchronise and ensure precedence of parallel calculations in a distributed system.

3.9.7
Models Based on Fuzzy Logic

The principles of fuzzy logic were established by Lotfi Zadeh in 1965 (Zadeh, 1965). Whereas regression allows one to fit a model to a collection of measurement data, fuzzy logic allows the fitting of a model to human sentiment and experience, and to extract useful numerical estimates from this ill-defined information.

Fuzzy logic is based on fuzzy set theory. This allows for *partial* membership of a set, as opposed to the traditional *bivalent* view in which an element either did belong (membership = 1) or did not belong (membership = 0). For example, an ambient temperature range might be described by *cold*, *mild* or *hot*. Whether or not 15 °C is *cold* or *mild* is subjective, and it is likely to have partial membership of both these sets (say, based on the views of a large population). The total membership must add up to 1. Various functional forms can be devised to represent these sets, but the use of straight-line segments as in Figure 3.48 is common. Here it is seen that 15 °C has a membership (or 'truth') of 0.5 for 'cold' and 0.5 for 'mild'.

So sets based on variable x have functional forms $A(x)$, $B(x)$ and so on, and operations defined as follows (Figure 3.49):

$$\text{complement}: \quad cA(x) = 1 - A(x) \tag{3.407}$$

$$\text{intersection ("AND")}: \quad (A \cap B)(x) = \min[A(x), B(x)] \tag{3.408}$$

$$\text{union ("OR")}: \quad (A \cup B)(x) = \max[A(x), B(x)] \tag{3.409}$$

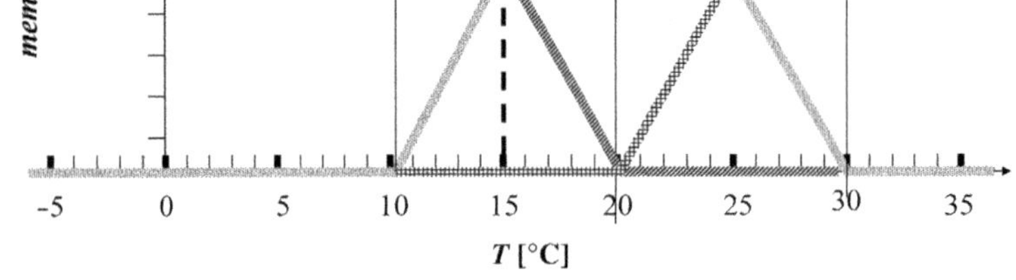

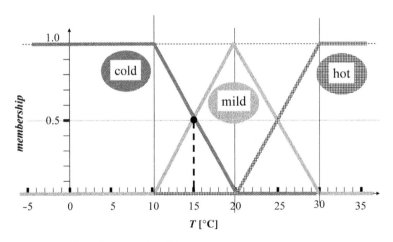

Figure 3.48 Three fuzzy sets describing a temperature range.

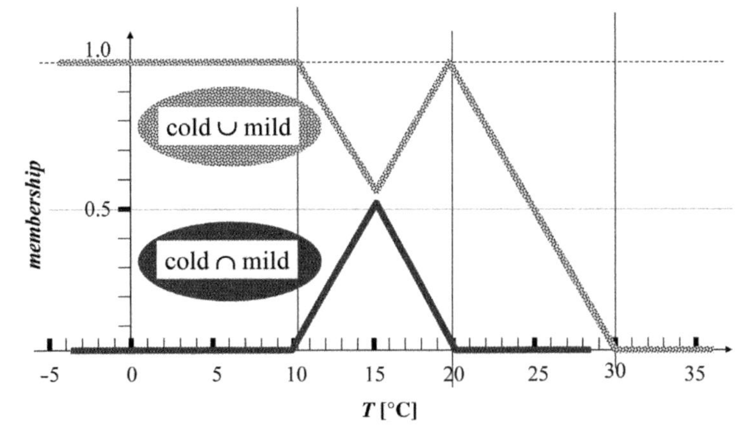

Figure 3.49 Union and intersection of two fuzzy sets shown in Figure 3.48.

The variables considered in fuzzy modelling are not often dynamically related – usually the equivalent of an algebraic representation is being sought. However, in keeping with this text's emphasis on dynamic modelling, an illustrative example will be based on the tank with two restricted outflows and a bursting disc (Figure 3.43). By predicting the *rate of change of level* in the tank, the result will lend itself directly to dynamic modelling.

Example 3.18

Fuzzy dynamic model of tank with two restricted outflows and a bursting disc (Figure 3.43).

An operator has been questioned about the behaviour of this system, voicing opinions such as

> When the inflow is high and the level is far below the overflow, the level rises fast, provided the disc has not burst, else it rises moderately.
>
> When the inflow is moderate and the level is just above overflow, the level remains static, provided the disc has not burst, else it drops moderately.
>
> And so on.

On this basis, the engineer has constructed the following fuzzy rule matrices (the ones shown only contain 'AND' requirements, but 'OR' requirements are also possible). Note that 'o/f' refers to the overflow level in the given observations.

(a) Disc intact ($b = 0$)

		Level variation (dh/dt)			
		Far <o/f	Just <o/f	Just >o/f	Far >o/f
Inflow f	Low	Slow rise	Static	Mod fall	Fast fall
	Moderate	Mod rise	Slow rise	Static	Mod fall
	High	Fast rise	Mod rise	Slow rise	Static

(continued)

(b) Disc burst ($b = 1$)

		Level variation (dh/dt)			
		Far <o/f	**Just <o/f**	**Just >o/f**	**Far >o/f**
Inflow f	Low	Static	Slow fall	Fast fall	Fast fall
	Moderate	Slow rise	Static	Mod fall	Fast fall
	High	Mod rise	Slow rise	Slow fall	Mod fall

Now the engineer questions further in order to attach some numerical meaning to the fuzzy descriptions used by the operator, and arrives at the sets in Figure 3.50. Notice the bivalent description of the bursting disc and overflow level.

The information required to evaluate the model is now complete. In general, *crisp* input values are going to be input for b, f and h, and one wants to get a *crisp* output of dh/dt which will be used for integration on that time step. By *crisp* is meant actual numerical value, instead of a *fuzzy* description. Consider the following case:

$$b = 1 \quad \{1.0 \times \text{burst} \tag{3.410}$$

$$f = 9\ \text{m}^3\ \text{h}^{-1} \quad \begin{cases} 0.75 \times \text{mod} \\ 0.25 \times \text{high} \end{cases} \tag{3.411}$$

$$h = 75\% \quad \begin{cases} 0.5 \times \text{just} > \text{o/f} \\ 0.5 \times \text{far} > \text{o/f} \end{cases} \tag{3.412}$$

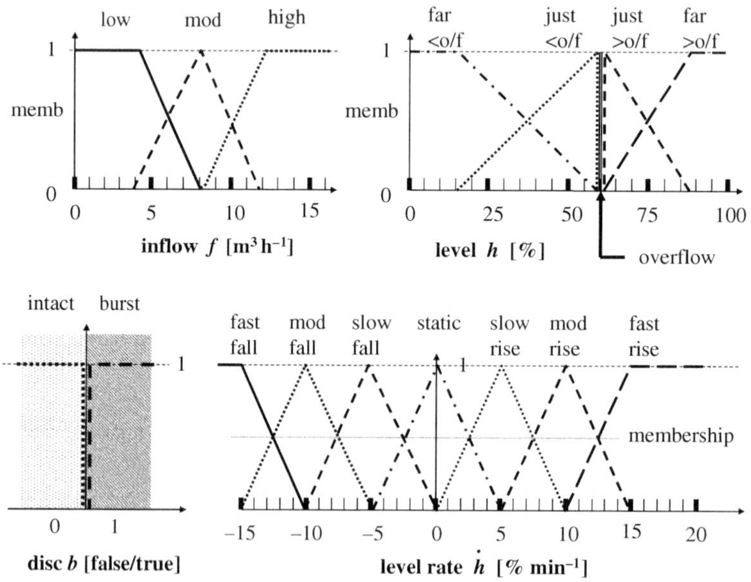

Figure 3.50 Fuzzy sets attached to operator's views of ranges.

From the rule matrices:

$$b(\text{burst})\ AND\ f(\text{mod})\ AND\ h(\text{just} > \text{o/f}) \rightarrow \begin{cases} \dot{h}(\text{mod fall}) \\ \text{weighted by} \\ b \cap f \cap h = \min(1, 0.75, 0.5) = 0.5 \end{cases} \quad (3.413)$$

$$b(\text{burst})\ AND\ f(\text{high})\ AND\ h(\text{just} > \text{o/f}) \rightarrow \begin{cases} \dot{h}(\text{slow fall}) \\ \text{weighted by} \\ b \cap f \cap h = \min(1, 0.25, 0.5) = 0.25 \end{cases} \quad (3.414)$$

$$b(\text{burst})\ AND\ f(\text{mod})\ AND\ h(\text{far} > \text{o/f}) \rightarrow \begin{cases} \dot{h}(\text{fast fall}) \\ \text{weighted by} \\ b \cap f \cap h = \min(1, 0.75, 0.5) = 0.5 \end{cases} \quad (3.415)$$

$$b(\text{burst})\ AND\ f(\text{high})\ AND\ h(\text{far} > \text{o/f}) \rightarrow \begin{cases} \dot{h}(\text{mod fall}) \\ \text{weighted by} \\ b \cap f \cap h = \min(1, 0.25, 0.5) = 0.25 \end{cases} \quad (3.416)$$

There are several defuzzification techniques. The general aim is to arrive at a point on the output axis (dh/dt) by weighting the above results according to their intersection values (0.5, 0.25, 0.5 and 0.25). The range and shape of each function need to be accounted for, so one way of doing it is by superimposing parts of each function up to the weighting fraction, and finding the centre of gravity of the resultant shape (Figure 3.51). In the illustration, a result $dh/dt = -11\%\ \text{min}^{-1}$ is obtained, allowing integration to find h for the next time step, when the model evaluation procedure must be repeated.

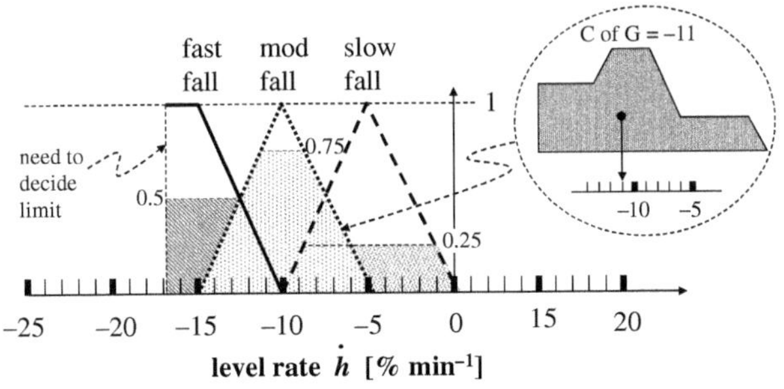

Figure 3.51 Possible defuzzification procedure.

There is some similarity between the concepts of fuzzy modelling, and *expert systems* in general. An expert system is preloaded with a *rule base* based on 'expert knowledge' in much the same way as the rules are constructed in fuzzy logic. However, there is generally no direct connection to numerical scales, with expert systems using an *inference engine* to navigate logical trees to obtain a conclusion of a discrete nature. The user input information can in cases be entirely preloaded, but because redundancy will occur with the narrowing of options as the search progresses, it is more efficient to enter data on demand.

3.10
Use of Random Variables in Modelling

The modelling discussed so far has been *deterministic* in nature, but one is often interested in *stochastic* (or *random*) behaviour. The area of *robustness* of control systems (Section 8.8) addresses the problem that the basis of model and controller development is usually some kind of simplistic statistical average, or perhaps even a single random sample. Then what if the real behaviour differs? In the open loop, to what extent should one expect the real output to differ from the model? In the closed loop, there can be serious implications for performance and stability.

Two issues are usually of interest: (a) the variation of input signals and disturbances and (b) the variation of the intrinsic system behaviour. Rather than embrace the fields of *stochastic differential equations* or even attempt to model the *moments* (*mean*, *variance*, etc.) of the outputs, one can learn a lot just by supplying the model with inputs and disturbances of a random characteristic nature, or varying the assumed model parameters in a random way.

Figure 3.52 shows two common *probability density functions* used to represent the range of values obtained in random sampling. Most natural processes produce samples which follow the Gaussian (or 'normal') distribution (a). The (pseudo-) random numbers available on most computer systems, however, follow a top-hat distribution (b) over a range $0<x<1$. In this range, one has uniformly $f(x)=1$, since the *cumulative probability* up to x

$$F(x) = \int_0^x f(x')\mathrm{d}x' \tag{3.417}$$

must reach 1 for $x=1$.

According to the *central limit theorem*, the Gaussian distribution can be approximated by creating individual samples by summing n top-hat samples – the larger the value of n, the closer the approximation.

$$y = \sum_{i=1}^{n} x_i \tag{3.418}$$

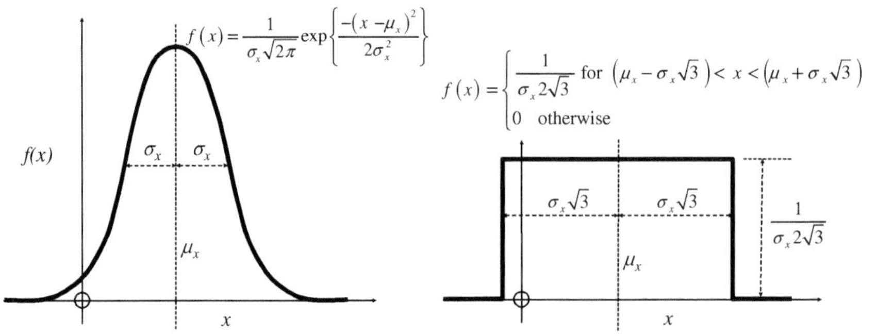

(a) Gaussian probability density function (b) Uniform probability density function

Figure 3.52 Gaussian (normal) and uniform (top-hat) probability density functions with mean μ_x and standard deviation σ_x.

$$\mu_y = \sum_{i=1}^{n} E\{x_i\} = n\mu_x \tag{3.419}$$

$$\sigma_y^2 = E\left\{\left(y - \mu_y\right)^2\right\} = E\left\{\left[\left(\sum_{i=1}^{n} x_i\right) - n\mu_x\right]^2\right\} \tag{3.420}$$

$$\sigma_y^2 = E\left\{\left[\sum_{i=1}^{n} (x_i - \mu_x)\right]^2\right\} = \sum_{i=1}^{n}\sum_{j=1}^{n} E\{(x_i - \mu_x)(x_j - \mu_x)\} \tag{3.421}$$

$$\sigma_y^2 = n^2\sigma_x^2 \tag{3.422}$$

$$\sigma_y = n\,\sigma_x \tag{3.423}$$

Consider the case where it is wished to use n samples from a {0, 1} uniform distribution to create a single random sample with a Gaussian distribution of mean μ_y and standard deviation σ_y. Firstly note that one requires

$$\mu_x = \frac{\mu_y}{n} \tag{3.424}$$

$$\sigma_x = \frac{\sigma_y}{n} \tag{3.425}$$

Since the samples z drawn from the {0, 1} distribution will have

$$\mu_z = 0.5 \tag{3.426}$$

$$\sigma_z = \frac{1}{2\sqrt{3}} \tag{3.427}$$

each z sample must be transformed according to

$$x = \left(z - \frac{1}{2}\right)\sigma_x 2\sqrt{3} + \mu_x \tag{3.428}$$

$$x = \frac{\left(z - \frac{1}{2}\right)\sigma_y 2\sqrt{3} + \mu_y}{n} \tag{3.429}$$

before the n samples x are summed. Even with n as low as 4, a good approximation to the shape of the Gaussian distribution is obtained, though it should be clear that no y sample could ever occur outside the range

$$\left(-\sigma_y 2\sqrt{3} + \mu_y\right) < y < \left(\sigma_y 2\sqrt{3} + \mu_y\right) \tag{3.430}$$

An alternative way to produce random normally distributed samples is through the cumulative probability function obtained for the Gaussian distribution as in Figure 3.53, after noting that

$$f(y) = \frac{1}{\sigma_y\sqrt{2\pi}}\exp\left\{\frac{-\left(y - \mu_y\right)^2}{2\sigma_y^2}\right\} \tag{3.431}$$

$$F(y) = \int_{-\infty}^{y} f(y')\mathrm{d}y' = 1 + \frac{1}{2}\operatorname{erf}\left(\frac{y - \mu_y}{\sigma_y\sqrt{2}}\right) \tag{3.432}$$

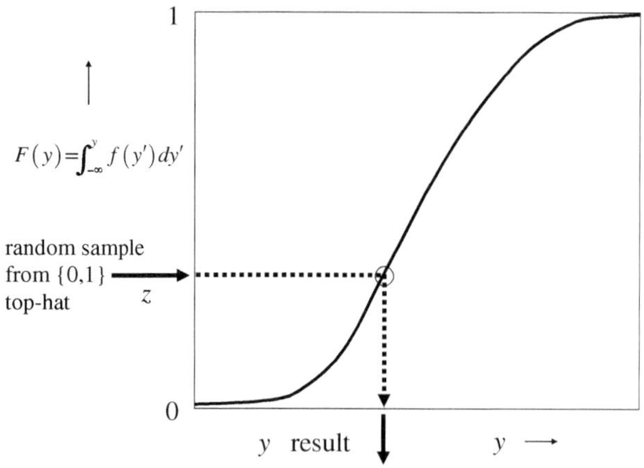

Figure 3.53 Inversion of a Gaussian cumulative probability function, with uniform {0, 1} samples as input, to get random Gaussian samples.

In practice, it is very unlikely that signal disturbances or model errors will vary in an uncorrelated way between sampling intervals – unless the sampling interval is large compared with the relevant time constants. So it is useful in simulation studies to prepare random variations of signals or model parameters which have a degree of *autocorrelation*. A simple way of doing this is by means of the *single-exponential filter* (Figure 3.54).

It is intended here to supply this filter with random samples x as input at the discrete time intervals. The filter will then produce a correlated random output y, making this a *Markov process* like Brownian motion. The input x will be unrelated samples, say from a Gaussian distribution as above. The output will also be distributed in the Gaussian sense, but will more importantly have *autocorrelation*, that is it will be correlated with itself across a time gap (Figure 3.55).

In the discrete system (Example 3.11), the z-domain transfer function for a first-order process with gain K, time constant τ and sampling interval T was

$$G(z) = \frac{z-1}{z} Z\left\{\frac{K}{s} - \frac{K}{s+\frac{1}{\tau}}\right\} \tag{3.433}$$

$$G(z) = K\left(\frac{z-1}{z}\right)\left(\frac{z}{z-1} - \frac{z}{z-e^{-T/\tau}}\right) \tag{3.434}$$

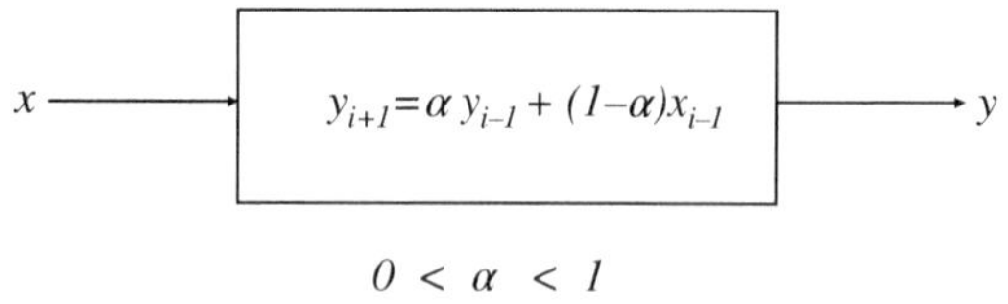

Figure 3.54 Single-exponential filter.

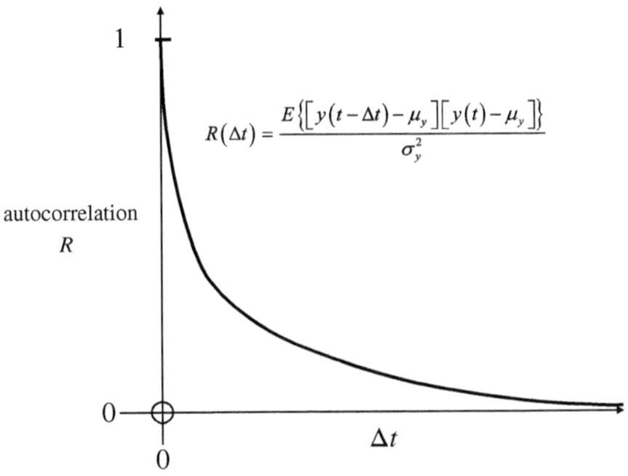

Figure 3.55 Typical autocorrelation function.

$$G(z) = K\left(1 - \frac{z-1}{z - e^{-T/\tau}}\right) \tag{3.435}$$

$$G(z) = K\left(\frac{1 - e^{-T/\tau}}{z - e^{-T/\tau}}\right) \tag{3.436}$$

The transfer function for the single-exponential filter in Figure 3.54 is

$$y_{i+1} = \alpha y_i + (1-\alpha)x_i \tag{3.437}$$

$$zy(z) = \alpha y(z) + (1-\alpha)x(z) \tag{3.438}$$

$$G(z) = \frac{y(z)}{x(z)} = \frac{1-\alpha}{z-\alpha} \tag{3.439}$$

Quite clearly

$$K = 1 \tag{3.440}$$

$$\alpha = e^{-T/\tau} \tag{3.441}$$

so the time constant τ of the single-exponential filter is

$$\tau = -\frac{T}{\ln(\alpha)} \tag{3.442}$$

Appropriate choice of α and T then allows one to set the degree of variation of the signal y between sample points. One notes from Equation 3.437 and Figure 3.55 that the autocorrelation function for this filter is

$$R(0) = 1 \tag{3.443}$$

$$R(1T) = \alpha = e^{-T/\tau} \tag{3.444}$$

$$R(2T) = \alpha^2 = e^{-2T/\tau} \tag{3.445}$$

$$\vdots$$

$$R(iT) = \alpha^i = e^{-iT/\tau} \tag{3.446}$$

Large α (near the maximum of 1) will give a slowly varying signal (T/τ small) and small α (near the minimum of 0) will give a fast-varying, almost uncorrelated signal (T/τ large).

In Example 3.19, a situation is represented in which a dynamic model of a process has been developed, and one wishes to provide a typically time-varying input signal, perhaps to check the performance of a controller, identifier or optimiser. Additionally, time-correlated random Gaussian errors are included in system parameters, representing model uncertainty and drift. This type of random perturbation of signals and parameters is referred to as *Monte Carlo* modelling, and it is often useful for revealing the effects of a large range of event combinations.

Example 3.19

Signal and model parameter variations for a second-order system.

Recall that *s*-domain transfer functions lying between sample points should be transformed *in entirety* to the *z*-domain. However, for convenience, approximate the non-interacting two-tank flow system of Figure 3.25a as having a second *sample and hold* between the two tanks, determining overall behaviour according to the product of the individual tank transfer functions:

$$G_1(z) = \left(\frac{1 - e^{-T/\tau_1}}{z - e^{-T/\tau_1}}\right) \tag{3.447}$$

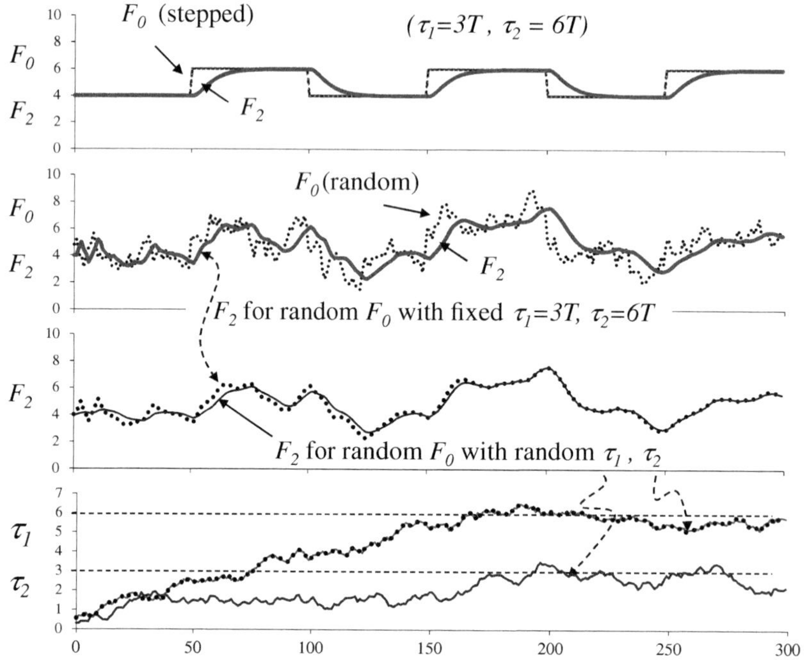

Figure 3.56 Second-order tank flow system response without and with time-correlated Gaussian random variations in feed F_0 and time constants τ_1 and τ_2.

$$G_2(z) = \left(\frac{1 - e^{-T/\tau_2}}{z - e^{-T/\tau_2}}\right) \tag{3.448}$$

$$G(z) = \frac{F_2(z)}{F_0(z)} = G_1(z)G_2(z) = \frac{\left(1 - e^{-T/\tau_1}\right)\left(1 - e^{-T/\tau_2}\right)}{z^2 - \left(e^{-T/\tau_1} + e^{-T/\tau_2}\right)z + e^{-(T/\tau_1 + T/\tau_2)}} \tag{3.449}$$

For $\tau_1 = 3T,\ \tau_2 = 6T$

$$F_2(z) = \left(e^{-1/3} + e^{-1/6}\right)z^{-1}F_2(z) - e^{-(1/3+1/6)}z^{-2}F_2(z) + \left(1 - e^{-1/3}\right)\left(1 - e^{-1/6}\right)z^{-2}F_0(z) \tag{3.450}$$

In Figure 3.56, responses are given for this system without and with time-correlated Gaussian random variations in the feed F_0 and the time constants τ_1 and τ_2.

3.11 Modelling of Closed Loops

The focus so far has been on *open-loop* system modelling. In later chapters, controllers, identifiers and optimisers will be developed which one normally tests *offline* using a model of the process. Effectively, these algorithms just add extra equations to the overall system description, so there is nothing different to what has already been discussed. The most important aspect of closing a loop is to be aware of what can be manipulated and what cannot. In the *open loop*, MVs and DVs were accessible to excite a system. In the *closed loop*, some or all MVs will be taken over by the algorithm. Instead, higher level variables can now be set as input, such as setpoints and constraints (Figure 3.57).

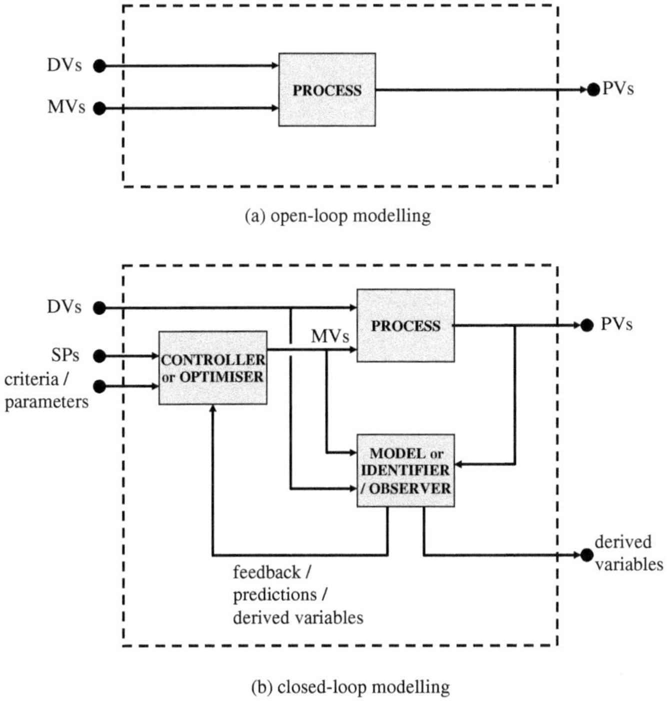

Figure 3.57 Net inputs and outputs for open-loop and closed-loop modelling.

References

Ascher, U.M. and Petzold, L.R. (1998) *Computer Methods for Ordinary Differential Equations and Differential–Algebraic Equations*, SIAM, Philadelphia, PA.

Becerra, V.M., Roberts, P.D. and Griffiths, G.W. (2001) Applying the extended Kalman filter to systems described by nonlinear differential–algebraic equations. *Control Engineering Practice*, **9** (3), 267–281.

Bemporad, A. and Morari, M. (1999) Control of systems integrating logic, dynamics and constraints. *Automatica*, **35**, 407–427.

Champagnat, R., Esteban, P., Pingaud, H. and Valette, R. (1998) Petri net based modelling of hybrid systems. *Computers in Industry*, **36**, 139–146.

Mao, G. and Petzold, L.R. (2002) Efficient integration over discontinuities for differential–algebraic systems. *Computers & Mathematics with Applications*, **43** (1–2), 65–79.

Petri, C.A. (1962) Kommunikation mit Automaten. Ph.D. thesis, Darmstadt University of Technology.

Zadeh, L.A. (1965) Fuzzy sets. *Information and Control*, **8**, 338–353.

4
Basic Elements Used in Plant Control Schemes

So far the focus has been on open-loop systems, that is how a system naturally behaves before any intelligence is added to it. Now simple signal manipulations and rerouting, including feedback, will be considered. The 'base layer' control scheme for a processing plant is generally constructed using signal interconnections between a number of standard elements. Configuration software for SCADA and distributed control systems allow one to do this graphically. This chapter focuses on the standard functions made available in these schemes.

4.1
Signal Filtering/Conditioning

Measurement signals can be altered at three points:

1) Some measurement devices (such as DP cells) have on-board electronic or digital signal-smoothing algorithms.
2) Signal conversion hardware such as analogue-to-digital converters can have accompanying electronics such as RC filters to smooth signals.
3) Once the measurement is accessible inside the control system software, it can be subjected to numerical calculations, typically smoothing, before it is inserted into the database.

Signals are subjected to a degree of random 'noise', either electrical interference, such as RF (radio-frequency) radiation, or natural variations, such as liquid motion in a vessel affecting a level measurement. This is despite such efforts by instrument technicians as opto-isolators or stilling chambers. Where these perturbations are short enough not to affect the control objectives, the control engineer usually wants to get rid of them. One reason is that the algorithm sampling cycle might coincide with unrepresentative outlying points. Another is that it is pointless to try to track rapid variations around a mean value when the rest of a scheme is too slow to react to them.

Figure 4.1 represents a typical signal-smoothing application, showing equivalent analogue and digital filters (single-exponential filter: Section 3.10). The value of the filter time constant τ has to be decided. As mentioned, this will be on the basis of what rates of variation are significant. In the figure, one notes that the signal time constant is around 10 s, whilst that of the filter is around 1 min. This can be judged in an approximate way from the length of the shortest oscillations. For the digital filter, a decision must be made about the sampling interval T. To obtain enough detail from the signal, and avoid aliasing (Figure 3.39), it will be short – less than 10 s in this case. Algorithms using the smoothed result h_m are likely to be operating on longer time intervals.

Applied Process Control: Essential Methods, First Edition. Michael Mulholland.

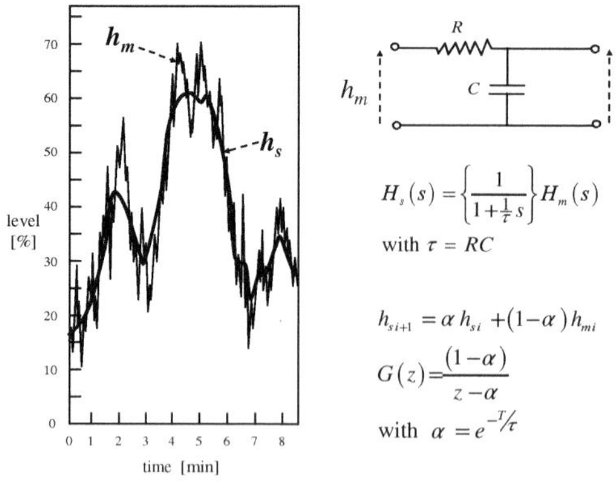

Figure 4.1 Electronic and digital smoothing filters.

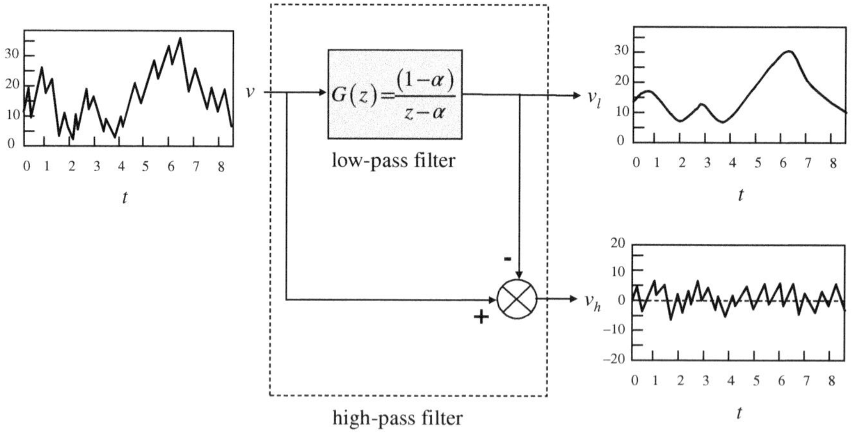

Figure 4.2 Low-pass and high-pass filters.

The filter arrangement in Figure 4.1 is sometimes referred to as a *low-pass filter*, because it will only pass the lower frequencies. In some applications, for example measurement of eddy turbulence, one is interested in isolating the higher frequencies. For this purpose, a *high-pass filter* can be constructed as in Figure 4.2 by subtracting the low-pass signal from the original signal.

Example 4.1

Filter to correct outlying data points.

In the handling of real-time data, a typical problem is to avoid reacting to outlying data points which all available evidence suggests cannot be valid. One way of doing this is by means of a 'fish-tail' filter. For this one would define a fixed maximum change ($+\Delta x$ and $-\Delta x$) that is to be allowed between successive points in time. Effectively, these ramp limits define a 'fish-tail' zone for future measurements, with the apex at the last accepted point. Any point occurring outside of the zone is brought back onto the fish-tail boundary.

One could make an even more sophisticated filter, which has limits that respond to the current variability of the signal. If the current standard deviation of the signal has been consistently high, then one would like to have wider limits, and vice versa. That is the subject of the present example, which is shown in Figure 4.3. Here one can 'tune' the filter using a factor γ to give a limit envelope proportional to a local standard deviation obtained from smoothing filters. An important aspect of this filter is that the adjusted data point *does* still enter the updating calculation of the standard deviation, with its original value, to allow the standard deviation to gradually increase when a new period of variability is entered.

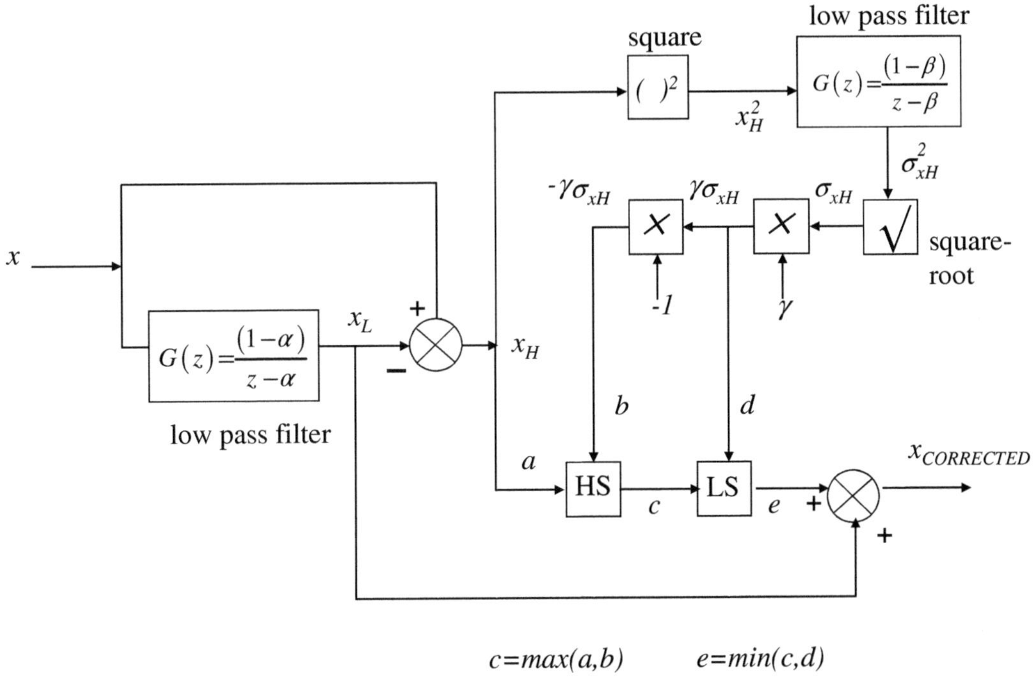

Figure 4.3 Example filter to correct outlying data points according to a running standard deviation.

In situations where high-integrity measurements are required, a *voting system* might be used in which several instruments taking the same reading are compared. A voting system might be used to ignore the most outlying reading(s) at any time. Alternatively, if an average is taken, large deviations might be used to trigger an alarm. An example of a two out of three voting system is given in Figure 4.4. Here the closest two measurements are averaged for the final result.

As far as the output signals for *manipulated variables* are concerned, only one adjustment comes to mind. Quite often the final control element has only two modes. For example, a refrigeration compressor motor may be either *on* or *off*, or a solenoid valve may be either *open* or *shut*. In these situations, it is sometimes useful to *time-slice* the actual signal to the final control element in order to obtain an intermediate effect. This would allow an algorithm to output an intermediate (i.e. analogue) signal, and have it interpreted by the signal manipulation as in Figure 4.5. Reflux ratios on small distillation columns are sometimes achieved like this. Clearly, the repeat interval T needs to be small enough not to create significant transients in the system, yet not so small that the mechanism is overworked or cannot respond properly.

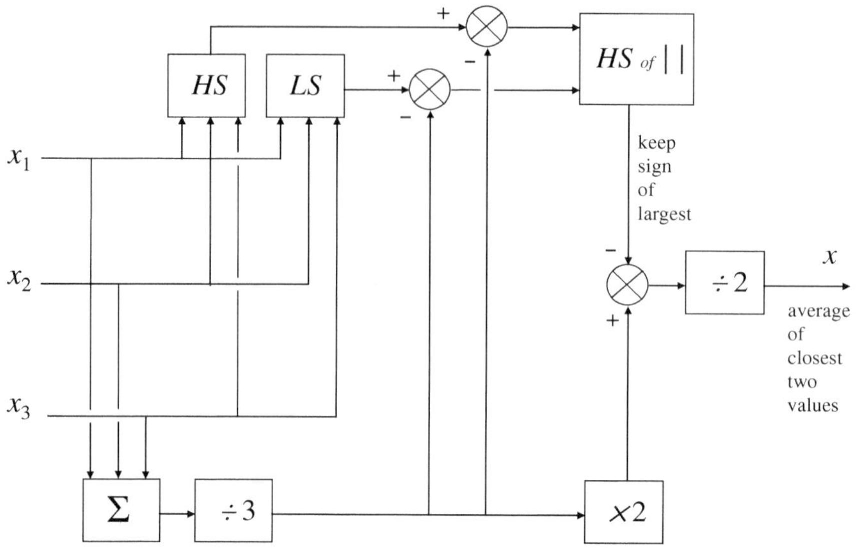

Figure 4.4 Example of two out of three voting system with averaging of closest readings (HS is a highest selector and LS is a lowest selector).

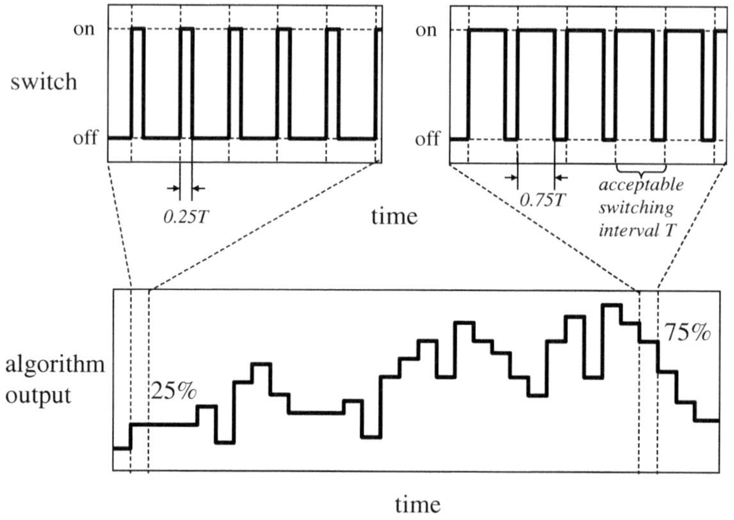

Figure 4.5 Time slicing of a discrete manipulated variable in order to achieve intermediate effects.

4.2 Basic SISO Controllers

With a basic single-input single-output (SISO) controller, one adds significant intelligence to the operation of a process. Here one is starting to ask the process to take care of itself. On the downside, the act of closing a feedback loop introduces complicated destabilising dynamic effects which need to be considered carefully, and these will be embraced later in Chapter 8. One gets an idea of the potential for information recycle to cause disruption from the howl in a microphone/loudspeaker feedback loop.

4.2.1 Block Diagram Representation of Control Loops

Although one is probably not working with a linear system in the *s*-domain, classical viewpoints based on Laplace transfer functions are useful for clarification of the information flow in control loops. Figure 4.6 illustrates some important conceptual paradigms using transfer functions. The additivity/subtractability of effects from parallel transfer functions at the summers/comparators definitely relies on linearity. More generally, for MIMO systems, 'bigger blocks' would need to receive several inputs and/or produce several outputs.

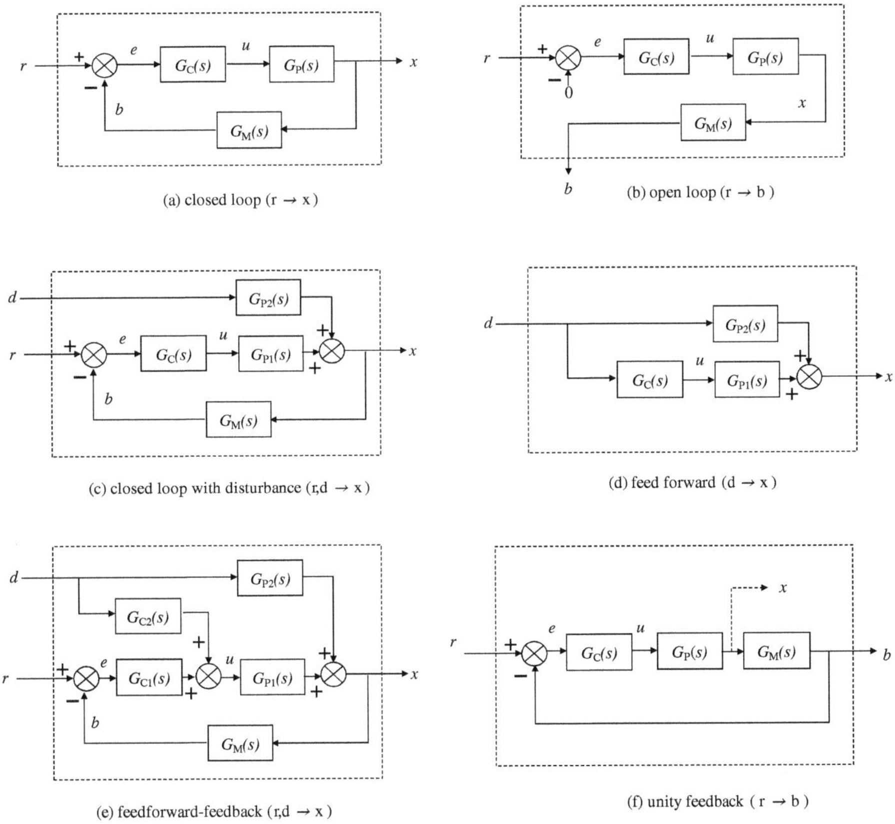

Figure 4.6 Conceptual control loop configurations from a transfer function viewpoint.

In Figure 4.6, G_P refers to the actual process, G_C is the control algorithm, r is the setpoint and d is a disturbance signal. G_M is the measurement device which prepares the feedback signal. There is a very subtle distinction between the *actual* PV, shown as x, and the signal fed back to the controller, shown as b. The variable x is the actual instantaneous variable – say, a level in a tank. Perhaps the closest way of knowing this is by looking at it – then the only lag in our observation is the time that light takes to travel to our eyes! However, all of the instruments used to obtain measurement signals on plants will have some sort of dynamic lag – and the purpose of showing G_M explicitly is to remind one of this problem – what is being seen in the control loop is not necessarily where the level is right now. Process engineers quickly ignore this aspect, and see G_M as part of G_P, similar to *unity feedback* in (f), and this generally does not cause any problems.

Now the relationship between *open-loop* and *closed-loop* transfer functions will be derived. Though the emphasis in this section is on SISO controllers, the equations will be developed in their more general multivariable form. Thus, care must be taken regarding the multiplication order, for example $\boldsymbol{G}_C(s)\boldsymbol{G}_P(s) \neq \boldsymbol{G}_P(s)\boldsymbol{G}_C(s)$.

The error $\boldsymbol{e}$ between setpoint and feedback is

$$\boldsymbol{E}(s) = \boldsymbol{R}(s) - \boldsymbol{B}(s) \tag{4.1}$$

and

$$\boldsymbol{U}(s) = \boldsymbol{G}_C(s)\boldsymbol{E}(s) \tag{4.2}$$

The closed-loop arrangement in Figure 4.6a gives

$$\boldsymbol{X}(s) = \boldsymbol{G}_P(s)\boldsymbol{U}(s) \tag{4.3}$$

$$\boldsymbol{X}(s) = \boldsymbol{G}_P(s)\boldsymbol{G}_C(s)\boldsymbol{E}(s) \tag{4.4}$$

Since

$$\boldsymbol{B}(s) = \boldsymbol{G}_M(s)\boldsymbol{X}(s) \tag{4.5}$$

$$\boldsymbol{X}(s) = \boldsymbol{G}_P(s)\boldsymbol{G}_C(s)[\boldsymbol{R}(s) - \boldsymbol{G}_M(s)\boldsymbol{X}(s)] \tag{4.6}$$

so

$$[\boldsymbol{I} + \boldsymbol{G}_P(s)\boldsymbol{G}_C(s)\boldsymbol{G}_M(s)]\,\boldsymbol{X}(s) = \boldsymbol{G}_P(s)\boldsymbol{G}_C(s)\boldsymbol{R}(s) \tag{4.7}$$

$$\boldsymbol{X}(s) = \left\{[\boldsymbol{I} + \boldsymbol{G}_P(s)\boldsymbol{G}_C(s)\boldsymbol{G}_M(s)]^{-1}\boldsymbol{G}_P(s)\boldsymbol{G}_C(s)\right\}\boldsymbol{R}(s) \tag{4.8}$$

So the *closed-loop* transfer function is

$$\boldsymbol{G}_{CL}(s) = \left\{[\boldsymbol{I} + \boldsymbol{G}_P(s)\boldsymbol{G}_C(s)\boldsymbol{G}_M(s)]^{-1}\boldsymbol{G}_P(s)\boldsymbol{G}_C(s)\right\} \tag{4.9}$$

Now it is necessary to embrace another subtle distinction in common use. Engineers on a plant will usually refer to the 'open-loop' process as what lies between the MV input and the CV output, for example from valve to tank level. In control loop analysis, however, the full sequence of transfer functions around the loop is termed the *open loop*, as in Figure 4.6b. The best practice is to qualify what is meant by *open loop* each time the term is used. Continuing with the analytical definition, the *open-loop transfer function* is

$$\boldsymbol{G}_O(s) = \boldsymbol{G}_P(s)\boldsymbol{G}_C(s)\boldsymbol{G}_M(s) \tag{4.10}$$

So the closed-loop transfer function is

$$G^*_{CL}(s) = \{[I + G_O(s)]^{-1} G_P(s) G_C(s)\} \tag{4.11}$$

$$G^*_{CL}(s) = \{[I + G_O(s)]^{-1} G_O(s)\}[G_M(s)]^{-1} \tag{4.12}$$

Usually one ignores the multiplier $[G_M(s)]^{-1}$, meaning that one's view is really the *unity feedback* form in Figure 4.6f. Since the only form of the controlled variable available is b, there is little interest in the true value x. Thus, one takes

$$G_{CL}(s) = \{[I + G_O(s)]^{-1} G_O(s)\} \tag{4.13}$$

so that

$$B(s) = G_{CL}(s) R(s) \tag{4.14}$$

For SISO systems, there is the common form

$$G_{CL}(s) = \frac{G_O(s)}{1 + G_O(s)} \tag{4.15}$$

The perceived relationship between *process* and *controller* is illustrated in Figure 4.7. Note that there are always three signals associated with a controller, which acts as a translator and a solver.

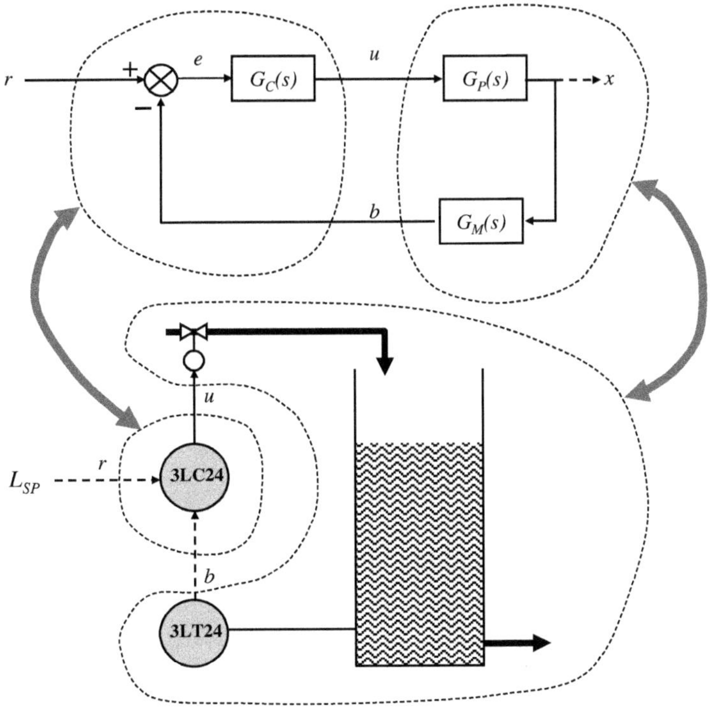

Figure 4.7 Perceived relationships between *process* and *controller.*

In the example, it is going to solve for such a valve setting, in the language *% open*, to cause the process output in the language *m* (or *% level*) to arrive at a value L_{SP}.

4.2.2
Proportional Controller

Figure 2.35 in Section 2.7 shows a centrifugal governor for rotational speed, and a ball float valve for level control. Both of these devices are *proportional controllers* – they give a control action which is linearly related to the deviation e of feedback from setpoint. It is usually not possible to ensure that the offset (or bias) control action u_0 exactly eliminates the error, so such controllers are prone to *offset*.

$$u(t) = K_C e(t) + u_0 \tag{4.16}$$

The parameter K_C is the *proportional gain*. It is sometimes expressed in inverse form as the *proportional band*:

$$\text{proportional band (\%)} = \frac{100}{K_C} \tag{4.17}$$

Clearly K_C would be dimensionless here in terms of fractions of instrument ranges. It is best not to dwell on whether the controller is *reverse acting* or *direct acting*. Suffice it to say that for the particular configurations shown, K_C will be *positive* if it is required to *increase* the MV to correct a *low* CV, and K_C with be *negative* if it is required to *decrease* the MV to correct a *low* CV.

As in Section 3.8, linearise Equation 4.16 by passing the Δ operator through it.

$$\Delta u(t) = K_C \Delta e(t) \tag{4.18}$$

and taking the Laplace transform

$$U(s) = K_C E(s) \tag{4.19}$$

so the transfer functions for a SISO *proportional* (P) controller are

$$G_C(s) = \frac{U(s)}{E(s)} = K_C \tag{4.20}$$

$$G_C(z) = \frac{u(z)}{e(z)} = K_C \tag{4.21}$$

The perturbation variable Δu is taken as the deviation from the initial value of u when the controller is switched on (i.e. u_0). The expression obviously relies on having $e_0 = 0$ for $u = u_0$, otherwise a high controller gain K_C would be required to reduce the steady $e \neq 0$. Even with $e_0 = 0$ for a steady $u = u_0$, a residual nonzero error e will be required to drive the control action to any other position $u \neq u_0$, for example to counter the effect of some 'load disturbance' in the system (i.e. caused by an unmeasured DV). In the case of a float valve, if a continuous flow is drawn from the tank, the level would have to stay below its equilibrium position ($u = u_0 = 0$, $e = e_0 = 0$), creating a steady-state error $e_{SS} \neq 0$ in order to cause the extra inflow through the valve. This highlights a typical problem of systems under proportional control – they experience an offset from setpoint under load disturbances or setpoint changes (Figure 4.8). The error is simply required to keep the valve in a new position. An exception to this is a system which is naturally integrating, for example boiler pressure control using the furnace

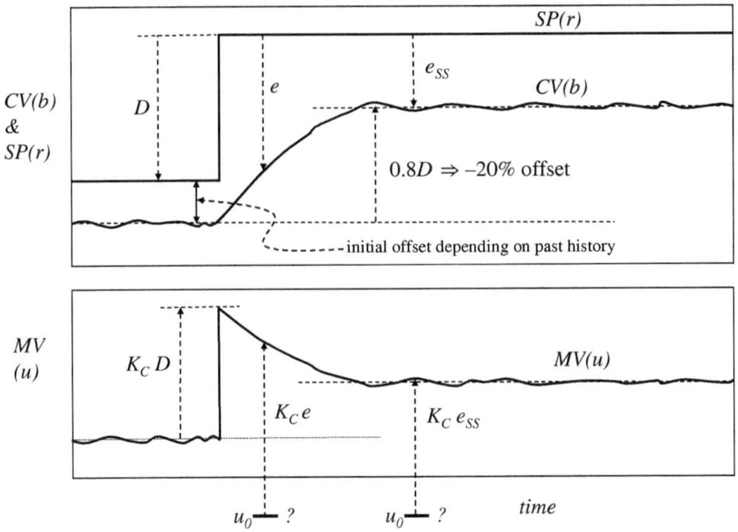

Figure 4.8 Proportional controller response to a setpoint step.

fuel valve. In any case, in many systems the exact value of the controlled variable is not of prime concern, and proportional control can be quite adequate. An example of this would be level control in a distillation column, where the liquid depth is acting largely as a vapour seal.

4.2.3 Proportional–Integral Controller

The problem of offset usually encountered with proportional controllers is dealt with by adding *integral action* to the *proportional action.*

$$u(t) = K_C\left[e(t) + \frac{1}{\tau_I}\int_0^t e(t')\mathrm{d}t'\right] + u_0 \tag{4.22}$$

The smaller the *integral time* τ_I is set, the stronger the integral action becomes in comparison with the proportional action. The integral time clearly corrects the units arising from the integration, and is sometimes expressed as its inverse (the frequency at which the proportional action is equated [*repeats per unit time*]). Integral action is sometimes called *reset.* A degree of proportional action is always necessary to dampen the oscillations caused by integral action. Linearise Equation 4.22, and then take the Laplace transform to obtain

$$U(s) = K_C\left[E(s) + \frac{1}{\tau_I}\left\{\frac{1}{s}\right\}E(s)\right] \tag{4.23}$$

so the transfer function for a *proportional–integral* (PI) controller is

$$G_C(s) = \frac{U(s)}{E(s)} = K_C\left[1 + \frac{1}{\tau_I s}\right] \tag{4.24}$$

A common derivation of the online discrete controller is based on a z-domain approximation for 's' as in Section 3.9.3.4. For example, using the Tustin approximation

$$s \approx \left(\frac{2}{T}\right)\frac{z-1}{z+1}$$

one has

$$G_C(z) = K_C\left[\frac{\left(\frac{T}{2\tau_I}+1\right)z+\left(\frac{T}{2\tau_I}-1\right)}{(z-1)}\right] \tag{4.25}$$

which would be used in the form

$$u_{i+1} = u_i + K_C\left(\frac{T}{2\tau_I}+1\right)e_{i+1} + K_C\left(\frac{T}{2\tau_I}-1\right)e_i \tag{4.26}$$

When this controller is switched on ($i = 0$), u_0 is obtained from the current MV value, and e_1 and e_0 are *both* set to the start-up error SP-CV. The use of a stack to update these values will be illustrated in Section 4.2.4.

Figure 4.9 depicts the reaction of a system under PI control to a step in setpoint. The immediate proportional response of the MV is followed by a further build-up of the integral action, until the CV starts to respond. The integral action will build up continuously until the error is eliminated. If it cannot be eliminated, one has the problem of integral *windup*, and the manner of dealing with this is discussed in Section 4.2.5.

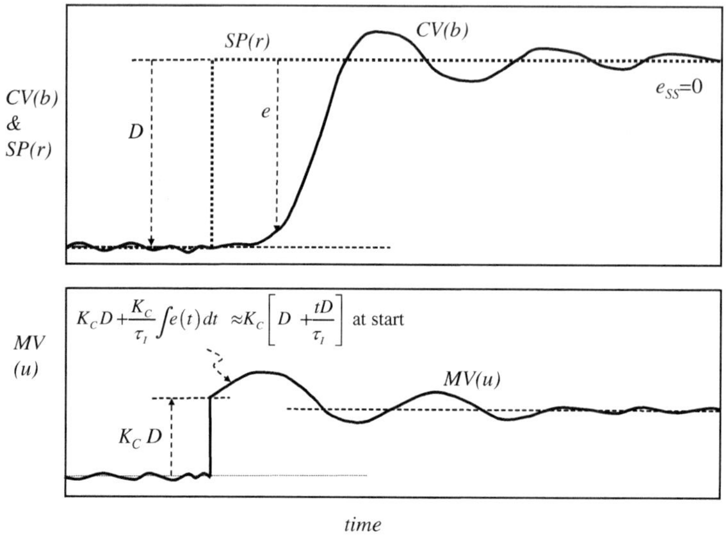

Figure 4.9 Proportional–integral controller response to a setpoint step.

4.2.4 Proportional–Integral–Derivative Controller

Although the PI controller effectively deals with offset, one notes that *integral action* is based on *past* errors, and is thus reacting after a time lag (a 90° phase lag as will be seen later). Moreover, most load disturbances entering a system do not cause an immediate response (i.e. unlike the setpoint steps discussed above). Rather, the effect builds up gradually, so the full correction from the *proportional action* only comes some time after the onset of the disturbance. The solution to these delays was to introduce a degree of *derivative action* into the controller. The controller would then react to the initial *slope* of the error – effectively using this slope to gauge the size of the full incoming error. The form of the *theoretical* proportional–integral–derivative (PID) controller is

$$u(t) = K_C \left[e(t) + \frac{1}{\tau_I} \int_0^t e(t')\mathrm{d}t' + \tau_D \frac{\mathrm{d}e(t)}{\mathrm{d}t} \right] + u_0 \tag{4.27}$$

The larger the *derivative time* τ_D, the larger the derivative action. Linearising Equation 4.27 and taking the Laplace transform, one obtains

$$U(s) = K_C \left[E(s) + \frac{1}{\tau_I} \left\{ \frac{1}{s} \right\} + \tau_D s E(s) \right] \tag{4.28}$$

so the transfer function for a theoretical PID controller is

$$G_C(s) = \frac{U(s)}{E(s)} = K_C \left[1 + \frac{1}{\tau_I s} + \tau_D s \right] \tag{4.29}$$

Consider the response of a system under control by such a theoretical PID controller to a step in the setpoint (Figure 4.10). Since the controller differentiates the step, it initially gives an infinite output. As discussed in Section 3.9.2.1, this theoretical controller is not *physically realisable.* This is apparent considering the order $m = 2$ of the numerator which is greater than $n = 1$ in the denominator of the form

$$G_C(s) = K_C \left[\frac{\tau_I \tau_D s^2 + \tau_I s + 1}{\tau_I s} \right] \tag{4.30}$$

In *real* PID controllers, the derivative action behaves as if it has passed through a first-order filter:

$$\tau_D s \to \tau_D s \times \left(\frac{1}{\alpha \tau_D s + 1} \right) \quad \text{with } \alpha \text{ small, e.g. } \alpha = 0.1 \tag{4.31}$$

So from Equation 4.29, a *real* PID controller has a transfer function of the form

$$G_C(s) = K_C \left[1 + \frac{1}{\tau_I s} + \frac{\tau_D s}{\alpha \tau_D s + 1} \right] \tag{4.32}$$

which has $m = n = 2$. Finally, it is not usual to apply the *derivative* action to $e =$ (SP-CV), but rather just to '-CV' $= -b$. So the differentiation of the setpoint SP is excluded, because manual setpoint

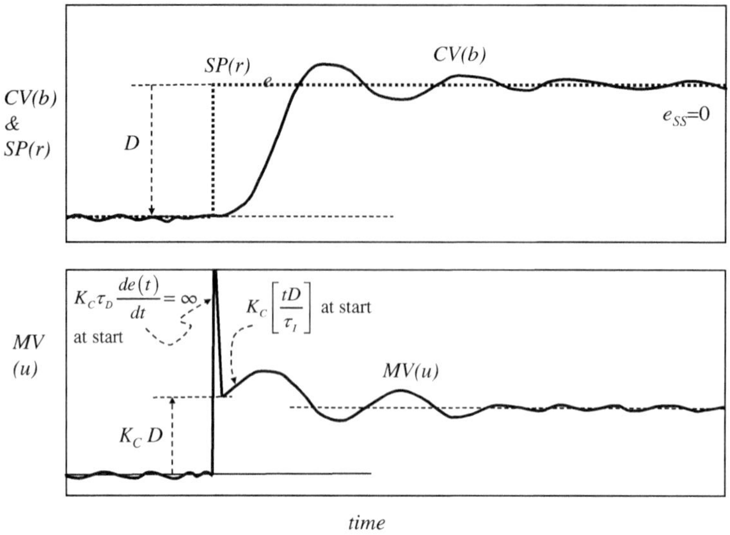

Figure 4.10 Theoretical response of a proportional–integral–derivative controller to a setpoint step (not physically realisable).

step adjustments will give impulses in the MV. In any case, in the processing industries, the main requirement is in *regulation*, so the SP seldom changes, and the control action MV will generally be responding just to '-CV', that is it will be responding to the load disturbances.

$$U(s) = \mathbf{G}_{\mathrm{C}}(s)\begin{pmatrix} E(s) \\ B(s) \end{pmatrix} \tag{4.33}$$

$$U(s) = K_{\mathrm{C}}\left[\left\{1 + \frac{1}{\tau_{\mathrm{I}}s}\right\} \quad \left\{\frac{-\tau_{\mathrm{D}}s}{\alpha\tau_{\mathrm{D}}s + 1}\right\}\right]\begin{pmatrix} E(s) \\ B(s) \end{pmatrix} \tag{4.34}$$

$$U(s) = K_{\mathrm{C}}\left[\left\{1 + \frac{1}{\tau_{\mathrm{I}}s}\right\} \quad \left\{-1 - \frac{1}{\tau_{\mathrm{I}}s} - \frac{\tau_{\mathrm{D}}s}{\alpha\tau_{\mathrm{D}}s + 1}\right\}\right]\begin{pmatrix} R(s) \\ B(s) \end{pmatrix} \tag{4.35}$$

Using the Tustin approximation for '*s*' as in Section 4.2.3 (see also Section 3.9.3.3), and setting

$$\beta = \frac{T}{2\tau_{\mathrm{I}}} \tag{4.36}$$

$$\gamma = \frac{T}{2\alpha\tau_{\mathrm{D}}} \tag{4.37}$$

$$\delta = K_{\mathrm{C}}\{\beta + 1\}\{\gamma + 1\} \tag{4.38}$$

$$\epsilon = K_{\mathrm{C}}\left[\{\beta - 1\}\{\gamma + 1\} + \{\beta + 1\}\{\gamma - 1\}\right] \tag{4.39}$$

$$\phi = K_{\mathrm{C}}\{\beta - 1\}\{\gamma - 1\} \tag{4.40}$$

$$\rho = \frac{K_{\mathrm{C}}}{\alpha} \tag{4.41}$$

then it is easily shown that the discrete form of a *real PID controller*, with setpoint removed from the derivative action, becomes

$$u_{i+1} = \left\{\frac{2}{\gamma+1}\right\}u_i + \left\{\frac{\gamma-1}{\gamma+1}\right\}u_{i-1} + \delta(r_{i+1} - b_{i+1}) + \epsilon(r_i - b_i) + \phi(r_{i-1} - b_{i-1}) - \rho b_{i+1} + 2\rho b_i - \rho b_{i-1} \tag{4.42}$$

A computer algorithm for a PID controller based on the form of Equation 4.42 is presented in Figure 4.11. Notice the use of stacks to cascade past values of r, b and u. The stacks must be updated whether or not the controller is operating (i.e. even if it is not on 'AUTO').

An alternative form of Equation 4.32 for a *real PID controller* arises from viewing it as a PI and a PD controller in series, the so-called 'series compensator':

$$G_C(s) = K_C\left[\frac{\tau_I s + 1}{\tau_I s}\right]\left[\frac{\tau_D s + 1}{\alpha\tau_D s + 1}\right] \tag{4.43}$$

for which a discrete form can be derived in the same way as above.

Derivative action theoretically adds up to 90° of *phase lead* to the system response, reducing the lag incurred by integral action. However, it is not extensively used, and the PI format tends to be more common. Derivative action will react strongly to even minor perturbations on the CV signal, if their rate of change is rapid (Figure 4.12). In many cases, feedback signals have high-frequency 'noise' superimposed on them, either from electrical interference or from natural phenomena, such as liquid motion in a vessel affecting a level measurement. With derivative action, this causes a rapid oscillation of the MV signal over a large range, leading to increased wear, and possibly introducing unwarranted variability in the process.

4.2.5 Integral Action Windup

A SISO control loop may, for some reason, find itself unable to reach the setpoint value. Typically, this happens when a control valve cannot travel any further – it has reached 100% open, or 0% open. If the controller has integral action, it will continue to 'wind up' with continuous integration of the residual error e. This phenomenon is also called *reset windup*. The main problem with windup is that if the setpoint subsequently becomes attainable, the controller will not take the CV to its setpoint, but will rather place the CV with an error of the opposite sign for a sustained period, in order to eliminate the original error integral (Figure 4.13).

Analogue controllers have a facility which stops the integration if the output signal does not match the calculated signal. This *anti-windup* feature is sometimes shown as an *external reset* which senses the output signal. As far as the discrete computer algorithms are concerned (Equation 4.26 for PI and Equation 4.42 for PID), these recursive calculations lend themselves to anti-windup naturally, just by 'clipping' the output u value at a limit if a value beyond the limit is calculated (see Figure 4.11).

4.2.6 Tuning of P, PI and PID Controllers

The methods provided in this section are 'black box' in the sense that they are based on measurable relationships between the process input and output. The few measurements involved cannot

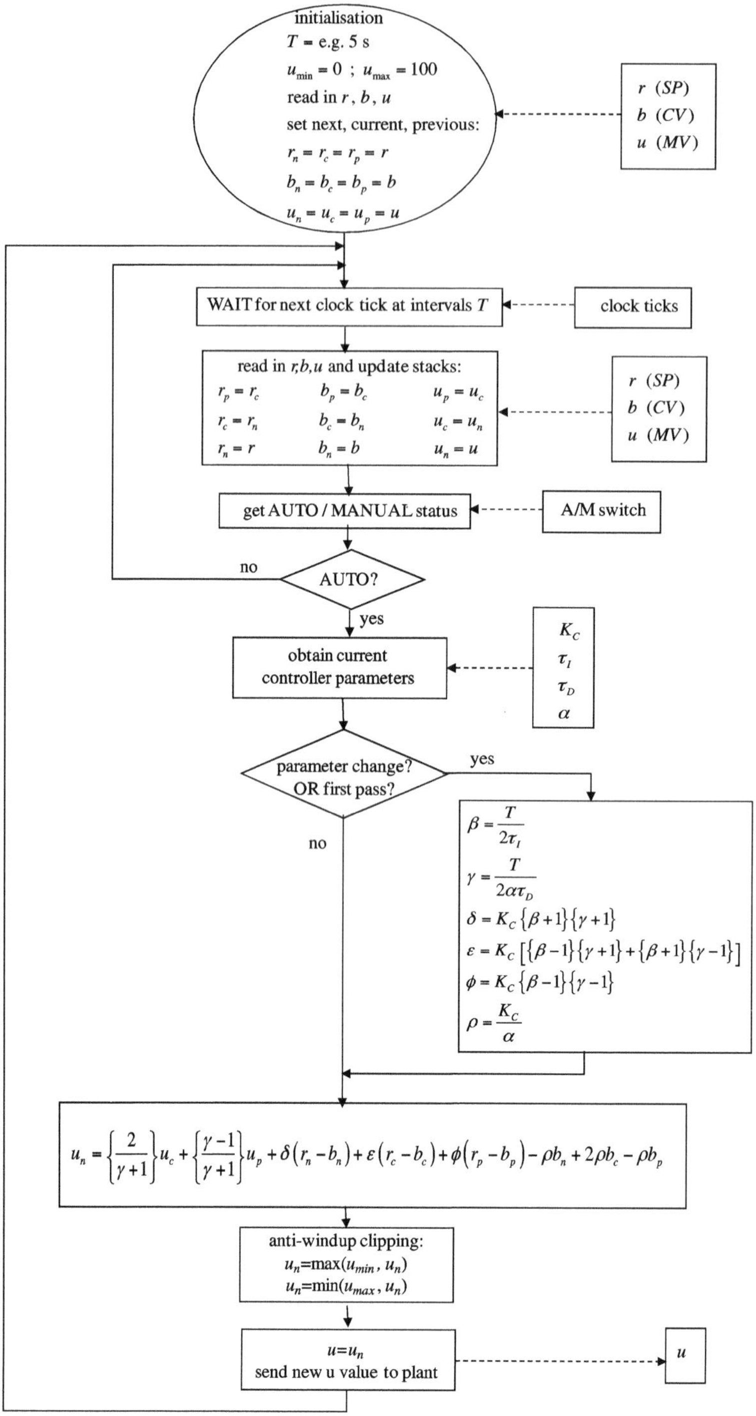

Figure 4.11 Algorithm for a real PID controller according to Equation 4.42 (note: '=' implies assignment).

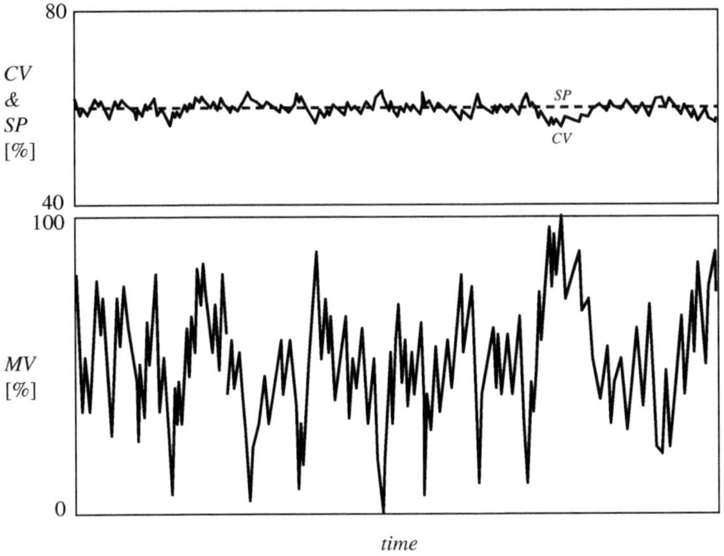

Figure 4.12 Rapid small perturbations on CV signal causing large MV oscillations under derivative action.

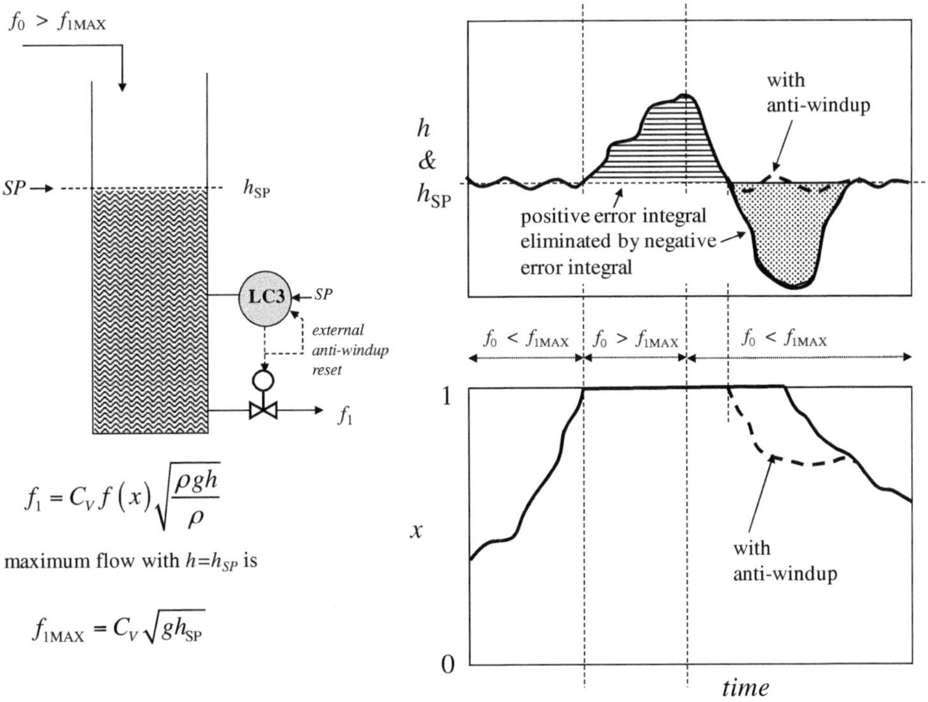

Figure 4.13 Level controller with and without anti-windup protection.

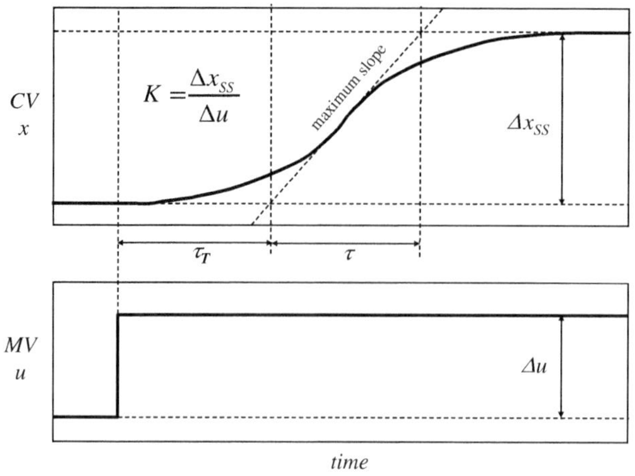

Figure 4.14 Step response reaction curve measurements viewed as arising from a transport lag and first-order dynamic lag.

completely describe anything except the simplest systems, so it is not surprising that the suggested controller settings cannot be considered 'optimal', but rather 'acceptable' for most common systems.

4.2.6.1 Step Response Controller Tuning

This method, also called the *reaction curve* method, effectively represents higher order responses as arising from a combination of transport lag (dead time) and first-order dynamic lag:

$$G_P(s) = \frac{K\,e^{-\tau_T s}}{\tau s + 1} \tag{4.44}$$

In Figure 4.14, it is seen how the three defining parameters of this process are fitted to a step response measurement. One notes that a real first-order dynamic lag would produce its steepest slope ($\Delta x_{SS}/\tau$) of its response to a step input at the time of the step (see Example 3.6). Higher order responses have a zero gradient at this time. Obviously, this method cannot be used for systems with integrating behaviour, such as filling a tank with no balancing exit flow. Ziegler and Nichols (1942) proposed 'good' controller settings based on these measurements as in Table 4.1.

Table 4.1 Ziegler–Nichols *reaction curve* controller settings.

	P	PI	PID
K_C	$\frac{\tau}{K\tau_T}$	$\frac{0.9\tau}{K\tau_T}$	$\frac{1.2\tau}{K\tau_T}$
τ_I	—	$\frac{\tau_T}{0.3}$	$\frac{\tau_T}{0.5}$
τ_D	—	—	$\frac{\tau_T}{2}$

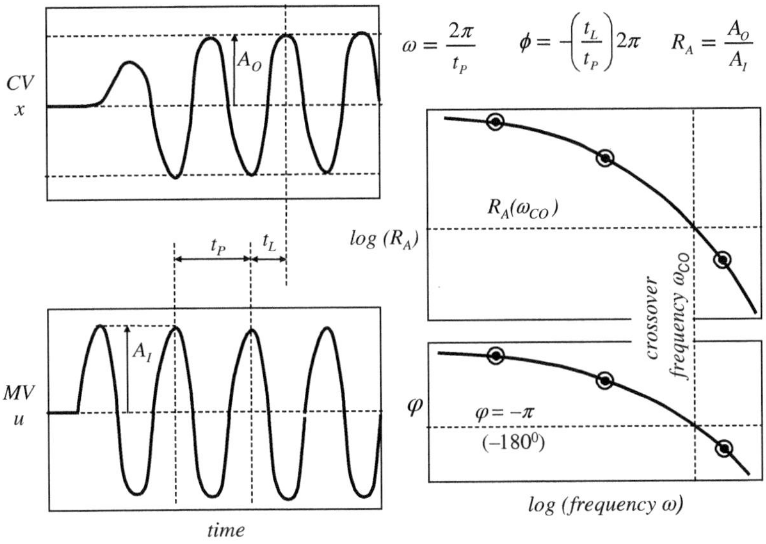

Figure 4.15 Process frequency response measurements for controller tuning.

4.2.6.2 Frequency Response Controller Tuning

In this method, the process input is cycled continuously over a range of frequencies, to obtain the dependence of output:input amplitude ratio R_A and phase angle ϕ on frequency ω as in Figure 4.15. Unlike the *Bode plots* for the *open loop* which will be considered in Chapter 8, the amplitude ratio R_A in this Bode plot will have units such as [$m^3\,h^{-1}$ (% open)$^{-1}$] because it represents G_P alone, not the entire open-loop G_O (which is dimensionless). *Ultimate gain* K_U and *ultimate period* P_U represent conditions at the *crossover frequency* ω_{CO} (where $\phi = -\pi$), and are obtained as follows:

$$K_U = \frac{1}{R_A(\omega_{CO})} \tag{4.45}$$

$$P_U = \frac{2\pi}{\omega_{CO}} \tag{4.46}$$

Based on these frequency response measurements, Ziegler and Nichols (1942) proposed 'good' controller settings as in Table 4.2.

Table 4.2 Ziegler–Nichols *frequency response* controller settings.

	P	PI	PID
K_C	$0.5K_U$	$0.45K_U$	$0.6K_U$
τ_I	—	$\frac{P_U}{1.2}$	$\frac{P_U}{2}$
τ_D	—	—	$\frac{P_U}{8}$

4.2.6.3 Closed-Loop Trial-and-Error Controller Tuning

Conservative tuning of a controller will give sluggish responses to setpoint changes or load disturbances. Most tuning strategies attempt to speed up these responses, yet maintain a safety margin from the point at which the loop becomes unstable. At the point of instability, disturbances will cause continuous cycling. In the *online closed-loop* method described below for a PID controller, poorly damped cycling is used to indicate this approach to instability. The method adapts to PI and PID by stopping a sequence of changes in K_C, τ_I and τ_D at an appropriate point with respect to the onset of undamped cycling:

1) Set τ_I high and τ_D low to remove integral and derivative action. Start with a low choice for K_C.
2) Double K_C and disturb the system slightly with a minor setpoint movement.
3) Repeat step 2 until an almost undamped oscillatory response is obtained. The gain at this point is effectively K_U, the *ultimate gain* (see Section 4.2.6.2), because that part of the feedback frequency spectrum at ω_{CO} will be 180° out of phase, but will invert at the comparator through subtraction, and be adjusted back to its original size by $K_C G_P$, so it will perpetually cycle around the loop. If K_C were larger than K_U, it would grow each time it went around, giving instability.
4) Set K_C at half of K_U (compare the entry in Table 4.2).
5) Now *halve* τ_I on successive steps until an almost undamped oscillatory response is obtained for a slight setpoint disturbance.
6) Set τ_I to twice the final value from step 5.
7) Now increase τ_D in successive steps. A load disturbance may be required to test responses, since derivative action usually acts only on the feedback variable. Seek a τ_D value which gives good control without amplifying variations in the feedback signal.
8) Now K_C can be increased in small steps (10%) until desirable performance is achieved.

4.2.7
Feedforward Control

Figure 4.6d shows a *feedforward* control arrangement. The distinguishing features are as follows:

- it requires input from some measurable disturbance;
- there is no measurement available of the actual variable being controlled;
- it has to be based on a 'model' that it uses to anticipate how to correct for the disturbance in order to maintain the predicted value of the controlled variable at the desired value.

The advantage of feedforward control is that it can be used to avoid deviation of the CV from its setpoint (at least under ideal circumstances), whereas feedback control *requires* a deviation before it takes action. In some situations, it is possible to use feedforward control alone (e.g. Figure 4.16). The problem is that the model and measurements can never be perfectly accurate, so such applications require a fair safety margin to be applied. Certainly, in integrating systems, such as tank level control, there would be a serious problem in using feedforward alone. One could not control the level in a tank by setting its exit flow rate equal to its measured inlet flow rate – a minor error would eventually result in the tank overflowing or emptying! Nevertheless, even in integrating systems, a combination of feedforward and feedback (Figure 4.6e) has great benefits. The feedforward eliminates most of the effect of a disturbance, leaving the feedback to perform a minor trim to correct a smaller deviation from setpoint.

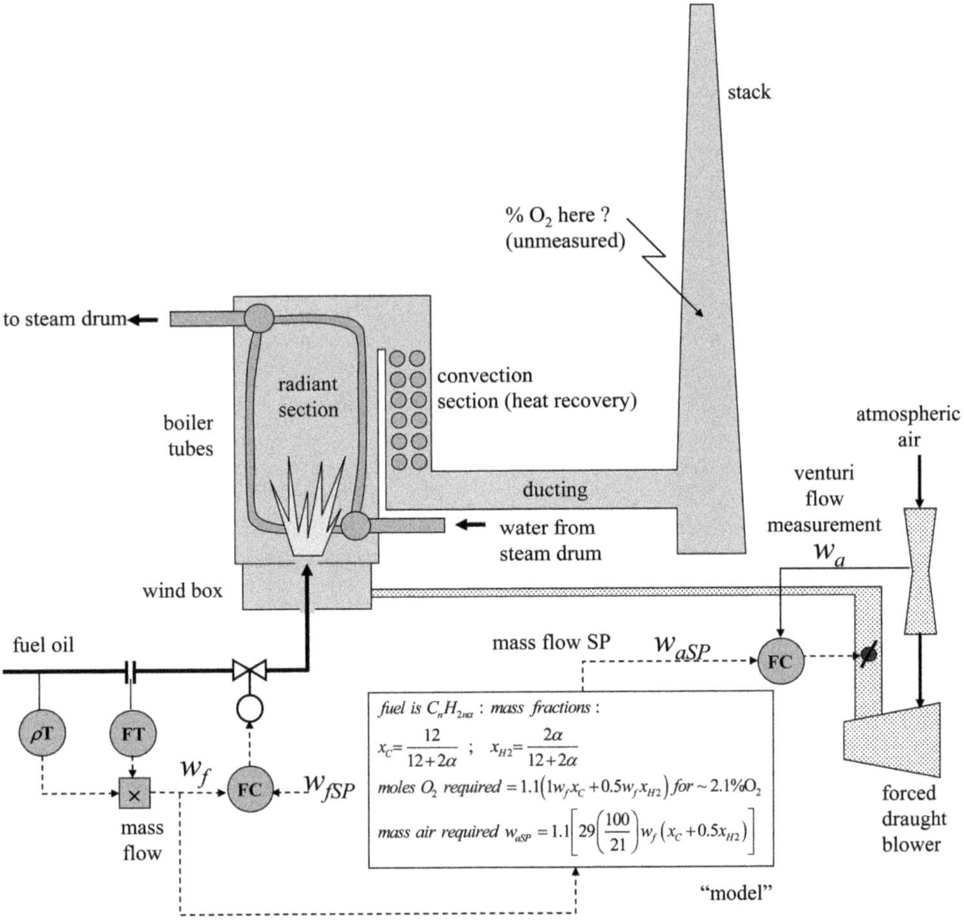

Figure 4.16 Boiler furnace feedforward control of flue gas $\%O_2$.

Figure 4.16 illustrates a feedforward control of the residual $\%O_2$ in the flue gas from a boiler furnace. It is important to keep residual O_2 reasonably above zero, for example around 3% on a molar basis (bearing in mind that the maximum is 21%, if huge amounts of air were fed). If O_2 drops to zero, it is likely that unburnt fuel will be accumulating in the furnace or ducting. If a source of ignition is subsequently encountered, the whole system could explode. Of course, too much air makes the process inefficient, because the entire mass (including 79% N_2) must be raised to the combustion temperature, and it is difficult to recover all of this heat in the economiser (convection section). Hence, operation is typically around 3%, and a purely feedforward scheme as shown in Figure 4.16 would probably be set at a greater excess than the 2.1% shown, to allow for inaccuracy in the assumed combustion stoichiometry.

One notes that a 'model' is required to anticipate the required adjustment of the air flow rate. Though the focus here is on the residual $\%O_2$, this really constitutes *ratio control* (Section 4.4) which is merely a particular form of feedforward control.

4.2.8 Other Simple Controllers

4.2.8.1 On/Off Deadband Control

Many SISO control loops involve control actions which have just two states: on and off. This may be due to the switching of a motor on or off, as for air-conditioning or refrigeration compressors, or switching a heating element, as for a water bath or geyser thermostat. As in continuous system control loops, these systems will be comparing a feedback measurement of the CV with a setpoint, and manipulating the control action MV accordingly. However, a state will soon be reached where the CV is oscillating closely around the setpoint, with very frequent switching of the MV. Frequent start-up surging of motors, or switching of relays, adds to wear and tear, and this is thus avoided in such systems by creation of a *deadband* in which no further action is taken. For example, a refrigeration compressor may only switch on once the temperature exceeds 1 °C above setpoint, and then off again once it drops below 1 °C below setpoint (Figure 4.17).

Without the deadband, operation would be similar to a normal continuous proportional controller with a very high gain K_C. As the feedback CV crosses the setpoint, the control action MV would go directly from one end of its span to the other. Continuous controllers like this which are seldom at an intermediate output are said to be in *bang-bang* mode. Though control may appear good, the control action mechanism is working hard, and resultant variations (e.g. in an exit flow) may well be upsetting downstream parts of the plant.

4.2.8.2 Simple Nonlinear and Adaptive Controllers

The computer age has brought many variations of controller algorithms to suit particular circumstances. In Figure 4.18, two simple *nonlinear* strategies are represented which also have *adaptive* features. In case (a), the controller gain K_C increases with the size of the deviation of

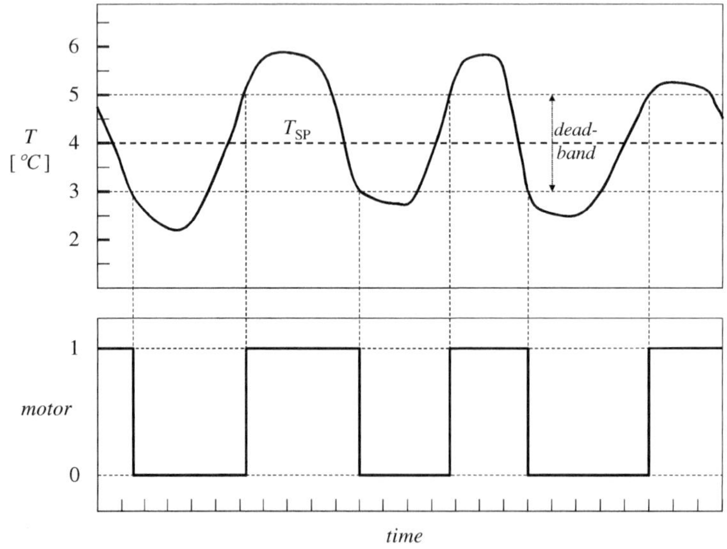

Figure 4.17 On/off temperature control with deadband using a refrigeration compressor.

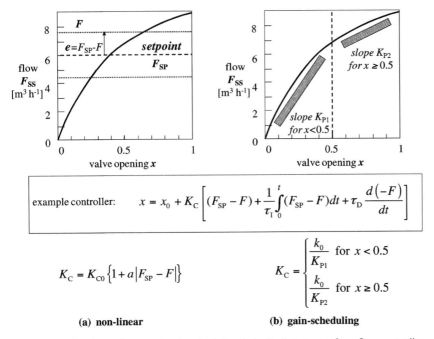

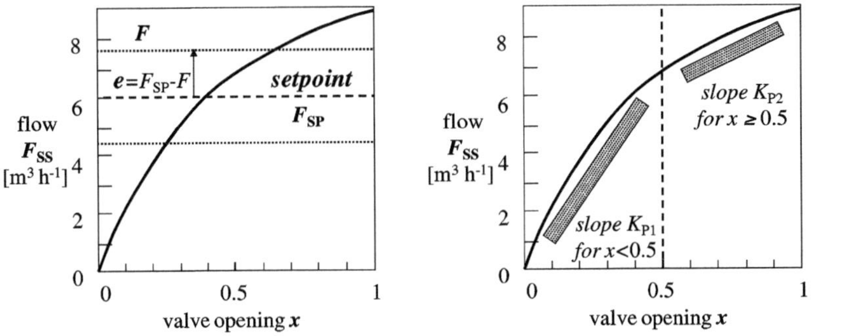

example controller: $x = x_0 + K_C\left[(F_{SP} - F) + \frac{1}{\tau_I}\int_0^t (F_{SP} - F)dt + \tau_D \frac{d(-F)}{dt}\right]$

$$K_C = K_{C0}\{1 + a|F_{SP} - F|\}$$

$$K_C = \begin{cases} \frac{k_0}{K_{P1}} & \text{for } x < 0.5 \\ \frac{k_0}{K_{P2}} & \text{for } x \geq 0.5 \end{cases}$$

(a) non-linear **(b) gain-scheduling**

Figure 4.18 Simple nonlinear and gain-scheduling (adaptive) strategies for a flow controller.

the feedback variable from its setpoint, that is the size of the control error *e*. The objective here is to take stronger control actions when the error is large, but not to upset the system unduly when the error is small. There is effectively a multiplication $|e|e$ which makes the strategy nonlinear. In case (b), there is an attempt to handle the nonlinear characteristic of the valve by dividing its range into two sections, each represented by its own process gain K_P. So the higher process gain K_{P1} on the lower end gives a smaller controller gain K_C, whilst K_{P2} for the upper end gives a larger K_C. This type of adjustment is often called *gain scheduling*. The dependence of K_C on x, and implicitly F, makes this controller *nonlinear*. Both cases (a) and (b) are *adaptive* in trying to adjust a controller parameter dependent on current circumstances.

4.3 Cascade Arrangement of Controllers

Quite often it is useful to isolate tasks in a control strategy by delegation. One controller can *supervise* another by passing its MV output to the *slave controller* as a remote setpoint. The slave of course has to be switched into 'remote' mode (Figure 2.36) in order to accept it. In Figure 4.19, the *direct* scheme (a) has the drawback that variations in fuel supply pressure will proceed to disturb the controlled temperature before corrective action is finally taken by the TC. In contrast, the *cascaded* scheme (b) isolates the problem of maintaining fuel flow at the level of the FC, so that the temperature should not need to deviate from setpoint in order to correct for a fuel pressure change.

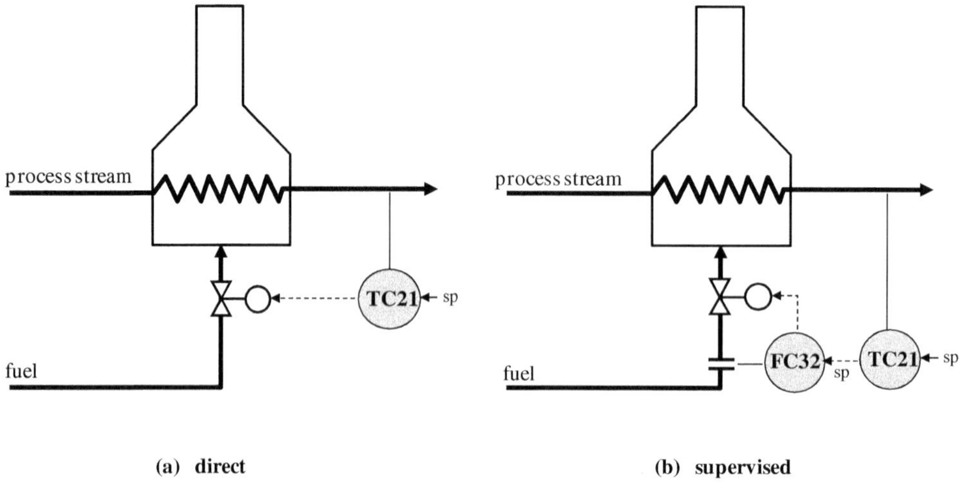

Figure 4.19 Furnace exit temperature control without and with a cascade to eliminate fuel pressure disturbances.

4.4
Ratio Control

In most cases, ratio control concerns the flow rates of two streams. By keeping the two flows in a fixed ratio, it is usually hoped that some downstream condition (e.g. flue $\%O_2$, pH) is met, and where this is not actually measured, the ratio control constitutes a feedforward control. Two arrangements are possible as depicted in Figure 4.20. Of these, (a) is not recommended, because the variable factor 1/F2 in the feedback loop means that the tuning of the RC cannot be fixed optimally.

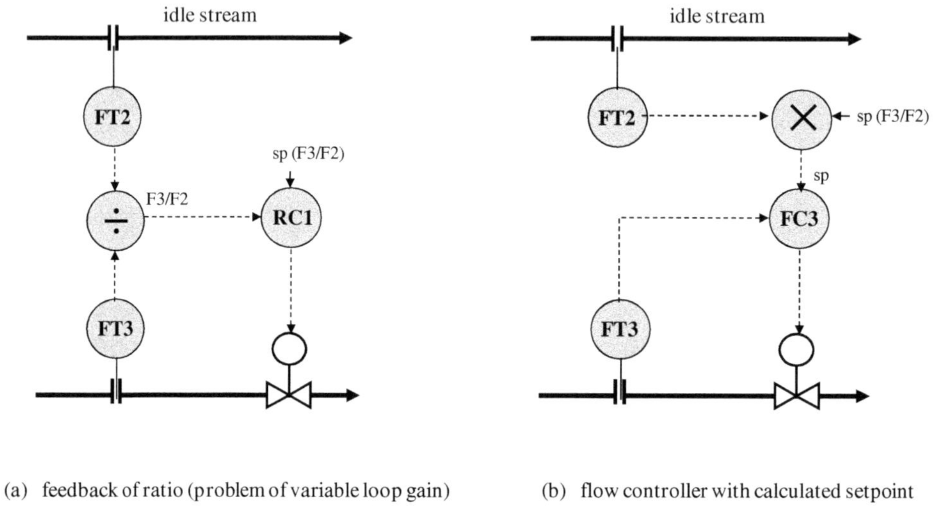

Figure 4.20 Ratio control configurations.

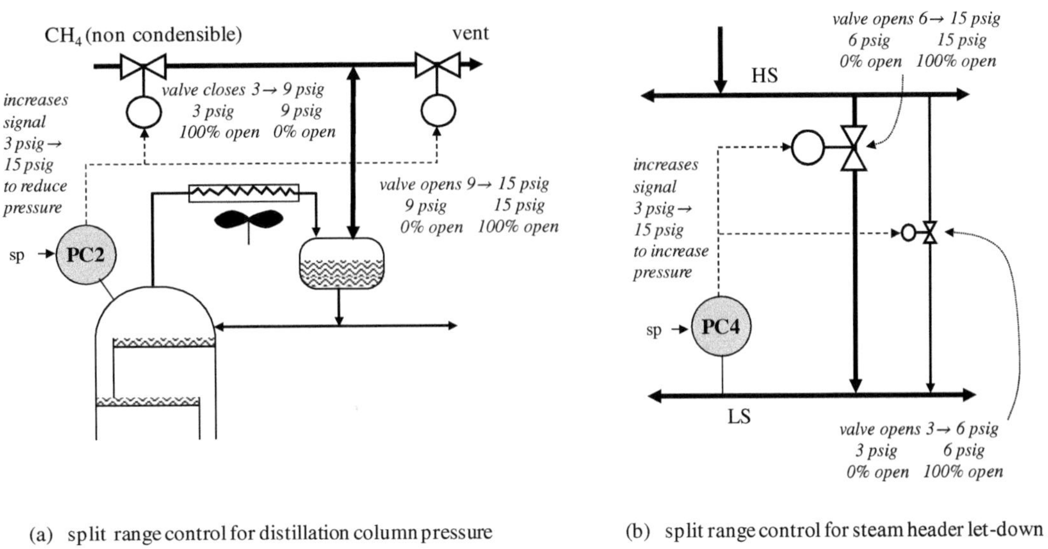

Figure 4.21 Split range control examples.

4.5 Split Range Control

The MV output of a controller is generally thought of as taking a single final control element, such as a valve, through its full range from 0 to 100% open. This continuous MV signal can, however, be reinterpreted in several ranges acting on different control elements. This is perhaps most easily understood using standard analogue signal ranges as in the examples of Figure 4.21. In case (a), a noncondensable gas is initially introduced if the pressure in the column is too low. The extra partial pressure of this gas is felt in the column through the open channelling of the condenser. Conversely, as the pressure rises, the noncondensable gas supply is cut off, and overhead vapours are vented instead. In case (b) of the steam header let-down, a big valve has been specified to cope with normal production rates. However, at low rates, a very poor quality of control would be achieved by this large valve as it 'bumps' off its seat. Rather, a smaller valve in parallel sees to the control in that range, with the large valve shut.

4.6 Control of a Calculated Variable

All feedback variables are manipulated mathematically in some way, but what is considered here is a more complicated derivation involving several measurements. The modern idea of *intelligent sensors* follows this pattern. One example of this is the mass flow control of a gas. It was noted in Equation 2.15 that orifice plate measurements require a correction

$$w_{\mathrm{ACTUAL}} = w_{\mathrm{INDICATED}}\sqrt{\frac{M_{\mathrm{ACTUAL}}}{M_{\mathrm{CALIBRATION}}}\cdot\frac{P_{\mathrm{ACTUAL}}}{P_{\mathrm{CALIBRATION}}}\cdot\frac{T_{\mathrm{CALIBRATION}}}{T_{\mathrm{ACTUAL}}}} \tag{4.47}$$

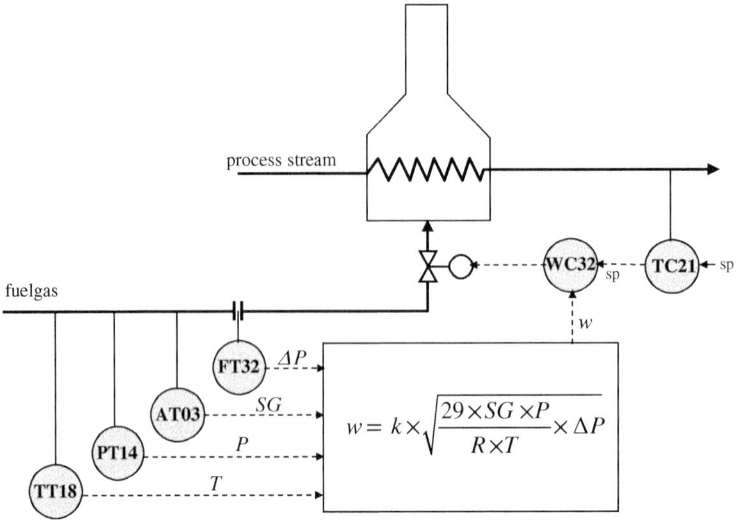

Figure 4.22 Fuel gas mass flow control used in furnace firing.

where M is molecular mass, P is absolute pressure and T is absolute temperature. Where it is important to regulate the mass flow, and these conditions vary, the appropriate measurements may be taken, and the calculation performed online. An example of this is the supply of a variable fuel gas quality to fire a furnace. The heating value is strongly related to the amount of carbon, which largely constitutes the mass, so mass flow control stabilises the firing duty. The example in Figure 4.22 is based on the type of instrument that gives the gas specific gravity relative to air ($SG = M_{gas}/M_{air}$).

The type of manipulation of variables used to produce a feedback signal might instead depend on process objectives, as in the *weighted average bed temperature* (WABT) control used in catalytic reforming of oil (Figure 4.23). Here the quality of the reformed product is strongly related to the WABT.

Another common example of calculated variable control is that of *pressure-compensated temperature* T_{PC} in distillation. In Section 2.4.5, it was noted that one usually seeks a more robust, accessible and indirect measurement of composition, rather than using a sophisticated device such an online gas chromatograph. In distillation, operators have got used to the idea of *temperature* at selected points in a column giving an indication of composition. For a *binary* mixture at a fixed pressure, this is fine – if the temperature is higher, then more of the less volatile component must be present (Figure 4.24). However, the one-to-one relationship between temperature and composition will vary with pressure. So Figure 4.24 shows how a temperature measurement at a different pressure can be corrected back to what it would be at the reference pressure P_{REF}, which would normally be chosen at mid-range for the operation.

$$T_{PC} = T - \left[\frac{\partial T}{\partial P}\right]_{x\ \text{const}} (P - P_{REF}) \tag{4.48}$$

An average slope might be used as indicated. T_{PC} will have a fixed relation to composition, and will allow operators to work with a familiar 'temperature' indication. But multicomponent mixtures will of course not have such a unique relationship. Fortunately, process feeds normally have a characteristic distribution of components, so the behaviour tends to be similar. It is not difficult to

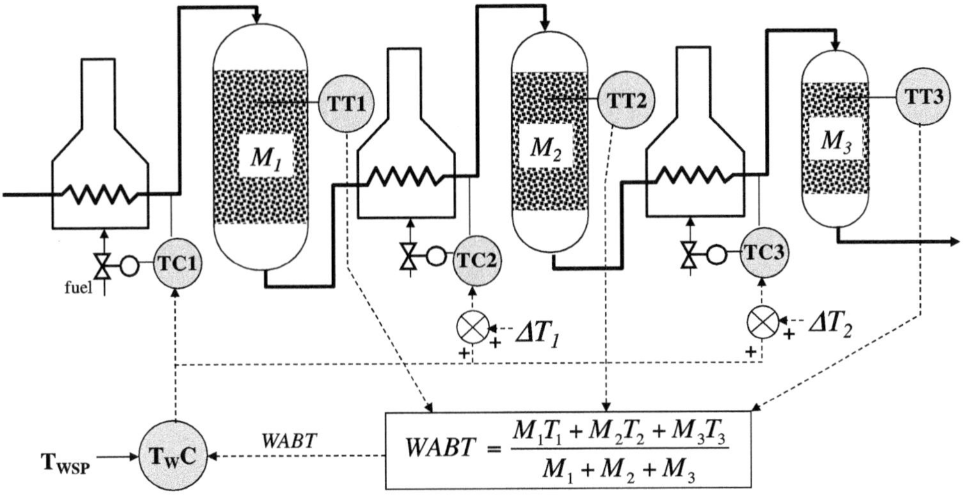

Figure 4.23 Calculated feedback of WABT in oil reforming.

understand that the most effective source of information for this type of measurement correction would be to draw samples for a range of P and T measurements, and have them analysed in the plant laboratory. A regression to find a functional form T_{PC} or liquid composition $x_A = f(T, P)$ then provides a useful online interpreter. Refineries also use this type of inference for more obscure composition indicators such as *octane number* and *cut point*. The interpretations discussed in this section are simple *observer models* which will be discussed in more detail in Chapter 6.

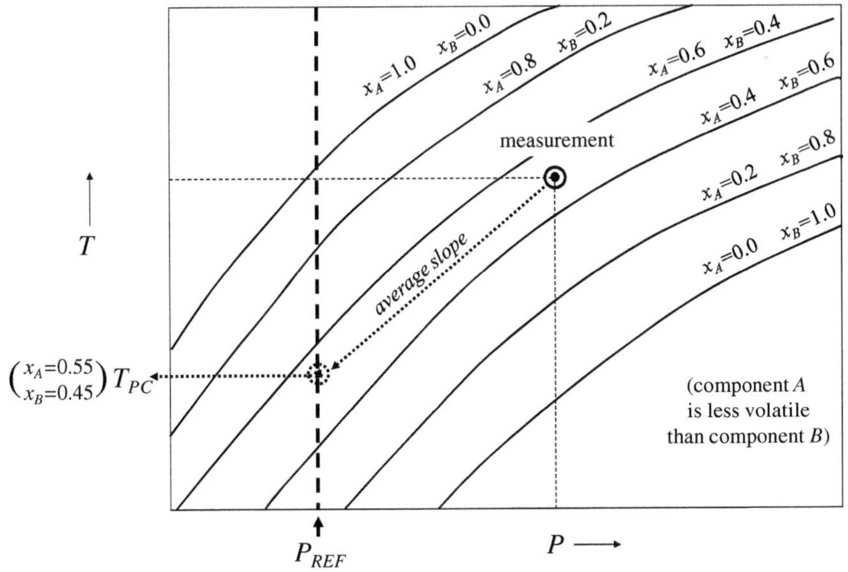

Figure 4.24 Adjustment of the temperature of an equilibrium binary mixture to that at a reference pressure for the same composition.

4.7
Use of High Selector or Low Selector on Measurement Signals

In Section 4.6, it was seen how several measurements might be combined to produce a single feedback signal. The voting system discussed in Section 4.1 is a particular case of this too. A more radical interference in the measurement signal path comes from the use of a *high selector* (HS) or *low selector* (LS). A typical example of this is control of a reactor according to the highest measured temperature amongst a number of probes distributed in a catalytic bed. In this way, one would be aiming to maximise rate, yet avoid catalyst damage (Figure 4.25).

4.8
Overrides: Use of High Selector or Low Selector on Control Action Signals

In comparison with the action of HS or LS on *measurement* signals above (Section 4.7), a more complicated situation arises from the competition of *control action* signals at an HS or LS. The situation being envisaged is where two different controllers might send competing signals to a final control element such as a valve (or even to a *slave* controller). It turns out that this is often a useful arrangement for compliance with a higher priority objective.

Figure 4.26 shows an override controller being used to protect the level at the bottom of a distillation column. One appreciates that it may be impossible to achieve a given temperature setpoint at the base of a distillation column, if the column feed for a period is too light to achieve it at the operating pressure. Without level protection, the reboiler would simply boil up all liquid at the column base. The column would cease to operate as a separator, and the bottom liquid seal would be lost, potentially allowing vapours to pass to the downstream equipment instead of liquid. To avoid this possibly hazardous

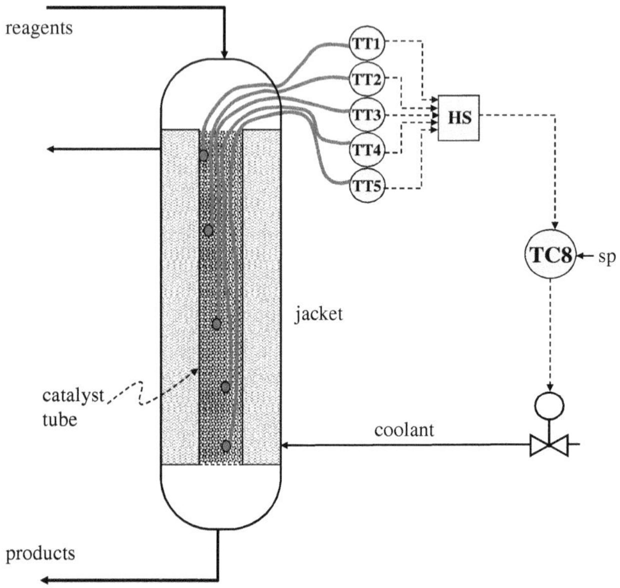

Figure 4.25 Control of maximum amongst five measured temperatures in a tubular catalytic reactor.

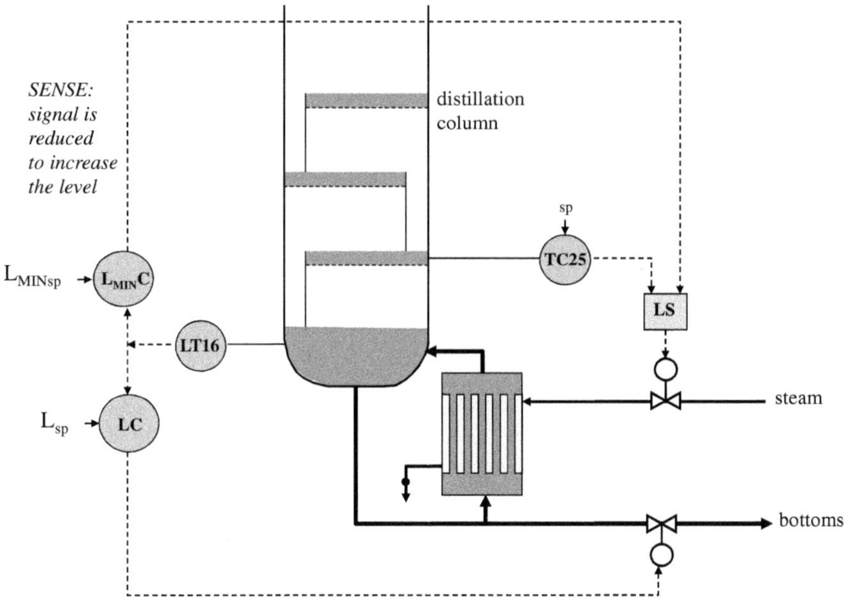

Figure 4.26 Protection of the bottom level of a distillation column using a minimum level override.

situation, it may be deemed preferable to allow the bottoms product to go off its temperature specification, by overriding the temperature controller action with the signal from a minimum level controller.

In Figure 4.26, the override device is shown as a controller. Here one is thinking of a proportional controller with a very large gain ('bang-bang' control). As the feedback signal crosses the setpoint, the output MV goes directly from one extreme to the other, for example 100% open to 0% open as the level drops below L_{MINsp}. Whilst its output has been above L_{MINsp}, the override controller has not been in contention at the LS, because its output has been the highest possible value. As it crosses to a level value below L_{MINsp}, the new override controller output of 0% open will of course dominate the result of the LS, and the steam valve will completely shut. As the level rises above L_{MINsp} again, the override output will switch back to 100% open, and control of the valve will be released to the TC again.

As mentioned in Section 4.2.5, a controller which is cut off from its final control element (say a valve) cannot achieve its setpoint, and will *wind up* its integral action (if present). *Analogue controllers* have an anti-windup feature which stops the integration if the actual valve signal does not match the controller's output signal. As mentioned, windup at the extremes of an output range is not a problem for *computer controller algorithms* which work recursively, since they just 'mark time' at 100% open or 0% open (see Figure 4.11). Since the valve should be at either extreme of its range when an override interferes with the original controller output, there should be no windup problem when a computer algorithm is overridden.

Although this text will continue to show override devices as controllers, one appreciates that such a device does not require many of the features of a full-fledged controller. As described above, it merely needs to switch its output from maximum to minimum (or vice versa) as the setpoint is crossed (Figure 4.27). Where override devices are included in control scheme diagrams, it is important to show the *sense* of the device as in Figure 4.26, because the logic of the arrangement is usually not self-explanatory.

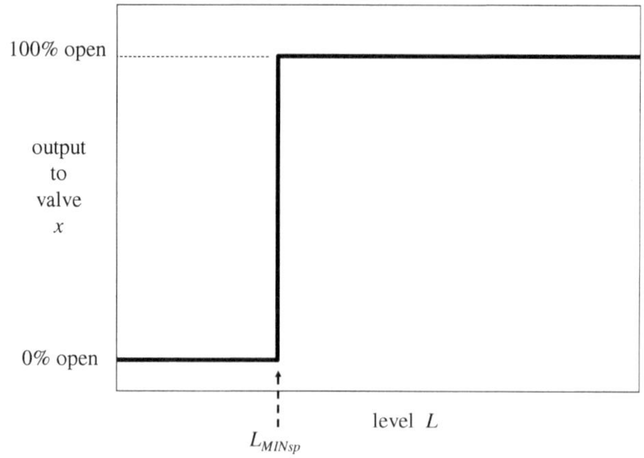

Figure 4.27 Simplified behaviour of the override controller $L_{MIN}C$ in Figure 4.26.

4.9 Clipping, Interlocks, Trips and Latching

The various control scheme elements discussed so far in this chapter enable one to communicate most instrumentation arrangements. Occasionally, interlocks or trips are required to complete a concept. In Section 2.8 and Figure 2.38, drawing representations were briefly considered for these, but at present a more self-explanatory representation is sought which is similar to typical commercial configuration environments. These are not standard, so for the meantime some ideas will be initiated which will help to proceed with exercises (Figure 4.28). More importantly, if a representation does not unambiguously define a scheme, elaborative notes should be included.

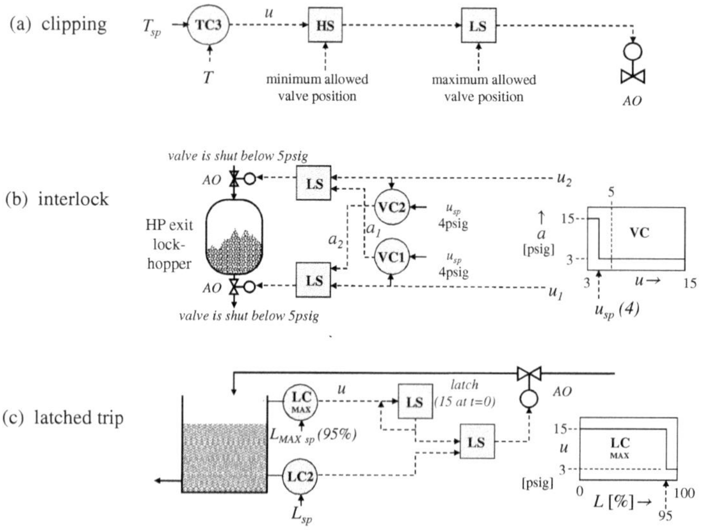

Figure 4.28 Possible representations of clipping, interlocking and a latched trip.

4.10 Valve Position Control

Valve position control is a simple form of optimisation which seeks to maximise or minimise the use of a process stream by positioning a valve near either end of its range, but not so close that it cannot respond to process disturbances. It involves an inner loop, taking care of the primary control task, and a nested outer loop which slowly manipulates the setpoint of the inner loop in order to cause the valve to move to the desired position.

The original idea was proposed by Shinskey (1976, 1978) in connection with his *floating pressure control* of distillation columns (Figure 4.29). Shinskey recognised that the vapour pressure ratios (relative volatility) of components in a mixture tended to expand over a larger range as temperature (and therefore pressure) dropped. The implication was that distillation columns would separate a mixture more efficiently at lower pressures, so why not just set the condenser to maximum cooling duty? This was not advisable, since the uncontrolled column pressure would respond sharply to disturbances in feed conditions, reboiler adjustments and the coolant temperature. A sudden pressure drop could, for example, cause flashing on the trays, loss of level control and flooding. So Shinskey's idea was that enough movement be left in the valve to counter such disturbances, but that a much slower control loop should gradually move the *setpoint* of the pressure controller to cause the control valve to move close to its maximum, but leaving a margin for control adjustments (say 95% open).

One appreciates that the outer VPC loop is attempting to control an algorithm – say a PID controller, and there is a real danger that this loop will go into unstable oscillation. As mentioned, it is important to make the VPC loop very slow, so that the process has plenty of time to adjust to the pressure setpoint variations. Other applications of VPC are generally based on economic criteria, and do not necessarily involve a floating setpoint. Figure 4.30 shows a dual-fuel furnace with a VPC arranged to maximise the use of cheap refinery off-gas, and minimise the use of expensive and reliable naphtha. The primary control of the temperature is by means of naphtha, but the intention

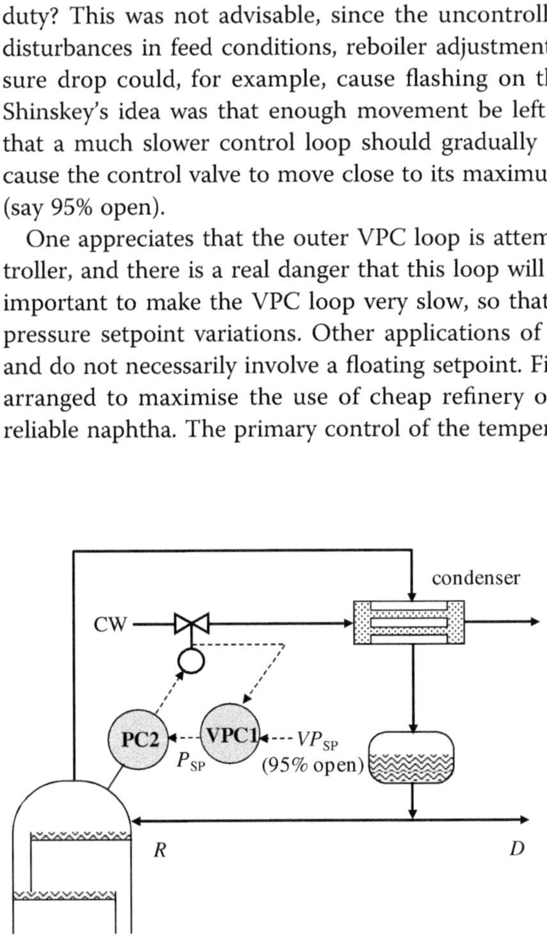

Figure 4.29 Floating pressure control – a form of valve position control.

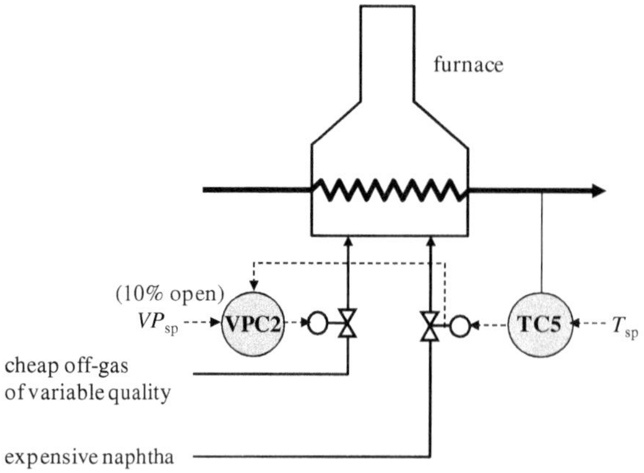

Figure 4.30 Valve position controller used to maximise the use of a cheaper furnace fuel.

is to supply most of the heat from the off-gas. More general process optimisation will be considered later in Chapter 9.

4.11 Advanced Level Control

Liquid levels in a system are generally maintained by manipulating the flow rates of arriving or departing streams. Exceptions include the use of heat to maintain a level, or the equivalent situation with a vapour, where the inventory may be represented by pressure instead of level. In all of these cases, flows in the system are being disturbed in order to maintain the inventory at some setpoint. Technicians would often tune ordinary level controllers to hold the level tightly at some point specified by the operating personnel. The result would be wildly varying flows in the plant, 'painting the chart' as the recording pen moved from side to side. The continuous disturbance of upstream or downstream operations (such as column 2 in Figure 4.31) meant that controllers there could not meet important process objectives. Eventually, it was realised that the level itself is usually not an important objective. It may simply act as a seal or buffer between two units, and could vary over a wide range yet still allow perfectly safe operation. Many process disturbances are cyclical in nature, so a level could be allowed to cycle up and down between limits with only occasional adjustment of the manipulated flow. Most processing plants now use this concept wherever they can, and even at the design phase, additional capacity is added to these intermediate storages to allow the absorption of fluctuations using *advanced level control.*

It is interesting to contemplate a possible algorithm for an advanced level controller. In practice, these can require several pages of programming code. In Figure 4.31, L_U and L_L are physical upper and lower limits constraining the allowed values of the actual level L. These might be determined by tray spacing and weir depth. A desired time margin τ_D has been specified as the minimum reaction time that must be maintained against transgression of either limit. On each controller time step, if the predicted time margin to transgression τ is shown to be less than τ_D, an adjustment is made in the flow setting F_S, otherwise not (Equations 4.49 and 4.50). A gradual and infrequent

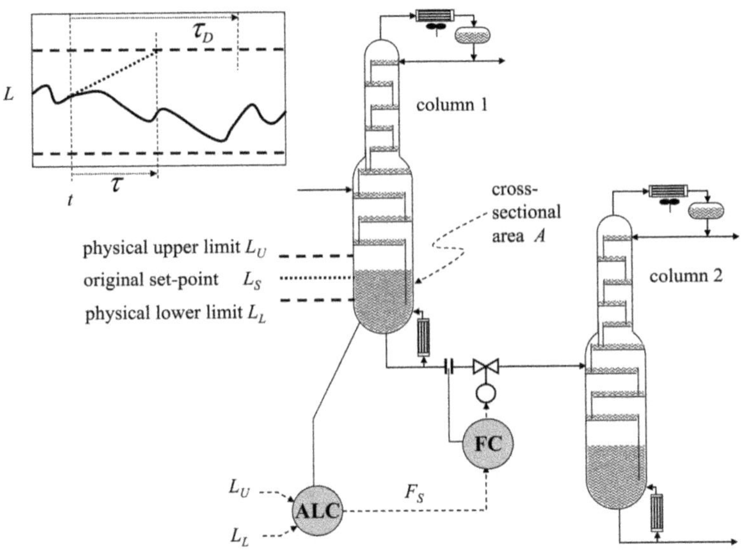

Figure 4.31 Advanced level control to minimise adjustments of a stream fed to a downstream unit.

adjustment might also be superimposed to reposition L in the centre of the range, to provide space for movement in either direction.

$$\tau = \begin{cases} \left(\dfrac{L_\mathrm{U} - L}{\left.\dfrac{\mathrm{d}L}{\mathrm{d}t}\right|_t}\right) & \text{for} \quad \left.\dfrac{\mathrm{d}L}{\mathrm{d}t}\right|_t > 0 \\ \left(\dfrac{L - L_\mathrm{L}}{-\left.\dfrac{\mathrm{d}L}{\mathrm{d}t}\right|_t}\right) & \text{for} \quad \left.\dfrac{\mathrm{d}L}{\mathrm{d}t}\right|_t < 0 \end{cases} \tag{4.49}$$

$$F_\mathrm{S} = \begin{cases} F_\mathrm{S} \quad \text{for} \quad \tau \geq \tau_\mathrm{D} \text{ (unchanged)} \\ F_\mathrm{S} + A(L_\mathrm{U} - L)\left(\dfrac{1}{\tau} - \dfrac{1}{\tau_\mathrm{D}}\right) & \text{for} \quad \tau < \tau_\mathrm{D} \quad \text{and} \quad \left.\dfrac{\mathrm{d}L}{\mathrm{d}t}\right|_t > 0 \\ F_\mathrm{S} - A(L - L_\mathrm{L})\left(\dfrac{1}{\tau} - \dfrac{1}{\tau_\mathrm{D}}\right) & \text{for} \quad \tau < \tau_\mathrm{D} \quad \text{and} \quad \left.\dfrac{\mathrm{d}L}{\mathrm{d}t}\right|_t < 0 \end{cases} \tag{4.50}$$

4.12
Calculation of Closed-Loop Responses: Process Model with Control Element

In Chapter 3, the simulation of the open-loop process was considered, using both numerical and analytical techniques. In Chapter 4, algorithms for some of the basic elements of plant control schemes have been addressed. It is common to combine these in an offline simulation before actual implementation on a plant, to detect problems and establish basic tuning settings. Here one is considering not just *controllers* in *closed loop*, but overall signal handling including such *open-loop* instances as filters, selectors and overrides. In this section, examples of numerical and analytical techniques for the combined simulation are developed. Often one is interested in the *robustness* of the technique. Thus, a control algorithm developed on the basis of one's best model of part of a process may in fact be used to control the same model with additional noise or parameter variations inserted into the model.

4.12.1 Closed-Loop Simulation by Numerical Techniques

Some numerical algorithms for open-loop modelling were developed in Sections 3.1–3.4. Here a general method will be presented by example, combining such a model with an adaptive PID controller based on the discrete form of Section 4.2.4.

Example 4.2

Adaptive PID control of the level in a conical tank.

Figure 4.32 shows a level control loop installed on a conical vessel. A nonlinear dynamic model is presented, and this will be used as an accurate representation of the system by numerical solution in a simple Euler integration. To compensate for the system nonlinearities, a PID controller will be *adapted* on each time step using the local open-loop gain and time constant, also derived in Figure 4.32. The basis of the adaptation will be the Ziegler–Nichols recommended reaction curve tuning parameters in Table 4.1, which it is seen are expressed in terms of an equivalent first-order system (K,τ) with transport lag τ_T. The system as described does not have a transport lag, so τ_T will be left free for adjustment to account for the inevitable higher order effects which will be found in practice. Figure 4.33 shows an arrangement of the algorithm suitable for computer implementation.

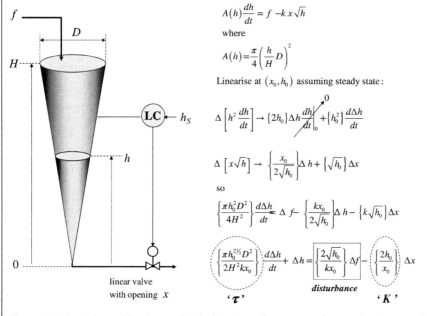

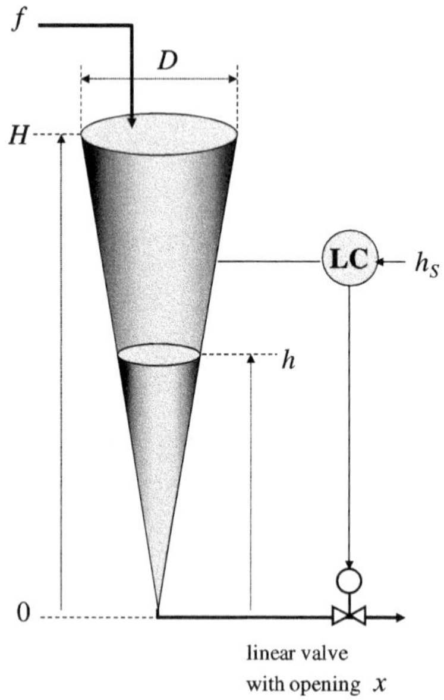

$$A(h)\frac{dh}{dt} = f - kx\sqrt{h}$$

where

$$A(h)=\frac{\pi}{4}\left(\frac{h}{H}D\right)^2$$

Linearise at (x_0,h_0) assuming steady state:

$$\Delta\left[h^2\frac{dh}{dt}\right] \rightarrow \{2h_0\}\Delta h\left.\frac{dh}{dt}\right|_0 + \{h_0^2\}\frac{d\Delta h}{dt} \qquad \left(\left.\frac{dh}{dt}\right|_0 = 0\right)$$

$$\Delta\left[x\sqrt{h}\right] \rightarrow \left\{\frac{x_0}{2\sqrt{h_0}}\right\}\Delta h + \left\{\sqrt{h_0}\right\}\Delta x$$

so

$$\left\{\frac{\pi h_0^2 D^2}{4H^2}\right\}\frac{d\Delta h}{dt} = \Delta f - \left\{\frac{kx_0}{2\sqrt{h_0}}\right\}\Delta h - \left\{k\sqrt{h_0}\right\}\Delta x$$

$$\underbrace{\left\{\frac{\pi h_0^{2\frac{1}{2}} D^2}{2H^2 k x_0}\right\}}_{\text{'}\tau\text{'}}\frac{d\Delta h}{dt} + \Delta h = \underbrace{\left\{\frac{2\sqrt{h_0}}{kx_0}\right\}\Delta f}_{\text{disturbance}} - \underbrace{\left\{\frac{2h_0}{x_0}\right\}}_{\text{'}K\text{'}}\Delta x$$

Figure 4.32 Conical vessel level control with derivation of parameters for an adaptive controller.

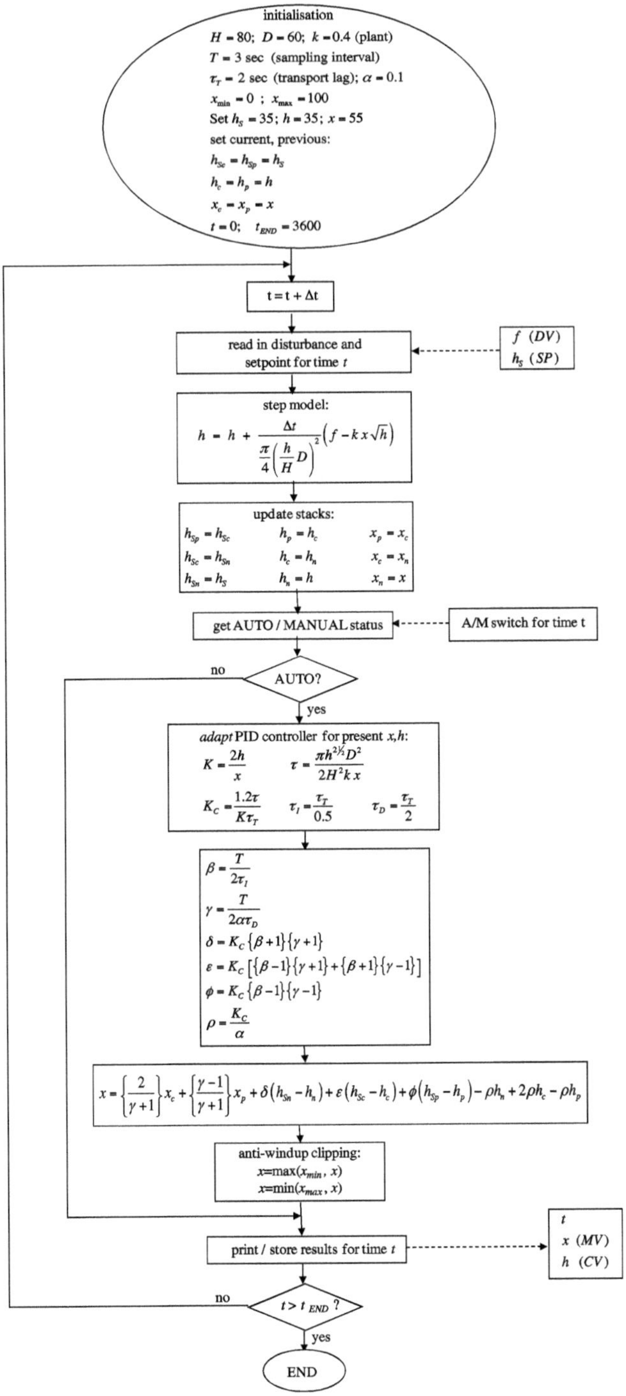

Figure 4.33 Algorithm for simulation of *adaptive* PID control of level in conical vessel of Figure 4.32 (note: '=' implies assignment).

4.12.2
Closed-Loop Simulation Using Laplace Transforms

Consider the two linearised systems in Figure 4.34. The open-loop behaviour of the tank with exit restriction (a) is such that it will gradually adjust to a new equilibrium level if the inflow is changed. The thermal mass of the furnace in (b) responds similarly to a fuel increase, reaching a new *radiation temperature*. This means that more heat must be transferred to create steam, but if the steam users are drawing fixed amounts (deviation variable = 0), the only possibility is that the additional steam must accumulate in the vapour space of the drum. The pure integral 1/s in the transfer function (b) accounts for the associated ramp build-up of pressure.

Consider control of each system in Figure 4.34 using a proportional controller with gain K_C (i.e. $G_C(s) = K_C$). Recall from Equation 4.15 that the closed-loop transfer function of a SISO process is

$$G_{CL}(s) = \frac{G_O(s)}{1 + G_O(s)} \quad \text{with} \quad G_O(s) = G_M(s)G_P(s)G_C(s) \tag{4.51}$$

So for the system (a)

$$G_{CL}(s) = \frac{B(s)}{R(s)} = \frac{K_C K}{\tau s + 1 + K_C K} = \frac{\left(\dfrac{K_C K}{1 + K_C K}\right)}{\left(\dfrac{\tau}{1 + K_C K}\right)s + 1} \tag{4.52}$$

$$G_p(s) = \frac{B(s)}{U(s)} = \frac{K}{\tau s + 1} \qquad\qquad G_p(s) = \frac{B(s)}{U(s)} = \frac{K}{s(\tau s + 1)}$$

(a) tank, valve and restriction with linear behaviour

(b) boiler and fuel valve with linear behaviour

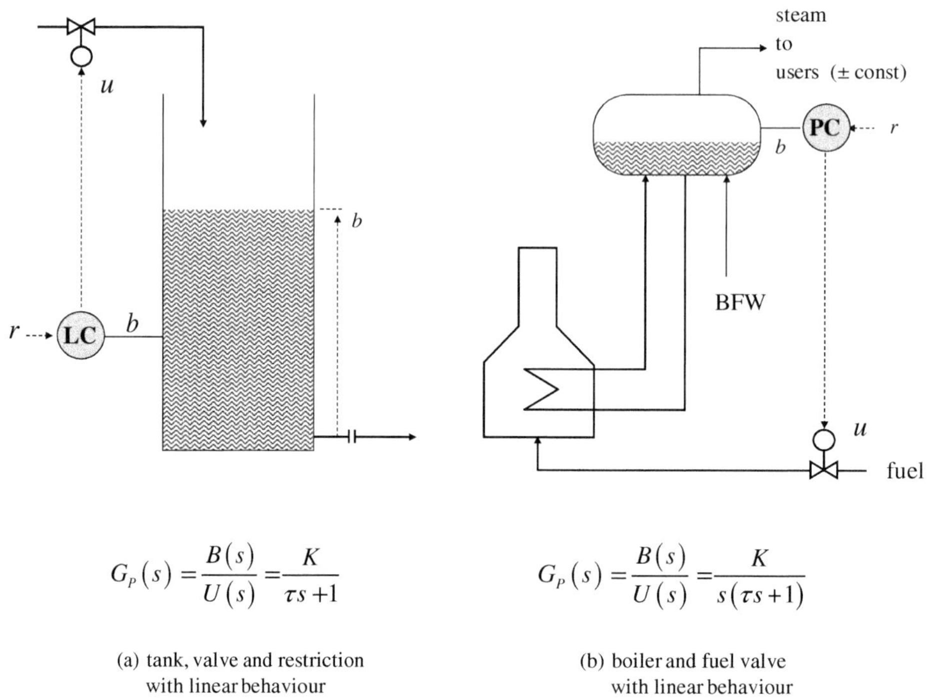

Figure 4.34 Control of non-integrating (a) and integrating (b) systems: open-loop transfer functions.

Again, this is a first-order process, and it is noted that the gain and time constant are

$$K_{\mathrm{CL}} = \left(\frac{K_{\mathrm{C}}K}{1 + K_{\mathrm{C}}K}\right) \tag{4.53}$$

$$\tau_{\mathrm{CL}} = \left(\frac{\tau}{1 + K_{\mathrm{C}}K}\right) \tag{4.54}$$

The gain is less than unity, so that a step in the setpoint r will result in a final change in the CV b which is smaller. For example, for a setpoint step of unity, this is easily seen using the *final value theorem* (Table 3.5):

$$B(s) = G_{\mathrm{CL}}(s)R(s) = \left\{\frac{K_{\mathrm{C}}K}{\tau s + 1 + K_{\mathrm{C}}K}\right\}\frac{1}{s} \tag{4.55}$$

$$\lim_{t\to\infty} b(t) = \lim_{s\to 0} sB(s) = \frac{K_{\mathrm{C}}K}{1 + K_{\mathrm{C}}K} \tag{4.56}$$

So for the non-interacting system (a), there is a fractional offset of $-1/(1 + K_{\mathrm{C}}K)$.

Applying the same procedure to the integrating system (b),

$$G_{\mathrm{CL}}(s) = \frac{B(s)}{R(s)} = \frac{K_{\mathrm{C}}K}{s(\tau s + 1) + K_{\mathrm{C}}K} \tag{4.57}$$

For a unit step in setpoint

$$B(s) = G_{\mathrm{CL}}(s)R(s) = \left\{\frac{K_{\mathrm{C}}K}{s(\tau s + 1) + K_{\mathrm{C}}K}\right\}\frac{1}{s} \tag{4.58}$$

$$\lim_{t\to\infty} b(t) = \lim_{s\to 0} sB(s) = 1$$

So for the integrating system (b) there is no offset, despite the use of a proportional controller.

If it is desired to obtain the full time response, $b(t)$ can be evaluated by inversion of $B(s)$ using a partial fraction expansion as discussed for open-loop modelling in Section 3.9.2.1.

References

Shinskey, F.G. (1976) Energy-conserving control systems for distillation units. *Chemical Engineering Progress*, **72**, 73.

Shinskey, F.G. (1978) Control systems can save energy. *Chemical Engineering Progress*, **74**, 43.

Ziegler, J.G. and Nichols, N.B. (1942) Optimum settings for automatic controllers. *Transactions of the American Society of Mechanical Engineers*, **64**, 759–768.

5
Control Strategy Design for Processing Plants

The piping and instrumentation diagram (Section 2.1) is the key document describing a processing plant, around which the rest of the design revolves. In the design and construction phases, it must be kept up-to-date, and in computerised documentation systems it will be found to link automatically to virtually every other design document. So at an early stage, it becomes necessary to specify the plant instrumentation scheme and control philosophy. The objective of the control engineer is initially that the plant 'should be able to take care of itself' under most circumstances. Once that objective is reached, he or she will seek to get the plant to 'take care of itself *optimally*'. This rests on the field of *advanced process control,* arising from the common use of digital computers. As industries have improved their safety, efficiency and product specification compliance through advanced approaches, so too have their competitors been obliged to develop advanced control capacity, in order to remain competitive.

In modern processing plants, the processing scheme itself is becoming more complex, owing to flow scheme optimisations such as 'pinch' technology which seeks to minimise energy demands or waste streams. These developments lead to a lot of interactions across the plant, for example heat recovery interchange after a reactor. It is understandable that this complicates plant control – how does one get a stream up to its reaction temperature when it requires the exothermic reaction heat to get there? At one time just a few units such as distillation columns were viewed as intrinsically interactive 'black boxes' requiring a coordinated control strategy for best performance. Now more interactive links are arising from the flow scheme itself, so where does the control engineer start with his or her design?

There are two extreme views in the control scheme design:

- The entire plant is an interactive black box – any one of the available manipulated variables (MV) potentially affects one or more of the controlled variables (CV), requiring overall MIMO control.
- With careful pairing of CVs and MVs, and careful tuning of these SISO controllers, it is possible to meet all of the control objectives of the plant. Interactions can be broken by wide separation of the speeds (time-constants) of the controllers.

Before the use of digital computers, plants were necessarily controlled using multiple SISO analogue controllers. An advantage here was that the control scheme was easy to comprehend and maintain. In modern plants, one finds a combination of MIMO and SISO controllers. The use of SISO controllers is extended as far as possible over the *base layer* (Figure 1.5). MIMO controllers will then be developed for more interactive subsections of the plant such as a distillation column. These are located in the *advanced process control* layer, and usually manipulate setpoints in the base layer.

Applied Process Control: Essential Methods, First Edition. Michael Mulholland.

5.1
General Guidelines to the Specification of an Overall Plant Control Scheme

A good understanding of a processing plant and its operation leads in an intuitive way to the control scheme specification. The mass-balance, for example, might start with an uncontrolled feed, which might have to be accommodated with level controls that manipulate the exit streams of subsequent plant items. Between plant items, one could not operate two valves in series, and so on. Usually the specification involves a number of SISO control loops, with occasional MIMO control tasks identified for highly interactive subsections of the plant, such as distillation columns or recycled reactors. Following Buckley (1964) and Luyben (1990), some useful ideas are listed below:

1) Keep the scheme as simple as possible. Operators are prone to switching off schemes which they cannot understand. Complicated schemes will require the recall of an expert should the behaviour become strange, for example due to an embedded faulty measurement.
2) Minimise instrumentation where possible, for example using overflows for level control, atmospheric connections (possibly through a condenser) or barometric legs for pressure control.
3) First specify the control loops required to maintain the various process flows and inventories (levels and pressures). Use proportional controllers where levels are not critical, to reduce MV variations.
4) Then specify the remaining loops (temperature, composition) bearing in mind the possibility of interactions. Make the inventory and flow loops 'slow' (e.g. low gain, high integral time) where necessary to break interactions, otherwise assess for a MIMO strategy.
5) Use feedforward control to reduce the effects of disturbances.
6) Use cascaded controllers to delegate tasks, with the aim of preventing disturbances from affecting CVs.
7) Avoid controlling across transport lags (see the Smith Predictor in Section 6.2).
8) In most situations, one avoids indirect 'nesting' of control loops, where some other controller has to operate effectively to cause the desired change. For example, if the first of two tanks in series has a level control valve on the exit, one would not arrange for the second tank's level controller to manipulate feed to the first tank.

5.2
Systematic Approaches to the Specification of an Overall Plant Control Scheme

The discussion in Chapter 4 has revealed that the 'control philosophy' will entail far more than just SISO or MIMO controllers tasked to manipulate MVs to achieve SPs. Elements such as summers, multipliers, dividers, selectors, limits and trips all will play a part. However, the discussion in this section will focus only on the task of achieving all of the control objectives (SPs) of the plant, given all of the available control actions (MVs). The increasingly interactive nature of plants means that several different MVs could be individually used to achieve a given SP. Moreover, it is sometimes not clear whether a particular SP *can* be achieved, so a formal method of detecting *uncontrollability* will be useful. Some quantitative methods to assess these aspects will be addressed later, but for the meantime it is useful to consider the more qualitative *structural approach*.

5.2.1
Structural Synthesis of the Plant Control Scheme

The *structural approach to the synthesis of control systems* follows the work of Johnston and Barton (1985a, 1985b), and Lin and Lim (1994). It is based on *structural matrices* which have fixed zeros in certain locations and arbitrary entries (denoted by ×) in the remaining locations instead of numerical values. The implication of a nonzero value is that the variable represented by the column affects the variable represented by the row. One recognises that the variables one wishes to control, in vector $\boldsymbol{y}$, are not necessarily individual states from the state vector $\boldsymbol{x}$, but may be functionally related to the states and/or inputs $\boldsymbol{u}$, such as in the case of *pressure-compensated temperature* (Section 4.6). So a representation of a general continuous system could be

$$\dot{\boldsymbol{x}} = \boldsymbol{f}(\boldsymbol{x}, \boldsymbol{u}) \tag{5.1}$$

$$\boldsymbol{y} = \boldsymbol{g}(\boldsymbol{x}, \boldsymbol{u}) \tag{5.2}$$

(Recalling Equations 3.6–3.9, one notes that in Equation 5.1 the discrete states ($\boldsymbol{w}$) are excluded, and the ancillary variables ($\boldsymbol{y}$) in Equation 3.8 substituted by an explicit rearrangement of Equation 3.9 as Equation 5.2). In this system, consider $\boldsymbol{x} \in \mathfrak{R}^{n}$, $\boldsymbol{u} \in \mathfrak{R}^{m}$, $\boldsymbol{y} \in \mathfrak{R}^{r}$. The equivalent form of a linear (*state-space*) system is

$$\dot{\boldsymbol{x}} = \boldsymbol{A}\boldsymbol{x}+\boldsymbol{B}\boldsymbol{u} \tag{5.3}$$

$$\boldsymbol{y} = \boldsymbol{C}\boldsymbol{x} + \boldsymbol{D}\boldsymbol{u} \tag{5.4}$$

A *structured system* is thus defined as

$$\boldsymbol{S} = \begin{bmatrix} \boldsymbol{A} & \boldsymbol{B} \\ \boldsymbol{C} & \boldsymbol{D} \end{bmatrix} \tag{5.5}$$

where the elements of the matrices are now × or 0 representing 'effect' or 'no-effect', and these would indeed match the presence of nonzero or zero entries if the system were linear. The individual structural matrices will be of the following sizes:

$$\boldsymbol{A} : n \times n$$

$$\boldsymbol{B} : n \times m$$

$$\boldsymbol{C} : r \times n$$

$$\boldsymbol{D} : r \times m$$

For a given structured system $\boldsymbol{S}$, an $r \times m$ *cause-and-effect matrix* (CEM) can be formulated using the algorithm of Lin, Tade and Newell (1991). Each column of the CEM represents an MV, and each row represents a CV (output). This structured matrix will clearly represent cross-interactions through off-diagonal values in the $\boldsymbol{A}$ matrix as well as summations in the $\boldsymbol{C}$ and $\boldsymbol{D}$ matrices. Such a CEM can of course be constructed intuitively through careful examination of a process.

The *generic rank* of a structured matrix is the highest rank that can be achieved by any structurally equivalent numerical matrix. It follows then that a system is *output structurally* controllable only if the generic rank of its CEM is r.

Figure 5.1 shows a plant for the catalytic hydrogenation of furfural (F) to furfuryl alcohol (FA). Liquid F feed is vaporised into a circulating hydrogen stream which is fed to a tubular reactor at about 10% v/v F. After the reactor, the F and FA are separated out by condensation, with unreacted

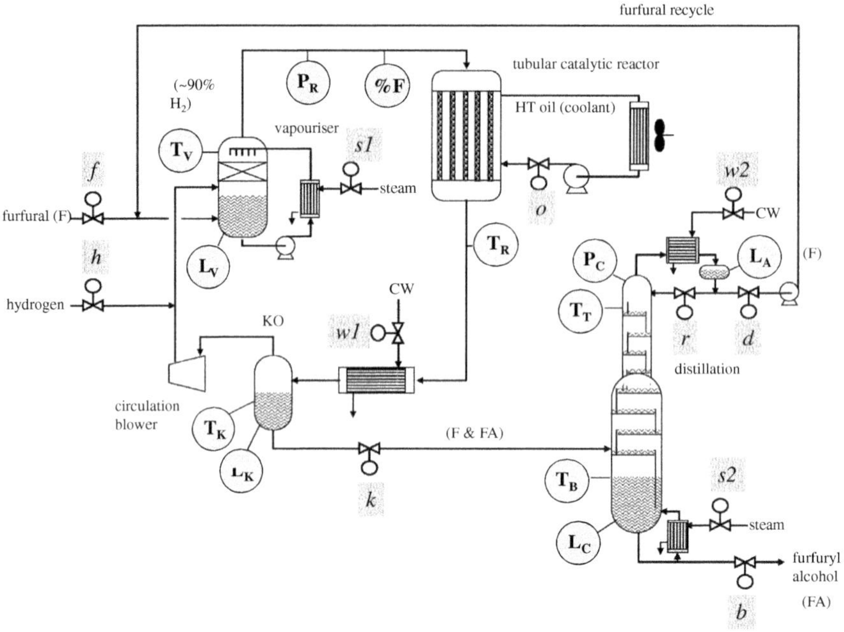

Figure 5.1 Furfuryl alcohol plant showing control objectives and available control actions.

F being recovered by distillation, and recycled back to the feed. A heat transfer oil is circulated on the jacket side of the reactor for heat removal. The first step of control scheme synthesis has been completed by identification of the important control variables of the plant (e.g. T_R and P_R) and the available manipulated variables (e.g. the valves *f*, *h*).

The first step in the structural synthesis of a control system is to construct a *cause-and-effect matrix* for the plant. In doing so, one needs to focus on the *open-loop* plant behaviour, that is what each MV affects whilst *all* control loops are open. In addition to using just ×'s, a *w* for weaker disturbances and *s* for slower responses will also be used following Lin and Lim (1994). Blank spaces have been left to represent zeros (Figure 5.2a).

Johnston and Barton (1984) give a method of row swopping and column swopping to achieve a block lower diagonal form (Figure 5.2b). In essence, one aims to clear the upper triangle as far as possible and further group the interactive variables near the diagonal. Nonzero entries above the diagonal cause multi-entry blocks, which need not be square. A non-square block moves the continued diagonal up or down.

The result is a sequence of feasible 'localisations' of control. In Figure 5.2b, one notes that ***s1*** is manipulated to regulate $\mathbf{T_V}$. Although ***s1*** also affects $\mathbf{L_V}$, the lower triangular form has allowed allocation of ***f*** to regulate $\mathbf{L_V}$ (and in so doing, it will be taking the necessary compensatory actions for the ***s1*** disturbances).

The blocks containing multiple entries may need MIMO strategies. Such a strategy may well make good use of excess MVs. Alternatively, if one is intent on finding reasonable SISO pairs even in these blocks, excess MVs must be eliminated. Likewise, excess CVs, indicating *generic rank deficiency* in a block, must be eliminated. Thus '**%F**' has been eliminated from the fifth block. In any

	f	*h*	*s1*	*w1*	*k*	*o*	*w2*	*r*	*d*	*s2*	*b*
T_V	*w*		×								
L_V	×		×						*w*		
T_K				×							
L_K				*w*	×						
P_R		×									
%F		×	×								
T_R		*w*				×					
P_C							×	×		×	
T_T							*w*	×		*s*	
T_B							*s*	*s*		×	
L_C							*s*	*s*		×	×
L_A							×	×	×	*s*	

(a)

	s1	*f*	*w1*	*k*	*h*	*o*	*w2*	*s2*	*r*	*b*	*d*
T_V	×	*w*									
L_V	×	×									*w*
T_K			×								
L_K			*w*	×							
P_R					×						
%F	×				×						
T_R					*w*	×					
P_C							×	×	×		
T_T							*w*	*s*	×		
T_B							*s*	×	*s*		
L_C							*s*	×	*s*	×	
L_A							×	*s*	×		×

(b)

	s1	*f*	*w1*	*k*	*h*	*o*	*w2*	*r*	*s2*	*b*	*d*
T_V	**X**	*w*									
L_V	×	**X**									*w*
T_K			**X**								
L_K			*w*	**X**							
P_R					**X**						
T_R					*w*	**X**					
P_C							**X**	×	×		
T_T							*w*	**X**	*s*		
T_B							*s*	*s*	**X**		
L_C							*s*	*s*	×	**X**	
L_A							×	×	*s*		**X**

(c)

Figure 5.2 Use of cause-and-effect matrix (CEM) to determine output control structure for the furfuryl alcohol plant in Figure 5.1. (a) Initial CEM. (b) CEM adjusted to (block-) lower diagonal form (two upper *w*'s ignored) and blocks marked off. (c) CEM after removal of '%F' control objective to correct generic rank deficiency in fifth block. Also, selected SISO pairings are shown with **X** (although seventh block could be MIMO instead).

case, the setting of the pressure and vaporiser temperature will fix the concentration of furfural (F) in the reactor feed. Note that the plant rate will be determined by this composition and the reactor pressure and temperature. The resultant controlled plant is shown in Figure 5.3.

Even in a small plant, the possibility that several alternative MVs could regulate a particular CV rapidly leads to numerous permutations of the multiple SISO control scheme. If one considers just the distillation column control block in Figure 5.2,

	w2	*r*	*s2*
P_C	×	×	×
T_T	*w*	×	*s*
T_B	*s*	*s*	×

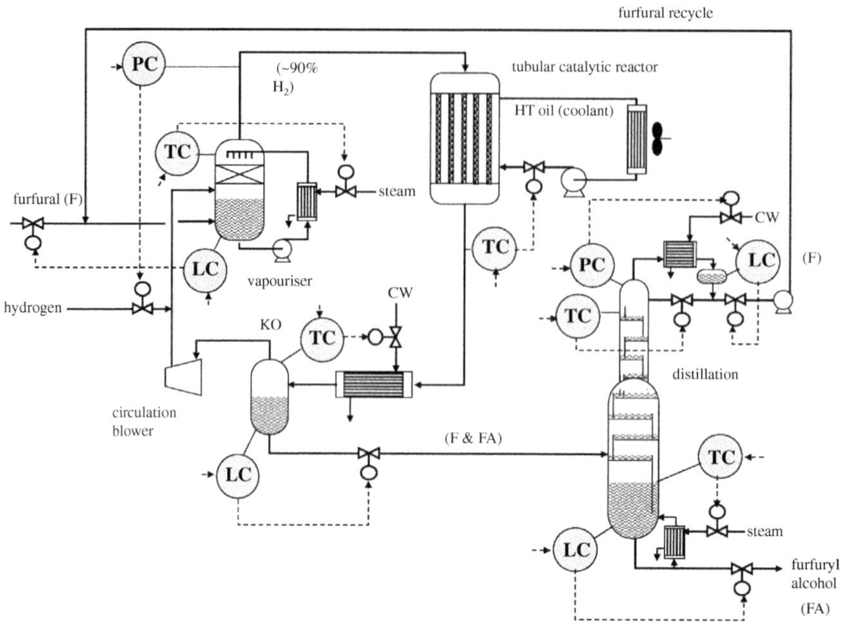

Figure 5.3 Furfuryl alcohol plant of Figure 5.1 showing SISO control scheme arising from *structural output control synthesis* of Figure 5.2.

it is noted that there are potentially $3! = 6$ different schemes:

e.g. first allocation to $\mathbf{P_C}$: choice of 3 MVs
$\times$ second allocation to $\mathbf{T_T}$: choice of 2 remaining MVs
$\times$ third allocation to $\mathbf{T_B}$: choice of 1 remaining MV
$= 3 \times 2 \times 1$

A binary distillation column need not only have dual-composition (temperature) objectives as shown. Other possible objectives are the actual flow rates of the bottom and distillate streams – for example, the flow and composition of the *bottom* stream could be important, regardless of the *distillate* flow and composition. Such specific objectives will determine one's choice of pairing in a multiple SISO control scheme. Stephanopoulos (1983) has an interesting example of a 4 MV, 4 CV case, leading to $4! = 24$ permutations. An important consideration in distillation control is always to avoid pairings where the MV has to change conditions right through the column (e.g. tray compositions/temperatures) in order to affect a CV at the opposite end. Such slow pairs are marked 's' in the CEM, and would normally not be considered for pairing in a SISO scheme.

5.2.2 Controllability and Observability

The structural synthesis of the control scheme discussed above can ensure that a configuration is *output structurally controllable*, but this is not enough for one to proceed. When one discusses *controllability*, one is usually referring to the *states* of a system, and it is noted that the development based on the CEM bypassed the states directly to the desired *outputs* $\boldsymbol{y}$ (i.e. CVs).

It is clearly possible that the outputs (of interest) are controllable, without all of the states defining the system being controllable. Moreover, the qualification is only *structural* (qualitative) and it will soon be obvious that the quantitative relationships between variables are required to assert *controllability*.

From Equation 5.1, consider again the general open-loop system:

$$\dot{\boldsymbol{x}} = \boldsymbol{f}(\boldsymbol{x}, \boldsymbol{u}) \tag{5.6}$$

Here one is usually thinking of some subsection of the plant which must be considered for MIMO control, such as the 3×3 distillation block in Figure 5.2, although there is no reason why the method cannot be applied to the entire plant, as a 'black box'.

> This system is said to be *controllable* if $\boldsymbol{u}(t)$ can be found to take $\boldsymbol{x}$ from $\boldsymbol{x}(0)$ to an arbitrary finite state $\boldsymbol{x}(t_f)$ in a finite time t_f.

For linear systems, there is a useful result to determine controllability. Local linearisations of nonlinear systems (Section 3.7) may also provide useful indications following this method. Firstly, consider a discrete multivariable system of the form of Equation 3.285:

$$\boldsymbol{x}_{i+1} = \boldsymbol{A}\boldsymbol{x}_i + \boldsymbol{B}\boldsymbol{u}_i \tag{5.7}$$

where the state vector $\boldsymbol{x}$ has n elements. If the matrix $\boldsymbol{B}$ were square and non-singular, the controllability condition would immediately be satisfied. More generally, however, noting that

$$\boldsymbol{x}_1 = \boldsymbol{A}\boldsymbol{x}_0 + \boldsymbol{B}\boldsymbol{u}_0 \tag{5.8}$$

$$\boldsymbol{x}_2 = \boldsymbol{A}(\boldsymbol{A}\boldsymbol{x}_0 + \boldsymbol{B}\boldsymbol{u}_0) + \boldsymbol{B}\boldsymbol{u}_1 \tag{5.9}$$

$$\vdots$$

$$\boldsymbol{x}_n = \boldsymbol{A}^n\boldsymbol{x}_0 + \boldsymbol{A}^{n-1}\boldsymbol{B}\boldsymbol{u}_0 + \boldsymbol{A}^{n-2}\boldsymbol{B}\boldsymbol{u}_1 + \cdots + \boldsymbol{B}\boldsymbol{u}_{n-1} \tag{5.10}$$

that is

$$\boldsymbol{x}_n - \boldsymbol{A}^n\boldsymbol{x}_0 = \begin{bmatrix}\boldsymbol{B} & \boldsymbol{A}\boldsymbol{B} & \cdots & \boldsymbol{A}^{n-2}\boldsymbol{B} & \boldsymbol{A}^{n-1}\boldsymbol{B}\end{bmatrix}\begin{pmatrix}\boldsymbol{u}_{n-1}\\ \boldsymbol{u}_{n-2}\\ \vdots\\ \boldsymbol{u}_1\\ \boldsymbol{u}_0\end{pmatrix} \tag{5.11}$$

it follows that the sequence of control vector values $\boldsymbol{u}_i$ required to take the state from $\boldsymbol{x}_0$ to $\boldsymbol{x}_n$ can be determined if the matrix on the right-hand-side has rank n. Note that the row rank will always equal the column rank. It can be shown that the requirement

> 'for a linear system having n states to be *controllable*, the controllability matrix $\begin{bmatrix}\boldsymbol{B} & \boldsymbol{A}\boldsymbol{B} & \cdots & \boldsymbol{A}^{n-2}\boldsymbol{B} & \boldsymbol{A}^{n-1}\boldsymbol{B}\end{bmatrix}$ must have rank n'

applies also to the *continuous* linear system $\dot{\boldsymbol{x}}(t) = \boldsymbol{A}\boldsymbol{x}(t) + \boldsymbol{B}\boldsymbol{u}(t)$.

There is a surprising implication in the above definition of *controllability* that, for example, a system having two states can be controlled by a single input. However, it is the *path* to the desired state that is manipulated to achieve the final objective. For example, a motor car travelling along a road has two states – speed and position, but may only have one input – the accelerator. However, with a little planning, one can always achieve a given speed at a given position and time.

> A system is said to be *observable* if the output measurements $\mathbf{y}(t)$, over a finite time 0 to t_f, contain the information which completely identifies the initial state $\mathbf{x}(0)$.

It follows that if any one state does not affect any of the output measurements, the system must be classified as unobservable. If one considers the particular case of a linear discrete or continuous system in which the available observations are related to the states by the possibly non-square or singular matrix C, that is

$$y = Cx \tag{5.12}$$

then a similar development to that above yields the result

'*for* a linear system having n states to be *observable*, the observability matrix

$$\begin{bmatrix} C \\ CA \\ \vdots \\ CA^{n-2} \\ CA^{n-1} \end{bmatrix} \text{ must have rank } n\text{'}$$

In practice, individual states can be

(*controllable, observable*)

(*uncontrollable, observable*)

(*controllable, unobservable*)

(*uncontrollable, unobservable*)

As will be seen later (Section 6.4), a Kalman filter 'observer' can be constructed to estimate the actual state values continuously in systems where some or all states are unobservable.

Example 5.1

Controllability and observability of furnace temperatures.

Figure 5.4 shows a furnace with two burners and two heated tubes which are symmetrically disposed relative to the burners. The objective is to determine whether the open-loop system below allows individual control of each temperature.

$$\begin{pmatrix} T_1 \\ T_2 \end{pmatrix}_{i+1} = \begin{bmatrix} -2 & 1 \\ 1 & -2 \end{bmatrix} \begin{pmatrix} T_1 \\ T_2 \end{pmatrix}_i + \begin{bmatrix} 2 & 1 \\ 2 & 1 \end{bmatrix} \begin{pmatrix} f_1 \\ f_2 \end{pmatrix}_i \tag{5.13}$$

$$\begin{bmatrix} B & AB \end{bmatrix} = \begin{bmatrix} 2 & 1 & -2 & -1 \\ 2 & 1 & -2 & -1 \end{bmatrix} \tag{5.14}$$

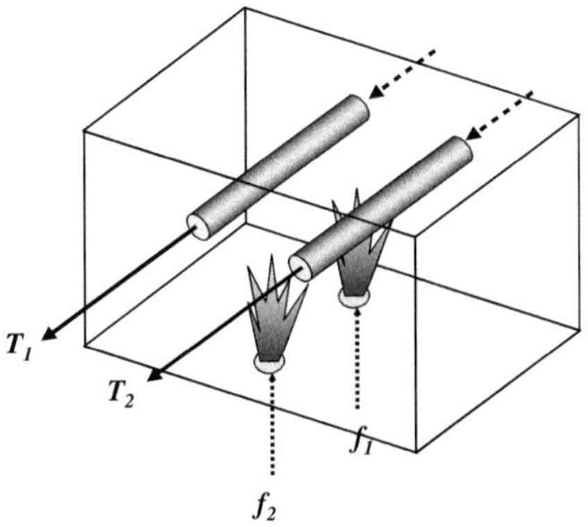

Figure 5.4 Furnace with symmetric firing.

The rank of the controllability matrix is 1 instead of the required 2, so the system is not controllable, as expected.

Now consider the case where the dynamic response of each tube to the firing differs slightly:

$$\begin{pmatrix} T_1 \\ T_2 \end{pmatrix}_{i+1} = \begin{bmatrix} -2 & 1 \\ 0 & -2 \end{bmatrix} \begin{pmatrix} T_1 \\ T_2 \end{pmatrix}_i + \begin{bmatrix} 2 & 1 \\ 2 & 1 \end{bmatrix} \begin{pmatrix} f_1 \\ f_2 \end{pmatrix}_i \tag{5.15}$$

$$\begin{bmatrix} B & AB \end{bmatrix} = \begin{bmatrix} 2 & 1 & -2 & -1 \\ 2 & 1 & -4 & -2 \end{bmatrix} \tag{5.16}$$

This restores the full rank 2 and controllability. Likewise, a slight relocation of one of the burners, nearer either tube, would cause the $\boldsymbol{B}$ matrix to lose its singularity, also restoring controllability.

Now consider that one hopes to 'observe' this system having available only a measured *average* of the two temperatures.

$$\begin{pmatrix} T_1 \\ T_2 \end{pmatrix}_{i+1} = \begin{bmatrix} -2 & 1 \\ 1 & -2 \end{bmatrix} \begin{pmatrix} T_1 \\ T_2 \end{pmatrix}_i + \cdots \tag{5.17}$$

$$\boldsymbol{C} = \begin{bmatrix} 1/2 & 1/2 \end{bmatrix} \tag{5.18}$$

$$\begin{bmatrix} \boldsymbol{C} \\ \boldsymbol{CA} \end{bmatrix} = \begin{bmatrix} 1/2 & 1/2 \\ -1/2 & -1/2 \end{bmatrix} \tag{5.19}$$

This observability matrix does not have the required rank of 2, so the system is not observable. On the other hand, for the dynamically asymmetric system,

$$\begin{pmatrix} T_1 \\ T_2 \end{pmatrix}_{i+1} = \begin{bmatrix} -2 & 1 \\ 0 & -2 \end{bmatrix} \begin{pmatrix} T_1 \\ T_2 \end{pmatrix}_i + \cdots \tag{5.20}$$

$$\boldsymbol{C} = \begin{bmatrix} 1/2 & 1/2 \end{bmatrix} \tag{5.21}$$

$$\begin{bmatrix} \boldsymbol{C} \\ \boldsymbol{CA} \end{bmatrix} = \begin{bmatrix} 1/2 & 1/2 \\ -1 & -1/2 \end{bmatrix} \tag{5.22}$$

the observability matrix now has its full rank, so the system is observable.

5.2.3 Morari Resiliency Index

Morari (1983) observed that process control effectively requires *inversion* of the open-loop model, that is 'find the input values required to achieve particular output values'. For a static linear n-input × n-output system, this will depend on whether the gain matrix relating inputs to outputs can be inverted. Morari took this idea further, to possibly non-square systems, and provided a measure of how close a system is to being 'un-invertible' based on the singular values.

In applying Morari's method, one is usually thinking of some subsection of the plant which is being considered for MIMO control, such as the 3 × 3 distillation block in Figure 5.2, although again there is no reason why the method cannot be applied to the whole plant as a 'black box'. The index is useful in a relative sense, for example 'is the resiliency improved by a different choice of variables?'

The ith singular value of a possibly non-square matrix $\boldsymbol{M}$ is the positive square root of the ith eigenvalue of $\boldsymbol{M}^T\boldsymbol{M}$:

$$\sigma_{i[\boldsymbol{M}]} = \sqrt{\lambda_{i[\boldsymbol{M}^T\boldsymbol{M}]}} \tag{5.23}$$

If the matrix $\boldsymbol{M}$ is square, then its determinant has the same magnitude as the product of its singular values. A zero singular value thus indicates that the square matrix cannot be inverted. For a non-square matrix $\boldsymbol{M}$, the implication of a zero singular value is that the matrix is *rank deficient.* In both instances, the implication is that the equation

$$\boldsymbol{x} = \boldsymbol{M}\boldsymbol{u} \tag{5.24}$$

cannot be solved (to get the $\boldsymbol{u}$ required to achieve a particular $\boldsymbol{x}$). As the *minimum* singular value $\sigma^{\min}_{[\boldsymbol{M}]}$ decreases towards zero, the equation draws closer to being insoluble.

Morari applied this idea to a gain matrix relating outputs to inputs at different frequencies. It will be seen in Chapter 8 that a multivariable transfer function, for example

$$\boldsymbol{G}(s) = \begin{bmatrix} \dfrac{1}{(s+1)} & 0 \\ \dfrac{2}{(s+1)(s+2)} & \dfrac{3}{(s+2)} \end{bmatrix} \quad \text{where } \boldsymbol{X}(s) = \boldsymbol{G}(s)\boldsymbol{U}(s) \tag{5.25}$$

converts a vector of steady input oscillations at frequency ω radians per unit time to a vector of steady output oscillations at the same frequency, but with the amplitudes altered according to the gain matrix:

$$\boldsymbol{K_P}(\omega) = |\boldsymbol{G}(j\omega)| = \begin{bmatrix} \left|\dfrac{1}{(j\omega+1)}\right| & 0 \\ \left|\dfrac{2}{(j\omega+1)(j\omega+2)}\right| & \left|\dfrac{3}{(j\omega+2)}\right| \end{bmatrix} \tag{5.26}$$

One notes that $\boldsymbol{K_P}$ need not be square, and that $\boldsymbol{K_P}(0) = \begin{bmatrix} 1 & 0 \\ 1 & 3/2 \end{bmatrix}$ is the steady-state gain matrix. If such a matrix cannot be inverted (or is rank-deficient), one cannot find the steady-state inputs required to achieve a particular steady-state output. This situation will vary with frequency, depending on the individual frequency responses, so the *Morari resiliency index* is defined for the whole frequency range as

$$\mathrm{MRI}(\omega) = \sigma^{\min}_{[\boldsymbol{K_P}(\omega)]} \tag{5.27}$$

As the MRI reduces towards zero for a particular frequency, it is understood that the system will become harder to control at that frequency. In the processing industries, one is generally interested in the steady-state gain $\boldsymbol{K_P}(0)$, rather than any other frequency.

The MRI value is sensitive to the units of the individual gain entries in $\boldsymbol{K_P}$. To allow comparison of this 'controllability' index as variables are changed, it is necessary to normalise K_{P11}, K_{P12}, K_{P21} and so on. A value such as 0.2 °C (%open)$^{-1}$ might, for example, be normalised to

$$0.2\,^{\circ}\mathrm{C}(\%\mathrm{open})^{-1} \rightarrow \frac{\text{temperature change as fraction of its range}}{\text{valve change as fraction of its range}} \rightarrow \frac{\dfrac{0.2}{(60-20)}}{\dfrac{1.0}{(100-20)}} = 0.4 \qquad (5.28)$$

if the normal range of operation is between 20 and 60 °C, and the valve is restricted between 20%open and 100%open.

Example 5.2

Morari resiliency index for the open-loop burner/temperature Example 5.1.

$$\begin{pmatrix} T_1 \\ T_2 \end{pmatrix}_{i+1} = \begin{bmatrix} -2 & 1 \\ 1 & -2 \end{bmatrix} \begin{pmatrix} T_1 \\ T_2 \end{pmatrix}_i + \begin{bmatrix} 2 & 1 \\ 2 & 1 \end{bmatrix} \begin{pmatrix} f_1 \\ f_2 \end{pmatrix}_i \qquad (5.29)$$

Steady-state ($\omega = 0$):

$$\begin{bmatrix} 1+2 & 0-1 \\ 0-1 & 1+2 \end{bmatrix} \begin{pmatrix} T_1 \\ T_2 \end{pmatrix} = \begin{bmatrix} 2 & 1 \\ 2 & 1 \end{bmatrix} \begin{pmatrix} f_1 \\ f_2 \end{pmatrix} \qquad (5.30)$$

$$\begin{pmatrix} T_1 \\ T_2 \end{pmatrix} = \begin{bmatrix} 3 & -1 \\ -1 & 3 \end{bmatrix}^{-1} \begin{bmatrix} 2 & 1 \\ 2 & 1 \end{bmatrix} \begin{pmatrix} f_1 \\ f_2 \end{pmatrix} \qquad (5.31)$$

$$= \frac{1}{8} \begin{bmatrix} 3 & 1 \\ 1 & 3 \end{bmatrix} \begin{bmatrix} 2 & 1 \\ 2 & 1 \end{bmatrix} \begin{pmatrix} f_1 \\ f_2 \end{pmatrix} \qquad (5.32)$$

$$= \begin{bmatrix} 1 & \frac{1}{2} \\ 1 & \frac{1}{2} \end{bmatrix} \begin{pmatrix} f_1 \\ f_2 \end{pmatrix} \qquad (5.33)$$

$$\text{So } \boldsymbol{K_P}(0) = \begin{bmatrix} 1 & \tfrac{1}{2} \\ 1 & \tfrac{1}{2} \end{bmatrix} \text{ and it is obvious that MRI} = \sigma^{\min}_{[\boldsymbol{K_P}(0)]} = 0 \qquad (5.34)$$

The next case considered in Example 5.1, where just the dynamic response of each tube temperature differs slightly, also gives MRI = 0, since the symmetry of the burner arrangement with respect to the tubes dictates an equal effect from each burner at steady state. So according to the MRI, this dynamically asymmetric system is uncontrollable, whereas the strict definition of *controllability* in Section 5.2.2 uses the differing dynamics to bring each state of the system to a specified setpoint at a specified time, making the system *controllable*!

Now consider a further system in which the apportionment of heat from each burner to the two tubes is asymmetric:

$$\begin{pmatrix} T_1 \\ T_2 \end{pmatrix}_{i+1} = \begin{bmatrix} -2 & 1 \\ 1 & -2 \end{bmatrix} \begin{pmatrix} T_1 \\ T_2 \end{pmatrix}_i + \begin{bmatrix} 2 & 1 \\ 2 & 0 \end{bmatrix} \begin{pmatrix} f_1 \\ f_2 \end{pmatrix}_i \tag{5.35}$$

At steady state,

$$\begin{pmatrix} T_1 \\ T_2 \end{pmatrix} = \begin{bmatrix} 3 & -1 \\ -1 & 3 \end{bmatrix}^{-1} \begin{bmatrix} 2 & 1 \\ 2 & 0 \end{bmatrix} \begin{pmatrix} f_1 \\ f_2 \end{pmatrix} \tag{5.36}$$

$$= \frac{1}{8} \begin{bmatrix} 3 & 1 \\ 1 & 3 \end{bmatrix} \begin{bmatrix} 2 & 1 \\ 2 & 0 \end{bmatrix} \begin{pmatrix} f_1 \\ f_2 \end{pmatrix} \tag{5.37}$$

$$= \frac{1}{8} \begin{bmatrix} 8 & 3 \\ 8 & 1 \end{bmatrix} \begin{pmatrix} f_1 \\ f_2 \end{pmatrix} \tag{5.38}$$

$$= \begin{bmatrix} 1 & \frac{3}{8} \\ 1 & \frac{1}{8} \end{bmatrix} \begin{pmatrix} f_1 \\ f_2 \end{pmatrix} \tag{5.39}$$

$$\text{So } \boldsymbol{K_P}(0) = \begin{bmatrix} 1 & \frac{3}{8} \\ 1 & \frac{1}{8} \end{bmatrix} \tag{5.40}$$

$$\boldsymbol{K}_P^T \boldsymbol{K}_P = \begin{bmatrix} 1 & 1 \\ \frac{3}{8} & \frac{1}{8} \end{bmatrix} \begin{bmatrix} 1 & \frac{3}{8} \\ 1 & \frac{1}{8} \end{bmatrix} = \begin{bmatrix} 2 & \frac{1}{2} \\ \frac{1}{2} & \frac{5}{32} \end{bmatrix} \tag{5.41}$$

For eigenvalues, solve

$$\det \begin{bmatrix} \lambda - 2 & -\frac{1}{2} \\ -\frac{1}{2} & \lambda - \frac{5}{32} \end{bmatrix} = 0 \tag{5.42}$$

$$\text{that is } (\lambda - 2)\left(\lambda - \frac{5}{32}\right) - \frac{1}{4} = 0 \tag{5.43}$$

$$\lambda^2 - \frac{69}{32}\lambda + \frac{1}{16} = 0$$

$$\lambda = \frac{\frac{69}{32} \pm \sqrt{\frac{4761}{1024} - \frac{1}{4}}}{2} = 3.18, 0.029 \tag{5.44}$$

$$\sigma_{[K_P(0)]}^{\min} = \sqrt{0.029} = 0.17 \tag{5.45}$$

Thus the MRI is 0.17, and this is the value that must me monitored for its closeness to zero, as changes are made in the physical system.

5.2.4 Relative Gain Array (Bristol Array)

In Sections 5.2.2 and 5.2.3, quantitative methods were considered to assess the controllability of part of a plant, based on the dynamic equations for the open-loop model relating inputs (proposed MVs) to outputs (proposed CVs), for example, Figure 5.5 (which is based on Figure 5.2). The methods could be applied to an entire plant, but it is likely that only difficult interactive subsections would be analysed in this way. The methods there gave no indication of how the control was to be performed.

Particularly before the widespread use of digital computers (but even nowadays as one aims for as simple a scheme as possible), it was necessary to break down interactive plant subsections into a number of SISO control loops. The *relative gain array* (RGA), sometimes called the *Bristol array*, was developed by Bristol (1966) to identify the pairings of MVs with CVs which would minimise undesirable interactions in a system. Even simple control tasks, such as in Figure 5.6, can be susceptible to loop interaction. In this example, if the supply line has appreciable resistance, FC1 can 'ride' FC2 in ever-increasing oscillations as each loop attempts to compensate for the pressure variation caused by the other.

Bristol recognised that if one focused on a particular control loop, an indication of its interaction with other controllers could be obtained by comparing its direct open-loop gain with

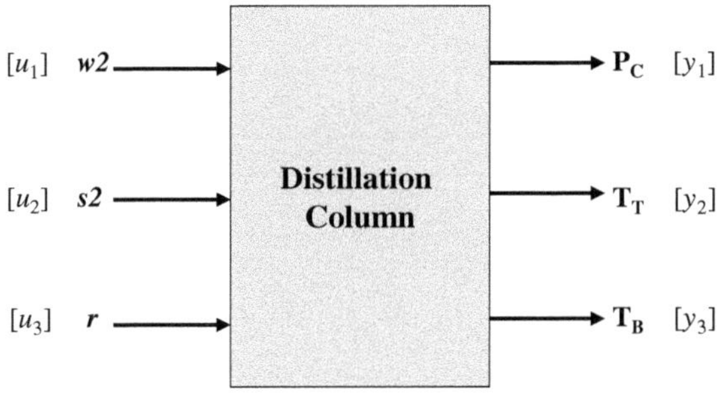

Figure 5.5 Section of a plant from Figure 5.2 to be assessed for pairing of MVs and CVs in SISO control loops.

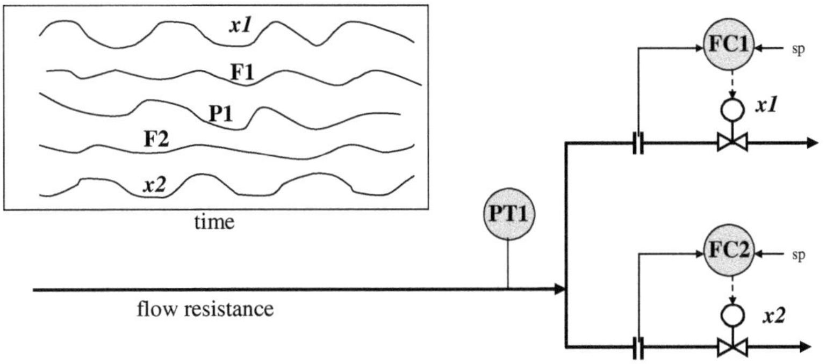

Figure 5.6 Flow control loops 'riding' each other due to their effects on a common supply pressure.

the gain achieved whilst all other controllers compensated. The *ij*th element of the RGA is thus the ratio

$$\beta_{ij} = \frac{\left[\frac{dy_i}{du_j}\right]_{\substack{\text{Steady state} \\ u_k=\text{const} \, for \, k \neq j}}}{\left[\frac{dy_i}{du_j}\right]_{\substack{\text{Steady state} \\ y_k=\text{const} \, for \, k \neq i}}} = \frac{\text{steady-state gain } CV_i/MV_j \text{ with other MVs constant}}{\text{steady-state gain } CV_i/MV_j \text{ with MVs adjusted to hold other CVs constant}} \tag{5.46}$$

The numerator term is clearly the *ij*th element of the steady-state gain matrix $\boldsymbol{K_P}$ (i.e. $\boldsymbol{K_P}(0)$ above). However, the denominator term must be derived from (a square, non-singular) $\boldsymbol{K_P}$ as follows:

At steady-state,

$$\begin{pmatrix} y_1 \\ y_2 \\ \vdots \\ y_n \end{pmatrix} = \boldsymbol{K_P} \begin{pmatrix} u_1 \\ u_2 \\ \vdots \\ u_n \end{pmatrix} = \begin{bmatrix} k_{11} & k_{12} & \cdots & k_{1n} \\ k_{21} & k_{22} & \cdots & k_{2n} \\ \vdots & \vdots & \ddots & \vdots \\ k_{n1} & k_{n2} & \cdots & k_{nn} \end{bmatrix} \begin{pmatrix} u_1 \\ u_2 \\ \vdots \\ u_n \end{pmatrix} \tag{5.47}$$

Allow a change $y_2 \rightarrow y_2 + \Delta y_2$ but no other y is to change:

$$\begin{pmatrix} u_1+\Delta u_1 \\ u_2+\Delta u_2 \\ \vdots \\ u_n+\Delta u_n \end{pmatrix} = \begin{bmatrix} k_{11} & k_{12} & \cdots & k_{1n} \\ k_{21} & k_{22} & \cdots & k_{2n} \\ \vdots & \vdots & \ddots & \vdots \\ k_{n1} & k_{n2} & \cdots & k_{nn} \end{bmatrix}^{-1} \begin{pmatrix} y_1 \\ y_2+\Delta y_2 \\ \vdots \\ y_n \end{pmatrix} = \begin{pmatrix} u_1 \\ u_2 \\ \vdots \\ u_n \end{pmatrix} + \begin{bmatrix} k_{11} & k_{12} & \cdots & k_{1n} \\ k_{21} & k_{22} & \cdots & k_{2n} \\ \vdots & \vdots & \ddots & \vdots \\ k_{n1} & k_{n2} & \cdots & k_{nn} \end{bmatrix}^{-1} \begin{pmatrix} 0 \\ \Delta y_2 \\ \vdots \\ 0 \end{pmatrix} \tag{5.48}$$

So (perhaps obviously)

$$\begin{pmatrix} \Delta u_1 \\ \Delta u_2 \\ \vdots \\ \Delta u_n \end{pmatrix} = \begin{bmatrix} k_{11} & k_{12} & \cdots & k_{1n} \\ k_{21} & k_{22} & \cdots & k_{2n} \\ \vdots & \vdots & \ddots & \vdots \\ k_{n1} & k_{n2} & \cdots & k_{nn} \end{bmatrix}^{-1} \begin{pmatrix} 0 \\ \Delta y_2 \\ \vdots \\ 0 \end{pmatrix} \tag{5.49}$$

Thus,

$$\left.\frac{\Delta y_2}{\Delta u_1}\right|_{\text{other } y \text{ const}} = \frac{1}{\text{element in the1st row, 2nd column of } \boldsymbol{K_P^{-1}}} \tag{5.50}$$

and more generally

$$\left.\frac{\Delta y_i}{\Delta u_j}\right|_{\text{other } y \text{ const}} = \frac{1}{\text{element in the } i\text{th row}, j\text{th column of } \left[\boldsymbol{K_P^{-1}}\right]^T} \tag{5.51}$$

Finally, it is clear that

$$\beta_{ij} = (\text{element in the } i\text{th row}, j\text{th column of } \boldsymbol{K_P}) \times \left(\text{element in the } i\text{th row, } j\text{th column of} \left[\boldsymbol{K_P^{-1}}\right]^T\right) \tag{5.52}$$

An element *ij* occurring in the RGA with a value of unity indicates that that particular pair (*i*th CV controlled using *j*th MV) would have no interaction at all with other loops. Smaller (e.g. 0.5) and larger (e.g. 2) values indicate progressively more interaction. A negative sign would imply that the sign of the controller gain has to be swopped if that loop is switched on at the same time as the other loops!

The idea of the RGA is that the control engineer would populate his or her plant with SISO loops, starting with the CV/MV pairings closest to unity. The worse pairings might still be used, but it would be sensible to make such loops progressively slower to avoid unstable interactions. Alternatively, nowadays, the engineer might deal with the more interactive combinations within a MIMO control strategy, rather than several SISO control loops.

As a caution, Luyben (1990) notes that the main requirement in the chemical processing industry is load rejection. Controllers usually regulate at fixed setpoints, and aim to minimise the effects of disturbances. Configuring a control system to minimise loop interaction does not necessarily give the best load rejection.

5.3
Control Schemes Involving More Complex Interconnections of Basic Elements

It is rare that the systematic approaches discussed in Section 5.2 are able to completely define an overall plant control scheme. Section 5.2 focussed only on the installation of SISO or MIMO controllers to meet the overall process objectives. No mention was made of the basic elements in Chapter 4 such as selectors, overrides, ratios, adders, feedforward controllers and valve position controllers which are common parts of control schemes. This Section 5.3 will illustrate how the basic elements are brought together in a few common industrial control schemes. The associated exercises, in the accompanying volume of solved problems, focus on the step-by-step interpretation of a set of instructions defining a plant control scheme. The process engineer needs a clear understanding of the process phenomena in order to properly execute these instructions. Of course, the skill one ultimately aims for is the ability to devise the overall scheme oneself, which should arise from sufficient experience with this type of specification.

5.3.1
Boiler Drum-Level Control

One of the earlier historical control schemes concerned the maintenance of a setpoint level in boiler steam drums. Figure 5.7 shows the well-known ‘three-element’ control scheme. The three elements are clearly feedback, feedforward and a supervised flow control loop.

As mentioned in Section 4.2.7, all feedforward controllers require a ‘model’, and in this case it is seen to be a very simple one, with 1 tonne of *steam drawn* translating exactly to 1 tonne of *BFW to be supplied*. Since the system integrates, and the flow measurements cannot be perfect, a feedback ‘trim’ is essential, if for nothing else, just to get the level to its initial setpoint!

The delegation of the task of maintaining a desired BFW flow rate, to a slave flow control loop, isolates the rest of the algorithm from such factors affecting the BFW flow as BFW and drum pressure fluctuations, and indeed nonlinearity of the valve itself. The direct summation of feedforward and feedback BFW demands may appear to require twice as much BFW as necessary, until one

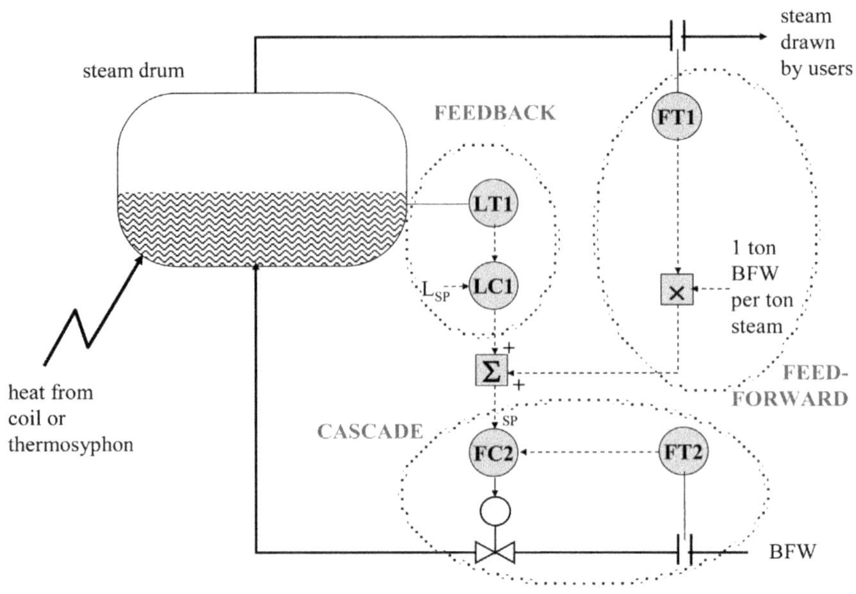

Figure 5.7 Three-element control of boiler steam drum level.

recalls that the entire algorithm is working on the basis of *deviations* from the initial 'switch on' condition.

In Section 4.2.1, it was noted that a controller acts as a 'translator', converting a requirement in one language (e.g. level %) into an action in another language (e.g. BFW flow t h^{-1}). In viewing a diagram like Figure 5.7, it important to identify the implied 'languages' of the various signals. In particular, note that the summation involves three signals which must represent 't BFW h^{-1}'.

5.3.1.1 Note on Boiler Drum-Level Inverse Response

Some processes exhibit *inverse response* in their open-loop behaviour. It is convenient to consider this phenomenon here, because the effect of boiler feedwater (BFW) flow on the boiler drum level is a classic example. Assume that a boiler is operating steadily, with the BFW exactly making up for the steam delivered to consumers, and the level is thus steady. The observed effects of stepping the BFW inflow up or down in this circumstance are depicted in Figure 5.8.

Boiler feedwater will usually be de-aerated and preheated, but can in general be expected to be below the enthalpy of the equilibrium liquid in the boiler drum. Thus, an increase in the BFW flow (a), all other variables remaining constant, will cause a drop in the average enthalpy of the liquid in the drum, so that some of the immersed bubbles collapse, releasing latent heat and establishing a new equilibrium. The bubble collapse temporarily reduces the *volume* in the drum, and thus the level. For the case (b), where the BFW flow is decreased, additional bubbles form, as the liquid contents no longer have to compensate as much for the lower enthalpy of the BFW inflow. Clearly, these level changes would not register on a differential-pressure measurement (Section 2.4.2.1), since the level and mean density of the bubbly contents will vary as inverses.

Inverse responses of this nature, where the variable of interest temporarily 'goes the wrong way' after a control action adjustment, require special care in process control. In the above case (a), a

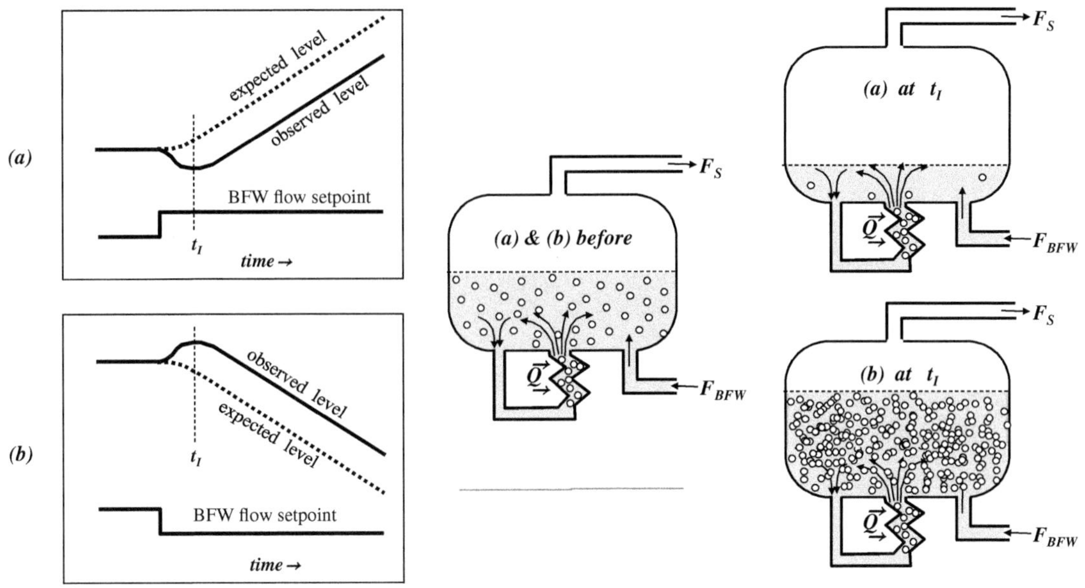

Figure 5.8 Shrink and swell of boiler drum level due to boiler feedwater flow variation.

controller trying to increase the level, seeing the initial drop, might try to add even more BFW, making the situation worse, and eventually over-shooting the target level. So simple controllers such as the PID (Section 4.2.4) need to be 'de-tuned', that is work with a lower gain K_C, in order to give the level more time to respond. A predictive controller like the *dynamic matrix controller* (Section 7.8.2) fully anticipates the initial inverse response, allowing much tighter tuning.

5.3.2
Furnace Full Metering Control with Oxygen Trim Control

In Section 4.2.7 the maintenance of oxygen in the flue gas from a furnace was used to illustrate a purely feedforward strategy. Mention was made there of the necessity to maintain flue gas $\%O_2$ above zero to ensure combustion, but not so high as to incur significant heat loss. About 3% on a molar basis (compared to a maximum of 21% in air) seems typical industrially. If there is no actual measurement of oxygen, then a larger margin may be necessary to ensure safety.

Figure 5.9 shows the full metering control with oxygen trim control presented by Smith and Corripio (1985). This is an air/fuel ratio control with an additive feedback trim from a flue gas oxygen controller. The implication at the summer is that the feedback signal from the AC will request a zero adjustment if the ratio controller, which is working with absolute flow rates, is already achieving the setpoint $\%O_2$. This feedforward–feedback arrangement will minimise $\%O_2$ deviations from setpoint. The air flow rate controller is a slave in the cascade, whilst the fuel flow setpoint will arrive from the operator, or another controller such as the process stream TC or a boiler PC. The scheme shown includes a high and a low 'clip' on the air flow rate setpoint. The low clip is a good safety measure, but the high clip could lead to incomplete combustion at high fuel demands. A slight variation of this scheme is sometimes encountered where instead of supplying an

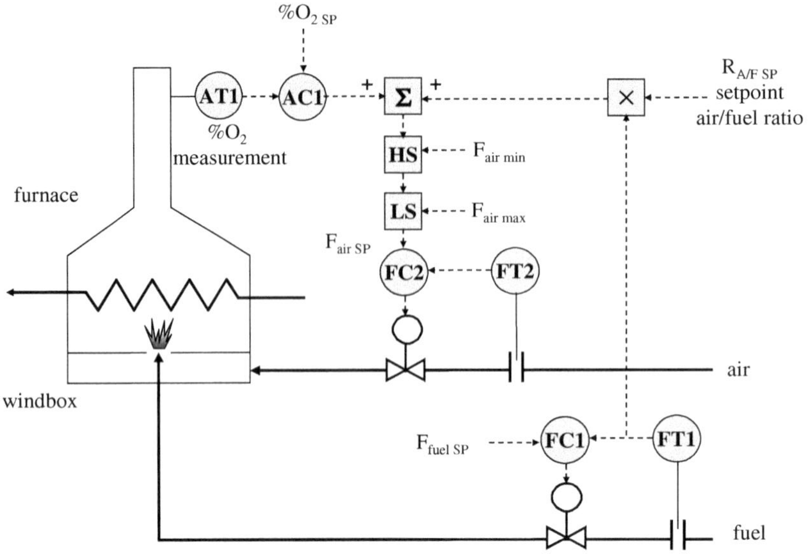

Figure 5.9 Furnace full metering control with oxygen trim control.

additive trim to the air flow rate setpoint, the oxygen controller manipulates the setpoint air/fuel ratio $R_{AF\ SP}$ directly.

The scheme in Figure 5.9 has another drawback to do with the system dynamics. One notes that the fuel flow rate will always change in advance of the air adjustments, simply because there will be dynamic lag of the actual air flow as the ratio controller moves the air flow setpoint in response to the measured fuel flow variations. This situation is described as *fuel leads air in* and *fuel leads air out*. The latter situation is safe, because air will temporarily be in excess. However, the *fuel leads air in* is dangerous, because the implication is that there will temporarily be a deficit of oxygen.

5.3.3
Furnace Cross-Limiting Control

The dynamic lag of the air flow controller in Section 5.3.2, Figure 5.9, was seen to cause a temporary drop in the air/fuel ratio when the fuel flow increased. The cross-limiting scheme in Figure 5.10 overcomes this problem by using both measured flows to ensure a minimum air/fuel ratio at all times.

The cross-limiting control scheme is best explained using an example. Consider the situation where the system is initially steady maintaining the correct air/fuel ratio. Now TC1 demands an increase in fuel in order to maintain its setpoint temperature. This demand will be ignored at the LS, because the *desired fuel* will be higher than the *allowed fuel* (which initially will be the *actual fuel*). However, the HS will pass this higher demand rather than the *actual fuel*, so after multiplication by the air/fuel ratio, the air flow will start to rise to match the *desired fuel*. As the actual air flow rises, following its controller setpoint, the cut-off limit (*allowed fuel*) arriving at the LS rises in proportion, gradually allowing the fuel setpoint to increase, and ultimately the *actual fuel* will rise to the *desired fuel*. One notes that *air leads fuel in*, which is a safe action. Conversely, if one follows

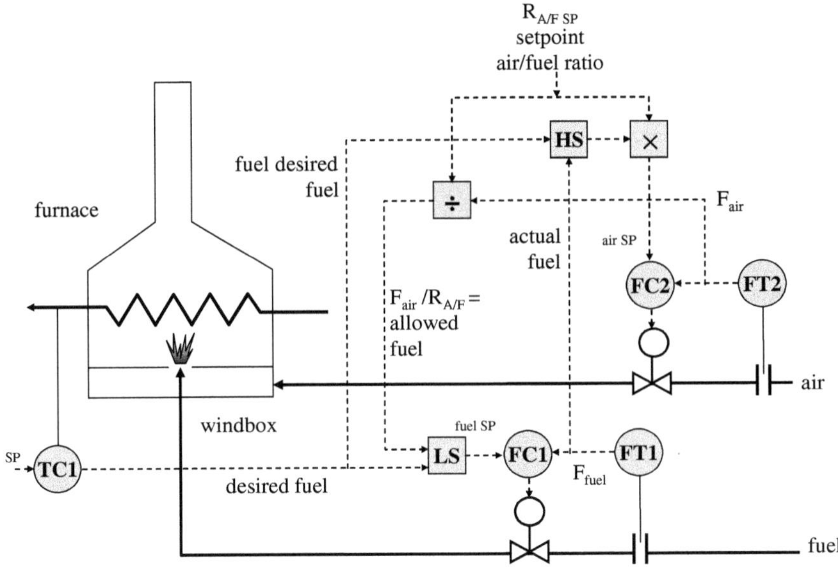

Figure 5.10 Furnace cross-limiting control to avoid decreases of air/fuel ratio below setpoint.

the sequence of events when *desired fuel* decreases, it will be found that *air follows fuel out*, which is again a safe action. So in any transient, there will temporarily be a safe excess of air, which will return to the correct ratio when the process settles down again.

Furnace control schemes as in Figures 5.9 and 5.10 may also have an additional safety feature to prevent *flame-out*. It is understandable that the fuel valve could swing to the shut position temporarily, depending on the controller gain, as it seeks to hold the flow setpoint. This would cause an irreversible situation where the flame is lost. As the valve moves open again, uncombusted fuel will accumulate in the furnace box and ducting, and may explode all at once should a source of ignition be found, for example at a neighbouring furnace sharing the same ducting. Thus, it is necessary to ensure a minimum fuel flow. This can be done using an override controller which senses the fuel pressure just before the burner nozzle. Figure 5.11 shows an arrangement where an override pressure controller maintains a minimum pressure, and implicitly a flow.

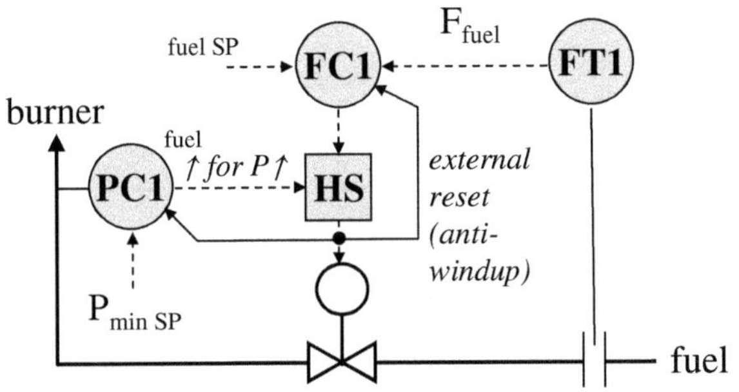

Figure 5.11 Fuel pressure override to prevent furnace 'flame-out'.

References

Bristol, E.H. (1966) On a new measure of interaction for multivariable process control. *IEEE Transactions on Automatic Control*, **11**, 133–134.

Buckley, P.S. (1964) *Techniques of Process Control*, John Wiley & Sons, Inc., New York.

Johnston, R.D. and Barton, G.W. (1984) Control objective reduction in single-input single-output control schemes. *International Journal of Control*, **40**, 265–270.

Johnston, R.D. and Barton, G.W. (1985a) Structural equivalence and model reduction. *International Journal of Control*, **41**, 1477–1491.

Johnston, R.D. and Barton, G.W. (1985b) Single input-single output control system synthesis. *Computers and Chemical Engineering*, **9** (6), 547–555.

Lin, R.M. and Lim, M.K. (1994) Analytical model updating and model reduction. *International Journal for Numerical Methods in Engineering*, **37**, 1881–1896.

Lin, X., Tade, M.O. and Newell, R.B. (1991) Output structural controllability condition for the synthesis of control systems for chemical processes. *International Journal of Systems Science*, **22** (1), 107–132.

Luyben, W.L. (1990) *Process Modeling, Simulation and Control for Chemical Engineers*, 2nd edn, McGraw-Hill, New York.

Morari, M. (1983) Design of resilient processing plants III: a general framework for the assessment of dynamic resilience. *Chemical Engineering Science*, **38**, 1881–1891.

Smith, C.A. and Corripio, A.B. (1985) *Principles and Practice of Automatic Process Control*, John Wiley & Sons, Inc., New York.

Stephanopoulos, G. (1983) *Chemical Process Control: An Introduction to Theory and Practice*, Prentice Hall.

6
Estimation of Variables and Model Parameters from Plant Data

In this chapter, the idea of 'observer models' will be developed. The use of digital computers in industry, and even computerised measurement instruments, has led to the notions of 'intelligent sensors', 'soft measurements' and 'data mining'. An example might be the varying temperature in the centre of cooling cast steel ingots, estimated from the surface temperature. Such calculated variables were considered briefly in Section 4.6, but the more complex issues of state and parameter estimation during dynamic variation, for multivariable systems, must be considered in more detail here owing to the growing importance of these techniques industrially.

Figure 6.1 shows the general case of control on the basis of an observer model output – a form of internal model control (IMC). Typical examples include state estimation and parameter estimation for adaptive control. However, such an estimator need not be part of a feedback control loop – other important applications include *fault detection* and *management information* such as process efficiency. An observer like this might also be used to reconcile process data, for example to provide a 'best-fit' mass balance as input to an optimiser, which might, or might not, operate in closed loop.

6.1
Estimation of Signal Properties

In Section 4.1, basic signal filtering was described. The objective was simply to remove random 'noise' superimposed on measurements by physical or electrical phenomena, so that filtered values became representative of the most likely current value of a variable. From a process control point of view, some further properties of signals could be of interest.

6.1.1
Calculation of Cross-Correlation and Autocorrelation

Some important aspects of the behaviour of a process can be determined from batch or real-time measurements of the input vector $\boldsymbol{u}$ and the output vector $\boldsymbol{y}$, by means of the time series *cross-correlation* between the elements u_i and y_j of $\boldsymbol{u}$ and $\boldsymbol{y}$.

$$R_{u_i y_j}(\tau) = \frac{E\left\{ \left[u_i(t) - \mu_{u_i}\right] \left[y_j(t+\tau) - \mu_{y_j}\right] \right\}}{\sigma_{u_i}\sigma_{y_j}} \tag{6.1}$$

Applied Process Control: Essential Methods, First Edition. Michael Mulholland.

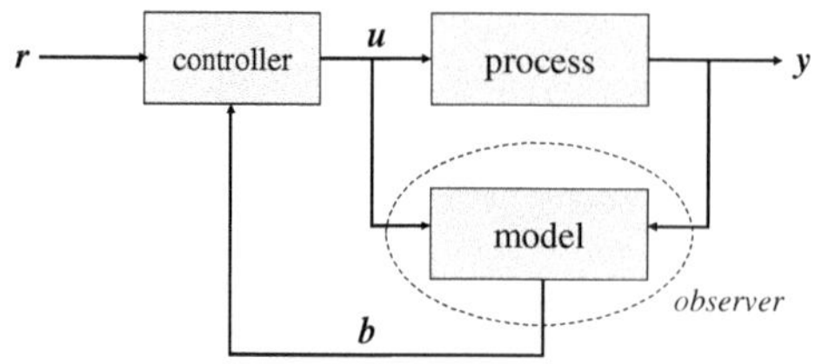

Figure 6.1 Control on the basis of an observer model: a form of internal model control.

Here $E\{\cdot\}$ is the *expectation*, that is the average. For sampled data, the cross-correlation across an n-interval gap can be estimated from $N+n$ measurement points using

$$R_{u_iy_j}(n) = \frac{\frac{1}{N}\sum_{k=1}^{N}\left[u_i(k) - \mu_{u_i}\right]\left[y_j(k+n) - \mu_{y_j}\right]}{\sigma_{u_i}\sigma_{y_j}} \tag{6.2}$$

Here μ and σ are the mean and standard deviation of the respective signals. A basic algorithm for the calculation of $R_{uy}(n)$ is given in Figure 6.2.

This normalised cross-correlation will give zero if the two variables are uncorrelated at interval τ or n, and +1 if they are exactly correlated at this interval (i.e. both up or both down). A −1 implies that they are negatively correlated at this interval time (i.e. one is up whilst the other is down). Pure plug flow would cause a short-lived spike of this nature, but in practice the *dynamic* lag (arising from mixing or accumulations) will spread the effect, smoothing such a cross-correlation spike. In Section 6.7.1, a similar method provides the *impulse response* model of the process. In this way, for example, the dead time or space-time of a vessel could be estimated in real time, and used to improve the performance of estimation and control algorithms.

Consider the real-time evaluation of $R_{uy}(n)$ for $1 \le n \le M$, based on a *moving window* of the last $N+M$ points as in Figure 6.3. The corresponding equations (Equations 6.3–6.7) offer a more efficient means of updating the statistical parameters on each time step, by introducing the new point and discarding the old point.

Mean updates:

$$\mu_{u,k} = \frac{(M+N)\mu_{u,k-1} + u_i - u_{k-M-N}}{M+N} \tag{6.3}$$

$$\mu_{y,k} = \frac{(M+N)\mu_{y,k-1} + y_i - y_{k-M-N}}{M+N} \tag{6.4}$$

Variance updates:

$$\sigma_{u,k}^2 = \frac{(M+N-1)\sigma_{u,k-1}^2 + \left(u_k - \mu_{u,k}\right)^2 - \left(u_{k-M-N} - \mu_{u,k-M-N}\right)^2}{M+N-1} \tag{6.5}$$

$$\sigma_{y,k}^2 = \frac{(M+N-1)\sigma_{y,k-1}^2 + \left(y_k - \mu_{y,k}\right)^2 - \left(y_{k-M-N} - \mu_{y,k-M-N}\right)^2}{M+N-1} \tag{6.6}$$

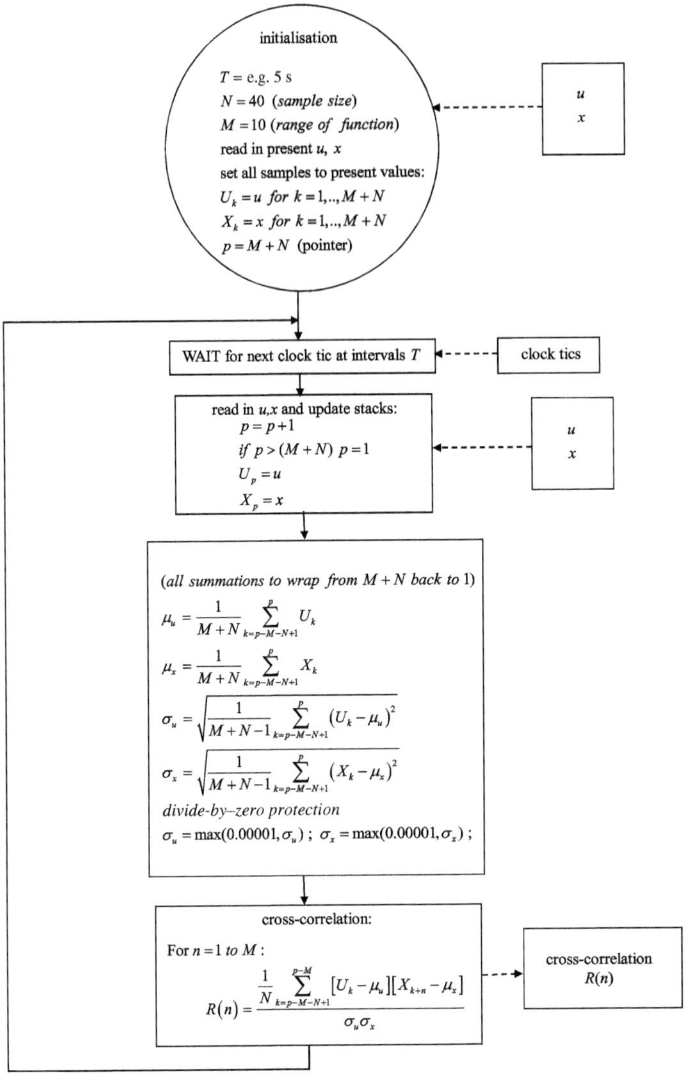

Figure 6.2 Algorithm for real-time estimation of input–output cross-correlation function.

Cross-correlation updates:

$$R_{uy,k}(n) = \frac{\sigma_{u,k-1}\sigma_{y,k-1}(N+M-n)R_{uy,k-1}(n) + \left(y_k - \mu_{y,k}\right)\left(u_{k-n} - \mu_{u,k}\right) - \left(y_{k-N-M+n} - \mu_{y,k-N-M+n}\right)\left(u_{k-N-M} - \mu_{u,k-N-M}\right)}{\sigma_{u,k}\sigma_{y,k}(N+M-n)} \tag{6.7}$$

When the above functions are applied to the same signal (e.g. y_i versus y_j instead of the separate signals u_i versus y_j), the result is the *autocorrelation* function, useful for establishing the time

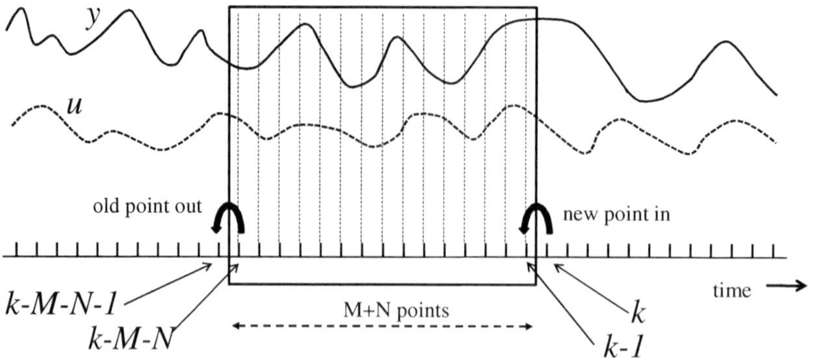

Figure 6.3 Moving window stack for $N+M$ data points.

constant and any periodicity in a single signal. Though obtainable similarly for cross-correlations, the *integral timescale* is usually considered in relation to autocorrelations as

$$\Gamma = \int_0^\infty R_{yy}(\tau)\mathrm{d}\tau \tag{6.8}$$

or

$$\Gamma \approx T \sum_{n=1}^{\infty} R_{yy}(n) \tag{6.9}$$

where T is the sampling interval.

6.1.2 Calculation of Frequency Spectrum

The *Fourier transform* allows identification of the extent to which a signal is made up of different frequencies. For example, if one transformed a long sequence of measurements of electricity consumption in a city, one would discover several peaks – representing variations at frequencies of day^{-1}, week^{-1} and year^{-1} (winter–summer). These spectral fingerprints are obviously useful in model identification, revealing the contributions of different inputs to the outputs, indicating the types of sampling frequencies required, and the time constants of the system.

The discrete Fourier transform (DFT) of a sequence of measurements $x_0, x_1, x_2, \ldots, x_{N-1}$ is given by

$$X_k = \sum_{n=0}^{N-1} x_n \,\mathrm{e}^{-j(2\pi/N)kn}, \quad k = 0, 1, \ldots, N-1 \tag{6.10}$$

where j is the complex value $\sqrt{-1}$. The original sequence of measurements can be reconstructed from these Fourier coefficients using the inverse discrete Fourier transform (IDFT):

$$x_n = \frac{1}{N}\sum_{k=0}^{N-1} X_k \,\mathrm{e}^{j(2\pi/N)kn}, \quad n = 0, 1, \ldots, N-1 \tag{6.11}$$

Noting that X_k will be complex,

$$X_k = a_k + jb_k = A_k\, e^{j\phi} \quad \text{where} \quad \begin{cases} A_k = \sqrt{a_k^2 + b_k^2} \\ \phi = \angle(a_k + jb_k) \end{cases} \tag{6.12}$$

each term of Equation 6.11 is

$$A_k\, e^{j\phi}\, e^{jk\theta} = A_k\, e^{j(k\theta+\phi)} = A_k\left[\cos(k\theta + \phi) + j\sin(k\theta + \phi)\right] \tag{6.13}$$

where

$$\theta = \frac{2\pi}{N}n \tag{6.14}$$

Bearing in mind the form of Equation 6.10, the magnitude A_k of the Fourier coefficient X_k can thus be interpreted as the amplitude of the 'k' frequency contribution needed to reconstruct the original measurements $x_0, x_1, x_2, \ldots, x_{N-1}$. *Power spectral density* is plotted in terms of A_k^2.

Various fast Fourier transform (FFT) algorithms have been developed to speed up the evaluation of the complex coefficients in Equation 6.10. A common form is the Cooley and Tukey (1965) method which requires an input sequence of $N = 2^M$ points. The coefficients are then constructed from M halvings of the number of data points. In a *real-time* analysis, a stack of N points could be updated as in the trailing window of Figure 6.3.

6.1.3 Calculation of Principal Components

Principal components analysis (PCA) offers another way of analysing process data with a view to reducing the dimensions of a problem, or detecting unusual behaviour, perhaps caused by a fault. The idea is quite straightforward, and is illustrated by the following problem.

Measurements of concentrations into and out of a flow reactor in which $\boldsymbol{A}$ is converted to $\boldsymbol{B}$ have been determined as in Figure 6.4. Firstly determine the *mean* of each of the three variables, and subtract it from the corresponding variable to obtain the 6×3 matrix of deviations $\boldsymbol{X}$.

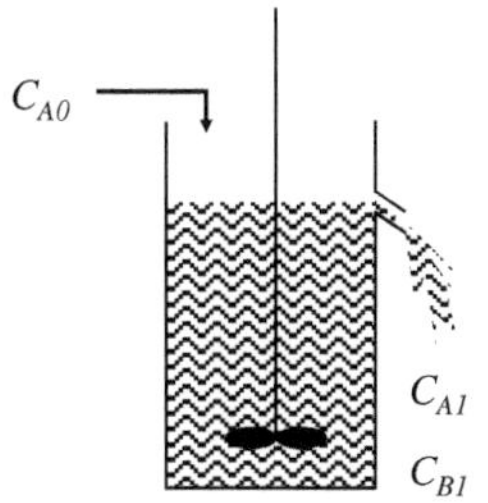

t	C_{A0}	C_{A1}	C_{B1}
1	20	4	15
2	21	8	12
3	32	8	23
4	12	7	6
5	14	3	12
6	26	11	13

Figure 6.4 Concentration measurements around a flow reactor for PCA.

$$\boldsymbol{X} = \begin{pmatrix} \boldsymbol{x}_1^{\mathrm{T}} \\ \boldsymbol{x}_2^{\mathrm{T}} \\ \boldsymbol{x}_3^{\mathrm{T}} \\ \boldsymbol{x}_4^{\mathrm{T}} \\ \boldsymbol{x}_5^{\mathrm{T}} \\ \boldsymbol{x}_6^{\mathrm{T}} \end{pmatrix} = \begin{pmatrix} C_{A01} - \overline{C}_{A0} & C_{A11} - \overline{C}_{A1} & C_{B11} - \overline{C}_{B1} \\ \vdots & \vdots & \vdots \\ \vdots & \vdots & \vdots \\ \vdots & \vdots & \vdots \\ \vdots & \vdots & \vdots \\ C_{A06} - \overline{C}_{A0} & C_{A16} - \overline{C}_{A1} & C_{B16} - \overline{C}_{B1} \end{pmatrix} = \begin{bmatrix} -1 & -3 & 3 \\ 0 & 1 & -3 \\ 12 & 2 & 9 \\ -9 & 0 & -7 \\ -7 & -4 & -2 \\ 5 & 4 & 0 \end{bmatrix} \tag{6.15}$$

The covariance matrix is thus obtained for the $n = 6$ data points as

$$\boldsymbol{C} = E\{\boldsymbol{x}\boldsymbol{x}^{\mathrm{T}}\} = \frac{1}{(n-1)} \sum_{i=1}^{n} \boldsymbol{x}_i \boldsymbol{x}_i^{\mathrm{T}} = \begin{bmatrix} 60 & 15 & 36.4 \\ 15 & 9.2 & 2.8 \\ 36.4 & 2.8 & 30.4 \end{bmatrix} \tag{6.16}$$

Arrange the eigenvectors of $\boldsymbol{C}$ as unit vectors in columns of matrix $\boldsymbol{V}$ in order of decreasing size of eigenvalue, so that the *principal component* is in the first column.

$$\lambda_1 = 87.03, \quad \lambda_2 = 12.253, \quad \lambda_3 = 0.317 \tag{6.17}$$

$$\boldsymbol{V} = \begin{bmatrix} 0.8237 & -0.2662 & -0.5006 \\ 0.1781 & -0.7167 & 0.6742 \\ 0.5383 & 0.6445 & 0.5430 \end{bmatrix} \tag{6.18}$$

The eigenvectors are orthogonal, and the eigenvalues λ_i represent the amounts of 'energy' (variance) in the original data lying in these three new coordinate directions. In this example, it is seen that the third component is contributing very little.

The data set can be represented by selected principal components by removing right-hand columns of $\boldsymbol{V}$ as required – say

$$\boldsymbol{V}' = \begin{bmatrix} 0.8237 & -0.2662 \\ 0.1781 & -0.7167 \\ 0.5383 & 0.6445 \end{bmatrix} \tag{6.19}$$

Then the measurements in the reduced orthogonal frame $\boldsymbol{Y}$ representing the original set $\boldsymbol{X}$ are

$$\boldsymbol{Y} = \boldsymbol{Y}\boldsymbol{V}' = \begin{bmatrix} -1 & -3 & 3 \\ 0 & 1 & -3 \\ 12 & 2 & 9 \\ -9 & 0 & -7 \\ -7 & -4 & -2 \\ 5 & 4 & 0 \end{bmatrix} \begin{bmatrix} 0.8237 & -0.2662 \\ 0.1781 & -0.7167 \\ 0.5383 & 0.6445 \end{bmatrix} = \begin{bmatrix} 0.2567 & 4.3500 \\ -1.4367 & -2.6503 \\ 15.0855 & 1.1730 \\ -11.1815 & -2.1160 \\ -7.5551 & 3.4412 \\ 4.8311 & -4.1979 \end{bmatrix} \tag{6.20}$$

It turns out that for a matrix with orthogonal vectors for its columns, its inverse (in the 'best-fit' sense) is given by its transpose, so $[\boldsymbol{V}']^{-1} = [\boldsymbol{V}']^{\mathrm{T}}$, allowing one to make a prediction $\boldsymbol{X}'$ of the original data set using the reduced set $\boldsymbol{Y}$ as follows:

$$X' = Y[V']^{\mathrm{T}} = \begin{bmatrix} 0.2567 & 4.3500 \\ -1.4367 & -2.6503 \\ 15.0855 & 1.1730 \\ -11.1815 & -2.1160 \\ -7.5551 & 3.4412 \\ 4.8311 & -4.1979 \end{bmatrix} \begin{bmatrix} 0.8237 & 0.1781 & 0.5383 \\ -0.2662 & -0.7167 & 0.6445 \end{bmatrix} = \begin{bmatrix} -0.9465 & -3.0721 & 2.9419 \\ -0.4780 & 1.6437 & -2.4816 \\ 12.1141 & 1.8463 & 8.8762 \\ -9.6473 & -0.4750 & -7.3826 \\ -7.1394 & -3.8122 & -1.8488 \\ 5.0970 & 3.8693 & -0.1052 \end{bmatrix} \tag{6.21}$$

Comparing X' with X, one notes that the errors are quite small, so just two variables in the new axis system have been proven adequate to represent these data. This result is understandable in this case. The reactor clearly has a variable feed rate, giving varying conversions of the feed concentration C_{A0} of A to B. The data suggest that any A not converted remains as A, implying that there are no other reactions. Thus, the three measurements are not independent, allowing the reduced representation by PCA.

The simple analysis above focuses on the steady state for the flow reactor. However, dynamically varying measurements are easily handled by considering also values of the same variable at several points in time, just as in a discrete dynamic model. In a real-time application, stacks would have to maintain the variables of interest for the n preceding data points, just as in Figure 6.3. The *clustering* of effects in a reduced number of variables clearly allows the use of a reduced model of the system, whether for control or identification.

Finally, it is noted that an important application of the PCA technique is to monitor the data pattern. For example, in the given case, a significant side reaction may suddenly develop due to impurities present. All of the original measurements would then become necessary, and the third eigenvalue would increase whilst the eigenvectors changed. A behaviour-monitoring system tracking these variables could then alarm regarding the unexpected behaviour change.

6.2 Real-Time Estimation of Variables for Which a Delayed Measurement Is Available for Correction

Online analysers in industry are often prohibitively expensive, and sometimes unreliable or intermittent in their operation. As a result, 'quality predictors' have become commonplace, and serve to provide a continuous 'measurement' for either feedback control or open-loop operator decisions.

The original idea was proposed by Smith (1957), and may be applied to any measurement which only becomes available some time after a control action has determined it. Such a transport lag (dead time) within a control loop seriously undermines the quality of control, so one seeks the result of the control action as soon as possible after the determining action. The 'Smith predictor' delays this predicted effect for the period of the transport lag, so that a prediction error can be determined from the actual process measurement. The error is fed back continuously for correction of an early prediction which excludes the transport lag (Figure 6.5).

Figure 6.6 elucidates the general concept of Figure 6.5 as a block diagram in IMC format. Note that the Smith predictor itself is merely the estimation device, and does not include the controller. In this figure, Laplace domain transfer functions are used to represent the functions of the blocks. The process and its approximating model could in reality have any desired behaviour – nonlinear,

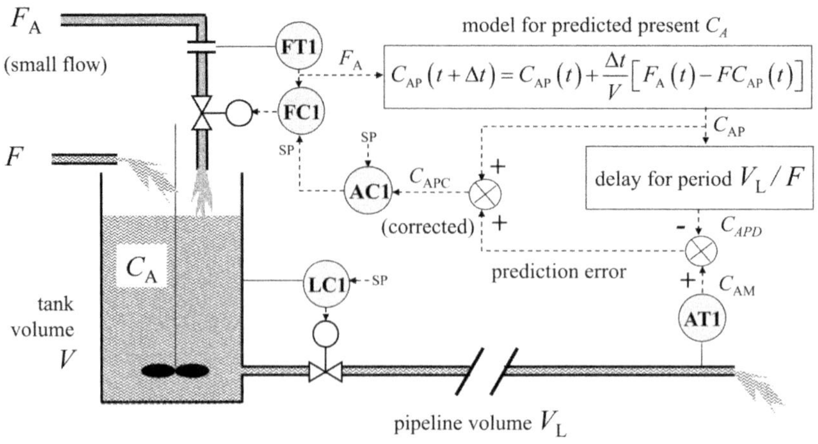

Figure 6.5 Smith predictor applied to a system with long measurement delay.

steady state and so on. The illustrative form is first order plus dead time. The important point is that an 'early' prediction is taken from the model, before the approximation of the dead time. The dead-time simulation, probably a computer cyclical 'stack' array as in Section 3.2, attempts to synchronise with the measurement coming from the process. This 'old' error is used to correct the current 'early' prediction. The intention here is to correct for such problems as slowly varying offset errors. The parameters in the model are shown with primes, representing estimated values that may differ from the true process values.

Differences in the dead time will manifest in the *predicted and corrected* result as a short positive or negative error disturbance. Figure 6.7 shows a particular case where the process is steady state,

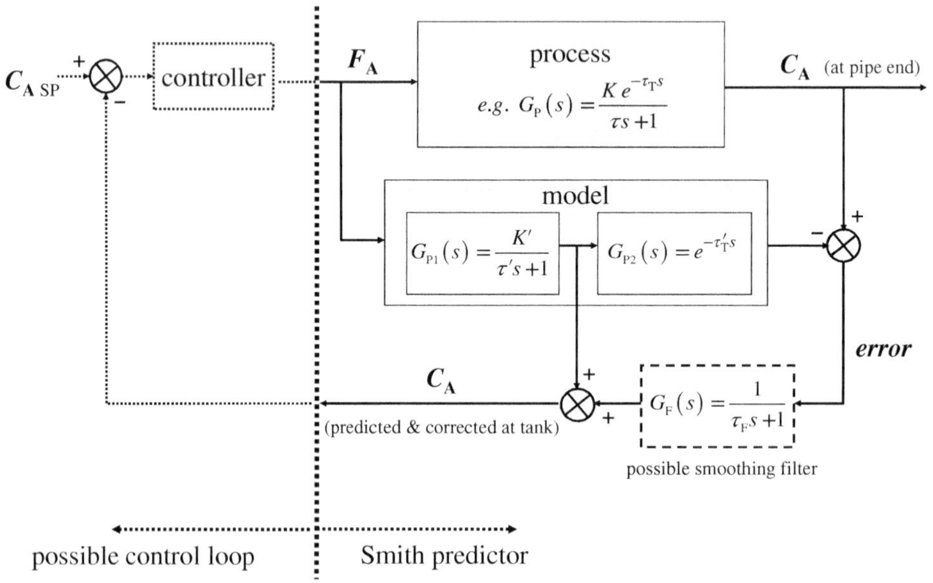

Figure 6.6 IMC format block diagram of a Smith predictor following Figure 6.5.

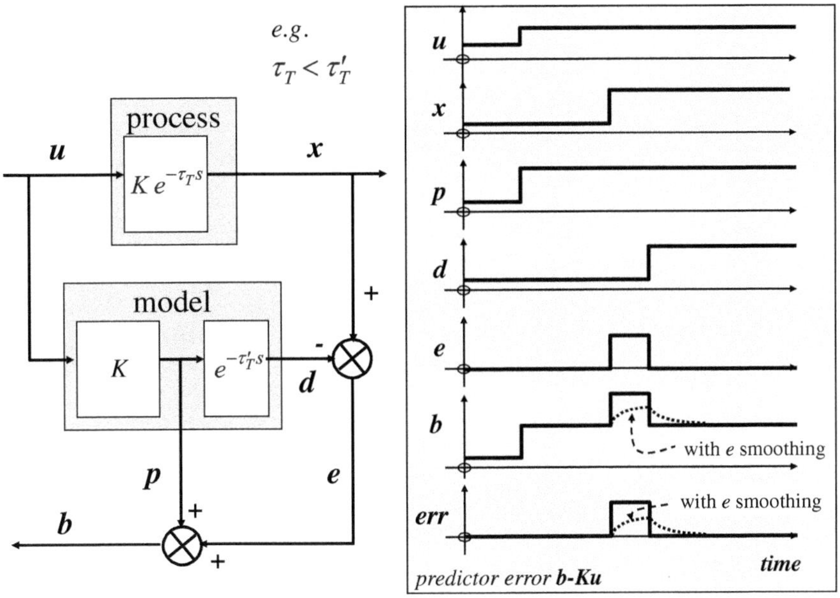

Figure 6.7 Rapid error disturbance in the output of a Smith predictor arising from mismatched dead time (transport lag).

the gain of which has been accurately determined for the model. However, the transport lag used in the model is shorter than that of the process. This type of error due to a dead-time phasing shift (or errors due to rapid signal noise on the output) can be reduced by insertion of a smoothing filter on the error feedback signal (Section 4.1).

The correction applied in the Figure 6.5 example is additive, which is by far the most common form. In rare situations (Giles and Gaines, 1977), there might be a benefit in using a multiplicative correction, the so-called gain-adaptive approach.

A refinement of the method used in Figures 6.5 and 6.6 would allow for a varying flow rate F, and thus a varying transport lag. An updated stack of the volumetric flow rate would need to be maintained. At any point in time, one would need to integrate backwards through the stack until the dead volume was reached, and this point would fix the time at which the stored early prediction must be extracted to be compared with the current measurement (Mulholland, 1988).

In some situations, there is little point in seeking the dynamic part of the model. In large plug flow reactors, such as pulp bleachers or digesters, the exit composition is dictated by feed rates or settings near the reactor inlet. The steady-state stoichiometry will be important, but the time constant of the dynamic response will be short compared with reactor space times such as 2–8 h. One is unlikely to get a better match with the measurement by including the ODE dynamics in such cases, and a steady-state early prediction will be quite adequate.

Having investigated the original Smith predictor format above, now is the opportunity to consider the modern *quality predictor* which has become prolific and valuable in industry. The principle is the same as the Smith predictor, but dynamics are usually omitted, there is usually no transport lag involved, measurements are intermittent and instantaneous, and the time delay of concern is to do with drawing a sample and taking it to the laboratory for analysis (Figure 6.8).

Quality predictors will seldom be based on a physical model of the process. Modern plant data historisers allow engineers to go back through the measurement record of part of the process to

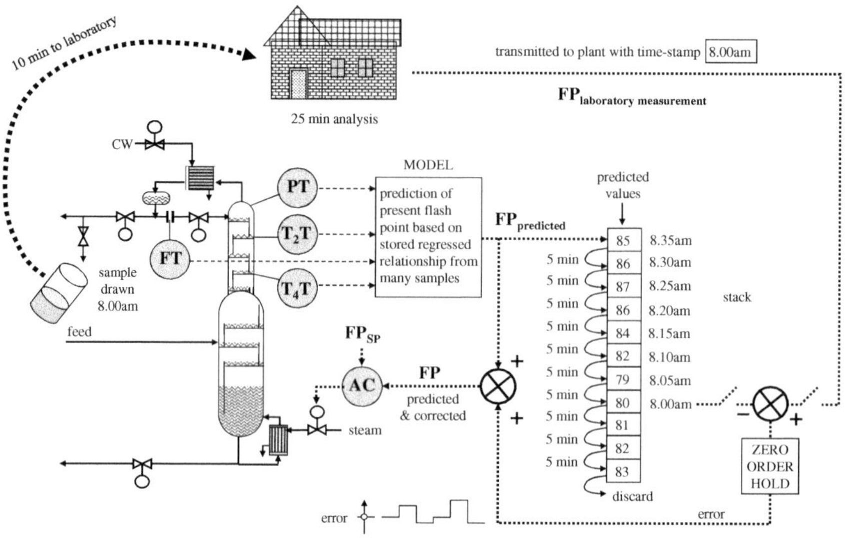

Figure 6.8 Implementation of an industrial *quality predictor* for overheads flashpoint.

extract a set of measurements at the time each laboratory sample was drawn. Simple regressions, perhaps aided by *analysis of variance*, rapidly identify useful correlations for the laboratory measurement, which can then be installed as real-time models. Thus, cumbersome measurements such as *octane*, *cut point* or *flash point* can appear to be coming from the process instantaneously. The rest of the information flow then concerns the correction of the model. The intermittent samples continue to be drawn, and are analysed in the plant testing station or laboratory. An integrated data capture system allows the chemist to enter the analysis with a 'time stamp' for the sampling time. An instantaneous model error is established by lookup in the historical stack maintained by the online model. This error is then used for model correction until the next result arrives. Sudden correction steps will have the same effect as jerking the composition controller setpoint, so again it is advisable to apply some smoothing to the stepped error signal.

In connection with the instantaneous error calculation determining a stepped error signal, a similar situation can arise in the case of online gas or liquid chromatographs. Such an analyser updates its measured compositions at the end of each injection/elutriation cycle, but these compositions remain constant (e.g. for 5 min) until the next update, which in any case will represent delayed values. In some situations, for example polyethylene manufacture, control benefits are possible by 'bridging' the updates and eliminating the dead-time lag with a process model. The effect and principle employed are similar to Figure 6.8, but with a fixed sampling interval.

6.3 Plant Data Reconciliation

In this section, consideration will be given to a particular problem where the data set representing a plant must be exactly consistent in terms of a mass or energy balance. This is typically where a

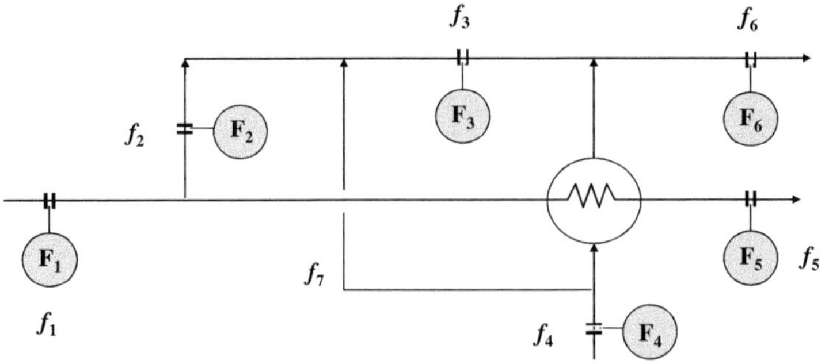

Figure 6.9 Measurements (*F*) available for data reconciliation and estimation, with desired reconciled values (*f*) shown in italics.

process optimiser requires a balanced picture of the process, and is quite a different problem to techniques considered later, such as the Kalman filter (Section 6.4), which seek a *compromise* solution – somewhere between a model and the actual measurements. Nowadays, versions of standard flowsheeting packages may be used online in order to get a parallel data set suitable for a real-time plant optimiser. Such optimisers, usually based on the process steady state, act as higher level controllers, supervising the setpoints of plant dynamic controllers (Figure 1.5).

An example of a data reconciliation problem is given in Figure 6.9, where a single liquid flows in all of the pipes. The available measurements $\boldsymbol{y}$ are shown in the instrument circles, whilst the desired reconciled variables ($\boldsymbol{x}$, $\boldsymbol{u}$) are indicated in italics. The $\boldsymbol{x}$ are the reconciled values to be found for the measurements $\boldsymbol{y}$, whilst the $\boldsymbol{u}$ are additional reconciled variables to be found. A steady-state model demands that

$$\boldsymbol{g}(\boldsymbol{x}, \boldsymbol{u}) = \boldsymbol{0} \tag{6.22}$$

Then a typical data reconciliation problem aims to minimise the weighted sum of squared deviations between $\boldsymbol{x}$ and $\boldsymbol{y}$, yet at the same time satisfy Equation 6.22. The weight used for each measurement in the $\boldsymbol{y}$ vector takes into account the units as well as the confidence that one has in that measurement.

$$w_i = \frac{1}{\sigma_{e_i}^2} \tag{6.23}$$

Here $\sigma_{e_i}^2$ is the variance of the error expected in the measurement y_i. If one cannot distinguish such confidence levels, one might just equate σ_{e_i} to the normal operating span of the ith measurement. A simple expression of the weighted sum of square deviations is facilitated by positioning the w_i in order down the diagonal of a diagonal matrix $\boldsymbol{W}$. A statement of the data reconciliation problem is then

Find $\boldsymbol{x}$ and $\boldsymbol{u}$ such that

$$\boldsymbol{g}(\boldsymbol{x}, \boldsymbol{u}) = \boldsymbol{0} \tag{6.24}$$

and

$$\phi(\boldsymbol{x}) = [\boldsymbol{y} - \boldsymbol{x}]^{\mathrm{T}} \boldsymbol{W} [\boldsymbol{y} - \boldsymbol{x}] \tag{6.25}$$

is minimised.

There are indeed solvers for this general *nonlinear* problem, even when one constrains the allowed ranges of the $\boldsymbol{x}$. However, it happens that the functions $\boldsymbol{g}$ for the example in Figure 6.9 are linear, readily allowing a solution for this particular case.

$$y = \begin{pmatrix} F_1 \\ F_2 \\ F_3 \\ F_4 \\ F_5 \\ F_6 \end{pmatrix} \tag{6.26}$$

$$\boldsymbol{x} = \begin{pmatrix} f_1 \\ f_2 \\ f_3 \\ f_4 \\ f_5 \\ f_6 \end{pmatrix} \tag{6.27}$$

$$\boldsymbol{u} = (f_7) \tag{6.28}$$

$$\boldsymbol{g}(\boldsymbol{x},\boldsymbol{u}) = \boldsymbol{Ax} + \boldsymbol{Bu} - \boldsymbol{b} = \begin{bmatrix} 0 & -1 & 1 & 0 & 0 & 0 \\ 0 & 0 & -1 & -1 & 0 & 1 \\ -1 & -1 & 0 & 0 & 1 & 0 \end{bmatrix} \begin{pmatrix} f_1 \\ f_2 \\ f_3 \\ f_4 \\ f_5 \\ f_6 \end{pmatrix} + \begin{bmatrix} -1 \\ 1 \\ 0 \end{bmatrix} (f_7) - \begin{pmatrix} 0 \\ 0 \\ 0 \end{pmatrix} \tag{6.29}$$

Some row operations are now necessary to obtain an identity matrix at, say, the bottom of the $\boldsymbol{B}$ matrix, so that the elements of $\boldsymbol{u}$ appear only in these rows.

$$\left[\boldsymbol{A} \,\vdots\, \boldsymbol{B}\right] \begin{pmatrix} \boldsymbol{x} \\ \hdashline \boldsymbol{u} \end{pmatrix} = \boldsymbol{b} \rightarrow \begin{bmatrix} \boldsymbol{A}_1 & \vdots & \boldsymbol{0} \\ \hdashline \boldsymbol{A}_2 & \vdots & \boldsymbol{I} \end{bmatrix} \begin{pmatrix} \boldsymbol{x} \\ \hdashline \boldsymbol{u} \end{pmatrix} = \begin{pmatrix} \boldsymbol{b}_1 \\ \hdashline \boldsymbol{b}_2 \end{pmatrix} \tag{6.30}$$

that is for $\left[\boldsymbol{A} \,\vdots\, \boldsymbol{B}\right], \boldsymbol{b}$:

$$\begin{bmatrix} 0 & -1 & 1 & 0 & 0 & 0 & \vdots & -1 \\ 0 & 0 & -1 & -1 & 0 & 1 & \vdots & 1 \\ -1 & -1 & 0 & 0 & 1 & 0 & \vdots & 0 \end{bmatrix}, \begin{pmatrix} 0 \\ 0 \\ 0 \end{pmatrix} \rightarrow \begin{bmatrix} 0 & -1 & 0 & -1 & 0 & 1 & \vdots & 0 \\ -1 & -1 & 0 & 0 & 1 & 0 & \vdots & 0 \\ \hdashline 0 & 0 & -1 & -1 & 0 & 1 & \vdots & 1 \end{bmatrix}, \begin{pmatrix} 0 \\ 0 \\ \hdashline 0 \end{pmatrix} \tag{6.31}$$

So the problem for this linear example is to minimise

$$\phi(\boldsymbol{x}) = [\boldsymbol{y} - \boldsymbol{x}]^{\mathrm{T}} \boldsymbol{W} [\boldsymbol{y} - \boldsymbol{x}] \tag{6.32}$$

subject to the constraint

$$\boldsymbol{A}_1 \boldsymbol{x} = \boldsymbol{b}_1 \tag{6.33}$$

The use of a *Lagrange multiplier* vector $\boldsymbol{\lambda}$ allows one to incorporate the constraint directly by finding instead $(\boldsymbol{x}, \boldsymbol{\lambda})$ which minimise a new objective function, called the *Lagrangian*,

$$J(\boldsymbol{x}, \boldsymbol{\lambda}) = [\boldsymbol{y} - \boldsymbol{x}]^{\mathrm{T}} \boldsymbol{W} [\boldsymbol{y} - \boldsymbol{x}] + \boldsymbol{\lambda}^{\mathrm{T}} (\boldsymbol{A}_1 \boldsymbol{x} - \boldsymbol{b}_1) \tag{6.34}$$

At the minimum for all values of $(\boldsymbol{x}, \boldsymbol{\lambda})$, the derivatives of J with respect to elements of $\boldsymbol{\lambda}$ must be zero, which it is seen forces the constraints to be satisfied. When the derivatives with respect to elements of $\boldsymbol{x}$ are zero, it means that simultaneously the constraint is parallel to the contours of $\phi(\boldsymbol{x}) = [\boldsymbol{y} - \boldsymbol{x}]^{\mathrm{T}} \boldsymbol{W} [\boldsymbol{y} - \boldsymbol{x}]$ as $\boldsymbol{x}$ varies, that is the constraint must be passing through the minimum of ϕ on its path.

Noting that the partial differentiation of a scalar with respect to each element of a vector yields the results

$$\frac{\partial}{\partial \boldsymbol{x}}\{\boldsymbol{x}^{\mathrm{T}} \boldsymbol{M} \boldsymbol{x}\} = 2\boldsymbol{M}\boldsymbol{x} \quad \text{and} \quad \frac{\partial}{\partial \boldsymbol{x}}\{\boldsymbol{x}^{\mathrm{T}} \boldsymbol{M} \boldsymbol{y}\} = \boldsymbol{M}\boldsymbol{y} \tag{6.35}$$

and using the fact that the scalar $\boldsymbol{\lambda}^{\mathrm{T}} \boldsymbol{A}_1 \boldsymbol{x} = \left(\boldsymbol{\lambda}^{\mathrm{T}} \boldsymbol{A}_1 \boldsymbol{x}\right)^{\mathrm{T}} = \boldsymbol{x}^{\mathrm{T}} \boldsymbol{A}_1^{\mathrm{T}} \boldsymbol{\lambda}$, it follows that

$$\frac{\partial}{\partial \boldsymbol{x}}\left\{[\boldsymbol{y} - \boldsymbol{x}]^{\mathrm{T}} \boldsymbol{W} [\boldsymbol{y} - \boldsymbol{x}] + \boldsymbol{\lambda}^{\mathrm{T}} (\boldsymbol{A}_1 \boldsymbol{x} - \boldsymbol{b}_1)\right\} = 2\boldsymbol{W}[\boldsymbol{y} - \boldsymbol{x}] + \boldsymbol{A}_1^{\mathrm{T}} \boldsymbol{\lambda} \tag{6.36}$$

and

$$\frac{\partial}{\partial \boldsymbol{\lambda}}\left\{[\boldsymbol{y} - \boldsymbol{x}]^{\mathrm{T}} \boldsymbol{W} [\boldsymbol{y} - \boldsymbol{x}] + \boldsymbol{\lambda}^{\mathrm{T}} (\boldsymbol{A}_1 \boldsymbol{x} - \boldsymbol{b}_1)\right\} = \boldsymbol{A}_1 \boldsymbol{x} - \boldsymbol{b}_1 \tag{6.37}$$

Since both Equations 6.36 and 6.37 must yield zero vectors at the minimum of J,

$$\begin{bmatrix} -2\boldsymbol{W} & \boldsymbol{A}_1^{\mathrm{T}} \\ \boldsymbol{A}_1 & \boldsymbol{0} \end{bmatrix} \begin{pmatrix} \boldsymbol{x} \\ \boldsymbol{\lambda} \end{pmatrix} = \begin{pmatrix} -2\boldsymbol{W}\boldsymbol{y} \\ \boldsymbol{b}_1 \end{pmatrix} \tag{6.38}$$

Solution by elimination of $\boldsymbol{\lambda}$ yields the desired reconciled fit of $\boldsymbol{x}$ to the measurements $\boldsymbol{y}$:

$$\boldsymbol{x} = \boldsymbol{y} + \boldsymbol{W}^{-1} \boldsymbol{A}_1^{\mathrm{T}} \left[\boldsymbol{A}_1 \boldsymbol{W}^{-1} \boldsymbol{A}_1^{\mathrm{T}}\right]^{-1} (\boldsymbol{b}_1 - \boldsymbol{A}_1 \boldsymbol{y}) \tag{6.39}$$

Finally, from Equation 6.30, the reconciled estimates of the unmeasured variables are given by

$$\boldsymbol{u} = \boldsymbol{b}_2 - \boldsymbol{A}_2 \boldsymbol{x} \tag{6.40}$$

In general, plant data reconciliation will account for all of the available balances, such as *energy* and *species* reaction stoichiometry. In Figure 6.9, the four temperatures around the heat exchanger might be measured, and might easily be incorporated as a check on the balances. One notes though that this leads to a constraint such as

$$\begin{aligned} g_4\left[\left(f_1, f_2, f_3, f_4, f_5, f_6, T_{1\mathrm{i}}, T_{1\mathrm{o}}, T_{4\mathrm{i}}, T_{4\mathrm{o}}\right), (f_7)\right] &= (f_1 - f_2) T_{1\mathrm{i}} + (f_4 - f_7) \\ \times T_{4\mathrm{i}} - f_5 T_{1\mathrm{o}} - (f_6 - f_3) T_{4\mathrm{o}} &= 0 \end{aligned} \tag{6.41}$$

which is nonlinear, requiring an iterative numerical solution. At this point, one has not even considered unsteady-state behaviour, caused by mass or energy accumulations in the system. Some methods have been developed for the *unsteady* case (e.g. Franks, 1982), but these are not widely used yet.

6.4 Recursive State Estimation

The *data reconciliation* approach considered in Section 6.3 assumed that the 'model' was perfectly accurate, and provided estimates of measured and unmeasured variables that gave a perfect fit to the model. Now another procedure will be considered where varying degrees of confidence are assigned to both the *model* and the *measurements*, so that the estimates lie somewhere between the two. The idea will be developed using the linear state equations (Equations 6.42 and 6.43), with an extension to nonlinear systems considered later. Here a set of measurements $\boldsymbol{y}'$ is available, corresponding to the variables $\boldsymbol{y}$ which arise as a linear combination of some, or all, of the states $\boldsymbol{x}$. (A minor adjustment of the following equations will incorporate dependence of $\boldsymbol{y}$ also on $\boldsymbol{u}$ if necessary.)

$$\frac{\mathrm{d}\boldsymbol{x}}{\mathrm{d}t} = \boldsymbol{A}\boldsymbol{x} + \boldsymbol{B}\boldsymbol{u} \tag{6.42}$$

$$\boldsymbol{y} = \boldsymbol{C}\boldsymbol{x} \tag{6.43}$$

A *Luenberger observer* (Luenberger, 1966) seeks estimates $\hat{\boldsymbol{x}}$ of the true state $\boldsymbol{x}$, given measurements $\boldsymbol{y}'$ and $\boldsymbol{u}'$, by adding a term onto the right-hand side of the differential equation based on the difference $(\boldsymbol{y}' - \hat{\boldsymbol{y}})$:

$$\hat{\boldsymbol{y}} = \boldsymbol{C}\hat{\boldsymbol{x}} \tag{6.44}$$

$$\frac{\mathrm{d}\hat{\boldsymbol{x}}}{\mathrm{d}t} = \boldsymbol{A}\hat{\boldsymbol{x}} + \boldsymbol{B}\boldsymbol{u}' + \boldsymbol{K}[\boldsymbol{y}' - \hat{\boldsymbol{y}}] \tag{6.45}$$

$$\frac{\mathrm{d}\hat{\boldsymbol{x}}}{\mathrm{d}t} = \boldsymbol{A}\hat{\boldsymbol{x}} + \boldsymbol{B}\boldsymbol{u}' + \boldsymbol{K}[\boldsymbol{y}' - \boldsymbol{C}\hat{\boldsymbol{x}}] \tag{6.46}$$

$$\frac{\mathrm{d}\hat{\boldsymbol{x}}}{\mathrm{d}t} = [\boldsymbol{A} - \boldsymbol{K}\boldsymbol{C}]\hat{\boldsymbol{x}} + \boldsymbol{B}\boldsymbol{u}' + \boldsymbol{K}\boldsymbol{y}' \tag{6.47}$$

So it is seen that the dynamics of this observer are determined by the matrix $[\boldsymbol{A} - \boldsymbol{K}\boldsymbol{C}]$ (compare with Section 3.9.2.2). The following example illustrates how this distortion of the original governing equation can be helpful in tracking the behaviour of a process.

For simplicity, consider that the exit flow from the tank in Figure 6.10 is proportional (α) to the level in the tank, so that the required observer is

$$\frac{\mathrm{d}\hat{h}}{\mathrm{d}t} = \left\{-\frac{\alpha}{A}\right\}\hat{h} + \left\{\frac{1}{A}\right\}f' + k\left[h' - \hat{h}\right] \tag{6.48}$$

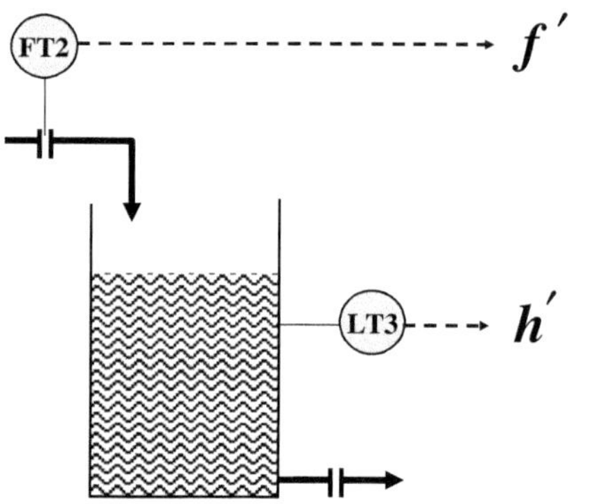

Figure 6.10 SISO example for a Luenberger observer.

where A is the liquid surface area. Imagine that this equation is being integrated in real time whilst the process is operating. If the estimate $\hat{h}$ falls below the measurement h', and the gain k happens to be positive, a positive quantity will feed into the integration 'cycle' until the error is eliminated. In this sense, the filter is acting like a proportional feedback controller, attempting to eliminate the difference between h' (as setpoint) and $\hat{h}$ (as feedback). The gain k need not be large – just gradually eliminating offsets. Then, if the error $[h' - \hat{h}]$ suddenly becomes larger, one can use this fact to confirm a fault in the system based on 'unexpected behaviour'. Moreover, the original measurement h' might be noisy, so that $\hat{h}$ will more closely represent its expected value.

A simple filter like this could be tuned by trial-and-error adjustment of k to get the desired behaviour. However, the determination of a 'good' gain matrix $\boldsymbol{K}$ for more complex MIMO systems is not that easy. The *Kalman–Bucy filter* (Kalman and Bucy, 1961) is the specific case of the continuous system in Equation 6.45 for which the gain matrix $\boldsymbol{K}$ has been determined to optimally compensate for random errors in the observation $\boldsymbol{y}'$ and in the model. In the following sections, the optimal $\boldsymbol{K}$ will first be developed for discrete linear systems, and then continuous linear systems.

6.4.1 Discrete Kalman Filter

It is easier to develop the ideas of optimal state estimation using discrete linear systems, before proceeding to continuous systems. Following from Equation 3.285, the discrete equations for the states and observations equivalent to Equations 6.42 and 6.43 are

$$\boldsymbol{x}_i = \boldsymbol{A}_i\boldsymbol{x}_{i-1} + \boldsymbol{B}_i\boldsymbol{u}'_{i-1} + \boldsymbol{\mu}_{i-1} \tag{6.49}$$

$$\boldsymbol{y}'_i = \boldsymbol{C}_i\boldsymbol{x}_i + \boldsymbol{\nu}_{i-1} \tag{6.50}$$

This form allows for the more general case where the coefficients might change in time. The model now includes a random variable vector $\boldsymbol{\mu}$ to reflect the fact that it cannot accurately predict the next true state from the present true state (i.e. there is model error), whilst the observation includes a random variable vector $\boldsymbol{\nu}$ reflecting that the actual measurement deviates from the correct values determined by the true state, owing to measurement inaccuracy. The *Kalman filter* is based on the following requirements with respect to the terms in these random error vectors:

1) $\boldsymbol{\mu}$ and $\boldsymbol{\nu}$ are uncorrelated;
2) $\boldsymbol{\mu}$ and $\boldsymbol{\nu}$ have Gaussian distributions;
3) $\boldsymbol{\mu}$ and $\boldsymbol{\nu}$ both have zero means ($E\{\boldsymbol{\mu}\} = \boldsymbol{0}$, $E\{\boldsymbol{\nu}\} = \boldsymbol{0}$);
4) estimates of the covariances of each error vector are stored as $\boldsymbol{Q} = E\{\boldsymbol{\mu}\boldsymbol{\mu}^{\mathrm{T}}\}$, $\boldsymbol{R} = E\{\boldsymbol{\nu}\boldsymbol{\nu}^{\mathrm{T}}\}$.

To develop the Kalman filter, one has to break each step of the calculation of the estimated state $\hat{\boldsymbol{x}}$ into two phases:

a) *Prediction of the present state based on the previous updated state estimate*:

$$\hat{\boldsymbol{x}}_{i|i-1} = \boldsymbol{A}_i\hat{\boldsymbol{x}}_{i-1|i-1} + \boldsymbol{B}_i\boldsymbol{u}'_{i-1} \tag{6.51}$$

b) *Updating of the prediction of the present state, using the present measurement error*:

$$\hat{\boldsymbol{x}}_{i|i} = \hat{\boldsymbol{x}}_{i|i-1} + \boldsymbol{K}_i\left[\boldsymbol{y}'_i - \boldsymbol{C}_i\hat{\boldsymbol{x}}_{i|i-1}\right] \tag{6.52}$$

Subtracting Equation 6.51 from Equation 6.49, one has the present predicted error based on the previous updated error:

$$\boldsymbol{e}_{i|i-1} = \boldsymbol{A}_i \boldsymbol{e}_{i-1|i-1} + \boldsymbol{\mu}_{i-1} \tag{6.53}$$

where $\boldsymbol{e}_{i|i-1} = \boldsymbol{x}_i - \hat{\boldsymbol{x}}_{i|i-1}$ and $\boldsymbol{e}_{i-1|i-1} = \boldsymbol{x}_{i-1} - \hat{\boldsymbol{x}}_{i-1|i-1}$.

Because the $\boldsymbol{\mu}$ are not correlated with the $\boldsymbol{e}$, the covariances of the terms in Equation 6.53 are additive, that is using expectations E:

$$E\left\{\left(\boldsymbol{e}_{i|i-1}\right)\left(\boldsymbol{e}_{i|i-1}\right)^{\mathrm{T}}\right\} = E\left\{\left(\boldsymbol{A}_i \boldsymbol{e}_{i-1|i-1}\right)\left(\boldsymbol{A}_i \boldsymbol{e}_{i-1|i-1}\right)^{\mathrm{T}}\right\} + E\left\{\boldsymbol{\mu}_{i-1}\boldsymbol{\mu}_{i-1}^{\mathrm{T}}\right\} \tag{6.54}$$

that is

$$\boldsymbol{P}_{i|i-1} = \boldsymbol{A}_i \boldsymbol{P}_{i-1|i-1} \boldsymbol{A}_i^{\mathrm{T}} + \boldsymbol{Q} \tag{6.55}$$

where

$$\boldsymbol{P}_{i|i-1} = E\left\{\boldsymbol{e}_{i|i-1}\boldsymbol{e}_{i|i-1}^{\mathrm{T}}\right\}, \quad \boldsymbol{P}_{i-1|i-1} = E\left\{\boldsymbol{e}_{i-1|i-1}\boldsymbol{e}_{i-1|i-1}^{\mathrm{T}}\right\} \tag{6.56}$$

Note that $\boldsymbol{P}_{i|i-1}$ is the estimate of the present covariance matrix of the model error, before updating. Now consider this matrix instead based on $\hat{\boldsymbol{x}}_{i|i}$, that is *after* the updating by Equation 6.52:

$$\boldsymbol{P}_{i|i} = E\left\{\boldsymbol{e}_{i|i}\boldsymbol{e}_{i|i}^{\mathrm{T}}\right\} \tag{6.57}$$

$$\boldsymbol{P}_{i|i} = E\left\{\left(\boldsymbol{x}_i - \hat{\boldsymbol{x}}_{i|i}\right)\left(\boldsymbol{x}_i - \hat{\boldsymbol{x}}_{i|i}\right)^{\mathrm{T}}\right\} \tag{6.58}$$

$$\boldsymbol{P}_{i|i} = E\left\{\left(\boldsymbol{x}_i - \hat{\boldsymbol{x}}_{i|i-1} - \boldsymbol{K}_i\left[\boldsymbol{y}_i - \boldsymbol{C}_i\hat{\boldsymbol{x}}_{i|i-1}\right]\right)\left(\boldsymbol{x}_i - \hat{\boldsymbol{x}}_{i|i-1} - \boldsymbol{K}_i\left[\boldsymbol{y}'_i - \boldsymbol{C}_i\hat{\boldsymbol{x}}_{i|i-1}\right]\right)^{\mathrm{T}}\right\} \tag{6.59}$$

Using Equation 6.50, this is

$$\boldsymbol{P}_{i|i} = E\left\{\left(\boldsymbol{x}_i - \hat{\boldsymbol{x}}_{i|i-1} - \boldsymbol{K}_i\left[\boldsymbol{C}_i\boldsymbol{x}_i + \boldsymbol{v}_{i-1} - \boldsymbol{C}_i\hat{\boldsymbol{x}}_{i|i-1}\right]\right)\left(\boldsymbol{x}_i - \hat{\boldsymbol{x}}_{i|i-1} - \boldsymbol{K}_i\left[\boldsymbol{C}_i\boldsymbol{x}_i + \boldsymbol{v}_{i-1} - \boldsymbol{C}_i\hat{\boldsymbol{x}}_{i|i-1}\right]\right)^{\mathrm{T}}\right\} \tag{6.60}$$

$$\boldsymbol{P}_{i|i} = E\left\{\left[(\boldsymbol{I} - \boldsymbol{K}_i\boldsymbol{C}_i)(\boldsymbol{x}_i - \hat{\boldsymbol{x}}_{i|i-1}) - \boldsymbol{K}_i\boldsymbol{v}_{i-1}\right]\left[\left(\boldsymbol{x}_i - \hat{\boldsymbol{x}}_{i|i-1}\right)^{\mathrm{T}}(\boldsymbol{I} - \boldsymbol{K}_i\boldsymbol{C}_i)^{\mathrm{T}} - \boldsymbol{v}_{i-1}^{\mathrm{T}}\boldsymbol{K}_i^{\mathrm{T}}\right]\right\} \tag{6.61}$$

$$\boldsymbol{P}_{i|i} = (\boldsymbol{I} - \boldsymbol{K}_i\boldsymbol{C}_i)E\left\{\left(\boldsymbol{x}_i - \hat{\boldsymbol{x}}_{i|i-1}\right)\left(\boldsymbol{x}_i - \hat{\boldsymbol{x}}_{i|i-1}\right)^{\mathrm{T}}\right\}(\boldsymbol{I} - \boldsymbol{K}_i\boldsymbol{C}_i)^{\mathrm{T}} + \boldsymbol{K}_i E\left\{\boldsymbol{v}_{i-1}\boldsymbol{v}_{i-1}^{\mathrm{T}}\right\}\boldsymbol{K}_i^{\mathrm{T}} \tag{6.62}$$

Because the cross-terms are uncorrelated, and thus have expectations of zero, they have disappeared. It follows that

$$\boldsymbol{P}_{i|i} = (\boldsymbol{I} - \boldsymbol{K}_i\boldsymbol{C}_i)\boldsymbol{P}_{i|i-1}(\boldsymbol{I} - \boldsymbol{K}_i\boldsymbol{C}_i)^{\mathrm{T}} + \boldsymbol{K}_i\boldsymbol{R}\boldsymbol{K}_i^{\mathrm{T}} \tag{6.63}$$

Expanding Equation 6.63, one obtains

$$\boldsymbol{P}_{i|i} = \boldsymbol{P}_{i|i-1} - \boldsymbol{K}_i\boldsymbol{C}_i\boldsymbol{P}_{i|i-1} - \boldsymbol{P}_{i|i-1}\boldsymbol{C}_i^{\mathrm{T}}\boldsymbol{K}_i^{\mathrm{T}} + \boldsymbol{K}_i\left(\boldsymbol{C}_i\boldsymbol{P}_{i|i-1}\boldsymbol{C}_i^{\mathrm{T}} + \boldsymbol{R}\right)\boldsymbol{K}_i^{\mathrm{T}} \tag{6.64}$$

Now one seeks the particular gain $\boldsymbol{K}_i$ that will minimise the sum of square deviations of the state estimates from their true values, that is minimise

$$J = E\left\{(\boldsymbol{x}_i - \hat{\boldsymbol{x}}_i)^{\mathrm{T}}(\boldsymbol{x}_i - \hat{\boldsymbol{x}}_i)\right\} \tag{6.65}$$

It happens that this is the sum of the terms on the diagonal of $\boldsymbol{P}_{i|i}$, that is the *trace* of $\boldsymbol{P}_{i|i}$. Some known results for differentiation of *traces* with respect to matrices are

$$\frac{\partial\,\mathrm{tr}(\boldsymbol{AB})}{\partial \boldsymbol{A}} = \frac{\partial\,\mathrm{tr}\left(\boldsymbol{B}^{\mathrm{T}}\boldsymbol{A}^{\mathrm{T}}\right)}{\partial \boldsymbol{A}} = \boldsymbol{B}^{\mathrm{T}} \quad \text{and} \quad \frac{\partial\,\mathrm{tr}\left(\boldsymbol{ABA}^{\mathrm{T}}\right)}{\partial \boldsymbol{A}} = \boldsymbol{A}\left(\boldsymbol{B}^{\mathrm{T}} + \boldsymbol{B}\right) \tag{6.66}$$

It follows that

$$\frac{\partial J}{\partial \boldsymbol{K}_i} = \frac{\partial\,\mathrm{tr}\left(\boldsymbol{P}_{i|i}\right)}{\partial \boldsymbol{K}_i} = -\boldsymbol{P}_{i|i-1}^{\mathrm{T}}\boldsymbol{C}_i^{\mathrm{T}} - \boldsymbol{P}_{i|i-1}\boldsymbol{C}_i^{\mathrm{T}} + \boldsymbol{K}_i\left[\left(\boldsymbol{C}_i\boldsymbol{P}_{i|i-1}^{\mathrm{T}}\boldsymbol{C}_i^{\mathrm{T}} + \boldsymbol{R}^{\mathrm{T}}\right) + \left(\boldsymbol{C}_i\boldsymbol{P}_{i|i-1}\boldsymbol{C}_i^{\mathrm{T}} + \boldsymbol{R}\right)\right] \tag{6.67}$$

It is clear from Equation 6.56 that $\boldsymbol{P}_{i|i-1}$ must be symmetric, and the same applies to $\boldsymbol{R}$. Thus, setting the result to the zero matrix in order to minimise J, one obtains

$$-\boldsymbol{P}_{i|i-1}\boldsymbol{C}_i^{\mathrm{T}} + \boldsymbol{K}_i\left[\boldsymbol{C}_i\boldsymbol{P}_{i|i-1}\boldsymbol{C}_i^{\mathrm{T}} + \boldsymbol{R}\right] = \boldsymbol{0} \tag{6.68}$$

so that the optimal gain matrix in the *Kalman filter* is given by

$$\boldsymbol{K}_i = \boldsymbol{P}_{i|i-1}\boldsymbol{C}_i^{\mathrm{T}}\left[\boldsymbol{C}_i\boldsymbol{P}_{i|i-1}\boldsymbol{C}_i^{\mathrm{T}} + \boldsymbol{R}\right]^{-1} \tag{6.69}$$

Substitution of this optimal gain matrix into Equation 6.64 yields

$$\boldsymbol{P}_{i|i} = \boldsymbol{P}_{i|i-1} - \boldsymbol{P}_{i|i-1}\boldsymbol{C}_i^{\mathrm{T}}\left[\boldsymbol{C}_i\boldsymbol{P}_{i|i-1}\boldsymbol{C}_i^{\mathrm{T}} + \boldsymbol{R}\right]^{-1}\boldsymbol{C}_i\boldsymbol{P}_{i|i-1} \tag{6.70}$$

and assuming Equation 6.69 has already been evaluated, computation can be reduced by back-substituting $\boldsymbol{K}_i$ to obtain

$$\boldsymbol{P}_{i|i} = \boldsymbol{P}_{i|i-1} - \boldsymbol{K}_i\boldsymbol{C}_i\boldsymbol{P}_{i|i-1} = [\boldsymbol{I} - \boldsymbol{K}_i\boldsymbol{C}_i]\boldsymbol{P}_{i|i-1} \tag{6.71}$$

In summary then, from Equations 6.55, 6.69, 6.22, 6.51 and 6.71, the execution of the *Kalman filter* on each time step i requires the following sequence of assignments:

$$\boxed{\boldsymbol{P}_{i|i-1} = \boldsymbol{A}_i\boldsymbol{P}_{i-1|i-1}\boldsymbol{A}_i^{\mathrm{T}} + \boldsymbol{Q} \quad \text{(predicted)}} \tag{6.72}$$

$$\boxed{\boldsymbol{K}_i = \boldsymbol{P}_{i|i-1}\boldsymbol{C}_i^{\mathrm{T}}\left[\boldsymbol{C}_i\boldsymbol{P}_{i|i-1}\boldsymbol{C}_i^{\mathrm{T}} + \boldsymbol{R}\right]^{-1}} \tag{6.73}$$

$$\boxed{\hat{\boldsymbol{x}}_{i|i-1} = \boldsymbol{A}_i\hat{\boldsymbol{x}}_{i-1|i-1} + \boldsymbol{B}_i\boldsymbol{u}'_{i-1} \quad \text{(predicted)}} \tag{6.74}$$

$$\boxed{\hat{\boldsymbol{x}}_{i|i} = \hat{\boldsymbol{x}}_{i|i-1} + \boldsymbol{K}_i\left[\boldsymbol{y}'_i - \boldsymbol{C}_i\hat{\boldsymbol{x}}_{i|i-1}\right] \quad \text{(updated)}} \tag{6.75}$$

$$\boxed{\boldsymbol{P}_{i|i} = [\boldsymbol{I} - \boldsymbol{K}_i\boldsymbol{C}_i]\boldsymbol{P}_{i|i-1} \quad \text{(updated)}} \tag{6.76}$$

which then cycles back to Equation 6.72 for the next i step (Figure 6.11).

Another way of viewing the $\boldsymbol{P}-\boldsymbol{K}$ recursion is by elimination of the intermediate term $\boldsymbol{P}_{i|i-1}$ to obtain an independent updating equation for $\boldsymbol{P}_{i|i}$:

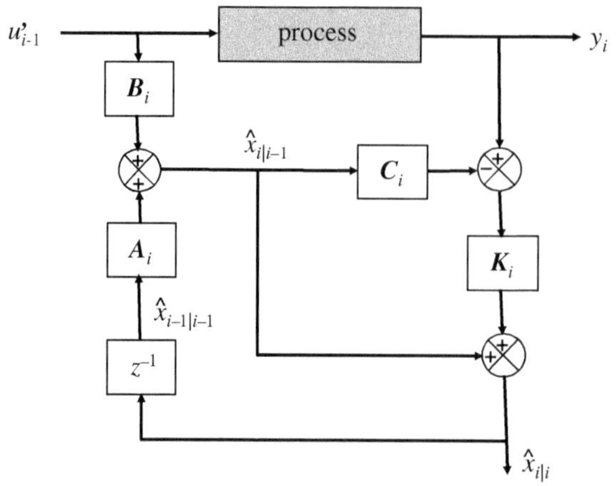

Figure 6.11 Block diagram representation of discrete Kalman filter.

$$\begin{aligned} P_{i|i} &= \left[A_i P_{i-1|i-1} A_i^{\mathrm{T}} + Q\right] - \left[A_i P_{i-1|i-1} A_i^{\mathrm{T}} + Q\right] C_i^{\mathrm{T}} \left\{C_i \left[A_i P_{i-1|i-1} A_i^{\mathrm{T}} + Q\right] C_i^{\mathrm{T}} + R\right\}^{-1} \\ &\quad \times C_i \left[A_i P_{i-1|i-1} A_i^{\mathrm{T}} + Q\right] \\ K_i &= \left[A_i P_{i-1|i-1} A_i^{\mathrm{T}} + Q\right] C_i^{\mathrm{T}} \left\{C_i \left[A_i P_{i-1|i-1} A_i^{\mathrm{T}} + Q\right] C_i^{\mathrm{T}} + R\right\}^{-1} \end{aligned} \tag{6.77}$$

More succinctly, letting

$$M_i = \left[A_i P_{i|i} A_i^{\mathrm{T}} + Q\right] \tag{6.78}$$

use

$$\boxed{K_i = M_{i-1} C_i^{\mathrm{T}} \left\{C_i M_{i-1} C_i^{\mathrm{T}} + R\right\}^{-1}} \tag{6.79}$$

$$\boxed{M_i = A_i \{I - K_i C_i\} M_{i-1} A_i^{\mathrm{T}} + Q} \tag{6.80}$$

For constant system matrices A and C, M (and thus P and K) will approach steady values determined by the *algebraic Riccati equation*

$$M = AM\left\{I - C^{\mathrm{T}}\left[CMC^{\mathrm{T}} + R\right]^{-1} CM\right\} A^{\mathrm{T}} + Q \tag{6.81}$$

The complete stepwise calculation sequence for the discrete *Kalman filter* is illustrated in Figure 6.12.

The model error covariance matrix $Q = E\{\mu\mu^{\mathrm{T}}\}$ and the measurement error covariance matrix $R = E\{\nu\nu^{\mathrm{T}}\}$ determine the 'tuning' of the Kalman filter. Large values in Q or small values in R imply that larger errors are expected in the model than the measurements, so the result $C\hat{x}$ should track the measurements y' more closely. Conversely, small Q values or large R values display a lack of confidence in the measurements, so the model will be satisfied more closely with poor tracking of measurements. The 'tracking gain' of the filter is thus determined by the 'ratio' Q/R. If it is required to start with a high gain K, P is initialised 'large', for example $1000I$ or conversely $0.1I$ for a low initial gain.

In practice, little is known about the cross-correlation of the terms in either error vector μ or ν, so it is common to take the matrices Q and R as diagonal, with an estimated variance of error for each term in order down the diagonal. At the outset, such terms will normalise by taking care of

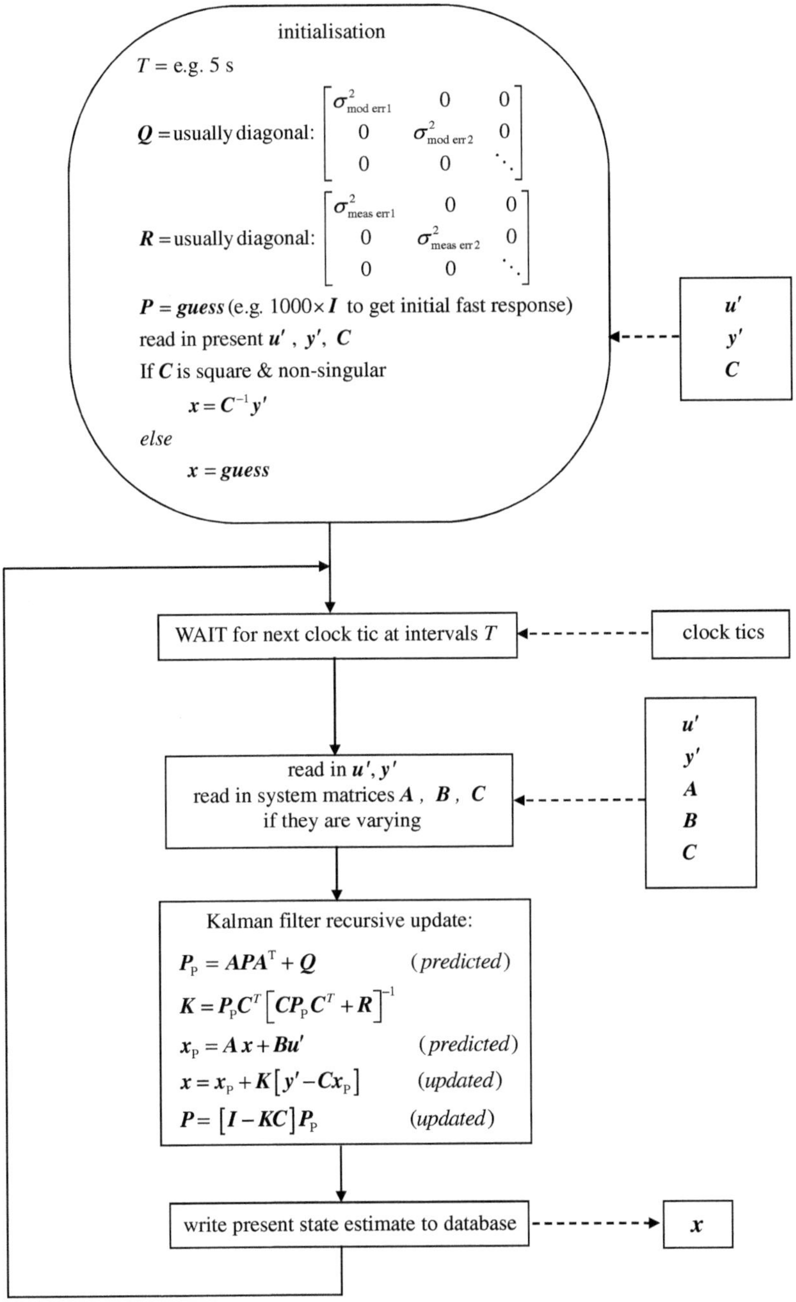

Figure 6.12 Real-time implementation of a Kalman filter allowing for varying system matrices (note: '=' implies assignment).

the different units of each variable, and may be sized simply to reflect the normal operating span, say $(35\,°\mathrm{C})^2$ for operation between 60 and 95 °C. Hopefully, more is known about the measurement (or model) error, say 5% of operating range, in which case a variance of $(0.05 \times 35\,°\mathrm{C})^2$ would be used instead.

Example 6.1

Kalman filter for observation of tank levels.

Two non-interacting tanks are arranged as in Figure 6.13. For simplicity, the exit flows are taken as proportional to level. Only the inlet flow and the level in the second tank are available as measurements.

Equations 6.49 and 6.50 have been determined as

$$\begin{pmatrix} h_1 \\ h_2 \end{pmatrix}_i = \begin{bmatrix} \frac{1}{2} & 0 \\ \frac{1}{4} & \frac{1}{2} \end{bmatrix} \begin{pmatrix} h_1 \\ h_2 \end{pmatrix}_{i-1} + \begin{bmatrix} 2 \\ 0 \end{bmatrix} f + \begin{pmatrix} \mu_1 \\ \mu_2 \end{pmatrix}_{i-1} \quad (6.82)$$

$$h_{2\mathrm{MEAS}\,i} = \begin{bmatrix} 0 & 1 \end{bmatrix} \begin{pmatrix} h_1 \\ h_2 \end{pmatrix}_i + (\nu)_{i-1} \quad (6.83)$$

Taking

$$\boldsymbol{Q} = \begin{bmatrix} 10 & 0 \\ 0 & 1 \end{bmatrix} \quad (6.84)$$

$$\boldsymbol{R} = [10] \quad (6.85)$$

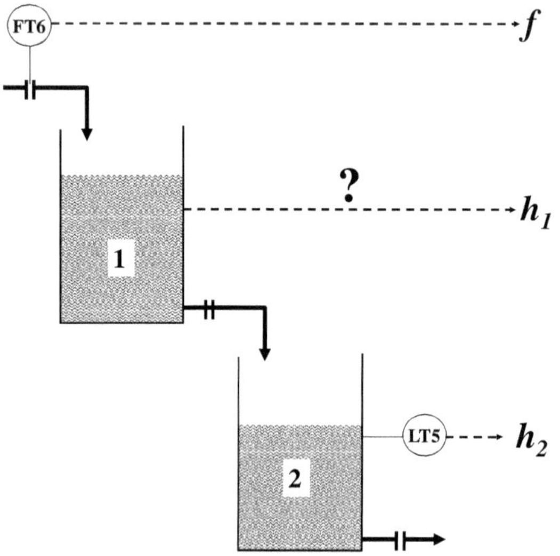

Figure 6.13 Flow through two non-interacting tanks for which only the feed and final levels are measured.

and following the assignments in Figure 6.12, the necessary filter equations are

$$\boldsymbol{P}_P = \begin{bmatrix} \frac{1}{2} & 0 \\ \frac{1}{4} & \frac{1}{2} \end{bmatrix} \boldsymbol{P} \begin{bmatrix} \frac{1}{2} & \frac{1}{4} \\ 0 & \frac{1}{2} \end{bmatrix} + \begin{bmatrix} 10 & 0 \\ 0 & 1 \end{bmatrix} \tag{6.86}$$

$$\boldsymbol{K} = \boldsymbol{P}_P \begin{bmatrix} 0 \\ 1 \end{bmatrix} \left[\begin{bmatrix} 0 & 1 \end{bmatrix} \boldsymbol{P}_P \begin{bmatrix} 0 \\ 1 \end{bmatrix} + 10 \right]^{-1} \tag{6.87}$$

$$\begin{pmatrix} h_1 \\ h_2 \end{pmatrix}_P = \begin{bmatrix} \frac{1}{2} & 0 \\ \frac{1}{4} & \frac{1}{2} \end{bmatrix} \begin{pmatrix} h_1 \\ h_2 \end{pmatrix} + \begin{bmatrix} 2 \\ 0 \end{bmatrix} f \tag{6.88}$$

$$\begin{pmatrix} h_1 \\ h_2 \end{pmatrix} = \begin{pmatrix} h_1 \\ h_2 \end{pmatrix}_P + \boldsymbol{K} \left[h_{2\text{MEAS}} - \begin{bmatrix} 0 & 1 \end{bmatrix} \begin{pmatrix} h_1 \\ h_2 \end{pmatrix}_P \right] \tag{6.89}$$

$$\boldsymbol{P} = \left[\boldsymbol{I} - \boldsymbol{K} \begin{bmatrix} 0 & 1 \end{bmatrix} \right] \boldsymbol{P}_P \tag{6.90}$$

and one could start with

$$\boldsymbol{P} = \begin{bmatrix} 0.0001 & 0 \\ 0 & 0.0001 \end{bmatrix} \tag{6.91}$$

The basic model

$$\begin{pmatrix} h_1 \\ h_2 \end{pmatrix} = \begin{bmatrix} \frac{1}{2} & 0 \\ \frac{1}{4} & \frac{1}{2} \end{bmatrix} \begin{pmatrix} h_1 \\ h_2 \end{pmatrix} + \begin{bmatrix} 2 \\ 0 \end{bmatrix} f \tag{6.92}$$

was run with a step in f from 50 to 70 at $t = 40$. Two zero-mean time-correlated random errors were added to this h_2 signal to yield the *measurement* $h_{2\text{MEAS}}$ according to Equation 6.83, the one error with a short time constant and the other with a long time constant (see Section 3.10). Using the above filtering equations, the behaviour in Figure 6.14 was obtained for three different settings of $\boldsymbol{R}$. As expected, reduction of $\boldsymbol{R}$ causes closer tracking of the measurement, simultaneously causing more sympathetic variation in the unmeasured state h_1 in order to maintain consistency within the model.

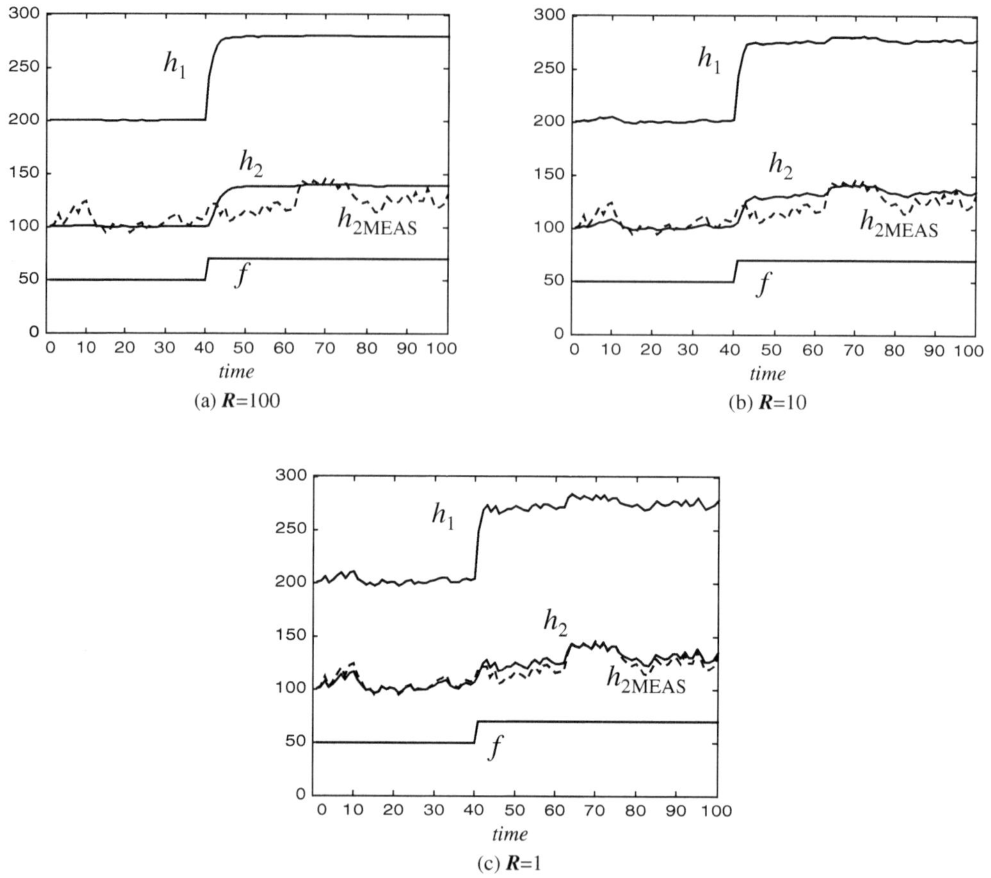

Figure 6.14 Tracking of h_2 measurement, and prediction of unmeasured h_1, for two-tank example in Figure 6.13, for decreasing measurement error variance $\boldsymbol{R}$.

6.4.2
Continuous Kalman–Bucy Filter

The above result for the discrete Kalman filter will be used in a plausible development of the continuous-time *Kalman–Bucy* filter below. Here one is interested in the linear model and measurement relationships of Equations 6.42 and 6.43:

$$\frac{\mathrm{d}\boldsymbol{x}(t)}{\mathrm{d}t} = \boldsymbol{A}(t)\boldsymbol{x}(t) + \boldsymbol{B}(t)\boldsymbol{u}'(t) + \boldsymbol{\mu}(t) \tag{6.93}$$

$$\boldsymbol{y}'(t) = \boldsymbol{C}(t)\boldsymbol{x}(t) + \boldsymbol{v}(t) \tag{6.94}$$

where the variables have the same meaning as in Section 6.4.1, except now in continuous time. Recall that $\boldsymbol{u}'$ and $\boldsymbol{y}'$ are the measured process input and output, respectively. One seeks an optimal estimation $\hat{\boldsymbol{x}}(t)$ of the true state $\boldsymbol{x}(t)$ using the form

$$\frac{\mathrm{d}\hat{\boldsymbol{x}}(t)}{\mathrm{d}t} = \boldsymbol{A}(t)\hat{\boldsymbol{x}}(t) + \boldsymbol{B}(t)\boldsymbol{u}'(t) + \boldsymbol{K}(t)[\boldsymbol{y}'(t) - \boldsymbol{C}(t)\hat{\boldsymbol{x}}(t)] \tag{6.95}$$

If a small step dt is considered, this is

$$\hat{\boldsymbol{x}}(t + \mathrm{d}t) = [\boldsymbol{I} + \mathrm{d}t\,\boldsymbol{A}(t)]\hat{\boldsymbol{x}}(t) + \mathrm{d}t\,\boldsymbol{B}(t)\boldsymbol{u}'(t) + \boldsymbol{K}(t)[\mathrm{d}t\,\boldsymbol{y}'(t) - \mathrm{d}t\,\boldsymbol{C}(t)\hat{\boldsymbol{x}}(t)] \tag{6.96}$$

which has the same form as the discrete Kalman filter (Equations 6.51 and 6.52). The error corresponding to the first two terms on the right now becomes dt $\boldsymbol{\mu}$ (from Equation 6.93), so the corresponding error covariance matrix becomes $(\mathrm{d}t)^{-2}$ times the discrete filter '$\boldsymbol{Q}$', whilst $\boldsymbol{R}$ remains the same.

Substitution of Equation 6.55 into Equation 6.69, and replacing the discrete matrices according to Equation 6.96, one obtains

$$\boldsymbol{K} = \left\{[\boldsymbol{I} + \mathrm{d}t\,\boldsymbol{A}]\boldsymbol{P}[\boldsymbol{I} + \mathrm{d}t\,\boldsymbol{A}]^{\mathrm{T}} + \mathrm{d}t\,\boldsymbol{Q}\right\}\mathrm{d}t\,\boldsymbol{C}^{\mathrm{T}}\left[(\mathrm{d}t)^2\boldsymbol{C}\left\{[\boldsymbol{I} + \mathrm{d}t\,\boldsymbol{A}]\boldsymbol{P}[\boldsymbol{I} + \mathrm{d}t\,\boldsymbol{A}]^{\mathrm{T}} + \mathrm{d}t\,\boldsymbol{Q}\right\}\boldsymbol{C}^{\mathrm{T}} + \mathrm{d}t\,\boldsymbol{R}\right]^{-1} \tag{6.97}$$

The terms in the curly brackets are dominated by $\boldsymbol{P}$ for small dt, whilst the $(\mathrm{d}t)^2$ term in the inverse is negligible, so this equation is

$$\boldsymbol{K} = \boldsymbol{P}\boldsymbol{C}^{\mathrm{T}}\boldsymbol{R}^{-1} \tag{6.98}$$

Equations 6.55 and 6.71 combine to give

$$\begin{aligned} \boldsymbol{P}_{t+\mathrm{d}t} &= [\boldsymbol{I} - \boldsymbol{K}\,\mathrm{d}t\,\boldsymbol{C}]\left\{[\boldsymbol{I} + \mathrm{d}t\,\boldsymbol{A}]\boldsymbol{P}_t[\boldsymbol{I} + \mathrm{d}t\,\boldsymbol{A}]^{\mathrm{T}} + \mathrm{d}t\,\boldsymbol{Q}\right\} \\ &= [\boldsymbol{I} - \boldsymbol{K}\,\mathrm{d}t\,\boldsymbol{C}]\left\{\boldsymbol{P}_t + \mathrm{d}t\,\boldsymbol{A}\,\boldsymbol{P}_t + \mathrm{d}t\,\boldsymbol{P}_t\boldsymbol{A}^{\mathrm{T}} + \mathrm{d}t\,\boldsymbol{Q}\right\} \end{aligned} \tag{6.99}$$

where the small $(\mathrm{d}t)^2$ term has been ignored. So the change in $\boldsymbol{P}$ is

$$\mathrm{d}\boldsymbol{P} = -\mathrm{d}t\,\boldsymbol{KCP} + [\boldsymbol{I} - \mathrm{d}t\,\boldsymbol{KC}]\left\{\mathrm{d}t\,\boldsymbol{AP} + \mathrm{d}t\,\boldsymbol{PA}^{\mathrm{T}} + \mathrm{d}t\,\boldsymbol{Q}\right\} \tag{6.100}$$

and again ignoring the small $(\mathrm{d}t)^2$ terms

$$\frac{\mathrm{d}\boldsymbol{P}}{\mathrm{d}t} = -\boldsymbol{KCP} + \boldsymbol{AP} + \boldsymbol{PA}^{\mathrm{T}} + \boldsymbol{Q} \tag{6.101}$$

Finally, using Equation 6.98, one has

$$\frac{\mathrm{d}\boldsymbol{P}}{\mathrm{d}t} = \boldsymbol{AP} + \boldsymbol{PA}^{\mathrm{T}} - \boldsymbol{PC}^{\mathrm{T}}\boldsymbol{R}^{-1}\boldsymbol{CP} + \boldsymbol{Q} \tag{6.102}$$

where it is understood that the error covariance matrix $\boldsymbol{P}$ is time-varying, and all of the other matrices may be time-varying if so desired. This is the well-known *Riccati differential equation*.

In summary then, the *Kalman–Bucy filter* requires simultaneous solution of the following equations:

$$\boxed{\frac{\mathrm{d}\boldsymbol{P}}{\mathrm{d}t} = \boldsymbol{AP} + \boldsymbol{PA}^{\mathrm{T}} - \boldsymbol{PC}^{\mathrm{T}}\boldsymbol{R}^{-1}\boldsymbol{CP} + \boldsymbol{Q}} \tag{6.103}$$

$$\boxed{\boldsymbol{K} = \boldsymbol{PC}^{\mathrm{T}}\boldsymbol{R}^{-1}} \tag{6.104}$$

$$\boxed{\frac{\mathrm{d}\hat{\boldsymbol{x}}}{\mathrm{d}t} = \boldsymbol{A}\hat{\boldsymbol{x}} + \boldsymbol{B}\boldsymbol{u}' + \boldsymbol{K}[\boldsymbol{y}' - \boldsymbol{C}\hat{\boldsymbol{x}}]} \tag{6.105}$$

where it is understood that the matrices $\boldsymbol{P}$ and $\boldsymbol{K}$ and the vectors $\hat{\boldsymbol{x}}$, $\boldsymbol{u}'$ and $\boldsymbol{y}'$ are time-varying, and all of the other matrices may be time-varying if so desired. The solution has to start with guesses of $\hat{\boldsymbol{x}}(0)$ and $\boldsymbol{P}(0)$. For $\boldsymbol{P}(0)$, a diagonal matrix with suitably large values on the axis may be used, to ensure a rapid response at start-up.

For constant $\boldsymbol{A}$, $\boldsymbol{C}$, $\boldsymbol{R}$ and $\boldsymbol{Q}$ matrices, the solution of the Riccati differential equation will reach a steady state determined by

$$\boldsymbol{AP} + \boldsymbol{PA}^{\mathrm{T}} - \boldsymbol{PC}^{\mathrm{T}}\boldsymbol{R}^{-1}\boldsymbol{CP} + \boldsymbol{Q} = \boldsymbol{0} \tag{6.106}$$

which is the *algebraic Riccati equation* and is observed to be quadratic in matrix $\boldsymbol{P}$. In Section 7.6.6, it will be seen how Riccati equations very similar to Equations 6.103 and 6.106 arise in connection with the *linear quadratic regulator* (LQR). A number of solution techniques have been developed for Equation 6.106. The recursive numerical method of Kleinman (1968) underlies several of these, and can be briefly explained as follows:

1) choose a $\boldsymbol{K}_0$ such that $[\boldsymbol{A} + \boldsymbol{K}_0\boldsymbol{C}]$ has all eigenvalues with negative real parts (stable filter);
2) define $\boldsymbol{P}_i$, $\boldsymbol{K}_i$ recursively as

$$[\boldsymbol{A} + \boldsymbol{K}_i\boldsymbol{C}]\boldsymbol{P}_i + \boldsymbol{P}_i[\boldsymbol{A} + \boldsymbol{K}_i\boldsymbol{C}]^{\mathrm{T}} = -\boldsymbol{K}_i\boldsymbol{R}\boldsymbol{K}_i^{\mathrm{T}} - \boldsymbol{Q} \tag{6.107}$$

$$\boldsymbol{K}_{i+1} = -\boldsymbol{P}_i\boldsymbol{C}^{\mathrm{T}}\boldsymbol{R}^{-1} \tag{6.108}$$

3) then $\|\boldsymbol{P}_{i+1}\| \le \|\boldsymbol{P}_i\|$ and $\|\boldsymbol{P}_{i+1} - \boldsymbol{P}\| \le c\|\boldsymbol{P}_i - \boldsymbol{P}\|^2$ where $c > 0$ (quadratic convergence based on, for example, the 1-norm: maximum sum of absolute values of any row), and

$$\lim_{i\to\infty} \boldsymbol{P}_i = \boldsymbol{P} \tag{6.109}$$

It is noted that if Equations 6.107 and 6.108 are simultaneously satisfied, then Equation 6.106 will also be satisfied. Although Equation 6.107 is linear in $\boldsymbol{P}_i$, the elements of $\boldsymbol{P}_i$ have to be individually extracted in the solution owing to the left and right multiplications. However, recall that $\boldsymbol{P}_i$ will be symmetric, so just over half of the elements need be solved for. Before embarking on more complex solutions like this, it is worthwhile attempting to simply integrate the $\boldsymbol{P}$ differential equation (Equation 6.103) with a small time step until steady state is achieved.

6.4.3 Extended Kalman Filter

The Kalman (discrete) and Kalman–Bucy (continuous) filters, developed above in terms of linear systems, have been successfully applied to many nonlinear systems by repeated online re-linearisation to find local estimates of the system coefficient matrices $\boldsymbol{A}$, $\boldsymbol{B}$ and $\boldsymbol{C}$. Effectively, one pretends that the variations experienced are time-dependent, namely $\boldsymbol{A}(t)$, $\boldsymbol{B}(t)$ and $\boldsymbol{C}(t)$, instead of state-dependent. This approximation is adequate in many situations, but will clearly be problematic in the case of strong nonlinearity.

The procedure is similar for discrete and continuous filters, so this discussion will just focus on the continuous filter. The general system of interest involves the state $\boldsymbol{x}$, input $\boldsymbol{u}$ and measurement $\boldsymbol{y}$ as follows:

$$\frac{\mathrm{d}\boldsymbol{x}}{\mathrm{d}t} = \boldsymbol{f}(\boldsymbol{x}, \boldsymbol{u}) \tag{6.110}$$

$$\boldsymbol{y} = \boldsymbol{g}(\boldsymbol{x}) \tag{6.111}$$

More complex measurement dependence (e.g. $\boldsymbol{g}(\boldsymbol{y}, \boldsymbol{x}, \boldsymbol{u}) = 0$) can be handled if necessary as in Section 3.7. Following that method, define the Jacobian matrices for a system with n states, m inputs and p measurements,

$$\boldsymbol{A}(\boldsymbol{x}, \boldsymbol{u}) = \begin{bmatrix} \frac{\partial f_1(\boldsymbol{x}, \boldsymbol{u})}{\partial x_1} & \frac{\partial f_1(\boldsymbol{x}, \boldsymbol{u})}{\partial x_2} & \cdots & \frac{\partial f_1(\boldsymbol{x}, \boldsymbol{u})}{\partial x_n} \\ \frac{\partial f_2(\boldsymbol{x}, \boldsymbol{u})}{\partial x_1} & \frac{\partial f_2(\boldsymbol{x}, \boldsymbol{u})}{\partial x_2} & \cdots & \frac{\partial f_2(\boldsymbol{x}, \boldsymbol{u})}{\partial x_n} \\ \vdots & \vdots & \ddots & \vdots \\ \frac{\partial f_n(\boldsymbol{x}, \boldsymbol{u})}{\partial x_1} & \frac{\partial f_n(\boldsymbol{x}, \boldsymbol{u})}{\partial x_2} & \cdots & \frac{\partial f_n(\boldsymbol{x}, \boldsymbol{u})}{\partial x_n} \end{bmatrix} \tag{6.112}$$

$$\boldsymbol{B}(\boldsymbol{x}, \boldsymbol{u}) = \begin{bmatrix} \frac{\partial f_1(\boldsymbol{x}, \boldsymbol{u})}{\partial u_1} & \frac{\partial f_1(\boldsymbol{x}, \boldsymbol{u})}{\partial u_2} & \cdots & \frac{\partial f_1(\boldsymbol{x}, \boldsymbol{u})}{\partial u_m} \\ \frac{\partial f_2(\boldsymbol{x}, \boldsymbol{u})}{\partial u_1} & \frac{\partial f_2(\boldsymbol{x}, \boldsymbol{u})}{\partial u_2} & \cdots & \frac{\partial f_2(\boldsymbol{x}, \boldsymbol{u})}{\partial u_m} \\ \vdots & \vdots & \ddots & \vdots \\ \frac{\partial f_n(\boldsymbol{x}, \boldsymbol{u})}{\partial u_1} & \frac{\partial f_n(\boldsymbol{x}, \boldsymbol{u})}{\partial u_2} & \cdots & \frac{\partial f_n(\boldsymbol{x}, \boldsymbol{u})}{\partial u_m} \end{bmatrix} \tag{6.113}$$

$$\boldsymbol{C}(\boldsymbol{x}) = \begin{bmatrix} \frac{\partial g_1(\boldsymbol{x})}{\partial x_1} & \frac{\partial g_1(\boldsymbol{x})}{\partial x_2} & \cdots & \frac{\partial g_1(\boldsymbol{x})}{\partial x_n} \\ \frac{\partial g_2(\boldsymbol{x})}{\partial x_1} & \frac{\partial g_2(\boldsymbol{x})}{\partial x_2} & \cdots & \frac{\partial g_2(\boldsymbol{x})}{\partial x_n} \\ \vdots & \vdots & \ddots & \vdots \\ \frac{\partial g_p(\boldsymbol{x})}{\partial x_1} & \frac{\partial g_p(\boldsymbol{x})}{\partial x_2} & \cdots & \frac{\partial g_p(\boldsymbol{x})}{\partial x_n} \end{bmatrix} \tag{6.114}$$

Assuming that small changes will occur in $\boldsymbol{x}$ and $\boldsymbol{u}$ over a 'small' time step $t \to t + \Delta t$, a Taylor expansion of $\boldsymbol{f}$ to the first term gives

$$\left.\frac{\mathrm{d}\boldsymbol{x}}{\mathrm{d}t}\right|_{t \to t+\Delta t} \approx \boldsymbol{A}(\boldsymbol{x}_t, \boldsymbol{u}_t)[\boldsymbol{x} - \boldsymbol{x}_t] + \boldsymbol{B}(\boldsymbol{x}_t, \boldsymbol{u}_t)[\boldsymbol{u} - \boldsymbol{u}_t] + \boldsymbol{f}(\boldsymbol{x}_t, \boldsymbol{u}_t) \tag{6.115}$$

Representing

$$\boldsymbol{A}_t = \boldsymbol{A}(\hat{\boldsymbol{x}}_t, \boldsymbol{u}'_t) \tag{6.116}$$

$$\boldsymbol{B}_t = \boldsymbol{B}(\hat{\boldsymbol{x}}_t, \boldsymbol{u}'_t) \tag{6.117}$$

$$\boldsymbol{d}_t = \boldsymbol{f}(\hat{\boldsymbol{x}}_t, \boldsymbol{u}'_t) - \boldsymbol{A}(\hat{\boldsymbol{x}}_t, \boldsymbol{u}'_t)\boldsymbol{x}_t - \boldsymbol{B}(\hat{\boldsymbol{x}}_t, \boldsymbol{u}'_t)\boldsymbol{u}'_t \tag{6.118}$$

and including the random error contributions, the behaviour in the interval $t \to t + \Delta t$ is thus

$$\left.\frac{d\boldsymbol{x}}{dt}\right|_{t \to t+\Delta t} \approx \boldsymbol{A}_t\boldsymbol{x} + \boldsymbol{B}_t\boldsymbol{u}' + \boldsymbol{d}_t + \boldsymbol{\mu} \tag{6.119}$$

A similar expansion of $\boldsymbol{g}$ gives

$$\boldsymbol{y}' \approx \boldsymbol{C}_t\boldsymbol{x} + \boldsymbol{e}_t + \boldsymbol{v} \tag{6.120}$$

where

$$\boldsymbol{C}_t = \boldsymbol{C}(\hat{\boldsymbol{x}}_t) \tag{6.121}$$

$$\boldsymbol{e}_t = \hat{\boldsymbol{y}}_t - \boldsymbol{C}(\hat{\boldsymbol{x}}_t)\hat{\boldsymbol{x}}_t \tag{6.122}$$

The *extended Kalman filter* (EKF) procedure now follows Equations 6.103–6.105, but on an interval-to-interval basis, by simultaneous integration as τ proceeds from t to $t + \Delta t$,

$$\boxed{\frac{d\boldsymbol{P}(\tau)}{d\tau} = \boldsymbol{A}_t\boldsymbol{P}(\tau) + \boldsymbol{P}(\tau)\boldsymbol{A}_t^{\mathrm{T}} - \boldsymbol{P}(\tau)\boldsymbol{C}_t^{\mathrm{T}}\boldsymbol{R}_t^{-1}\boldsymbol{C}_t\boldsymbol{P}(\tau) + \boldsymbol{Q}_t} \tag{6.123}$$

$$\boxed{\boldsymbol{K}(\tau) = \boldsymbol{P}(\tau)\boldsymbol{C}_t^{\mathrm{T}}\boldsymbol{R}_t^{-1}} \tag{6.124}$$

$$\boxed{\frac{d\hat{\boldsymbol{x}}(\tau)}{d\tau} = \boldsymbol{f}(\hat{\boldsymbol{x}}(\tau), \boldsymbol{u}'(\tau)) + \boldsymbol{K}(\tau)[\boldsymbol{y}'(\tau) - \boldsymbol{C}_t\hat{\boldsymbol{x}}(\tau) - \boldsymbol{e}_t]} \tag{6.125}$$

As each interval is completed, Equations 6.116–6.118 and 6.121–6.122 are re-evaluated at the new state estimate $\hat{\boldsymbol{x}}$, input $\boldsymbol{u}'$ and measurement $\boldsymbol{y}'$, in this way maintaining more appropriate matrices $\boldsymbol{A}$, $\boldsymbol{B}$ and $\boldsymbol{C}$, and vectors $\boldsymbol{d}$ and $\boldsymbol{e}$. An integration of the original nonlinear $\boldsymbol{f}$ functions is indicated, but there may be some benefit in using the approximate form of Equation 6.119.

The interval size to be used between re-evaluations of $\boldsymbol{A}$, $\boldsymbol{B}$, $\boldsymbol{C}$, $\boldsymbol{d}$ and $\boldsymbol{e}$ will depend on how quickly the operating point moves to a region of distinctly different behaviour. So far nothing has been said about the forms of $\boldsymbol{f}$ and $\boldsymbol{g}$. Detailed equations may be available, or a real-time identification may be updating model coefficients or physical parameters, as in Section 6.5.

As far as the *discrete* Kalman filter is concerned (Section 6.4.1), it lends itself directly to the above procedure by periodic updating of $\boldsymbol{A}$, $\boldsymbol{B}$ and $\boldsymbol{C}$ (and $\boldsymbol{d}$ and $\boldsymbol{e}$ for an *absolute* representation) depending on the operating point. If the basis of the discrete model is a continuous model, for example using the matrix exponential as in Section 3.9.3.2, then parameters such as heat transfer coefficients may change with the operating point, requiring re-evaluation of the discrete matrices $\boldsymbol{A}$ and $\boldsymbol{B}$. Alternatively, elements in these matrices may be determined directly by real-time identification (Section 6.5).

6.5 Identification of the Parameters of a Process Model

If one is aware of the general *structure* of a mathematical model describing a process, one may wish to use the process measurements directly to evaluate some or all of the parameters in the model. To draw a parallel, in the consideration of the Kalman filter in Section 6.4, the focus was on estimating the *state variables*. Now consideration will be given to estimating the *coefficients* of the actual model itself, in a real-time fashion, from plant data. Such estimations need not be complex, for example a valve 'gain' ($m^3\ h^{-1}$ [% open]$^{-1}$) might be estimated from the slope of a line connecting the last two different steady-state operating points.

Consider a dynamic model expressed as a system of first-order ODEs involving the states $\boldsymbol{x}$, with a functional representation $\boldsymbol{y}$ available as measurements:

$$\frac{d\boldsymbol{x}(t)}{dt} = \boldsymbol{f}(\boldsymbol{x}(t), \boldsymbol{u}(t)) \tag{6.126}$$

$$\boldsymbol{y}(t) = \boldsymbol{g}(\boldsymbol{x}(t), \boldsymbol{u}(t)) \tag{6.127}$$

There is no guarantee that the model connecting these variables *can* be identified. For example, if the process is operating at steady state, it will be impossible to establish the time constants. If the selection of measurements $\boldsymbol{y}$ is too limited, it may be difficult to extract model details. (Consider *observability* in Section 5.2.2.) Conversely, it is possible that the measurement of a *single* variable could allow estimation of *several* model parameters – if the measurement is rich in information (e.g. strong variability). So one is in general at the mercy of the 'information content' of the measurements.

As written, Equations 6.126 and 6.127 include the possibility of a nonlinear model. Given a batch of measurements of $\boldsymbol{y}(t)$ and $\boldsymbol{u}(t)$ at a sequence of points in time, this constitutes a nonlinear optimisation problem, where one typically wishes to minimise the deviation between model output and actual measurements thereof, in order to determine a 'best-fit' model. A number of algorithms, commercial and otherwise, are available for this job. In this section, the focus will instead be on some of the useful simple results in less complex situations.

There are three main aims in real-time estimation of model parameters:

1) provide an updated model for adjustment of a control or optimisation algorithm (see adaptive control in Section 4.2.8.2 and Chapter 7);
2) provide an updated model for adjustment of a state estimator (e.g. EKF, Section 6.4.3);
3) monitor equipment parameters (e.g. heat transfer coefficient) to detect failures (e.g. fouling).

As will be seen, identification will usually lead to values for lumped terms, involving combinations of actual physical parameters such as transfer coefficients. So some further calculation will normally be required to extract actual physical parameter values in the case of (3) above.

Considerations of model parameter identification will be limited to two important types of discrete linear models:

a) *State-space model*:

$$\boldsymbol{x}_i = \boldsymbol{A}'\boldsymbol{x}_{i-1} + \boldsymbol{B}'\boldsymbol{u}_{i-1} \tag{6.128}$$

$$\boldsymbol{y}_i = \boldsymbol{C}'\boldsymbol{x}_i \tag{6.129}$$

b) *Input–output model*:

$$\mathbf{y}_i = \mathbf{\Lambda}_1 \mathbf{y}_{i-1} + \mathbf{\Lambda}_2 \mathbf{y}_{i-2} + \cdots + \mathbf{\Lambda}_n \mathbf{y}_{i-n} + \boldsymbol{\beta}_1 \mathbf{u}_{i-1} + \boldsymbol{\beta}_2 \mathbf{u}_{i-2} + \cdots + \boldsymbol{\beta}_m \mathbf{u}_{i-m} \tag{6.130}$$

In the input–output model, the matrices $\mathbf{\Lambda}_i$ acting on the earlier values of the output can be arranged to be diagonal. This may be seen by multiplying each row of Equation 6.131 by the common denominator of that row of the matrix. Unlike the left-multiplying scalar $D(z^{-1})$ in state systems, however, the coefficients will vary along each diagonal. So each output element is seen to arise from a limited series of its *own* preceding values, plus a linear combination of *all* input elements at a limited series of preceding times (i.e. the $\boldsymbol{\beta}_i$ are potentially fully populated matrices).

$$\mathbf{y}(z) = \begin{bmatrix} q^{-d_{11}} \dfrac{b_{111}q^{-1} + \cdots + b_{11m_{11}}q^{-m_{11}}}{1 + a_{111}q^{-1} + \cdots + a_{11n_{11}}q^{-n_{11}}} & q^{-d_{12}} \dfrac{b_{121}q^{-1} + \cdots + b_{12m_{12}}q^{-m_{12}}}{1 + a_{121}q^{-1} + \cdots + a_{12n_{12}}q^{-n_{12}}} & \cdots & \cdots \\ q^{-d_{21}} \dfrac{b_{211}q^{-1} + \cdots + b_{21m_{21}}q^{-m_{21}}}{1 + a_{211}q^{-1} + \cdots + a_{21n_{21}}q^{-n_{21}}} & q^{-d_{22}} \dfrac{b_{221}q^{-1} + \cdots + b_{22m_{22}}q^{-m_{22}}}{1 + a_{221}q^{-1} + \cdots + a_{22n_{22}}q^{-n_{22}}} & \cdots & \cdots \\ \vdots & \vdots & \ddots & \cdots \\ \vdots & \vdots & \vdots & \ddots \end{bmatrix} \mathbf{u}(z) \tag{6.131}$$

Input–output models arise by identification of the relationship between each input element and each output element, for example in step tests. Each input–output pair is thus described by the corresponding transfer function element in Equation 6.131. Here the dead-time lag q^{-d} (with d integer) is explicitly extracted as a factor to reduce the number of terms in each dynamic transfer function. Equation 6.130, with a diagonal matrix applied to the output vector at each preceding time, arises when each row is multiplied by its common denominator. The identification of the transfer function in each element of the matrix in Equation 6.131 is seen to be straightforward in controlled tests where the inputs $\mathbf{u}$ can be individually manipulated. Conversely, real-time plant data will yield the *combined* contributions of all inputs, requiring the simultaneous identification of far more coefficients.

For the state-space model (Equation 6.128), one requires measurements of the chosen states $\mathbf{x}_i$ (i.e. $\mathbf{C}' = \mathbf{I}$, or alternatively, $\mathbf{C}'^{-1}$ must exist). Both the state-space model (a) and input–output model (b) can then be represented by an equation of the form

$$\mathbf{A}(q)\mathbf{y}(t) = \mathbf{B}(q)\mathbf{u}(t) \tag{6.132}$$

with

$$\left.\begin{array}{ll} \text{(a) state-space :} & \mathbf{A}(q) = \mathbf{I} - \left[\mathbf{C}'\mathbf{A}'\mathbf{C}'^{-1}\right]q^{-1} \\ & \mathbf{B}(q) = \mathbf{C}'\mathbf{B}'q^{-1} \\ \text{(b) input–output :} & \mathbf{A}(q) = \mathbf{I} - \mathbf{\Lambda}_1 q^{-1} - \cdots - \mathbf{\Lambda}_n q^{-n} \\ & \mathbf{B}(q) = \boldsymbol{\beta}_1 q^{-1} + \cdots + \boldsymbol{\beta}_m q^{-m} \end{array}\right\} \begin{array}{l} \mathbf{A}(q) = \mathbf{I} - \mathbf{A}_1 q^{-1} - \mathbf{A}_2 q^{-2} - \cdots - \mathbf{A}_n q^{-n} \\ \mathbf{B}(q) = \mathbf{B}_1 q^{-1} + \mathbf{B}_2 q^{-2} + \cdots + \mathbf{B}_m q^{-m} \end{array}$$

Recall that q^{-n} represents a backward shift of n intervals in the sampled data sequence. In case (a), only the $\mathbf{A}_1$ and $\mathbf{B}_1$ matrices are required, with potentially all of their elements being nonzero. Conversely, in case (b), the $\mathbf{A}$ matrices are diagonal, but the $\mathbf{A}$ and $\mathbf{B}$ matrices are required for a series of previous $\mathbf{y}$ and $\mathbf{u}$ values.

At time i, both the present measurement $\mathbf{y}_i$ and the previous measurements $\mathbf{y}_{i-1}, \mathbf{y}_{i-2}, \ldots, \mathbf{u}_{i-1}, \mathbf{u}_{i-2}$ are available. Use these to construct partitioned matrices and vectors as follows:

$$\overline{\boldsymbol{A}} = \begin{bmatrix} \boldsymbol{A}_1 & \vdots & \boldsymbol{A}_2 & \vdots & \cdots & \vdots & \boldsymbol{A}_\text{n} \end{bmatrix} \tag{6.133}$$

$$\overline{\boldsymbol{B}} = \begin{bmatrix} \boldsymbol{B}_1 & \vdots & \boldsymbol{B}_2 & \vdots & \cdots & \vdots & \boldsymbol{B}_\text{m} \end{bmatrix} \tag{6.134}$$

$$\overline{\boldsymbol{y}}_{i-1} = \begin{pmatrix} \boldsymbol{y}_{i-1} \\ \hdashline \boldsymbol{y}_{i-2} \\ \hdashline \vdots \\ \hdashline \boldsymbol{y}_{i-n} \end{pmatrix} \tag{6.135}$$

$$\overline{\boldsymbol{u}}_{i-1} = \begin{pmatrix} \boldsymbol{u}_{i-1} \\ \hdashline \boldsymbol{u}_{i-2} \\ \hdashline \vdots \\ \hdashline \boldsymbol{u}_{i-m} \end{pmatrix} \tag{6.136}$$

Now assuming that measurements are available of present (i) and previous ($i-1$) $\overline{\boldsymbol{y}}$ and $\overline{\boldsymbol{u}}$ values, the output prediction error at time i is thus

$$\boldsymbol{e}_i = \boldsymbol{y}_i - \overline{\boldsymbol{A}}\,\overline{\boldsymbol{y}}_{i-1} - \overline{\boldsymbol{B}}\,\overline{\boldsymbol{u}}_{i-1} \tag{6.137}$$

Focusing on a single row of this system, that is a single output,

$$e_i = y_i - \overline{\boldsymbol{a}}^\text{T}\overline{\boldsymbol{y}}_{i-1} - \overline{\boldsymbol{b}}^\text{T}\overline{\boldsymbol{u}}_{i-1} \tag{6.138}$$

where the vectors $\overline{\boldsymbol{a}}$ and $\overline{\boldsymbol{b}}$ are clearly constructed from the appropriate row of $\overline{\boldsymbol{A}}$ and $\overline{\boldsymbol{B}}$. Now for any one row, depending on the form (a) or (b) above, various elements of vector $\overline{\boldsymbol{a}}$ and vector $\overline{\boldsymbol{b}}$ are known to be zero, such as in the case of diagonal matrices. Packing *only the potentially nonzero* coefficients of $\overline{\boldsymbol{a}}$ and $\overline{\boldsymbol{b}}$ into vector $\boldsymbol{\theta}$, and the corresponding measurements from the $\overline{\boldsymbol{y}}_{i-1}$ and $\overline{\boldsymbol{u}}_{i-1}$ vectors into vector $\boldsymbol{\phi}_{\text{i}-1}$, the output prediction error is

$$e_i = y_i - \boldsymbol{\theta}^\text{T}\boldsymbol{\phi}_{i-1} = y_i - \boldsymbol{\phi}_{i-1}^\text{T}\boldsymbol{\theta} \tag{6.139}$$

6.5.1 Model Identification by Least-Squares Fitting to a Batch of Measurements

From Equations 6.133–6.136, the data required to evaluate a single error e_i in Equation 6.139 spans a maximum number of sample times (of $\boldsymbol{y}$ and $\boldsymbol{u}$) stretching back over $m+1$ or $n+1$ points, whichever is greater. Call this number N_{yu}. From Equation 6.132, the number of unknown $\boldsymbol{A}$ and $\boldsymbol{B}$ coefficients in $\boldsymbol{\theta}$ will be $N_\text{O}+N_\text{I}$ for the state-space form (a), and $n+(m\times N_\text{I})$ for the input–output form (b), where N_O and N_I are the numbers of output and input variables of the system, that is the dimension of the matrix in Equation 6.131. Hence, the number of sample times required to start minimising $|e_i|$ by choice of the coefficients in $\boldsymbol{\theta}$ is determined by

a) state-space form: $N\geq N_{yu}+N_\text{O}+N_\text{I}-1$;
b) input–output form: $N\geq N_{yu}+n+(m\times N_\text{I})-1$.

As the number of data sample times increases above these thresholds, one is in a position to improve the estimates of coefficients by 'fitting'. When such a batch of N measurement points is available, a straightforward *least-squares (LS) fit* can be used in the identification of the parameters in $\boldsymbol{\theta}$ above. In the *offline* case, one set of coefficients will be established for the whole record. In the *real-time* case, one may trail a *window* of width N behind the current time, with the new measurement being added to the stack at one end and the oldest falling off at the other end. In this latter case, the identified parameters can of course vary from step to step. One notes further that a separate parameter vector $\boldsymbol{\theta}$ must be found for each output of a multi-output system, but that the form of Equation 6.132 allows these solutions to be performed independently.

One seeks $\boldsymbol{\theta}$ which minimises

$$J(\boldsymbol{\theta}) = \sum_{i=n+1}^{N} e_i^2 = \sum_{i=n+1}^{N} \left[y_i - \boldsymbol{\phi}_{i-1}^{\mathrm{T}}\boldsymbol{\theta}\right]^2 \tag{6.140}$$

Differentiating with respect to each element in $\boldsymbol{\theta}$ and setting to zero for the minimum,

$$\frac{\mathrm{d}J(\boldsymbol{\theta})}{\mathrm{d}\boldsymbol{\theta}} = \sum_{i=n+1}^{N} 2(-\boldsymbol{\phi}_{i-1})\left[y_i - \boldsymbol{\phi}_{i-1}^{\mathrm{T}}\boldsymbol{\theta}\right] = \mathbf{0} \tag{6.141}$$

So

$$\left[\sum_{i=n+1}^{N} \boldsymbol{\phi}_{i-1}\boldsymbol{\phi}_{i-1}^{\mathrm{T}}\right]\boldsymbol{\theta} = \left(\sum_{i=n+1}^{N} y_i\boldsymbol{\phi}_{i-1}\right) \tag{6.142}$$

that is

$$\begin{bmatrix} \sum_{i=n+1}^{N} \phi_{1,i-1}\phi_{1,i-1} & \sum_{i=n+1}^{N} \phi_{2,i-1}\phi_{1,i-1} & \sum_{i=n+1}^{N} \phi_{3,i-1}\phi_{1,i-1} & \cdots \\ \sum_{i=n+1}^{N} \phi_{1,i-1}\phi_{2,i-1} & \sum_{i=n+1}^{N} \phi_{2,i-1}\phi_{2,i-1} & \sum_{i=n+1}^{N} \phi_{3,i-1}\phi_{2,i-1} & \cdots \\ \sum_{i=n+1}^{N} \phi_{1,i-1}\phi_{3,i-1} & \sum_{i=n+1}^{N} \phi_{2,i-1}\phi_{3,i-1} & \sum_{i=n+1}^{N} \phi_{3,i-1}\phi_{3,i-1} & \cdots \\ \vdots & \vdots & \vdots & \ddots \end{bmatrix} \begin{pmatrix} \theta_1 \\ \theta_2 \\ \theta_3 \\ \vdots \end{pmatrix} = \begin{pmatrix} \sum_{i=n+1}^{N} y_i\phi_{1,i-1} \\ \sum_{i=n+1}^{N} y_i\phi_{2,i-1} \\ \sum_{i=n+1}^{N} y_i\phi_{3,i-1} \\ \vdots \end{pmatrix} \tag{6.143}$$

which is solved for $\boldsymbol{\theta}$.

In the case of a moving window trailing N observation samples behind present time i, one may weight the contributions of observations in a decreasing fashion with age, producing an updated $\boldsymbol{\theta}_i$ on each time step i. Thus, one would seek to minimise

$$J(\boldsymbol{\theta}_i) = \sum_{k=i-N+1}^{i} w_{i-k+1}e_k^2 = \sum_{k=i-N+1}^{i} w_{i-k+1}\left[y_k - \boldsymbol{\phi}_{k-1}^{\mathrm{T}}\boldsymbol{\theta}_i\right]^2 \tag{6.144}$$

with the result

$$\left[\sum_{k=i-N+1}^{i} w_{i-k+1}\boldsymbol{\phi}_{k-1}\boldsymbol{\phi}_{k-1}^{\mathrm{T}}\right]\boldsymbol{\theta} = \left(\sum_{k=i-N+1}^{i} w_{i-k+1}y_k\boldsymbol{\phi}_{k-1}\right) \tag{6.145}$$

and appropriate alteration of Equation 6.143. The amount of computation required on each time step may be reduced as in Equations 6.3–6.7, by attending only to the new measurement point, and discarding terms involving the oldest point.

6.5.2
Model Identification Using Recursive Least Squares on Measurements

The discrete Kalman filter of Section 6.4 can be configured as a recursive model parameter estimator. Recall the basic form of the equations for estimation of the state $\boldsymbol{x}$:

System with random, zero-mean errors on model and measurements:

$$\boldsymbol{x}_i = \boldsymbol{A}_i\boldsymbol{x}_{i-1} + \boldsymbol{B}_i\boldsymbol{u}_{i-1} + \boldsymbol{\mu}_{i-1} \tag{6.146}$$

$$\boldsymbol{y}_i = \boldsymbol{C}_i\boldsymbol{x}_i + \boldsymbol{\nu}_{i-1} \tag{6.147}$$

Recursive estimates $\hat{\boldsymbol{x}}$ of $\boldsymbol{x}$:

$$\boldsymbol{P}_{i|i-1} = \boldsymbol{A}_i\boldsymbol{P}_{i-1|i-1}\boldsymbol{A}_i^{\mathrm{T}} + \boldsymbol{Q} \tag{6.148}$$

$$\boldsymbol{K}_i = \boldsymbol{P}_{i|i-1}\boldsymbol{C}_i^{\mathrm{T}}\left[\boldsymbol{C}_i\boldsymbol{P}_{i|i-1}\boldsymbol{C}_i^{\mathrm{T}} + \boldsymbol{R}\right]^{-1} \tag{6.149}$$

$$\hat{\boldsymbol{x}}_{i|i-1} = \boldsymbol{A}_i\hat{\boldsymbol{x}}_{i-1|i-1} + \boldsymbol{B}_i\boldsymbol{u}_{i-1} \tag{6.150}$$

$$\hat{\boldsymbol{x}}_{i|i} = \hat{\boldsymbol{x}}_{i|i-1} + \boldsymbol{K}_i\left[\boldsymbol{y}_i - \boldsymbol{C}_i\hat{\boldsymbol{x}}_{i|i-1}\right] \tag{6.151}$$

Imagining Equation 6.139 to be evaluated at each time step i, one's belief is that the parameters $\boldsymbol{\theta}$ should remain constant. Thus, $\boldsymbol{\theta}$ replaces $\boldsymbol{x}$ in the model equations (Equations 6.49 and 6.50), with $\boldsymbol{A}_i \to \boldsymbol{I}$ and $\boldsymbol{B}_i \to [\boldsymbol{0}]$. In the observation equation (Equation 6.50), $\boldsymbol{C}_i \to \boldsymbol{\Phi}_{i-1}^{\mathrm{T}}$. So the system becomes

$$\boldsymbol{\theta}_i = \boldsymbol{\theta}_{i-1} + \boldsymbol{\mu}_{i-1} \tag{6.152}$$

$$y_i = \boldsymbol{\Phi}_{i-1}^{\mathrm{T}}\boldsymbol{\theta}_i + \nu_{i-1} \tag{6.153}$$

with recursive estimates $\hat{\boldsymbol{\theta}}$ of $\boldsymbol{\theta}$ from

$$\boldsymbol{P}_{i|i-1} = \boldsymbol{P}_{i-1|i-1} + \boldsymbol{Q} \tag{6.154}$$

$$\boldsymbol{K}_i = \boldsymbol{P}_{i|i-1}\boldsymbol{\Phi}_{i-1}\left[\boldsymbol{\Phi}_{i-1}^{\mathrm{T}}\boldsymbol{P}_{i|i-1}\boldsymbol{\Phi}_{i-1} + R\right]^{-1} \tag{6.155}$$

$$\hat{\boldsymbol{\theta}}_{i|i} = \hat{\boldsymbol{\theta}}_{i-1|i-1} + \boldsymbol{K}_i\left[y_i - \boldsymbol{\Phi}_{i-1}^{\mathrm{T}}\hat{\boldsymbol{\theta}}_{i-1|i-1}\right] \tag{6.156}$$

$$\boldsymbol{P}_{i|i} = \left[\boldsymbol{I} - \boldsymbol{K}_i\boldsymbol{\Phi}_{i-1}^{\mathrm{T}}\right]\boldsymbol{P}_{i|i-1} \tag{6.157}$$

Recall that

$$\boldsymbol{Q} = E\{\boldsymbol{\mu}\boldsymbol{\mu}^{\mathrm{T}}\} \text{ where } \boldsymbol{\mu} \text{ is the error in the assumption of a constant } \boldsymbol{\theta} \tag{6.158}$$

$$R = E\{\nu\nu^{\mathrm{T}}\} = E\{\nu^2\} \text{ where } \nu \text{ is the measurement error for } y \tag{6.159}$$

Large $\boldsymbol{Q}$ terms compared with the R term will cause rapid variation of $\boldsymbol{\theta}$ in an attempt to fit the measurements closely from step to step. Conversely, a large R term compared with the $\boldsymbol{Q}$ terms will cause slow variation of $\boldsymbol{\theta}$. As noted in Section 6.4.1, if it is required to start with a high gain $\boldsymbol{K}$, $\boldsymbol{P}$ is initialised 'large', for example $1000\boldsymbol{I}$, or conversely $0.1\boldsymbol{I}$ for a low initial gain.

Example 6.2

Comparison of (a) least squares and (b) recursive least squares for identification of a two-input, two-output discrete system.

A system with measurable states is thought to behave as

$$\begin{pmatrix} x_1 \\ x_2 \end{pmatrix}_i = \begin{bmatrix} a_{11} & a_{12} \\ a_{21} & a_{22} \end{bmatrix} \begin{pmatrix} x_1 \\ x_2 \end{pmatrix}_{i-1} + \begin{bmatrix} b_{11} & b_{12} \\ b_{21} & b_{22} \end{bmatrix} \begin{pmatrix} u_1 \\ u_2 \end{pmatrix}_{i-1} \tag{6.160}$$

Measurements of $\boldsymbol{x}$ and $\boldsymbol{u}$ at a series of points in time will be used for real-time identification of suitable a_{ij} and b_{ij}. One notes that the longest backward time shift in this system is $n = 1$.

In this comparison, both identifiers are provided with the input signal $\begin{pmatrix} u_1 \\ u_2 \end{pmatrix}_i$ and the resultant output signal $\begin{pmatrix} x_1 \\ x_2 \end{pmatrix}_i$ which are constructed using a vector of 'top-hat' random samples r in the range $0 \leq r \leq 1$, according to the initialisation:

$$\begin{pmatrix} u_1 \\ u_2 \end{pmatrix}_0 = \begin{pmatrix} 50 \\ 30 \end{pmatrix} \tag{6.161}$$

$$\begin{pmatrix} \Delta u_1 \\ \Delta u_2 \end{pmatrix}_0 = \begin{pmatrix} 0 \\ 0 \end{pmatrix} \qquad \alpha_{\Delta u} = 0.95 \tag{6.162}$$

$$\begin{pmatrix} x_1 \\ x_2 \end{pmatrix}_0 = \begin{pmatrix} 105 \\ 100 \end{pmatrix} \tag{6.163}$$

$$\begin{pmatrix} \Delta x_1 \\ \Delta x_2 \end{pmatrix}_0 = \begin{pmatrix} 0 \\ 0 \end{pmatrix} \qquad \alpha_{\Delta x} = 0.95 \tag{6.164}$$

and the following assignments repeated on each step i:

$$\begin{pmatrix} \Delta u_1 \\ \Delta u_2 \end{pmatrix}_i = \alpha_{\Delta u} \begin{pmatrix} \Delta u_1 \\ \Delta u_2 \end{pmatrix}_{i-1} + [1 - \alpha_{\Delta u}]1000 \left[\begin{pmatrix} r_a \\ r_b \end{pmatrix}_{i-1} - \begin{pmatrix} \frac{1}{2} \\ \frac{1}{2} \end{pmatrix} \right] \tag{6.165}$$

$$\begin{pmatrix} u_1 \\ u_2 \end{pmatrix}_i = \begin{pmatrix} u_1 \\ u_2 \end{pmatrix}_0 + \begin{pmatrix} \Delta u_1 \\ \Delta u_2 \end{pmatrix}_i \tag{6.166}$$

$$\begin{pmatrix} \Delta x_1 \\ \Delta x_2 \end{pmatrix}_i = \alpha_{\Delta x} \begin{pmatrix} \Delta x_1 \\ \Delta x_2 \end{pmatrix}_{i-1} + [1 - \alpha_{\Delta x}]5 \left[\begin{pmatrix} r_e \\ r_f \end{pmatrix}_{i-1} - \begin{pmatrix} \frac{1}{2} \\ \frac{1}{2} \end{pmatrix} \right] \tag{6.167}$$

$$\begin{pmatrix} x_1 \\ x_2 \end{pmatrix}_i = \begin{bmatrix} 0.4 & 0.1 \\ 0.2 & 0.3 \end{bmatrix} \begin{pmatrix} x_1 \\ x_2 \end{pmatrix}_{i-1} + \begin{bmatrix} 0.7 & 0.6 \\ 0.5 & 0.8 \end{bmatrix}_{i-1} \begin{pmatrix} u_1 \\ u_2 \end{pmatrix}_i + \begin{pmatrix} \Delta x_1 \\ \Delta x_2 \end{pmatrix}_i \tag{6.168}$$

Thus, it is seen that the input is the sum of a steady signal at (50, 50), and a zero-mean, time-correlated first-order signal starting at (0, 0). A fixed model is used to compute an intermediate output, which is then combined with an error arising from a similar zero-mean first-order system starting at (0, 0), to provide the final measured output.

a. *Least-squares identification (batch)*: Using the symbols x_{1i}, x_{2i}, u_{1i}, and u_{2i} to represent the actual measurements available for time i, Equation 6.139

$$e_i = y_i - \boldsymbol{\Phi}_{i-1}^{\mathrm{T}}\boldsymbol{\theta} \tag{6.169}$$

can be rewritten as

$$e_{1i} = x_{1i} - \begin{bmatrix} x_{1i-1} & x_{2i-1} & u_{1i-1} & u_{2i-1} \end{bmatrix} \begin{pmatrix} a_{11} \\ a_{12} \\ b_{11} \\ b_{12} \end{pmatrix} \tag{6.170}$$

$$e_{2i} = x_{2i} - \begin{bmatrix} x_{1i-1} & x_{2i-1} & u_{1i-1} & u_{2i-1} \end{bmatrix} \begin{pmatrix} a_{21} \\ a_{22} \\ b_{21} \\ b_{22} \end{pmatrix} \tag{6.171}$$

Following Equation 6.142, the LS fit for the coefficients affecting x_1, over the recent N samples, is given by

$$\begin{pmatrix} a_{11} \\ a_{12} \\ b_{11} \\ b_{12} \end{pmatrix} = \begin{bmatrix} \sum_{i=2}^{N} x_{1,i-1}x_{1,i-1} & \sum_{i=2}^{N} x_{2,i-1}x_{1,i-1} & \sum_{i=2}^{N} u_{1,i-1}x_{1,i-1} & \sum_{i=2}^{N} u_{2,i-1}x_{1,i-1} \\ \sum_{i=2}^{N} x_{1,i-1}x_{2,i-1} & \sum_{i=2}^{N} x_{2,i-1}x_{2,i-1} & \sum_{i=2}^{N} u_{1,i-1}x_{2,i-1} & \sum_{i=2}^{N} u_{2,i-1}x_{2,i-1} \\ \sum_{i=2}^{N} x_{1,i-1}u_{1,i-1} & \sum_{i=2}^{N} x_{2,i-1}u_{1,i-1} & \sum_{i=2}^{N} u_{1,i-1}u_{1,i-1} & \sum_{i=2}^{N} u_{2,i-1}u_{1,i-1} \\ \sum_{i=2}^{N} x_{1,i-1}u_{2,i-1} & \sum_{i=2}^{N} x_{2,i-1}u_{2,i-1} & \sum_{i=2}^{N} u_{1,i-1}u_{2,i-1} & \sum_{i=2}^{N} u_{2,i-1}u_{2,i-1} \end{bmatrix}^{-1} \begin{pmatrix} \sum_{i=2}^{N} x_{1,i}x_{1,i-1} \\ \sum_{i=2}^{N} x_{1,i}x_{2,i-1} \\ \sum_{i=2}^{N} x_{1,i}u_{1,i-1} \\ \sum_{i=2}^{N} x_{1,i}u_{2,i-1} \end{pmatrix} \tag{6.172}$$

and for the coefficients affecting x_2:

$$\begin{pmatrix} a_{21} \\ a_{22} \\ b_{21} \\ b_{22} \end{pmatrix} = \begin{bmatrix} \sum_{i=2}^{N} x_{1,i-1}x_{1,i-1} & \sum_{i=2}^{N} x_{2,i-1}x_{1,i-1} & \sum_{i=2}^{N} u_{1,i-1}x_{1,i-1} & \sum_{i=2}^{N} u_{2,i-1}x_{1,i-1} \\ \sum_{i=2}^{N} x_{1,i-1}x_{2,i-1} & \sum_{i=2}^{N} x_{2,i-1}x_{2,i-1} & \sum_{i=2}^{N} u_{1,i-1}x_{2,i-1} & \sum_{i=2}^{N} u_{2,i-1}x_{2,i-1} \\ \sum_{i=2}^{N} x_{1,i-1}u_{1,i-1} & \sum_{i=2}^{N} x_{2,i-1}u_{1,i-1} & \sum_{i=2}^{N} u_{1,i-1}u_{1,i-1} & \sum_{i=2}^{N} u_{2,i-1}u_{1,i-1} \\ \sum_{i=2}^{N} x_{1,i-1}u_{2,i-1} & \sum_{i=2}^{N} x_{2,i-1}u_{2,i-1} & \sum_{i=2}^{N} u_{1,i-1}u_{2,i-1} & \sum_{i=2}^{N} u_{2,i-1}u_{2,i-1} \end{bmatrix}^{-1} \begin{pmatrix} \sum_{i=2}^{N} x_{2,i}x_{1,i-1} \\ \sum_{i=2}^{N} x_{2,i}x_{2,i-1} \\ \sum_{i=2}^{N} x_{2,i}u_{1,i-1} \\ \sum_{i=2}^{N} x_{2,i}u_{2,i-1} \end{pmatrix} \tag{6.173}$$

b. *Recursive least-squares (RLS) identification*: Again, the identifications are performed independently for the coefficients associated with each output. Focusing on the first output,

Equation 6.170 gave

$$e_{1i} = x_{1i} - \left[x_{1i-1} \quad x_{2i-1} \quad u_{1i-1} \quad u_{2i-1} \right] \begin{pmatrix} a_{11} \\ a_{12} \\ b_{11} \\ b_{12} \end{pmatrix} \tag{6.174}$$

that is

$$y_{1i} = x_{1i} \tag{6.175}$$

$$\mathbf{\Phi}_{1i-1} = \begin{pmatrix} x_{1i-1} \\ x_{2i-1} \\ u_{1i-1} \\ u_{2i-1} \end{pmatrix} \tag{6.176}$$

$$\boldsymbol{\theta} = \begin{pmatrix} a_{11} \\ a_{12} \\ b_{11} \\ b_{12} \end{pmatrix} \tag{6.177}$$

Following Equations 6.154–6.157, the following *assignments* are required:

$$\boldsymbol{P}_{i|i-1} = \boldsymbol{P}_{i-1|i-1} + \boldsymbol{Q} \quad \text{(predicted)} \tag{6.178}$$

$$\boldsymbol{K}_i = \boldsymbol{P}_{i|i-1} \begin{bmatrix} x_1 \\ x_2 \\ u_1 \\ u_2 \end{bmatrix}_{i-1} \left[\left[x_1 \quad x_2 \quad u_1 \quad u_2 \right]_{i-1} \boldsymbol{P}_{i|i-1} \begin{bmatrix} x_1 \\ x_2 \\ u_1 \\ u_2 \end{bmatrix}_{i-1} + \boldsymbol{R} \right]^{-1} \tag{6.179}$$

$$\begin{pmatrix} a_{11} \\ a_{12} \\ b_{11} \\ b_{12} \end{pmatrix}_i = \begin{pmatrix} a_{11} \\ a_{12} \\ b_{11} \\ b_{12} \end{pmatrix}_{i-1} + \boldsymbol{K}_i \left[(x_1)_i - \left[x_1 \quad x_2 \quad u_1 \quad u_2 \right]_{i-1} \begin{pmatrix} a_{11} \\ a_{12} \\ b_{11} \\ b_{12} \end{pmatrix}_{i-1} \right] \tag{6.180}$$

$$\boldsymbol{P}_{i|i} = \left[\boldsymbol{I} - \boldsymbol{K}_i \left[x_1 \quad x_2 \quad u_1 \quad u_2 \right]_{i-1} \right] \boldsymbol{P}_{i|i-1} \quad \text{(updated)} \tag{6.181}$$

The error covariance matrix $\boldsymbol{Q}$ for the model (which assumes constant a_{ij} and b_{ij}) is set as

$$\boldsymbol{Q} = \begin{bmatrix} 1 & 0 & 0 & 0 \\ 0 & 1 & 0 & 0 \\ 0 & 0 & 1 & 0 \\ 0 & 0 & 0 & 1 \end{bmatrix} \tag{6.182}$$

and the error covariance matrix for the measurement $(x_1)_i$ in Equation 6.180 is set as

$$\boldsymbol{R} = [1] \tag{6.183}$$

In this example, the covariance matrix was initialised as $\boldsymbol{P} = \boldsymbol{0}$. Similar assignments to Equations 6.178–6.183 are used for the second output $y_{2i} = x_{2i}$.

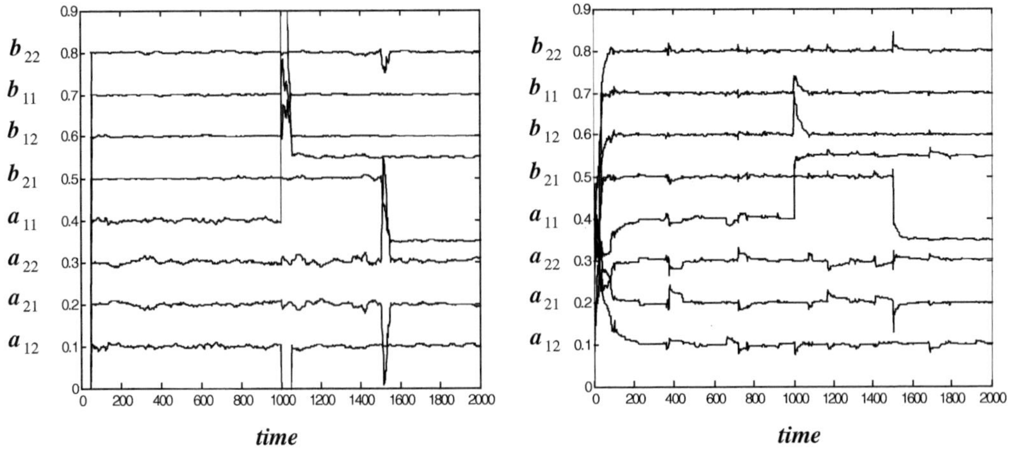

(a) Least-squares moving window

(b) Recursive least squares

Figure 6.15 LS based on a 50-point trailing window compared with RLS for identification of a two-input, two-output linear system (Example 6.2). The true value of a_{11} is stepped from 0.4 to 0.55 at $t = 1000$, and that of b_{21} is stepped from 0.5 to 0.35 at $t = 1500$.

The results of the two parallel parameter identifications based on the measurements y_{1i} and y_{2i} are shown together in Figure 6.15. Here the identified coefficients in $\boldsymbol{A}$ and $\boldsymbol{B}$ have been started at zero. The identification by least squares in a 50-point trailing window is compared with that of the above recursive least-squares settings. The algorithm for this simulation is shown schematically in Figure 6.16. For convenience, the parameter identification is done simultaneously for both output measurements, but in practice, since they are independent, one would seek to reduce array sizes by dealing with them separately.

6.5.3 Some Considerations in Model Identification

6.5.3.1 Type of Model

Following from Equation 6.132, *equation error* $\mathbf{e}(t)$ is described by

$$\boldsymbol{A}(q)\boldsymbol{y}(t) = \boldsymbol{B}(q)\boldsymbol{u}(t) + \mathbf{e}(t) \tag{6.184}$$

with

$$\boldsymbol{A}(q) = \boldsymbol{I} - \boldsymbol{A}_1 q^{-1} - \boldsymbol{A}_2 q^{-2} - \cdots - \boldsymbol{A}_n q^{-n} \tag{6.185}$$

$$\boldsymbol{B}(q) = \boldsymbol{B}_1 q^{-1} + \boldsymbol{B}_2 q^{-2} + \cdots + \boldsymbol{B}_m q^{-m} \tag{6.186}$$

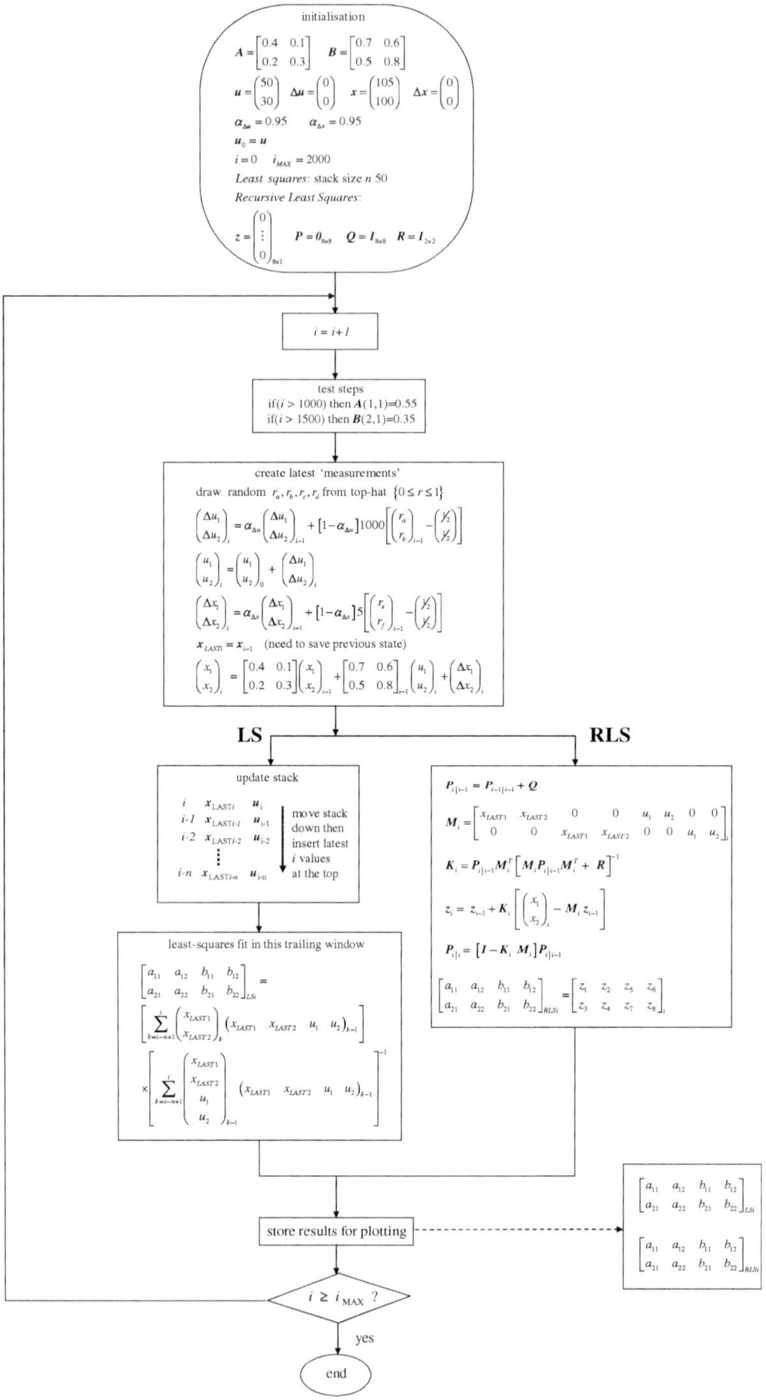

Figure 6.16 Algorithm for the comparative LS and RLS identification simulation of Example 6.2, yielding results in Figure 6.15 (note: '=' implies assignment). The identification for both measurements is done simultaneously. However, since they are independent, array sizes may be reduced by dealing with them separately.

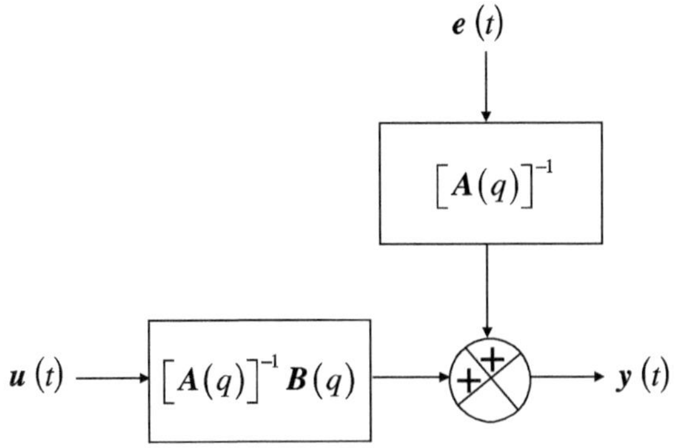

Figure 6.17 ARX model.

Thus,

$$y(t) = [A(q)]^{-1} B(q) u(t) + [A(q)]^{-1} e(t) \tag{6.187}$$

where the vector $e(t)$ consists of as yet unknown random error terms. From this point of view, the error finally affecting $y(t)$ includes dynamics associated with $A(q)$ (Figure 6.17).

The assumed behaviour of this system, that the outputs y are recursively related to their previous values, and that there are measurable exogenous inputs u, constitutes the ARX form of an *equation error* model. It is interesting to review some variations of this idea that have been used in identification. Noting a general form of the *equation error* model in Figure 6.18, some particular cases are listed in Table 6.1.

The *integration* feature of the ARIMAX model is seen to recognise only *changes* $(1 - q^{-1})$ in the measurements as determining the A and B matrices. Thus, it will eliminate the problem of offset in the measurements which one expects in the linearisation procedure of Section 3.7.

The *moving average* type of model equation (e.g. ARMAX, ARIMAX) allows one to anticipate the behaviour of the error affecting the identification. This feature is useful for time-correlated errors such as drift. Knowledge of the future behaviour of an extraneous *persistence* error allows one to isolate this effect from the identification of the intrinsic system. As will be seen, the penalty for this advantage is that the elements of the unknown C matrices determining the error behaviour *also*

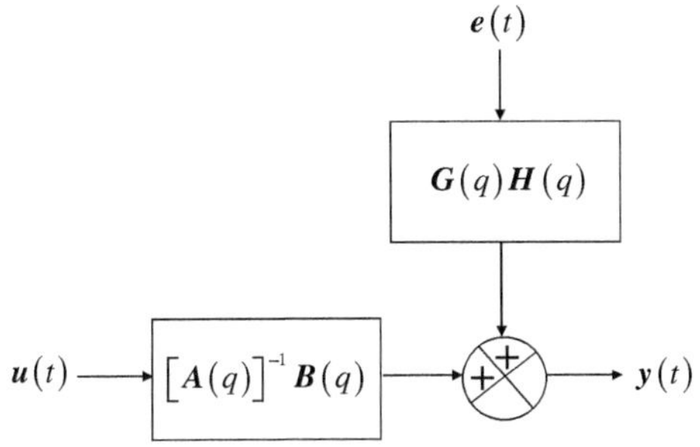

Figure 6.18 General form of *equation error* model.

Table 6.1 Some types of models used for identification.

Name	G(q)	H(q)	y(t)	Discrete calculation
ARX (autoregressive exogenous)	$[\boldsymbol{A}(q)]^{-1}$	$\boldsymbol{I}$	$y(t) = [\boldsymbol{A}(q)]^{-1}\boldsymbol{B}(q)\boldsymbol{u}(t) + [\boldsymbol{A}(q)]^{-1}\boldsymbol{e}(t)$	$\boldsymbol{y}_i = \boldsymbol{A}_1\boldsymbol{y}_{i-1} + \boldsymbol{A}_2\boldsymbol{y}_{i-2} + \cdots + \boldsymbol{B}_1\boldsymbol{u}_{i-1} + \boldsymbol{B}_2\boldsymbol{u}_{i-2} + \cdots + \boldsymbol{e}_i$
ARMAX (autoregressive moving average exogenous)	$[\boldsymbol{A}(q)]^{-1}$	$\boldsymbol{I} + C_1 q^{-1} + C_2 q^{-2} + \cdots$	$y(t) = [\boldsymbol{A}(q)]^{-1}\boldsymbol{B}(q)\boldsymbol{u}(t) + [\boldsymbol{A}(q)]^{-1}$ $[\boldsymbol{I} + \boldsymbol{C}_1 q^{-1} + \boldsymbol{C}_2 q^{-2} + \cdots]\boldsymbol{e}(t)$	$\boldsymbol{y}_i = \boldsymbol{A}_1\boldsymbol{y}_{i-1} + \boldsymbol{A}_2\boldsymbol{y}_{i-2} + \cdots + \boldsymbol{B}_1\boldsymbol{u}_{i-1} + \boldsymbol{B}_2\boldsymbol{u}_{i-2} + \cdots +$ $\boldsymbol{e}_i + \boldsymbol{C}_1\boldsymbol{e}_{i-1} + \boldsymbol{C}_2\boldsymbol{e}_{i-2} + \cdots$
ARIMAX (autoregressive integrated moving average exogenous)	$(1-q^{-1})$ $[\boldsymbol{A}(q)]^{-1}$	$\boldsymbol{I} + C_1 q^{-1} + C_2 q^{-2} + \cdots$	$y(t) = [\boldsymbol{A}(q)]^{-1}\boldsymbol{B}(q)\boldsymbol{u}(t) + [(1-q^{-1})\boldsymbol{A}(q)]^{-1}$ $[\boldsymbol{I} + \boldsymbol{C}_1 q^{-1} + \boldsymbol{C}_2 q^{-2} + \cdots]\boldsymbol{e}(t)$	$[\boldsymbol{y}_i - \boldsymbol{y}_{i-1}] = \boldsymbol{A}_1[\boldsymbol{y}_{i-1} - \boldsymbol{y}_{i-2}] + \boldsymbol{A}_2[\boldsymbol{y}_{i-2} - \boldsymbol{y}_{i-3}] + \cdots +$ $\boldsymbol{B}_1[\boldsymbol{u}_{i-1} - \boldsymbol{u}_{i-2}] + \boldsymbol{B}_2[\boldsymbol{u}_{i-2} - \boldsymbol{u}_{i-3}] + \cdots +$ $\boldsymbol{e}_i + \boldsymbol{C}_1\boldsymbol{e}_{i-1} + \boldsymbol{C}_2\boldsymbol{e}_{i-2} + \cdots$
Output error form	$\boldsymbol{I}$	$\boldsymbol{I}$	$y(t) = [\boldsymbol{A}(q)]^{-1}\boldsymbol{B}(q)\boldsymbol{u}(t) + \boldsymbol{e}(t)$	$[\boldsymbol{y}_i - \boldsymbol{e}_i] = \boldsymbol{A}_1[\boldsymbol{y}_{i-1} - \boldsymbol{e}_{i-1}] + \boldsymbol{A}_2[\boldsymbol{y}_{i-2} - \boldsymbol{e}_{i-2}] + \cdots +$ $\boldsymbol{B}_1\boldsymbol{u}_{i-1} + \boldsymbol{B}_2\boldsymbol{u}_{i-2} + \cdots$

have to be identified. $\boldsymbol{e}(t)$ is considered to be zero-mean *white noise* (i.e. uncorrelated random samples). The existence of a $\boldsymbol{C}_1$ matrix implies that a *final autocorrelated error* is present in the output $\boldsymbol{y}(t)$. Rearranging to express the *original zero-mean white noise error,*

$$\boldsymbol{e}(t) = \left[\boldsymbol{I} + \boldsymbol{C}_1 q^{-1} + \cdots\right]^{-1} \boldsymbol{A}(q)(1 - q^{-1})\boldsymbol{y}(t) - \left[\boldsymbol{I} + \boldsymbol{C}_1 q^{-1} + \cdots\right]^{-1} \boldsymbol{B}(q)(1 - q^{-1})\boldsymbol{u}(t) \quad (6.188)$$

The basis of the identification is then to choose the elements of the matrices $\boldsymbol{A}_1, \boldsymbol{A}_2, \ldots, \boldsymbol{B}_1, \boldsymbol{B}_2, \ldots, \boldsymbol{C}_1, \boldsymbol{C}_2, \ldots$ in such a way as to minimise the 'size' of the original random white noise error $\boldsymbol{e}(t)$ which is required to account for the observed measurements $\boldsymbol{y}(t)$. This is most easily illustrated with a SISO example, taking only up to the first term in each series:

$$e(t) = \frac{\left[1 + aq^{-1}\right]}{\left[1 + cq^{-1}\right]} \left[1 - q^{-1}\right] y(t) - \frac{\left[bq^{-1}\right]}{\left[1 + cq^{-1}\right]} \left[1 - q^{-1}\right] u(t) \quad (6.189)$$

The ratio terms can be expanded by long division as their impulse response series

$$e(t) = \left[1 + \alpha_1 q^{-1} + \alpha_2 q^{-2} + \cdots\right] \left(1 - q^{-1}\right) y(t) - \left[\beta_1 q^{-1} + \beta_2 q^{-2} + \cdots\right] \left(1 - q^{-1}\right) u(t) \quad (6.190)$$

with

$$\alpha_1 = (a - c) \quad (6.191)$$

$$\alpha_2 = -c(a - c) \quad (6.192)$$

$$\vdots$$

$$\beta_1 = b \quad (6.193)$$

$$\beta_2 = -bc \quad (6.194)$$

$$\vdots$$

The problem of identifying a and b is clearly much simpler in the case where $c = 0$ (e.g. ARX), which would not entail the 'dividing' series. However, the presence of the $\boldsymbol{C}_i$ series in MA problems generally requires resolution by a more complicated iterative procedure (Chatfield, 2004; Box and Jenkins, 1970).

To continue with the present simple SISO problem, note that *stationarity* of the process requires $|a| < 1$ and $|c| < 1$, so the terms in the series progressively reduce in magnitude. In practice this can be assured by further differencing of Equation 6.189. Consider a case where only insignificant terms occur after the first two terms in each series, and define

$$\boldsymbol{\theta} = \begin{pmatrix} \alpha_1 \\ \alpha_2 \\ \beta_1 \\ \beta_2 \end{pmatrix} \quad (6.195)$$

$$\boldsymbol{\Phi}_{i-1} = \begin{pmatrix} -y_{i-1} + y_{i-2} \\ -y_{i-2} + y_{i-3} \\ u_{i-1} - u_{i-2} \\ u_{i-2} - u_{i-3} \end{pmatrix} \quad (6.196)$$

Then

$$e_i \approx \Delta y_i - \boldsymbol{\Phi}_{i-1}^{\mathrm{T}} \boldsymbol{\theta} \tag{6.197}$$

where

$$\Delta y_i = y_i - y_{i-1} \tag{6.198}$$

i) For a *batch least-squares fit* to a series of measurements y_i, u_i over $i = 1, \ldots, N$ intervals, proceed as follows:

$$e_i^2 = \Delta y_i^2 - 2\Delta y_i \boldsymbol{\theta}^{\mathrm{T}} \boldsymbol{\Phi}_{i-1} + \boldsymbol{\theta}^{\mathrm{T}} \boldsymbol{\Phi}_{i-1} \boldsymbol{\Phi}_{i-1}^{\mathrm{T}} \boldsymbol{\theta} \tag{6.199}$$

$$J(\boldsymbol{\theta}) = \sum_{i=1}^{N} e_i^2 \tag{6.200}$$

$$J(\boldsymbol{\theta}) = \left\{ \sum_{i=1}^{N} \Delta y_i^2 \right\} - 2\boldsymbol{\theta}^{\mathrm{T}} \left\{ \sum_{i=1}^{N} \Delta y_i \boldsymbol{\Phi}_{i-1} \right\} + \boldsymbol{\theta}^{\mathrm{T}} \left\{ \sum_{i=1}^{N} \boldsymbol{\Phi}_{i-1} \boldsymbol{\Phi}_{i-1}^{\mathrm{T}} \right\} \boldsymbol{\theta} \tag{6.201}$$

Minimise with respect to $\boldsymbol{\theta}$:

$$\frac{\mathrm{d}J(\boldsymbol{\theta})}{\mathrm{d}\boldsymbol{\theta}} = 2\left\{ \sum_{i=1}^{N} \Delta y_i \boldsymbol{\Phi}_{i-1} \right\} - 2\left\{ \sum_{i=1}^{N} \boldsymbol{\Phi}_{i-1} \boldsymbol{\Phi}_{i-1}^{\mathrm{T}} \right\} \boldsymbol{\theta} = \boldsymbol{0} \tag{6.202}$$

$$\boldsymbol{\theta} = \left\{ \sum_{i=1}^{N} \boldsymbol{\Phi}_{i-1} \boldsymbol{\Phi}_{i-1}^{\mathrm{T}} \right\}^{-1} \left\{ \sum_{i=1}^{N} \Delta y_i \boldsymbol{\Phi}_{i-1} \right\} \tag{6.203}$$

ii) For a *recursive least-squares fit* one wishes for $e_i = 0$, that is for the predicted change to approach the actual change:

$$\Delta \hat{y}_i \to \Delta y_i \tag{6.204}$$

where

$$\Delta \hat{y}_i = \boldsymbol{\Phi}_{i-1}^{\mathrm{T}} \boldsymbol{\theta}_{i-1} \tag{6.205}$$

So adapting the Kalman filter as in Section 6.5.2,

$$\boldsymbol{\theta}_i = \boldsymbol{\theta}_{i-1} + \boldsymbol{\mu}_{i-1} \tag{6.206}$$

$$\Delta y_i = \boldsymbol{\Phi}_{i-1}^{\mathrm{T}} \boldsymbol{\theta}_{i-1} + \nu_{i-1} \tag{6.207}$$

with recursive estimates $\hat{\boldsymbol{\theta}}$ of $\boldsymbol{\theta}$ from

$$\boldsymbol{P}_{i|i-1} = \boldsymbol{P}_{i-1|i-1} + \boldsymbol{Q} \tag{6.208}$$

$$\boldsymbol{K}_i = \boldsymbol{P}_{i|i-1} \boldsymbol{\Phi}_{i-1} \left[\boldsymbol{\Phi}_{i-1}^{\mathrm{T}} \boldsymbol{P}_{i|i-1} \boldsymbol{\Phi}_{i-1} + R \right]^{-1} \tag{6.209}$$

$$\hat{\boldsymbol{\theta}}_{i|i} = \hat{\boldsymbol{\theta}}_{i-1|i-1} + \boldsymbol{K}_i \left[\Delta y_i - \boldsymbol{\Phi}_{i-1}^{\mathrm{T}} \hat{\boldsymbol{\theta}}_{i-1|i-1} \right] \tag{6.210}$$

$$\boldsymbol{P}_{i|i} = \left[\boldsymbol{I} - \boldsymbol{K}_i \boldsymbol{\Phi}_{i-1}^{\mathrm{T}} \right] \boldsymbol{P}_{i|i-1} \tag{6.211}$$

In the method of Box and Jenkins (1970), $\boldsymbol{u}(t)$ is initially 'prewhitened' by differencing, with the same transform applied to $\boldsymbol{y}(t)$, allowing the $\boldsymbol{C}_i$ series to be ignored initially, then fitted back later to the residual $\boldsymbol{e}(t)$.

6.5.3.2 Forgetting Factor

In real-time process identification, an important consideration is the extent to which the relevance of older measurements should decrease as time passes. The most recent measurements are, of course, very relevant, but they could arise from a short disturbance. On the other hand, if the measurements far back in time still have too strong an influence, the identifier will only respond slowly to a valid, recent change. An impression of the relative importance of the measurements up until present time is given in Figure 6.19.

In the formulation of the discrete Kalman filter as a parameter identifier in Section 6.5.2, it is noted that high $\boldsymbol{Q}$ values or low $\boldsymbol{R}$ values act to decrease the memory (higher gain $\boldsymbol{K}$) and vice versa. Retention of the two matrices in this formulation gives the user the opportunity to determine the rates at which individual parameters in vector $\boldsymbol{\theta}_i$ will adjust ($\boldsymbol{Q}$) and the recognition given to individual measurements in the $\boldsymbol{y}_i$ vector ($\boldsymbol{R}$).

In some formulations of the RLS identifier, a scalar *forgetting factor* λ is used instead of $\boldsymbol{Q}$ and $\boldsymbol{R}$. For example, following the approach of Ydstie, Kershenbaum and Sargent (1985), the assignments to be performed on each step i in the ARX case of Equations 6.146–6.147 become

$$\boldsymbol{P}_{i|i-1} = \frac{1}{\lambda_{i-1}}\left[\boldsymbol{P}_{i-1|i-1} + \boldsymbol{Q}\right] \quad \text{(predicted)} \tag{6.212}$$

$$\boldsymbol{K}_i = \boldsymbol{P}_{i|i-1}\boldsymbol{\Phi}_{i-1}\left[\boldsymbol{\Phi}_{i-1}^{\mathrm{T}}\boldsymbol{P}_{i|i-1}\boldsymbol{\Phi}_{i-1} + \lambda_{i-1}\boldsymbol{I}\right]^{-1} \tag{6.213}$$

$$\hat{\boldsymbol{\theta}}_{i|i} = \hat{\boldsymbol{\theta}}_{i-1|i-1} + \boldsymbol{K}_i\left[y_i - \boldsymbol{\Phi}_{i-1}^{\mathrm{T}}\hat{\boldsymbol{\theta}}_{i-1|i-1}\right] \tag{6.214}$$

$$\boldsymbol{P}_{i|i} = \left[\boldsymbol{I} - \boldsymbol{K}_i\boldsymbol{\Phi}_{i-1}^{\mathrm{T}}\right]\boldsymbol{P}_{i|i-1} \quad \text{(updated)} \tag{6.215}$$

Indeed, the retention here of the positive definite weighting matrix $\boldsymbol{Q}$ in Equation 6.212, as proposed by Young (1969), avoids the tendency for $\boldsymbol{P}$ to decrease. Forms also exist without the multiplier λ_{i-1} in Equation 6.213.

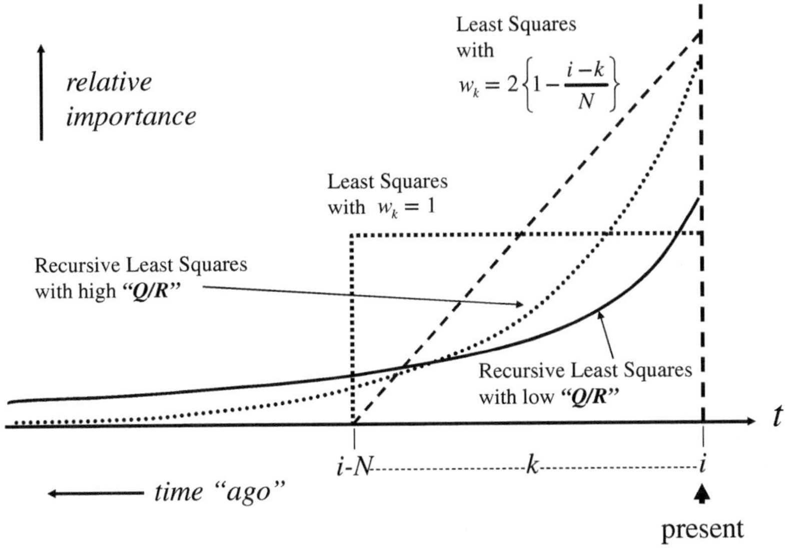

Figure 6.19 Relative importance of past measurements for process model identification.

For stability, $0 < \lambda < 1$. As $\lambda \to 0$, the gain increases and $\boldsymbol{\theta}$ will vary more rapidly to match the measurements. As $\lambda \to 1$, the gain decreases and $\boldsymbol{\theta}$ will vary more slowly, rather responding to measurements over a longer period. In this context, consider the following formulation of a *variable forgetting factor*:

$$\lambda_{i-1} = \frac{M}{M + \boldsymbol{e}_{i-1}^{\mathrm{T}} \left[\boldsymbol{\Phi}_{i-1}^{\mathrm{T}} \boldsymbol{P}_{i-1|i-1} \boldsymbol{\Phi}_{i-1} + \boldsymbol{I} \right]^{-1} \boldsymbol{e}_{i-1}} \tag{6.216}$$

A review of Equation 6.72 shows that the matrix in the denominator is a representation of the *present* inverse predictive error covariance. The scalar product of the *past* predictive error $\boldsymbol{e}_{i-1} = \boldsymbol{y}_{i-1} - \boldsymbol{\Phi}_{i-2}^{\mathrm{T}} \hat{\boldsymbol{\theta}}_{i-2|i-2}$ across this inverse matrix gives a ratio measure of *past/present* predictive errors. In Equation 6.216, M is a *memory length* chosen by the user, which will depend on the units of $\boldsymbol{e}$. Thus, if predictive errors are *reducing*, λ will be *larger* (slower forgetting), and if predictive errors are *increasing*, λ will be *smaller* (faster forgetting). In this way, the rate of adjustment of $\boldsymbol{\theta}$ is automatically determined by changes in the information content of the measurements, which is judged in terms of how *badly* the current model fits.

A number of formulations of the forgetting factor have been proposed by other workers. Equations 6.212–6.215 do not permit individual recognition of the confidence in equations and measurements. However, it is not difficult to superimpose the selective features of the $\boldsymbol{Q}$ and $\boldsymbol{R}$ matrices to deal with this.

6.5.3.3 Steady-State Offset

The general input–output equation (Equation 6.132)

$$\boldsymbol{A}(q)\boldsymbol{y}(t) = \boldsymbol{B}(q)\boldsymbol{u}(t) \tag{6.217}$$

arises in terms of *perturbation variables*, that is the deviations of $\boldsymbol{u}$ and $\boldsymbol{y}$ from some initial steady state. Following Section 3.7, a linearised difference equation using *absolute* values of $\boldsymbol{y}$ and $\boldsymbol{u}$ would involve a constant offset vector $\boldsymbol{z}$:

$$\boldsymbol{A}(q)\boldsymbol{y}(t) = \boldsymbol{B}(q)\boldsymbol{u}(t) + \boldsymbol{z} \tag{6.218}$$

For example, consider the state system with input $\boldsymbol{u}$ and output $\boldsymbol{y}$:

$$\dot{\boldsymbol{x}} = \boldsymbol{f}(\boldsymbol{x}, \boldsymbol{u}) \tag{6.219}$$

$$\boldsymbol{y} = \boldsymbol{g}(\boldsymbol{x}, \boldsymbol{u}) \tag{6.220}$$

Linearise as in Section 3.7 at the operating point $(\boldsymbol{x}_0, \boldsymbol{u}_0)$ to yield

$$\dot{\boldsymbol{x}} \approx \boldsymbol{A}'\boldsymbol{x} + \boldsymbol{B}'\boldsymbol{u} + \boldsymbol{w}_0 \tag{6.221}$$

$$\boldsymbol{y} \approx \boldsymbol{C}'\boldsymbol{x} + \boldsymbol{D}'\boldsymbol{u} + \boldsymbol{v}_0 \tag{6.222}$$

where

$$\boldsymbol{w}_0 = \boldsymbol{f}(\boldsymbol{x}_0, \boldsymbol{u}_0) - \boldsymbol{A}'\boldsymbol{x}_0 - \boldsymbol{B}'\boldsymbol{u}_0 \tag{6.223}$$

$$\boldsymbol{v}_0 = \boldsymbol{g}(\boldsymbol{x}_0, \boldsymbol{u}_0) - \boldsymbol{C}'\boldsymbol{x}_0 - \boldsymbol{D}'\boldsymbol{u}_0 \tag{6.224}$$

$$\boldsymbol{A}' = \left[\frac{\partial \boldsymbol{f}}{\partial \boldsymbol{x}}\right]_0 \tag{6.225}$$

$$\boldsymbol{B}' = \left[\frac{\partial \boldsymbol{f}}{\partial \boldsymbol{u}}\right]_0 \tag{6.226}$$

$$\boldsymbol{C}' = \left[\frac{\partial \boldsymbol{g}}{\partial \boldsymbol{x}}\right]_0 \tag{6.227}$$

$$\boldsymbol{D}' = \left[\frac{\partial \boldsymbol{g}}{\partial \boldsymbol{u}}\right]_0 \tag{6.228}$$

If the input $\boldsymbol{u}$ is only adjusted at intervals T, Equation 3.285 shows that the state variation is represented discretely by

$$\boldsymbol{x}_i = \boldsymbol{A}''\boldsymbol{x}_{i-1} + \boldsymbol{B}''\boldsymbol{u}_{i-1} + \boldsymbol{p} \tag{6.229}$$

where

$$\boldsymbol{A}'' = \boldsymbol{e}^{\boldsymbol{A}'T} \tag{6.230}$$

$$\boldsymbol{B}'' = \left[\boldsymbol{e}^{\boldsymbol{A}'T} - \boldsymbol{I}\right]\boldsymbol{A}'^{-1}\boldsymbol{B}' \tag{6.231}$$

$$\boldsymbol{p} = \left[\boldsymbol{e}^{\boldsymbol{A}'T} - \boldsymbol{I}\right]\boldsymbol{A}'^{-1}\boldsymbol{w}_0 \tag{6.232}$$

As in Equation 6.132, if $\boldsymbol{C}'$ is square and nonsingular, Equation 6.222 gives

$$\boldsymbol{x}_i = \boldsymbol{C}'^{-1}\boldsymbol{y}_i - \boldsymbol{C}'^{-1}\boldsymbol{D}'\boldsymbol{u}_i - \boldsymbol{C}'^{-1}\boldsymbol{v}_0 \tag{6.233}$$

whence the input–output form (Equation 6.218) is obtained for this state-space system as

$$\boldsymbol{y}_i - \boldsymbol{C}'\boldsymbol{A}''\boldsymbol{C}'^{-1}\boldsymbol{y}_{i-1} = \boldsymbol{D}'\boldsymbol{u}_i + \left[\boldsymbol{C}'\boldsymbol{B}'' - \boldsymbol{C}'\boldsymbol{A}''\boldsymbol{C}'^{-1}\boldsymbol{D}'\right]\boldsymbol{u}_{i-1} + \boldsymbol{z} \tag{6.234}$$

where the offset is seen to be

$$\boldsymbol{z} = \boldsymbol{C}'\boldsymbol{p} + \left[\boldsymbol{I} - \boldsymbol{C}'\boldsymbol{A}''\boldsymbol{C}'^{-1}\right]\boldsymbol{v}_0 \tag{6.235}$$

In the case of *integrated moving average* models such as ARIMAX, the offset $\boldsymbol{z}$ will be excluded in the identification. However, other forms will treat it as part of the *equation error*, unless the additional constant vector is explicitly identified along with the elements of the polynomial $\boldsymbol{A}$ and $\boldsymbol{B}$ matrices in Equation 6.218.

6.5.3.4 Extraction of Physical Parameters

Consider the case where the discrete system coefficient matrices $\boldsymbol{A}$ and $\boldsymbol{B}$ have been identified. These may be used online for adaptation of a controller, also based on the discrete form, so no further processing of the coefficients may be required. On the other hand, it may be that an estimate is required of some physical parameter imbedded in the original continuous system matrices $\boldsymbol{A}'$ or $\boldsymbol{B}'$ in Equation 6.221.

Assuming a constant flow rate f, the dynamics of the heated mixed flow vessel in Figure 6.20 are described by

$$\rho c_P V \frac{\mathrm{d}T}{\mathrm{d}t} = \rho c_P f[T_0 - T] + UA[T_S - T] \tag{6.236}$$

$$\frac{\mathrm{d}T}{\mathrm{d}t} = -\left\{\frac{f}{V} + \frac{UA}{\rho c_P V}\right\}T + \left\{\frac{UA}{\rho c_P V}\right\}T_S + \left\{\frac{f}{V}\right\}T_0 \tag{6.237}$$

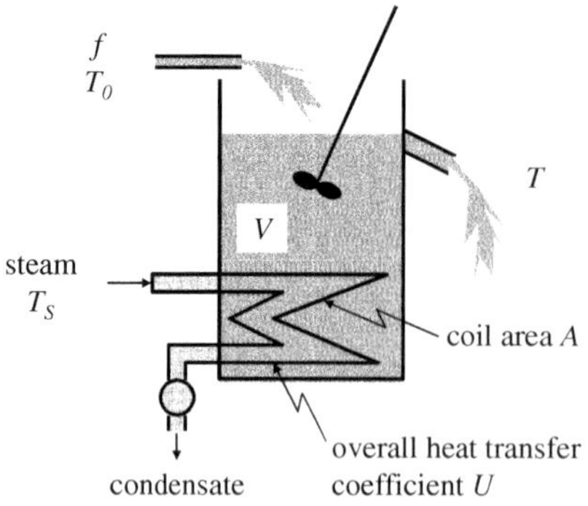

Figure 6.20 Parameter extraction in identification: heated mixed flow vessel.

Thus,

$$\frac{\mathrm{d}\boldsymbol{x}}{\mathrm{d}t} = \boldsymbol{A}'\boldsymbol{x} + \boldsymbol{B}'\boldsymbol{u} \tag{6.238}$$

where

$$\boldsymbol{x} = (T) \tag{6.239}$$

$$\boldsymbol{u} = \begin{pmatrix} T_S \\ T_0 \end{pmatrix} \tag{6.240}$$

$$\boldsymbol{A}' = \left[\left\{-\frac{f}{V} - \frac{UA}{\rho c_P V}\right\}\right] \quad (= [a'_{11}]) \tag{6.241}$$

$$\boldsymbol{B}' = \left[\left\{\frac{UA}{\rho c_P V}\right\} \quad \left\{\frac{f}{V}\right\}\right] \quad (= [b'_{11} \quad b'_{12}]) \tag{6.242}$$

A typical concern here would be to monitor fouling of the heat transfer coil, so one would need to extract the OHTC U from the original matrix coefficients. Basing the parameter identification on the discrete form of Equation 6.238, one needs as in Section 3.9.3.2 the matrices

$$\boldsymbol{A}'' = \boldsymbol{e}^{\boldsymbol{A}'\Delta t} \quad (= [a''_{11}]) \tag{6.243}$$

$$\boldsymbol{B}'' = \left[\boldsymbol{e}^{\boldsymbol{A}'\Delta t} - \boldsymbol{I}\right]\boldsymbol{A}'^{-1}\boldsymbol{B}' \quad \left(= \left[b''_{11} \quad b''_{12}\right]\right) \tag{6.244}$$

which are available in MATLAB® using 'expm', or can be computed as in Section 3.9.3.3. In this case, the state is observed directly ($\boldsymbol{y} = \boldsymbol{x}$), so the input–output equation as in Equation 6.132 is

$$\left[\boldsymbol{I}q^0 - \boldsymbol{A}''q^{-1}\right]T(t) = \left[\boldsymbol{B}''q^{-1}\right]\begin{pmatrix} T_S(t) \\ T_0(t) \end{pmatrix} \tag{6.245}$$

Thus, it is the elements of $\boldsymbol{A}''$ and $\boldsymbol{B}''$ that will be identified. In order to extract a physical parameter (U in this case), one must revert to the $\boldsymbol{A}'$ and $\boldsymbol{B}'$ of Equation 6.238. A possibility is the MATLAB®

'logm' function. Alternatively, one could use the approximate relations

$$e^{A'\Delta t} \approx I + A'\Delta t \tag{6.246}$$

whence

$$A' \approx \frac{1}{\Delta t}[A'' - I] \tag{6.247}$$

and

$$B' \approx \frac{1}{\Delta t} B'' \tag{6.248}$$

so that

$$U = \left\{\frac{\rho c_P V}{A}\right\} b'_{11} \approx \left\{\frac{\rho c_P V}{\Delta t A}\right\} b''_{11} \tag{6.249}$$

6.5.3.5 Transport Lag (Dead Time)

The heated mixed flow vessel in Section 6.5.3.4 might have been subject to a transport lag, for example if the temperature T were only available some distance downstream on the exit pipeline. If this dead time were known to be $p \times \Delta t$, then the system could be described by

$$\begin{pmatrix} T \\ T_{+p-1} \\ \vdots \\ T_{+2} \\ T_{+1} \\ T_{+0} \end{pmatrix}_i = \begin{bmatrix} 0 & 1 & 0 & \cdots & 0 & 0 \\ 0 & 0 & 1 & \vdots & 0 & 0 \\ \vdots & \vdots & \cdots & \ddots & \vdots & \vdots \\ 0 & 0 & 0 & \cdots & 1 & 0 \\ 0 & 0 & 0 & \cdots & 0 & 1 \\ 0 & 0 & 0 & \cdots & 0 & a_{11} \end{bmatrix} \begin{pmatrix} T \\ T_{+p-1} \\ T_{+p-2} \\ \vdots \\ T_{+1} \\ T_{+0} \end{pmatrix}_{i-1} + \begin{pmatrix} 0 & 0 \\ 0 & 0 \\ 0 & 0 \\ 0 & 0 \\ 0 & 0 \\ b_{11} & b_{12} \end{pmatrix} \begin{pmatrix} T_S \\ T_0 \end{pmatrix}_{i-1} \tag{6.250}$$

where a_{11}, b_{11} and b_{12} have the same meaning as in Section 6.5.3.4. However, this is a cumbersome form for identification of just a_{11}, b_{11} and b_{12}, so the system equation is more succinctly expressed as

$$x_i = Ax_{i-1} + Bu_{i-1-p} \tag{6.251}$$

and the term $\begin{pmatrix} x \\ \hdashline u \end{pmatrix}_{i-1}$ in equations like Equation 6.176 or 6.179 obviously becomes $\begin{pmatrix} x_{i-1} \\ \hdashline u_{i-1-p} \end{pmatrix}$.

The most general identification case, entailing a different delay acting on each input and output, requires solution for elements of the matrices $A_1, A_2, \ldots, B_1, B_2, \ldots$ in

$$x_i = A_1 x_{i-1} + A_2 x_{i-2} + \cdots + B_1 u_{i-1} + B_2 u_{i-2} + \cdots \tag{6.252}$$

6.6 Combined State and Parameter Observation Based on a System of Differential and Algebraic Equations

Following Equations 3.8 and 3.9, consider the system of first-order differential and algebraic equations (DAEs) with continuous states and inputs x and u, continuous ancillary variables z (such as flows) and continuous parameters p (such as rate constants and transfer coefficients):

$$\frac{\mathrm{d}\boldsymbol{x}}{\mathrm{d}t} = \boldsymbol{f}(\boldsymbol{x}, \boldsymbol{u}, \boldsymbol{z}, \boldsymbol{p}) \tag{6.253}$$

$$\boldsymbol{0} = \boldsymbol{g}(\boldsymbol{x}, \boldsymbol{u}, \boldsymbol{z}, \boldsymbol{p}) \tag{6.254}$$

Defining the Jacobians

$$\boldsymbol{A} = \frac{\partial \boldsymbol{f}}{\partial \boldsymbol{x}} = \begin{bmatrix} \frac{\partial f_1}{\partial x_1} & \frac{\partial f_1}{\partial x_2} & \cdots & \frac{\partial f_1}{\partial x_n} \\ \frac{\partial f_2}{\partial x_1} & \frac{\partial f_2}{\partial x_2} & \cdots & \frac{\partial f_2}{\partial x_n} \\ \vdots & \vdots & \ddots & \vdots \\ \frac{\partial f_n}{\partial x_1} & \frac{\partial f_n}{\partial x_2} & \cdots & \frac{\partial f_n}{\partial x_n} \end{bmatrix} \tag{6.255}$$

and

$$\boldsymbol{E} = \frac{\partial \boldsymbol{g}}{\partial \boldsymbol{x}} = \begin{bmatrix} \frac{\partial g_1}{\partial x_1} & \frac{\partial g_1}{\partial x_2} & \cdots & \frac{\partial g_1}{\partial x_n} \\ \frac{\partial g_2}{\partial x_1} & \frac{\partial g_2}{\partial x_2} & \cdots & \frac{\partial g_2}{\partial x_n} \\ \vdots & \vdots & \ddots & \vdots \\ \frac{\partial g_m}{\partial x_1} & \frac{\partial g_m}{\partial x_2} & \cdots & \frac{\partial g_m}{\partial x_n} \end{bmatrix} \tag{6.256}$$

and similarly

$$\boldsymbol{B} = \frac{\partial \boldsymbol{f}}{\partial \boldsymbol{u}} \tag{6.257}$$

and

$$\boldsymbol{F} = \frac{\partial \boldsymbol{g}}{\partial \boldsymbol{u}} \tag{6.258}$$

$$\boldsymbol{C} = \frac{\partial \boldsymbol{f}}{\partial \boldsymbol{z}} \tag{6.259}$$

and

$$\boldsymbol{G} = \frac{\partial \boldsymbol{g}}{\partial \boldsymbol{z}} \tag{6.260}$$

$$\boldsymbol{D} = \frac{\partial \boldsymbol{f}}{\partial \boldsymbol{p}} \tag{6.261}$$

and

$$\boldsymbol{H} = \frac{\partial \boldsymbol{g}}{\partial \boldsymbol{p}} \tag{6.262}$$

Then, evaluating the coefficients and offsets at the start of the interval, t_{i-1},

$$\left(\begin{array}{c}\dot{\boldsymbol{x}}\\ \hline \boldsymbol{0}\end{array}\right) \approx \left[\begin{array}{c|c|c|c}\boldsymbol{A} & \boldsymbol{B} & \boldsymbol{C} & \boldsymbol{D}\\ \hline \boldsymbol{E} & \boldsymbol{F} & \boldsymbol{G} & \boldsymbol{H}\end{array}\right]_{i-1} \left(\begin{array}{c}\boldsymbol{x}\\ \hline \boldsymbol{u}\\ \hline \boldsymbol{z}\\ \hline \boldsymbol{p}\end{array}\right) + \left(\begin{array}{c}\boldsymbol{v}_{i-1}\\ \hline \boldsymbol{w}_{i-1}\end{array}\right) \tag{6.263}$$

where

$$\left(\begin{array}{c}\boldsymbol{v}_{i-1}\\ \boldsymbol{w}_{i-1}\end{array}\right) = \left(\begin{array}{c}\boldsymbol{f}(\boldsymbol{x}_{i-1}, \boldsymbol{u}_{i-1}, \boldsymbol{z}_{i-1}, \boldsymbol{p}_{i-1})\\ \boldsymbol{g}(\boldsymbol{x}_{i-1}, \boldsymbol{u}_{i-1}, \boldsymbol{z}_{i-1}, \boldsymbol{p}_{i-1})\end{array}\right) - \left[\begin{array}{c|c|c|c}\boldsymbol{A} & \boldsymbol{B} & \boldsymbol{C} & \boldsymbol{D}\\ \hline \boldsymbol{E} & \boldsymbol{F} & \boldsymbol{G} & \boldsymbol{H}\end{array}\right]_{i-1} \left(\begin{array}{c}\boldsymbol{x}_{i-1}\\ \hline \boldsymbol{u}_{i-1}\\ \hline \boldsymbol{z}_{i-1}\\ \hline \boldsymbol{p}_{i-1}\end{array}\right) \tag{6.264}$$

Postulate that the ancillary variables and parameters have a tendency to drift towards 'measured values' $\boldsymbol{z}_{\mathrm{m}i-1}$ and $\boldsymbol{p}_{\mathrm{m}i-1}$ according to

$$\left(\begin{array}{c}\dot{\boldsymbol{z}}\\ \hline \dot{\boldsymbol{p}}\end{array}\right) \approx \left[\begin{array}{c|c}-(1/\tau)\boldsymbol{I} & \boldsymbol{0}\\ \hline \boldsymbol{0} & -(1/\tau)\boldsymbol{I}\end{array}\right] \left(\begin{array}{c}\boldsymbol{z}\\ \hline \boldsymbol{p}\end{array}\right) + \left[\begin{array}{c|c}(1/\tau)\boldsymbol{I} & \boldsymbol{0}\\ \hline \boldsymbol{0} & (1/\tau)\boldsymbol{I}\end{array}\right] \left(\begin{array}{c}\boldsymbol{z}_{\mathrm{m}i-1}\\ \hline \boldsymbol{p}_{\mathrm{m}i-1}\end{array}\right) \tag{6.265}$$

Then the continuous system, in an interval t_{i-1} to t_i, has the requirements

$$\left(\begin{array}{c}\dot{\boldsymbol{x}}\\ \hline \dot{\boldsymbol{z}}\\ \hline \dot{\boldsymbol{p}}\end{array}\right) \approx \left[\begin{array}{c|c|c}\boldsymbol{A} & \boldsymbol{C} & \boldsymbol{D}\\ \hline \boldsymbol{0} & -(1/\tau)\boldsymbol{I} & \boldsymbol{0}\\ \hline \boldsymbol{0} & \boldsymbol{0} & -(1/\tau)\boldsymbol{I}\end{array}\right]_{i-1} \left(\begin{array}{c}\boldsymbol{x}\\ \hline \boldsymbol{z}\\ \hline \boldsymbol{p}\end{array}\right) + \left[\begin{array}{c}\boldsymbol{B}\\ \hline \boldsymbol{0}\\ \hline \boldsymbol{0}\end{array}\right]_{i-1} \boldsymbol{u} + \left(\begin{array}{c}\boldsymbol{v}\\ \hline (1/\tau)\boldsymbol{z}_{\mathrm{m}}\\ \hline (1/\tau)\boldsymbol{p}_{\mathrm{m}}\end{array}\right)_{i-1} \tag{6.266}$$

$$-\boldsymbol{w}_{i-1} = \left[\begin{array}{c|c|c}\boldsymbol{E} & \boldsymbol{G} & \boldsymbol{H}\end{array}\right]_{i-1} \left(\begin{array}{c}\boldsymbol{x}\\ \hline \boldsymbol{z}\\ \hline \boldsymbol{p}\end{array}\right) + [\boldsymbol{F}]_{i-1}\boldsymbol{u} \tag{6.267}$$

Comparing the forms in Equations 6.266 and 6.267 with Equations 6.93 and 6.94, one has the 'state' model and an observation that must be matched, which is the basis of the linear Kalman–Bucy filter. The extra term $[\boldsymbol{F}]_{i-1}\boldsymbol{u}$ in the observation equation is easily included in Equations 6.93 and 6.94, so Equations 6.266 and 6.267 allow identification of $\boldsymbol{x}$, $\boldsymbol{z}$ and $\boldsymbol{p}$ in the interval $t_{i-1} < t < t_i$ using the continuous Kalman–Bucy filter of Section 6.4.2. Alternatively, if the $\boldsymbol{u}$ is piecewise constant at value $\boldsymbol{u}_{i-1}$ in the interval $t_{i-1} < t < t_i$, then Equation 6.266 may be converted into the discrete form using the matrix exponential according to Equations 3.282–3.285. Then, with Equation 6.267 set to

$$-\boldsymbol{w}_{i-1} = \left[\begin{array}{c|c|c}\boldsymbol{E} & \boldsymbol{G} & \boldsymbol{H}\end{array}\right]_{i-1} \left(\begin{array}{c}\boldsymbol{x}\\ \hline \boldsymbol{z}\\ \hline \boldsymbol{p}\end{array}\right)_i + [\boldsymbol{F}]_{i-1}\boldsymbol{u}_{i-1} \tag{6.268}$$

the discrete Kalman filter of Section 6.4.1 allows identification of $\boldsymbol{x}$, $\boldsymbol{z}$ and $\boldsymbol{p}$ on each step i. In this mode, the 'measurements' $\boldsymbol{z}_{\mathrm{m}i}$ and $\boldsymbol{p}_{\mathrm{m}i}$ can be updated at the end of each step to the new $\boldsymbol{z}_i$ and $\boldsymbol{p}_i$ values, allowing a larger range of movement.

With the re-evaluation of coefficients and offsets on each step, it is noted that both these filters would be operating in *extended* form (Section 6.4.3). Generally speaking, the measurements need to be very 'rich' in information for such a general identification to succeed. Nevertheless, convergence can be obtained by restricting the movement of as many $\boldsymbol{z}$ and $\boldsymbol{p}$ values as necessary to their starting values, by the choice of small values for the appropriate $\boldsymbol{Q}$ matrix terms for Equation 6.266 or its discrete equivalent.

6.7 Nonparametric Identification

Indirect identifications such as the cross-correlation function in Section 6.1.1, the frequency spectrum in Section 6.1.2 and the principal components in Section 6.1.3 are termed *nonparametric* because they do not seek the actual model parameters. Two further cases are now considered, starting from the linear input–output parametric model of Equation 6.132.

6.7.1 Impulse Response Coefficients by Cross-Correlation

$$A(q)\mathbf{y}(t) = \mathbf{B}(q)\mathbf{u}(t) \tag{6.269}$$

with

$$A(q) = \mathbf{I} - \mathbf{A}_1 q^{-1} - \mathbf{A}_2 q^{-2} - \cdots - \mathbf{A}_n q^{-n} \tag{6.270}$$

$$\mathbf{B}(q) = \mathbf{B}_1 q^{-1} + \mathbf{B}_2 q^{-2} + \cdots + \mathbf{B}_m q^{-m} \tag{6.271}$$

Writing

$$\mathbf{y}(t) = [A(q)]^{-1}\mathbf{B}(q)\mathbf{u}(t) \tag{6.272}$$

one notes that the vectors $\mathbf{y}$ and $\mathbf{u}$ are related by a matrix of transfer functions $g_{ij}(q)$

$$g_{ij}(q) = \frac{b_{ij1}q^{-1} + \cdots + b_{ijm_{ij}}q^{-m_{ij}}}{1 + a_{ij1}q^{-1} + \cdots + a_{ijn_{ij}}q^{-n_{ij}}} \tag{6.273}$$

Long division yields an infinite series

$$g_{ij}(q) = c_{ij1}q^{-1} + c_{ij2}q^{-2} + c_{ij3}q^{-3} + \cdots \tag{6.274}$$

This form is called the (infinite) *impulse response* (IIR). One notes that a single unit impulse at $t=0$ will give subsequent output values equal to the series of coefficients. As in Section 6.1.1, the cross-correlation analysis for a chosen input–output pair (y_i, u_j) is

$$R_{y_iu_j}(n) = \frac{E\left\{\left[y_i(t) - \mu_{y_i}\right]\left[u_j(t - n\Delta t) - \mu_{u_j}\right]\right\}}{\sigma_{y_i}\sigma_{u_j}} \tag{6.275}$$

$$R_{y_iu_j}(n) = \frac{E\left\{\left[\sum_{k=1}^{\infty} c_{ijk}\left(u_j(t - k\Delta t) - \mu_{u_j}\right)\right]\left[u_j(t - n\Delta t) - \mu_{u_j}\right]\right\}}{\sigma_{y_i}\sigma_{u_j}} \tag{6.276}$$

$$R_{y_iu_j}(n) = \frac{\sigma_{u_j}}{\sigma_{y_i}}\sum_{k=1}^{\infty} c_{ijk}R_{u_ju_j}(n-k) \tag{6.277}$$

Thus, the coefficients c_{ijk} can be determined by comparison of the cross-correlation function $R_{y_iu_j}(n)$ with earlier points $(n-k)$ in the autocorrelation function $R_{u_ju_j}(n)$. In ergodic, non-integrating systems, the coefficients c_{ijk} decrease rapidly as k increases, so a truncated series is sufficient. This also means that the achievable values of y become bounded, so one speaks of the *finite impulse response* (FIR) form. Although this method does not reveal the actual coefficients in $\boldsymbol{A}$ and $\boldsymbol{B}$, it could be used to check deviations from the c coefficients expected in Equation 6.274, giving some indication of the validity of the model for each input–output pair.

Dynamic matrix controllers (Section 7.8.2) use either the *finite impulse response* or its integral, the *step response*, as the basis of their predictive models. Equations 6.275–6.277 thus offer a direct means of estimating the necessary coefficients from process data, for example in real-time adaptive control applications. Where the model basis is presented in the form of Equation 6.273, the model validity can alternatively be checked by comparison of the implied c_{ijk} coefficients using Equations 6.274 and 6.277.

Example 6.3

Determination of impulse response coefficients from measured process data for a first-order system with dead time.

The SISO discrete system

$$y_i = 0.9y_{i-1} + 0.1u_{i-20} \tag{6.278}$$

is used to generate a long (u, y) data record using a correlated random u input generated by passing random top-hat samples through a second-order filter. The auto- and cross-correlations R_{uu} and R_{yu} are found, and used to estimate c_k as in Equations 6.275–6.277 (see Figure 6.21). The solution for this example is given in Figure 6.22.

6.7.2 Direct RLS Identification of a Dynamic Matrix (Step Response)

Another common form of discrete linear model is that based on the dynamic matrix of Section 3.9.5.1. In the Nth line of Equation 3.387, one has

$$\boldsymbol{y}_{(N)} = \boldsymbol{y}_{(0)} + \boldsymbol{b}_{(N)}\Delta\boldsymbol{u}_{(0)} + \boldsymbol{b}_{(N-1)}\Delta\boldsymbol{u}_{(1)} + \cdots + \boldsymbol{b}_{(2)}\Delta\boldsymbol{u}_{(N-2)} + \boldsymbol{b}_{(1)}\Delta\boldsymbol{u}_{(N-1)} \tag{6.279}$$

There was an assumption in this equation that the system was initially at steady state at $\boldsymbol{y}_{(0)}$, and that there were no earlier moves $\Delta\boldsymbol{u}$ which could affect the response. Now allow for the possibility of earlier moves than $t = (0)$. For non-integrating responses that reach equilibrium at $t < (N)$, only a fixed steady offset $\boldsymbol{y}_{\text{off}}$ will be contributed at $t \geq (N)$,

$$\boldsymbol{y}_{(N)} = \boldsymbol{y}_{(0)} + \boldsymbol{b}_{(N)}\Delta\boldsymbol{u}_{(0)} + \boldsymbol{b}_{(N-1)}\Delta\boldsymbol{u}_{(1)} + \cdots + \boldsymbol{b}_{(2)}\Delta\boldsymbol{u}_{(N-2)} + \boldsymbol{b}_{(1)}\Delta\boldsymbol{u}_{(N-1)} + \boldsymbol{y}_{\text{off}} \tag{6.280}$$

Using the same starting point to predict the $(N+i)$th $\boldsymbol{y}$ output, the full sequence is

$$\boldsymbol{y}_{(N)} = \boldsymbol{y}_{(0)} + \boldsymbol{b}_{(N)}\Delta\boldsymbol{u}_{(0)} + \boldsymbol{b}_{(N-1)}\Delta\boldsymbol{u}_{(1)} + \cdots + \boldsymbol{b}_{(2)}\Delta\mathbf{u}_{(N-2)} + \boldsymbol{b}_{(1)}\Delta\boldsymbol{u}_{(N-1)} + \boldsymbol{y}_{\text{off}} \tag{6.281}$$

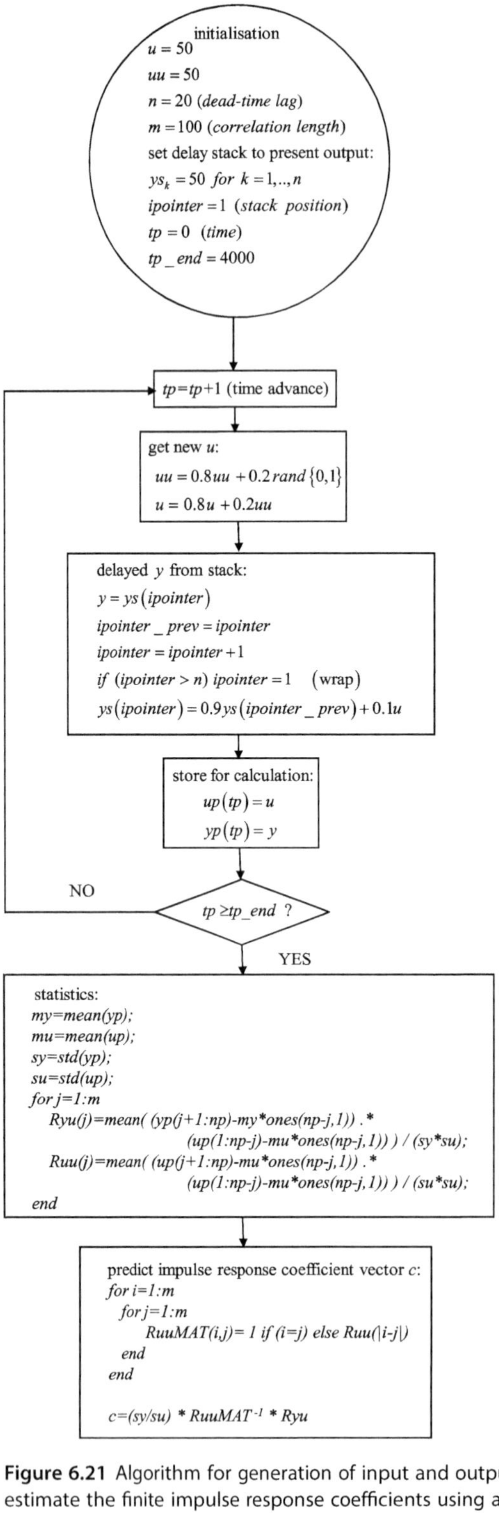

Figure 6.21 Algorithm for generation of input and output data for a first-order-plus-dead-time process, and to estimate the finite impulse response coefficients using auto- and cross-correlation functions. Refer to Example 6.3.

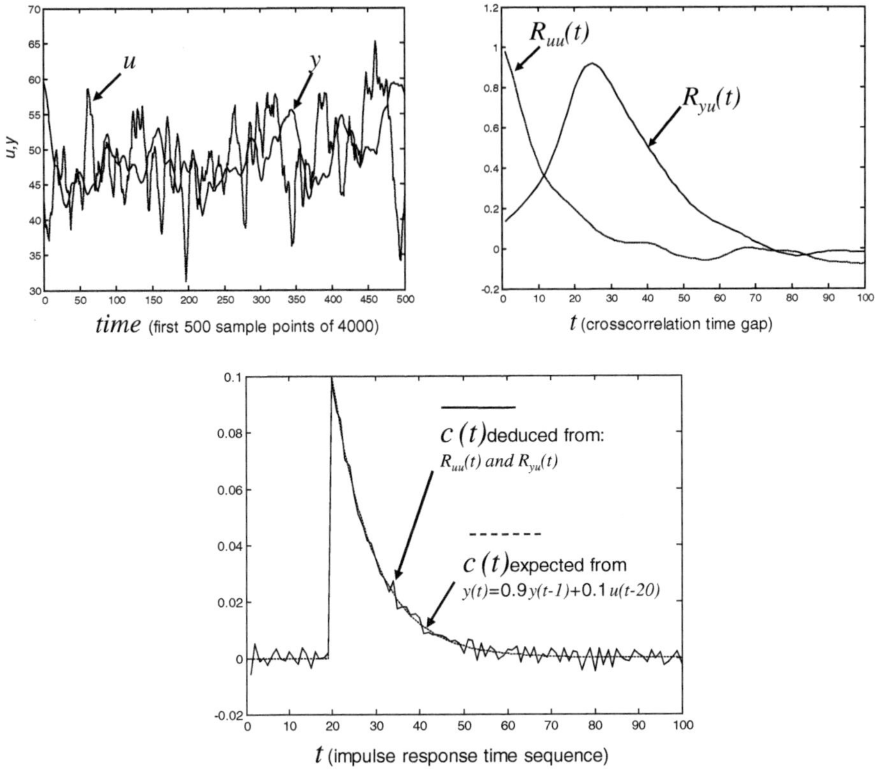

Figure 6.22 Solution using the algorithm in Figure 6.21: estimation of finite impulse response coefficients using moving-window autocorrelation $R_{uu}(t)$ and cross-correlation $R_{uu}(t)$. Comparison with theoretical finite impulse response coefficients.

$$\boldsymbol{y}_{(N+1)} = \boldsymbol{y}_{(0)} + \boldsymbol{b}_{(N)}\Delta\boldsymbol{u}_{(0)} + \boldsymbol{b}_{(N)}\Delta\boldsymbol{u}_{(1)} + \boldsymbol{b}_{(N-1)}\Delta\boldsymbol{u}_{(2)} + \cdots + \boldsymbol{b}_{(2)}\Delta\boldsymbol{u}_{(N-1)} + \boldsymbol{b}_{(1)}\Delta\boldsymbol{u}_{(N)} + \boldsymbol{y}_{\text{off}} \tag{6.282}$$

$$\boldsymbol{y}_{(N+2)} = \boldsymbol{y}_{(0)} + \boldsymbol{b}_{(N)}\Delta\boldsymbol{u}_{(0)} + \boldsymbol{b}_{(N)}\Delta\boldsymbol{u}_{(1)} + \boldsymbol{b}_{(N)}\Delta\boldsymbol{u}_{(2)} + \boldsymbol{b}_{(N-1)}\Delta\boldsymbol{u}_{(3)} + \cdots + \boldsymbol{b}_{(2)}\Delta\boldsymbol{u}_{(N)} + \boldsymbol{b}_{(1)}\Delta\boldsymbol{u}_{(N+1)} + \boldsymbol{y}_{\text{off}} \tag{6.283}$$

$$\vdots$$

Arranging for the output changes over the interval $N+1$ to $2N$:

$$\begin{bmatrix} (\boldsymbol{y}_{N+1}-\boldsymbol{y}_N)^{\mathrm{T}} \\ (\boldsymbol{y}_{N+2}-\boldsymbol{y}_N)^{\mathrm{T}} \\ \vdots \\ (\boldsymbol{y}_{2N-1}-\boldsymbol{y}_N)^{\mathrm{T}} \\ (\boldsymbol{y}_{2N}-\boldsymbol{y}_N)^{\mathrm{T}} \end{bmatrix} = \begin{bmatrix} (\Delta\boldsymbol{u}_{(N)}-\Delta\boldsymbol{u}_{(N-1)})^{\mathrm{T}} & (\Delta\boldsymbol{u}_{(N-1)}-\Delta\boldsymbol{u}_{(N-2)})^{\mathrm{T}} & \cdots & (\Delta\boldsymbol{u}_{(2)}-\Delta\boldsymbol{u}_{(1)})^{\mathrm{T}} & \Delta\boldsymbol{u}_{(1)}^{\mathrm{T}} \\ (\Delta\boldsymbol{u}_{(N+1)}-\Delta\boldsymbol{u}_{(N-1)})^{\mathrm{T}} & (\Delta\boldsymbol{u}_{(N)}-\Delta\boldsymbol{u}_{(N-2)})^{\mathrm{T}} & \cdots & (\Delta\boldsymbol{u}_{(3)}-\Delta\boldsymbol{u}_{(1)})^{\mathrm{T}} & \Delta\boldsymbol{u}_{(1)}^{\mathrm{T}}+\Delta\boldsymbol{u}_{(2)}^{\mathrm{T}} \\ \vdots & \vdots & \ddots & \vdots & \vdots \\ (\Delta\boldsymbol{u}_{(2N-2)}-\Delta\boldsymbol{u}_{(N-1)})^{\mathrm{T}} & (\Delta\boldsymbol{u}_{(2N-3)}-\Delta\boldsymbol{u}_{(N-2)})^{\mathrm{T}} & \cdots & (\Delta\boldsymbol{u}_{(N)}-\Delta\boldsymbol{u}_{(1)})^{\mathrm{T}} & \Delta\boldsymbol{u}_{(1)}^{\mathrm{T}}+\cdots+\Delta\boldsymbol{u}_{(N-1)}^{\mathrm{T}} \\ (\Delta\boldsymbol{u}_{(2N-1)}-\Delta\boldsymbol{u}_{(N-1)})^{\mathrm{T}} & (\Delta\boldsymbol{u}_{(2N-2)}-\Delta\boldsymbol{u}_{(N-2)})^{\mathrm{T}} & \cdots & (\Delta\boldsymbol{u}_{(N+1)}-\Delta\boldsymbol{u}_{(1)})^{\mathrm{T}} & \Delta\boldsymbol{u}_{(1)}^{\mathrm{T}}+\cdots+\Delta\boldsymbol{u}_{(N)}^{\mathrm{T}} \end{bmatrix} \begin{bmatrix} \boldsymbol{b}_{(1)}^{\mathrm{T}} \\ \boldsymbol{b}_{(2)}^{\mathrm{T}} \\ \vdots \\ \boldsymbol{b}_{(N-1)}^{\mathrm{T}} \\ \boldsymbol{b}_{(N)}^{\mathrm{T}} \end{bmatrix} \tag{6.284}$$

For recursive identification, let the $2N$ point be current time i in a moving window:

$$\underbrace{\begin{bmatrix} (y_{i-N+1}-y_{i-N})^{\mathrm{T}} \\ (y_{i-N+2}-y_{i-N})^{\mathrm{T}} \\ \vdots \\ (y_{i-1}-y_{i-N})^{\mathrm{T}} \\ (y_i-y_{i-N})^{\mathrm{T}} \end{bmatrix}}_{Y_i} = \underbrace{\begin{bmatrix} (\Delta u_{(i-N)}-\Delta u_{(i-N-1)})^{\mathrm{T}} & (\Delta u_{(i-N-1)}-\Delta u_{(i-N-2)})^{\mathrm{T}} & \cdots & (\Delta u_{(i-2N+2)}-\Delta u_{(i-2N+1)})^{\mathrm{T}} & \Delta u_{(i-2N+1)}^{\mathrm{T}} \\ (\Delta u_{(i-N+1)}-\Delta u_{(i-N-1)})^{\mathrm{T}} & (\Delta u_{(i-N)}-\Delta u_{(i-N-2)})^{\mathrm{T}} & \cdots & (\Delta u_{(i-2N+3)}-\Delta u_{(i-2N+1)})^{\mathrm{T}} & \Delta u_{(i-2N+1)}^{\mathrm{T}}+\Delta u_{(i-2N+2)}^{\mathrm{T}} \\ \vdots & \vdots & \ddots & \vdots & \vdots \\ (\Delta u_{(i-2)}-\Delta u_{(i-N-1)})^{\mathrm{T}} & (\Delta u_{(i-3)}-\Delta u_{(i-N-2)})^{\mathrm{T}} & \cdots & (\Delta u_{(i-N)}-\Delta u_{(i-2N+1)})^{\mathrm{T}} & \Delta u_{(i-2N+1)}^{\mathrm{T}}+\cdots+\Delta u_{(i-N-1)}^{\mathrm{T}} \\ (\Delta u_{(i-1)}-\Delta u_{(i-N-1)})^{\mathrm{T}} & (\Delta u_{(i-2)}-\Delta u_{(i-N-2)})^{\mathrm{T}} & \cdots & (\Delta u_{(i-N+1)}-\Delta u_{(i-2N+1)})^{\mathrm{T}} & \Delta u_{(i-2N+1)}^{\mathrm{T}}+\cdots+\Delta u_{(i-N)}^{\mathrm{T}} \end{bmatrix}}_{U_i} \underbrace{\begin{bmatrix} b_{(1)}^{\mathrm{T}} \\ b_{(2)}^{\mathrm{T}} \\ \vdots \\ b_{(N-1)}^{\mathrm{T}} \\ b_{(N)}^{\mathrm{T}} \end{bmatrix}}_{\beta_i} \quad (6.285)$$

The RLS procedure then requires on each time step i the following sequence of assignments:

$$\boxed{P_{i|i-1} = P_{i-1|i-1} + Q \quad \text{(predicted)}} \quad (6.286)$$

$$\boxed{K_i = P_{i|i-1} U_i^{\mathrm{T}} \left[U_i P_{i|i-1} U_i^{\mathrm{T}} + R \right]^{-1}} \quad (6.287)$$

$$\boxed{\beta_i = \beta_{i-1} + K_i \left[Y_{\mathrm{i}} - U_i \beta_{i-1} \right]} \quad (6.288)$$

$$\boxed{P_{i|i} = [I - K_i U_i] P_{i|i-1} \quad \text{(updated)}} \quad (6.289)$$

As written, for the general MIMO case where the $b_{(i)}$ are matrices, Y_i and β_i in Equation 6.285 are matrices. The implication is that P, Q and R will be fourth-order tensors in Equations 6.286–6.289. To avoid this, Equation 6.285 should be rearranged such that the Y_i and β_i elements are in vector format.

References

Box, G.E.P. and Jenkins, G.M. (1970) *Time series analysis forecasting and control*, Holden-Day, San Francisco.

Chatfield, C. (2004) *The Analysis of Time Series: An Introduction*, 6th edn, Chapman & Hall/CRC Press, New York.

Cooley, J.W. and Tukey, J.W. (1965) An algorithm for the machine calculation of complex Fourier series. *Mathematics of Computation*, **19**, 297–301.

Franks, R.G.E. (1982) DYFLO update: DYFLO2. Summer Computer Simulation Conference, Society for Computer Simulation, La Jolla, CA, pp. 507–513.

Giles, R.F. and Gaines, L.D. (1977) Controlling inerts in ammonia synthesis. *Instrumentation Technology*, **24**, 41–45.

Kalman, R.E. and Bucy, R.S. (1961) New results in linear filtering and prediction theory. *Journal of Basic Engineering, Transactions of the American Society of Mechanical Engineers, Series D*, **83**, 95–108.

Kleinman, D.L. (1968) On the iterative technique for Riccati equation computations. *IEEE Transactions on Automatic Control*, **13**, 114–115.

Luenberger, D.G. (1966) Observers for multivariable systems. *IEEE Transactions on Automatic Control*, **11**, 190–197.

Mulholland, M. (1988) Application of minimum-variance control strategies on an ammonia plant. *Computers & Chemical Engineering*, **12**, 297–302.

Smith, O.J.M. (1957) Closer control of loops with dead time. *Chemical Engineering Progress*, **53** (5), 217–219.

Ydstie, B.E., Kershenbaum, L.S. and Sargent, R.W.H. (1985) Theory and application of a self-tuning controller. *AIChE Journal*, **31** (11), 1771–1780.

Young, P.C. (1969) Applying parameter estimation to dynamic systems. Part I. *Control Engineering*, **16** (10), 119–125.

7
Advanced Control Algorithms

This chapter focuses largely on multi-input multi-output (MIMO) controllers. The same equations can be simplified where necessary for the SISO case.

7.1
Discrete z-Domain Minimal Prototype Controllers

This is a sampled data controller based on a model of the process. The design procedure seeks to achieve a stipulated sampled response to a stipulated disturbance to the system. This may not always be achievable, so an alternative desired response may have to be given. Figure 7.1 shows a typical installation in which the manipulated variable values are updated by comparison of feedback with setpoint only at fixed intervals T. The aim will be to achieve a particular form of measured output $\boldsymbol{y}$ in response to particular forms of setpoint variation $\boldsymbol{y}_\mathrm{s}$ and/or load disturbance variations $\boldsymbol{d}$.

Recalling from Section 3.9.3.1 that the analogue sections must be combined between sampling points, before transforming to 'impulse-modulated' (z) form, and being careful to preserve the transfer function multiplication order so that MIMO systems are included, the system shown in Figure 7.1 is seen to have a measured output determined by

$$\boldsymbol{y}(z) = [\boldsymbol{G}_\mathrm{M}\boldsymbol{G}_\mathrm{L}\boldsymbol{d}](z) + [\boldsymbol{G}_\mathrm{M}\boldsymbol{G}_\mathrm{P}\boldsymbol{G}_\mathrm{H}](z)\boldsymbol{G}_\mathrm{C}(z)\{\boldsymbol{y}_\mathrm{s}(z) - \boldsymbol{y}(z)\} \tag{7.1}$$

that is

$$\boldsymbol{y}(z) = \{\boldsymbol{I} + [\boldsymbol{G}_\mathrm{M}\boldsymbol{G}_\mathrm{P}\boldsymbol{G}_\mathrm{H}](z)\boldsymbol{G}_\mathrm{C}(z)\}^{-1}\{[\boldsymbol{G}_\mathrm{M}\boldsymbol{G}_\mathrm{L}\boldsymbol{d}](z) + [\boldsymbol{G}_\mathrm{M}\boldsymbol{G}_\mathrm{P}\boldsymbol{G}_\mathrm{H}](z)\boldsymbol{G}_\mathrm{C}(z)\boldsymbol{y}_\mathrm{s}(z)\} \tag{7.2}$$

7.1.1
Setpoint Tracking Discrete Minimal Prototype Controller

Minimal prototype controllers will differ for setpoint or load disturbance, and for the particular type of disturbance chosen, for example step versus ramp. In Examples 7.1 and 7.2, control of a first-order system, with and without dead time, is considered, based on specifications for setpoint tracking.

Applied Process Control: Essential Methods, First Edition. Michael Mulholland.

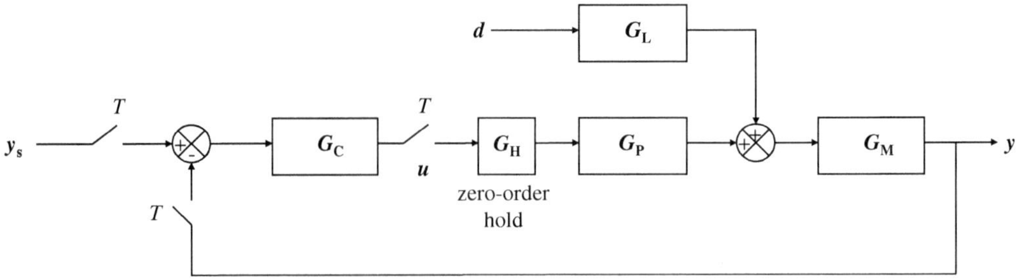

Figure 7.1 Discrete feedback control of process G_P with load disturbance.

Example 7.1

Minimal prototype controller for a first-order SISO system based on a setpoint step.

In the absence of a load disturbance,

$$\mathbf{G}_M\mathbf{G}_L\mathbf{d}(z) = \mathbf{0} \tag{7.3}$$

and it is known from Example 3.11 that the first-order system without measurement dynamics is represented by

$$[\mathbf{G}_M\mathbf{G}_P\mathbf{G}_H](z) = \frac{K\left(1 - e^{aT}\right)}{z - e^{aT}} \tag{7.4}$$

Consider a unit setpoint step

$$\mathbf{y}_s = \frac{z}{z-1} \tag{7.5}$$

from Table 3.7. One already knows that y cannot follow this immediately, but propose that it is able to achieve the setpoint value one step later, and thereafter is able to equal the setpoint at each sampling instant.

$$\mathbf{y} = \frac{1}{z-1} \tag{7.6}$$

Substitute in Equation 7.2 to obtain

$$\frac{1}{z-1} = \left\{1 + \frac{K\left(1 - e^{aT}\right)}{z - e^{aT}}\mathbf{G}_C(z)\right\}^{-1}\left\{\frac{K\left(1 - e^{aT}\right)}{z - e^{aT}}\mathbf{G}_C(z)\frac{z}{z-1}\right\} \tag{7.7}$$

whence

$$\mathbf{G}_C(z) = \frac{\mathbf{u}(z)}{\mathbf{y}_s(z) - \mathbf{y}(z)} = \frac{1}{K(1 - e^{aT})}\left(\frac{z - e^{aT}}{z-1}\right) \tag{7.8}$$

which is physically realisable since the order of the numerator polynomial is not greater than that of the denominator polynomial. The closed-loop pole is determined by the *characteristic equation*

$$\det\{\boldsymbol{I} + [\boldsymbol{G}_M \boldsymbol{G}_P \boldsymbol{G}_H](z)\boldsymbol{G}_C(z)\} = 0 \tag{7.9}$$

which in this case is

$$\frac{z}{z-1} = 0, \quad \text{i.e.} \quad z = 0 \Rightarrow s = -\infty \quad (\zeta = 1, \text{ critically damped}) \tag{7.10}$$

so there is no overshoot (Figure 7.2). The particular behaviour of a controller like this, which guarantees to strike the setpoint at the sampling instants, is described as *dead-beat control*. One notes that no information is available about the behaviour of the feedback signal *between* the sampling instants!

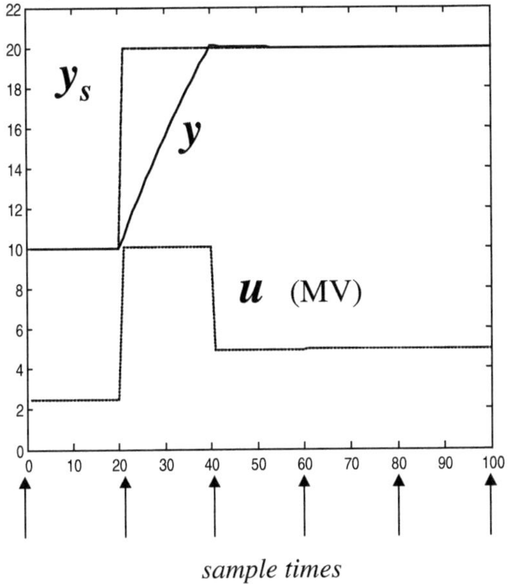

Figure 7.2 Setpoint step response for controller of Example 7.1 with $K = 4$, $a = -0.02$ and $T = 20$.

Example 7.2

Minimal prototype controller: Dahlin algorithm for a first-order SISO system with dead time, based on a setpoint step.

For a transport lag of k intervals,

$$[\boldsymbol{G}_M \boldsymbol{G}_P \boldsymbol{G}_H](z) = \frac{z^{-k} K (1 - e^{aT})}{z - e^{aT}} \tag{7.11}$$

For a unit setpoint step, Dahlin (1968) stipulated a *first-order plus dead-time* response (regardless of the form of the open-loop system), which Equation 3.254 shows to be

$$y = z^{-k}\left(\frac{z}{z-1} - \frac{z}{z-e^{bT}}\right) \tag{7.12}$$

and it is noted that b will differ from a in the example SISO system equation (Equation 7.11). As in Example 7.1, substitute in Equation 7.2 to obtain

$$G_C(z) = \frac{(1-e^{bT})}{K(1-e^{aT})}\left\{\frac{(z-e^{aT})}{(z-e^{bT}) - z^{-k}(1-e^{bT})}\right\} \tag{7.13}$$

$$G_C(z) = \frac{\boldsymbol{u}(z)}{\boldsymbol{y}_s(z) - \boldsymbol{y}(z)} = \frac{(1-e^{bT})}{K(1-e^{aT})}\left\{\frac{(z-e^{aT})}{(z-e^{bT}) - z^{-k}(1-e^{bT})}\right\} \tag{7.14}$$

that is

$$\boldsymbol{u}_i = e^{bT}\boldsymbol{u}_{i-1} + (1-e^{bT})\boldsymbol{u}_{i-k-1} + \frac{(1-e^{bT})}{K(1-e^{aT})}\{(\boldsymbol{y}_{si} - \boldsymbol{y}_i) - e^{aT}(\boldsymbol{y}_{si-1} - \boldsymbol{y}_{i-1})\} \tag{7.15}$$

The parameter $b = -1/\tau$ for the desired response is used for tuning purposes. Figure 7.3 shows the closed-loop response of this controller to a setpoint step. The Dahlin algorithm, also found independently by Higham (1968), was the first practical feedback algorithm with dead-time compensation. Adapted versions can include elements of a PID controller within the expression.

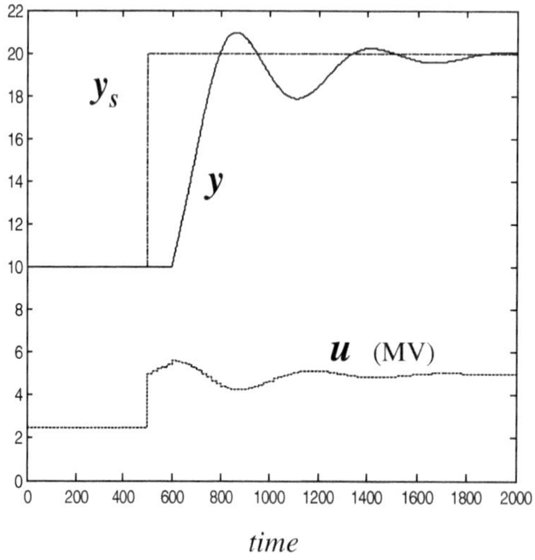

Figure 7.3 Setpoint step response for dead-time compensating Dahlin controller of Example 7.2, with dead time = 100, $K = 4$, $a = -0.005$, $T = 20$ and tuning set as $b = -0.0025$.

7.1.2
Setpoint Tracking and Load Disturbance Suppression with a Discrete Minimal Prototype Controller (Two-Degree-of-Freedom Controller)

A similar procedure to Example 7.1 is used to design the controller on the basis of a *load disturbance* alone. For example, for a unit step, one might take

$$[\boldsymbol{G}_{\mathrm{M}}\boldsymbol{G}_{\mathrm{L}}\boldsymbol{d}](z) = \frac{z}{z-1} \tag{7.16}$$

One would like to demand in this case that the output $\boldsymbol{y}$ remains unchanged ($\boldsymbol{y}=\boldsymbol{y}_{\mathrm{s}}$) at the subsequent sampling points, but of course this is impossible, because the controller will only react once it is aware of the deviation of $\boldsymbol{y}$ after the first interval, and could only achieve the setpoint again at the end of the second interval. So the stipulation of the $\boldsymbol{y}$ output must give the *open-loop* response to the load disturbance at the end of the first interval, and the setpoint value thereafter.

In practice, it is often desirable for a controller to handle both setpoint tracking and disturbance suppression well. This requires a *two-degree-of-freedom controller* set-up as in Figure 7.4.

For the arrangement in Figure 7.4, Equation 7.2 becomes

$$\boldsymbol{y}(z) = \{\boldsymbol{I} + [\boldsymbol{G}_{\mathrm{M}}\boldsymbol{G}_{\mathrm{P}}\boldsymbol{G}_{\mathrm{H}}](z)\boldsymbol{G}_{\mathrm{C1}}(z)\boldsymbol{G}_{\mathrm{C3}}(z)\}^{-1}\{[\boldsymbol{G}_{\mathrm{M}}\boldsymbol{G}_{\mathrm{L}}\boldsymbol{d}](z) + [\boldsymbol{G}_{\mathrm{M}}\boldsymbol{G}_{\mathrm{P}}\boldsymbol{G}_{\mathrm{H}}](z)\boldsymbol{G}_{\mathrm{C1}}(z)\boldsymbol{G}_{\mathrm{C2}}(z)\boldsymbol{y}_{\mathrm{s}}(z)\} \tag{7.17}$$

For a SISO system, this is

$$y = \left(\frac{1}{1 + G_{\mathrm{MPH}}G_{\mathrm{C1}}G_{\mathrm{C3}}}\right)g_{\mathrm{MLd}} + \left(\frac{G_{\mathrm{MPH}}G_{\mathrm{C1}}G_{\mathrm{C2}}}{1 + G_{\mathrm{MPH}}G_{\mathrm{C1}}G_{\mathrm{C3}}}\right)y_{\mathrm{s}} \tag{7.18}$$

where

$$G_{\mathrm{MPH}} = [G_{\mathrm{M}}G_{\mathrm{P}}G_{\mathrm{H}}](z) \tag{7.19}$$

and

$$g_{\mathrm{MLd}} = [G_{\mathrm{M}}G_{\mathrm{L}}d](z) \tag{7.20}$$

The design procedure is to first choose $\boldsymbol{G}_{\mathrm{C1}}$ and $\boldsymbol{G}_{\mathrm{C3}}$ to achieve the desirable *disturbance suppression* as described above in this section. Once this is done, the *prefilter* $\boldsymbol{G}_{\mathrm{C2}}$ is specified as in Example 7.1 or 7.2 to achieve the desirable *setpoint tracking*. A 'dual algorithm' of this nature has been proposed by Shunta and Luyben (1972).

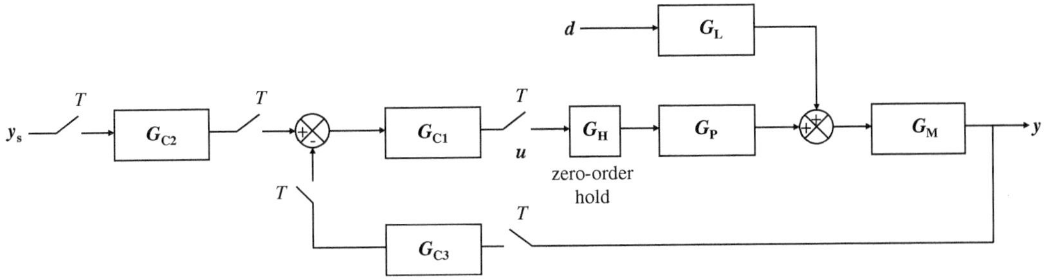

Figure 7.4 Two-degree-of-freedom controller arrangement.

7.2
Continuous s-Domain MIMO Controller Decoupling Design by Inverse Nyquist Array

Consider a continuous MIMO system equivalent to the discrete system in Figure 7.1, with a *decoupler* $\boldsymbol{G}_\mathrm{D}$ inserted between the controller and the process (Figure 7.5). Here one is considering a system with an equal number n of inputs and outputs, to offer the possibility of individually driving each output to a setpoint.

The closed-loop behaviour is determined by

$$\boldsymbol{y} = \{\boldsymbol{I} + \boldsymbol{G}_\mathrm{M}\boldsymbol{G}_\mathrm{P}\boldsymbol{G}_\mathrm{D}\boldsymbol{G}_\mathrm{C}\}^{-1}\{\boldsymbol{G}_\mathrm{M}\boldsymbol{G}_\mathrm{L}\boldsymbol{d} + \boldsymbol{G}_\mathrm{M}\boldsymbol{G}_\mathrm{P}\boldsymbol{G}_\mathrm{D}\boldsymbol{G}_\mathrm{C}\boldsymbol{y}_\mathrm{s}\} \tag{7.21}$$

where it is understood that the vectors and matrices are functions in the s-domain. From a setpoint tracking point of view, one desires to obtain a closed-loop transfer function

$$\boldsymbol{G} = \{\boldsymbol{I} + \boldsymbol{G}_\mathrm{M}\boldsymbol{G}_\mathrm{P}\boldsymbol{G}_\mathrm{D}\boldsymbol{G}_\mathrm{C}\}^{-1}\boldsymbol{G}_\mathrm{M}\boldsymbol{G}_\mathrm{P}\boldsymbol{G}_\mathrm{D}\boldsymbol{G}_\mathrm{C} \tag{7.22}$$

which is as 'diagonal' as possible, to minimise cross-disturbances if a setpoint is changed. This effectively means that the open-loop transfer function $\boldsymbol{G}_\mathrm{M}\boldsymbol{G}_\mathrm{P}\boldsymbol{G}_\mathrm{D}\boldsymbol{G}_\mathrm{C}$ should be as diagonal as possible, creating a canonical system in which the loops operate independently.

Quite a lot of work has been done on the design of a controller $\boldsymbol{G}_\mathrm{C}$ that does this without needing $\boldsymbol{G}_\mathrm{D}$, or alternatively the design of a decoupler $\boldsymbol{G}_\mathrm{D}$ which allows $\boldsymbol{G}_\mathrm{C}$ to have a simple diagonal form, for example with a PID controller in each diagonal position. However, in the processing industries, the most important task is *load disturbance suppression*, rather than *setpoint tracking*, so there are not many applications of the diagonalisation technique, which will thus be outlined only briefly below.

Rosenbrock (1974) proposed a simple approximate method to visualise the problems of interaction. Considering the open-loop transfer function

$$\boldsymbol{G}_\mathrm{O}(s) = \boldsymbol{G}_\mathrm{M}(s)\boldsymbol{G}_\mathrm{P}(s)\boldsymbol{G}_\mathrm{D}(s)\boldsymbol{G}_\mathrm{C}(s) \tag{7.23}$$

his method is based on the inverse

$$\widehat{\boldsymbol{Q}}(s) = \boldsymbol{G}_\mathrm{O}^{-1}(s) = \boldsymbol{G}_\mathrm{C}^{-1}(s)\boldsymbol{G}_\mathrm{D}^{-1}(s)\boldsymbol{G}_\mathrm{P}^{-1}(s)\boldsymbol{G}_\mathrm{M}^{-1}(s) \tag{7.24}$$

As illustrated in Figure 7.6, this transfer function will convert a vector of output changes $\Delta\boldsymbol{y}$ into the vector of error changes $\Delta\boldsymbol{e}$ at the comparator in Figure 7.5 *which would be needed to cause the*

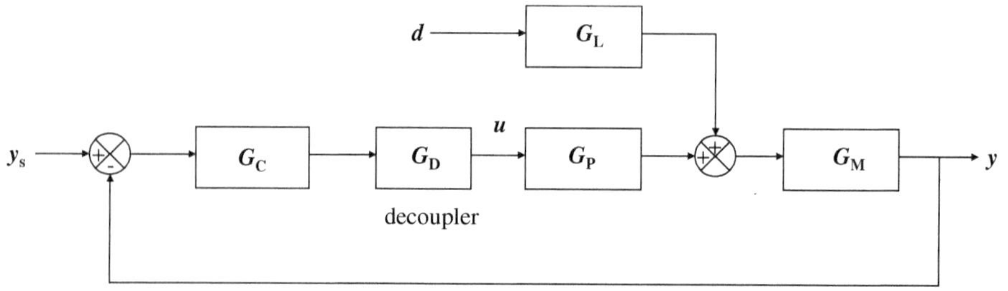

Figure 7.5 Continuous MIMO system with controller and decoupler.

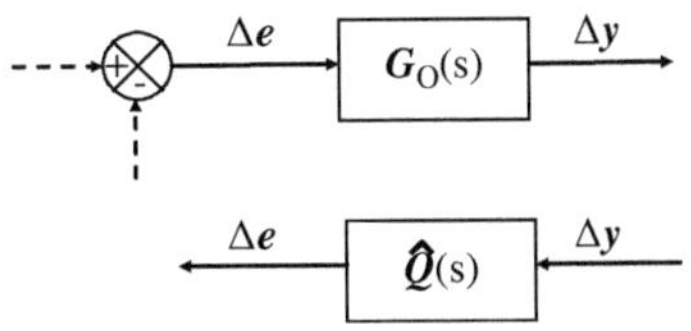

Figure 7.6 Conceptual view of inverse system considered by Rosenbrock (1974).

output changes. Just as one seeks a diagonal $\mathbf{G}_O(s)$, so one equally seeks a diagonal $\widehat{\mathbf{Q}}(s)$ to minimise cross-interactions between the $\Delta\mathbf{y}$ and the $\Delta\mathbf{e}$, and therefore between individual loops that would otherwise be operating in parallel.

The *frequency response* (Section 8.6.1) for this inverse system is obtained by substituting $j\omega$ for s in the transfer function

$$\widehat{\mathbf{Q}}(j\omega) = \mathbf{A}(\omega) + j\mathbf{B}(\omega) \tag{7.25}$$

A particular frequency ω (radians/time) determines a particular matrix of points $a_{ij}+jb_{ij}$ on the complex plane, with real values $\mathbf{A}(\omega)$ and imaginary values $\mathbf{B}(\omega)$. As ω varies from 0 to $+\infty$, these complex values trace out curves on the complex plane known as *inverse Nyquist plots* (see Section 8.6.4 for the *Nyquist plot* of element g_{ij} of $\mathbf{G}_O$). In terms of polar coordinates, the ijth curve of the inverse is determined by

$$R_{Aij}(\omega)e^{j\phi_{ij}(\omega)} \quad \text{or} \quad R_{Aij}(\omega)\angle\phi_{ij}(\omega) \quad \text{for } \omega \text{ proceeding from 0 to } +\infty \tag{7.26}$$

$$\text{MAGNITUDE:} \quad R_{Aij}(\omega) = \sqrt{a_{ij}^2(\omega) + b_{ij}^2(\omega)} \quad \text{(amplitude ratio)} \tag{7.27}$$

$$\text{ANGLE:} \quad \phi_{ij}(\omega) = \angle\left(a_{ij}(\omega) + jb_{ij}(\omega)\right) \quad \text{(phase angle)} \tag{7.28}$$

The effect of inverting the original system is illustrated for a particular element ij in Figure 7.7. Here the Nyquist plot is shown for the original $g_{ij(j\omega)}$ and the inverted system $1/g_{ij(j\omega)}$. The inverse magnitude $R_{Aij}(\omega) = 1/R_{Agij}(\omega)$ and the inverse angle $\phi_{ij}(\omega) = -\phi_{gij}(\omega)$. In this case, one notes that the original system is third order with a stable input–output relationship, since the origin is approached at $-3\times\pi/2$ and the point $-1+j0$ lies *outside* of the curve (see Section 8.6.4). Note that the related *inverse* plot must *enclose* the $-1+j0$ point at the *crossover frequency* ω_{CO} for stability.

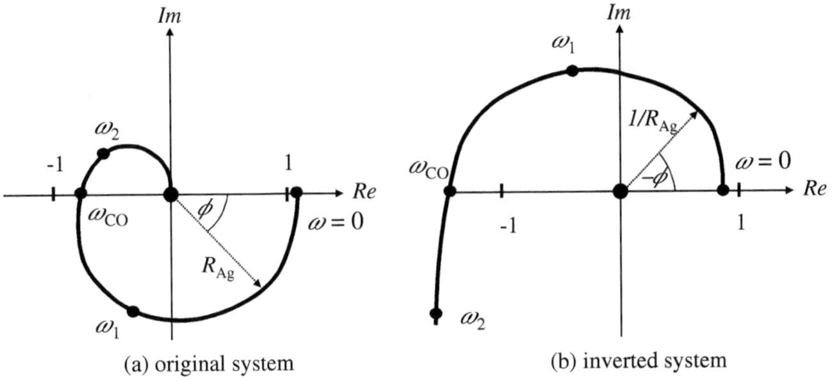

Figure 7.7 Nyquist plots for a SISO system and its inverse.

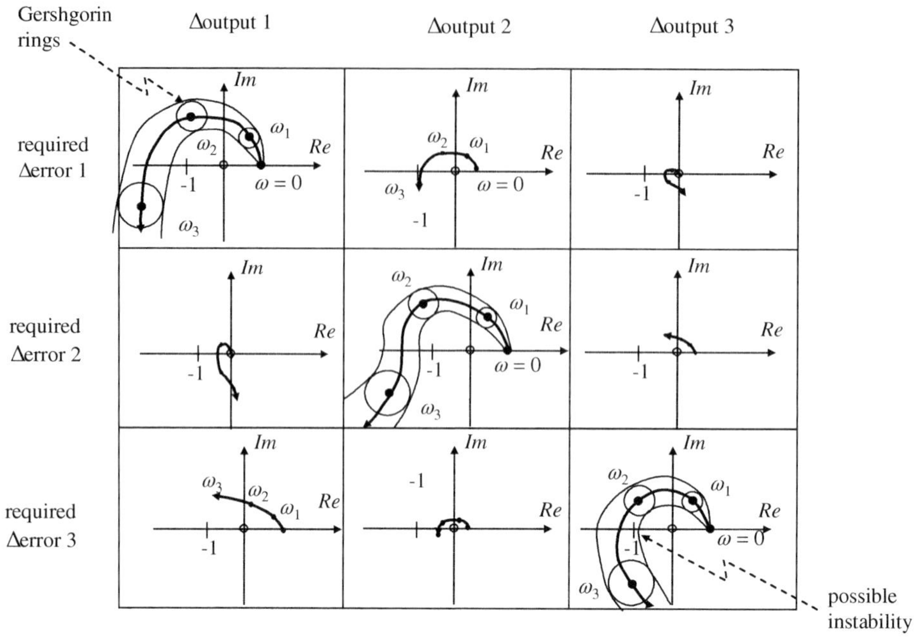

Figure 7.8 Inverse Nyquist array for a three-input three-output system.

The *inverse Nyquist array* (INA) is obtained by arranging the Nyquist plot for each input–output transfer function in the inverted system $\widehat{Q}(s)$ as a matrix of plots, so that a 3×3 system would appear as in Figure 7.8. The importance of frequency response plots like this lies in the fact that the often random signals passing through a system can be decomposed into a range of sinusoidal curves of differing frequency and amplitude. For each frequency, a corresponding point on the plot can be found that shows how the amplitude and phase of that frequency signal will be altered as it passes through (in this case, for conversion of the output $\boldsymbol{y}$ back to the error $\boldsymbol{e}$, amplitude is multiplied by the plot magnitude, and phase is shifted by the plot angle). In the INA, one is particularly interested in the 'interactive' off-diagonal terms, where the magnitudes indicate the degree of interaction as frequency varies. An effective decoupling thus needs to account for this whole frequency range, and will obviously aim to minimise the magnitudes of the off-diagonal plots.

Rosenbrock (1974) proposed a simple approximate method to visualise the problems of interaction. In the present context, one can use this visualisation on a trial-and-error basis to specify a matrix of controllers $\boldsymbol{G}_\mathrm{C}$ which minimises interaction, or alternatively, if the controllers are to be separated onto the diagonal of $\boldsymbol{G}_\mathrm{C}$, then it can be used to specify the decoupler $\boldsymbol{G}_\mathrm{D}$, which in its simplest form could be a matrix of constants.

At the outset, one needs to scale the units of the variables being controlled (and thus their matching errors), say as fractions of their normal operating ranges, so that the plot magnitudes have equal significance. Now one needs to manipulate the elements of $\boldsymbol{G}_\mathrm{C}(s)$ or $\boldsymbol{G}_\mathrm{D}(s)$ in order to make the magnitudes of the off-diagonal plots as small as possible, and this is done on a trial-and-error basis.

A technique has been developed to estimate the impact of the residual off-diagonal terms using *Gershgorin rings*. At each frequency, the magnitudes of the off-diagonal plots in a row are summed, and this value is used as the radius of a circle centred on that frequency on the diagonal plot of that row (see Figure 7.8). The successive circles thus define an outer envelope on each diagonal plot, representing the uncertainty of the relationship between an output and its error. As noted regarding stability in Figure 7.7, a prime concern (for the inverse system) is that there must be no possibility of the plot crossing the negative real axis at a magnitude of less than 1, which could indicate instability.

7.3 Continuous s-Domain MIMO Controller Design Based on Characteristic Loci

From Equation 7.22, the closed-loop transfer function for a general system is

$$\boldsymbol{G}(s) = [\boldsymbol{I} + \boldsymbol{G}_{\mathrm{M}}(s)\boldsymbol{G}_{\mathrm{P}}(s)\boldsymbol{G}_{\mathrm{C}}(s)]^{-1}\boldsymbol{G}_{\mathrm{M}}(s)\boldsymbol{G}_{\mathrm{P}}(s)\boldsymbol{G}_{\mathrm{D}}(s)\boldsymbol{G}_{\mathrm{C}}(s) \quad (7.29)$$

$$\boldsymbol{G}(s) = [\boldsymbol{I} + \boldsymbol{G}_{\mathrm{O}}(s)]^{-1}\boldsymbol{G}_{\mathrm{O}}(s) \quad (7.30)$$

$$\boldsymbol{G}(s) = \frac{1}{\det[\boldsymbol{I} + \boldsymbol{G}_{\mathrm{O}}(s)]}\left[\mathrm{adj}\{\boldsymbol{I} + \boldsymbol{G}_{\mathrm{O}}(s)\}\right]\boldsymbol{G}_{\mathrm{O}}(s) \quad (7.31)$$

where the determinant is a scalar dependent on s and the adjoint matrix is a matrix of functions of s (Section 3.9.2.2).

The *poles* of the closed loop are the particular values of s on the complex plane which cause the determinant to equal zero, and thus $\boldsymbol{G}(s)$ to become infinite. In general, the elements of $\boldsymbol{G}_{\mathrm{O}}(s)$ will be ratios of polynomials in s, for example

$$\boldsymbol{G}_{\mathrm{O}}(s) = \begin{bmatrix} \dfrac{s+2}{2s^2+3s+1} & \dfrac{4}{2s+1} \\ \dfrac{2}{s} & \dfrac{5}{s(6s+1)} \end{bmatrix} \quad (7.32)$$

so the characteristic equation

$$\det[\boldsymbol{I} + \boldsymbol{G}_{\mathrm{O}}(s)] = 0 \quad (7.33)$$

will have m roots (with $m = 4$ in this example). Thus, it can be re-expressed as

$$\left(s - p_1\right)\left(s - p_2\right)\left(s - p_3\right)\cdots\left(s - p_m\right) = 0 \quad (7.34)$$

From Section 3.9.2.2, one is aware that the factors $\left(s - p_1\right), \left(s - p_2\right), \ldots$ will appear in the denominators of any response, so one requires that all of the closed-loop poles p_i must lie in the left-hand side of the complex plane (i.e. $\mathrm{Re}(p_i) < 0$) for the closed-loop system to be stable. Indeed, one can simply solve for these pole locations for any choice of $\boldsymbol{G}_{\mathrm{C}}(s)$, and manipulate this controller to obtain suitably damped and responsive closed-loop pole locations (Figure 3.28). However, more useful insight can be obtained by examining the behaviour of the system as signal frequency varies, and this is what is proposed below.

It will be seen in Section 8.6.4 that if one plots the single complex value of $\det[\boldsymbol{I}+\boldsymbol{G}_O(s)]$ as s follows a (Nyquist) contour in a positive angular direction (anticlockwise) around the entire right-hand complex plane, then the resultant trace will encircle the origin $Z-P$ times in a positive angular direction, where Z is the number of zeros of the determinant in the right-hand complex plane and P is the number of poles of the determinant in the right-hand complex plane. For an open-loop stable system, $\boldsymbol{G}_O(s)$ will have no poles there, so the number of encirclements Z of $0+j0$ will equal the number of zeros of the determinant lying there, that is the number of *poles of the closed loop* (out of the m) which lie in the right-hand complex plane. Any such pole indicates closed-loop instability, and a design procedure would aim to eliminate this possibility with a reasonable safety margin.

As can be imagined, the many influences on the single Nyquist plot above lead to a complex curve with many loops which can be difficult to interpret. Often it is easier to base a design on the *characteristic loci* instead. Here one considers the n factors making up the $n\times \text{n}$ system $\boldsymbol{G}_O(s)$ separately (e.g. $n=2$ in the example of Equation 7.21). A particular choice of s in the complex plane will result in a particular matrix of complex numbers $\boldsymbol{G}_O(s)$. The determinant of this matrix can be expressed as the product of its eigenvalues, which will also depend on s.

$$\det[\boldsymbol{G}_O(s)] = \lambda_1(s)\lambda_2(s)\cdots\lambda_n(s) \tag{7.35}$$

But if $\lambda_i(s)$ is an eigenvalue of $\boldsymbol{G}_O(s)$, then $1+\lambda_i(s)$ must be an eigenvalue of $[\boldsymbol{I}+\boldsymbol{G}_O(s)]$ so that

$$\det[\boldsymbol{I}+\boldsymbol{G}_O(s)] = \{1+\lambda_1(s)\}\{1+\lambda_2(s)\}\cdots\{1+\lambda_n(s)\} \tag{7.35}$$

As s moves positively around the Nyquist contour, positive rotations of each factor around its origin will combine to give the total number of positive rotations of $\det[\boldsymbol{I}+\boldsymbol{G}_O(s)]$ around *its* origin. If $\boldsymbol{G}_O(s)$ is stable, negative rotations will represent the number of zeros of $\det[\boldsymbol{I}+\boldsymbol{G}_O(s)]$ and thus the number of poles of the closed loop which lie in the right-hand complex plane. (If $\boldsymbol{G}_O(s)$ were *unstable*, one would need to subtract the number of negative rotations caused by its poles, which will also be poles of $[\boldsymbol{I}+\boldsymbol{G}_O(s)]$.)

One notes that a rotation of $\{1+\lambda_i(s)\}$ around its origin is the same as a rotation of $\lambda_i(s)$ around the point $-1+j0$. Moreover, since the system $\boldsymbol{G}_O(s)$ is physically realisable (higher power of s in the numerator than denominator), the $\lambda_i(s)$ will be small as s follows the large values at the extremes of the complex plane. As a result, one only needs to consider s as it follows finite values on the imaginary axis. Thus, n curves are obtained, one for each $\lambda_i(j\omega)$ in the range $0<\omega<+\infty$ (positive direction is infinity to zero). One understands (but does not plot) the remainder of this curve, which is the complex conjugate in the range $-\infty<\omega<0$. These are the *characteristic loci*. The design procedure is based on how close any of these curves comes to encircling the point $-1+j0$, allowing such measures as *gain margin* and *phase margin* to be manipulated (Section 8.6.4). A controller design procedure is described by Maciejowski (1989) using three compensators in series, for the high-, medium- and low-frequency ranges. The latter two controllers tend towards identity matrices as frequency increases.

7.4
Continuous *s*-Domain MIMO Controller Design Based on Largest Modulus

As noted in Section 7.3, if one plots the single complex value of $\det[\boldsymbol{I}+\boldsymbol{G}_O(s)]$ as s follows a (Nyquist) contour in a positive angular direction (anticlockwise) around the entire right-hand complex plane, then the resultant trace will encircle the origin $Z-P$ times in a positive angular direction, where Z is the number of zeros of the determinant in the right-hand complex plane and P is the number of poles of the determinant in the right-hand complex plane. For an open-loop stable system, the number of encirclements of $0+j0$ will equal the number of zeros of $\det[\boldsymbol{I} + \boldsymbol{G}_{\mathrm{O}}(s)]$ lying in the right-half complex plane, which would imply closed-loop instability. To create an analogy to the SISO system Nyquist analysis (Section 8.6.4), one can shift the plot of $\det[\boldsymbol{I} + \boldsymbol{G}_{\mathrm{O}}(s)]$ by -1 and rather look for encirclements of $-1+j0$. Moreover, it has been noted that it is only necessary to look at one of the complex conjugates of this curve, namely for $s=j\omega$, $0<\omega<+\infty$ (positive direction is infinity to zero). So the *largest modulus* method (Luyben, 1986) considers the approach of the scalar function

$$w(j\omega) = -1 + \det\left[\boldsymbol{I} + \boldsymbol{G}_{\mathrm{O}}(j\omega)\right] \tag{7.36}$$

to the point $-1+j0$. As noted, it has to pass to the right of that point for the controlled MIMO system to be stable. (This at least applies to less complex plots – the *full* Nyquist contour can of course cause w to pass on the left and right, and still not *encircle* the point.)

The multivariable system *modulus* is defined as

$$L(j\omega) = \left|\frac{w(j\omega)}{1 + w(j\omega)}\right| \tag{7.37}$$

so the largest modulus will occur at the closest approach of w to $-1+j0$. In a typical design procedure, one might occupy just n positions in the $n\times n$ matrix $\boldsymbol{G}_{\mathrm{C}}(s)$ with PI controllers $G_{\mathrm{C}i}(s)$ for chosen pairs of inputs and outputs. A common detuning parameter α in the range 1.5–4 is introduced as follows:

$$G_{\mathrm{C}i}(s) = \frac{K_{\mathrm{C}i}}{\alpha}\left[1 + \frac{1}{\alpha\tau_{\mathrm{I}i}s}\right] \tag{7.38}$$

and α is varied to obtain the desired *largest modulus* value. A value of $2n$ dB would be equivalent to n identical loops each having a largest modulus of 2 dB, a typical tuning goal.

7.5 MIMO Controller Design Based on Pole Placement

7.5.1 Continuous *s*-Domain MIMO Controller Design Based on Pole Placement

So far the impact of the closed-loop poles on the controlled system performance has been considered passively in the controller design. Controllers have been adjusted, whereafter the closed-loop pole locations have been evaluated, or their influence detected in frequency response analysis. Pole locations as in Figure 3.28 have a profound impact on the response of a system, including its stability, so the idea of *setting* the closed-loop pole locations to arrive at a suitable controller is attractive.

In Figure 7.9, consider a two-degree-of-freedom controller in 'RST' format similar to Figure 7.4.

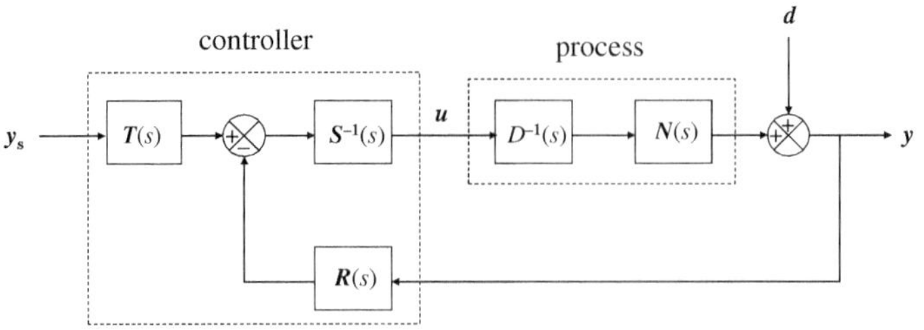

Figure 7.9 Two-degree-of-freedom controller arrangement for a continuous MIMO system.

The solution is facilitated by defining the s-domain matrices $\boldsymbol{N}(s)$, $\boldsymbol{R}(s)$, $\boldsymbol{T}(s)$ and $\boldsymbol{S}(s)$ as containing only polynomials in s (no ratios). In general, $D(s)$ is scalar, acting as a common denominator for the process transfer function elements.

The closed-loop transfer function is obtained as follows:

$$\boldsymbol{u} = \boldsymbol{S}^{-1}\{\boldsymbol{T}\boldsymbol{y}_{\mathrm{S}} - \boldsymbol{R}\boldsymbol{y}\} \tag{7.39}$$

$$\boldsymbol{y} = D^{-1}\boldsymbol{N}\boldsymbol{u} + \boldsymbol{d} \tag{7.40}$$

$$\boldsymbol{y} = D^{-1}\boldsymbol{N}\boldsymbol{S}^{-1}\{\boldsymbol{T}\boldsymbol{y}_{\mathrm{S}} - \boldsymbol{R}\boldsymbol{y}\} + \boldsymbol{d} \tag{7.41}$$

$$[\boldsymbol{I} + D^{-1}\boldsymbol{N}\boldsymbol{S}^{-1}\boldsymbol{R}]\boldsymbol{y} = D^{-1}\boldsymbol{N}\boldsymbol{S}^{-1}\boldsymbol{T}\boldsymbol{y}_{\mathrm{S}} + \boldsymbol{d} \tag{7.42}$$

$$[\boldsymbol{N}^{-1}D + \boldsymbol{S}^{-1}\boldsymbol{R}]\boldsymbol{y} = \boldsymbol{S}^{-1}\boldsymbol{T}\boldsymbol{y}_{\mathrm{S}} + \boldsymbol{N}^{-1}D\boldsymbol{d} \tag{7.43}$$

Since $D(s)$ is scalar, multiplication by $\boldsymbol{S}^{-1}$ from the left and $\boldsymbol{N}^{-1}$ from the right allows

$$[\boldsymbol{N}^{-1}D + \boldsymbol{S}^{-1}\boldsymbol{R}] = \boldsymbol{S}^{-1}[\boldsymbol{S}\boldsymbol{N}^{-1}D + \boldsymbol{R}] \tag{7.43}$$

$$[\boldsymbol{N}^{-1}D + \boldsymbol{S}^{-1}\boldsymbol{R}] = \boldsymbol{S}^{-1}[\boldsymbol{S}D + \boldsymbol{R}\boldsymbol{N}]\boldsymbol{N}^{-1} \tag{7.44}$$

so

$$[\boldsymbol{N}^{-1}D + \boldsymbol{S}^{-1}\boldsymbol{R}]^{-1} = \boldsymbol{N}[\boldsymbol{S}D + \boldsymbol{R}\boldsymbol{N}]^{-1}\boldsymbol{S} \tag{7.45}$$

and

$$\boldsymbol{y} = \boldsymbol{N}[\boldsymbol{S}D + \boldsymbol{R}\boldsymbol{N}]^{-1}\{\boldsymbol{T}\boldsymbol{y}_{\mathrm{S}} + \boldsymbol{S}\boldsymbol{N}^{-1}D\boldsymbol{d}\} \tag{7.46}$$

Since the coefficients of the $D(s)$ and $\boldsymbol{N}(s)$ polynomials describing the process are known, the objective is now to find those in $\boldsymbol{R}(s)$, $\boldsymbol{S}(s)$ and $\boldsymbol{T}(s)$ required to cause the desired closed-loop pole (and zero) locations. Following the procedure of Aström and Wittenmark (1989), let

$$\boldsymbol{P}(s) = D(s)\boldsymbol{S}(s) + \boldsymbol{R}(s)\boldsymbol{N}(s) \tag{7.47}$$

which is a Diophantine equation known as the Bezout equation. Then the closed-loop transfer function is

$$\boldsymbol{G}(s) = \boldsymbol{N}_{\mathrm{CL}}(s)D_{\mathrm{CL}}^{-1}(s) = \boldsymbol{N}(s)\boldsymbol{P}(s)^{-1}\boldsymbol{T}(s) \tag{7.48}$$

Because $\boldsymbol{N}_{\mathrm{CL}}(s)$ is a matrix of polynomials (with no denominators), choosing the closed-loop poles sets the coefficients of the closed-loop denominator polynomial $D_{\mathrm{CL}}(s)$ in terms of these poles, that is

$$D_{CL}(s) = (s - \lambda_{CL1})(s - \lambda_{CL2}) \cdots (s - \lambda_{CLN}) \tag{7.49}$$

which *monic* form arises by accommodating any premultiplying constant in $\boldsymbol{N}_{CL}(s)$. It is also recommended to cancel the 'good' well-damped and stable zeros of $\boldsymbol{N}(s)$ of the process, but the 'bad' unstable, poorly damped and slow zeros must be kept in the closed-loop transfer function. The method is based on factoring $\boldsymbol{N}(s)$ into two polynomial matrices – usually only feasible for a set of SISO systems when $\boldsymbol{N}(s)$ is diagonal.

$$\boldsymbol{N}(s) \to \boldsymbol{N}_b(s)\boldsymbol{N}_g(s) \tag{7.50}$$

de Larminat (1993) proposes an alternative strategy that will avoid this factorisation, by improving the bad zeros so that they can be included in $\boldsymbol{P}(s)$ in 'modified' form: Unstable zeros should have their real parts multiplied by -1, and the poorly damped zeros should be shifted slightly to better damped positions. Slow zeros should be speeded up. If all of the zeros are good, or have been improved ($\boldsymbol{N} \to \boldsymbol{N}'$) like this, one can take $\boldsymbol{N}_g(s) = \boldsymbol{N}'(s)/\beta$ and $\boldsymbol{N}_b(s) = \beta$ (scalar). Alternatively, if abandoning zero cancellation altogether, take $\boldsymbol{N}_g(s) = \boldsymbol{I}$ and $\boldsymbol{N}_b(s) = \boldsymbol{N}(s)$. The desired poles and the 'good' zero cancellation are then achieved by considering $\boldsymbol{P}(s)$ to be

$$\boldsymbol{P}(s) = D_{CL}(s)\boldsymbol{D}_0(s)\boldsymbol{N}_g(s) \tag{7.51}$$

where $\boldsymbol{D}_0(s)$ is specified to be stable and fast, for example $(s+\alpha)\boldsymbol{I}$ with α large, representing the 'observer dynamics', since $\boldsymbol{P}(s)$ multiplies on the left-hand side of Equation 7.46. Since the three terms on the right-hand side of Equation 7.51 have been specified so far, the coefficients of $\boldsymbol{P}(s)$ can now be evaluated. Thus, the coefficients of $\boldsymbol{R}(s)$ and $\boldsymbol{S}(s)$ can be evaluated by matching the s^k coefficients for $k = 0, 1, \ldots$ in the equation

$$\boldsymbol{P}(s) = D(s)\boldsymbol{S}(s) + \boldsymbol{R}(s)\boldsymbol{N}(s) \tag{7.52}$$

namely

$$\begin{aligned}
&(d_0 + d_1 s + \cdots + d_n s^n) \begin{bmatrix} (s_{110} + s_{111}s + s_{112}s^2 + \cdots) & (s_{120} + s_{121}s + s_{122}s^2 + \cdots) & \cdots \\ (s_{210} + s_{211}s + s_{212}s^2 + \cdots) & (s_{220} + s_{221}s + s_{222}s^2 + \cdots) & \cdots \\ \vdots & \vdots & \ddots \end{bmatrix} \\
&+ \begin{bmatrix} (r_{110} + r_{111}s + r_{112}s^2 + \cdots) & (r_{120} + r_{121}s + r_{122}s^2 + \cdots) & \cdots \\ (r_{210} + r_{211}s + r_{212}s^2 + \cdots) & (r_{220} + r_{221}s + r_{222}s^2 + \cdots) & \cdots \\ \vdots & \vdots & \ddots \end{bmatrix} \\
&\times \begin{bmatrix} (n_{110} + n_{111}s + n_{112}s^2 + \cdots) & (n_{120} + n_{121}s + n_{122}s^2 + \cdots) & \cdots \\ (n_{210} + n_{211}s + n_{212}s^2 + \cdots) & (n_{220} + n_{221}s + n_{222}s^2 + \cdots) & \cdots \\ \vdots & \vdots & \ddots \end{bmatrix} \\
&= \begin{bmatrix} (p_{110} + p_{111}s + p_{112}s^2 + \cdots) & (p_{120} + p_{121}s + p_{122}s^2 + \cdots) & \cdots \\ (p_{210} + p_{211}s + p_{212}s^2 + \cdots) & (p_{220} + p_{221}s + p_{222}s^2 + \cdots) & \cdots \\ \vdots & \vdots & \ddots \end{bmatrix}
\end{aligned} \tag{7.53}$$

and solving the resultant linear system. (For a SISO system, this would involve the $2n \times 2n$ Sylvester matrix.)

Now $N_{CL}(s)$ is constructed as a matrix of polynomials by suitable manipulation $N_*(s)$, if required, of $N_b(s)$:

$$N_{CL}(s) = N_b(s)N_*(s) \tag{7.54}$$

Using Equations 7.48 and 7.50, and recalling that $D_{CL}(s)$ is scalar, the prefilter matrix of polynomials $T(s)$ can then be evaluated as follows:

$$\{N_b(s)N_*(s)\}D_{CL}^{-1}(s) = N(s)\left\{N_g^{-1}(s)D_0^{-1}(s)D_{CL}^{-1}(s)\right\}T(s) \tag{7.55}$$

that is

$$N_*(s) = D_0^{-1}(s)T(s) \tag{7.56}$$

$$T(s) = D_0(s)N_*(s) \tag{7.57}$$

The matching of the coefficients of powers of s in this result again gives the necessary coefficients for T, completing the specification of the RST controller.

In constructing the necessary transfer functions (and ultimately the actual algorithm) along the above lines, some practical issues are noted. For modelling purposes, one moves $S^{-1}(s)$ from the output of the comparator to both of its inputs. Where necessary to obtain physically realisable transfer functions, the denominator could be extended, for example by multiplying by a fast first-order term $(0.1s+1)$. The $T(s)$ function must have its gain adjusted (by applying a common factor to all coefficients), in order to match the steady-state gain of the $R(s)$ function at the comparator.

7.5.2
Discrete z-Domain MIMO Controller Design Based on Pole Placement

Equation 3.295 for a discrete process gives

$$\left(z^n + d_{n-1}z^{n-1} + \cdots + d_1 z + d_0\right)y(z) = \left[N_{n-1}z^{n-1} + N_{n-2}z^{n-2} + \cdots + N_1 z + N_0\right]u(z) \tag{7.58}$$

that is

$$D(z)y(z) = N(z)u(z) \tag{7.59}$$

where $D(z)$ is a scalar polynomial in z and $N(z)$ is a matrix polynomial in z using the matrices of coefficients N_i, $i = 0, \ldots, n-1$. For state processes, n will be the number of states, but for an arbitrary input-output system, n will increase (Section 7.8.1). As in Section 7.5.1, choosing $R(z)$, $S(z)$ and $T(z)$ to be similar matrix polynomials in z, one arranges the *two-degree-of-freedom* controller as in Figure 7.10.

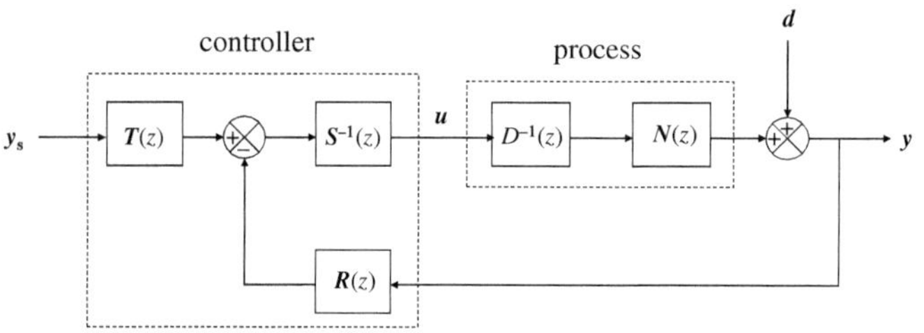

Figure 7.10 Two-degree-of-freedom controller arrangement for a discrete MIMO system.

The procedure is now the same as for the continuous system in Equations 7.43–7.49, allowing specification of the eigenvalues (poles) of the closed-loop transfer function according to

$$D_{\mathrm{CL}}(z) = (z - \lambda_{\mathrm{CL1}})(z - \lambda_{\mathrm{CL2}}) \cdots (z - \lambda_{\mathrm{CL}N}) \tag{7.60}$$

Recalling that the closed-loop poles must lie in the unit circle for stability (Figure 3.33) and that

$$z = \mathrm{e}^{Ts} \quad \text{where } s = a + jb \tag{7.61}$$

so

$$z = \exp\{aT + jbT\} \tag{7.62}$$

$$z = \mathrm{e}^{aT}\,\mathrm{e}^{jbT} \begin{cases} \text{magnitude:} & \mathrm{e}^{aT} \\ \text{angle:} & bT \end{cases} \tag{7.63}$$

the choice of discrete pole locations will determine closed-loop behaviour according to Figure 7.11.

The procedure of Aström and Wittenmark (1989) can again be used to eliminate 'good' process zeros by specifying the denominator polynomial matrix

$$\boldsymbol{P}(z) = D_{\mathrm{CL}}(z)\boldsymbol{D}_0(z)\boldsymbol{N}_{\mathrm{g}}(z) \tag{7.64}$$

Once the coefficients of $\boldsymbol{P}(z)$ are known, the Bezout equation

$$\boldsymbol{P}(z) = D(z)\boldsymbol{S}(z) + \boldsymbol{R}(z)\boldsymbol{N}(z) \tag{7.65}$$

allows solution for the coefficients of $\boldsymbol{R}(z)$ and $\boldsymbol{S}(z)$, and then $\boldsymbol{T}(z)$ may be obtained as in Section 7.5.1. Additionally, one may ensure that there is no steady-state offset by stipulating an integrating factor $(z-1)$ in the $\boldsymbol{S}(z)$ terms, whilst a 'robustness filter' on the feedback signals can be incorporated as a factor $(z-\alpha)/(1-\alpha)$ in the $\boldsymbol{R}(z)$ terms (Corriou, 2004). Dividing Equation 7.58 by z^n shows as expected that the most recent input affecting the output is one interval T ago determined by matrix $\boldsymbol{N}_{n-1}$. An additional dead-time lag of kT would cause $\boldsymbol{N}_{n-1}, \boldsymbol{N}_{n-2}, \ldots, \boldsymbol{N}_{n-k}$ to be zero matrices.

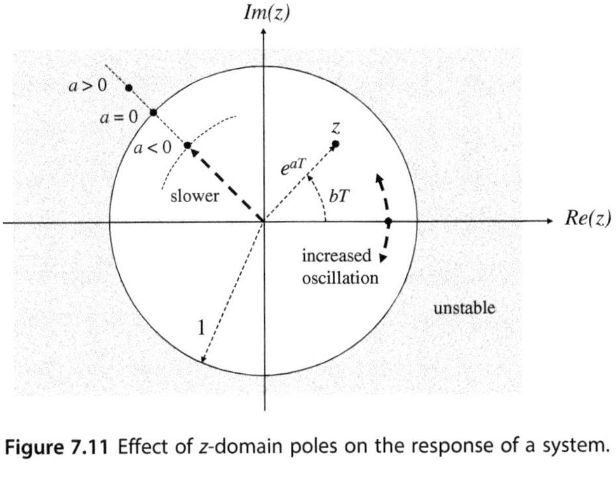

Figure 7.11 Effect of z-domain poles on the response of a system.

7.6 State-Space MIMO Controller Design

The controllers discussed here rely on the full state of the system $\boldsymbol{x}$ being available. If this is not the case, an estimate of the state must be provided as in Section 6.4. Note that *integral* and *derivative* action is easily included in these systems by adding additional equations to define the relationships of these new states to the existing states (Problems 7.7 and 7.8 in the accompanying book).

7.6.1 Continuous State-Space MIMO Modal Control: Proportional Feedback

Here one is considering the linear state equation

$$\frac{\mathrm{d}x(t)}{\mathrm{d}t} = Ax(t) + Bu(t) \tag{7.66}$$

In Sections 3.9.2.2 and 3.9.2.3, it was noted that the eigenvalues of $\boldsymbol{A}$ dominate the behaviour of this system, and must lie in the left-hand complex plane for stability. These determine the 'modes' of the system, that is the underlying speeds of response and damping as illustrated in Figure 3.7. Under proportional feedback control

$$\boldsymbol{u}(t) = \boldsymbol{K}\{\boldsymbol{x}_{\mathrm{S}}(t) - \boldsymbol{x}(t)\} \tag{7.67}$$

so that the closed-loop behaviour is determined by

$$\frac{\mathrm{d}\boldsymbol{x}(t)}{\mathrm{d}t} = [\boldsymbol{A} - \boldsymbol{BK}]\boldsymbol{x}(t) + \boldsymbol{BK}\boldsymbol{x}_{\mathrm{S}}(t) \tag{7.68}$$

and the eigenvalues of $[\boldsymbol{A} - \boldsymbol{BK}]$ now determine the closed-loop behaviour. In the *modal* design approach, one diagonalises this system matrix, making the relative adjustment of the modes from open loop to closed loop transparent. In this way, one aims to speed up the slower modes (eigenvalues closest to the imaginary axis) to get a desired closed-loop performance.

Recall that if one constructs a matrix $\boldsymbol{E}$ which contains the eigenvectors of $\boldsymbol{A}$ as columns, then

$$\boldsymbol{\Lambda} = \boldsymbol{E}^{-1}\boldsymbol{AE} \tag{7.69}$$

will be diagonal with the eigenvalues of $\boldsymbol{A}$ on its axis. Thus, defining a transformed state

$$\boldsymbol{z} = \boldsymbol{E}^{-1}\boldsymbol{x} \tag{7.70}$$

Equation 7.68 becomes

$$\frac{\mathrm{d}\boldsymbol{z}(t)}{\mathrm{d}t} = \left[\boldsymbol{\Lambda} - \boldsymbol{E}^{-1}\boldsymbol{BKE}\right]\boldsymbol{z}(t) + \boldsymbol{E}^{-1}\boldsymbol{BK}\boldsymbol{x}_{\mathrm{S}}(t) \tag{7.71}$$

Choosing $\boldsymbol{\Lambda}_{\mathrm{DESIRED}} = \boldsymbol{\Lambda} - \boldsymbol{E}^{-1}\boldsymbol{BKE}$ to be diagonal, with the adjusted eigenvalues on its axis, then allows solution for the necessary proportional gain matrix

$$\boldsymbol{K} = \boldsymbol{B}^{-1}\boldsymbol{E}[\boldsymbol{\Lambda} - \boldsymbol{\Lambda}_{\mathrm{DESIRED}}]\boldsymbol{E}^{-1} \tag{7.72}$$

provided that matrix $\boldsymbol{B}$ is square and nonsingular. As noted, integral and derivative action may easily be included by extending the state vector in Equation 7.66.

7.6.2 Discrete State-Space MIMO Modal Control: Proportional Feedback

A linear discrete closed-loop under proportional state feedback is governed by the equations

$$z\boldsymbol{x}(z) = \boldsymbol{A}\boldsymbol{x}(z) + \boldsymbol{B}\boldsymbol{u}(z) \tag{7.73}$$

$$\boldsymbol{u}(z) = \boldsymbol{K}[\boldsymbol{x}_S(z) - \boldsymbol{x}(z)] \tag{7.74}$$

so that

$$\boldsymbol{x}(z) = \{z\boldsymbol{I} - [\boldsymbol{A} - \boldsymbol{B}\boldsymbol{K}]\}^{-1}\boldsymbol{B}\boldsymbol{K}\boldsymbol{x}_S(z) \tag{7.75}$$

As in the continuous system (Section 7.6.1), the closed-loop behaviour will depend on the eigenvalues of $[\boldsymbol{A} - \boldsymbol{B}\boldsymbol{K}]$, except that the stability, damping and speed of these closed-loop poles are now determined as in Figure 7.11. As in Equation 7.72, a proportional gain matrix is determined according to the desired shift in eigenvalues relative to the open-loop values on the diagonal of $\boldsymbol{\Lambda} = \boldsymbol{E}^{-1}\boldsymbol{A}\boldsymbol{E}$, that is

$$\boldsymbol{K} = \boldsymbol{B}^{-1}\boldsymbol{E}[\boldsymbol{\Lambda} - \boldsymbol{\Lambda}_{DESIRED}]\boldsymbol{E}^{-1} \tag{7.76}$$

provided that matrix $\boldsymbol{B}$ is square and nonsingular. Integral and derivative action is easily included by extending the state vector in Equation 7.73.

7.6.3 Continuous State-Space MIMO Controller Design Based on 'Controllable System' Pole Placement

The system of first-order equations describing the state evolution in a linear continuous system is

$$\frac{\mathrm{d}\boldsymbol{x}(t)}{\mathrm{d}t} = \boldsymbol{A}\boldsymbol{x}(t) + \boldsymbol{B}\boldsymbol{u}(t) \tag{7.77}$$

Taking the Laplace transform

$$\boldsymbol{X}(s) = [s\boldsymbol{I} - \boldsymbol{A}]^{-1}\boldsymbol{B}\boldsymbol{U}(s) \tag{7.78}$$

$$\boldsymbol{X}(s) = D^{-1}(s)\boldsymbol{N}(s)\boldsymbol{U}(s) \tag{7.79}$$

where Equation 3.163 gives for $\boldsymbol{A}$ $n \times n$

$$D(s) = s^n + d_{n-1}s^{n-1} + \cdots + d_1 s + d_0 \quad \text{(scalar polynomial)} \tag{7.80}$$

$$\boldsymbol{N}(s) = \boldsymbol{N}_{n-1}s^{n-1} + \boldsymbol{N}_{n-2}s^{n-2} + \cdots + \boldsymbol{N}_1 s + \boldsymbol{N}_0 \quad \text{(matrix of polynomials)} \tag{7.81}$$

As in Figure 7.12, introduce a *partial state* $\boldsymbol{w}(t)$ such that

$$\boldsymbol{W}(s) = D^{-1}(s)\boldsymbol{U}(s) \tag{7.82}$$

then

$$\boldsymbol{X}(s) = \boldsymbol{N}(s)\boldsymbol{W}(s) \tag{7.83}$$

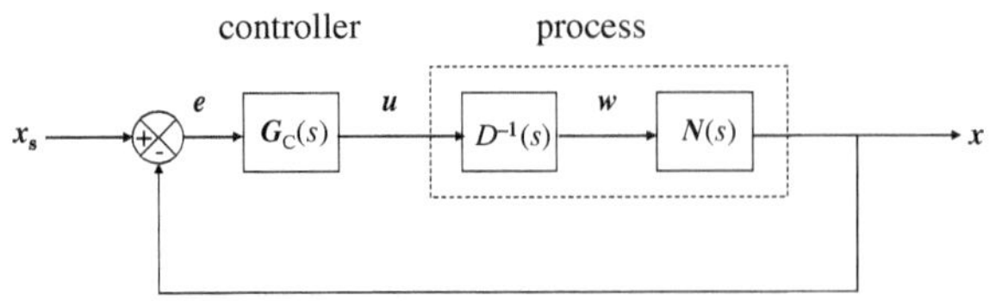

Figure 7.12 Concept of partial state $\boldsymbol{w}$ to be used in continuous pole placement controller.

Defining the vector of vectors

$$\widehat{\boldsymbol{w}}(t) = \begin{pmatrix} \dfrac{d^{n-1}\boldsymbol{w}(t)}{dt^{n-1}} \\ \dfrac{d^{n-2}\boldsymbol{w}(t)}{dt^{n-2}} \\ \vdots \\ \dfrac{d\boldsymbol{w}(t)}{dt} \\ \boldsymbol{w}(t) \end{pmatrix} \quad \text{so that} \quad \widehat{\boldsymbol{W}}(s) = \begin{pmatrix} s^{n-1}\boldsymbol{I} \\ s^{n-2}\boldsymbol{I} \\ \vdots \\ s\boldsymbol{I} \\ \boldsymbol{I} \end{pmatrix} \boldsymbol{W}(s) \tag{7.84}$$

the evolution of the partial states is described by

$$s\widehat{W}(s) = \begin{pmatrix} s^{n}\boldsymbol{I} \\ s^{n-1}\boldsymbol{I} \\ \vdots \\ s^{2}\boldsymbol{I} \\ s\boldsymbol{I} \end{pmatrix} \boldsymbol{W}(s) = \underbrace{\begin{bmatrix} -d_{n-1}\boldsymbol{I} & -d_{n-2}\boldsymbol{I} & \cdots & -d_1\boldsymbol{I} & -d_0\boldsymbol{I} \\ \boldsymbol{I} & \boldsymbol{0} & \cdots & \boldsymbol{0} & \boldsymbol{0} \\ \boldsymbol{0} & \boldsymbol{I} & \cdots & \boldsymbol{0} & \boldsymbol{0} \\ \vdots & \vdots & \ddots & \vdots & \vdots \\ \boldsymbol{0} & \boldsymbol{0} & \cdots & \boldsymbol{I} & \boldsymbol{0} \end{bmatrix}}_{\widehat{A}} \underbrace{\begin{pmatrix} s^{n-1}\boldsymbol{I} \\ s^{n-2}\boldsymbol{I} \\ \vdots \\ s\boldsymbol{I} \\ \boldsymbol{I} \end{pmatrix}}_{\widehat{W}(s)} \boldsymbol{W}(s) + \underbrace{\begin{bmatrix} \boldsymbol{I} \\ \boldsymbol{0} \\ \vdots \\ \boldsymbol{0} \\ \boldsymbol{0} \end{bmatrix}}_{\widehat{B}} \boldsymbol{U}(s) \tag{7.85}$$

that is

$$\frac{d\widehat{\boldsymbol{w}}(t)}{dt} = \widehat{\boldsymbol{A}}\widehat{\boldsymbol{w}}(t) + \widehat{\boldsymbol{B}}\boldsymbol{u}(t) \tag{7.86}$$

which is now in the *standard controllable form*. The first row expresses Equation 7.82 which is a canonical set of equations with identical coefficients for each of the partial states in $\boldsymbol{w}(t)$.

$$\frac{d^{n}\boldsymbol{w}(t)}{dt^{n}} + d_{n-1}\frac{d^{n-1}\boldsymbol{w}(t)}{dt^{n-1}} + \cdots + d_1\frac{d\boldsymbol{w}(t)}{dt} + d_0\boldsymbol{w}(t) = \boldsymbol{u}(t) \tag{7.87}$$

Now consider proportional state feedback control of the standard form (Equation 7.86)

$$\boldsymbol{u}(t) = \widehat{\boldsymbol{K}}[\widehat{\boldsymbol{w}}_S(t) - \widehat{\boldsymbol{w}}(t)] \tag{7.88}$$

$$\frac{d\widehat{\boldsymbol{w}}(t)}{dt} = \left[\widehat{\boldsymbol{A}} - \widehat{\boldsymbol{B}}\widehat{\boldsymbol{K}}\right]\widehat{\boldsymbol{w}}(t) + \widehat{\boldsymbol{B}}\widehat{\boldsymbol{K}}\widehat{\boldsymbol{w}}_S(t) \tag{7.89}$$

Noting that

$$\widehat{\boldsymbol{K}} = \left[\widehat{k}_{n-1}\boldsymbol{I} \quad \widehat{k}_{n-2}\boldsymbol{I} \quad \cdots \quad \widehat{k}_1\boldsymbol{I} \quad \widehat{k}_0\boldsymbol{I}\right] \tag{7.90}$$

because the elements in each subvector of $\widehat{\boldsymbol{w}}(t)$ have the same behaviour and no interaction (Equation 7.87), the dynamics are now determined by

$$\widehat{\boldsymbol{A}} - \widehat{\boldsymbol{B}}\widehat{\boldsymbol{K}} = \begin{bmatrix} -\left(d_{n-1}+\widehat{k}_{n-1}\right)\boldsymbol{I} & -\left(d_{n-2}+\widehat{k}_{n-2}\right)\boldsymbol{I} & \cdots & -\left(d_1+\widehat{k}_1\right)\boldsymbol{I} & -\left(d_0+\widehat{k}_0\right)\boldsymbol{I} \\ \boldsymbol{I} & \boldsymbol{0} & \cdots & \boldsymbol{0} & \boldsymbol{0} \\ \boldsymbol{0} & \boldsymbol{I} & \cdots & \boldsymbol{0} & \boldsymbol{0} \\ \vdots & \vdots & \ddots & \vdots & \vdots \\ \boldsymbol{0} & \boldsymbol{0} & \cdots & \boldsymbol{I} & \boldsymbol{0} \end{bmatrix} \tag{7.91}$$

Moreover, taking the derivatives of $\boldsymbol{w}_{\mathrm{S}}(t)$ to be zero,

$$\widehat{\boldsymbol{B}}\widehat{\boldsymbol{K}}\widehat{\boldsymbol{w}}_{\mathrm{S}}(t) = \begin{pmatrix} \widehat{k}_0\boldsymbol{w}_{\mathrm{S}}(t) \\ \boldsymbol{0} \\ \vdots \\ \boldsymbol{0} \end{pmatrix} \tag{7.92}$$

the first row of the system gives the closed-loop form of Equation 7.87:

$$\frac{\mathrm{d}^n\boldsymbol{w}(t)}{\mathrm{d}t^n} + \left(d_{n-1}+\widehat{k}_{n-1}\right)\frac{\mathrm{d}^{n-1}\boldsymbol{w}(t)}{\mathrm{d}t^{n-1}} + \cdots + \left(d_1+\widehat{k}_1\right)\frac{\mathrm{d}\boldsymbol{w}(t)}{\mathrm{d}t} + \left(d_0+\widehat{k}_0\right)\boldsymbol{w}(t) = \widehat{k}_0\boldsymbol{w}_{\mathrm{S}}(t) \tag{7.93}$$

The open-loop poles determine

$$D(s) = s^n + d_{n-1}s^{n-1} + \cdots + d_1 s + d_0 = (s-\lambda_1)(s-\lambda_2)\cdots(s-\lambda_n) = 0 \tag{7.94}$$

Choosing closed-loop poles $\lambda_{\mathrm{CL}i}$ relative to these, evaluate the coefficients $d_{\mathrm{CL}i}$ in

$$D_{\mathrm{CL}}(s) = s^n + d_{\mathrm{CL}n-1}s^{n-1} + \cdots + d_{\mathrm{CL}1}s + d_{\mathrm{CL}0} = (s-\lambda_{\mathrm{CL}1})(s-\lambda_{\mathrm{CL}2})\cdots(s-\lambda_{\mathrm{CL}n}) = 0 \tag{7.95}$$

whence $\hat{k}$ in Equation 7.90 is easily evaluated using

$$\widehat{k}_{\mathrm{CL}i} = d_{\mathrm{CL}i} - d_i \quad \text{for} \quad i = 0, \ldots, n-1 \tag{7.96}$$

To implement this controller, it is necessary to create an observer for $\widehat{\boldsymbol{w}}(t)$. From Equation 7.83

$$\boldsymbol{X}(s) = \boldsymbol{N}(s)\boldsymbol{W}(s) \tag{7.97}$$

$$\boldsymbol{X}(s) = \left[\boldsymbol{N}_{n-1} \quad \boldsymbol{N}_{n-2} \quad \cdots \quad \boldsymbol{N}_1 \quad \boldsymbol{N}_0\right]\widehat{\boldsymbol{W}}(s) \tag{7.98}$$

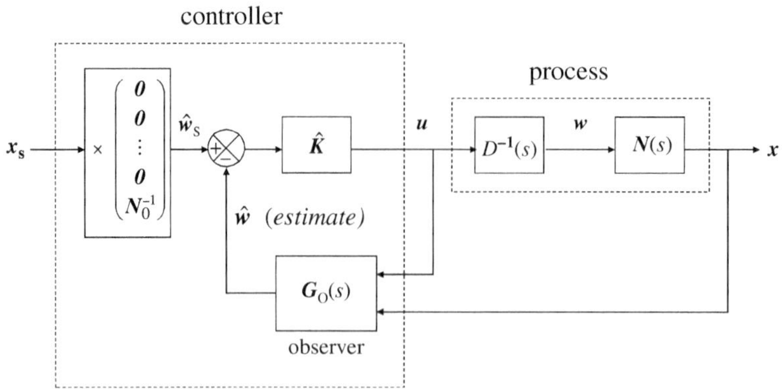

Figure 7.13 Continuous state-space MIMO pole placement controller showing dynamic observer and proportional gain.

so that

$$x(t) = \begin{bmatrix} N_{n-1} & N_{n-2} & \cdots & N_1 & N_0 \end{bmatrix} \hat{w}(t) \tag{7.99}$$

$$x(t) = \hat{N}\hat{w}(t) \tag{7.100}$$

Then, with the understanding that the $\hat{w}(t)$ are now *estimates* and that $x(t)$ is the *measurement* (or perhaps a similar estimate) of the state, a Luenberger observer can be based on Equation 7.86 as

$$\frac{d\hat{w}(t)}{dt} = \hat{A}\hat{w}(t) + \hat{B}u(t) + L(t)\left[x(t) - \hat{N}\hat{w}(t)\right] \tag{7.101}$$

where $L(t)$ can be a Kalman–Bucy gain matrix (Section 6.4.2). At steady state, Equation 7.99 shows

$$x_{SS} = \begin{bmatrix} N_{n-1} & N_{n-2} & \cdots & N_1 & N_0 \end{bmatrix} \begin{pmatrix} 0 \\ 0 \\ \vdots \\ 0 \\ w_{SS} \end{pmatrix} \quad \text{so take} \quad \hat{w}_S(t) = \begin{pmatrix} 0 \\ 0 \\ \vdots \\ 0 \\ N_0^{-1} x_S \end{pmatrix} \tag{7.102}$$

Finally, the necessary control action vector is given by Equation 7.88:

$$u(t) = \hat{K}[\hat{w}_S(t) - \hat{w}(t)] \tag{7.103}$$

A more realistic representation of this controller is then as in Figure 7.13.

Note that the original state input matrix B in Equation 7.77 need not be square, allowing, for example, for the inclusion of integral action (see Problems 7.7 and 7.8 in the accompanying book).

7.6.4
Discrete State-Space MIMO Controller Design Based on 'Controllable System' Pole Placement

The system of first-order equations describing the state evolution in a linear discrete system is

$$x_{i+1} = Ax_i + Bu_i \tag{7.104}$$

In the z-domain, this is

$$x(z) = [zI - A]^{-1}Bu(z) \tag{7.105}$$

$$x(z) = D^{-1}(z)N(z)u(z) \tag{7.106}$$

where for A $n \times n$, Equations 3.291 and 3.293 give

$$D(z) = z^n + d_{n-1}z^{n-1} + \cdots + d_1 z + d_0 \quad \text{(scalar polynomial)} \tag{7.107}$$

$$N(z) = N_{n-1}z^{n-1} + N_{n-2}z^{n-2} + \cdots + N_1 z + N_0 \quad \text{(matrix of polynomials)} \tag{7.108}$$

The format is just the same as Equation 7.81 for continuous systems, with z replacing s. Thus, following the same procedure as in Section 7.6.3, with the understanding that the $\widehat{w}_i$ are now *estimates* and that x_i is the *measurement* (or perhaps a similar estimate) of the state, the Luenberger observer following Equation 7.101 is

$$\widehat{w}_{i+1} = \widehat{A}\widehat{w}_i + \widehat{B}u_i + L_i\left[x_i - \widehat{N}\widehat{w}_i\right] \tag{7.109}$$

where L_i can be a Kalman gain matrix (Section 6.4.1). At steady state, Equation 7.99 shows

$$x_{SS} = \begin{bmatrix} N_{n-1} & N_{n-2} & \cdots & N_1 & N_0 \end{bmatrix} \begin{pmatrix} 0 \\ 0 \\ \vdots \\ 0 \\ w_{SS} \end{pmatrix} \quad \text{so take} \quad \widehat{w}_{Si} = \begin{pmatrix} 0 \\ 0 \\ \vdots \\ 0 \\ N_0^{-1}x_{Si} \end{pmatrix} \tag{7.110}$$

Finally, the necessary control action vector follows from Equation 7.88:

$$u_i = \widehat{K}[\widehat{w}_{Si} - \widehat{w}_i] \tag{7.111}$$

where

$$\widehat{K} = \begin{bmatrix} \widehat{k}_{n-1}I & \widehat{k}_{n-2}I & \cdots & \widehat{k}_1 I & \widehat{k}_0 I \end{bmatrix} \tag{7.112}$$

and the $\widehat{k}_i$ are used to cause the required shifts in the closed-loop poles following Equations 7.94, 7.95 and 7.96. For the discrete system, the closed-loop pole properties are determined as in Figure 7.11. The discrete state-space pole placement controller may thus be represented as in Figure 7.14.

As in Section 7.6.3, note that the original state input matrix B in Equation 7.104 need not be square, allowing, for example, for the inclusion of integral action (see Problems 7.7 and 7.8 in the accompanying book).

7.6.5
Discrete State-Space MIMO Controller Design Using the Linear Quadratic Regulator Approach

This type of controller is also referred to as a 'minimum variance controller'. It is based on minimising the sum of both weighted square setpoint deviations and weighted square control actions over a defined control horizon. In this way, the tightness of regulation can be played off against the control effort in a multivariable sense. This controller is based on feedback of the full state, possibly requiring an observer as in Section 6.4. The aim in LQR (linear quadratic regulator) is the 'dual problem' of the Kalman filter (Section 6.4), which was instead based on the weighted sum of square errors in the measurements and model. In this sense, the Kalman filter is an ideal observer for the LQR, and the combination, when used, is referred to as linear quadratic Gaussian (LQG) control.

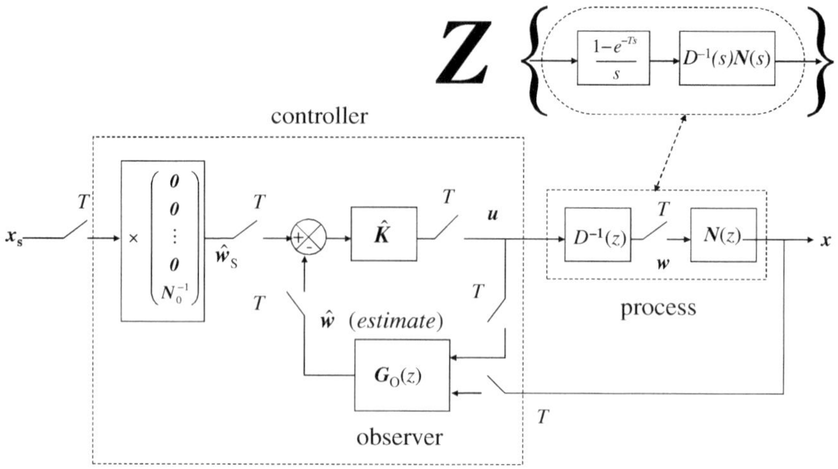

Figure 7.14 Discrete state-space MIMO pole placement controller showing dynamic observer and proportional gain.

A general state system described by

$$\frac{\mathrm{d}\boldsymbol{x}}{\mathrm{d}t} = \boldsymbol{f}(\boldsymbol{x}, \boldsymbol{u}) \tag{7.113}$$

may be linearised as

$$\frac{\mathrm{d}\boldsymbol{x}}{\mathrm{d}t} = \boldsymbol{A}'\boldsymbol{x} + \boldsymbol{B}'\boldsymbol{u} + \boldsymbol{w}_0 \tag{7.114}$$

where

$$w_0 = f(x_0, u_0) - A'x_0 - B'u_0 \tag{7.115}$$

$$\boldsymbol{A}' = \left.\frac{\partial \boldsymbol{f}}{\partial \boldsymbol{x}}\right|_0 \quad \text{Jacobian} \tag{7.116}$$

$$\boldsymbol{B}' = \left.\frac{\partial \boldsymbol{f}}{\partial \boldsymbol{x}}\right|_0 \quad \text{Jacobian} \tag{7.117}$$

Application of the methods in Section 3.9.3.2 leads to a discrete form over intervals T:

$$\boldsymbol{x}_{i+1} = \boldsymbol{A}\boldsymbol{x}_i + \boldsymbol{B}\boldsymbol{u}_i + T\boldsymbol{w}_0 \tag{7.118}$$

so at steady state

$$\boldsymbol{x}_{\mathrm{SS}} = \boldsymbol{A}\boldsymbol{x}_{\mathrm{SS}} + \boldsymbol{B}\boldsymbol{u}_{\mathrm{SS}} + T\boldsymbol{w}_0 \tag{7.119}$$

and this is the condition which is assumed to exist when the controller is switched on ($t = 0$, $i = 0$), with the setpoint set equal to the initial state, that is $\boldsymbol{x}_{\mathrm{S}i} = \boldsymbol{x}_{\mathrm{SS}}$, at least for $i = 0$. Indeed, in the

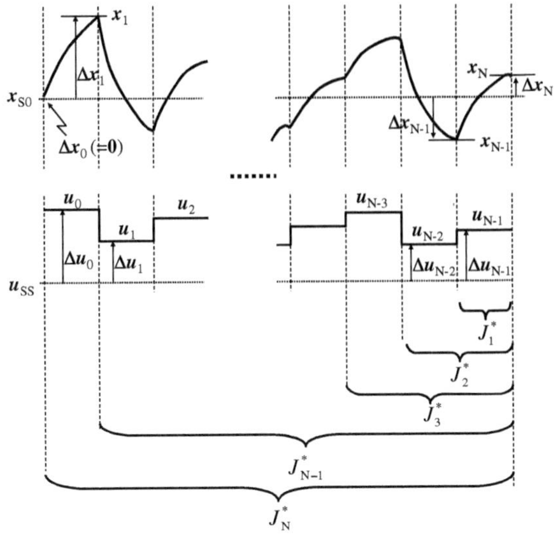

Figure 7.15 Determination of optimal control policy by dynamic programming.

processing industries, it is not expected to change much thereafter. Defining the setpoint deviation and control effort

$$\Delta \boldsymbol{x}_i = \boldsymbol{x}_i - \boldsymbol{x}_{Si} \tag{7.120}$$

$$\Delta \boldsymbol{u}_i = \boldsymbol{u}_i - \boldsymbol{u}_{SS} \tag{7.121}$$

subtraction of Equation 7.119 from Equation 7.118 yields

$$\Delta \boldsymbol{x}_{i+1} = \boldsymbol{A}\Delta \boldsymbol{x}_i + \boldsymbol{B}\Delta \boldsymbol{u}_i \tag{7.122}$$

In the LQR approach, only regulation is considered, not setpoint tracking, so the method is based entirely on $\Delta \boldsymbol{x}$ and $\Delta \boldsymbol{u}$. In Figure 7.15, consider the trajectory of $\boldsymbol{x}$ over N steps as a result of the control action policy $\Delta \boldsymbol{u}_0, \Delta \boldsymbol{u}_1, \ldots, \Delta \boldsymbol{u}_{N-1}$.

Define a performance index for the N steps starting with $\Delta \boldsymbol{x} = \Delta \boldsymbol{x}_0$:

$$J_N(\Delta \boldsymbol{x}_0) = \sum_{i=1}^{N} \Delta \boldsymbol{x}_i^{\mathrm{T}} \boldsymbol{Q} \Delta \boldsymbol{x}_i + \Delta \boldsymbol{u}_{i-1}^{\mathrm{T}} \boldsymbol{R} \Delta \boldsymbol{u}_{i-1} \tag{7.123}$$

where the weighting matrices $\boldsymbol{Q}$ and $\boldsymbol{R}$ are symmetric and positive definite, and in most situations can be taken as diagonal. The aim is to minimise this by choice of an optimum control action sequence $\Delta \boldsymbol{u}_0, \Delta \boldsymbol{u}_1, \ldots, \Delta \boldsymbol{u}_{N-1}$. However, this is seen to be an Augean task with the choice in any interval affecting the choice in any other interval. Fortunately, the *principle of optimality* originated by Bellman (1957) as a basis for *dynamic programming* brings a huge simplification to the calculation:

> An optimal policy has the property that whatever the initial state and initial decision are, the remaining decisions must constitute an optimal policy with regard to the state resulting from the first decision

To use this idea, it is necessary to work backwards from the last step, where it is required to find $\Delta\boldsymbol{u}_{N-1}$ that minimises

$$J_1(\Delta\boldsymbol{x}_{N-1}) = \Delta\boldsymbol{x}_N^{\mathrm{T}}\boldsymbol{Q}\Delta\boldsymbol{x}_N + \Delta\boldsymbol{u}_{N-1}^{\mathrm{T}}\boldsymbol{R}\Delta\boldsymbol{u}_{N-1} \tag{7.124}$$

that is to solve

$$J_1^*(\Delta\boldsymbol{x}_{N-1}) = \min_{\Delta\boldsymbol{u}_{N-1}} \left\{(\boldsymbol{A}\Delta\boldsymbol{x}_{N-1} + \boldsymbol{B}\Delta\boldsymbol{u}_{N-1})^{\mathrm{T}}\boldsymbol{Q}(\boldsymbol{A}\Delta\boldsymbol{x}_{N-1} + \boldsymbol{B}\Delta\boldsymbol{u}_{N-1}) + \Delta\boldsymbol{u}_{N-1}^{\mathrm{T}}\boldsymbol{R}\Delta\boldsymbol{u}_{N-1}\right\} \tag{7.125}$$

Recalling that

$$\frac{\partial}{\partial\boldsymbol{x}}\left\{\boldsymbol{x}^{\mathrm{T}}\boldsymbol{M}\boldsymbol{x}\right\} = 2\boldsymbol{M}\boldsymbol{x} \quad \text{and} \quad \frac{\partial}{\partial\boldsymbol{x}}\left\{\boldsymbol{x}^{\mathrm{T}}\boldsymbol{M}\boldsymbol{y}\right\} = \boldsymbol{M}\boldsymbol{y} \tag{7.126}$$

differentiate the scalar J_1 with respect to $\Delta\boldsymbol{u}_{N-1}$:

$$\frac{\partial}{\partial\Delta\boldsymbol{u}_{N-1}}\{J_1(\Delta\boldsymbol{x}_{N-1})\} = 2\boldsymbol{B}^{\mathrm{T}}\boldsymbol{Q}(\boldsymbol{A}\Delta\boldsymbol{x}_{N-1} + \boldsymbol{B}\Delta\boldsymbol{u}_{N-1}) + 2\boldsymbol{R}\Delta\boldsymbol{u}_{N-1} \tag{7.127}$$

For $J_1(\Delta\boldsymbol{x}_{N-1})$ to be minimised with respect to the choice of all the elements of $\Delta\boldsymbol{u}_{N-1}$, Equation 7.127 must yield a zero vector, so that

$$\Delta\boldsymbol{u}_{N-1} = -\left[\boldsymbol{B}^{\mathrm{T}}\boldsymbol{Q}\boldsymbol{B} + \boldsymbol{R}\right]^{-1}\boldsymbol{B}^{\mathrm{T}}\boldsymbol{Q}\boldsymbol{A}\Delta\boldsymbol{x}_{N-1} \tag{7.128}$$

Thus, the minimum value of J_1, starting at point $\Delta\boldsymbol{x}_{N-1}$, is obtained using Equation 7.125 as

$$J_1^*(\Delta\boldsymbol{x}_{N-1}) = \Delta\boldsymbol{x}_{N-1}^{\mathrm{T}}\boldsymbol{P}_1\Delta\boldsymbol{x}_{N-1} \tag{7.129}$$

where

$$\boldsymbol{P}_1 = (\boldsymbol{A} - \boldsymbol{B}\boldsymbol{K}_0)^{\mathrm{T}}\boldsymbol{Q}(\boldsymbol{A} - \boldsymbol{B}\boldsymbol{K}_0) + \boldsymbol{K}_0^{\mathrm{T}}\boldsymbol{R}\boldsymbol{K}_0 \tag{7.130}$$

$$\boldsymbol{K}_0 = \left[\boldsymbol{B}^{\mathrm{T}}\boldsymbol{Q}\boldsymbol{B} + \boldsymbol{R}\right]^{-1}\boldsymbol{B}^{\mathrm{T}}\boldsymbol{Q}\boldsymbol{A} \tag{7.131}$$

From Equation 7.123

$$J_2(\Delta\boldsymbol{x}_{N-2}) = \left[\Delta\boldsymbol{x}_{N-1}^{\mathrm{T}}\boldsymbol{Q}\Delta\boldsymbol{x}_{N-1} + \Delta\boldsymbol{u}_{N-2}^{\mathrm{T}}\boldsymbol{R}\Delta\boldsymbol{u}_{N-2}\right] + \left[\Delta\boldsymbol{x}_N^{\mathrm{T}}\boldsymbol{Q}\Delta\boldsymbol{x}_N + \Delta\boldsymbol{u}_{N-1}^{\mathrm{T}}\boldsymbol{R}\Delta\boldsymbol{u}_{N-1}\right] \tag{7.132}$$

$$J_2(\Delta\boldsymbol{x}_{N-2}) = \left[\Delta\boldsymbol{x}_{N-1}^{\mathrm{T}}\boldsymbol{Q}\Delta\boldsymbol{x}_{N-1} + \Delta\boldsymbol{u}_{N-2}^{\mathrm{T}}\boldsymbol{R}\Delta\boldsymbol{u}_{N-2}\right] + J_1(\Delta\boldsymbol{x}_{N-1}) \tag{7.133}$$

According to the *principle of optimality*, an overall optimal path from $\Delta\boldsymbol{x}_{N-2}$ to $\Delta\boldsymbol{x}_N$ which happens to pass through $\Delta\boldsymbol{x}_{N-1}$ must continue from $\Delta\boldsymbol{x}_{N-1}$ on the original optimal path established from that point, that is the contribution to the minimum value $J_2^*(\Delta\boldsymbol{x}_{N-2})$ for that section must be $J_1^*(\Delta\boldsymbol{x}_{N-1})$, so that

$$J_2^*(\Delta\boldsymbol{x}_{N-2}) = \min_{\Delta\boldsymbol{u}_{N-2}} \left[\Delta\boldsymbol{x}_{N-1}^{\mathrm{T}}\boldsymbol{Q}\Delta\boldsymbol{x}_{N-1} + \Delta\boldsymbol{u}_{N-2}^{\mathrm{T}}\boldsymbol{R}\Delta\boldsymbol{u}_{N-2}\right] + J_1^*(\Delta\boldsymbol{x}_{N-1}) \tag{7.134}$$

$$J_2^*(\Delta\boldsymbol{x}_{N-2}) = \min_{\Delta\boldsymbol{u}_{N-2}} \left\{\Delta\boldsymbol{x}_{N-1}^{\mathrm{T}}\boldsymbol{Q}\Delta\boldsymbol{x}_{N-1} + \Delta\boldsymbol{u}_{N-2}^{\mathrm{T}}\boldsymbol{R}\Delta\boldsymbol{u}_{N-2} + \Delta\boldsymbol{x}_{N-1}^{\mathrm{T}}\boldsymbol{P}_1\Delta\boldsymbol{x}_{N-1}\right\} \tag{7.135}$$

$$J_2^*(\Delta\boldsymbol{x}_{N-2}) = \min_{\Delta\boldsymbol{u}_{N-2}} \left\{\Delta\boldsymbol{x}_{N-1}^{\mathrm{T}}[\boldsymbol{Q} + \boldsymbol{P}_1]\Delta\boldsymbol{x}_{N-1} + \Delta\boldsymbol{u}_{N-2}^{\mathrm{T}}\boldsymbol{R}\Delta\boldsymbol{u}_{N-2}\right\} \tag{7.136}$$

$$J_2^*(\Delta\boldsymbol{x}_{N-2}) = \min_{\Delta\boldsymbol{u}_{N-2}} \left\{(\boldsymbol{A}\Delta\boldsymbol{x}_{N-2} + \boldsymbol{B}\Delta\boldsymbol{u}_{N-2})^{\mathrm{T}}[\boldsymbol{Q} + \boldsymbol{P}_1](\boldsymbol{A}\Delta\boldsymbol{x}_{N-2} + \boldsymbol{B}\Delta\boldsymbol{u}_{N-2}) + \Delta\boldsymbol{u}_{N-2}^{\mathrm{T}}\boldsymbol{R}\Delta\boldsymbol{u}_{N-2}\right\} \tag{7.137}$$

which yields

$$\Delta u_{N-2} = -\{B^{\mathrm{T}}[Q + P_1]B + R\}^{-1} B^{\mathrm{T}}[Q + P_1]A\Delta x_{N-2} \tag{7.138}$$

Here one sees the benefit of the *dynamic programming* method – each step backwards extends the forward optimal paths back to that point, so the individual control settings Δu_i in those paths need not be considered again. Equation 7.137 is the same form as Equation 7.125, with $Q + P_1$ substituted for Q. Thus, on the basis of Equations 7.128, 7.129 and 7.137, one can write the simple recurrence relations

$$K_i = \{B^{\mathrm{T}}[Q + P_i]B + R\}^{-1} B^{\mathrm{T}}[Q + P_i]A \tag{7.139}$$

$$P_{i+1} = (A - BK_i)^{\mathrm{T}}[Q + P_i](A - BK_i) + K_i^{\mathrm{T}} R K_i \tag{7.140}$$

Starting with $P_0 = 0$, this recurrence must be run *backwards in time* as i increases from 0 to $N-1$. Only then can the successive values of the state be obtained by running *forwards in time* as k increases from 0 to $N-1$ in

$$\Delta u_k = -K_{N-1-k}\Delta x_k \tag{7.141}$$

$$\Delta x_{k+1} = A\Delta x_k + B\Delta u_k \tag{7.142}$$

Following Equation 7.129, it will be found that

$$J_N^*(\Delta x_0) = \Delta x_0^{\mathrm{T}} P_N \Delta x_0 \tag{7.143}$$

To obtain an independent updating equation for P, substitute K in Equation 7.140 to obtain

$$P_{i+1} = A^{\mathrm{T}}[Q + P_i]A - A^{\mathrm{T}}[Q + P_i]B\{B^{\mathrm{T}}[Q + P_i]B + R\}^{-1} B^{\mathrm{T}}[Q + P_i]A \tag{7.144}$$

Letting

$$M_i = P_i + Q \tag{7.145}$$

this gives the more succinct recursive relations

$$M_{i+1} = A^{\mathrm{T}} M_i \left\{ I - B\left[B^{\mathrm{T}} M_i B + R\right]^{-1} B^{\mathrm{T}} M_i \right\} A + Q \tag{7.146}$$

$$K_i = \{B^{\mathrm{T}} M_i B + R\}^{-1} B^{\mathrm{T}} M_i A \tag{7.147}$$

Comparing Equations 6.78, 6.79 and 6.80 for the *discrete Kalman filter*, it is interesting to note the correspondence of terms in Table 7.1.

As the control horizon is lengthened (larger N), it is found that the P_i in Equation 7.144 and thus the M_i in Equation 7.146 approach constant values determined by the *algebraic Riccati equation*

$$M = A^{\mathrm{T}} M \left\{ I - B\left[B^{\mathrm{T}} M B + R\right]^{-1} B^{\mathrm{T}} M \right\} A + Q \tag{7.148}$$

Although Equations 7.146 and 7.147 should run *backwards* in time, they are useful in adaptive control where updated estimates of the system matrices A_i and B_i are available. Slow variations in

Table 7.1 Variable correspondence in the discrete Kalman filter and LQR.

Kalman filter	LQR controller
$M_i = [AP_iA^T + Q]$	$M_i = P_i + Q$
A	A^T
C	B^T
Q	Q
R	R
$K_i = M_{i-1}C^T\{CM_{i-1}C^T + R\}^{-1}$	$K_i = \{B^TM_iB + R\}^{-1}B^TM_iA$

these parameters, or the tuning matrices $\boldsymbol{Q}_i$ and $\boldsymbol{R}_i$, may be accommodated by continually running the recursion of these two equations with the updated values.

The LQR has been developed in terms of the setpoint deviation vector $\Delta\boldsymbol{x}$ and the control action deviation vector $\Delta\boldsymbol{u}$ from its initial setting. The weighted square sum of the elements of each vector is determined by the matrices $\boldsymbol{Q}$ and $\boldsymbol{R}$, respectively, which are usually diagonal. In minimising this total, the LQR thus has a 'gain' which will depend on the relative sizes of the elements of $\boldsymbol{Q}$ and $\boldsymbol{R}$. Large $\boldsymbol{Q}$ terms relative to $\boldsymbol{R}$ will give strong control actions, and vice versa. This clearly allows individual attention to each input and output in a MIMO system, and must take care of the units of measurement and ranges.

Although the system has been optimised in terms of *regulation*, setpoint variations are dealt with by defining $\Delta\boldsymbol{x}$ as the deviation from setpoint. A change in setpoint will thus appear as a disturbance, thus requiring $\Delta\boldsymbol{u}$ to 'stretch' further away from some arbitrary starting point determined by $\Delta\boldsymbol{u} = \boldsymbol{0}$ for $\boldsymbol{u} = \boldsymbol{u}_0 = \boldsymbol{u}_{SS}$. This offset value of $\boldsymbol{u}$ might thus be defined by the user to rather be some otherwise desirable control action setting from which it is wished to resist movement. *Integral action* is easily incorporated in the method by augmenting the original system (Equation 7.122) as follows:

$$\begin{pmatrix} \Delta\boldsymbol{x} \\ \Delta\boldsymbol{x}_I \end{pmatrix}_{i+1} = \begin{bmatrix} \boldsymbol{A} & \boldsymbol{0} \\ \boldsymbol{\psi} & \boldsymbol{I} \end{bmatrix} \begin{pmatrix} \Delta\boldsymbol{x} \\ \Delta\boldsymbol{x}_I \end{pmatrix}_i + \begin{bmatrix} \boldsymbol{B} \\ \boldsymbol{0} \end{bmatrix} \Delta\boldsymbol{u}_i \tag{7.149}$$

where

$$\boldsymbol{\psi} = \begin{bmatrix} T/\tau_{I1} & 0 & \cdots \\ 0 & T/\tau_{I2} & \cdots \\ \vdots & \vdots & \ddots \end{bmatrix} \tag{7.150}$$

If it is not desired to change the basis model or tuning of the LQR online, the steady-state gain $\boldsymbol{K}$ can be precalculated. In this sense, it is a simple MIMO controller with few computational requirements. A disadvantage of LQR is that there is no means of handling constraints optimally – the best that can be done is to just 'clip' computed control actions that fall outside of a desirable range. In Section 7.8, fully optimal controllers that use a similar objective function will be discussed. These handle constraints on both control actions and controlled variables in an optimal fashion, but have a large real-time computational requirement.

7.6.6 Continuous State-Space MIMO Controller Design Using the Linear Quadratic Regulator Approach

This is the continuous version of the 'minimum variance' controller. Taking the viewpoint used in Section 7.6.5, the system of first-order equations describing the state evolution in a linear continuous system is

$$\frac{\mathrm{d}\Delta\boldsymbol{x}(t)}{\mathrm{d}t} = \boldsymbol{A}\Delta\boldsymbol{x}(t) + \boldsymbol{B}\Delta\boldsymbol{u}(t) \tag{7.151}$$

As in Section 7.6.5, $\Delta\boldsymbol{x}(t)$ will ultimately be considered as the deviation of the state vector from its setpoint (suggesting regulation around a value of $\boldsymbol{0}$), and $\Delta\boldsymbol{u}(t)$ will be considered the deviation of the control action vector from the initial steady-state condition. For a small interval $\mathrm{d}t$, consider

$$\Delta\boldsymbol{x}(t+\mathrm{d}t) = [\boldsymbol{I} + \boldsymbol{A}\,\mathrm{d}t]\Delta\boldsymbol{x}(t) + \boldsymbol{B}\,\mathrm{d}t\,\Delta\boldsymbol{u}(t) \tag{7.152}$$

It is desired to find a control policy $\Delta\boldsymbol{u}(t)$ which minimises the objective function

$$J[\Delta\boldsymbol{x}(0)] = \int_0^{\tau} \left\{\Delta\boldsymbol{x}^{\mathrm{T}}(t)\boldsymbol{Q}\Delta\boldsymbol{x}(t) + \Delta\boldsymbol{u}^{\mathrm{T}}(t)\boldsymbol{R}\Delta\boldsymbol{u}(t)\right\}\mathrm{d}t \tag{7.153}$$

where τ is a defined control horizon. This problem can be solved by variational methods, but instead an explanation here will be based on the discrete LQR development in Section 7.6.5.

In the last $\mathrm{d}t$ step of the objective integration,

$$J_{\mathrm{d}t}[\Delta\boldsymbol{x}(\tau-\mathrm{d}t)] \approx \left\{\Delta\boldsymbol{x}^{\mathrm{T}}(\tau)\boldsymbol{Q}\Delta\boldsymbol{x}(\tau) + \Delta\boldsymbol{u}^{\mathrm{T}}(\tau-\mathrm{d}t)\boldsymbol{R}\Delta\boldsymbol{u}(\tau-\mathrm{d}t)\right\}\mathrm{d}t \tag{7.154}$$

Substituting for $\Delta\boldsymbol{x}(\tau)$ from Equation 7.152, it is required to solve

$$J_{\mathrm{d}t}[\Delta\boldsymbol{x}(\tau-\mathrm{d}t)] = \min_{\Delta\boldsymbol{u}(\tau-\mathrm{d}t)} \left\{ \begin{array}{l} [\{\boldsymbol{I}+\boldsymbol{A}\,\mathrm{d}t\}\Delta\boldsymbol{x}(\tau-\mathrm{d}t) + \boldsymbol{B}\,\mathrm{d}t\,\Delta\boldsymbol{u}(\tau-\mathrm{d}t)]^{\mathrm{T}}\mathrm{d}t\,\boldsymbol{Q}\left[\{\boldsymbol{I}+\boldsymbol{A}\,\mathrm{d}t\}\Delta\boldsymbol{x}(\tau-\mathrm{d}t) + \boldsymbol{B}\,\mathrm{d}t\,\Delta\boldsymbol{u}(\tau-\mathrm{d}t)\right] \\ +\,\Delta\boldsymbol{u}^{\mathrm{T}}(\tau-\mathrm{d}t)\mathrm{d}t\,\boldsymbol{R}\Delta\boldsymbol{u}(\tau-\mathrm{d}t) \end{array} \right\} \tag{7.155}$$

Solving by differentiation as in Equations 7.124–7.128 obtain

$$\Delta\boldsymbol{u}(\tau-\mathrm{d}t) = -\left[(\mathrm{d}t)^2\boldsymbol{B}^{\mathrm{T}}\,\mathrm{d}t\,\boldsymbol{Q}\boldsymbol{B} + \mathrm{d}t\,\boldsymbol{R}\right]^{-1}\mathrm{d}t\,\boldsymbol{B}^{\mathrm{T}}\,\mathrm{d}t\,\boldsymbol{Q}[\boldsymbol{I}+\boldsymbol{A}\,\mathrm{d}t]\Delta\boldsymbol{x}(\tau-\mathrm{d}t) \tag{7.156}$$

yielding a minimum value

$$J^*_{\mathrm{d}t}[\Delta\boldsymbol{x}(\tau-\mathrm{d}t)] = \Delta\boldsymbol{x}^{\mathrm{T}}(\tau-\mathrm{d}t)\boldsymbol{P}_{\mathrm{d}t}\Delta\boldsymbol{x}(\tau-\mathrm{d}t) \tag{7.157}$$

where

$$\boldsymbol{P}_{\mathrm{d}t} = [\boldsymbol{I}+\boldsymbol{A}\,\mathrm{d}t - \mathrm{d}t\,\boldsymbol{B}\boldsymbol{K}_0]^{\mathrm{T}}\mathrm{d}t\,\boldsymbol{Q}[\boldsymbol{I}+\boldsymbol{A}\,\mathrm{d}t - \mathrm{d}t\,\boldsymbol{B}\boldsymbol{K}_0] + \boldsymbol{K}_0^{\mathrm{T}}\,\mathrm{d}t\,\boldsymbol{R}\boldsymbol{K}_0 \tag{7.158}$$

$$\boldsymbol{K}_0 = \left[(\mathrm{d}t)^2\boldsymbol{B}^{\mathrm{T}}\,\mathrm{d}t\,\boldsymbol{Q}\boldsymbol{B} + \mathrm{d}t\,\boldsymbol{R}\right]^{-1}\mathrm{d}t\,\boldsymbol{B}^{\mathrm{T}}\,\mathrm{d}t\,\boldsymbol{Q}[\boldsymbol{I}+\boldsymbol{A}\,\mathrm{d}t] \tag{7.159}$$

As in Equations 7.134–7.138, optimising from $\tau - 2\mathrm{d}t$ on yields

$$J^*_{2\mathrm{d}t}[\Delta\boldsymbol{x}(\tau-2\mathrm{d}t)] = \min_{\Delta\boldsymbol{u}(\tau-2\mathrm{d}t)} \left[\Delta\boldsymbol{x}^{\mathrm{T}}(\tau-\mathrm{d}t)\boldsymbol{Q}\Delta\boldsymbol{x}(\tau-\mathrm{d}t) + \Delta\boldsymbol{u}^{\mathrm{T}}(\tau-2\mathrm{d}t)\boldsymbol{R}\Delta\boldsymbol{u}(\tau-2\mathrm{d}t)\right]\mathrm{d}t + J^*_{\mathrm{d}t}[\Delta\boldsymbol{x}(\tau-\mathrm{d}t)] \quad (7.160)$$

$$J^*_{2\mathrm{d}t}[\Delta\boldsymbol{x}(\tau-2\mathrm{d}t)] = \min_{\Delta\boldsymbol{u}(\tau-2\mathrm{d}t)} \left\{ \begin{array}{l} \left[\{\boldsymbol{I}+\boldsymbol{A}\,\mathrm{d}t\}\Delta\boldsymbol{u}(\tau-2\mathrm{d}t) + \boldsymbol{B}\,\mathrm{d}t\,\Delta\boldsymbol{u}(\tau-2\mathrm{d}t)\right]^{\mathrm{T}}[\mathrm{d}t\,\boldsymbol{Q}+\boldsymbol{P}_{\mathrm{d}t}]\left[\{\boldsymbol{I}+\boldsymbol{A}\,\mathrm{d}t\}\Delta\boldsymbol{u}(\tau-2\mathrm{d}t) + \boldsymbol{B}\,\mathrm{d}t\,\Delta\boldsymbol{u}(\tau-2\mathrm{d}t)\right] \\ +\Delta\boldsymbol{u}^{\mathrm{T}}(\tau-2\mathrm{d}t)\mathrm{d}t\,\boldsymbol{R}\Delta\boldsymbol{u}(\tau-2\mathrm{d}t) \end{array} \right\} \quad (7.161)$$

which yields

$$\Delta\boldsymbol{u}(\tau-2\mathrm{d}t) = -\left\{(\mathrm{d}t)^2\boldsymbol{B}^{\mathrm{T}}[\mathrm{d}t\,\boldsymbol{Q}+\boldsymbol{P}_{\mathrm{d}t}]\boldsymbol{B} + \mathrm{d}t\,\boldsymbol{R}\right\}^{-1}\mathrm{d}t\,\boldsymbol{B}^{\mathrm{T}}[\mathrm{d}t\,\boldsymbol{Q}+\boldsymbol{P}_{\mathrm{d}t}][\boldsymbol{I}+\boldsymbol{A}\,\mathrm{d}t]\Delta\boldsymbol{x}(\tau-2\mathrm{d}t) \quad (7.162)$$

Equation 7.161 is the same form as Equation 7.155, with $\mathrm{d}t\ \boldsymbol{Q}+\boldsymbol{P}_{\mathrm{d}t}$ substituted for $\mathrm{d}t\ \boldsymbol{Q}$. Thus, on the basis of Equations 7.157–7.162, one can write the recurrence relations

$$\boldsymbol{K}_{i\,\mathrm{d}t} = \left\{(\mathrm{d}t)^2\boldsymbol{B}^{\mathrm{T}}[\mathrm{d}t\,\boldsymbol{Q}+\boldsymbol{P}_{i\,\mathrm{d}t}]\boldsymbol{B} + \mathrm{d}t\,\boldsymbol{R}\right\}^{-1}\mathrm{d}t\,\boldsymbol{B}^{\mathrm{T}}[\mathrm{d}t\,\boldsymbol{Q}+\boldsymbol{P}_{i\,\mathrm{d}t}][\boldsymbol{I}+\mathrm{d}t\,\boldsymbol{A}] \quad (7.163)$$

$$\boldsymbol{P}_{(i+1)\mathrm{d}t} = (\boldsymbol{I}+\mathrm{d}t\,\boldsymbol{A}-\mathrm{d}t\,\boldsymbol{B}\boldsymbol{K}_{i\,\mathrm{d}t})^{\mathrm{T}}[\mathrm{d}t\,\boldsymbol{Q}+\boldsymbol{P}_{i\,\mathrm{d}t}](\boldsymbol{I}+\mathrm{d}t\,\boldsymbol{A}-\mathrm{d}t\,\boldsymbol{B}\boldsymbol{K}_{i\,\mathrm{d}t}) + \boldsymbol{K}^{\mathrm{T}}_{i\,\mathrm{d}t}\,\mathrm{d}t\,\boldsymbol{R}\boldsymbol{K}_{i\,\mathrm{d}t} \quad (7.164)$$

Noting that $\boldsymbol{P}$, $\boldsymbol{Q}$ and $\boldsymbol{R}$ are symmetric, and neglecting terms with the small factor $(\mathrm{d}t)^2$,

$$\boldsymbol{K}_{i\,\mathrm{d}t} = \boldsymbol{R}^{-1}\boldsymbol{B}^{\mathrm{T}}\{\boldsymbol{P}_{i\,\mathrm{d}t} + \mathrm{d}t[\boldsymbol{Q}+\boldsymbol{P}_{i\,\mathrm{d}t}\boldsymbol{A}]\} \quad (7.165)$$

$$\boldsymbol{P}_{(i+1)\mathrm{d}t} = [\mathrm{d}t\,\boldsymbol{Q}+\boldsymbol{P}_{i\,\mathrm{d}t}] + \mathrm{d}t\,\boldsymbol{A}^{\mathrm{T}}\boldsymbol{P}_{i\,\mathrm{d}t} + \mathrm{d}t\,\boldsymbol{P}_{i\,\mathrm{d}t}\boldsymbol{A} - \mathrm{d}t\,\boldsymbol{P}_{i\,\mathrm{d}t}\boldsymbol{B}\boldsymbol{R}^{-1}\boldsymbol{B}^{\mathrm{T}}\boldsymbol{P}_{i\,\mathrm{d}t} \quad (7.166)$$

Letting $t = i\ \mathrm{d}t$ (but beware that it is running *backwards* from $t = \tau$), dividing by $\mathrm{d}t$ and taking the limit as $\mathrm{d}t \to 0$, one finally has the *differential Riccati equation*

$$\frac{\mathrm{d}\boldsymbol{P}(t)}{\mathrm{d}t} = \boldsymbol{A}^{\mathrm{T}}\boldsymbol{P}(t) + \boldsymbol{P}(t)\boldsymbol{A} - \boldsymbol{P}(t)\boldsymbol{B}\boldsymbol{R}^{-1}\boldsymbol{B}^{\mathrm{T}}\boldsymbol{P}(t) + \boldsymbol{Q} \quad (7.167)$$

and the required proportional state feedback gain for LQR control

$$\boldsymbol{K}(t) = \boldsymbol{R}^{-1}\boldsymbol{B}^{\mathrm{T}}\boldsymbol{P}(t) \quad (7.168)$$

Comparing Equations 6.103–6.105 for the *continuous Kalman–Bucy filter*, it is interesting to note the correspondence of terms in Table 7.2.

Table 7.2 Variable correspondence in the continuous Kalman–Bucy filter and LQR.

Kalman filter	LQR controller
$\boldsymbol{A}$	$\boldsymbol{A}^{\mathrm{T}}$
$\boldsymbol{C}$	$\boldsymbol{B}^{\mathrm{T}}$
$\boldsymbol{Q}$	$\boldsymbol{Q}$
$\boldsymbol{R}$	$\boldsymbol{R}$
$\boldsymbol{K} = \boldsymbol{P}\boldsymbol{C}^{\mathrm{T}}\boldsymbol{R}^{-1}$	$\boldsymbol{K} = \boldsymbol{R}^{-1}\boldsymbol{B}^{\mathrm{T}}\boldsymbol{P}$

As the control horizon τ is lengthened, it is found that the $\boldsymbol{P}(t)$ in Equation 7.167 approaches a constant value determined by the *algebraic Riccati equation*

$$\boldsymbol{A}^{\mathrm{T}}\boldsymbol{P}(t) + \boldsymbol{P}(t)\boldsymbol{A} - \boldsymbol{P}(t)\boldsymbol{B}\boldsymbol{R}^{-1}\boldsymbol{B}^{\mathrm{T}}\boldsymbol{P}(t) = \boldsymbol{0} \tag{7.169}$$

Solutions for this equation are mentioned in Section 6.4.2. The solution of Equation 7.167 starts with a typical choice $\boldsymbol{P}(0) = \boldsymbol{0}$, and runs from $t=0$ to $t=\tau$. Actually, this is running backwards from the horizon, and the optimal gain matrix $\boldsymbol{K}(t)$ will really apply at $(\tau - t)$ in finite horizon problems. The differential equation form (Equation 7.167) can nevertheless be useful in an adaptive sense for slow variation of the parameters $\boldsymbol{A}_i$, $\boldsymbol{B}_i$, $\boldsymbol{Q}_i$ and $\boldsymbol{R}_i$.

The comments at the end of Section 7.6.5 regarding the discrete form of the LQR apply also to the continuous LQR in this section. Note that *integral action* is easily incorporated into the method by augmenting the original system equation (Equation 7.151) as in Equation 7.170. The matrix $\boldsymbol{\psi}$ could be an identity matrix if it desired to perform all of the tuning by means of the corresponding diagonal elements of $\boldsymbol{Q}$.

$$\frac{\mathrm{d}}{\mathrm{d}t}\begin{pmatrix}\Delta\boldsymbol{x}(t)\\ \Delta\boldsymbol{x}_{\mathrm{I}}(t)\end{pmatrix} = \begin{bmatrix}\boldsymbol{A} & \boldsymbol{0}\\ \boldsymbol{\psi} & \boldsymbol{I}\end{bmatrix}\begin{pmatrix}\Delta\boldsymbol{x}(t)\\ \Delta\boldsymbol{x}_{\mathrm{I}}(t)\end{pmatrix} + \begin{bmatrix}\boldsymbol{B}\\ \boldsymbol{0}\end{bmatrix}\Delta\boldsymbol{u}(t) \tag{7.170}$$

where

$$\boldsymbol{\psi} = \begin{bmatrix} 1/\tau_{\mathrm{I}1} & 0 & \cdots \\ 0 & 1/\tau_{\mathrm{I}2} & \cdots \\ \vdots & \vdots & \ddots \end{bmatrix} \tag{7.171}$$

7.7
Concept of Internal Model Control

A number of approaches to process control fall naturally into the class of *internal model control* (IMC). A model representation of the real process is run in parallel with the process in real time, using available input and output information, and this provides some form of feedback for control purposes (Figure 7.16). Examples of such internal models used in closed loops are the Smith predictor (Section 6.2), data reconcilers (Section 6.3), state estimators (Section 6.4) and model parameter estimators (Section 6.5) for adaptive control. This section serves only to highlight the concept,

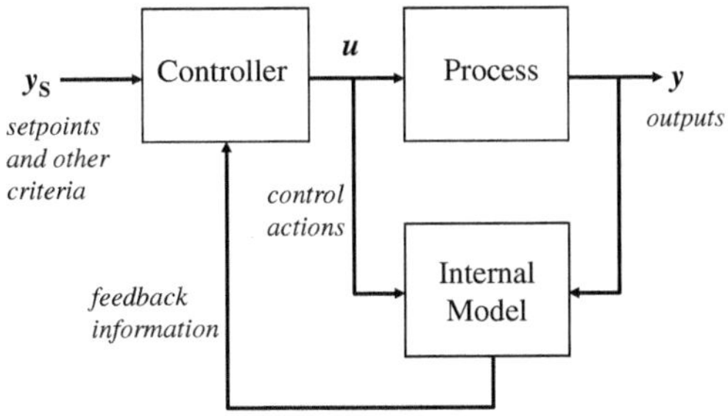

Figure 7.16 General IMC structure.

which will be adopted in some advanced controllers later in this chapter. A simple direct design procedure is noted in the following subsection.

7.7.1
A General MIMO Controller Design Approach Based on IMC

Figure 7.17 represents an IMC structure in which transfer functions $\boldsymbol{G}(z)$ (Equation 3.291) or $\boldsymbol{G}(s)$ (Equation 3.161) are used as the basis for a controller design. In general, for the *z-domain* or *s-domain*, one expects the model of the process to be determined by

$$\hat{\boldsymbol{y}}(z) = \hat{\boldsymbol{G}}_{\mathrm{P}}(z)\boldsymbol{u}(z) \quad \text{or} \quad \hat{\boldsymbol{y}}(s) = \hat{\boldsymbol{G}}_{\mathrm{P}}(s)\boldsymbol{u}(s) \tag{7.172}$$

In Sections 3.9.2.2 and 3.9.3.2, it was found that state systems lead to the forms

$$\hat{\boldsymbol{G}}_{\mathrm{P}}(z) = [D(z)]^{-1}\boldsymbol{N}(z) \quad \text{or} \quad \hat{\boldsymbol{G}}_{\mathrm{P}}(s) = [D(s)]^{-1}\boldsymbol{N}(s) \tag{7.173}$$

and the system is realisable for $m < n$ in

$$\begin{aligned} D(z) &= z^n + d_1 z^{n-1} + d_2 z^{n-2} + \cdots + d_{n-1}z + d_n \quad \text{or} \\ D(s) &= s^n + d_1 s^{n-1} + d_2 s^{n-2} + \cdots + d_{n-1}s + d_n \end{aligned} \tag{7.174}$$

$$\begin{aligned} \boldsymbol{N}(z) &= \boldsymbol{N}_0 z^m + \boldsymbol{N}_1 z^{m-1} + \boldsymbol{N}_2 z^{m-2} + \cdots + \boldsymbol{N}_{m-1}z + \boldsymbol{N}_m \quad \text{or} \\ \boldsymbol{N}(s) &= \boldsymbol{N}_0 s^m + \boldsymbol{N}_1 s^{m-1} + \boldsymbol{N}_2 s^{m-2} + \cdots + \boldsymbol{N}_{m-1}s + \boldsymbol{N}_m \end{aligned} \tag{7.175}$$

where the d_i are scalars and the $\boldsymbol{N}_i$ are constant matrices. It turns out that the $\boldsymbol{N}$ and D polynomials need to be much larger to accommodate the general input-output case (see Section 7.8.1), but the same principles apply. The process is said to be *minimum phase* if the poles of both

$$\hat{\boldsymbol{G}}_{\mathrm{P}} = [D]^{-1}\boldsymbol{N} \tag{7.176}$$

and

$$\hat{\boldsymbol{G}}_{\mathrm{P}}^{-1} = [\boldsymbol{N}]^{-1}D \tag{7.177}$$

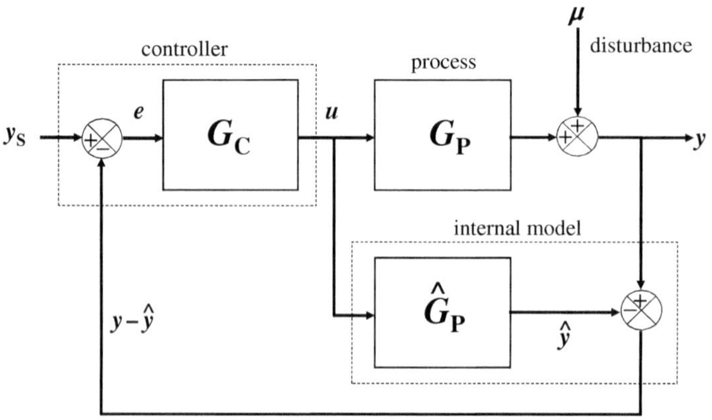

Figure 7.17 Transfer function-based design of an IMC.

are stable (i.e. poles z within the unit circle, or s on the left of the complex plane). A minimum-phase system thus has a stable inverse, so a controller can be found directly as this inverse. The closed-loop transfer function in Figure 7.17 is obtained as follows:

$$\boldsymbol{u} = \boldsymbol{G}_\mathrm{C}\left[\boldsymbol{y}_\mathrm{S} - \boldsymbol{y} + \widehat{\boldsymbol{y}}\right] \tag{7.178}$$

$$\boldsymbol{u} = \boldsymbol{G}_\mathrm{C}\left[\boldsymbol{y}_\mathrm{S} - \boldsymbol{G}_\mathrm{P}\boldsymbol{u} - \boldsymbol{\mu} + \widehat{\boldsymbol{G}}_\mathrm{P}\boldsymbol{u}\right] \tag{7.179}$$

$$\boldsymbol{u} = \boldsymbol{G}_\mathrm{C}\left[\boldsymbol{y}_\mathrm{S} - \boldsymbol{\mu}\right] - \boldsymbol{G}_\mathrm{C}\left[\widehat{\boldsymbol{G}}_\mathrm{P} - \boldsymbol{G}_\mathrm{P}\right]\boldsymbol{u} \tag{7.180}$$

$$\boldsymbol{u} = \left\{\boldsymbol{I} + \boldsymbol{G}_\mathrm{C}\left[\widehat{\boldsymbol{G}}_\mathrm{P} - \boldsymbol{G}_\mathrm{P}\right]\right\}^{-1}\boldsymbol{G}_\mathrm{C}\left[\boldsymbol{y}_\mathrm{S} - \boldsymbol{\mu}\right] \tag{7.181}$$

$$\boldsymbol{y} = \boldsymbol{G}_\mathrm{P}\left\{\boldsymbol{I} + \boldsymbol{G}_\mathrm{C}\left[\widehat{\boldsymbol{G}}_\mathrm{P} - \boldsymbol{G}_\mathrm{P}\right]\right\}^{-1}\boldsymbol{G}_\mathrm{C}\left[\boldsymbol{y}_\mathrm{S} - \boldsymbol{\mu}\right] + \boldsymbol{\mu} \tag{7.182}$$

Thus, if $\widehat{\boldsymbol{G}}_\mathrm{P}$ is *minimum phase*, and $m = n$, one would clearly choose

$$\boldsymbol{G}_\mathrm{C} = \left[\widehat{\boldsymbol{G}}_\mathrm{P}\right]^{-1} \tag{7.183}$$

giving

$$\boldsymbol{y} = \boldsymbol{G}_\mathrm{P}\left\{\boldsymbol{I} + \left[\boldsymbol{I} - \widehat{\boldsymbol{G}}_\mathrm{P}^{-1}\boldsymbol{G}_\mathrm{P}\right]\right\}^{-1}\widehat{\boldsymbol{G}}_\mathrm{P}^{-1}\left[\boldsymbol{y}_\mathrm{S} - \boldsymbol{\mu}\right] + \boldsymbol{\mu} \tag{7.184}$$

which would transmit the setpoint directly through the system if the model is accurate and there is no disturbance.

Even if the above procedure were possible, one would need to avoid slow and poorly damped zeros in the $\widehat{\boldsymbol{G}}_\mathrm{P}$, which it is seen become poles in the closed loop. In general, such zeros of $\widehat{\boldsymbol{G}}_\mathrm{P}$, and any unstable zeros or dead-time lags, must be factored into a 'bad' transfer function $\widehat{\boldsymbol{G}}_\mathrm{Pb}$ and the rest into a 'good' transfer function $\widehat{\boldsymbol{G}}_\mathrm{Pg}$ such that

$$\widehat{\boldsymbol{G}}_\mathrm{P} = \widehat{\boldsymbol{G}}_\mathrm{Pg}\widehat{\boldsymbol{G}}_\mathrm{Pb} \tag{7.185}$$

Then, choosing

$$\boldsymbol{G}_\mathrm{C} = \left[\widehat{\boldsymbol{G}}_\mathrm{Pg}\right]^{-1} \tag{7.186}$$

the closed-loop equation becomes

$$\boldsymbol{y} = \boldsymbol{G}_\mathrm{Pg}\boldsymbol{G}_\mathrm{Pb}\left\{\boldsymbol{I} + \left[\widehat{\boldsymbol{G}}_\mathrm{Pb} - \widehat{\boldsymbol{G}}_\mathrm{Pg}^{-1}\boldsymbol{G}_\mathrm{Pg}\boldsymbol{G}_\mathrm{Pb}\right]\right\}^{-1}\widehat{\boldsymbol{G}}_\mathrm{Pg}^{-1}\left[\boldsymbol{y}_\mathrm{S} - \boldsymbol{\mu}\right] \tag{7.187}$$

which if the model is accurate and there is no disturbance gives

$$\boldsymbol{y} = \boldsymbol{G}_\mathrm{Pg}\boldsymbol{G}_\mathrm{Pb}\widehat{\boldsymbol{G}}_\mathrm{Pg}^{-1}\boldsymbol{y}_\mathrm{S} \tag{7.188}$$

Thus, one of the complications in separating $\widehat{\boldsymbol{G}}_\mathrm{Pg}$ and $\widehat{\boldsymbol{G}}_\mathrm{Pb}$ will be to apportion the steady-state gains so that this closed-loop transfer function has a unity steady-state gain from each setpoint to each output. Assuming this is done, consider, for example, the case where it has been possible to separate out all of the dead-time lags onto the axis of a diagonal $\widehat{\boldsymbol{G}}_\mathrm{Pb}$. Then the ideal closed loop in

Equation 7.188 would perfectly track any setpoint change, but with the corresponding dead-time lag in each case.

7.8 Predictive Control

Most of the controller design techniques discussed so far in this chapter have relied on some kind of inversion of the original process, or use of the model parameters to derive new mathematical forms. One exception has been the LQR in Sections 7.6.5 and 7.6.6, which was seen to choose its control actions by effectively viewing the forward trajectory from the present time. In the LQR, a closed analytical solution is obtained for the best control action, bearing in mind the future control strategy up to a finite or infinite horizon. So the question to be asked is – why not perform trial-and-error tests up to some horizon, for *any* objective function, for *any* form of model? Indeed, this is the basis nowadays of many successful implementations in industry. As the open-loop model is run forward from the present time t, one has the opportunity to ensure that both PV and MV constraints are not violated. One is optimising a sequence of future choices $\boldsymbol{u}_{i+1}, \boldsymbol{u}_{i+2}, \ldots, \boldsymbol{u}_{i+N}$, or alternatively continuous functions $\boldsymbol{u}(t')$, $t < t' < t+\tau$, for the MVs. The problem of optimising *functions* like this is dealt with in the *calculus of variations*. For state systems, it turns out to be more efficient to work *backwards* from the target condition, so that established optimal path sections may be reused (Bellman, 1957).

The various predictive control approaches differ mainly in the form of the model, and the technique of the trajectory optimisation. A common view taken is that the future *closed-loop response* differs from the future *open-loop response* only because the control actions are not left static at their present values (Figures 7.18 and 7.19). On each time step, the optimal future control strategy is

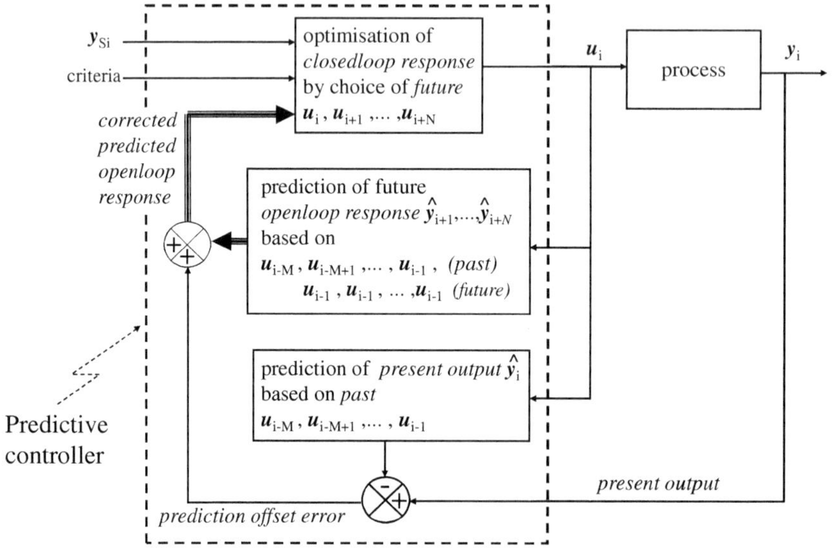

Figure 7.18 Concept of a predictive controller based on optimising the *closed-loop response* relative to the *open-loop response*.

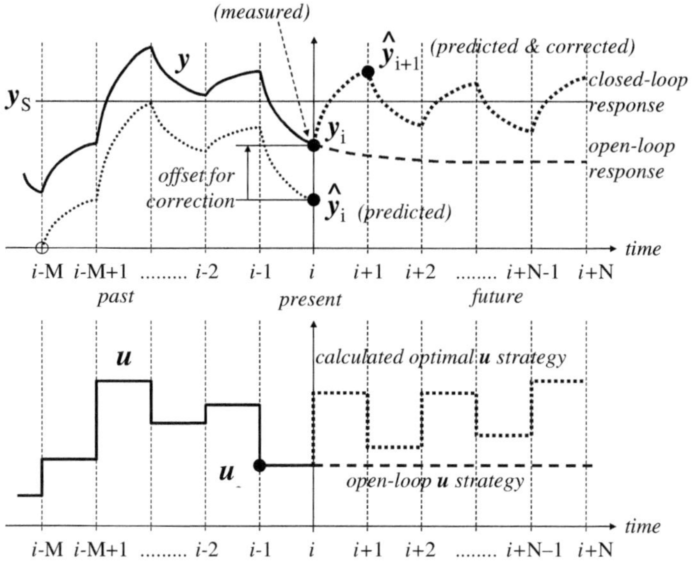

Figure 7.19 Typical discrete system predictive control concept based on *open-loop* and *closed-loop* responses.

computed, but it is only the first control action in that sequence that is implemented. Moving forward one time step, the prediction horizon also *recedes* one step, maintaining the length of the future period under consideration, and the entire calculation is repeated.

7.8.1 Generalised Predictive Control for a Discrete *z*-Domain MIMO System

Before considering some of the popular specific forms of *model predictive control* (MPC), it is worthwhile reviewing the *generalised predictive control* (GPC) framework of Clarke, Mohtadi and Tuffs (1987).

A general version of the discrete linear dynamic system is represented by the *input–output* form

$$\boldsymbol{A}(z^{-1})\boldsymbol{y}(z) = \boldsymbol{B}(z^{-1})\boldsymbol{u}(z) \tag{7.189}$$

with

$$\boldsymbol{A}(z^{-1}) = \boldsymbol{A}_0 + \boldsymbol{A}_1 z^{-1} + \cdots + \boldsymbol{A}_n z^{-n} \tag{7.190}$$

$$\boldsymbol{B}(z^{-1}) = \boldsymbol{B}_0 + \boldsymbol{B}_1 z^{-1} + \cdots + \boldsymbol{B}_m z^{-m} \tag{7.191}$$

and $\boldsymbol{A}_i$ and $\boldsymbol{B}_i$ constant matrices. Then

$$\boldsymbol{y}(z) = \frac{\operatorname{adj}\left[\boldsymbol{A}(z^{-1})\right]\boldsymbol{B}(z^{-1})\boldsymbol{u}(z)}{\det[\boldsymbol{A}(z^{-1})]} \tag{7.192}$$

For N elements in $\boldsymbol{y}$ and M elements in $\boldsymbol{u}$, the dividing scalar determinant will have the form

$$\det\left[\boldsymbol{A}\left(z^{-1}\right)\right] = a_0 + a_1 z^{-1} + \cdots + a_{N\times n} z^{-N\times n} \tag{7.193}$$

so setting

$$\boldsymbol{N}(z^{-1}) = \frac{1}{a_0}\operatorname{adj}\left[\boldsymbol{A}(z^{-1})\right]\boldsymbol{B}(z^{-1}) \tag{7.194}$$

$$d_i = \frac{a_i}{a_0}, \quad i = 1, \ldots, N\times n \tag{7.195}$$

obtain the general representation

$$D(z^{-1})\boldsymbol{y}(z) = \boldsymbol{N}(z^{-1})\boldsymbol{u}(z) \tag{7.196}$$

with

$$D(z^{-1}) = 1 + d_1 z^{-1} + \cdots + d_{N\times n} z^{-N\times n} \tag{7.197}$$

$$\boldsymbol{N}(z^{-1}) = \boldsymbol{N}_0 + \boldsymbol{N}_1 z^{-1} + \cdots + \boldsymbol{N}_{(N-1)\times n+m} z^{-(N-1)\times n+m} \tag{7.198}$$

and it is necessary that $\boldsymbol{N}_0 = \boldsymbol{0}$ because the output cannot react instantly to the input. A similar large number of D and $\boldsymbol{N}$ terms arises if an arbitrary transfer function $\boldsymbol{G}$ (Equation 6.131) is specified, and is multiplied through by the common denominator of all denominators.

In passing, one notes that for the special case of the linear discrete state system (Section 3.9.3.2),

$$z\boldsymbol{x}(z) = \boldsymbol{A}'\boldsymbol{x}(z) + \boldsymbol{B}'\boldsymbol{u}(z) \tag{7.199}$$

where only a selection or combination of the states is available according to the linear equation

$$\boldsymbol{y}(z) = \boldsymbol{H}'\boldsymbol{x}(z) \tag{7.200}$$

one would find

$$\boldsymbol{N}(z^{-1}) = z^{-n}\boldsymbol{H}' \operatorname{adj}[z\boldsymbol{I} - \boldsymbol{A}']\boldsymbol{B}' \tag{7.201}$$

$$D(z^{-1}) = z^{-n} \det[z\boldsymbol{I} - \boldsymbol{A}'] \tag{7.202}$$

7.8.1.1 GPC for a Discrete MIMO System Represented by z-Domain Polynomials (Input–Output Form)

In the approach of Clarke, Mohtadi and Tuffs (1987), the process is represented by a CARIMA model using the 'difference equation' polynomials in Equations 7.196–7.198. These forms may be established by the identification methods discussed in Chapter 6, either offline or online. In the latter case, with the model undergoing periodic updating, the GPC technique described here will amount to *adaptive control*. The various models used in *process identification* are listed in Table 6.1 in Section 6.5.3.1, including the ARIMAX model. The CARIMA model is the same as the ARIMAX, except that the mnemonic recognises that the incoming control action $\boldsymbol{u}$ will now emanate from a *controller*, rather than *exogenously*. Using the discrete forms in Equations 7.196–7.198, the controlled autoregressive integrated moving average model is thus

$$D\left(z^{-1}\right)\boldsymbol{y}(z) = \boldsymbol{N}\left(z^{-1}\right)\boldsymbol{u}(z) + \frac{\boldsymbol{C}\left(z^{-1}\right)\boldsymbol{\mu}(z)}{[1 - z^{-1}]} \tag{7.203}$$

where

$$C(z^{-1}) = C_0 + C_1 z^{-1} + C_2 z^{-2} + \cdots + C_{p-1} z^{-p+1} + C_p z^{-p} \tag{7.204}$$

Recall that $\boldsymbol{\mu}$ is a vector of random and uncorrelated variables, and that the function $\boldsymbol{C}(z)$ is used to create a degree of smoothing and cross-correlation in the disturbances ultimately affecting the output, so that such correlated behaviour is not erroneously attributed to $D(z)$ or $\boldsymbol{N}(z)$ in the course of an identification. Division by the differencing function $[1 - z^{-1}]$ moreover integrates the disturbance, to allow for the changes in offset that typically affect identification.

For a j-step-ahead predictor, Clarke, Mohtadi and Tuffs (1987) define uniquely matching functions $\boldsymbol{E}_j(z)$ and $\boldsymbol{F}_j(z)$ by the requirement that they satisfy the Diophantine equation

$$\boldsymbol{I} = \left[1 - z^{-1}\right]\boldsymbol{E}_j\left(z^{-1}\right)D\left(z^{-1}\right) + z^{-j}\boldsymbol{F}_j\left(z^{-1}\right) \tag{7.205}$$

The purpose of this construction is to create a gap in the $\boldsymbol{y}$-dependence between the present and desired future predictions, rather than a step-by-step dependence. More discussion on this will follow later. Multiplying through Equation 7.203 by $\left[1 - z^{-1}\right]\boldsymbol{E}_j\left(z^{-1}\right)$, one has

$$\left[1 - z^{-1}\right]\boldsymbol{E}_j\left(z^{-1}\right)D\left(z^{-1}\right)\boldsymbol{y}(z) = \left[1 - z^{-1}\right]\boldsymbol{E}_j\left(z^{-1}\right)\boldsymbol{N}\left(z^{-1}\right)\boldsymbol{u}(z) + \boldsymbol{E}_j\left(z^{-1}\right)\boldsymbol{C}\left(z^{-1}\right)\boldsymbol{\mu}(z) \tag{7.206}$$

and using Equation 7.205:

$$\left[\boldsymbol{I} - z^{-j}\boldsymbol{F}_j\left(z^{-1}\right)\right]\boldsymbol{y}(z) = \left[1 - z^{-1}\right]\boldsymbol{E}_j\left(z^{-1}\right)\boldsymbol{N}\left(z^{-1}\right)\boldsymbol{u}(z) + \boldsymbol{E}_j\left(z^{-1}\right)\boldsymbol{C}\left(z^{-1}\right)\boldsymbol{\mu}(z) \tag{7.207}$$

$$\times z^{+j}: \quad z^j\boldsymbol{y}(z) = \boldsymbol{F}_j\left(z^{-1}\right)\boldsymbol{y}(z) + \left[z^j - z^{j-1}\right]\boldsymbol{E}_j\left(z^{-1}\right)\boldsymbol{N}\left(z^{-1}\right)\boldsymbol{u}(z) + z^j\boldsymbol{E}_j\left(z^{-1}\right)\boldsymbol{C}\left(z^{-1}\right)\boldsymbol{\mu}(z) \tag{7.208}$$

which has opened a gap between the *present/past* $\boldsymbol{y}$ values and the desired *future* value j steps ahead. Because $\boldsymbol{E}_j\left(z^{-1}\right)$ has a maximum lag factor of z^{-j+1}, at least in the simplest case $\boldsymbol{C}\left(z^{-1}\right) = \boldsymbol{I}$ the disturbance contributions all occur in the future and thus must have an expectation of zero in the predictor. Thus, setting

$$\boldsymbol{G}_j\left(z^{-1}\right) = \boldsymbol{E}_j\left(z^{-1}\right)\boldsymbol{N}\left(z^{-1}\right), \tag{7.209}$$

one has

$$z^j\boldsymbol{y}(z) = \boldsymbol{F}_j\left(z^{-1}\right)\boldsymbol{y}(z) + \boldsymbol{G}_j\left(z^{-1}\right)\left[z^j - z^{j-1}\right]\boldsymbol{u}(z) \tag{7.210}$$

$$z^j\boldsymbol{y}(z) = \boldsymbol{F}_j\left(z^{-1}\right)\boldsymbol{y}(z) + \boldsymbol{G}_j\left(z^{-1}\right)z^j\Delta\boldsymbol{u}(z) \tag{7.211}$$

where clearly

$$\Delta\boldsymbol{u}_i = \boldsymbol{u}_i - \boldsymbol{u}_{i-1} \quad \text{(the control ‘move’)} \tag{7.212}$$

From Equation 7.198, the smallest lag in $\boldsymbol{N}\left(z^{-1}\right)$ is z^{-1}, which by Equation 7.209 must then also be the case for $\boldsymbol{G}_j\left(z^{-1}\right)$. So one can represent

$$\boldsymbol{F}_j\left(z^{-1}\right) = \boldsymbol{F}_{j,0} + \boldsymbol{F}_{j,1}z^{-1} + \boldsymbol{F}_{j,2}z^{-2} + \cdots + \boldsymbol{F}_{j,(n-2)}z^{-n+2} + \boldsymbol{F}_{j,(n-1)}z^{-n+1} \tag{7.213}$$

$$\boldsymbol{G}_j\left(z^{-1}\right) = \boldsymbol{0} + \boldsymbol{G}_{j,1}z^{-1} + \boldsymbol{G}_{j,2}z^{-2} + \cdots + \boldsymbol{G}_{j,(n+j-2)}z^{-n-j+2} + \boldsymbol{G}_{j,(n+j-1)}z^{-n-j+1} \tag{7.214}$$

The predictive control algorithm will require estimates of the vector $\boldsymbol{y}$ over some interval ahead of the present time – say from $j=1$ to $j=N$:

$$\underbrace{\begin{pmatrix} z\boldsymbol{I} \\ z^2\boldsymbol{I} \\ \vdots \\ z^{N-1}\boldsymbol{I} \\ z^N\boldsymbol{I} \end{pmatrix} \boldsymbol{y}(z)}_{\overline{\boldsymbol{y}}_\mathrm{F}} = \underbrace{\begin{bmatrix} \boldsymbol{F}_{1,n-1} & \boldsymbol{F}_{1,n-2} & \cdots & \boldsymbol{F}_{1,1} & \boldsymbol{F}_{1,0} \\ \boldsymbol{F}_{2,n-1} & \boldsymbol{F}_{2,n-2} & \cdots & \boldsymbol{F}_{2,1} & \boldsymbol{F}_{2,0} \\ \vdots & \vdots & \vdots & \vdots & \vdots \\ \boldsymbol{F}_{N-1,n-1} & \boldsymbol{F}_{N-1,n-2} & \cdots & \boldsymbol{F}_{N-1,1} & \boldsymbol{F}_{N-1,0} \\ \boldsymbol{F}_{N,n-1} & \boldsymbol{F}_{N,n-2} & \cdots & \boldsymbol{F}_{N,1} & \boldsymbol{F}_{N,0} \end{bmatrix}}_{\overline{\boldsymbol{F}}} \underbrace{\begin{pmatrix} z^{-n+1}\boldsymbol{I} \\ z^{-n+2}\boldsymbol{I} \\ \vdots \\ z^{-1}\boldsymbol{I} \\ \boldsymbol{I} \end{pmatrix} \boldsymbol{y}(z)}_{\overline{\boldsymbol{y}}_\mathrm{P}}$$

$$+ \underbrace{\begin{bmatrix} \boldsymbol{G}_{1,n} & \boldsymbol{G}_{1,n-1} & \cdots & \boldsymbol{G}_{1,3} & \boldsymbol{G}_{1,2} \\ \boldsymbol{G}_{2,n+1} & \boldsymbol{G}_{2,n} & \cdots & \boldsymbol{G}_{2,4} & \boldsymbol{G}_{2,3} \\ \ddots & \ddots & \ddots & \ddots & \ddots \\ \boldsymbol{G}_{N-1,n+N-2} & \boldsymbol{G}_{N-1,n+N-3} & \cdots & \boldsymbol{G}_{N-1,N+1} & \boldsymbol{G}_{N-1,N} \\ \boldsymbol{G}_{N,n+N-1} & \boldsymbol{G}_{N,n+N-2} & \cdots & \boldsymbol{G}_{N,N+2} & \boldsymbol{G}_{N,N+1} \end{bmatrix}}_{\overline{\boldsymbol{G}}_\mathrm{P}} \underbrace{\begin{pmatrix} z^{-n+1}\boldsymbol{I} \\ z^{-n+2}\boldsymbol{I} \\ \vdots \\ z^{-2}\boldsymbol{I} \\ z^{-1}\boldsymbol{I} \end{pmatrix} \Delta\boldsymbol{u}(z)}_{\overline{\Delta\boldsymbol{u}_\mathrm{P}}}$$

$$+ \underbrace{\begin{bmatrix} \boldsymbol{G}_{1,1} & \boldsymbol{0} & \boldsymbol{0} & \cdots & \boldsymbol{0} & \boldsymbol{0} \\ \boldsymbol{G}_{2,2} & \boldsymbol{G}_{2,1} & \boldsymbol{0} & \cdots & \boldsymbol{0} & \boldsymbol{0} \\ \ddots & \ddots & \ddots & \ddots & \ddots & \ddots \\ \boldsymbol{G}_{N-1,N-1} & \boldsymbol{G}_{N-1,N-2} & \boldsymbol{G}_{N-1,N-3} & \cdots & \boldsymbol{G}_{N-1,1} & \boldsymbol{0} \\ \boldsymbol{G}_{N,N} & \boldsymbol{G}_{N,N-1} & \boldsymbol{G}_{N,N-2} & \cdots & \boldsymbol{G}_{N,2} & \boldsymbol{G}_{N,1} \end{bmatrix}}_{\overline{\boldsymbol{G}}_\mathrm{F}} \underbrace{\begin{pmatrix} \boldsymbol{I} \\ z\boldsymbol{I} \\ z^2\boldsymbol{I} \\ \vdots \\ z^{N-2}\boldsymbol{I} \\ z^{N-1}\boldsymbol{I} \end{pmatrix} \Delta\boldsymbol{u}(z)}_{\overline{\Delta\boldsymbol{u}_\mathrm{F}}} \tag{7.215}$$

The form here is

$$\overline{\boldsymbol{y}}_\mathrm{F} = \overline{\boldsymbol{F}}\,\overline{\boldsymbol{y}}_\mathrm{P} + \overline{\boldsymbol{G}}_\mathrm{P}\,\overline{\Delta\boldsymbol{u}_\mathrm{P}} + \overline{\boldsymbol{G}}_\mathrm{F}\,\overline{\Delta\boldsymbol{u}_\mathrm{F}} \tag{7.216}$$

where $\overline{\boldsymbol{y}}_\mathrm{P}$ is a partitioned vector containing the *past* output vectors for $n-1$ to 1 step ago, $\overline{\boldsymbol{y}}_\mathrm{F}$ is a partitioned vector containing the *future* output vectors for 1 to N steps ahead, $\overline{\Delta\boldsymbol{u}_\mathrm{P}}$ is a partitioned vector containing the *past* input *move* vectors from $n-1$ to 1 step ago, $\overline{\Delta\boldsymbol{u}_\mathrm{F}}$ is a partitioned vector containing the *future* input *move* vectors from *now* until $N-1$ steps ahead, $\overline{\boldsymbol{F}}$ is a partitioned matrix for the contribution of the *past* output from $n-1$ steps ago to the present, $\overline{\boldsymbol{G}}_\mathrm{P}$ is a partitioned matrix for the contribution of the *past* input *move* vectors $\overline{\Delta\boldsymbol{u}_\mathrm{P}}$ and $\overline{\boldsymbol{G}}_\mathrm{F}$ is a partitioned matrix for the contribution of the *future* input *move* vectors $\overline{\Delta\boldsymbol{u}_\mathrm{F}}$.

One notes that $\overline{\boldsymbol{G}}_\mathrm{F}$ is simply a continuation to the right of $\overline{\boldsymbol{G}}_\mathrm{P}$. The predictive control problem is thus to find suitable $\overline{\Delta\boldsymbol{u}_\mathrm{F}}$ according to criteria such as minimising the deviation of $\overline{\boldsymbol{y}}_\mathrm{F}$ from the future setpoint trajectory $\overline{\boldsymbol{y}}_\mathrm{FS}$, and minimising the control effort $\overline{\Delta\boldsymbol{u}_\mathrm{F}}$. Usually it is sufficient to optimise just the first M control moves of the available N moves (with $M=1$, 2 or 3). The remaining moves are left at zero, freezing the control action after the Mth move. So truncated versions of $\overline{\Delta \boldsymbol{u}_{\mathrm{F}}}$ and $\overline{\boldsymbol{G}}_{\mathrm{F}}$ are defined as

$$\overline{\boldsymbol{G}}_{\mathrm{F}}^{*}=\begin{bmatrix} \boldsymbol{G}_{1,1} & \boldsymbol{0} & \ddots & \boldsymbol{0} & \boldsymbol{0} \\ \boldsymbol{G}_{2,2} & \boldsymbol{G}_{2,1} & \ddots & \boldsymbol{0} & \boldsymbol{0} \\ \boldsymbol{G}_{3,3} & \boldsymbol{G}_{3,2} & \ddots & \boldsymbol{0} & \boldsymbol{0} \\ \boldsymbol{G}_{4,4} & \boldsymbol{G}_{4,3} & \ddots & \boldsymbol{G}_{M-1,1} & \boldsymbol{0} \\ \boldsymbol{G}_{5,5} & \boldsymbol{G}_{5,4} & \ddots & \boldsymbol{G}_{M,2} & \boldsymbol{G}_{M,1} \\ \vdots & \vdots & \vdots & \vdots & \vdots \\ \boldsymbol{G}_{N,N} & \boldsymbol{G}_{N,N-1} & \ddots & \boldsymbol{G}_{N,N-M+2} & \boldsymbol{G}_{N,N-M+1} \end{bmatrix} \quad \text{and} \quad \overline{\Delta \boldsymbol{u}_{\mathrm{F}}^{*}}=\begin{pmatrix} \boldsymbol{I} \\ z\boldsymbol{I} \\ \vdots \\ z^{M-2}\boldsymbol{I} \\ z^{M-1}\boldsymbol{I} \end{pmatrix} \Delta \boldsymbol{u}(z) \tag{7.217}$$

An objective function that is often used is the quadratic form

$$J\left(\overline{\Delta \boldsymbol{u}_{\mathrm{F}}^{*}}\right)=\left(\overline{\boldsymbol{y}}_{\mathrm{F}}-\overline{\boldsymbol{y}}_{\mathrm{FS}}\right)^{\mathrm{T}} \overline{\boldsymbol{Q}}\left(\overline{\boldsymbol{y}}_{\mathrm{F}}-\overline{\boldsymbol{y}}_{\mathrm{FS}}\right)+\left(\overline{\Delta \boldsymbol{u}_{\mathrm{F}}^{*}}\right)^{\mathrm{T}} \overline{\boldsymbol{R}}\left(\overline{\Delta \boldsymbol{u}_{\mathrm{F}}^{*}}\right) \tag{7.218}$$

Here the partitioned penalty matrices are typically diagonal:

$$\overline{\boldsymbol{Q}}=\begin{bmatrix} \boldsymbol{Q}_{1} & \boldsymbol{0} & \ddots \\ \boldsymbol{0} & \ddots & \boldsymbol{0} \\ \ddots & \boldsymbol{0} & \boldsymbol{Q}_{N} \end{bmatrix} \quad \text{and} \quad \overline{\boldsymbol{R}}=\begin{bmatrix} \boldsymbol{R}_{1} & \boldsymbol{0} & \ddots \\ \boldsymbol{0} & \ddots & \boldsymbol{0} \\ \ddots & \boldsymbol{0} & \boldsymbol{R}_{M} \end{bmatrix} \tag{7.219}$$

with the submatrices $\boldsymbol{Q}_i$ and $\boldsymbol{R}_i$ dealing with individual variables on each time step also usually diagonal. The weighting terms in $\boldsymbol{Q}_i$ are sometimes increased as i approaches N, to force final values of the controlled variables towards their setpoints. Defining

$$\overline{\boldsymbol{y}}_{\mathrm{OL}}=\overline{\boldsymbol{F}}\, \overline{\boldsymbol{y}}_{\mathrm{P}}+\overline{\boldsymbol{G}}_{\mathrm{P}}\, \overline{\Delta \boldsymbol{u}_{\mathrm{P}}} \quad \text{('open-loop' response)} \tag{7.220}$$

$$\overline{\boldsymbol{e}}_{\mathrm{OL}}=\overline{\boldsymbol{y}}_{\mathrm{OL}}-\overline{\boldsymbol{y}}_{\mathrm{FS}} \quad \text{('open-loop' error)} \tag{7.221}$$

then

$$J\left(\overline{\Delta \boldsymbol{u}_{\mathrm{F}}^{*}}\right)=\left(\overline{\boldsymbol{e}}_{\mathrm{OL}}+\overline{\boldsymbol{G}}_{\mathrm{F}}^{*} \overline{\Delta \boldsymbol{u}_{\mathrm{F}}^{*}}\right)^{\mathrm{T}} \boldsymbol{Q}\left(\overline{\boldsymbol{e}}_{\mathrm{OL}}+\overline{\boldsymbol{G}}_{\mathrm{F}}^{*} \overline{\Delta \boldsymbol{u}_{\mathrm{F}}^{*}}\right)+\left(\overline{\Delta \boldsymbol{u}_{\mathrm{F}}^{*}}\right)^{\mathrm{T}} \boldsymbol{R}\left(\overline{\Delta \boldsymbol{u}_{\mathrm{F}}^{*}}\right) \tag{7.222}$$

The optimal sequence of control vector values $\left(\overline{\Delta \boldsymbol{u}_{\mathrm{F}}^{*}}\right)_{\mathrm{OPT}}$ is that which minimises the index J. Industrially, one usually seeks to perform this optimisation within constraints determined by the future absolute values of the elements of both the output $\boldsymbol{y}$ and the input $\boldsymbol{u}$. Additionally, the rate of change of the inputs may be restricted by placing constraints on $\Delta \boldsymbol{u}$ (i.e. 'ramp' constraints). In this general case, techniques such as FSQP (feasible sequential quadratic programming) are required. In the meantime, it is instructive to consider the simple case of an unconstrained system. Recalling that

$$\frac{\partial}{\partial \boldsymbol{x}}\left\{\boldsymbol{x}^{\mathrm{T}} \boldsymbol{M} \boldsymbol{x}\right\}=2 \boldsymbol{M} \boldsymbol{x} \quad \text{and} \quad \frac{\partial}{\partial \boldsymbol{x}}\left\{\boldsymbol{x}^{\mathrm{T}} \boldsymbol{M} \boldsymbol{y}\right\}=\boldsymbol{M} \boldsymbol{y} \tag{7.223}$$

differentiate Equation 7.222 with respect to $\overline{\Delta \boldsymbol{u}_{\mathrm{F}}^{*}}$ to obtain

$$\frac{\partial J}{\partial \overline{\Delta \boldsymbol{u}_{\mathrm{F}}^{*}}}=2 \overline{\boldsymbol{G}}_{\mathrm{F}}^{* \mathrm{T}} \boldsymbol{Q}\left(\overline{\boldsymbol{e}}_{\mathrm{OL}}+\overline{\boldsymbol{G}}_{\mathrm{F}}^{*} \overline{\Delta \boldsymbol{u}_{\mathrm{F}}^{*}}\right)+2 \boldsymbol{R} \overline{\Delta \boldsymbol{u}_{\mathrm{F}}^{*}}=\boldsymbol{0} \quad \text{for minimum } J \tag{7.224}$$

Thus,

$$\left(\overline{\Delta \boldsymbol{u}_{\mathrm{F}}^{*}}\right)_{\mathrm{OPT}} = -\left[\overline{\boldsymbol{G}_{\mathrm{F}}^{*}}^{\mathrm{T}} \boldsymbol{Q} \overline{\boldsymbol{G}_{\mathrm{F}}^{*}} + \boldsymbol{R}\right]^{-1} \overline{\boldsymbol{G}_{\mathrm{F}}^{*}}^{\mathrm{T}} \boldsymbol{Q} \overline{\boldsymbol{e}}_{\mathrm{OL}} \tag{7.225}$$

Only the first one of the M control moves will be used. The entire calculation is repeated on the next time step. One notes that the resultant values of $\boldsymbol{u}$ itself will arise as an integral of the open-loop error (by summation of the $\left(\overline{\Delta \boldsymbol{u}_{\mathrm{F}}^{*}}\right)_{\mathrm{OPT}}$), so this controller has a natural integral action.

In the next section, the same predictive control problem will be considered instead in the *state space*, where it will be seen that the difficulty of Equation 7.205 does not arise. Recall that the solutions $\boldsymbol{E}_j$ and $\boldsymbol{F}_j$ of the equation

$$\boldsymbol{I} = \left[1 - z^{-1}\right] \boldsymbol{E}_j\left(z^{-1}\right) D\left(z^{-1}\right) + z^{-j} \boldsymbol{F}_j\left(z^{-1}\right) \tag{7.226}$$

were required to provide nonrecursive predictions of the output vector at a range of times in the future. Example 7.3 considers this calculation for a two-input, two-output system. Note that Clarke, Mohtadi and Tuffs (1987) provide a simple *recursive* technique for this calculation. A computer algorithm for this method is presented in Problem 7.10 of the accompanying book.

Example 7.3

Calculation of nonrecursive predictor polynomials $E_j(z^{-1})$ and $\boldsymbol{F}_j(z^{-1})$.

A certain two-input, two-output system is represented by

$$\boldsymbol{x}(z) = D^{-1}(z) \boldsymbol{N}(z) \boldsymbol{u}(z) \tag{7.227}$$

with

$$D(z) = \left(z - \frac{1}{2}\right)^2 = z^2 - z + \frac{1}{2} \tag{7.228}$$

$$\boldsymbol{N}(z) = \begin{bmatrix} z - \frac{1}{2} & 2z - 1 \\ -\frac{1}{2} & z + \frac{1}{2} \end{bmatrix} = \begin{bmatrix} 1 & 2 \\ 0 & 1 \end{bmatrix} z + \begin{bmatrix} -\frac{1}{2} & -1 \\ -\frac{1}{2} & +\frac{1}{2} \end{bmatrix} \tag{7.229}$$

$$\div z^2: \quad D\left(z^{-1}\right) = 1 - z^{-1} + \frac{1}{2} z^{-2} \tag{7.230}$$

$$\boldsymbol{N}\left(z^{-1}\right) = \begin{bmatrix} 1 & 2 \\ 0 & 1 \end{bmatrix} z^{-1} + \begin{bmatrix} -\frac{1}{2} & -1 \\ -\frac{1}{2} & +\frac{1}{2} \end{bmatrix} z^{-2} \tag{7.231}$$

Recalling that $\boldsymbol{E}_j\left(z^{-1}\right)$ has a maximum lag factor of z^{j+1}, consider firstly $j = 1$:

$$1 = \left[1 - z^{-1}\right]\left\{E_{10}\right\}\left(1 - z^{-1} + \frac{1}{2} z^{-2}\right) + z^{-1}\left\{F_{10} + F_{11} z^{-1} + F_{12} z^{-2}\right\} \tag{7.232}$$

$$1 = E_{10}\left[1 - 2z^{-1} + \frac{3}{2}z^{-2} - \frac{1}{2}z^{-3}\right] + F_{10}z^{-1} + F_{11}z^{-2} + F_{12}z^{-3} \tag{7.233}$$

Matching coefficients

$$1 = E_{10} \tag{7.234}$$

$$0 = -2E_{10} + F_{10}, \quad \text{so} \quad F_{10} = +2 \tag{7.235}$$

$$0 = +\frac{3}{2}E_{10} + F_{11}, \quad \text{so} \quad F_{11} = -\frac{3}{2} \tag{7.236}$$

$$0 = -\frac{1}{2}E_{10} + F_{12}, \quad \text{so} \quad F_{12} = +\frac{1}{2} \tag{7.237}$$

Now for $j = 2$:

$$\begin{aligned} 1 &= \left[1 - z^{-1}\right]\left\{E_{20} + E_{21}z^{-1}\right\}\left(1 - z^{-1} + \frac{1}{2}z^{-2}\right) + z^{-2}\left\{F_{20} + F_{21}z^{-1} + F_{22}z^{-2}\right\} \\ &= E_{20}\left[1 - 2z^{-1} + \frac{3}{2}z^{-2} - \frac{1}{2}z^{-3}\right] + E_{21}\left[z^{-1} - 2z^{-2} + \frac{3}{2}z^{-3} - \frac{1}{2}z^{-4}\right] + F_{20}z^{-2} + F_{21}z^{-3} + F_{22}z^{-4} \end{aligned} \tag{7.238}$$

Matching coefficients

$$1 = E_{20} \tag{7.239}$$

$$0 = -2E_{20} + E_{21}, \quad \text{so} \quad E_{21} = +2 \tag{7.240}$$

$$0 = +\frac{3}{2}E_{20} - 2E_{21} + F_{20}, \quad \text{so} \quad F_{20} = +2\frac{1}{2} \tag{7.241}$$

$$0 = -\frac{1}{2}E_{20} + \frac{3}{2}E_{21} + F_{21}, \quad \text{so} \quad F_{21} = -2\frac{1}{2} \tag{7.242}$$

$$0 = -\frac{1}{2}E_{21} + F_{22}, \quad \text{so} \quad F_{22} = +1 \tag{7.243}$$

$$\vdots$$

This process is continued for $j = 3$ before $\boldsymbol{G}_j(z^{-1})$ can be constructed according to Equation 7.209 for $j = 1, 2, 3$. Note the simple recursive procedure of Clarke, Mohtadi and Tuffs (1987) for calculation of the coefficients in $\boldsymbol{E}_j(z^{-1})$ and $\boldsymbol{F}_j(z^{-1})$ – see Problem 7.10 in the accompanying book.

7.8.1.2 Predictive Control for a Discrete MIMO System Represented in the State Space

It is interesting to obtain the equivalent of Equation 7.216 from a state-space viewpoint, assuming that measurements or estimates of all of the states exist. Following from Equation 7.199,

$$z\boldsymbol{x}(z) = \boldsymbol{A}\boldsymbol{x}(z) + \boldsymbol{B}\boldsymbol{u}(z) \tag{7.244}$$

$$z^2\boldsymbol{x}(z) = \boldsymbol{A}z\boldsymbol{x}(z) + \boldsymbol{B}z\boldsymbol{u}(z) \tag{7.245}$$

$$z^2\boldsymbol{x}(z) = \boldsymbol{A}[\boldsymbol{A}\boldsymbol{x}(z) + \boldsymbol{B}\boldsymbol{u}(z)] + \boldsymbol{B}z\boldsymbol{u}(z) \tag{7.246}$$

$$z^2\boldsymbol{x}(z) = \boldsymbol{A}^2\boldsymbol{x}(z) + [\boldsymbol{A}\boldsymbol{B} + \boldsymbol{B}z]\boldsymbol{u}(z) \tag{7.247}$$

$$z^3\boldsymbol{x}(z) = \boldsymbol{A}^2z\boldsymbol{x}(z) + [\boldsymbol{A}\boldsymbol{B} + \boldsymbol{B}z]z\boldsymbol{u}(z) \tag{7.248}$$

$$z^3\boldsymbol{x}(z) = \boldsymbol{A}^2[\boldsymbol{A}\boldsymbol{x}(z) + \boldsymbol{B}\boldsymbol{u}(z)] + \left[\boldsymbol{A}\boldsymbol{B}z + \boldsymbol{B}z^2\right]\boldsymbol{u}(z) \tag{7.249}$$

$$z^3\boldsymbol{x}(z) = \boldsymbol{A}^3\boldsymbol{x}(z) + \left[\boldsymbol{A}^2\boldsymbol{B} + \boldsymbol{A}\boldsymbol{B}z + \boldsymbol{B}z^2\right]\boldsymbol{u}(z) \tag{7.250}$$

$$\vdots$$

$$z^N\boldsymbol{x}(z) = \boldsymbol{A}^N\boldsymbol{x}(z) + \left[\boldsymbol{A}^{N-1}\boldsymbol{B} + \boldsymbol{A}^{N-2}\boldsymbol{B}z + \boldsymbol{A}^{N-3}\boldsymbol{B}z^2 + \cdots + \boldsymbol{A}\boldsymbol{B}z^{N-2} + \boldsymbol{B}z^{N-1}\right]\boldsymbol{u}(z) \tag{7.251}$$

Thus,

$$\begin{pmatrix} z\boldsymbol{I} \\ z^2\boldsymbol{I} \\ \vdots \\ z^{N-1}\boldsymbol{I} \\ z^N\boldsymbol{I} \end{pmatrix}\boldsymbol{x}(z) = \begin{bmatrix} \boldsymbol{A} \\ \boldsymbol{A}^2 \\ \vdots \\ \boldsymbol{A}^{N-1} \\ \boldsymbol{A}^N \end{bmatrix}\boldsymbol{x}(z) + \begin{bmatrix} \boldsymbol{I} & \boldsymbol{0} & \cdots & \boldsymbol{0} & \boldsymbol{0} \\ \boldsymbol{A} & \boldsymbol{I} & \cdots & \boldsymbol{0} & \boldsymbol{0} \\ \ddots & \ddots & \ddots & \ddots & \ddots \\ \boldsymbol{A}^{N-2} & \boldsymbol{A}^{N-3} & \cdots & \boldsymbol{I} & \boldsymbol{0} \\ \boldsymbol{A}^{N-1} & \boldsymbol{A}^{N-2} & \cdots & \boldsymbol{A} & \boldsymbol{I} \end{bmatrix}\overline{\boldsymbol{B}}\begin{pmatrix} \boldsymbol{I} \\ z\boldsymbol{I} \\ \vdots \\ z^{N-2}\boldsymbol{I} \\ z^{N-1}\boldsymbol{I} \end{pmatrix}\boldsymbol{u}(z) \tag{7.252}$$

where the new $\overline{\boldsymbol{B}}$ matrix is block-diagonal with $\boldsymbol{B}$'s on its diagonal. Differentiate Equation 7.252 by multiplication by $[1 - z^{-1}]$ in order to eliminate the effect of steady offsets

$$\begin{pmatrix} z\boldsymbol{I} \\ z^2\boldsymbol{I} \\ \vdots \\ z^{N-1}\boldsymbol{I} \\ z^N\boldsymbol{I} \end{pmatrix}\Delta\boldsymbol{x}(z) = \underbrace{\begin{bmatrix} \boldsymbol{A} \\ \boldsymbol{A}^2 \\ \vdots \\ \boldsymbol{A}^{N-1} \\ \boldsymbol{A}^N \end{bmatrix}}_{\overline{\boldsymbol{R}}}\left[\boldsymbol{x}(z) - z^{-1}\boldsymbol{x}(z)\right] + \underbrace{\begin{bmatrix} \boldsymbol{I} & \boldsymbol{0} & \cdots & \boldsymbol{0} & \boldsymbol{0} \\ \boldsymbol{A} & \boldsymbol{I} & \cdots & \boldsymbol{0} & \boldsymbol{0} \\ \ddots & \ddots & \ddots & \ddots & \ddots \\ \boldsymbol{A}^{N-2} & \boldsymbol{A}^{N-3} & \cdots & \boldsymbol{I} & \boldsymbol{0} \\ \boldsymbol{A}^{N-1} & \boldsymbol{A}^{N-2} & \cdots & \boldsymbol{A} & \boldsymbol{I} \end{bmatrix}}_{\overline{\boldsymbol{S}}}\overline{\boldsymbol{B}}\underbrace{\begin{pmatrix} \boldsymbol{I} \\ z\boldsymbol{I} \\ \vdots \\ z^{N-2}\boldsymbol{I} \\ z^{N-1}\boldsymbol{I} \end{pmatrix}\Delta\boldsymbol{u}(z)}_{\overline{\Delta\boldsymbol{u}_{\mathrm{F}}}} \tag{7.253}$$

Noting that

$$\underbrace{\begin{bmatrix} \boldsymbol{I} & \boldsymbol{0} & \cdots & \boldsymbol{0} & \boldsymbol{0} \\ \boldsymbol{I} & \boldsymbol{I} & \cdots & \boldsymbol{0} & \boldsymbol{0} \\ \ddots & \ddots & \ddots & \ddots & \ddots \\ \boldsymbol{I} & \boldsymbol{I} & \cdots & \boldsymbol{I} & \boldsymbol{0} \\ \boldsymbol{I} & \boldsymbol{I} & \cdots & \boldsymbol{I} & \boldsymbol{I} \end{bmatrix}}_{\overline{\boldsymbol{L}}}\begin{pmatrix} z\boldsymbol{I} \\ z^2\boldsymbol{I} \\ \vdots \\ z^{N-1}\boldsymbol{I} \\ z^N\boldsymbol{I} \end{pmatrix}\Delta\boldsymbol{x}(z) = \underbrace{\begin{pmatrix} z\boldsymbol{I} \\ z^2\boldsymbol{I} \\ \vdots \\ z^{N-1}\boldsymbol{I} \\ z^N\boldsymbol{I} \end{pmatrix}\boldsymbol{x}(z)}_{\overline{\boldsymbol{x}}_{\mathrm{F}}} - \underbrace{\begin{bmatrix} \boldsymbol{I} \\ \boldsymbol{I} \\ \vdots \\ \boldsymbol{I} \\ \boldsymbol{I} \end{bmatrix}}_{\overline{\boldsymbol{I}}}\boldsymbol{x}(z) \tag{7.254}$$

then multiplication of Equation 7.253 by $\overline{\boldsymbol{L}}$ yields

$$\overline{\boldsymbol{x}}_{\mathrm{F}}(z) = \left[\overline{\boldsymbol{L}}\,\overline{\boldsymbol{R}} - \overline{\boldsymbol{I}}\right]\boldsymbol{x}(z) - \overline{\boldsymbol{L}}\,\overline{\boldsymbol{R}}z^{-1}\boldsymbol{x}(z) + \overline{\boldsymbol{L}}\,\overline{\boldsymbol{S}}\,\overline{\boldsymbol{B}}\,\overline{\Delta\boldsymbol{u}_{\mathrm{F}}} \tag{7.255}$$

Defining

$$\overline{\boldsymbol{x}}_{\mathrm{P}}(z) = \begin{pmatrix} z^{-1}\boldsymbol{I} \\ \boldsymbol{I} \end{pmatrix} \boldsymbol{x}(z) \tag{7.256}$$

$$\overline{\boldsymbol{F}} = \left[-\overline{\boldsymbol{L}}\overline{\boldsymbol{R}} \quad \overline{\boldsymbol{L}}\overline{\boldsymbol{R}} - \overline{\boldsymbol{I}} \right] \tag{7.257}$$

$$\overline{\boldsymbol{G}}_{\mathrm{F}} = \overline{\boldsymbol{L}}\,\overline{\boldsymbol{S}}\,\overline{\boldsymbol{B}} \tag{7.258}$$

Equation 7.255 becomes

$$\overline{\boldsymbol{x}}_{\mathrm{F}}(z) = \overline{\boldsymbol{F}}\,\overline{\boldsymbol{x}}_{\mathrm{P}}(z) + \overline{\boldsymbol{G}}_{\mathrm{F}}\,\overline{\Delta\boldsymbol{u}_{\mathrm{F}}} \tag{7.259}$$

which is analogous to Equation 7.216 except that $\overline{\boldsymbol{G}}_{\mathrm{P}} = \boldsymbol{0}$, because in the state space only the present state and future inputs are required to predict the future outputs. As mentioned earlier, if only a selection or combination of the states are available according to

$$\boldsymbol{y}(z) = \boldsymbol{H}\boldsymbol{x}(z) \tag{7.260}$$

then the one-step prediction form used in Equation 7.244 is not possible, and the forms

$$D(z) = \det[z\boldsymbol{I} - \boldsymbol{A}] = z^n + d_1 z^{n-1} + \cdots + d_{n-1}z + d_n \quad \text{(scalar)} \tag{7.261}$$

$$\boldsymbol{N}(z) = \boldsymbol{H}\,\mathrm{adj}[z\boldsymbol{I} - \boldsymbol{A}]\boldsymbol{B} = \boldsymbol{N}_1 z^{n-1} + \boldsymbol{N}_2 z^{n-2} + \cdots + \boldsymbol{N}_{n-1}z + \boldsymbol{N}_n \tag{7.262}$$

must be used as in Section 7.8.1.1.

7.8.2
Dynamic Matrix Control

Dynamic matrix control (DMC) is a form of model predictive control which uses a step response convolution model, or the impulse version in Section 7.8.2.5, for prediction of the effect of possible control actions. Since the early work of Cutler and Ramaker (1979) and Garcia and Morshedi (1984), these controllers, particularly DMC, have proven their worth in many industrial applications.

In Section 3.9.5.1, the behaviour of a process was modelled using measured step responses. The relationship developed envisaged a system initially at steady state, with no history of previous input *moves*. In this situation, a *functional description* of the output sequence $\overline{\boldsymbol{y}}$ at future time intervals, in response to the sequence of future values $\overline{\Delta\boldsymbol{u}}$ of the move vector, yielded

$$\overline{\boldsymbol{y}} = \overline{\boldsymbol{y}}_0 + \overline{\boldsymbol{B}}\,\overline{\Delta\boldsymbol{u}} \tag{7.263}$$

with

$$\overline{\boldsymbol{y}} = \begin{pmatrix} y_{(1)} \\ y_{(2)} \\ y_{(3)} \\ y_{(4)} \\ y_{(5)} \\ y_{(6)} \\ y_{(7)} \\ \vdots \end{pmatrix} \tag{7.264}$$

$$\bar{\boldsymbol{y}}_0 = \begin{pmatrix} \boldsymbol{y}_{(0)} \\ \boldsymbol{y}_{(0)} \\ \boldsymbol{y}_{(0)} \\ \boldsymbol{y}_{(0)} \\ \boldsymbol{y}_{(0)} \\ \boldsymbol{y}_{(0)} \\ \boldsymbol{y}_{(0)} \\ \vdots \end{pmatrix} \tag{7.265}$$

$$\overline{\Delta \boldsymbol{u}} = \begin{pmatrix} \Delta \boldsymbol{u}_{(0)} \\ \Delta \boldsymbol{u}_{(1)} \\ \Delta \boldsymbol{u}_{(2)} \\ \Delta \boldsymbol{u}_{(3)} \\ \Delta \boldsymbol{u}_{(4)} \\ \Delta \boldsymbol{u}_{(5)} \\ \Delta \boldsymbol{u}_{(6)} \\ \vdots \end{pmatrix} \tag{7.266}$$

$$\overline{\boldsymbol{B}} = \begin{bmatrix} \boldsymbol{b}_{(1)} & \boldsymbol{0} & \boldsymbol{0} & \boldsymbol{0} & \boldsymbol{0} & \boldsymbol{0} & \boldsymbol{0} & \cdots \\ \boldsymbol{b}_{(2)} & \boldsymbol{b}_{(1)} & \boldsymbol{0} & \boldsymbol{0} & \boldsymbol{0} & \boldsymbol{0} & \boldsymbol{0} & \cdots \\ \vdots & \boldsymbol{b}_{(2)} & \boldsymbol{b}_{(1)} & \boldsymbol{0} & \boldsymbol{0} & \boldsymbol{0} & \boldsymbol{0} & \cdots \\ \boldsymbol{b}_{(N-1)} & \vdots & \boldsymbol{b}_{(2)} & \boldsymbol{b}_{(1)} & \boldsymbol{0} & \boldsymbol{0} & \boldsymbol{0} & \cdots \\ \boldsymbol{b}_{(N)} & \boldsymbol{b}_{(N-1)} & \vdots & \boldsymbol{b}_{(2)} & \boldsymbol{b}_{(1)} & \boldsymbol{0} & \boldsymbol{0} & \cdots \\ \boldsymbol{b}_{(N)} & \boldsymbol{b}_{(N)} & \boldsymbol{b}_{(N-1)} & \vdots & \boldsymbol{b}_{(2)} & \boldsymbol{b}_{(1)} & \boldsymbol{0} & \cdots \\ \boldsymbol{b}_{(N)} & \boldsymbol{b}_{(N)} & \boldsymbol{b}_{(N)} & \boldsymbol{b}_{(N-1)} & \vdots & \boldsymbol{b}_{(2)} & \boldsymbol{b}_{(1)} & \cdots \\ \vdots & \vdots & \vdots & \vdots & \vdots & \vdots & \vdots & \ddots \end{bmatrix} \tag{7.267}$$

where $\overline{\boldsymbol{B}}$ is a partitioned matrix of matrices called the *dynamic matrix*. In order to proceed, one needs to recast this result into a moving frame of reference, so that the point in time '(0)' represents the present time. Now one has to expect that moves made before the present will both determine the present output and contribute to the future output. Initially, only systems which reach a steady state will be considered (i.e. not integrating systems). In Figure 7.20, it is seen that moves occurring before some time NT ago will not contribute any *variation* to the future response ($t+T$ to $t+NT$), only a steady-state offset.

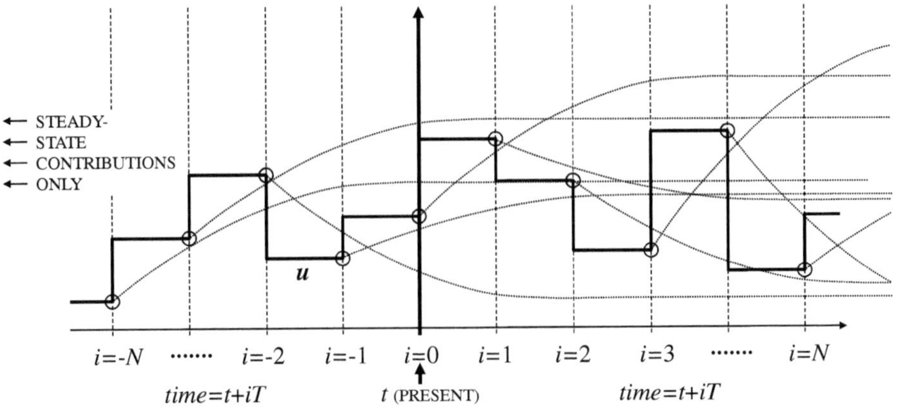

Figure 7.20 Schematic representation of the contributions of past and future moves to the future response.

The varying contributions to the future response can be included by extending matrix $\overline{\boldsymbol{B}}$ backwards as follows:

$$\boldsymbol{y}_0 = \Big[\overbrace{\boldsymbol{b}_{(N)} \;\; \boldsymbol{b}_{(N-1)} \;\; \cdots \;\; \boldsymbol{b}_{(3)} \;\; \boldsymbol{b}_{(2)} \;\; \boldsymbol{b}_{(1)}}^{\overline{\boldsymbol{B}}_0} \;\vdots\; \boldsymbol{0} \;\; \boldsymbol{0} \;\; \boldsymbol{0} \;\; \boldsymbol{0} \;\; \boldsymbol{0} \;\; \boldsymbol{0}\Big]$$

$$\overline{\boldsymbol{y}}_F = \Bigg[\underbrace{\begin{matrix} \boldsymbol{b}_{(N)} & \boldsymbol{b}_{(N)} & \boldsymbol{b}_{(N-1)} & \vdots & \boldsymbol{b}_{(3)} & \boldsymbol{b}_{(2)} \\ \boldsymbol{b}_{(N)} & \boldsymbol{b}_{(N)} & \boldsymbol{b}_{(N)} & \boldsymbol{b}_{(N-1)} & \vdots & \boldsymbol{b}_{(3)} \\ \boldsymbol{b}_{(N)} & \boldsymbol{b}_{(N)} & \boldsymbol{b}_{(N)} & \boldsymbol{b}_{(N)} & \boldsymbol{b}_{(N-1)} & \vdots \\ \boldsymbol{b}_{(N)} & \boldsymbol{b}_{(N)} & \boldsymbol{b}_{(N)} & \boldsymbol{b}_{(N)} & \boldsymbol{b}_{(N)} & \boldsymbol{b}_{(N-1)} \\ \boldsymbol{b}_{(N)} & \boldsymbol{b}_{(N)} & \boldsymbol{b}_{(N)} & \boldsymbol{b}_{(N)} & \boldsymbol{b}_{(N)} & \boldsymbol{b}_{(N)} \\ \boldsymbol{b}_{(N)} & \boldsymbol{b}_{(N)} & \boldsymbol{b}_{(N)} & \boldsymbol{b}_{(N)} & \boldsymbol{b}_{(N)} & \boldsymbol{b}_{(N)} \end{matrix}}_{\overline{\boldsymbol{B}}_P} \;\vdots\; \underbrace{\begin{matrix} \boldsymbol{b}_{(1)} & \boldsymbol{0} & \boldsymbol{0} & \boldsymbol{0} & \boldsymbol{0} & \boldsymbol{0} \\ \boldsymbol{b}_{(2)} & \boldsymbol{b}_{(1)} & \boldsymbol{0} & \boldsymbol{0} & \boldsymbol{0} & \boldsymbol{0} \\ \boldsymbol{b}_{(3)} & \boldsymbol{b}_{(2)} & \boldsymbol{b}_{(1)} & \boldsymbol{0} & \boldsymbol{0} & \boldsymbol{0} \\ \vdots & \boldsymbol{b}_{(3)} & \boldsymbol{b}_{(2)} & \boldsymbol{b}_{(1)} & \boldsymbol{0} & \boldsymbol{0} \\ \boldsymbol{b}_{(N-1)} & \vdots & \boldsymbol{b}_{(3)} & \boldsymbol{b}_{(2)} & \boldsymbol{b}_{(1)} & \boldsymbol{0} \\ \boldsymbol{b}_{(N)} & \boldsymbol{b}_{(N-1)} & \vdots & \boldsymbol{b}_{(3)} & \boldsymbol{b}_{(2)} & \boldsymbol{b}_{(1)} \end{matrix}}_{\overline{\boldsymbol{B}}_F}\Bigg] \begin{pmatrix} \left.\begin{matrix} \Delta\boldsymbol{u}_{(-N)} \\ \Delta\boldsymbol{u}_{(-N+1)} \\ \vdots \\ \Delta\boldsymbol{u}_{(-3)} \\ \Delta\boldsymbol{u}_{(-2)} \\ \Delta\boldsymbol{u}_{(-1)} \end{matrix}\right\} \overline{\Delta\boldsymbol{u}_P} \\ \cdots\cdots \\ \left.\begin{matrix} \Delta\boldsymbol{u}_{(0)} \\ \Delta\boldsymbol{u}_{(1)} \\ \Delta\boldsymbol{u}_{(2)} \\ \vdots \\ \Delta\boldsymbol{u}_{(N-2)} \\ \Delta\boldsymbol{u}_{(N-1)} \end{matrix}\right\} \overline{\Delta\boldsymbol{u}_F} \end{pmatrix} + \overline{\boldsymbol{y}}_{SS} \tag{7.268}$$

The only way to establish the steady-state offset $\overline{\boldsymbol{y}}_{SS}$ seems to be to integrate the effect of all moves up to the present, and add this to an initial value. In fact, a simple trick is used to establish the offset, by predicting the present output vector as

$$\boldsymbol{y}_0 = \overline{\boldsymbol{B}}_0 \overline{\Delta\boldsymbol{u}_P} \tag{7.269}$$

and subtracting it from a measurement of the present measured output $\boldsymbol{y}$. Thus,

$$\overline{\boldsymbol{y}}_F = \overline{\boldsymbol{I}}\left[\boldsymbol{y} - \overline{\boldsymbol{B}}_0 \overline{\Delta\boldsymbol{u}_P}\right] + \overline{\boldsymbol{B}}_P \overline{\Delta\boldsymbol{u}_P} + \overline{\boldsymbol{B}}_F \overline{\Delta\boldsymbol{u}_F} \tag{7.270}$$

$$\overline{\boldsymbol{y}}_F = \overline{\boldsymbol{I}}\boldsymbol{y} + \left[\overline{\boldsymbol{B}}_P - \overline{\boldsymbol{I}}\,\overline{\boldsymbol{B}}_0\right] \overline{\Delta\boldsymbol{u}_P} + \overline{\boldsymbol{B}}_F \overline{\Delta\boldsymbol{u}_F} \tag{7.271}$$

where

$$\bar{I} = \begin{bmatrix} I \\ I \\ \vdots \\ I \end{bmatrix} \tag{7.272}$$

is used to expand to N copies of the offset prediction. Referring to Figure 7.19, the predicted *open-loop response* is obviously

$$\bar{y}_{\mathrm{OL}} = \bar{I}y + \left[\bar{B}_{\mathrm{P}} - \bar{I}\,\bar{B}_{0}\right]\overline{\Delta u_{\mathrm{P}}} \tag{7.273}$$

and the predicted future values of the output are thus

$$\bar{y}_{\mathrm{F}} = \bar{y}_{\mathrm{OL}} + \bar{B}_{\mathrm{F}}\overline{\Delta u_{\mathrm{F}}} \tag{7.274}$$

On each time step, the future open-loop response $\bar{y}_{\mathrm{OL}}$ can be calculated based on past inputs and the present output. Thus, the control problem to achieve the desired trajectory $\bar{y}_{\mathrm{CL}}$ amounts to finding suitable $\overline{\Delta u_{\mathrm{F}}}$ as in Figure 7.21, where q^{-1} has been used to represent the backward shift.

The predicted open-loop and closed-loop error trajectories are determined by the deviation between the predicted future output and the desired setpoint trajectory:

$$\bar{e}_{\mathrm{OL}} = \bar{y}_{\mathrm{OL}} - \bar{y}_{\mathrm{FS}} \quad \text{('open-loop' error)} \tag{7.275}$$

$$\bar{e}_{\mathrm{CL}} = \bar{y}_{\mathrm{F}} - \bar{y}_{\mathrm{FS}} \quad \text{('closed-loop' error)} \tag{7.276}$$

$$\bar{e}_{\mathrm{CL}} = \bar{e}_{\mathrm{OL}} + \bar{B}_{\mathrm{F}}\overline{\Delta u_{\mathrm{F}}} \tag{7.277}$$

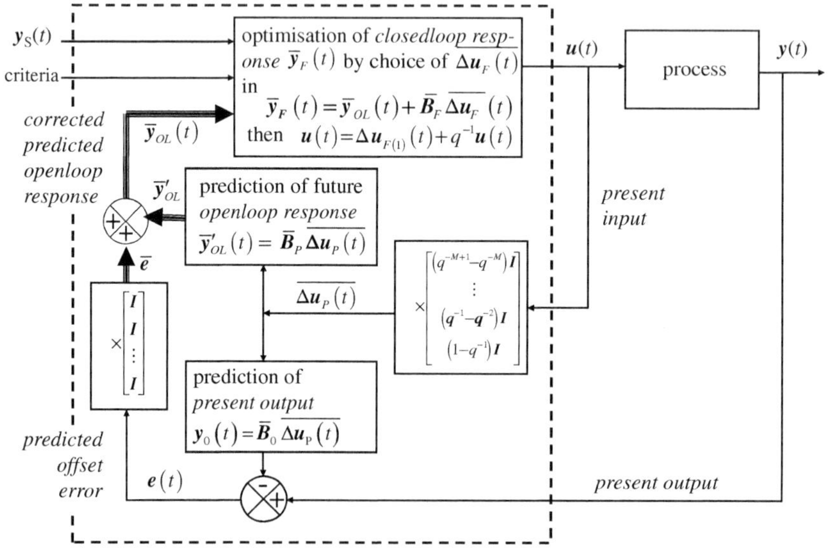

Figure 7.21 Dynamic matrix control.

This result is analogous to that for generalised predictive control in Equations 7.216 and 7.220. As mentioned in Section 7.8.1.1 on GPC, it is usually only necessary to optimise $M = 1$, 2 or at most 3 future moves, so one can truncate $\overline{\boldsymbol{B}}_{\mathrm{F}}$ by removing its rightmost columns, and $\overline{\Delta \boldsymbol{u}_{\mathrm{F}}}$ by removing its elements after M. Then a typical quadratic objective function is defined as

$$J\left(\overline{\Delta \boldsymbol{u}_{\mathrm{F}}^{*}}\right) = \left(\overline{\boldsymbol{e}}_{\mathrm{OL}} + \overline{\boldsymbol{B}}_{\mathrm{F}}^{*}\overline{\Delta \boldsymbol{u}_{\mathrm{F}}^{*}}\right)^{\mathrm{T}} \overline{\boldsymbol{Q}}\left(\overline{\boldsymbol{e}}_{\mathrm{OL}} + \overline{\boldsymbol{B}}_{\mathrm{F}}^{*}\overline{\Delta \boldsymbol{u}_{\mathrm{F}}^{*}}\right) + \left(\overline{\Delta \boldsymbol{u}_{\mathrm{F}}^{*}}\right)^{\mathrm{T}} \overline{\boldsymbol{R}}\left(\overline{\Delta \boldsymbol{u}_{\mathrm{F}}^{*}}\right) \tag{7.278}$$

for which one seeks the truncated strategy $\overline{\Delta \boldsymbol{u}_{\mathrm{F}}^{*}}$ which minimises J. Here the asterisk indicates the truncated versions to be used in the optimisation. The weighting matrices (of matrices) $\overline{\boldsymbol{Q}}$ and $\overline{\boldsymbol{R}}$ are usually diagonal, penalising setpoint deviations and control effort (step sizes), respectively. The second term is often referred to as *move suppression*. In the *unconstrained* case, there is an analytical solution to this optimisation, which was shown in Section 7.8.1.1 to be

$$\left(\overline{\Delta \boldsymbol{u}_{\mathrm{F}}^{*}}\right)_{\mathrm{UQO}} = -\left[\overline{\boldsymbol{B}}_{\mathrm{F}}^{*\mathrm{T}} \overline{\boldsymbol{Q}}\, \overline{\boldsymbol{B}}_{\mathrm{F}}^{*} + \overline{\boldsymbol{R}}\right]^{-1} \overline{\boldsymbol{B}}_{\mathrm{F}}^{*\mathrm{T}} \overline{\boldsymbol{Q}}\, \overline{\boldsymbol{e}}_{\mathrm{OL}} \tag{7.279}$$

Here the subscript UQO refers to the 'unconstrained quadratic optimum'. Only the first one of the M optimised control moves is actually used, with the entire calculation being repeated for the next time step. One notes that the value of $\boldsymbol{u}$ itself will arise as an integral of the open-loop error, so that this controller has a natural integral action.

Industrially, it is usually important to comply with constraints on both the inputs and the outputs. The mere 'clipping' of input control actions $\boldsymbol{u}$ at the allowed extremes of their ranges would render the control suboptimal. Moreover, the model predictive control approach lends itself to anticipation and avoidance of transgressing output $\boldsymbol{y}$ constraints. Another typical constraint restricts the speed of variation of the process' input control actions $\boldsymbol{u}$. Thus, the constrained problem is typically formulated as

Choose $\overline{\Delta \boldsymbol{u}_{\mathrm{F}}^{*}}$ to minimise

$$J = \left(\overline{\boldsymbol{e}}_{\mathrm{OL}} + \overline{\boldsymbol{B}}_{\mathrm{F}}^{*}\overline{\Delta \boldsymbol{u}_{\mathrm{F}}^{*}}\right)^{\mathrm{T}} \overline{\boldsymbol{Q}}\left(\overline{\boldsymbol{e}}_{\mathrm{OL}} + \overline{\boldsymbol{B}}_{\mathrm{F}}^{*}\overline{\Delta \boldsymbol{u}_{\mathrm{F}}^{*}}\right) + \left(\overline{\Delta \boldsymbol{u}_{\mathrm{F}}^{*}}\right)^{\mathrm{T}} \overline{\boldsymbol{R}}\left(\overline{\Delta \boldsymbol{u}_{\mathrm{F}}^{*}}\right) \tag{7.280}$$

such that

$$\overline{\boldsymbol{y}}_{\mathrm{MIN}} \leq \overline{\boldsymbol{y}}_{\mathrm{F}} \leq \overline{\boldsymbol{y}}_{\mathrm{MAX}} \tag{7.281}$$

$$\overline{\boldsymbol{u}}_{\mathrm{MIN}}^{*} \leq \overline{\boldsymbol{u}}_{\mathrm{F}}^{*} \leq \overline{\boldsymbol{u}}_{\mathrm{MAX}}^{*} \tag{7.282}$$

$$-\overline{\Delta \boldsymbol{u}_{\mathrm{MAX}}^{*}} \leq \overline{\Delta \boldsymbol{u}_{\mathrm{F}}^{*}} \leq +\overline{\Delta \boldsymbol{u}_{\mathrm{MAX}}^{*}} \tag{7.283}$$

where it is noted that

$$\overline{\boldsymbol{e}}_{\mathrm{OL}} = \overline{\boldsymbol{I}}\left[\boldsymbol{y} - \boldsymbol{y}_{\mathrm{SP}}\right] + \left[\overline{\boldsymbol{B}}_{\mathrm{P}} - \overline{\boldsymbol{I}}\, \overline{\boldsymbol{B}}_{0}\right] \overline{\Delta \boldsymbol{u}_{\mathrm{P}}} \tag{7.284}$$

$$\overline{\boldsymbol{u}}_{\mathrm{F}}^{*} = \overline{\boldsymbol{L}}\, \overline{\Delta \boldsymbol{u}_{\mathrm{F}}^{*}} + \overline{\boldsymbol{I}}\, \boldsymbol{u} \tag{7.285}$$

where $\boldsymbol{u}$ is the present absolute control action vector, $\overline{\boldsymbol{I}}$ truncated as required and the reduced

$$\overline{\boldsymbol{L}} = \begin{bmatrix} \boldsymbol{I} & \boldsymbol{0} & \ddots & \boldsymbol{0} \\ \boldsymbol{I} & \boldsymbol{I} & \ddots & \boldsymbol{0} \\ \ddots & \ddots & \ddots & \ddots \\ \boldsymbol{I} & \boldsymbol{I} & \ddots & \boldsymbol{I} \end{bmatrix} \tag{7.286}$$

7.8.2.1 Linear Dynamic Matrix Control

Linear dynamic matrix control (LDMC) is based on an approximate *linear programming* solution of the constrained optimisation problem in Equations 7.280–7.283, as proposed by Chang and Seborg (1983) and Morshedi, Cutler and Skrovanek (1985). The advantage of the method is that it requires smaller computational resources than an accurate solution. In it one seeks that combination of control moves which will get *as close as possible* to $\left(\overline{\Delta\boldsymbol{u}_{\mathrm{F}}^{*}}\right)_{\mathrm{UQO}}$, yet keeping within the constraints. This redefinition of the problem then allows the use of *linear programming* to handle the constraints. Although it does not guarantee the *quadratic optimum*, one expects to be close to it (and identical to it if it lies within the constraints). To distinguish this solution call it $\left(\overline{\Delta\boldsymbol{u}_{\mathrm{F}}^{*}}\right)_{\mathrm{BQO}}$, where the subscript BQO refers to 'bounded quadratic optimum'. Define the vector of residuals

$$\boldsymbol{r} = \left(\overline{\Delta\boldsymbol{u}_{\mathrm{F}}^{*}}\right)_{\mathrm{BQO}} - \left(\overline{\Delta\boldsymbol{u}_{\mathrm{F}}^{*}}\right)_{\mathrm{UQO}} \tag{7.287}$$

where the unbounded quadratic optimum arises from Equation 7.279. In order to allow minimisation of the *absolute* differences, $\boldsymbol{r}$ is represented using two non-negative quantities (one of which will be forced to zero in the LP solution):

$$\boldsymbol{r} = \boldsymbol{r}^{+} - \boldsymbol{r}^{-} \tag{7.288}$$

$$\left(\overline{\Delta\boldsymbol{u}_{\mathrm{F}}^{*}}\right)_{\mathrm{BQO}} = \left(\overline{\Delta\boldsymbol{u}_{\mathrm{F}}^{*}}\right)_{\mathrm{UQO}} + \left[\boldsymbol{r}^{+} - \boldsymbol{r}^{-}\right] \tag{7.289}$$

Multiply by $\overline{\boldsymbol{L}}$ and use Equation 7.285 to get

$$\left(\overline{\boldsymbol{u}}_{\mathrm{F}}^{*}\right)_{\mathrm{BQO}} = \overline{\boldsymbol{L}}\left(\overline{\Delta\boldsymbol{u}_{\mathrm{F}}^{*}}\right)_{\mathrm{UQO}} + \overline{\boldsymbol{I}}\,\boldsymbol{u} + \overline{\boldsymbol{L}}\left[r^{+} - r^{-}\right] \tag{7.290}$$

Then the linear programming problem is formulated as follows:

$$\text{Objective function:}\quad \boldsymbol{w}^{\mathrm{T}}\left[\boldsymbol{r}^{+} + \boldsymbol{r}^{-}\right] \quad \text{(to be minimised)} \tag{7.291}$$

where $\boldsymbol{w}$ is a weighting vector possibly chosen to improve the approach to $\left(\overline{\Delta\boldsymbol{u}_{\mathrm{F}}^{*}}\right)_{\mathrm{UQO}}$ by following the steepest descent of J (Mulholland and Prosser, 1997) subject to the constraints:

Input limits:

$$\left(\overline{\boldsymbol{u}}_{\mathrm{F}}^{*}\right)_{\mathrm{BQO}} \geq \overline{\boldsymbol{u}}_{\mathrm{MIN}}^{*} \Rightarrow \overline{\boldsymbol{L}}\left[\boldsymbol{r}^{+} - \boldsymbol{r}^{-}\right] \geq \overline{\boldsymbol{u}}_{\mathrm{MIN}}^{*} - \overline{\boldsymbol{L}}\left(\overline{\Delta\boldsymbol{u}_{\mathrm{F}}^{*}}\right)_{\mathrm{UQO}} - \overline{\boldsymbol{I}}\,\boldsymbol{u} \tag{7.292}$$

$$\left(\overline{\boldsymbol{u}}_{\mathrm{F}}^{*}\right)_{\mathrm{BQO}} \leq \overline{\boldsymbol{u}}_{\mathrm{MAX}}^{*} \Rightarrow \overline{\boldsymbol{L}}\left[\boldsymbol{r}^{+} - \boldsymbol{r}^{-}\right] \leq \overline{\boldsymbol{u}}_{\mathrm{MAX}}^{*} - \overline{\boldsymbol{L}}\left(\overline{\Delta\boldsymbol{u}_{\mathrm{F}}^{*}}\right)_{\mathrm{UQO}} - \overline{\boldsymbol{I}}\,\boldsymbol{u} \tag{7.293}$$

Input ramp limits:

$$\left(\overline{\Delta\boldsymbol{u}_{\mathrm{F}}^{*}}\right)_{\mathrm{BQO}} \geq -\overline{\Delta\boldsymbol{u}_{\mathrm{MAX}}^{*}} \Rightarrow \left[\boldsymbol{r}^{+} - \boldsymbol{r}^{-}\right] \geq -\overline{\Delta\boldsymbol{u}_{\mathrm{MAX}}^{*}} - \left(\overline{\Delta\boldsymbol{u}_{\mathrm{F}}^{*}}\right)_{\mathrm{UQO}} \tag{7.294}$$

$$\left(\overline{\Delta\boldsymbol{u}_{\mathrm{F}}^{*}}\right)_{\mathrm{BQO}} \leq +\overline{\Delta\boldsymbol{u}_{\mathrm{MAX}}^{*}} \Rightarrow \left[\boldsymbol{r}^{+} - \boldsymbol{r}^{-}\right] \leq +\overline{\Delta\boldsymbol{u}_{\mathrm{MAX}}^{*}} - \left(\overline{\Delta\boldsymbol{u}_{\mathrm{F}}^{*}}\right)_{\mathrm{UQO}} \tag{7.295}$$

Output limits:

$$\bar{y}_F \geq \bar{y}_{MIN} \Rightarrow \bar{B}_F^*[r^+ - r^-] \geq \bar{y}_{MIN} - \bar{y}_{OL} - \bar{B}_F^*(\overline{\Delta u_F^*})_{UQO} \tag{7.296}$$

$$\bar{y}_F \leq \bar{y}_{MAX} \Rightarrow \bar{B}_F^*[r^+ - r^-] \leq \bar{y}_{MAX} - \bar{y}_{OL} - \bar{B}_F^*(\overline{\Delta u_F^*})_{UQO} \tag{7.297}$$

In solving for r^+ and r^-, it will be found that one element of each matching pair will be zero. They have been arranged in this way to take advantage of the uniform minimum of zero in linear programming (or set a non-negative constraint explicitly if necessary). Since the difference between each corresponding element in r^+ and r^- moreover constitutes a single positive or negative number, the objective equation (Equation 7.291) will therefore ensure that one of each pair must be zero. The necessary input control actions are then obtained from Equation 7.289:

$$(\overline{\Delta u_F^*})_{BQO} = (\overline{\Delta u_F^*})_{UQO} + [r^+ - r^-] \tag{7.298}$$

Only the first step $\left(\Delta u_{F(1)}^*\right)_{BQO}$ of this solution sequence is actually used, before the entire optimisation process is repeated on the next time step. The effect of optimising more than one step is that the first step can be severe (overshooting), with subsequent steps aiming to correct the steady-state response (but never actually used). Increasing the number of optimised steps thus increases the severity of the control action.

Figure 7.22 illustrates the concept of the LDMC procedure for a SISO case where two successive control input steps are being optimised. Three coordinate systems are involved, but fortunately they are linearly related. The LP solution will always give a result at an apex. Actually, there are more apices than it appears, because the intersections of the r^+, r^- system axes with the constraints create further apices owing to the separation of the positive and negative ranges of r. Although this helps, the curvature of the contours of the original objective function, which led to $(\overline{\Delta u_F^*})_{UQO}$,

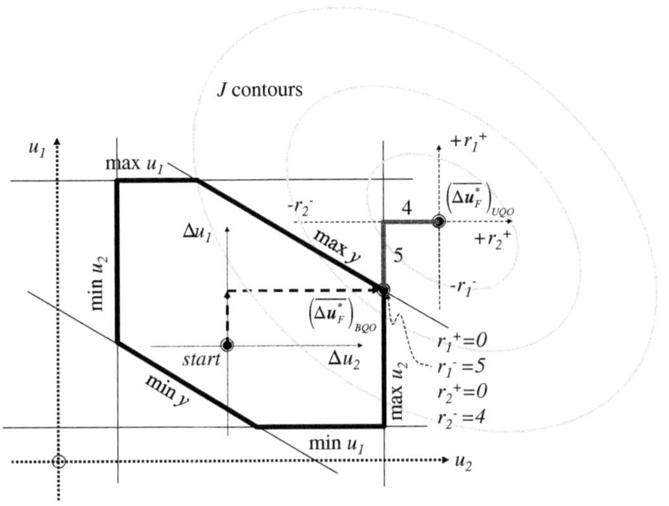

Figure 7.22 SISO control with two optimised moves: LDMC coordinate systems.

would in reality determine true optimal points at intermediate positions on a constraint, so the method is approximate. In the example, the minimum (weighted) sum of the distances (4 + 5) determines the optimal combination of moves. Problem 7.22 in the accompanying book is an exercise in programming the LDMC and testing its operation under constrained conditions.

7.8.2.2 Quadratic Dynamic Matrix Control in Industry

Industrial DMC applications are generally based on a *quadratic programming* solution of the constrained model predictive control problem (Garcia and Morshedi, 1984). This is a competitive area for vendors of distributed control hardware and advanced process control (APC) technology. Many features have been added to the versions available. Objective functions can include additional economic criteria, and constraints can be either *hard* or *soft*. In the preceding sections, only *hard* constraints were considered. These can present a problem when an algorithm finds itself beyond the constraints, and/or cannot find a solution within the constraints on its next step. Soft constraints are often useful in such instances, invoking a high asymmetric penalty on the outside of the constraint. The commercial packages include user-friendly software for the capturing and processing of plant step responses, and for setting up and tuning the online controller. Current examples include Honeywell's RMPCT®, Shell and Yokogawa's SMOC® and Pavilion/Neuralware/Aspentech's NEUCOP®. The last-named package is based on an underlying linear model, with an artificial neural net superimposed to model the residual nonlinearity.

7.8.2.3 Recursive Representation of the Future Output

From Equation 7.268, for only the first move $\Delta\boldsymbol{u}_{(0)}$, one extends $\overline{\boldsymbol{B}}_{\mathrm{P}}$ using the first column of $\overline{\boldsymbol{B}}_{\mathrm{F}}$ to obtain a prediction of the future output sequence as follows:

$$\overline{y}_{\mathrm{F}}(t) = \begin{pmatrix} y_{\mathrm{F}(1)}(t) \\ y_{\mathrm{F}(2)}(t) \\ y_{\mathrm{F}(3)}(t) \\ \vdots \\ y_{\mathrm{F}(N-1)}(t) \\ y_{\mathrm{F}(N)}(t) \end{pmatrix} = \begin{bmatrix} \cdots & \cdots & \boldsymbol{b}_{(5)} & \boldsymbol{b}_{(4)} & \boldsymbol{b}_{(3)} & \boldsymbol{b}_{(2)} & \boldsymbol{b}_{(1)} \\ \cdots & \cdots & \boldsymbol{b}_{(6)} & \boldsymbol{b}_{(5)} & \boldsymbol{b}_{(4)} & \boldsymbol{b}_{(3)} & \boldsymbol{b}_{(2)} \\ \cdots & \cdots & \boldsymbol{b}_{(7)} & \boldsymbol{b}_{(6)} & \boldsymbol{b}_{(5)} & \boldsymbol{b}_{(4)} & \boldsymbol{b}_{(3)} \\ \cdots & \cdots & \ddots & \ddots & \ddots & \ddots & \ddots \\ \cdots & \cdots & \boldsymbol{b}_{(N)} & \boldsymbol{b}_{(N)} & \boldsymbol{b}_{(N)} & \boldsymbol{b}_{(N)} & \boldsymbol{b}_{(N-1)} \\ \cdots & \cdots & \boldsymbol{b}_{(N)} & \boldsymbol{b}_{(N)} & \boldsymbol{b}_{(N)} & \boldsymbol{b}_{(N)} & \boldsymbol{b}_{(N)} \end{bmatrix} \begin{pmatrix} \vdots \\ \vdots \\ \Delta\boldsymbol{u}(t-4T) \\ \Delta\boldsymbol{u}(t-3T) \\ \Delta\boldsymbol{u}(t-2T) \\ \Delta\boldsymbol{u}(t-T) \\ \Delta\boldsymbol{u}(t) \end{pmatrix} + \overline{y}_{\mathrm{SS}} \tag{7.299}$$

$$\overline{y}_{\mathrm{F}}(t) = \begin{pmatrix} \cdots & \cdots & +\boldsymbol{b}_{(5)}\Delta\boldsymbol{u}(t-4T) & +\boldsymbol{b}_{(4)}\Delta\boldsymbol{u}(t-3T) & +\boldsymbol{b}_{(3)}\Delta\boldsymbol{u}(t-2T) & +\boldsymbol{b}_{(2)}\Delta\boldsymbol{u}(t-T) & +\boldsymbol{b}_{(1)}\Delta\boldsymbol{u}(t) \\ \cdots & \cdots & +\boldsymbol{b}_{(6)}\Delta\boldsymbol{u}(t-4T) & +\boldsymbol{b}_{(5)}\Delta\boldsymbol{u}(t-3T) & +\boldsymbol{b}_{(4)}\Delta\boldsymbol{u}(t-2T) & +\boldsymbol{b}_{(3)}\Delta\boldsymbol{u}(t-T) & +\boldsymbol{b}_{(2)}\Delta\boldsymbol{u}(t) \\ \cdots & \cdots & +\boldsymbol{b}_{(7)}\Delta\boldsymbol{u}(t-4T) & +\boldsymbol{b}_{(6)}\Delta\boldsymbol{u}(t-3T) & +\boldsymbol{b}_{(5)}\Delta\boldsymbol{u}(t-2T) & +\boldsymbol{b}_{(4)}\Delta\boldsymbol{u}(t-T) & +\boldsymbol{b}_{(3)}\Delta\boldsymbol{u}(t) \\ \cdots & \cdots & \ddots & \ddots & \ddots & \ddots & \ddots \\ \cdots & \cdots & +\boldsymbol{b}_{(N)}\Delta\boldsymbol{u}(t-4T) & +\boldsymbol{b}_{(N)}\Delta\boldsymbol{u}(t-3T) & +\boldsymbol{b}_{(N)}\Delta\boldsymbol{u}(t-2T) & +\boldsymbol{b}_{(N)}\Delta\boldsymbol{u}(t-T) & +\boldsymbol{b}_{(N-1)}\Delta\boldsymbol{u}(t) \\ \cdots & \cdots & +\boldsymbol{b}_{(N)}\Delta\boldsymbol{u}(t-4T) & +\boldsymbol{b}_{(N)}\Delta\boldsymbol{u}(t-3T) & +\boldsymbol{b}_{(N)}\Delta\boldsymbol{u}(t-2T) & +\boldsymbol{b}_{(N)}\Delta\boldsymbol{u}(t-T) & +\boldsymbol{b}_{(N)}\Delta\boldsymbol{u}(t) \end{pmatrix} + \overline{y}_{\mathrm{SS}} \tag{7.300}$$

Here one is acknowledging that all nonzero moves from the past are included in the partitioned move vector, and that the steady-state output $\bar{y}_{SS}$ existed before the first nonzero move. Note that, for example, $\boldsymbol{y}_{F(3)}(t)$ is a prediction of the output vector three steps ahead from the *present*, based on the *present* move $\Delta\boldsymbol{u}(t)$ and *all past moves.* As time moves forward by T, this system predicts another set of N future outputs effectively by adding a further column on the left of the matrix and a further row at the bottom of the vector:

$$\bar{y}_F(t+T) = \begin{pmatrix} \boldsymbol{y}_{F(1)}(t+T) \\ \boldsymbol{y}_{F(2)}(t+T) \\ \boldsymbol{y}_{F(3)}(t+T) \\ \vdots \\ \boldsymbol{y}_{F(N-1)}(t+T) \\ \boldsymbol{y}_{F(N)}(t+T) \end{pmatrix}$$

$$= \begin{bmatrix} \cdots & \cdots & \boldsymbol{b}_{(6)} & \boldsymbol{b}_{(5)} & \boldsymbol{b}_{(4)} & \boldsymbol{b}_{(3)} & \boldsymbol{b}_{(2)} & \boldsymbol{b}_{(1)} \\ \cdots & \cdots & \boldsymbol{b}_{(7)} & \boldsymbol{b}_{(6)} & \boldsymbol{b}_{(5)} & \boldsymbol{b}_{(4)} & \boldsymbol{b}_{(3)} & \boldsymbol{b}_{(2)} \\ \cdots & \cdots & \boldsymbol{b}_{(8)} & \boldsymbol{b}_{(7)} & \boldsymbol{b}_{(6)} & \boldsymbol{b}_{(5)} & \boldsymbol{b}_{(4)} & \boldsymbol{b}_{(3)} \\ \cdots & \cdots & \ddots & \ddots & \ddots & \ddots & \ddots & \ddots \\ \cdots & \cdots & \boldsymbol{b}_{(N)} & \boldsymbol{b}_{(N)} & \boldsymbol{b}_{(N)} & \boldsymbol{b}_{(N)} & \boldsymbol{b}_{(N)} & \boldsymbol{b}_{(N-1)} \\ \cdots & \cdots & \boldsymbol{b}_{(N)} & \boldsymbol{b}_{(N)} & \boldsymbol{b}_{(N)} & \boldsymbol{b}_{(N)} & \boldsymbol{b}_{(N)} & \boldsymbol{b}_{(N)} \end{bmatrix} \begin{pmatrix} \vdots \\ \vdots \\ \Delta\boldsymbol{u}(t-4T) \\ \Delta\boldsymbol{u}(t-3T) \\ \Delta\boldsymbol{u}(t-2T) \\ \Delta\boldsymbol{u}(t-T) \\ \Delta\boldsymbol{u}(t) \\ \Delta\boldsymbol{u}(t+T) \end{pmatrix} + \bar{y}_{SS} \quad (7.301)$$

$$\bar{y}_F(t+T) = \begin{pmatrix} \cdots & \cdots & +\boldsymbol{b}_{(6)}\Delta\boldsymbol{u}(t-4T) & +\boldsymbol{b}_{(5)}\Delta\boldsymbol{u}(t-3T) & +\boldsymbol{b}_{(4)}\Delta\boldsymbol{u}(t-2T) & +\boldsymbol{b}_{(3)}\Delta\boldsymbol{u}(t-T) & +\boldsymbol{b}_{(2)}\Delta\boldsymbol{u}(t) & +\boldsymbol{b}_{(1)}\Delta\boldsymbol{u}(t+T) \\ \cdots & \cdots & +\boldsymbol{b}_{(7)}\Delta\boldsymbol{u}(t-4T) & +\boldsymbol{b}_{(6)}\Delta\boldsymbol{u}(t-3T) & +\boldsymbol{b}_{(5)}\Delta\boldsymbol{u}(t-2T) & +\boldsymbol{b}_{(4)}\Delta\boldsymbol{u}(t-T) & +\boldsymbol{b}_{(3)}\Delta\boldsymbol{u}(t) & +\boldsymbol{b}_{(2)}\Delta\boldsymbol{u}(t+T) \\ \cdots & \cdots & +\boldsymbol{b}_{(8)}\Delta\boldsymbol{u}(t-4T) & +\boldsymbol{b}_{(7)}\Delta\boldsymbol{u}(t-3T) & +\boldsymbol{b}_{(6)}\Delta\boldsymbol{u}(t-2T) & +\boldsymbol{b}_{(5)}\Delta\boldsymbol{u}(t-T) & +\boldsymbol{b}_{(4)}\Delta\boldsymbol{u}(t) & +\boldsymbol{b}_{(3)}\Delta\boldsymbol{u}(t+T) \\ \cdots & \cdots & \ddots & \ddots & \ddots & \ddots & \ddots & \ddots \\ \cdots & \cdots & +\boldsymbol{b}_{(N)}\Delta\boldsymbol{u}(t-4T) & +\boldsymbol{b}_{(N)}\Delta\boldsymbol{u}(t-3T) & +\boldsymbol{b}_{(N)}\Delta\boldsymbol{u}(t-2T) & +\boldsymbol{b}_{(N)}\Delta\boldsymbol{u}(t-T) & +\boldsymbol{b}_{(N)}\Delta\boldsymbol{u}(t) & +\boldsymbol{b}_{(N-1)}\Delta\boldsymbol{u}(t+T) \\ \cdots & \cdots & +\boldsymbol{b}_{(N)}\Delta\boldsymbol{u}(t-4T) & +\boldsymbol{b}_{(N)}\Delta\boldsymbol{u}(t-3T) & +\boldsymbol{b}_{(N)}\Delta\boldsymbol{u}(t-2T) & +\boldsymbol{b}_{(N)}\Delta\boldsymbol{u}(t-T) & +\boldsymbol{b}_{(N)}\Delta\boldsymbol{u}(t) & +\boldsymbol{b}_{(N)}\Delta\boldsymbol{u}(t+T) \end{pmatrix} + \bar{y}_{SS} \quad (7.302)$$

Comparison of the terms in Equations 7.300 and 7.302 shows that

$$\boldsymbol{y}_{F(1)}(t+T) = \boldsymbol{y}_{F(2)}(t) + \boldsymbol{b}_{(1)}\Delta\boldsymbol{u}(t+T) \quad (7.303)$$

$$\boldsymbol{y}_{F(2)}(t+T) = \boldsymbol{y}_{F(3)}(t) + \boldsymbol{b}_{(2)}\Delta\boldsymbol{u}(t+T) \quad (7.304)$$

$$\boldsymbol{y}_{F(3)}(t+T) = \boldsymbol{y}_{F(4)}(t) + \boldsymbol{b}_{(3)}\Delta\boldsymbol{u}(t+T) \quad (7.305)$$

$$\vdots$$

$$\boldsymbol{y}_{F(N-1)}(t+T) = \boldsymbol{y}_{F(N)}(t) + \boldsymbol{b}_{(N-1)}\Delta\boldsymbol{u}(t+T) \quad (7.306)$$

$$\boldsymbol{y}_{F(N)}(t+T) = \boldsymbol{y}_{F(N)}(t) + \boldsymbol{b}_{(N)}\Delta\boldsymbol{u}(t+T) \quad (7.307)$$

that is

$$\bar{y}_{\rm F}(t+T) = \underbrace{\begin{bmatrix} 0 & I & 0 & \ddots & 0 & 0 \\ 0 & 0 & I & \ddots & 0 & 0 \\ \ddots & \ddots & \ddots & \ddots & \ddots & \ddots \\ 0 & 0 & 0 & \ddots & I & 0 \\ 0 & 0 & 0 & \ddots & 0 & I \\ 0 & 0 & 0 & \ddots & 0 & I \end{bmatrix}}_{A} \bar{y}_{\rm F}(t) + \underbrace{\begin{bmatrix} b_{(1)} \\ b_{(2)} \\ b_{(3)} \\ \vdots \\ b_{(N-1)} \\ b_{(N)} \end{bmatrix}}_{B} \Delta u(t+T) \tag{7.308}$$

which is in the autoregressive format, giving the set of N predicted future outputs at any time based on moves so far. Only in appearance is this a state equation, since previous information is aggregated in the move vector. The "future" outputs clearly need to be initialised, and the incoming output is given the final value as the last arrival. The one-step form is obtained by selection of the output

$$y_{{\rm F}(1)}(t+T) = C\,\bar{y}_{\rm F}(t+T) \quad \text{where} \quad C = \begin{bmatrix} I & 0 & \cdots & 0 & 0 & 0 \end{bmatrix} \tag{7.309}$$

7.8.2.4 Dynamic Matrix Control of an Integrating System

In a typical chemical processing plant, much of the equipment has an integrating property. Figure 7.23 shows step responses for a two-input, two-output system of this nature. Tanks receiving and dispensing liquid can quite easily overflow or empty – there is no equilibrium level. Likewise boiler pressures can rise or fall indefinitely depending on the energy and mass balances. Positive displacement pumps can build up line and vessel pressures almost indefinitely. If one seeks to use dynamic matrix control in these instances, a problem arises due to the continued dynamic response to all moves made since the last steady state. It happens that DMC nevertheless functions quite well

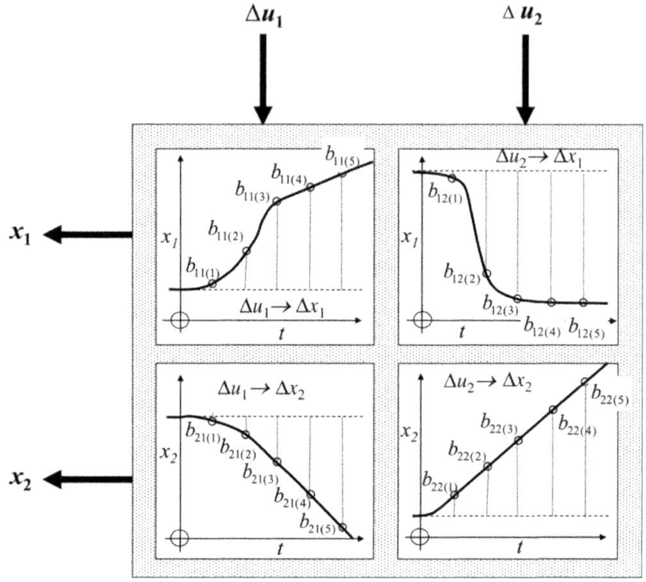

Figure 7.23 Step responses of a two-input, two-output system exhibiting integrating behaviour.

in these circumstances, using measured step response curves which will obviously be truncated before equilibrium. However, some improvements in performance can be obtained by adjusting the algorithm to recognise integration (Guiamba and Mulholland, 2004).

Several integrating process control studies have appeared in the literature. Lee, Morari and Garcia (1994) described a MPC technique based on step response parameters for systems of stable and/or integrating dynamics, developed using state-space estimation techniques. A Kalman filter identified ramp disturbances for inclusion in future predictions, within the context of known step response parameters.

Gupta (1998) presents an alternative approach to eliminate the steady-state offsets that are encountered when dealing with integrating process units. It can be implemented directly in the step response formulation of the DMC algorithm. This scheme takes advantage of the fact that the predicted response due to past inputs is a straight line passing through the output at the current control instant. Thus, the slope of the predicted response is determined from the slope of the output trajectory between the current and the previous control instants. Note that this slope includes the effect of unmeasured disturbances and any model mismatch that may be present. This approach allows the consideration of all inputs, outputs and constraints in a single optimisation problem, but appears to be best suited to processes that are purely integrating, without any other transient portion in the responses.

Dougherty and Cooper (2003) approximate the integrating process responses with first-order plus dead-time transfer functions. They provide recommendations for prediction and control horizons, and move suppression, using the parameters of these functions, and then implement a standard DMC controller on this basis.

Guiamba and Mulholland (2004) extended the $\overline{\boldsymbol{B}}_{\mathrm{P}}$ matrix in Equation 7.268 backwards to allow for the continuing integration

$$\underbrace{\begin{bmatrix} \boldsymbol{b}_{(N)}+\Delta\boldsymbol{b} & \boldsymbol{b}_{(N)} & \vdots & \boldsymbol{b}_{(2)} \\ \vdots & \boldsymbol{b}_{(N)}+\Delta\boldsymbol{b} & \boldsymbol{b}_{(N)} & \vdots \\ \boldsymbol{b}_{(N)}+(N-1)\Delta\boldsymbol{b} & \vdots & \boldsymbol{b}_{(N)}+\Delta\boldsymbol{b} & \boldsymbol{b}_{(N)} \\ \boldsymbol{b}_{(N)}+N\Delta\boldsymbol{b} & \boldsymbol{b}_{(N)}+(N-1)\Delta\boldsymbol{b} & \vdots & \boldsymbol{b}_{(N)}+\Delta\boldsymbol{b} \end{bmatrix}}_{\overline{\boldsymbol{B}}_{\mathrm{P}}} \vdots \underbrace{\begin{bmatrix} \boldsymbol{b}_{(1)} & \boldsymbol{0} & \boldsymbol{0} & \boldsymbol{0} \\ \boldsymbol{b}_{(2)} & \boldsymbol{b}_{(1)} & \boldsymbol{0} & \boldsymbol{0} \\ \vdots & \boldsymbol{b}_{(2)} & \boldsymbol{b}_{(1)} & \boldsymbol{0} \\ \boldsymbol{b}_{(N)} & \vdots & \boldsymbol{b}_{(2)} & \boldsymbol{b}_{(1)} \end{bmatrix}}_{\overline{\boldsymbol{B}}_{\mathrm{F}}} \tag{7.310}$$

and additionally included a residual gradient correction $\overline{\boldsymbol{g}}(t)$ in Equation 7.271:

$$\overline{\boldsymbol{y}}_{\mathrm{F}}(t) = \overline{\boldsymbol{I}}\,\boldsymbol{y}(t) + \left[\overline{\boldsymbol{B}}_{\mathrm{P}} - \overline{\boldsymbol{I}}\,\overline{\boldsymbol{B}}_{0}\right]\overline{\Delta\boldsymbol{u}_{\mathrm{P}}(t)} + \overline{\boldsymbol{B}}_{\mathrm{F}}\overline{\Delta\boldsymbol{u}_{\mathrm{F}}(t)} + \overline{\boldsymbol{g}}(t) \tag{7.311}$$

Equation 7.310 is presented in symmetric form, that is the 'past' horizon is equal to the 'future' horizon, though this need not be the case. It is assumed that the process step responses have been measured up to N intervals, at which point all responses had steady or zero gradients. The incremental gradient per step thereafter, based on the last two step response points, is stored in the submatrix $\Delta\boldsymbol{b}$. In the example of Figure 7.23, this would be

$$\Delta\boldsymbol{b} = \boldsymbol{b}_{(N)} - \boldsymbol{b}_{(N-1)} = \begin{bmatrix} b_{11(5)} - b_{11(4)} & b_{12(5)} - b_{12(4)} \\ b_{21(5)} - b_{21(4)} & b_{22(5)} - b_{22(4)} \end{bmatrix} \tag{7.312}$$

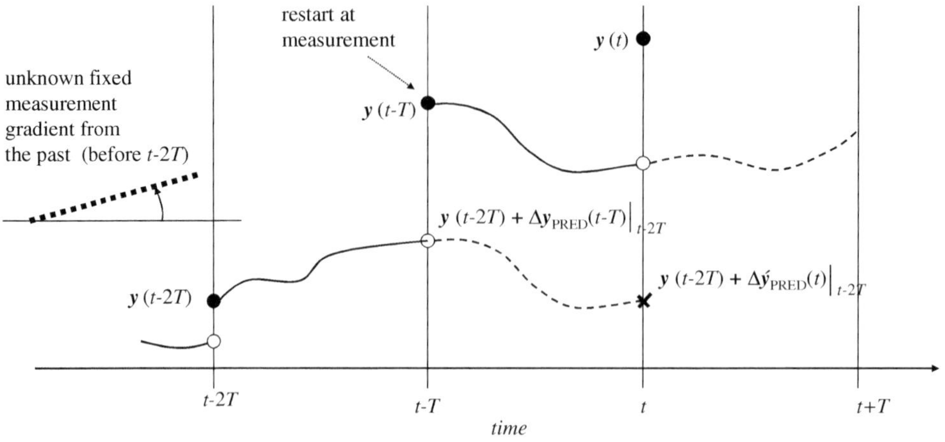

Figure 7.24 Juxtaposed N-step predictions with correction shown only every Nth step.

In Equation 7.310, the $\boldsymbol{b}_N$ terms are thus extended backwards with this gradient up to N steps before the present. Input moves before N steps ago will only contribute constant gradients to the future outputs.

Figure 7.24 illustrates how Equation 7.313 can be used to predict the output (solid line) in N-step jumps (i.e. full interval T apart). The starting point is corrected to the measurement at the start of each jump.

$$\Delta \boldsymbol{y}_{\mathrm{PRED}}(t-T)\big|_{t-2T} = \begin{bmatrix} \boldsymbol{b}_{(N)} & \boldsymbol{b}_{(N-1)} & \cdots & \boldsymbol{b}_{(1)} \end{bmatrix} \begin{pmatrix} \Delta \boldsymbol{u}(t-2NT) \\ \vdots \\ \Delta \boldsymbol{u}(t-\{N+2\}T) \\ \Delta \boldsymbol{u}(t-\{N+1\}T) \end{pmatrix} \tag{7.313}$$

The prediction using Equation 7.313 will of course have an offset error due to the moves *up until* $t-2T$. Also based on the measurement $\boldsymbol{y}(t-2T)$, a double-interval prediction (dashed line) can be done using

$$\Delta \boldsymbol{y}'_{\mathrm{PRED}}(t)\big|_{t-2T} = \begin{bmatrix} \boldsymbol{b}_{(N)} + N\Delta \boldsymbol{b} & \boldsymbol{b}_{(N)} + (N-1)\Delta \boldsymbol{b} & \cdots & \boldsymbol{b}_{(N)} + \Delta \boldsymbol{b} & \boldsymbol{b}_{(N)} & \boldsymbol{b}_{(N-1)} & \cdots & \boldsymbol{b}_{(1)} \end{bmatrix} \times \begin{pmatrix} \Delta \boldsymbol{u}(t-NT) \\ \vdots \\ \Delta \boldsymbol{u}(t-2T) \\ \Delta \boldsymbol{u}(t-T) \\ \Delta \boldsymbol{u}(t) \\ \Delta \boldsymbol{u}(t+T) \\ \vdots \\ \Delta \boldsymbol{u}(t+\{N-1\}T) \end{pmatrix} \tag{7.314}$$

The use of Equation 7.314 will be subject to the *same* offset error as Equation 7.313. Thus, if the gradient between $\Delta \boldsymbol{y}_{\mathrm{PRED}}(t-T)$ and $\Delta \boldsymbol{y}'_{\mathrm{PRED}}(t)$ differs from that between the measurements

$y(t-T)$ and $y(t)$, the error must be caused by an unknown constant gradient from before $t-2T$. The necessary gradient correction in Equation 7.311 is thus

$$\bar{g}(t) = \begin{bmatrix} (1/N)\boldsymbol{I} \\ (2/N)\boldsymbol{I} \\ \vdots \\ \boldsymbol{I} \end{bmatrix} \left\{ [y(t) - y(t-T)] - \left[y'_{\mathrm{PRED}}(t)\big|_{t-2T} - y_{\mathrm{PRED}}(t-T)\big|_{t-2T} \right] \right\} \tag{7.315}$$

It may be necessary to smooth this estimate by filtering, in order to reduce the effect of short disturbances.

7.8.2.5 Dynamic Matrix Control Based on a Finite Impulse Response

Equation 7.263 was

$$\bar{y} = \bar{y}_0 + \overline{\boldsymbol{B}}\,\overline{\Delta \boldsymbol{u}} \tag{7.316}$$

with

$$\bar{y} = \begin{pmatrix} y_{(1)} \\ y_{(2)} \\ y_{(3)} \\ \vdots \end{pmatrix}, \quad \bar{y}_0 = \begin{pmatrix} y_{(0)} \\ y_{(0)} \\ y_{(0)} \\ \vdots \end{pmatrix}, \quad \overline{\Delta \boldsymbol{u}} = \begin{pmatrix} \Delta \boldsymbol{u}_{(0)} \\ \Delta \boldsymbol{u}_{(1)} \\ \Delta \boldsymbol{u}_{(2)} \\ \vdots \end{pmatrix} \quad \text{and} \quad \overline{\boldsymbol{B}} = \begin{bmatrix} \boldsymbol{b}_{(1)} & \boldsymbol{0} & \boldsymbol{0} & \cdots \\ \boldsymbol{b}_{(2)} & \boldsymbol{b}_{(1)} & \boldsymbol{0} & \cdots \\ \boldsymbol{b}_{(3)} & \boldsymbol{b}_{(2)} & \boldsymbol{b}_{(1)} & \cdots \\ \vdots & \vdots & \vdots & \ddots \end{bmatrix} \tag{7.317}$$

Multiplying through by a differencing partitioned matrix

$$\overline{\boldsymbol{D}} = \begin{bmatrix} \boldsymbol{I} & \boldsymbol{0} & \boldsymbol{0} & \boldsymbol{0} & \boldsymbol{0} \\ -\boldsymbol{I} & \boldsymbol{I} & \boldsymbol{0} & \boldsymbol{0} & \ddots \\ \boldsymbol{0} & -\boldsymbol{I} & \boldsymbol{I} & \ddots & \boldsymbol{0} \\ \boldsymbol{0} & \boldsymbol{0} & \ddots & \boldsymbol{I} & \boldsymbol{0} \\ \boldsymbol{0} & \ddots & \boldsymbol{0} & -\boldsymbol{I} & \boldsymbol{I} \end{bmatrix} \tag{7.318}$$

obtain

$$\begin{pmatrix} y_{(1)} \\ y_{(2)} - y_{(1)} \\ y_{(3)} - y_{(2)} \\ \vdots \end{pmatrix} = \begin{pmatrix} y_{(0)} \\ \boldsymbol{0} \\ \boldsymbol{0} \\ \vdots \end{pmatrix} + \begin{bmatrix} \boldsymbol{b}_{(1)} & \boldsymbol{0} & \boldsymbol{0} & \cdots \\ \boldsymbol{b}_{(2)} - \boldsymbol{b}_{(1)} & \boldsymbol{b}_{(1)} & \boldsymbol{0} & \cdots \\ \boldsymbol{b}_{(3)} - \boldsymbol{b}_{(2)} & \boldsymbol{b}_{(2)} - \boldsymbol{b}_{(1)} & \boldsymbol{b}_{(1)} & \cdots \\ \vdots & \vdots & \vdots & \ddots \end{bmatrix} \begin{pmatrix} \Delta \boldsymbol{u}_{(0)} \\ \Delta \boldsymbol{u}_{(1)} \\ \Delta \boldsymbol{u}_{(2)} \\ \vdots \end{pmatrix} \tag{7.319}$$

that is

$$\begin{pmatrix} \Delta y_{(1)} \\ \Delta y_{(2)} \\ \Delta y_{(3)} \\ \vdots \end{pmatrix} = \begin{bmatrix} \boldsymbol{a}_{(1)} & \boldsymbol{0} & \boldsymbol{0} & \cdots \\ \boldsymbol{a}_{(2)} & \boldsymbol{a}_{(1)} & \boldsymbol{0} & \cdots \\ \boldsymbol{a}_{(3)} & \boldsymbol{a}_{(2)} & \boldsymbol{a}_{(1)} & \cdots \\ \vdots & \vdots & \vdots & \ddots \end{bmatrix} \begin{pmatrix} \Delta \boldsymbol{u}_{(0)} \\ \Delta \boldsymbol{u}_{(1)} \\ \Delta \boldsymbol{u}_{(2)} \\ \vdots \end{pmatrix} \tag{7.320}$$

where the submatrices $\boldsymbol{a}_{(i)}$ are the *impulse response* coefficients which can be obtained as the derivative of a step response as in Figure 7.25.

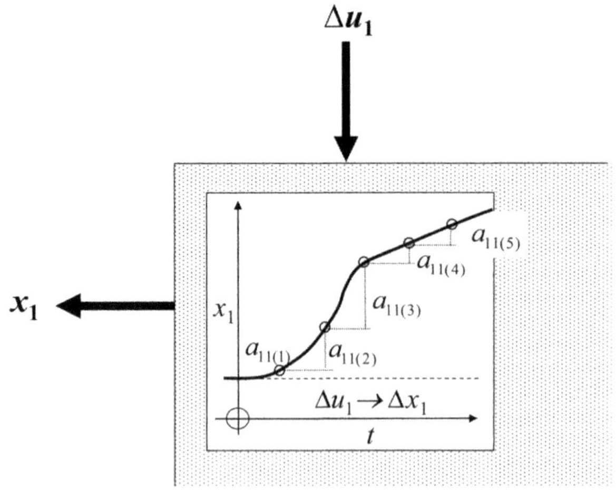

Figure 7.25 Impulse response parameters obtained from a unit step response.

From Equation 7.271, the closed-loop equation for future output predictions based on step responses was

$$\bar{\boldsymbol{y}}_{\mathrm{F}} = \bar{\boldsymbol{I}}\,\boldsymbol{y} + \left[\bar{\boldsymbol{B}}_{\mathrm{P}} - \bar{\boldsymbol{I}}\,\bar{\boldsymbol{B}}_{0}\right]\overline{\Delta \boldsymbol{u}_{\mathrm{P}}} + \bar{\boldsymbol{B}}_{\mathrm{F}}\overline{\Delta \boldsymbol{u}_{\mathrm{F}}} \tag{7.321}$$

Again, multiplying through by $\bar{\boldsymbol{D}}$ obtain

$$\overline{\Delta \boldsymbol{y}_{\mathrm{F}}} = \bar{\boldsymbol{A}}_{\mathrm{P}}\overline{\Delta \boldsymbol{u}_{\mathrm{P}}} + \bar{\boldsymbol{A}}_{\mathrm{F}}\overline{\Delta \boldsymbol{u}_{\mathrm{F}}} \tag{7.322}$$

that is

$$\underbrace{\begin{pmatrix} \Delta \boldsymbol{y}_{\mathrm{F}(1)} \\ \Delta \boldsymbol{y}_{\mathrm{F}(2)} \\ \vdots \\ \Delta \boldsymbol{y}_{\mathrm{F}(N-1)} \\ \Delta \boldsymbol{y}_{\mathrm{F}(N)} \end{pmatrix}}_{\overline{\Delta \boldsymbol{y}_{\mathrm{F}}}} = \underbrace{\begin{bmatrix} \boldsymbol{a}_{(N)} & \boldsymbol{a}_{(N)} & \cdots & \boldsymbol{a}_{(3)} & \boldsymbol{a}_{(2)} \\ \boldsymbol{a}_{(N)} & \boldsymbol{a}_{(N)} & \boldsymbol{a}_{(N)} & \cdots & \boldsymbol{a}_{(3)} \\ \cdots & \boldsymbol{a}_{(N)} & \boldsymbol{a}_{(N)} & \boldsymbol{a}_{(N)} & \cdots \\ \boldsymbol{a}_{(N)} & \cdots & \boldsymbol{a}_{(N)} & \boldsymbol{a}_{(N)} & \boldsymbol{a}_{(N)} \\ \boldsymbol{a}_{(N)} & \boldsymbol{a}_{(N)} & \cdots & \boldsymbol{a}_{(N)} & \boldsymbol{a}_{(N)} \end{bmatrix}}_{\bar{\boldsymbol{A}}_{\mathrm{P}}} \underbrace{\begin{pmatrix} \Delta \boldsymbol{u}_{(-N)} \\ \Delta \boldsymbol{u}_{(-N+1)} \\ \vdots \\ \Delta \boldsymbol{u}_{(-2)} \\ \Delta \boldsymbol{u}_{(-1)} \end{pmatrix}}_{\overline{\Delta \boldsymbol{u}_{\mathrm{P}}}} + \underbrace{\begin{bmatrix} \boldsymbol{a}_{(1)} & \boldsymbol{0} & \boldsymbol{0} & \cdots & \boldsymbol{0} \\ \boldsymbol{a}_{(2)} & \boldsymbol{a}_{(1)} & \boldsymbol{0} & \boldsymbol{0} & \cdots \\ \boldsymbol{a}_{(3)} & \boldsymbol{a}_{(2)} & \boldsymbol{a}_{(1)} & \boldsymbol{0} & \boldsymbol{0} \\ \cdots & \boldsymbol{a}_{(3)} & \boldsymbol{a}_{(2)} & \boldsymbol{a}_{(1)} & \boldsymbol{0} \\ \boldsymbol{a}_{(N)} & \cdots & \boldsymbol{a}_{(3)} & \boldsymbol{a}_{(2)} & \boldsymbol{a}_{(1)} \end{bmatrix}}_{\bar{\boldsymbol{A}}_{\mathrm{F}}} \underbrace{\begin{pmatrix} \Delta \boldsymbol{u}_{(0)} \\ \Delta \boldsymbol{u}_{(1)} \\ \vdots \\ \Delta \boldsymbol{u}_{(N-2)} \\ \Delta \boldsymbol{u}_{(N-1)} \end{pmatrix}}_{\overline{\Delta \boldsymbol{u}_{\mathrm{F}}}} \tag{7.323}$$

Here provision has been made for an integrating system (see Section 7.8.2.4), by extending constant $\boldsymbol{a}_{(N)}$ in the lower block triangle of $\bar{\boldsymbol{A}}_{\mathrm{P}}$. This portion would otherwise be occupied by zero matrices. To complete the correction for an integrating system, compare single-interval and double-interval predictions again as in Section 7.8.2.4.

$$\Delta\left\{\overline{\Delta \boldsymbol{y}_{\mathrm{F}}(t-T)}\right\}\Big|_{t-2T} = \bar{\boldsymbol{A}}_{\mathrm{F}}\overline{\Delta \boldsymbol{u}_{\mathrm{F}}(t-2T)} \tag{7.324}$$

$$\Delta\left\{\overline{\Delta \boldsymbol{y}_{\mathrm{F}}(t)}\right\}'\Big|_{t-2T} = \bar{\boldsymbol{A}}_{\mathrm{P}}\overline{\Delta \boldsymbol{u}_{\mathrm{P}}(t-T)} + \bar{\boldsymbol{A}}_{\mathrm{F}}\overline{\Delta \boldsymbol{u}_{\mathrm{F}}(t-T)} \tag{7.325}$$

where

$$\overline{\Delta \boldsymbol{u}_{\mathrm{P}}(t-T)} = \overline{\Delta \boldsymbol{u}_{\mathrm{F}}(t-2T)} \tag{7.326}$$

Then Equation 7.322 can be corrected for residual gradients in existence before $t-2T$ by

$$\overline{\Delta \boldsymbol{y}_{\mathrm{F}}(t)} = \overline{\boldsymbol{A}}_{\mathrm{P}} \overline{\Delta \boldsymbol{u}_{\mathrm{P}}(t)} + \overline{\boldsymbol{A}}_{\mathrm{F}} \overline{\Delta \boldsymbol{u}_{\mathrm{F}}(t)} + \overline{\boldsymbol{g}}(t) \tag{7.327}$$

where

$$\overline{\boldsymbol{g}}(t) = \frac{1}{N} \begin{bmatrix} \boldsymbol{I} \\ \boldsymbol{I} \\ \vdots \\ \boldsymbol{I} \end{bmatrix} \left\{ \sum_{i=1}^{N} \left[\Delta \left\{ \Delta \boldsymbol{y}_{\mathrm{F}(i)}(t-T) \right\} \Big|_{t-2T} - \Delta \left\{ \Delta \boldsymbol{y}_{\mathrm{F}(i)}(t) \right\}' \Big|_{t-2T} \right] - [\boldsymbol{y}(t) - \boldsymbol{y}(t-T)] \right\} \tag{7.328}$$

Defining the block lower triangular system

$$\overline{\boldsymbol{L}} = \begin{bmatrix} \boldsymbol{I} & \boldsymbol{0} & \ddots & \boldsymbol{0} \\ \boldsymbol{I} & \boldsymbol{I} & \ddots & \boldsymbol{0} \\ \ddots & \ddots & \ddots & \ddots \\ \boldsymbol{I} & \boldsymbol{I} & \ddots & \boldsymbol{I} \end{bmatrix} \tag{7.329}$$

then the equivalent of Equation 7.311 is

$$\overline{\boldsymbol{y}}_{\mathrm{F}} = \overline{\boldsymbol{I}}\,\boldsymbol{y} + \overline{\boldsymbol{L}}\,\overline{\boldsymbol{A}}_{\mathrm{P}} \overline{\Delta \boldsymbol{u}_{\mathrm{P}}} + \overline{\boldsymbol{L}}\,\overline{\boldsymbol{A}}_{\mathrm{F}} \overline{\Delta \boldsymbol{u}_{\mathrm{F}}} + \overline{\boldsymbol{L}}\,\overline{\boldsymbol{g}}(t) \tag{7.330}$$

which has a coefficient-by-coefficient equivalence as expected since

$$\overline{\boldsymbol{L}}\,\overline{\boldsymbol{D}} = \overline{\boldsymbol{D}}\,\overline{\boldsymbol{L}} = \overline{\boldsymbol{I}}. \tag{7.331}$$

An early pioneer of model-based predictive control (MBPC) was Jacques Richalet. Various forms were created, for example 'model predictive heuristic control' and 'model algorithmic control' (MAC). IDCOM (Richalet *et al.*, 1978) combined model identification with control, and was based on an impulse response model as above.

7.8.3
Approaches to the Optimisation of Control Action Trajectories

The various formulations of model predictive control all have in common the need of an algorithm which will find the optimal combination of the input variation and resultant output trajectory in the future. Actually, this problem of determining a future control sequence is not restricted to closed-loop predictive control. The developments to be discussed in this section also have relevance in open-loop situations where a future strategy is predetermined, possibly offline, for implementation in such situations as optimal batch reactor sequencing, or transitions in feed quality or product grade.

The model forms used in predictive control vary greatly (e.g. nonlinear DAE, artificial neural networks, state space, step response), and the objective function and constraints used in the optimisation can be equally varied. For MPC, the optimisation solution must execute reliably and robustly in real time. It must also suit the form of model used. The unconstrained MPC in Sections 7.8.1

and 7.8.2, based on linear models, had explicit solutions for the optimal control trajectory. For the constrained linear system in Section 7.8.2.1, a solution close to the optimal solution was achieved through linear programming (LP). The effectiveness of this solution has been examined by Rao and Rawlings (2000). A lot of attention has been paid to the path optimisation for MPC. Some of the developments are outlined below.

7.8.3.1 Some Concepts Used in Predictive Control Optimisation

So far only discrete predictive control problems have been considered. It is useful now to take a more general view of the problem. In the most general case, the model might be presented in the form of a set of implicit relationships

$$\boldsymbol{f}(t, \dot{\boldsymbol{x}}(t), \boldsymbol{x}(t), \boldsymbol{y}(t), \boldsymbol{u}(t)) = \boldsymbol{0} \tag{7.332}$$

where $\boldsymbol{x}$ is a vector of states, $\boldsymbol{u}$ the vector of control action inputs and $\boldsymbol{y}$ algebraic or output variables related to the states. Indeed, the model might only be available in a 'black box' form, such as a commercial dynamic flowsheeting package. A semi-explicit form of the relationships in Equation 7.332 is the set of differential and algebraic equations (DAEs)

$$\dot{\boldsymbol{x}}(t) = \boldsymbol{f}(\boldsymbol{x}(t), \boldsymbol{y}(t), \boldsymbol{u}(t)) \tag{7.333}$$

$$\boldsymbol{g}(\boldsymbol{x}(t), \boldsymbol{y}(t), \boldsymbol{u}(t)) = \boldsymbol{0} \tag{7.334}$$

Nonlinear equations in the DAE form are the typical result of modelling chemical processes, and thus most workers aim to accommodate these in their approaches to control trajectory optimisation. A general objective function for the period $t_0 < t < t_f$ has the form

$$J(\boldsymbol{u}(\bullet)) = E(\boldsymbol{x}(t_f)) + \int_{t_0}^{t_f} F(\boldsymbol{x}(t), \boldsymbol{u}(t))\mathrm{d}t \tag{7.335}$$

where $\boldsymbol{u}(\bullet)$ represents the chosen functional variation of $\boldsymbol{u}$ in the interval $t_0 < t < t_f$. The extra terminal penalty term $E(\boldsymbol{x}(t_f))$ effectively creates an upper bound of the finite horizon cost, provided the final state $\boldsymbol{x}(t_f)$ is constrained, say using a set of bounds $\boldsymbol{r}(\boldsymbol{x}(t_f)) \geq \boldsymbol{0}$ (Findeisen *et al.*, 2002b; Biegler, 2000; Chen and Allgöwer, 1998). This has the advantage that shorter horizons can be used without degrading performance. The optimisation problem for the control trajectory in the interval $t_0 < t < t_f$ can then be described by

$$\min_{\boldsymbol{u}(\bullet)} J(\boldsymbol{u}(\bullet)) \tag{7.336}$$

such that

$$\dot{\boldsymbol{x}}(t) = \boldsymbol{f}(\boldsymbol{x}(t), \boldsymbol{y}(t), \boldsymbol{u}(t)) \tag{7.337}$$

and the following equality and inequality constraints are obeyed:

$$\boldsymbol{g}(\boldsymbol{x}(t), \boldsymbol{y}(t), \boldsymbol{u}(t)) = \boldsymbol{0} \tag{7.338}$$

$$\boldsymbol{h}(\boldsymbol{x}(t), \boldsymbol{y}(t), \boldsymbol{u}(t)) \geq \boldsymbol{0} \tag{7.339}$$

$$\boldsymbol{r}(\boldsymbol{x}(t_f)) \geq \boldsymbol{0} \tag{7.340}$$

In the simplest case, the inequality constraints are box constraints of the form

$$\boldsymbol{x}_{\min} \leq \boldsymbol{x}(t) \leq \boldsymbol{x}_{\max} \tag{7.341}$$

$$\boldsymbol{y}_{\min} \leq \boldsymbol{y}(t) \leq \boldsymbol{y}_{\max} \tag{7.342}$$

$$\boldsymbol{u}_{\min} \leq \boldsymbol{u}(t) \leq \boldsymbol{u}_{\max} \tag{7.343}$$

$$x_{\mathrm{f\,min}} \leq x(t_{\mathrm{f}}) \leq x_{\mathrm{f\,max}} \tag{7.344}$$

and typical forms of functions E and F are quadratic functions with diagonal weighting matrices

$$E(\boldsymbol{x}(t_{\mathrm{f}})) = \{\boldsymbol{x}(t_{\mathrm{f}}) - \boldsymbol{x}_{\mathrm{S}}\}^{\mathrm{T}}\boldsymbol{Q}_{\mathrm{f}}\{\boldsymbol{x}(t_{\mathrm{f}}) - \boldsymbol{x}_{\mathrm{S}}\} \tag{7.345}$$

$$F(\boldsymbol{x}(t), \boldsymbol{u}(t)) = \{\boldsymbol{x}(t) - \boldsymbol{x}_{\mathrm{S}}\}^{\mathrm{T}}\boldsymbol{Q}\{\boldsymbol{x}(t) - \boldsymbol{x}_{\mathrm{S}}\} + \{\boldsymbol{u}(t) - \boldsymbol{u}_0\}^{\mathrm{T}}\boldsymbol{R}\{\boldsymbol{u}(t) - \boldsymbol{u}_0\} \tag{7.346}$$

By defining an extra state variable with the equation

$$\frac{\mathrm{d}x_{n+1}}{\mathrm{d}t} = F(x(t), u(t)) \tag{7.347}$$

and initialising

$$x_{n+1}(t_0) = 0 \tag{7.348}$$

the objective function in Equation 7.335 becomes

$$J(\boldsymbol{u}(\bullet)) = E(\boldsymbol{x}(t_{\mathrm{f}})) + x_{n+1}(t_{\mathrm{f}}) \tag{7.349}$$

With this inclusion of x_{n+1}, Equations 7.336–7.344 of the optimisation problem are now in 'standard form'. Solutions are possible following the *maximum principle* of Pontryagin *et al.* (1961, 1969) by solving a two-point boundary value problem which requires integration of differential equations for both the state and the adjoint vector. However, this chapter will follow the common numerical approaches for the nonlinear problem which are based on dynamic programming. In order to proceed with these common approaches, one needs to give some further description to the function $\boldsymbol{u}(t)$. This is done by *control variable parameterisation* (CVP). Three examples of CVP are given in Figure 7.26.

$$\text{polynomial:}\quad \boldsymbol{u}(t) = \boldsymbol{u}_0 + \boldsymbol{u}_1 t + \boldsymbol{u}_2 t^2 + \cdots + \boldsymbol{u}_N t^N \quad \text{for} \quad t_0 < t < t_{\mathrm{f}} \tag{7.350}$$

$$\text{linear:}\quad \boldsymbol{u}(t) = \boldsymbol{u}_{i-1} + \frac{(t - t_0 - [i-1]T)}{T}[\boldsymbol{u}_i - \boldsymbol{u}_{i-1}] \quad \text{for} \quad [i-1]T \leq t - t_0 \leq iT, \quad i = 1, \ldots, N \tag{7.351}$$

$$\text{step:}\quad \boldsymbol{u}(t) = \boldsymbol{u}_{i-1} \quad \text{for} \quad [i-1]T \leq t - t_0 < iT, \quad i = 1, \ldots, N \tag{7.352}$$

In parameterisations such as Equations 7.350–7.352, finding $\boldsymbol{u}(t)$ amounts to finding the coefficient vectors $\boldsymbol{u}_0$, $\boldsymbol{u}_1$, $\boldsymbol{u}_2$, . . . , $\boldsymbol{u}_N$. In scheduling problems, where a series of switch points must be found, parameterisation can instead be based on the variable intervals *between* switchings. For example, Fikar, Chachuat and Latifi (2005) avoided an *integer programming* solution for on/off values at a sequence of time instants, by rather solving for the optimal time interval sizes.

Assuming that CVP has been used to define $\boldsymbol{u}(t)$ by means of, say, $N \times m$ coefficients (where m is the size of the control vector), then the control trajectory optimisation problem of Equations 7.336–7.344 amounts to finding these coefficients for the interval $t_0 < t < t_{\mathrm{f}}$. The differential

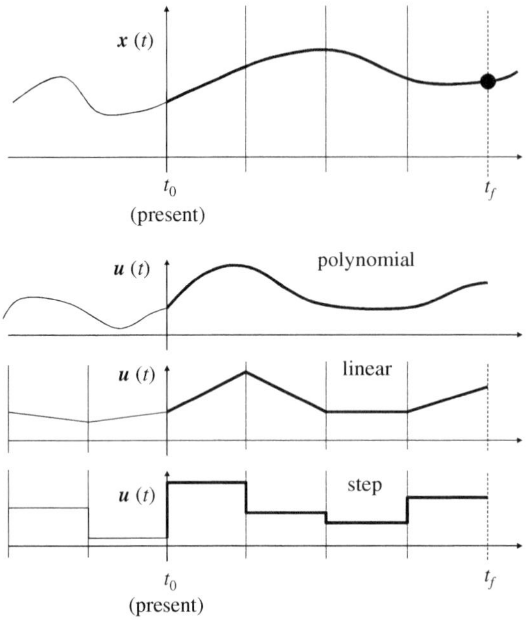

Figure 7.26 Examples of control variable parameterisation.

equation (Equation 7.337) could be written as a further equality constraint relating the state values at some small interval Δt. The entire set of equations, with all of the unknown intermediate values of $\boldsymbol{x}$ and $\boldsymbol{y}$, could then be loaded into a commercial *nonlinear programming* (NLP) solver (e.g. CPLEX®), and the $N \times m$ coefficients thus obtained that minimise J. This is referred to as the *simultaneous approach*, and will clearly result in a very large constrained NLP problem (Biegler, Cervantes and Wächter, 2002). However, more efficient solutions can be developed by taking advantage of the structure of the problem. In particular, a stage-wise solution, the so-called *sequential approach*, could make use of *dynamic programming* techniques.

The CVPs based on a series of intervals of size T are very common, because the sequence of 'stages' so defined, and the definition of the state at each stage, lends itself easily to path optimisation using Bellman's *principle of optimality* (Section 7.6.5). This will be the focus of the next sections.

In order to use Bellman's principle, namely that an optimal path that intersects another optimal path must continue along the latter, some thought needs to be given to the integration of the nonlinear differential equations (Equation 7.337). In general, the stage-wise solution requires one to work backwards from the final stage, evaluating the best path from any state value in that stage, to the final stage. Figure 7.27 represents a SISO case where it is desired to find an optimal sequence of input $\boldsymbol{u}$ values. In this example, the $\boldsymbol{u}$ values can be considered to be held constant in each interval, as in the CVP equation (Equation 7.352). The differential equation is solved in the forward direction across the interval. One can start at a chosen state $\boldsymbol{x}$ value, but one obviously has no control over the destination. Instead, one experiments with a range of input $\boldsymbol{u}$ values, to get a range of intermediate destinations. In the figure, only two choices of $\boldsymbol{u}$ are shown in each case. The net increase in the penalty incurred by each path is shown in lighter print on each trajectory. This idea

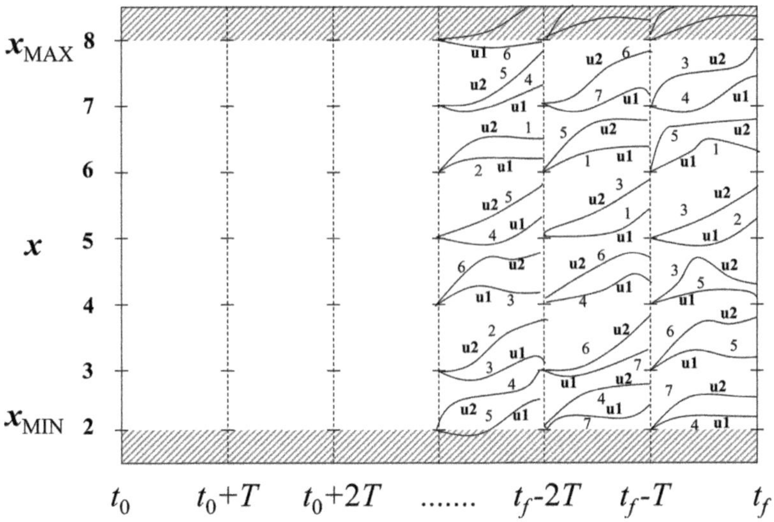

Figure 7.27 Multiple shooting of interval trajectories by integrating the state equation.

of launching test trajectories from chosen state values at chosen time intervals is called *multiple shooting*. In the next sections, it will be seen to be useful in the search for the optimal control trajectory. A complication is that an arbitrarily chosen starting state has to comply with the algebraic equality in Equation 7.334.

7.8.3.2 Direct Multiple Shooting

Findeisen *et al.* (2002a) following Bock *et al.* (2000) describe a formulation for the input to a NLP solver which they call 'direct multiple shooting'. It will be seen that this formulation is a combination of the *simultaneous* and *sequential* approaches referred to in Section 7.8.3.1. They decouple the solution of the DAE from the optimisation by defining variable vectors $\boldsymbol{X}_i$ and $\boldsymbol{Y}_i$ for the states and algebraic variables at the end of each stage i (Figure 7.28). Thus, $\boldsymbol{X}_0$ and $\boldsymbol{Y}_0$ are values at t_0 and $\boldsymbol{X}_N$ and $\boldsymbol{Y}_N$ are values at t_f. It is understood in the following that all variables represent modelling and optimisation values, except for $\boldsymbol{x}_P(t_0)$ and $\boldsymbol{y}_P(t_0)$ which are the actual process values at t_0. During each stage i, a stepwise constant control vector $\boldsymbol{U}_i$ is applied. The relaxed $[\boldsymbol{\alpha}_i(t)\rightarrow\boldsymbol{I}]$ DAE formulation is thus

$$\boldsymbol{x}(iT) = \boldsymbol{X}_i \qquad \text{for} \quad i = 1, \ldots, N \tag{7.353}$$

$$\boldsymbol{y}(iT) = \boldsymbol{Y}_i \qquad \text{for} \quad i = 1, \ldots, N \tag{7.354}$$

$$\dot{\boldsymbol{x}}(t) = \boldsymbol{f}(\boldsymbol{x}(t), \boldsymbol{y}(t), \boldsymbol{U}_i) \qquad \text{for} \quad i = 1, \ldots, N \quad \text{and} \quad (i-1)T \leq t \leq iT \tag{7.355}$$

$$\boldsymbol{0} = \boldsymbol{g}(\boldsymbol{x}(t), \boldsymbol{y}(t), \boldsymbol{U}_i) - \boldsymbol{\alpha}_i(t)\boldsymbol{g}(\boldsymbol{X}_i, \boldsymbol{Y}_i, \boldsymbol{U}_i) \quad \text{for} \quad i = 1, \ldots, N \quad \text{and} \quad (i-1)T \leq t \leq iT \tag{7.356}$$

$$J_i(\boldsymbol{X}_i, \boldsymbol{Y}_i, \boldsymbol{U}_i) = \int_{t_0+(i-1)T}^{t_0+iT} F(\boldsymbol{x}(t), \boldsymbol{U}_i)\mathrm{d}t \quad \text{for} \quad i = 1, \ldots, N \tag{7.357}$$

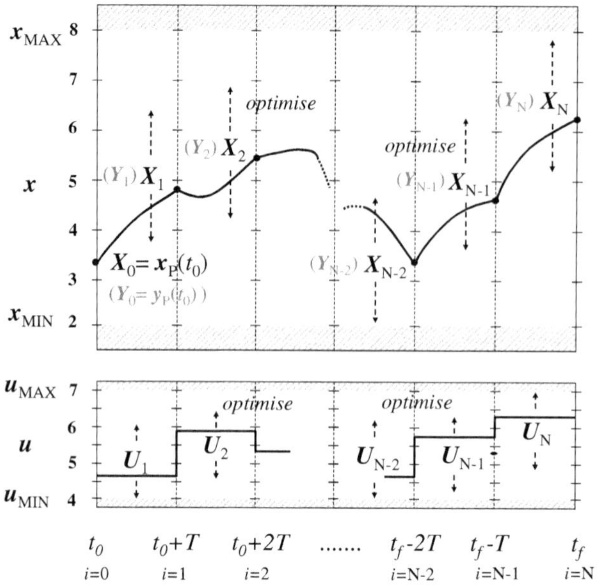

Figure 7.28 Posing of NLP problem in terms of X_i, Y_i and U_i by decoupling DAEs in shooting intervals.

The requirement is

$$\min_{X_i, Y_i, U_i, i=1,\ldots,N} \left\{ \sum_{i=1}^{N} J_i(X_i, Y_i, U_i) + E(X_N) \right\} \tag{7.358}$$

such that

$$x(t_0) = X_0 = x_P(t_0) \tag{7.359}$$

$$y(t_0) = Y_0 = y_P(t_0) \tag{7.360}$$

$$g(X_i, Y_i, U_i) = 0 \quad \text{for} \quad i = 1,\ldots,N \tag{7.361}$$

$$h(X_i, Y_i, U_i) \geq 0 \quad \text{for} \quad i = 1,\ldots,N \tag{7.362}$$

$$r(X_i) \geq 0 \quad \text{for} \quad i = 1,\ldots,N \tag{7.363}$$

In Equations 7.353–7.357, it is noted that the algebraic relations are not properly satisfied until the final solution is converged to satisfy Equations 7.359–7.360. This mechanism was used by Bock *et al.* (2000) to provide initial values in the DAE solution. The DAEs are clearly only being solved as 'shootings' inside the stage segments, for the iterative choices of X, Y on the left stage boundary and U. The integration equation is obviously inserted by the user, for example a small-step Euler. A specially tailored *partially reduced* SQP algorithm was used to solve this NLP problem for X_i, Y_i, U_i, $i = 1, \ldots, N$. It is noted that the constraints are only complied with at the stage nodes, that is there is a minor risk of noncompliance between nodes. Findeisen *et al.* (2002a) note that the repeated solution in an MPC application can be made very efficient by reuse of most results from the solution on the previous time step, for example the previous trajectory, offset by one step, can be used to create the initial values for the NLP iteration.

7.8.3.3 Interior Point Method and Barrier Functions

In a numerical solution for the optimal control trajectory, the presence of constraints on state, algebraic or control variables adds a significant difficulty. It is not difficult to program, say, a gradient search method for the minimum value of an objective function $J(\boldsymbol{u}(\bullet))$, but special measures must be taken once the search reaches a constraint. For the *simultaneous* approach, in which the coefficients of a control trajectory are simply variables in the NLP problem specification, some success has been achieved by replacing the inequality constraints with *barrier functions.* As a search approaches such a constraint, penalties increase dramatically, so the search path hopefully veers away in a more productive direction. Clearly, the search must start at an *interior point.* There are a number of refinements of the method, detailed by Biegler, Cervantes and Wächter (2002) or Bell and Sargent (2000), but only the basic idea will be outlined here.

Recall the problem specification in Equations 7.336–7.349:

$$J(\boldsymbol{u}(\bullet)) = E(\boldsymbol{x}(t_\mathrm{f}) + x_{n+1}(t_\mathrm{f})) \tag{7.364}$$

where

$$\frac{\mathrm{d}x_{n+1}}{\mathrm{d}t} = F(\boldsymbol{x}(t), \boldsymbol{u}(t)) \tag{7.365}$$

$$x_{n+1}(t_0) = 0 \tag{7.366}$$

and it is required to find

$$\min_{\boldsymbol{u}(\bullet)} J(\boldsymbol{u}(\bullet)) \tag{7.367}$$

such that

$$\dot{\boldsymbol{x}}(t) = \boldsymbol{f}(\boldsymbol{x}(t), \boldsymbol{y}(t), \boldsymbol{u}(t)) \tag{7.368}$$

$$\boldsymbol{g}(\boldsymbol{x}(t), \boldsymbol{y}(t), \boldsymbol{u}(t)) = \boldsymbol{0} \tag{7.369}$$

$$\boldsymbol{h}(\boldsymbol{x}(t), \boldsymbol{y}(t), \boldsymbol{u}(t)) \geq \boldsymbol{0} \tag{7.370}$$

$$\boldsymbol{r}(\boldsymbol{x}(t_\mathrm{f})) \geq \boldsymbol{0} \tag{7.371}$$

The inequality constraints are now transformed using slack variables $\boldsymbol{H}(t)$ and $\boldsymbol{R}_\mathrm{f}$:

$$\boldsymbol{h}(\boldsymbol{x}(t), \boldsymbol{y}(t), \boldsymbol{u}(t)) - \boldsymbol{H}(t) = \boldsymbol{0} \tag{7.372}$$

$$\boldsymbol{r}(\boldsymbol{x}(t_\mathrm{f})) - \boldsymbol{R}_\mathrm{f} = \boldsymbol{0} \tag{7.373}$$

Then, if, for example, the same type of trajectory discretisation as in Figure 7.28 were used, the values of the slack variable vectors at their nodes are $\boldsymbol{H}_1, \boldsymbol{H}_2, \ldots, \boldsymbol{H}_N$ and $\boldsymbol{R}_N$. The objective function (Equation 7.364) is then augmented as

$$J(\boldsymbol{U}_i, i = 1, N) = E(\boldsymbol{X}_N) + X_{N,n+1} - \sum_{i=1}^{N} \sum_{j=1}^{n_H} \mu \ln\{H_{ij}\} - \sum_{j=1}^{n_R} \mu \ln\{R_{Nj}\} \tag{7.374}$$

in which the first index on the scalars refers to the stage, and the second index refers to the element of the corresponding vector. As any H_{ij} or R_{Nj} approaches zero, the logarithms become increasingly large negative values, giving J a sharp increase. Thus, the search will be driven away from the bound (where H_{ij} or R_{Nj} would be zero). The parameter μ is an important search control variable, as it determines how far into the feasible region (the *interior*) the slope of the logarithm will extend. A search is started with a 'larger value' which is then successively reduced on each iteration. In the limit, for a very small μ value, it should be possible to find an optimal solution lying arbitrarily close to a constraint.

For the purpose of the interior search, Lagrange multipliers $\boldsymbol{\lambda}$ can be used as in Section 6.3 in order to incorporate the remaining equality constraints. Let

$$\boldsymbol{G}_i = \boldsymbol{g}(\boldsymbol{X}_i, \boldsymbol{Y}_i, \boldsymbol{U}_i) \tag{7.375}$$

Then it is required to minimise the objective function

$$J'(\boldsymbol{U}_i, \boldsymbol{\lambda}_{Gi}, \boldsymbol{\lambda}_{Hi}, i = 1, \ldots, N, \boldsymbol{\lambda}_R) = J(\boldsymbol{U}_i, i = 1, N) + \sum_{i=1}^{N} \boldsymbol{\lambda}_{Gi}^{\mathrm{T}} \boldsymbol{G}_i + \sum_{i=1}^{N} \boldsymbol{\lambda}_{Hi}^{\mathrm{T}} \boldsymbol{H}_i + \boldsymbol{\lambda}_R^{\mathrm{T}} \boldsymbol{R}_N \tag{7.376}$$

Thus, the elements of all of the vectors $\boldsymbol{U}_i$, $\boldsymbol{\lambda}_{Gi}$, $\boldsymbol{\lambda}_{Hi}$, $i = 1,\ldots, N$, $\boldsymbol{\lambda}_R$ must be found that minimise J'. At this point, the derivative of J' with respect to each element of the $\boldsymbol{\lambda}$ vectors must be zero, which means the associated constraint term is zero, that is one is *on* the equality constraint. Bearing in mind that the $\boldsymbol{G}_i$, $\boldsymbol{H}_i$ and $\boldsymbol{R}_N$ vectors will depend on the $\boldsymbol{U}_i$ choices, minimisation with respect to the $\boldsymbol{U}_i$ elements ensures that the constraint is tangential to the J contour, that is it is passing through the minimum of J on its route.

7.8.3.4 Iterative Dynamic Programming

Luus (1990) set out to solve for the optimal control trajectory by dynamic programming with the aim of overcoming the following problems:

- Poor resolution in a grid such as Figure 7.27, and the huge size of NLP problems if the resolution is improved.
- The problem that solutions often arrive at local optima and the user is left unaware of a better globally optimal solution.
- The need to solve the system equations (such as DAEs) in the forward direction, with stage terminations that will in general not arrive at state grid points.

In earlier work (de Tremblay and Luus, 1989), the method used to deal with the lack of an established optimal path from the exact landing point of a stage shooting trajectory was to simply take the 'nearest' optimal path from there on. This idea is illustrated in the extract in Figure 7.29 from the *fixed grid* shootings in Figure 7.27, where only two different control settings are tried at each $\boldsymbol{x}$ node X_{ij}. Working backwards from time N, the best path (lowest total cost) from each $\boldsymbol{x}$ position is marked in a heavier solid line. The total cost of the best path from the $N-2$ positions is shown in parentheses. Only one optimal path is shown from the $N-3$ positions, with its total cost in double parentheses. That requires the sequence $\boldsymbol{u1}$, $\boldsymbol{u1}$, $\boldsymbol{u1}$ in the last three stages. It happens in this case that there is an equal optimal path $\boldsymbol{u2}$, $\boldsymbol{u1}$, $\boldsymbol{u1}$ also with a total cost of ((7)). The choice of the 'nearest' grid point for multiple-value states (i.e. state vectors) is typically based on the Euclidean norm distance

$$d = \sqrt{\left(\boldsymbol{x}(iT) - \overline{\overline{\boldsymbol{X}}}_i\right)^{\mathrm{T}} \boldsymbol{W} \left(\boldsymbol{x}(iT) - \overline{\overline{\boldsymbol{X}}}_i\right)} \tag{7.377}$$

which includes a diagonal weighting matrix $\boldsymbol{W}$ to give comparative scales to the state variables. In Equation 7.377, $\boldsymbol{x}(iT)$ is the end of a shooting trajectory across stage i and $\overline{\overline{\boldsymbol{X}}}_i$ represents the

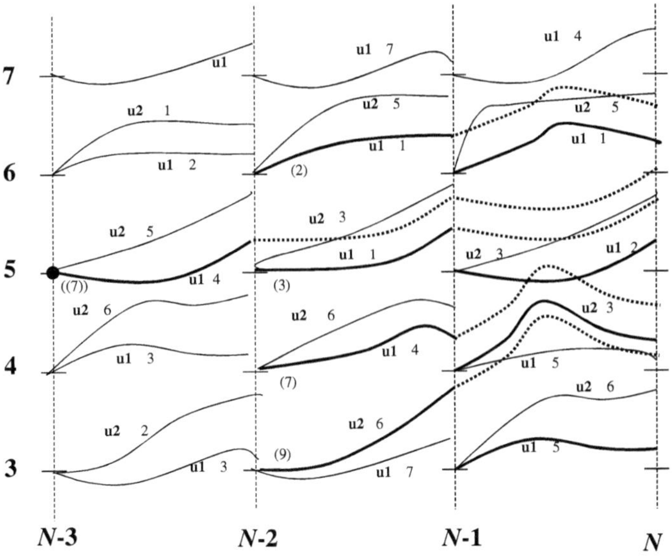

Figure 7.29 Optimal paths based on later optimal paths in a fixed grid by choice of 'nearest' position to the landing point.

collection of state grid points for stage i. So if there are n state variables, and each is discretised into 10 values for the grid, then $\overline{\overline{X}}_i$ represents 10^n individual points.

It is obvious that a multivariable problem will require a huge number of trajectory evaluations, because the control variable vector, size m, must also be discretised to provide the experimental shootings. Say that these variables were also discretised into 10 values. Then the total number of shootings required is $N \times 10^n \times 10^m$.

The approach used by Luus (1990) to overcome this *curse of dimensionality* was to restrict the region of attention closer to the operating range, so that the number of points in the grids could be reduced but resolution improved. Grids $\overline{\overline{X}}_i$ and $\overline{\overline{U}}_i$ for the terminal state X_i and control $\boldsymbol{U}_i$ for each stage i were *centred* on the trajectory of the last solution or iteration. Thus, the optimal control trajectory from the last controller time step is the middle reference point for an individual grid at each stage i (Figure 7.30). In this way, the solution could start with coarser grids spanning a large range, and hopefully embracing the global optimum. Then with each iteration the scale of the grids is shrunk so that the definition of the optimal path is refined.

If each of the n state variables x_i is discretised into p values covering the full range of interest, then they can be represented as the p deviations from a mean $x'_{ij} = x_{ij} - \overline{x}_i$, $j = 1, \ldots, p$. For the full set of states, this will define p^n points which are held in the tensor $\overline{\overline{X}}$ (multidimensional array). Likewise the q^m values representing discrete deviations from the means of the $\boldsymbol{u}$ elements are held in $\overline{\overline{\boldsymbol{U}}}$. For the first solution, the N grids $\overline{\overline{X}}_i$, $i = 1, \ldots, N$, and the N grids $\overline{\overline{\boldsymbol{U}}}_i$, $i = 1, \ldots, N$, are obtained by simply adding back the mean of the range:

$$\overline{\overline{X}}_i = \boldsymbol{I}\,\overline{\boldsymbol{x}} + \overline{\overline{X}}, \quad i = 1, \ldots, N \tag{7.378}$$

$$\overline{\overline{\boldsymbol{U}}}_i = \boldsymbol{I}\,\overline{\boldsymbol{u}} + \overline{\overline{\boldsymbol{U}}}, \quad i = 1, \ldots, N \tag{7.379}$$

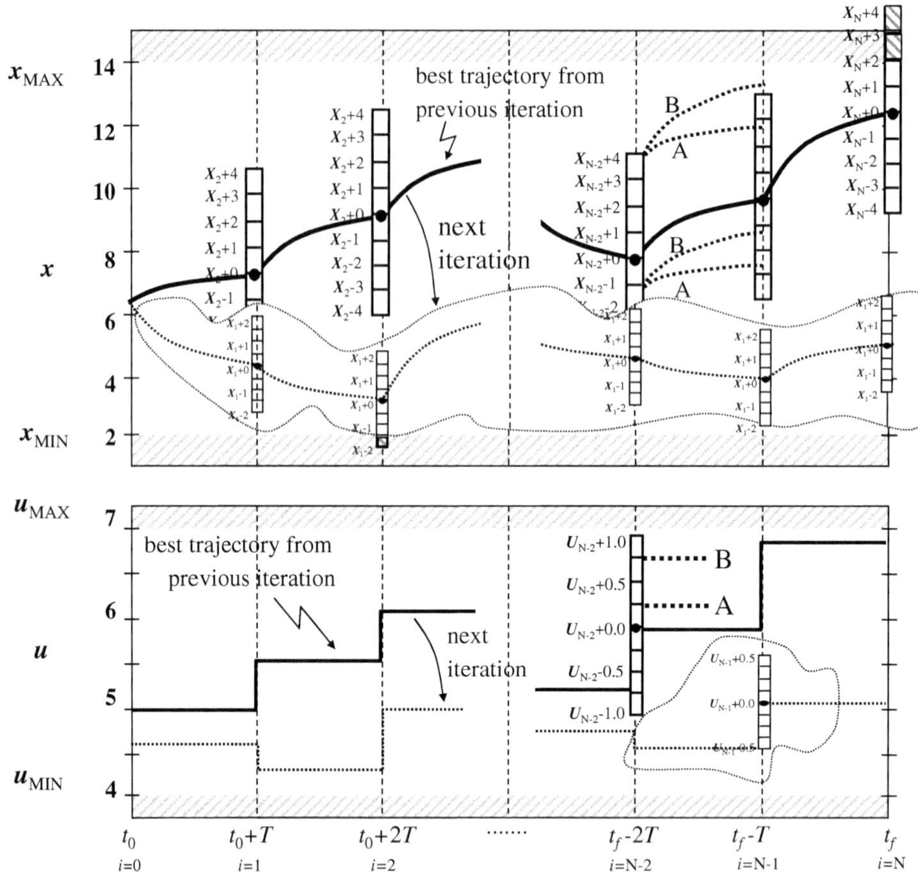

Figure 7.30 Grids centred on last optimal trajectory and shrinking with iterations for IDP.

so that the whole range of interest is covered. After the first solution, a sequence of points for the preceding optimal state trajectory X_i, $i = 1, \ldots, N$, and a sequence for the matching optimal control trajectory U_i, $i = 1, \ldots, N$, will be available on each iteration, so the grids are progressively shrunk around the trajectories as follows:

$$\overline{\overline{X}}_i = IX_i + (1 - \varepsilon)\overline{\overline{X}}, \quad i = 1, \ldots, N \tag{7.380}$$

$$\overline{\overline{U}}_i = IU_i + (1 - \varepsilon)\overline{\overline{U}}, \quad i = 1, \ldots, N \tag{7.381}$$

where ε is, say, 0.1 for a 10% shrinkage of the span on each pass. For every new definition of the grids, it is of course necessary to recalculate all of the $N \times p^n \times q^m$ shootings so that the optimal path can be established in the same way as Figure 7.29 by working backwards from the last stage, bearing in mind that the grids will now be (slightly) displaced from each other.

Bojkov and Luus (1994) propose a variation of the iterative dynamic programming (IDP) method for *minimisation of the time* required to bring the state close to a desired point, continuing with a method *suitable for high-dimensional* systems (Bojkov and Luus, 1995). Both of these approaches generate alternative shootings in a stage using randomly generated control variations rather than a grid of control values. Moreover, these shootings are done only *at one state point* rather than a grid of state points. This is an important saving (and approximation) when there are a large number of states. This method effectively uses the control trajectory variations to provide some exploration of the state space. It would need to start with a reasonable estimate of the control trajectory (Figure 7.31).

Figure 7.31 depicts the IDP method of Luus (1996) for high-dimensional systems. Instead of using the step CVP, Luus uses the linear interpolation across each stage according to Equation 7.351. Once the first control trajectory $\boldsymbol{U}_i$, $i = 1, \ldots, N$, has been guessed, a series of 'passes' is done, and within each pass a series of iterations is done.

On each iteration, the range of the control variable is reduced to a smaller region around the previous trajectory (darkest line) as in the original method above. Starting with the stage $N-1$ to N, random points are generated for $\boldsymbol{u}$ at the beginning and end of the last stage to give a series of interpolated $\boldsymbol{u}$ variations across the stage. The one with the least cost (A:1) is noted and used to replace both $\boldsymbol{U}_{N-1}$ and $\boldsymbol{U}_N$. Moving back one stage, random controls are generated

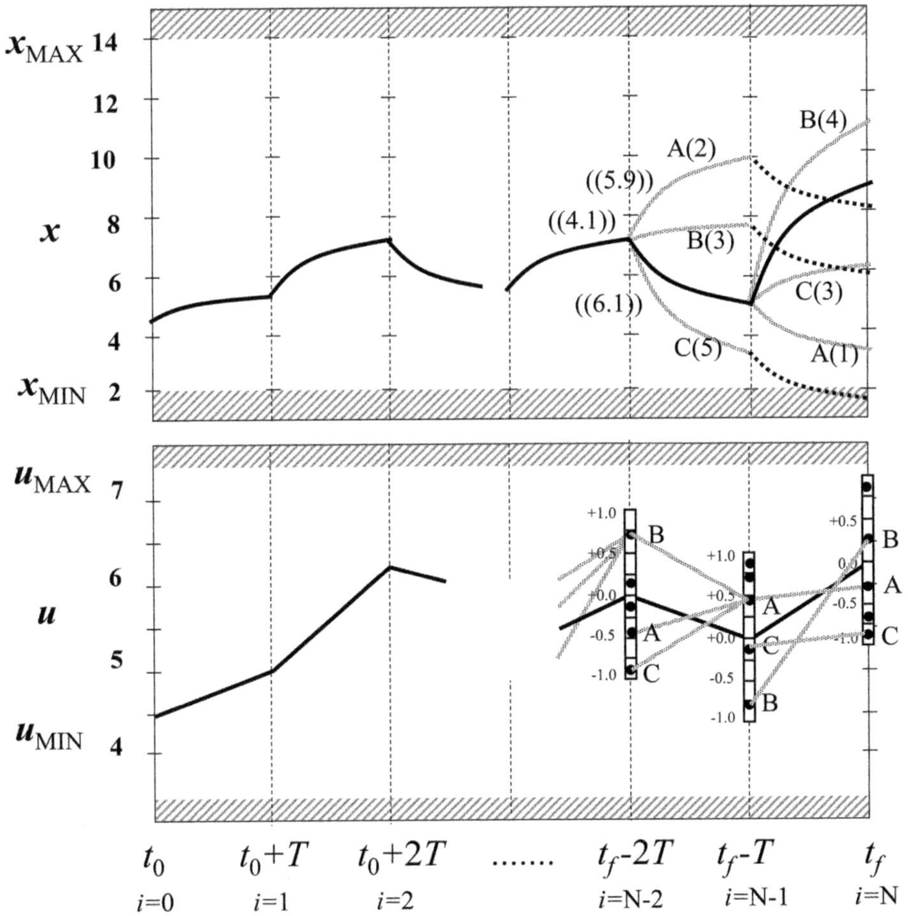

Figure 7.31 IDP method for larger number of states with shootings based on random control generation.

around U_{N-2} and interpolated to the recently updated U_{N-1} value for a series of shootings. The shootings do not stop at the end of the stage, but carry on to N using the recently updated U_i values in the last stage. The best overall path (B:4.1) is noted, and the U_{N-2} value is updated accordingly. Note that this path cost will not be exactly $3+1$ because the integration in the subsequent stage does not start at the original state. This procedure is repeated until the first stage is reached and a new U_0 is chosen. Now one replaces all of the X_i, $i=1, \ldots, N$, state values with those that arise using the fully updated U_i, $i=1, \ldots, N$. Then the current range in which the u values are randomised is reduced (say, by multiplication by 0.95), and the next iteration is conducted.

The iterations are repeated a set number of times before another 'pass' is initiated. This begins with the latest updated U and X trajectories, but not with the *original* range for u randomisation. Instead, the starting range for each pass is a progressively reduced (say, by multiplication by 0.90) version of the first pass range. This effectively re-expands the u randomisation range in effect at the end of the last pass. Then the set number of iterations is repeated with range reduction again. The U and X trajectories would have moved by now, so one hopes to detect other influences on the optimal trajectory in this way. The passes and iterations are continued thus until convergence.

7.8.3.5 Forward Iterative Dynamic Programming

One notes that the *dynamic programming* approaches used so far have started at the last stage $i=N$, and have ensured each time the preceding stage $i-1$ is considered that a record is kept of the best trajectory 'to the end' from positions at the current stage. In a *fixed grid* approach as depicted in Figures 7.27 and 7.29, it would not matter in which order all of the shootings are conducted. Once that is done, it is irrelevant in which direction time runs. One could start at the first stage in assembling optimal trajectories and proceed to the last. However, in order to take advantage of Bellman's principle, one would need in the same way to keep a record of the best trajectories up to all of the state positions at the current stage. A slight disadvantage if one moved in the *forward* direction is that divergence of the discrete paths might make it harder to address an important final state penalty $E(\boldsymbol{x}(t_f))$ as in Equation 7.335.

The *(backward) iterative dynamic programming* discussed in Section 7.8.3.4 does not appear to offer the possibility of such a *forward* solution. A single state position is used for each stage, and the integrations for all of the trial control values are pursued along an updated control trajectory right to the end. However, Lin and Hwang (1996) introduced the idea of *forward iterative dynamic programming* (FIDP) claiming two advantages:

a) (B)IDP cannot recognise control $\boldsymbol{u}$ or state $\boldsymbol{x}$ *time-delay* effects arising from before the stage under consideration (Figure 7.32), whereas FIDP can properly integrate these all of the way from $\boldsymbol{x}(t_0)$. Those from before t_0 could be obtained from the preceding controller time step and likewise included.
b) In working backwards, (B)IDP is unaware whether the new state positions it is generating in its shootings in a stage (i.e. at the end of that stage) are indeed accessible from $\boldsymbol{x}(t_0)$. Conversely, FIDP can ensure this by always integrating from t_0, so no computational effort is wasted.

Lin and Hwang (1998) presented an improved method referred to as 'forward–backward' dynamic programming which is outlined below. Their method is based on a piecewise-constant (step) form of the control trajectory. The procedure regarding iterations and reduction of the control grid is similar to that for IDP in Section 7.8.3.4. However, in order to have available at each

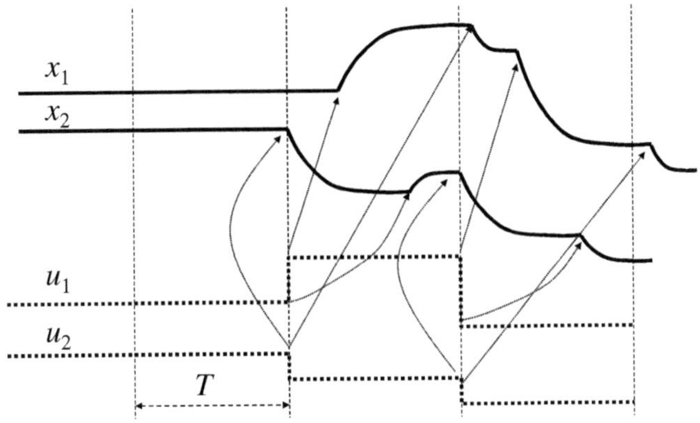

$$\frac{dx_1(t)}{dt} = -x_1(t) + 0.1x_2(t-0.8T) + 0.5u_1(t-0.3T) - 0.3u_2(t-1.1T)$$

$$\frac{dx_2(t)}{dt} = -0.05x_1(t-0.2T) - 0.8x_2(t) + 0.2u_1(t-0.7T) + 1.4u_2(t)$$

Figure 7.32 Example of a process with input and state time delays.

new state node the delayed control and state contributions, all state trajectories have to be built from the single node $\boldsymbol{x}(t_0)$. The illustration in Figure 7.33 shows an evenly distributed control grid, but these could equally be random samples from the present range as in Section 7.8.3.4.

Starting from $\boldsymbol{x}(t_0)$, the control samples A, B, C and D produce the state shootings shown in the first stage. For *each of the state grid points nearest these landings*, a further set of state shootings can be done in the second stage, using the control samples A, B, C and D in the second stage. This is possible because the preceding delayed state and control contribution is available from the nearest landing itself. This procedure is repeated until the final stage. There will obviously be a considerable economy in that shootings from neighbouring state grid points at a stage will share many of the same landing grid points, and only grid points nearest landings need further trajectories, the rest being eliminated due to *inaccessibility*. Once all of the possible path segments are identified, the optimal control path from any state grid point is determined by working backwards from the last stage, as in Figure 7.29 and as in the method of Luus (1990) in Figure 7.30.

In Figure 7.33, control D would have been determined as the best control from the −2 position at $i = N-1$. For all of the possible paths from +2 at $i = N-2$, in combination with the optimal paths from the $N-1$ state positions, it happens that DD is the best strategy, so this is noted against the +2 state at $N-2$. Likewise, it happens that DDD is the best strategy from the +4 state at $i = 1$. The other state grid positions at $i = 1$ will have other 'best' strategies noted against them, such as BAC and so on. However, in this example the route from the single position at $i = 0$ through the +4 state at $i = 1$ has the lowest penalty, resulting in the overall solution ADDD. Once an optimal path is chosen for a state grid position at a particular stage, it is of course wise to re-evaluate the overall penalty for that path by integration to the last stage.

The FIDP method as described does not strictly need a moving $\boldsymbol{x}$ grid centred on the previous state trajectory, since only state grid points nearest the final state shootings from the preceding stage are continued – these could be in a fixed grid. As mentioned, a course grid will reduce the number of shootings to be performed in the next stage. On the other hand, it will be necessary to increase the resolution of the state grid as the solution progresses, but it could remain in a fixed

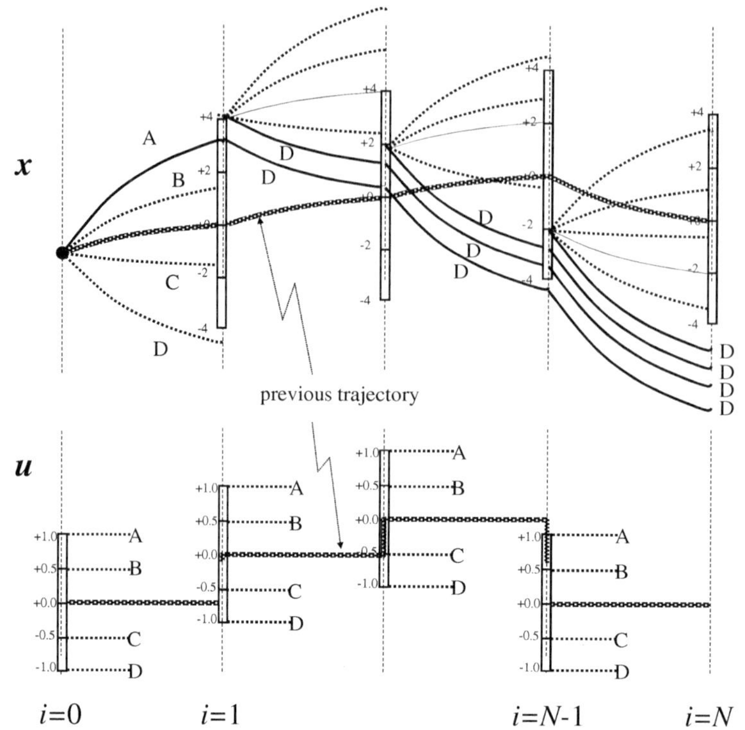

Figure 7.33 Depiction of FIDP method. State grid can be in a fixed position – only used points need be stored. State and control grid intervals are reduced with iterations to improve resolution.

position. Obviously, only *used* state positions need be stored. In summary then, the FIDP method of Lin and Hwang (1998) is similar to the IDP method of Luus (1990) depicted in Figure 7.30, except that the used multiple state positions at each stage are 'marked' by a range of control shootings in the preceding stage (during the *forward* pass), rather than being uniformly distributed on a shrinking, centred state grid. As in IDP, the control grid is centred on the previous trajectory, and is reduced on each iteration. All control grid points (or samples) must be considered for the shootings from every *used* state point at a stage.

7.8.3.6 Iterative Dynamic Programming Based on a Discrete Input–Output Model Instead of a State-Space Model

In closed-loop model predictive control, one will frequently aim to control a process in which not all of the states are observed. As noted, *dynamic programming* techniques require all of the states to be available in order to determine the future system behaviour based only on the future control. The control engineer has the option to construct a state observer. An alternative is to base the predictive control on a model relating the observable outputs directly to the inputs – a so-called *input–output* model. Such models are usually available in discrete form. An example of this might be an artificial neural network (ANN) model, or some other form arising from a 'black box' fitting technique (Section 6.5). Such models arise in process identification and are thus important in *adaptive control.* What they have in common is that the output vector $y(t)$ at the present time must be expressed in terms of a series of previous values of itself $\boldsymbol{y}(t-\Delta t)$, $\boldsymbol{y}(t-2\Delta t)$, ... as well as a series of previous values of the process input vector $\boldsymbol{u}(t-\Delta t)$, $\boldsymbol{u}(t-2\Delta t)$, ...

$$\boldsymbol{y}(t) = \boldsymbol{f}\{\boldsymbol{y}(t-\Delta t), \boldsymbol{y}(t-2\Delta t), \ldots, \boldsymbol{u}(t-\Delta t), \boldsymbol{u}(t-2\Delta t), \ldots\} \tag{7.382}$$

Note that Δt need not necessarily be the same as T, the controller (and stage) time step, and may be smaller. There is a strong connection here to the so-called *state-space models with input and/or output delays* considered in Section 7.8.3.5 and Figure 7.32 as candidates for FIDP by Lin and Hwang (1996). In fact, the additional delays there simply meant those models had missing states, namely the delayed contributions. A method for dealing with systems described by the general form of Equation 7.382 has been proposed by Rusnák et al. (2001), and this will be outlined below. The scheme is similar to the FIDP method described in Section 7.8.3.5 and Figure 7.33, but predefining, for each iteration, M variations of the control trajectory (and matching state trajectories), instead of a single control trajectory. It will be recalled that whole control and 'state' trajectories had to be available at the start of each iteration as a source of 'earlier inputs and "states" (outputs)' for predictions in subsequent stages.

The method proposed by Rusnák *et al.* (2001) is illustrated in Figure 7.34. The first iteration must start with a guessed control trajectory. The parameters of the control trajectory are perturbed by drawing random numbers from a range $\boldsymbol{r}$ centred on the last trajectory, to create a set of complete alternative control trajectories. All of these control trajectories are then integrated (actually 'recursed') to determine a set of output values $\boldsymbol{y}$ at the end on each stage. Then, starting in the last

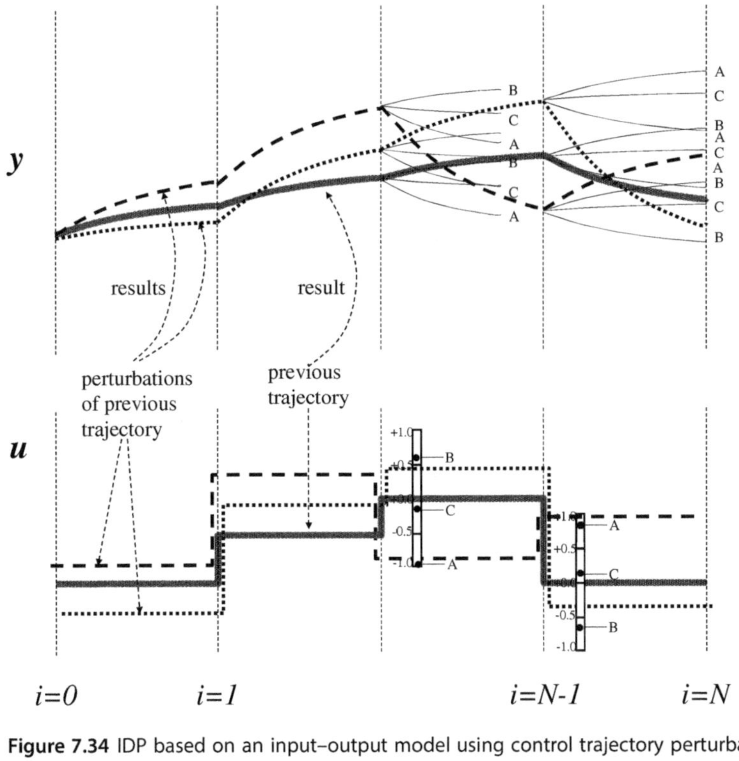

Figure 7.34 IDP based on an input–output model using control trajectory perturbations to create a distribution of state points.

stage, M variations of the last control are obtained by drawing random samples from a uniform distribution (again of range $\boldsymbol{r}$) centred on the last control of the previous iteration. For each output point at the $i=N-1$ stage, shootings are done using all M of the control random samples. The best control for each output point is noted. Now one moves back to the $i=N-2$ stage and repeats the process. However, the shootings are carried on to the end of the last stage. The control value used in the last stage is the previously determined optimal value for the nearest output point at which a shooting arrives at $i=N-1$. The 'distances' are determined by a (normalised) Euclidean norm as in Equation 7.377. The cost of each complete trajectory is obtained, and the best leaving each output point at $i=N-2$ is noted. In this way, the procedure is repeated moving back one stage each time until the best full trajectory is obtained from $i=0$ to $i=N$. This completes the iteration. Now the control randomisation range $\boldsymbol{r}$ is reduced (say by multiplying by 0.9), and the next iteration is started. Iterations are continued until the control trajectory converges within a desired tolerance.

The significant feature of the method of Rusnák *et al.* (2001) is that the individual output starting points at each stage have available their own histories $\boldsymbol{y}(\tau-\Delta t)$, $\boldsymbol{y}(\tau-2\Delta t)$, . . . , $\boldsymbol{u}(\tau-\Delta t)$, $\boldsymbol{u}(\tau-2\Delta t)$, . . . where τ is the effective stage time. These are of course obtained from the original set of complete randomised control trajectories obtained at the start of each iteration. In the case of $i=0$, these values are obtained from the preceding controller time steps.

Concerning the handling of constraints in this method, the following should be noted. Certainly, the randomly chosen input values $\boldsymbol{u}$ are just discarded and resampled should they lie outside of the $\boldsymbol{u}$ constraints. Likewise, the $\boldsymbol{u}$ values can be resampled should a resulting $\boldsymbol{y}$ value lie outside of the output constraints. But the presence of dead time renders this ineffective. Additionally, shrinking $\boldsymbol{u}$ sample ranges start to limit the choices. As a last resort then, $\boldsymbol{y}$ trajectory excursions outside of constraints have to be accepted, but can be heavily penalised to remove them from optimisation contention.

7.9 Control of Time-Delay Systems

In Section 7.1.1, Example 7.2, the minimal prototype Dahlin controller was developed as a SISO discrete setpoint step tracking controller which was able to compensate for (constant) dead time. In other sections, notably Sections 7.5 and 7.8, MIMO models were encountered in the discrete *input–output* format

$$\boldsymbol{y}(t)=\boldsymbol{f}\{\boldsymbol{y}(t-T),\boldsymbol{y}(t-2T),\ldots,\boldsymbol{u}(t-T),\boldsymbol{u}(t-2T),\ldots\} \tag{7.383}$$

or

$$\boldsymbol{y}(z)=\boldsymbol{G}(z)\boldsymbol{u}(z) \tag{7.384}$$

with

$$\boldsymbol{G}(z)=D^{-1}\left(z^{-1}\right)\boldsymbol{N}\left(z^{-1}\right) \tag{7.385}$$

$$D\left(z^{-1}\right)=1+d_1z^{-1}+d_2z^{-2}+d_3z^{-3}+\cdots \tag{7.386}$$

$$\boldsymbol{N}\left(z^{-1}\right)=\boldsymbol{0}+\boldsymbol{N}_1z^{-1}+\boldsymbol{N}_2z^{-2}+\boldsymbol{N}_3z^{-3}+\cdots \tag{7.387}$$

and these proved able to represent dead-time lags explicitly – as long as enough terms were added. Likewise, step response models based on the dynamic matrix (Section 3.9.5.1)

$$\overline{y} = \overline{y}_0 + \overline{B}\,\overline{\Delta u} \tag{7.388}$$

were able to represent arbitrary combinations of dynamics and dead time. The controllers based on these forms then automatically anticipate the effects of (constant) dead time, as seen in Section 7.5.2 (pole placement), Section 7.8.1 (GPC), Section 7.8.2 (DMC), Section 7.8.3.5 (FIDP) and Section 7.8.3.6 (IDP for input–output models).

Nevertheless, process dead time (transport lag) is problematic in many situations. The lack of immediate feedback usually requires sluggish controller tuning and can cause instability. Thus, it is worthwhile to consider ways of dealing with problems such as multiple dead-time delays in MIMO systems and time-variable dead time. The latter is a particular difficulty to be expected in the processing industries as the flow rates in pipes and vessels, as well as levels, vary.

7.9.1 MIMO Closed-Loop Control Using a Smith Predictor

In Section 6.2, the Smith predictor was introduced in terms of a SISO transfer function

$$\widehat{G}(s) = \widehat{G}^*(s)\mathrm{e}^{-\tau s} \tag{7.389}$$

which had been derived to represent the behaviour of the real process $G(s)$, which transforms the input to the output according to

$$y(s) = G(s)u(s) \tag{7.390}$$

Here it was possible to separate the model transfer function into a dynamic part without a dead-time delay $\widehat{G}^*(s)$ and a single delay $\mathrm{e}^{-\tau s}$. In a MIMO system, the output vector elements will be subject to a range of delays, arising from combinations of both state or output delays and input delays. Thus, a general model representation for a linear MIMO system with dead-time delays would be

$$\widehat{G}(s) = \begin{bmatrix} \widehat{G}^*_{11}(s)\mathrm{e}^{-\tau_{11}s} & \widehat{G}^*_{12}(s)\mathrm{e}^{-\tau_{12}s} & \cdots & \widehat{G}^*_{1m}(s)\mathrm{e}^{-\tau_{1m}s} \\ \widehat{G}^*_{21}(s)\mathrm{e}^{-\tau_{21}s} & \widehat{G}^*_{22}(s)\mathrm{e}^{-\tau_{22}s} & \cdots & \widehat{G}^*_{2m}(s)\mathrm{e}^{-\tau_{2m}s} \\ \vdots & \vdots & \ddots & \vdots \\ \widehat{G}^*_{n1}(s)\mathrm{e}^{-\tau_{n1}s} & \widehat{G}^*_{n2}(s)\mathrm{e}^{-\tau_{n2}s} & \cdots & \widehat{G}^*_{nm}(s)\mathrm{e}^{-\tau_{nm}s} \end{bmatrix} \tag{7.391}$$

Clearly, here one cannot separate out the dead-time delays. Only in the special case where an entire row has the same dead-time delay (i.e. an output delay) would it be possible to make the factorisation

$$\widehat{G}(s) = \boldsymbol{D}(s)\widehat{G}^*(s) \tag{7.392}$$

where $\boldsymbol{D}(s)$ is a diagonal matrix with the corresponding dead-time lags on its axis. In this special case, the original approach in Section 6.2 can be used to construct the Smith predictor

$$\widehat{y}(s) = \widehat{G}^*(s)\boldsymbol{u}(s) + \left\{y(s) - \boldsymbol{D}(s)\widehat{G}^*(s)\boldsymbol{u}(s)\right\} \tag{7.393}$$

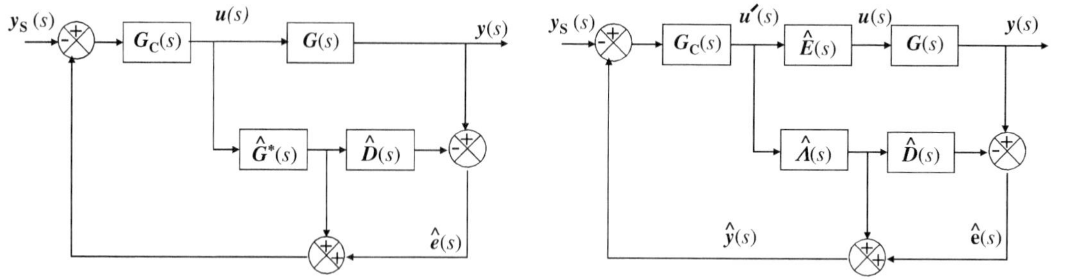

(a) MIMO Smith predictor for output-side delays only (b) MIMO Smith predictor using input-side diagonalisation

Figure 7.35 MIMO Smith predictors for output delays only (a) and mixed delays (b).

where $\widehat{y}(s)$ is the prediction of the output without the dead-time lags and the term in brackets is seen to be the correction based on the present process output $\boldsymbol{y}(s)$. A controller will compare the early prediction $\widehat{y}(s)$ with the desired setpoint vector, and use this error to deduce the required input vector $\boldsymbol{u}(s)$ as in Figure 7.35a.

Unfortunately, the situation where the lags τ_{ij} may all differ offers some difficulties to Smith prediction. For this case, Wang, Zou and Zhang (2000) propose a diagonalisation of the estimated process transfer function $\widehat{G}(s)$ as follows:

$$\widehat{\Lambda}(s) = \widehat{G}(s)\widehat{E}(s) \tag{7.394}$$

which predicts each output as early as possible based on a new input $\boldsymbol{u}'(s)$. Thus, $\widehat{\Lambda}(s)$ becomes the process that is actually being controlled by $\boldsymbol{G}_C(s)$ as in Figure 7.35b. For correction, each output is delayed the rest of the time until its normal availability by the diagonal delay matrix $\widehat{D}(s)$, and the error is fed back in the normal way. The diagonalisation is based on columns $\widehat{e}_i(s)$ of $\widehat{E}(s)$ needing the property

$$\widehat{e}_i(s) = -\left[\widehat{G}(s)\right]^{-1}\widehat{g}_i(s)\widehat{E}_{ii}(s) \tag{7.395}$$

where $\widehat{g}_i$ is the corresponding column in $\widehat{G}(s)$ and $\widehat{E}_{ii}(s)$ is the diagonal element of $\widehat{E}(s)$ which is therefore in $\widehat{e}_i(s)$ itself. Actually, $\widehat{E}_{ii}(s)$ is free for user specification, and is set as a long enough lag $e^{-\tau_i s}$ to just prevent any term in $\widehat{e}_i(s)$ from seeking a *future* value of $\boldsymbol{u}'(s)$ in the multiplication

$$\boldsymbol{u}(s) = \widehat{E}(s)\boldsymbol{u}'(s) \tag{7.396}$$

To make the solution of Equation 7.395 tractable, it is often necessary to find reduced-order approximations of the elements of $\widehat{G}(s)$, for example by the fitting of a second-order transfer function with dead-time lag at several frequency points.

7.9.2
Closed-Loop Control in the Presence of Variable Dead Time

As mentioned, dead time or transport lag in the processing industries is usually caused by the hold-up of fluids in vessel and pipe volumes. Some ideas for dealing with this problem are given in Section 6.2. If measurements exist for the flow rates and levels, then it is normally possible to adjust the controller in real time to accommodate the residence time variations, for example in the Dahlin controller where the z^{-k} term in Equation 7.11 represents the dead-time lag for k intervals of size T.

For controllers based on an input–output model with a numerator polynomial $\boldsymbol{N}(z^{-1}) = \boldsymbol{0} + \boldsymbol{N}_1 z^{-1} + \boldsymbol{N}_2 z^{-2} + \boldsymbol{N}_3 z^{-3} + \cdots$ as in Equations 7.385 and 7.387, this can be extended, or begun later ($\boldsymbol{N}_1 = \boldsymbol{0}, \boldsymbol{N}_2 = \boldsymbol{0}, \ldots$) to accommodate the current dead-time lag. When the individual inputs and outputs are subject to a range of dead-time lags, then the elements of the $\boldsymbol{N}_i$ matrices would need to be evaluated individually. It is possible that this could be done by real-time identification and adaptation of the controller. However, Dumont, Elnaggar and Elshafei (1993) remark that the high-order $\boldsymbol{N}(z^{-1})$ polynomial can be problematic due to the large number of coefficients to be identified, and possible cancellation of factors in the denominator $D(z^{-1})$. They considered SISO cases in which the process was modelled instead using an orthonormal set of Laguerre functions. The timescale in these functions can be altered explicitly using a single parameter. The model based on these functions is arranged in discrete j-step-ahead predictor form, and a minimal prototype controller (Section 7.1) developed based on a first-order reference trajectory.

7.10
A Note on Adaptive Control and Gain Scheduling

Following the discussion of observation and identification in Chapter 6, a general IMC format as in Figure 7.36 seems to accommodate all of the techniques. Indeed, Section 6.6 showed that the state and parameters might be 'observed' using a single algorithm. The idea of *adaptive control* is very general, and concerns really any form of real-time parameter estimation that is simultaneously used to update a control algorithm in real time. In this text then, it concerns the use of the identification techniques in Chapter 6 in conjunction with the controller design techniques in Chapter 7, *in real time*. Isermann (1982) reviewed parameter adaptive control algorithms based on input–output model identifications as in Table 6.1. Another large survey is presented by Seborg, Edgar and Shah (1986).

Section 4.2.8.2 considered a slightly different form of adaptive control which was referred to as *gain scheduling*. This was an explicit technique based on the current operating point of the process, and required no implicit or recursive extraction of parameters. The method was similar to *feedforward* or *open-loop* control in the sense that the controller was adjusted to compensate for current conditions. The method is not restricted to the controller gain – the entire controller specification could be altered in real time using optimal design relationships or lookup tables.

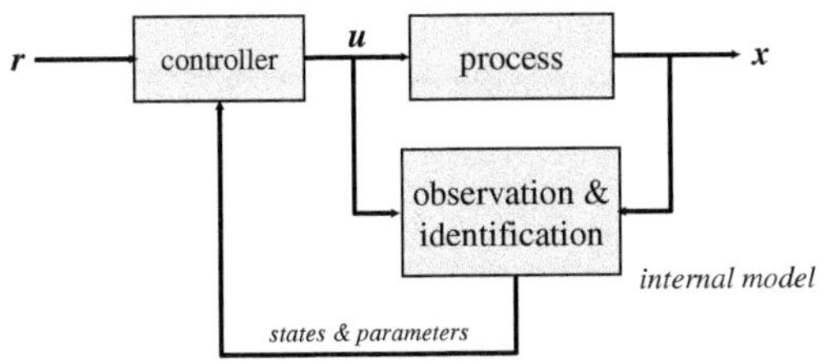

Figure 7.36 Possible simultaneous observation and identification in an IMC format.

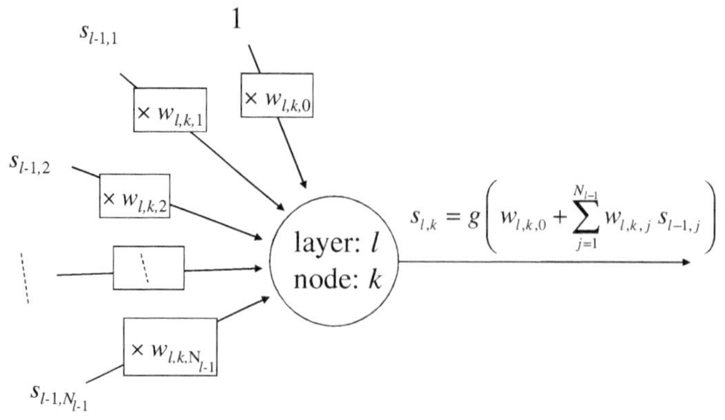

Figure 7.37 Calculation at an artificial neural network node.

7.11 Control Using Artificial Neural Networks

In Section 3.9.5, Example 3.17, it was noted that an artificial neural network could be trained to represent the relationships in a *nonlinear* discrete input–output system as follows:

$$\boldsymbol{y}_i = \boldsymbol{f}(\boldsymbol{y}_{i-1}, \boldsymbol{y}_{i-2}, \ldots \boldsymbol{u}_{i-1}, \boldsymbol{u}_{i-2}, \ldots) \tag{7.397}$$

The ability of the ANN to represent nonlinear processes arises from the nonlinear activation function located in its hidden layer and output layer nodes (Figure 7.37). This is typically a sigmoidal function:

$$g(x) = \frac{1}{1 - e^{-x}} \tag{7.398}$$

where x is the weighted sum of all of the inputs to a node (including a constant bias).

The weights are multipliers on all of the interconnections *within* the ANN, and the ANN itself is tuned by adjustment of these weights. A common type of ANN used in process control is the simple *feedforward* network in which all of the information moves in the forward direction. Figure 7.38 shows how this can be arranged to represent the dynamic equation (Equation 7.397) of a process. It is quite common to use just a single hidden layer, and it is important not to specify too many hidden nodes as this could result in 'overfitting' and conferment of nonexistent complex behaviour.

7.11.1 Back-propagation Training of an ANN

The feedforward ANN is trained by the *back-propagation* technique. In the case of Figure 7.37, the model is presented with measurements of the vector pair $\boldsymbol{u}_i$, $\boldsymbol{y}_i$ at a sequence of points in time i at intervals T. This can be done in real time, or offline using recorded measurements. Initially, all of the interconnection weights are given random values. Then iterations i can be repeated through the available data until the weights converge acceptably, or simply repeated on each time step i in real time. Weights are adjusted according to

$$w_{lkj(i)} = w_{lkj(i-1)} + \Delta w_{lkj(i)} \tag{7.399}$$

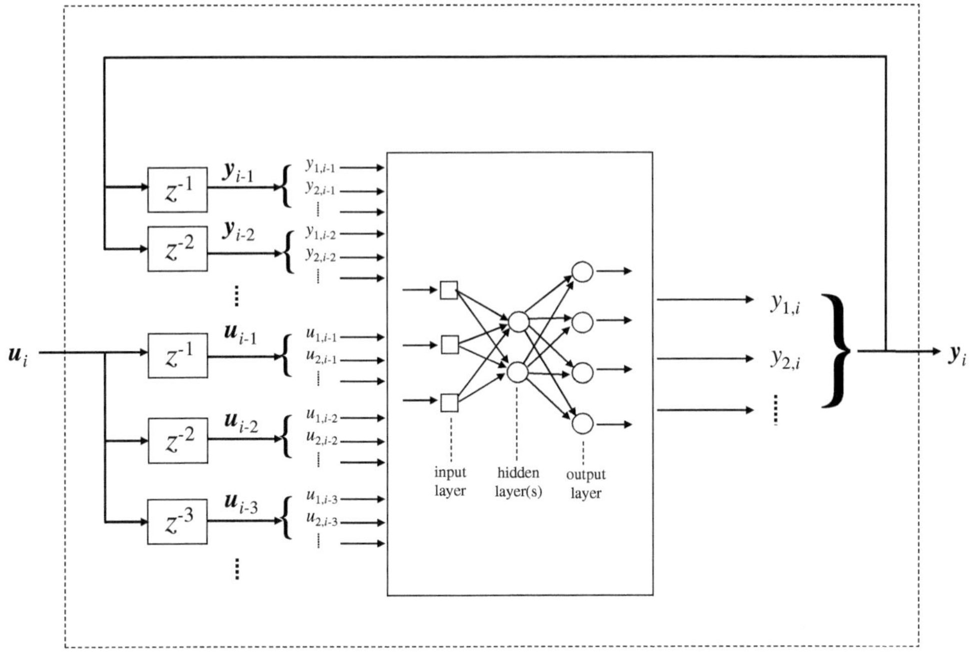

Figure 7.38 Feedforward artificial neural network used to model a dynamic process.

in which l is the layer, k is the node in that layer and j is the node in the preceding layer from which the signal is coming. Rumelhart and McClelland (1986) proposed that a weight should change in proportion to the sensitivity to that weight of the square error at the final output nodes. For a network with just three layers, one will be comparing $s_{31(i)}$ to a measured target value $y_{1(i)}$ at the first output node, $s_{32(i)}$ to $y_{2(i)}$ at the second and so on. So, noting that

$$\frac{\mathrm{d}\left[y_{k(i)} - s_{3k(i)}\right]^2}{\mathrm{d}\left[w_{3kj(i)}\right]} = -2\left[y_{k(i)} - s_{3k(i)}\right]\frac{\mathrm{d}\left[s_{3k(i)}\right]}{\mathrm{d}\left[w_{3kj(i)}\right]} \tag{7.400}$$

the weights feeding to the output layer are changed according to

$$\Delta w_{3kj(i)} = \eta\left[y_{k(i)} - s_{3k(i)}\right]\frac{\mathrm{d}\left[s_{3k(i)}\right]}{\mathrm{d}\left[w_{3kj(i)}\right]} \tag{7.401}$$

since each weight only affects the indicated output. Here η is the chosen *learning rate coefficient.* The derivative is easily evaluated as

$$\frac{\mathrm{d}\left[s_{3k(i)}\right]}{\mathrm{d}\left[w_{3kj(i)}\right]} = \left[\frac{\mathrm{d}g(x)}{\mathrm{d}x}\frac{\mathrm{d}x}{\mathrm{d}\left[w_{3kj(i)}\right]}\right] \tag{7.402}$$

where

$$x = w_{3k0(i)} + \sum_{p=1}^{N_2} w_{3kp(i)}s_{2p(i)} \tag{7.403}$$

and for the choice of the sigmoidal activation function in Equation 7.398:

$$g'(x) = \frac{-e^{-x}}{(1-e^{-x})^2} = g(x)\left[1-g(x)\right] \tag{7.404}$$

Thus,

$$\frac{d\left[s_{3k(i)}\right]}{d\left[w_{3kj(i)}\right]} = g'\left(w_{3k0(i)} + \sum_{p=1}^{N_2} w_{3kp(i)}s_{2p(i)}\right)s_{2j(i)} \tag{7.405}$$

Moving back to the hidden layer, the weights $w_{2kj(i)}$ now affect all of the output nodes, so

$$\Delta w_{2kj(i)} = \eta \sum_{p=1}^{N_3} \left[y_{p(i)} - s_{3p(i)}\right] \frac{d\left[s_{3p(i)}\right]}{d\left[w_{2kj(i)}\right]} \tag{7.406}$$

$$\Delta w_{2kj(i)} = \eta \sum_{p=1}^{N_3} \left[y_{p(i)} - s_{3p(i)}\right] \frac{d\left[s_{3p(i)}\right]}{d\left[s_{2k(i)}\right]} \frac{d\left[s_{2k(i)}\right]}{d\left[w_{2kj(i)}\right]} \tag{7.407}$$

$$\Delta w_{2kj(i)} = \eta \sum_{p=1}^{N_3} \left[y_{p(i)} - s_{3p(i)}\right] \left\{g'\left(w_{3p0(i)} + \sum_{q=1}^{N_2} w_{3pq(i)}s_{2q(i)}\right)w_{3pk(i)}\right\}\left\{g'\left(w_{2k0(i)} + \sum_{r=1}^{N_1} w_{2kr(i)}s_{1r(i)}\right)s_{1j(i)}\right\} \tag{7.408}$$

and in Figure 7.38, $s_{1j(i)}$, $j = 1, \ldots, N_1$, are the inputs $y_{1(i-1)}, y_{2(i-1)}, \ldots, y_{1(i-2)}, y_{2(i-2)}, \ldots, u_{1(i-1)}, u_{2(i-1)}, \ldots, u_{1(i-2)}, u_{2(i-2)}, \ldots$. In this procedure, further control over the process could be obtained by weighting the output node deviations individually, that is $W_k\left[y_{k(i)} - s_{3k(i)}\right]$ to handle range and unit variations. The risk of an algebraic mistake in obtaining the derivatives could be avoided by perturbing the internal weights individually to obtain the error sensitivities.

7.11.2
Process Control Arrangements Using ANNs

There is effectively no limit – except for the size of the ANN – to how long a sequence of output values $\boldsymbol{y}_i, \boldsymbol{y}_{i+1}, \boldsymbol{y}_{i+2}, \ldots$ one might wish to predict. This would of course require more measurements of the process input $\boldsymbol{u}_{i-1}, \boldsymbol{u}_i, \boldsymbol{u}_{i+1}, \ldots$, but the potential for predictive control is obvious. Actually, it will be seen soon that the smaller original ANN can be chained in sequence to achieve this. However, even this seems like unnecessary effort when it is noticed that the ANN could be arranged to produce information in any order. It offers the possibility of direct inversion to produce a controller. The ANN could be set up rather in the form of Figure 7.39, and trained using process measurements to provide the present control action required to achieve the desired next output value. This *inverse ANN* is therefore a controller. It too can be extended to give a sequence of future control settings $\boldsymbol{u}_i, \boldsymbol{u}_{i+1}, \boldsymbol{u}_{i+2}, \ldots$, necessary to achieve a desired future sequence of output values $\boldsymbol{y}_{Si+1}, \boldsymbol{y}_{Si+2}, \boldsymbol{y}_{Si+3}, \ldots$. It is not always feasible to obtain a reliable inverse like this, and some workers have used numerical techniques (e.g. Newton) to obtain inverse results from the forward model (Hussain, 1999).

If it is desired to use the ANN as the model basis of a receding-horizon model predictive controller, it has been mentioned that the one-step predictor form could be chained in sequence to provide the necessary predictions of the future output trajectory. This idea is illustrated in a SISO form in Figure 7.40.

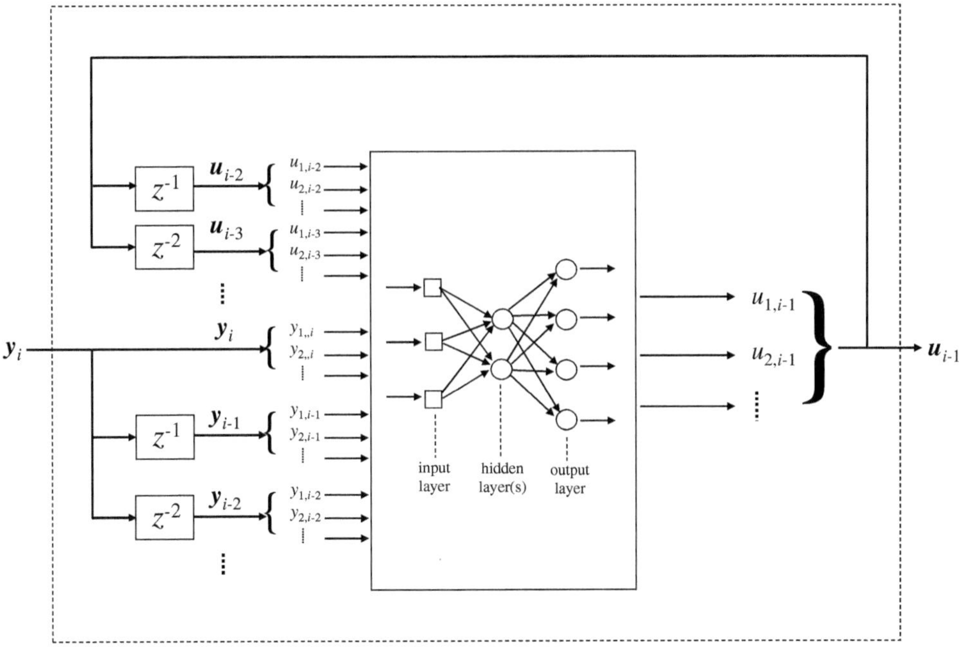

Figure 7.39 ANN trained to provide a control action setting necessary for the next desired output value ('inverse ANN' as a controller).

The ANN output prediction models such as in Figures 7.40 and 7.38 and the 'inverse' form for prediction of an input in Figure 7.39 all have in common the problem that they must work with absolute values of the variables in order to account for nonlinear behaviour. To handle problems of model offset error, it is thus wise to configure an IMC format as in Figure 7.41. A smoothing filter on the feedback error improves robustness.

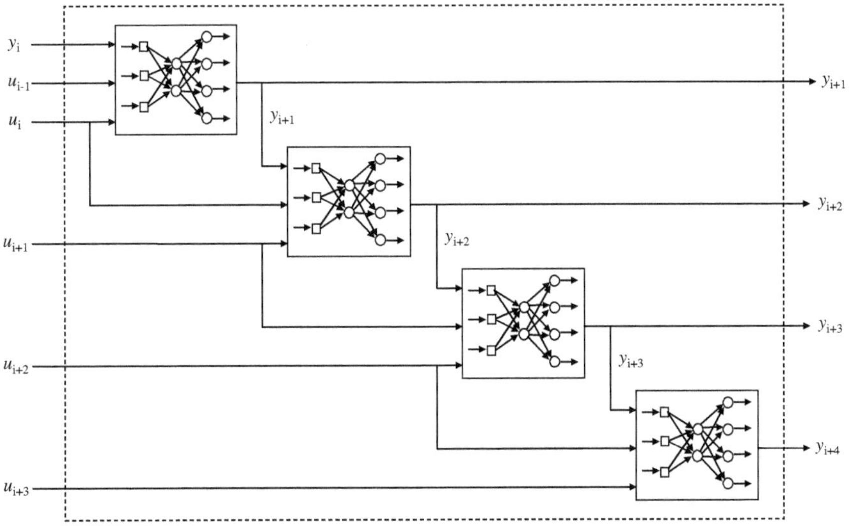

Figure 7.40 One-step predictor ANNs chained to provide a longer prediction horizon.

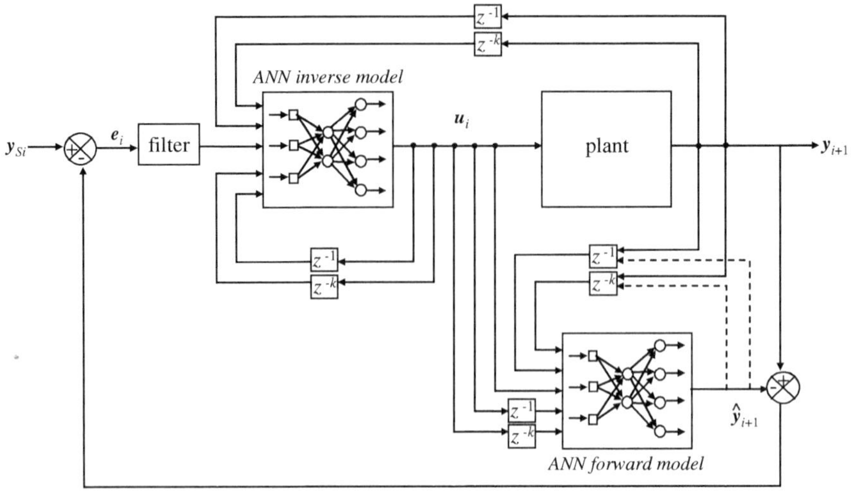

Figure 7.41 IMC arrangement of ANNs to correct for modelling error (Hussain, 1999).

The real-time training of ANNs offers the possibility of *adaptive control.* In the *indirect approach,* the weights in a forward model are updated online (e.g. by back-propagation training), and made available continuously to algorithms such as MPC. An example is the Neuralware/Pavilion Neucop® controller, which is based on a linear model with an ANN superimposed to model residual non-linearity. The *direct approach* on the other hand continuously updates the controller itself in order to meet various objectives. An obvious application on the direct approach would thus be to update an inverse model, to improve its accuracy to predict the control input $\boldsymbol{u}$ required to achieve any given output $\boldsymbol{y}$. Other objectives might be to improve stability, reduce overshoot and so on. These objectives would need to be formulated as variations of the objective function, which as seen in Section 7.11.1 is based on the sum of square output deviations in normal back-propagation training.

7.12 Control Based on Fuzzy Logic

The open-loop modelling of processes using fuzzy logic has been discussed in Section 3.9.7. It remains now to see how these ideas can be converted into a real-time control algorithm. Recall that 'crisp' variables (normal scalars) exist as fuzzy variables in terms of their degree of *membership* of fuzzy sets. These can also be thought of as the *probability* that a certain value belongs to that set. For example, using triangular membership functions for just two sets describing a temperature range in Figure 7.42, a temperature of 18 °C has a membership 30% of HOT and 70% of COLD. (Actually, if only two sets A and B are used to describe a whole range, one set can be made redundant since $\mu_A = 1 - \mu_B$.)

The full range of a crisp variable is considered to be the *universe of discourse* of that variable. Parts of this range will have nonzero membership of particular sets, referred to as the *support* of that set. A set which only has support at a particular value (where membership will be 1.0) is called a *fuzzy singleton.* In engineering, one does tend to work with crisp values anyway, so what is the use of *fuzzy modelling*? As mentioned in Section 3.9.7, it is a powerful means to capture 'experience',

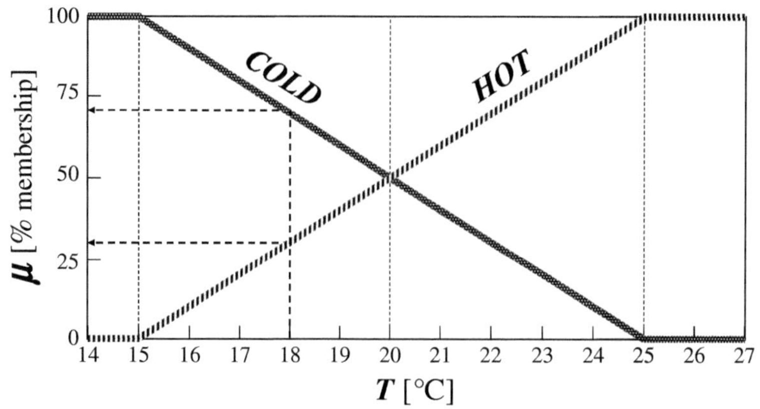

Figure 7.42 A temperature range described with just two fuzzy sets.

whether this involves operating a processing plant or catching a flying ball. To quantify these strategies entirely in crisp equations is virtually impossible, yet the human beings doing it still manage somehow. One captures the model as a series of linguistic rules:

IF (*antecedent*) THEN (*consequent*)

For example

IF (*it is HOT*) THEN (*your ice cream will MELT QUICKLY*)

From a *process control* point of view, one could also design *control rules* by observing operators at work:

IF {(*reactor is HOT*) AND (*feed rate is RAPID*)} THEN (*set cooling water valve HIGH*)

This is an example of a MISO rule: multi-input, single output. If there was a second consequent, then to stay in the MISO format one would operate the second rule in parallel, for example

IF {(*reactor is HOT*) AND (*feed rate is RAPID*)} THEN (*set catalyst feed GENTLE*)

The first fuzzy logic controllers were built like this in the 1970s, for example Mamdani (1974). However, it is an arduous and expensive exercise to construct a complete and consistent set of control rules which has adequate resolution – simply on the basis of operator interrogation.

So far what has been described appears to be an 'expert engine'. Actually, *fuzzy logic* grew around the idea of using the *IF (. . .) THEN (. . .)* logical structure to capture information, but then embedding it in such a way that engineers could continue to work with crisp numbers on the outside. This meant that scalar measurements going into the model had to be *fuzzified* (into their membership of sets), the logical rules had to be evaluated, and then the output set memberships had to be *defuzzified* back into crisp numbers that could be used elsewhere. An example of the procedure is given in Section 3.9.7. Consider an *open-loop* model of the reactor that was the subject of control above, for time intervals $i = 0, 1, \ldots$. Here $\Delta T_{i+1} = T_{i+1} - T_i$ is the predicted temperature change over the next time interval:

IF {(T_i *is HOT*) AND (F_i *is RAPID*)} THEN (ΔT_{i+1} *will be a LARGE INCREASE*)

IF {(T_i *is COLD*) AND (F_i *is RAPID*)} THEN (ΔT_{i+1} *will be a MODERATE INCREASE*)

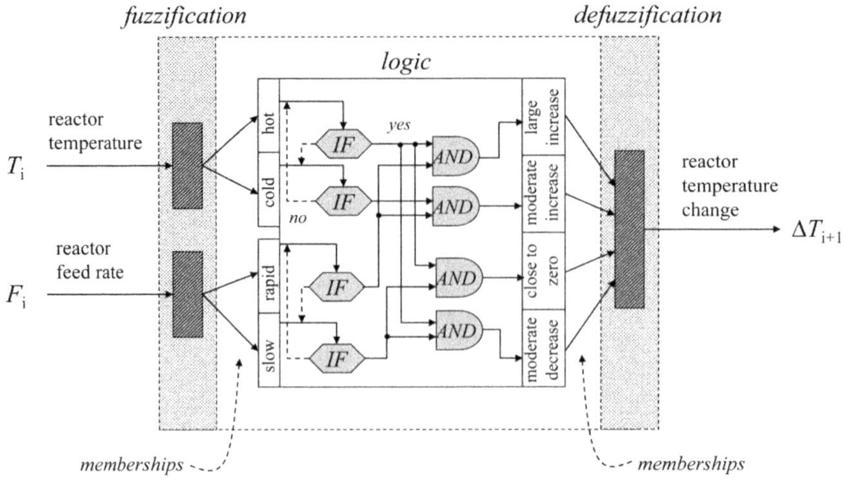

Figure 7.43 Open-loop fuzzy model of a reactor showing embedded logic.

IF {(T_i *is HOT*) AND (F_i *is SLOW*)} THEN (ΔT_{i+1} *will be CLOSE TO ZERO*)
IF {(T_i *is COLD*) AND (F_i *is SLOW*)} THEN (ΔT_{i+1} *will be a MODERATE DECREASE*)

Each input variable T_i and F_i has two membership sets and the output variable $\Delta T_{i+1} = T_{i+1} - T_i$ has four membership sets. Assuming that membership functions are devised (in consultation with the operator) to represent these sets, the model can be arranged for numerical input and output as in Figure 7.43. Thus, the model has crisp input and output, but is able to make use of operator experience in its embedded logic. Recall from Section 3.9.7 that the memberships (probabilities), sometimes called 'certainties', pass through the logical structure with Zadeh's rules:

'AND' ⇒ MINIMUM
'OR' ⇒ MAXIMUM

7.12.1
Fuzzy Relational Model

A dynamic model (e.g. a one-step predictor) based on embedded logic as in Figure 7.43 requires evaluation of the logical tree on each step. The designer needs to be careful that it has no 'gaps' and is consistent. As mentioned, it needs a very thorough interrogation of experience. Fortunately, there is an alternative to this type of model, called the *fuzzy relational model* (FRM), which can be constructed directly from process measurements, adaptively in real time if required. Sing and Postlethwaite (1996) remark that the FRM provides a complete model with a smooth prediction surface, and is able to make a reasonable prediction when the set definitions of ranges are relatively sparse.

Focusing now on an open-loop dynamic model, the FRM is expressed as

$$\boldsymbol{Y}(k+1) = \boldsymbol{R} \circ Y_1(k) \circ Y_2(k) \circ \cdots \circ Y_n(k) \circ U_1(k) \circ U_2(k) \circ \cdots \circ U_m(k) \tag{7.409}$$

Here k is the time interval, $\boldsymbol{Y}$ contains the memberships μ of the n output variables and $\boldsymbol{U}$ contains the memberships μ of the m input variables (after any dead-time delays, if applicable).

$$\boldsymbol{Y}(k) = \begin{pmatrix} Y_1(k) \\ Y_2(k) \\ \vdots \\ Y_n(k) \end{pmatrix} = \begin{pmatrix} \begin{pmatrix} \mu_{Y_11}(k) \\ \vdots \\ \mu_{Y_1N_1}(k) \end{pmatrix} \\ \begin{pmatrix} \mu_{Y_21}(k) \\ \vdots \\ \mu_{Y_2N_2}(k) \end{pmatrix} \\ \vdots \\ \begin{pmatrix} \mu_{Y_n1}(k) \\ \vdots \\ \mu_{Y_nN_n}(k) \end{pmatrix} \end{pmatrix} \tag{7.410}$$

and

$$\boldsymbol{U}(k) = \begin{pmatrix} \boldsymbol{U}_1(k) \\ \boldsymbol{U}_2(k) \\ \vdots \\ \boldsymbol{U}_m(k) \end{pmatrix} = \begin{pmatrix} \begin{pmatrix} \mu_{U_11}(k) \\ \vdots \\ \mu_{U_1M_1}(k) \end{pmatrix} \\ \begin{pmatrix} \mu_{U_21}(k) \\ \vdots \\ \mu_{U_2M_2}(k) \end{pmatrix} \\ \vdots \\ \begin{pmatrix} \mu_{U_m1}(k) \\ \vdots \\ \mu_{U_MM_m}(k) \end{pmatrix} \end{pmatrix} \tag{7.411}$$

The symbol '∘' is the *fuzzy compositional operator* and $\boldsymbol{R}$ is the *fuzzy relational array*. Both of these will be discussed further below. Because one is working in the fuzzy domain, the m input and n output variables *each* must be represented by memberships of several sets, with membership functions covering the respective ranges. The information flow through the fuzzy relational model is thus as depicted in Figure 7.44.

The multiplication involving the fuzzy compositional operator (∘) needs to be carefully defined. There are in fact alternative proposals of how this should be done, for example based on 'distance' or 'average'. However, this text will continue with the common *max–min* multiplication which is analogous to Zadeh's OR(+)–AND(×). Another aspect of this multiplication is that $\boldsymbol{R}$ is a high-order *tensor* (e.g. a *matrix* is a tensor of order 2). Say every one of the input variables was represented by M set memberships, and every output variable by N set memberships. Then $\boldsymbol{R}$ would need to combine $M^m \times N^n$ 'AND' (cross) probabilities (memberships μ) to produce the $N \times n$ output

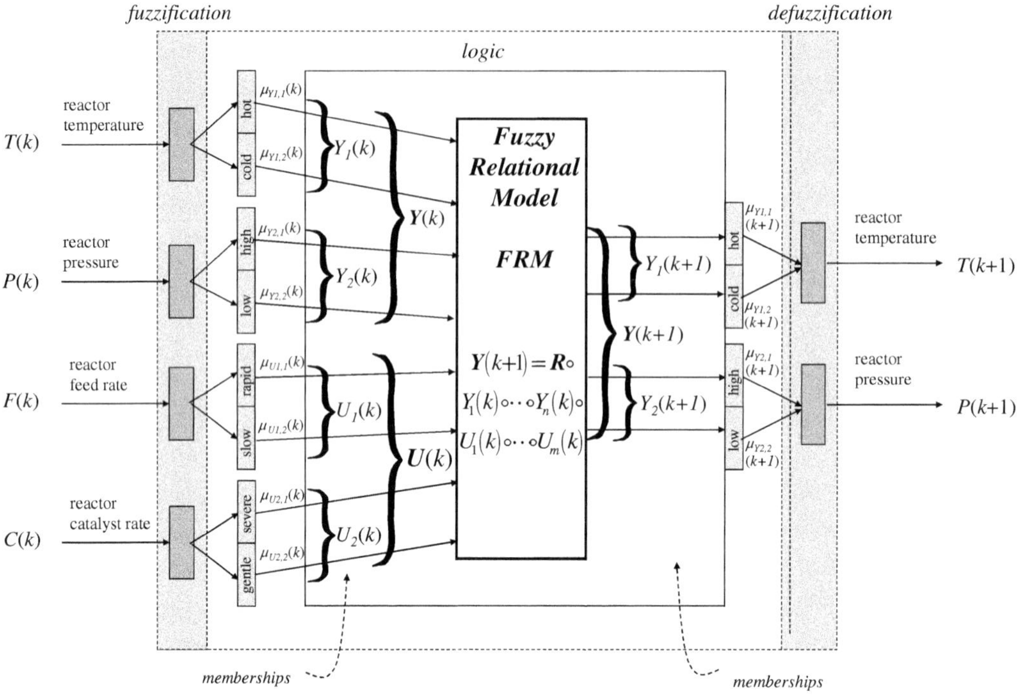

Figure 7.44 Inputs and outputs of a fuzzy relational model for a one-step dynamic predictor.

probabilities. Thus, $\boldsymbol{R}$ would have dimension $(N\times n)\times N^n\times M^m$ and its order is $1+n+m$. In the example of Figure 7.44, $\boldsymbol{R}$ is thus $4\times2\times2\times2\times2$, that is it has 64 elements and it has an order of 5, so individual elements would have to be subscripted r_{ijpqr}. More generally, with different numbers of sets N_j defining each output y_j and different numbers of sets M_p defining each input u_p, an element of $\boldsymbol{R}$ is

$$r_{i,\, j_1,\, j_2,\, \ldots,\, j_n,\, p_1,\, p_2,\, \ldots,\, p_m} \tag{7.412}$$

$p_m=1,\ldots,M_m$

$p_2=1,\ldots,M_2$ for $u_2(k)$

$p_1=1,\ldots,M_1$ for $u_1(k)$

$j_n=1,\ldots,N_n$

$j_2=1,\ldots,N_2$ for $y_2(k)$

$j_1=1,\ldots,N_1$ for $y_1(k)$

$i=1,\ldots,(N_1+N_2+\ldots\ldots+N_n)$ for $y_1(\mathrm{k}+1),\ldots\ldots, y_n(\mathrm{k}+1)$

Before going on, the determination of $\boldsymbol{R}$ must be considered. This is commonly done using actual process measurements, following the *RSK method* proposed by Ridley, Shaw and Kruger (1988). Postlethwaite (1994) considers this method to be more robust and superior to several others. Here one has recorded a sequence $k=1, 2, \ldots$ of process *measurements* $\mathbf{y}(k)$ and $\mathbf{u}(k)$. Using the agreed membership functions, these are converted into the sequence of memberships $\boldsymbol{Y}(k)$ and $\boldsymbol{U}(k)$. For each point in time, a tensor $\boldsymbol{F}$ is created by direct multiplication of the membership value vectors of the individual variables as follows:

$$\boldsymbol{f}(k) = Y_1(k) \otimes Y_2(k) \otimes \cdots \otimes Y_n(k) \otimes U_1(k) \otimes U_2(k) \otimes \cdots \otimes U_m(k) \tag{7.413}$$

where $\otimes$ represents the Kronecker tensor products – every element on the left multiplies every element on the right, just as a vector cross-product creates a matrix. On an element-by-element basis, this is

$$f_{j_1,j_2,\ldots,j_n,p_1,p_2,\ldots,p_m}(k) = \mu_{Y_1j_1}(k)\mu_{Y_2j_2}(k)\cdots\mu_{Y_nj_n}(k)\mu_{U_1p_1}(k)\mu_{U_2p_2}(k)\cdots\mu_{U_mp_m}(k) \tag{7.414}$$

$$f_{j_1,j_2,\ldots,j_n,p_1,p_2,\ldots,p_m}(k) = \left\{\prod_{r=1}^{n}\mu_{Y_rj_r}(k)\right\}\left\{\prod_{s=1}^{m}\mu_{U_sp_s}(k)\right\} \tag{7.415}$$

It is noted that the dimension of $\boldsymbol{F}$ follows that of $\boldsymbol{R}$ in Equation 7.412, except for the first subscript. Then the fuzzy relational array is updated at each time k by

$$r_{i,j_1,j_2,\ldots,j_n,p_1,p_2,\ldots,p_m}(k) = \frac{\left\{f_{j_1,j_2,\ldots,j_n,p_1,p_2,\ldots,p_m}(k)\right\}\mu_{Y,i}(k+1) + \left\{r_{i,j_1,j_2,\ldots,j_n,p_1,p_2,\ldots,p_m}(k-1)\right\}\left\{\sum_{q=1}^{k-1} f_{j_1,j_2,\ldots,j_n,p_1,p_2,\ldots,p_m}(q)\right\}}{\left\{\sum_{q=1}^{k} f_{j_1,j_2,\ldots,j_n,p_1,p_2,\ldots,p_m}(q)\right\}} \tag{7.416}$$

The calculation in Equation 7.416 must of course be repeated for the ranges of all of the subscripts of r. Note that the variable $\mu_{Y,i}(k+1)$ is the membership of the ith set where the membership vectors $Y_1, \ldots, Y_n$ of all measured outputs $y_1, \ldots, y_n$ have been concatenated in sequence in a single vector representing the values at the *next* time instant $(k+1)$, namely $\boldsymbol{Y}(k+1)$. In Equation 7.416, it is seen that the elements of $\boldsymbol{F}$ act to weight the increments of $\boldsymbol{R}$ according to the probability of each set combination. In practice, $\boldsymbol{R}$ may be determined offline using plant data records, and is likely to converge, allowing the use of a precalculated fixed model in the online control.

Finally, having seen the construction of $\boldsymbol{R}$, it is a little easier to understand how the multiplication according to the fuzzy compositional operator in Equation 7.403 is able to predict the memberships of the outputs at the next time instant. The equivalent of

$$\boldsymbol{Y}(k+1) = \boldsymbol{R} \circ Y_1(k) \circ Y_2(k) \circ \cdots \circ Y_n(k) \circ U_1(k) \circ U_2(k) \circ \cdots \circ U_m(k) \tag{7.417}$$

on an element-by-element basis is

$$\mu_{Y,i}(k+1) = \bigcup_{j_1=1}^{N_1}\left\{\cdots\left\{\bigcup_{j_n=1}^{N_n}\left\{\bigcup_{p_1=1}^{M_1}\left\{\cdots\left\{\bigcup_{p_m=1}^{M_m} r_{i,j_1,\ldots,j_n,p_1,\ldots,p_m} \cap \mu_{U_mp_m}(k)\right\}\cdots\right\}\cap\mu_{U_1p_1}(k)\right\}\cap\mu_{Y_nj_n}(k)\right\}\cdots\right\}\cap\mu_{Y_1j_1}(k) \tag{7.418}$$

where $\mu_{Y_ij_i}$ is the membership of the ith output y_i of its own j_ith set, and $\mu_{U_ip_i}$ is the membership of the ith input u_i of its own p_ith set. In Equation 7.418, the intersections ($\cap$) are performed in the fuzzy sense, that is the MINIMUM of the two probability (membership) values is taken, and

```
% Find fuzzy relational array R and use it
% DATA:
nout=2;
nin1=2;
nin2=2;
% [good, bad] <- [cold, hot], [wet, dry]
D=[ .9 .1 .2 .8 .1 .9 ;
    .5 .5 .4 .6 .6 .4 ;
    .2 .8 .7 .3 .6 .4 ;
    .7 .3 .3 .7 .4 .6 ;
    .5 .5 .9 .1 .1 .9 ];

% Train R
fsum_to_last=zeros(nin1,nin2);
R=zeros(nout,nin1,nin2);
for i=1:size(D,1)
  for j=1:nin1
    for k=1:nin2
      fjk=D(i,nout+j)*D(i,nout+nin1+k);
      fsumjk=fsum_to_last(j,k)+fjk;
      if (fsumjk~=0)
        for m=1:nout    % RSK method
          R(m,j,k)=
            (fjk*D(i,m)+R(m,j,k)*fsum_to_last(j,k))/fsumjk;
        end
      end
      fsum_to_last(j,k)=fsumjk;
    end
  end
end
```

RESULTS:

	MEAS out(1)	PRED out(1)	MEAS out(2)	PRED out(2)	MEAS sum out	PRED sum out
ans =	0.9000	0.7113	0.1000	0.2887	1.0000	1.0000
ans =	0.5000	0.5374	0.5000	0.4626	1.0000	1.0000
ans =	0.2000	0.4000	0.8000	0.6000	1.0000	1.0000
ans =	0.7000	0.6000	0.3000	0.4000	1.0000	1.0000
ans =	0.5000	0.5149	0.5000	0.4851	1.0000	1.0000

```
% Test R using original data
R1=zeros(nout,nin2); % for first multiplication: R1 = R o input1
for i=1:size(D,1)
  Pred=zeros(nout,1);
  for m=1:nout
    for k=1:nin2    % max-min multiplication
      max=0;
      for j=1:nin1
        min=1;
        if(R(m,j,k)<min)
          min=R(m,j,k);
        end
        if(D(i,nout+j)<min)
          min=D(i,nout+j);
        end
        if (min>max)
          max=min;
        end
      end
      R1(m,k)=max;
    end
  end

  for m=1:nout    % second multiplication: output = R1 o input2
    max=0;
    for k=1:nin2    % max-min multiplication
      min=1;
      if(R1(m,k)<min)
        min=R1(m,k);
      end
      if(D(i,nout+nin1+k)<min)
        min=D(i,nout+nin1+k);
      end
      if (min>max)
        max=min;
      end
    end
    Pred(m)=max;
  end
  % print result
  [ D(i,1) Pred(1) D(i,2) Pred(2) D(i,1)+D(i,2) Pred(1)+Pred(2) ]
end
```

Figure 7.45 MATLAB® code for determination and testing of a *fuzzy relational array* of a two-input, one-output system where each variable is represented by membership of two sets.

likewise the union (∪) gives the MAXIMUM of its argument terms – hence the description *max–min multiplication.* Figure 7.45 gives MATLAB® code for the determination of a fuzzy relational array $\boldsymbol{R}$ for a system with two inputs and one output, where each of these three variables is represented by memberships of two sets. The data are to do with good or bad weather on the basis of temperature and rain. $\boldsymbol{R}$ is determined from just five data points using the RSK method.

7.12.2 Fuzzy Relational Model-Based Control

Postlethwaite (1994) proposed an internal model control (Section 7.8) structure to make use of the FRM in closed-loop control. Recognising that operator experience could be captured to create the 'inverse fuzzy model', that is one which gives the control actions appropriate for present and desired output, he noted that it is difficult and expensive to capture a comprehensive control strategy by interrogation of operators. He proposed rather to obtain a fuzzy model of the open-loop process itself, for which supporting information could easily be obtained from plant data records. To make use of this forward-running model, a predictive control IMC structure was necessary (Figure 7.46).

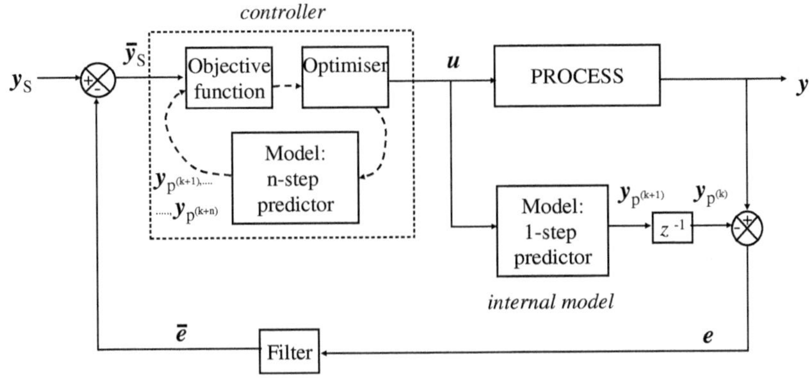

Figure 7.46 IMC structure used by Postlethwaite (1994) for predictive control using a *fuzzy relational model.*

Postlethwaite (1994) formulated the controller as follows:

$$
\begin{array}{lll}
\text{Last measurement:} & \boldsymbol{y}\,(k-1) & [\text{FRM}: \boldsymbol{y}\,(k-1) \rightarrow Y_1(k-1), Y_2(k-1), \ldots, Y_n(k-1)] \\
\text{Present measurement:} & \boldsymbol{y}\,(k) & [\text{FRM}: \boldsymbol{y}\,(k) \rightarrow Y_1(k), Y_2(k), \ldots, Y_n(k)] \\
\text{One-step prediction:} & \boldsymbol{y}_{\text{p}}(k) & [\text{FRM}: \boldsymbol{Y}_{\text{p}}(k) = \boldsymbol{R} \circ Y_1(k-1) \circ \cdots \circ Y_n(k-1) \circ U_1(k-1) \circ \cdots \circ U_m(k-1)] \\
\text{Modelling error:} & \boldsymbol{e}(k) & = \boldsymbol{y}\,(\text{k}) - \boldsymbol{y}_{\text{p}}(\text{k}) \quad [\text{FRM}: Y_{\text{P1}}(k), Y_{\text{P2}}(k), \ldots, Y_{\text{P}n}(k) \rightarrow \boldsymbol{y}_{\text{p}}(k)] \\
\text{Smoothed model error:} & \bar{\boldsymbol{e}}(k) & = \alpha \bar{\boldsymbol{e}}(k-1) + [1-\alpha]\boldsymbol{e}(k) \quad 0 < \alpha < 1 \\
\text{Offset setpoint:} & \bar{\boldsymbol{y}}_{\text{S}}(k) & = \boldsymbol{y}_{\text{S}}(k) - \bar{\boldsymbol{e}}(k)
\end{array}
$$

Predictions to n-step horizon:

$$
\begin{array}{ll}
\boldsymbol{y}_{\text{p}}(k+1) & [\text{FRM}: \boldsymbol{Y}_{\text{p}}(k+1) = \boldsymbol{R} \circ Y_1(k) \circ \cdots \circ Y_n(k) \circ U_1(k) \circ \cdots \circ U_m(k)] \\
\boldsymbol{y}_{\text{p}}(k+2) & [\text{FRM}: \boldsymbol{Y}_{\text{p}}(k+2) = \boldsymbol{R} \circ Y_{\text{p1}}(k+1) \circ \cdots \circ Y_{\text{p}n}(k+1) \circ U_1(k+1) \circ \cdots \circ U_m(k+1)] \\
\vdots & \\
\boldsymbol{y}_{\text{p}}(k+n) & [\text{FRM}: \boldsymbol{Y}_{\text{p}}(k+n) = \boldsymbol{R} \circ Y_{\text{p1}}(k+n-1) \circ \cdots \circ Y_{\text{p}n}(k+n-1) \circ U_1(k+n-1) \circ \cdots \circ U_m(k+n-1)]
\end{array}
$$

$$
\begin{array}{lll}
\text{where the choices are} & \boldsymbol{u}\,(k) & [\text{FRM}: \boldsymbol{u}\,(k) \rightarrow U_1(k), U_2(k), \ldots, U_m(k)] \\
 & \boldsymbol{u}\,(k+1) & [\text{FRM}: \boldsymbol{u}\,(k+1) \rightarrow U_1(k+1), U_2(k+1), \ldots, U_m(k+1)] \\
 & & \vdots \\
 & \boldsymbol{u}\,(k+n-1) & [\text{FRM}: \boldsymbol{u}\,(k+n-1) \rightarrow U_1(k+n-1), U_2(k+n-1), \ldots, U_m(k+n-1)]
\end{array}
\tag{7.419}
$$

As a basis for the choice of the optimal future control actions $\boldsymbol{u}(k)$, $\boldsymbol{u}(k+1), \ldots, \boldsymbol{u}(k+s-1)$, $s \leq n$, an objective function to be minimised can be defined as

$$
J(\boldsymbol{u}(k), \boldsymbol{u}(k+1), \ldots, \boldsymbol{u}(k+s)) = \sum_{j=1}^{n} \left[y(k+j) - \bar{y}_{\text{S}} \right]^{\text{T}} \boldsymbol{W} \left[y(k+j) - \bar{y}_{\text{S}} \right] + \sum_{i=0}^{s-1} [\boldsymbol{u}(k+i) - \boldsymbol{u}(k+i-1)]^{\text{T}} \boldsymbol{\Lambda} [\boldsymbol{u}(k+i) - \boldsymbol{u}(k+i-1)]
\tag{7.420}
$$

where $\boldsymbol{W}$ and $\boldsymbol{\Lambda}$ are diagonal weighting matrices to reduce setpoint deviations and for move suppression. Here the first s 'moves' are to be optimised (see Sections 7.8.1 and 7.8.2). Postlethwaite

(1994) performed a direct numerical 'Fibonacci' search (Beveridge and Schechter, 1970) for the best control actions. As usual, only the first control action is implemented before the entire calculation is repeated on the next controller time step. The effect of computing optimal actions at $s > 1$ step is to effectively increase the controller gain, as does increasing the ratio of elements in $\boldsymbol{W}$ relative to Λ.

The effect of basing an open-loop model on the fuzzy relational array as in Figure 7.44 is to create a patchwork response 'surface' over which behaviour is dominated by the relevant local combination of sets (e.g. temperature is HOT and feed is RAPID). If triangular membership functions are used as in Figure 3.47 or 7.42, then the responses $\boldsymbol{y}$ are determined as a series of interconnected straight-line (linear) segments as the set memberships μ of the inputs $\boldsymbol{u}$ vary. Postlethwaite and Edgar (2000) note that one seeks a 'prediction surface' for the outputs $\boldsymbol{y}$ which is *monotonic* as the memberships μ of the inputs $\boldsymbol{u}$ vary. If this is not achieved, then the objective function surface will become more complex, and there is a serious risk of finding only a local optimum J instead of a single global optimum for the future inputs $\boldsymbol{u}$.

To deal with the above problem, to reduce the size of the relational array and to eliminate the nonlinear solution required when multiple outputs are involved, Postlethwaite and Edgar (2000) reformulated their *fuzzy internal model controller* (FIMC) in a form that was linear in the memberships μ of $\boldsymbol{u}$. Instead of the form in Equation 7.418,

$$\boldsymbol{Y}(k+1) = \boldsymbol{R} \circ Y_1(k) \circ Y_2(k) \circ \cdots \circ Y_n(k) \circ U_1(k) \circ U_2(k) \circ \cdots \circ U_m(k) \tag{7.421}$$

they proposed a *superposition model*

$$\begin{aligned} \boldsymbol{y}(k+1) = {} & \boldsymbol{R}_1[Y_1(k) \otimes Y_2(k) \otimes \cdots \otimes Y_n(k) \otimes U_1(k)] \\ & + \boldsymbol{R}_2[Y_1(k) \otimes Y_2(k) \otimes \cdots \otimes Y_n(k) \otimes U_2(k)] \\ & + \cdots \\ & + \boldsymbol{R}_m[Y_1(k) \otimes Y_2(k) \otimes \cdots \otimes Y_n(k) \otimes U_m(k)] \end{aligned} \tag{7.422}$$

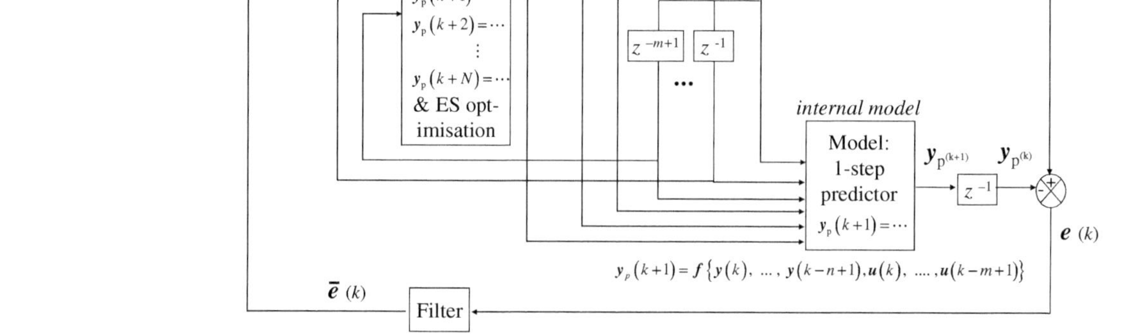

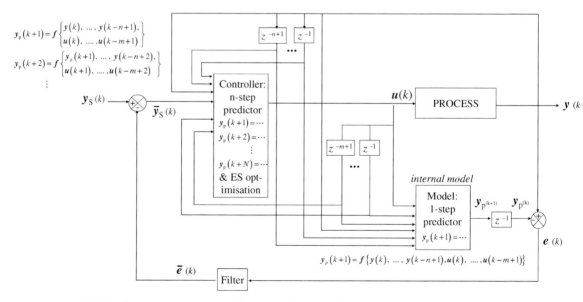

Figure 7.47 MPC using evolutionary strategy based on arbitrary nonlinear input–output model.

where ⊗ represents the Kronecker tensor product (see above). Apart from a very different arithmetic to Equation 7.418, this form obviously neglects the cross-product nonlinearities amongst the inputs. To obtain the $\boldsymbol{R}_i$, the method uses $k = 1, \ldots, N$ sets of measurements $\boldsymbol{u}(k)$, $\boldsymbol{y}(k)$ in an LP solution which minimises

$$J = \sum_{k=1}^{N} \mathbf{1}^{\mathrm{T}} \boldsymbol{e}_{+}(k) + \mathbf{1}^{\mathrm{T}} \boldsymbol{e}_{-}(k) \tag{7.423}$$

(i.e. sum of the elements of $\boldsymbol{e}_{+}$ and $\boldsymbol{e}_{-}$ for all of the measurements) with the N equivalence constraints

$$\begin{aligned} \boldsymbol{y}(k+1) = {} & \boldsymbol{R}_1[Y_1(k) \otimes Y_2(k) \otimes \cdots \otimes Y_n(k) \otimes U_1(k)] \\ & + \boldsymbol{R}_2[Y_1(k) \otimes Y_2(k) \otimes \cdots \otimes Y_n(k) \otimes U_2(k)] \\ & + \cdots \\ & + \boldsymbol{R}_m[Y_1(k) \otimes Y_2(k) \otimes \cdots \otimes Y_n(k) \otimes U_m(k)] + \boldsymbol{e}_{+}(k) - \boldsymbol{e}_{-}(k), \quad k = 1, \ldots, N \end{aligned} \tag{7.424}$$

by suitable choice of the elements of the $\boldsymbol{R}_i$, $i = 1, \ldots, m$.

The division of the error term into the additive and subtractive components merely respects the lower bound of zero in LP solutions, yet allows positive and negative deviations. If an element in $\boldsymbol{e}_{+}$ is nonzero, then the corresponding element in $\boldsymbol{e}_{-}$ must be zero, and vice versa. Thus, the LP solution will seek to minimise the sum of the *magnitudes* (moduli) of deviations from Equation 7.422, that is it will cause a best fit of the equations in this sense.

It can be seen in Equation 7.423 that the elements of the predicted crisp output vector $\boldsymbol{y}(k+1)$ will vary linearly with the *membership values* of $\boldsymbol{u}(k)$. To facilitate a globally optimal solution, these relationships were made monotonic by Postlethwaite and Edgar (2000) by the inclusion of appropriate extra inequality constraints for the LP on the elements of the $\boldsymbol{R}_i$, $i = 1, \ldots, m$. In this way, the slopes of the y_i with respect to the μ_j were retained between physically justifiable bounds.

7.13 Predictive Control Using Evolutionary Strategies

An introduction to *evolutionary strategies* is provided in Section 9.8. These are optimisation techniques based on a continuously updated *population* of points in the multidimensional search space, as opposed to a *single* point which is being moved around the search space in an attempt to locate the optimum. One particular evolutionary strategy that will be focused on is the *genetic algorithm.* Here each member of the population is effectively represented by a gene sequence, that is a collection of values in its different attribute categories. These attribute values can accommodate far more than traditional coordinate values, for example binary/discrete values, colours, item selections and so on. On each step of a genetic algorithm, the 'fitness' of each member of the population is determined. The fittest are *selected* to parent children by *recombination* – a process in which some of the attribute values are swopped to form new individuals. Some of these new individuals are then subjected to *mutation* – a random variation of some attributes. The process is then repeated, with the least fit individuals being weeded out of the population on each step. In this section, the focus will be on an *evolutionary strategy* approach arranged for the handling of continuous real numbers, with an application in model predictive control.

In Section 7.8, it was noted that predictive control requires the optimisation of the future control action and output trajectories, by suitable choice of the sequence of future control inputs. This optimisation can in general be required for an arbitrary nonlinear model, and it may have to comply with constraints. In this situation, one notes the potential of *evolutionary algorithms* to deal with both discrete and continuous variables. An important aspect of these techniques is their *flexibility* – so there is nothing prescriptive, and the literature reveals several different interpretations of the basic ideas of *recombination, mutation* and *selection.* The discussion that follows is based loosely on Schwefel (1965) and Lipták (2003).

Consider the MIMO input–output system described by

$$\boldsymbol{y}(k) = \boldsymbol{f}\{\boldsymbol{y}(k-1), \ldots, \boldsymbol{y}(k-n), \boldsymbol{u}(k-1), \ldots, \boldsymbol{u}(k-m)\} \tag{7.425}$$

where k refers to discrete time intervals. As in Section 7.12.2, arrange a predictive controller in IMC format to provide for the model offset correction (Figure 7.47). Predictions of the future output up to an N-step horizon make use of *measurements* as far as possible, otherwise *predictions*:

$$\boldsymbol{y}_{\mathrm{p}}(k+1) = \boldsymbol{f}\{\boldsymbol{y}(k), \boldsymbol{y}(k-1), \ldots, \boldsymbol{y}(k-n+1), \boldsymbol{u}(k), \ldots, \boldsymbol{u}(k-m+1)\} \tag{7.426}$$

$$\boldsymbol{y}_{\mathrm{p}}(k+2) = \boldsymbol{f}\left\{\boldsymbol{y}_{\mathrm{p}}(k+1), \boldsymbol{y}(k), \ldots, \boldsymbol{y}(k-n+2), \boldsymbol{u}(k+1), \ldots, \boldsymbol{u}(k-m+2)\right\} \tag{7.427}$$

$$\boldsymbol{y}_{\mathrm{p}}(k+3) = \boldsymbol{f}\left\{\boldsymbol{y}_{\mathrm{p}}(k+2), \boldsymbol{y}_{\mathrm{p}}(k+1), \ldots, \boldsymbol{y}(k-n+3), \boldsymbol{u}(k+2), \ldots, \boldsymbol{u}(k-m+3)\right\} \tag{7.428}$$

$$\vdots$$

$$\boldsymbol{y}_{\mathrm{p}}(k+N) = \boldsymbol{f}\left\{\boldsymbol{y}_{\mathrm{p}}(k+N-1), \boldsymbol{y}_{\mathrm{p}}(k+N-2), \ldots, \boldsymbol{y}(k-n+N), \boldsymbol{u}(k+N-1), \ldots, \boldsymbol{u}(k-m+N)\right\} \tag{7.429}$$

In these equations, the unknowns are the control vectors $\boldsymbol{u}(k)$, $\boldsymbol{u}(k+1)$, $\boldsymbol{u}(k+2)$, . . . , $\boldsymbol{u}(k+N-1)$. Probably only the first M (e.g. $M=1$ or 2) of these will be optimised, with an assumption of no further control changes thereafter, so

$$\boldsymbol{u}(k+M) = \boldsymbol{u}(k+M-1) \tag{7.430}$$

$$\boldsymbol{u}(k+M+1) = \boldsymbol{u}(k+M-1) \tag{7.431}$$

$$\vdots$$

It is thus the elements of $\boldsymbol{u}(k)$, $\boldsymbol{u}(k+1)$, . . . , $\boldsymbol{u}(k+\mathrm{M}-1)$ which must be solved for in the optimisation. An arbitrary 'fitness' measure J_j can be defined for an individual j selection of 'attributes' $\boldsymbol{u}_j(k)$, $\boldsymbol{u}_j(k+1)$, . . . , $\boldsymbol{u}_j(k+M-1)$, which causes the output response $\boldsymbol{y}_{\mathrm{p}j}(k+1)$, $\boldsymbol{y}_{\mathrm{p}j}(k+2)$, $\boldsymbol{y}_{\mathrm{p}j}(k+3)$, . . . , $\boldsymbol{y}_{\mathrm{p}j}(k+N)$, for example

$$\begin{aligned} J_j = J\left(u_j(k), \ldots, u_j(k+M-1)\right) = &-\sum_{i=k+1}^{k+N} \left[y_{\mathrm{p}j}(i) - \overline{y}_{\mathrm{S}}(i)\right]^{\mathrm{T}} W \left[y_{\mathrm{p}j}(i) - \overline{y}_{\mathrm{S}}(i)\right] \quad \text{(setpoint deviation penalty)} \\ &- \sum_{i=k}^{k+M-1} \left[u_j(i) - u_j(i-1)\right]^{\mathrm{T}} \Lambda \left[u_j(i) - u_j(i-1)\right] \quad \text{(move suppression)} \end{aligned} \tag{7.432}$$

where the weighting matrices $\boldsymbol{W}$ and $\boldsymbol{\Lambda}$ are typically diagonal.

Concatenate all of the unknown elements of the future control vectors into a single attribute U_{ji} (genes $i = 1, \ldots, P$) vector for the individual (trial) j:

$$\boldsymbol{U}_j(k) = \begin{pmatrix} \boldsymbol{u}_j(k) \\ \boldsymbol{u}_j(k+1) \\ \vdots \\ \boldsymbol{u}_j(k+M-1) \end{pmatrix} = \begin{pmatrix} \begin{pmatrix} u_{j1}(k) \\ \vdots \\ u_{jm}(k) \end{pmatrix} \\ \begin{pmatrix} u_{j1}(k+1) \\ \vdots \\ u_{jm}(k+1) \end{pmatrix} \\ \vdots \end{pmatrix} = \begin{pmatrix} U_{j1} \\ U_{j2} \\ \vdots \\ U_{j,P-1} \\ U_{j,P} \end{pmatrix}_k \quad \text{with} \quad P = M \times m \tag{7.433}$$

In the evolutionary strategy, an individual is considered to consist of two parts: an *object variable* (in this case $\boldsymbol{U}_j$) and a *strategy variable* $\boldsymbol{\sigma}_j$.

$$\boldsymbol{a}_j = (\boldsymbol{U}_j, \boldsymbol{\sigma}_j) = \left(\begin{pmatrix} U_{j1} \\ U_{j2} \\ \vdots \\ U_{j,P-1} \\ U_{j,P} \end{pmatrix}, \begin{pmatrix} \sigma_{j1} \\ \sigma_{j2} \\ \vdots \\ \sigma_{j,P-1} \\ \sigma_{j,P} \end{pmatrix} \right) \tag{7.434}$$

Say that there are L individuals in the population. Then each σ_{ji}, $j = 1, \ldots, L$, $i = 1, \ldots, P$, is going to act as a standard deviation for the random mutation of the corresponding U_{ji}. However, since the σ_{ji} must also vary to access both near and far values of the attribute i, they must first be subjected to their own 'mutation', and this is done using a multiplicative normally distributed random process

$$\sigma_{ji} = \sigma_{ji} \times \exp\{\tau' \boldsymbol{N}[0,1] + \tau \boldsymbol{N}_i[0,1]\} \quad \text{(assignment, per iteration)} \tag{7.435}$$

Here $\boldsymbol{N}[0,1]$ is a single random sample drawn from a normal distribution with a mean of 0 and a standard deviation of 1. The first term in the exponential is the same for all of the attributes i, whilst the second term (subscript i) gives individual variations per attribute. The parameters τ' and τ can be thought of as global learning rates. Schwefel (1965) suggests the values

$$\tau' = \frac{1}{\sqrt{2P}} \tag{7.436}$$

$$\tau = \frac{1}{\sqrt{2\sqrt{P}}} \tag{7.437}$$

Consider now the sequence of events as the controller arrives at time k and a decision is required regarding what control settings $\boldsymbol{u}(k)$ are to be implemented for the next time step:

1) $j = 1, \ldots, L$ copies of $\boldsymbol{U}_j$ are made (i.e. L individuals), containing all of the attributes (control vector elements), by random or uniform choice within the ranges of the elements, possibly centred on the $\boldsymbol{u}(k)$ values computed on the last controller step.

2) RECOMBINATION: The best μ individuals from the population (lowest J_j by Equation 7.432) are selected for this. Recombination can be done in pairs (male & female: $j=\mathrm{M}$ & F) or could involve more of the μ individuals in each assignment. Typical possibilities are to pass on attributes from the μ parents $(\boldsymbol{U},\boldsymbol{\sigma})$ to λ offspring $(\boldsymbol{U}',\boldsymbol{\sigma}')$ as follows:

$$U'_{ji} = U_{ji} \quad \text{for all } i \quad \text{(unchanged – no recombination)} \tag{7.438}$$

$$U'_{ji} = U_{\mathrm{F}i} \quad \text{for some of } i \quad \text{(discrete)} \tag{7.439}$$

$$U'_{j+1,i} = U_{\mathrm{M}i} \quad \text{for some of } i \quad \text{(discrete)} \tag{7.440}$$

$$U'_{ji} = U_{\mathrm{M}i} \quad \text{for rest of } i \quad \text{(discrete)} \tag{7.441}$$

$$U'_{j+1,i} = U_{\mathrm{F}i} \quad \text{for rest of } i \quad \text{(discrete)} \tag{7.442}$$

$$U'_{ji} = \frac{U_{\mathrm{F}i} + U_{\mathrm{M}i}}{2} \quad \text{for all } i \quad \text{(intermediate)} \tag{7.443}$$

$$U'_{ji} = \frac{1}{\mu}\sum_{j=1}^{\mu} U_{ji} \quad \text{for all } i \quad \text{(global average)} \tag{7.444}$$

$$\sigma'_{ji} = \sigma_{ji} \quad \text{for all } i \quad \text{(unchanged – no recombination)} \tag{7.445}$$

$$\sigma'_{ji} = \sigma_{\mathrm{F}i} \quad \text{for some of } i \quad \text{(discrete)} \tag{7.446}$$

$$\sigma'_{j+1,i} = \sigma_{\mathrm{M}i} \quad \text{for some of } i \quad \text{(discrete)} \tag{7.447}$$

$$\sigma'_{ji} = \sigma_{\mathrm{M}i} \quad \text{for rest of } i \quad \text{(discrete)} \tag{7.448}$$

$$\sigma'_{j+1,i} = \sigma_{\mathrm{F}i} \quad \text{for rest of } i \quad \text{(discrete)} \tag{7.449}$$

$$\sigma'_{ji} = \frac{\sigma_{\mathrm{F}i} + \sigma_{\mathrm{M}i}}{2} \quad \text{for all } i \quad \text{(intermediate)} \tag{7.450}$$

$$\sigma'_{ji} = \frac{1}{\mu}\sum_{j=1}^{\mu} \sigma_{ji} \quad \text{for all } i \quad \text{(global average)} \tag{7.451}$$

When these λ offspring have been created, there are two possibilities. The λ offspring can *replace* the selected parents – represented by the standard notation $(\mu,\ \lambda)$ – so after eliminating unselected individuals one has $L=\lambda$. Alternatively, the μ parents can be *re-inserted together* with their λ offspring – represented by the standard notation $(\mu+\lambda)$ – so after eliminating unselected individuals one has $L=\mu+\lambda$.

3) MUTATION: Mutate the $L\times P$ strategy variables according to Equations 7.435–7.437:

$$\sigma_{ji} = \sigma_{ji} \times \exp\{\tau' \boldsymbol{N}[0,1] + \tau \boldsymbol{N}_i[0,1]\} \quad \text{(assignment)} \tag{7.452}$$

Now mutate the $L\times P$ objective variables according to

$$U_{ji} = U_{ji} + \boldsymbol{N}\left[0, \sigma_{ji}\right] \quad \text{(assignment)} \tag{7.453}$$

4) Go back to step 2 and repeat iterations until there is sufficient convergence of the U_{ji}. Once converged, proceed to step 5.

5) Reverse the assignment in Equation 7.433:

$$\begin{pmatrix} u(k) \\ u(k+1) \\ \vdots \\ u(k+M-1) \end{pmatrix} = \begin{matrix} \text{average or} \\ \text{best } j \end{matrix} \begin{pmatrix} U_{j1} \\ U_{j2} \\ \vdots \\ U_{j,M\times m-1} \\ U_{j,M\times m} \end{pmatrix} \tag{7.454}$$

and implement the control $u(k)$ for the next controller time step. Wait until the end of that time step and then repeat from step 1.

7.14 Control of Hybrid Systems

A system is *hybrid* if it involves both continuous and discrete variables. A motor car is hybrid in the sense that its speed is a continuous variable in response to discrete settings of the gears, and continuous brake and accelerator inputs. Air conditioners and fridges operate in cycles as their compressors switch on and off. The discrete variables discussed so far are all input *controls*. However, there is another type of discrete behaviour referred to as *autonomous*, in which there are distinct changes in a system's behaviour as its state varies. An example would be a tank with exits at two levels as depicted in Figure 3.43. For a constant feed rate, the level will rise more slowly after it passes the second overflow exit. In Section 3.9.6, both *hybrid Petri nets* and *hybrid automata* were considered as possible *open-loop* modelling environments for these general continuous and discrete systems. Indeed, these two environments originally grew around the representation of only discrete (event-driven) systems, and have subsequently been adapted to include some representation of continuous dynamics.

Many important industrial processes involve discrete inputs, switching sequences, grade changes, and state-dependent behavioural changes. Some of these are seen as *scheduling* problems, where the problem seems to be the choice of points in time to make switches. Actually, these problems are all generic, and are not handled by the large established body of continuous control theory, so there have been strenuous efforts recently to develop useful frameworks for the control of such processes. Of course, the same ideas can be used to *extend* existing controls, for example by supervision of start-ups, shutdowns, batch operations and rate changes.

Optimal control of a system usually requires some form of iterative search for the best policy. Two of the few exceptions were seen to be the LQR (Section 7.6.5) and unbounded GPC or DMC (Section 7.8) in which the optimal policy could be stated explicitly. But for the general problem one aims for a statement of the problem that gives the iterative search the best chance of converging to the correct solution. In hybrid systems, the inclusion of discrete choices of inputs, or autonomous mode changes, renders solutions much more difficult because the normal gradient search techniques cannot be used. For example, if it were necessary to plan a binary control setting over the next 10 steps, 2^{10} different permutations would have to be tested. Fortunately, there are more efficient techniques than such exhaustive trials for conducting discrete searches (such as the *branch-and-bound* technique), but the fact remains that discrete systems offer very difficult optimisation problems.

The combination of continuous and discrete variables in hybrid system optimisation problems can be dealt with using modern software solutions in MILP and MINLP (mixed integer (non)linear programming), such as CPLEX. As problems get larger, these solutions easily fail if they are not posed in an appropriate and simple way. For example, a nonlinear product of two search variables can render the search impossible. On the other hand, one could take advantage of the physical structure of the problem, for example Fikar, Chachuat and Latifi (2005) eliminated switching decision variables by rather searching for the time interval sizes between a fixed number of switches.

In this section, some different ways of perceiving the problem of hybrid system optimal control are considered, giving one some choice in structuring the search problem. The potential of Petri net and automaton control representations will be considered, followed by a unifying framework which reposes the discrete–continuous control problem in a form suitable for computer solution.

7.14.1 Process Control Representation Using Hybrid Petri Nets

The use of Petri nets in Section 3.9.6 revealed their usefulness in the synchronisation of events in subsystems of a process model. They are able to model and supervise concurrent behaviour in a distributed system. With the inclusion of parallel continuous models in the *hybrid* version, it was seen that the discrete logical structure of the Petri net was able to interface and coordinate different continuous models. It will be seen that this is the overall significance of *hybrid systems* – the requirement that different dynamic models need to be interfaced in one structure by a series of logical decisions. The hybrid Petri net is indeed a useful means of visualisation and execution of such a structure.

Figure 7.48 compares two representations of Petri nets performing the same task. The distribution of tokens (black dots) in the system at any time is referred to as the *marking*. Version (a) employs the hybrid representation, with continuous variables shown using double circles and wide blank bars. The continuous transition supplying a continuous place 'fires' continuously at a fixed rate (f or q here) whilst its associated discrete place has a token. So in an interval $\mathrm{d}t$ a volume of $f \times \mathrm{d}t$ will flow into the tank. The discrete place loses its token when its exit transition condition is met. This standard is restricted to the transfer and accumulation of extensive properties such as volume, mass and heat (Champagnat *et al.*, 1998).

In version (b), a traditional discrete Petri net merely supervises the sequence of *transitions* (T) and *places* (P). These are described in each case by a separate set of equations which can have arbitrary complexity. Here such issues as motor speeds and reactor conversions would not be problematic. The sequence shown is fairly 'linear', but one appreciates that even in the approach (b) the Petri net representation will be very useful in distributed systems with multiple concurrent activities and repetitions. Recall from Section 3.9.6 that it is the presence of a *token* in a place, and the loss of such a token when all transition conditions are met, that forces the synchronisation and precedence in the system. From a programming point of view, whether it be for modelling *or* real-time control, the testing for arrival of any token, and testing for release of all tokens present, will obviously allow separate and asynchronous coding of all of the activities in the system.

Figure 7.49 depicts the problem of coal feed to a pressurised gasifier through twin lock hoppers. For operational reasons, the lock hoppers are not allowed to discharge into the gasifier simultaneously. Figure 7.49a shows the open-loop subsystem cycles and Figure 7.49 shows possible additions to create a closed-loop gasifier level control that meets the objectives.

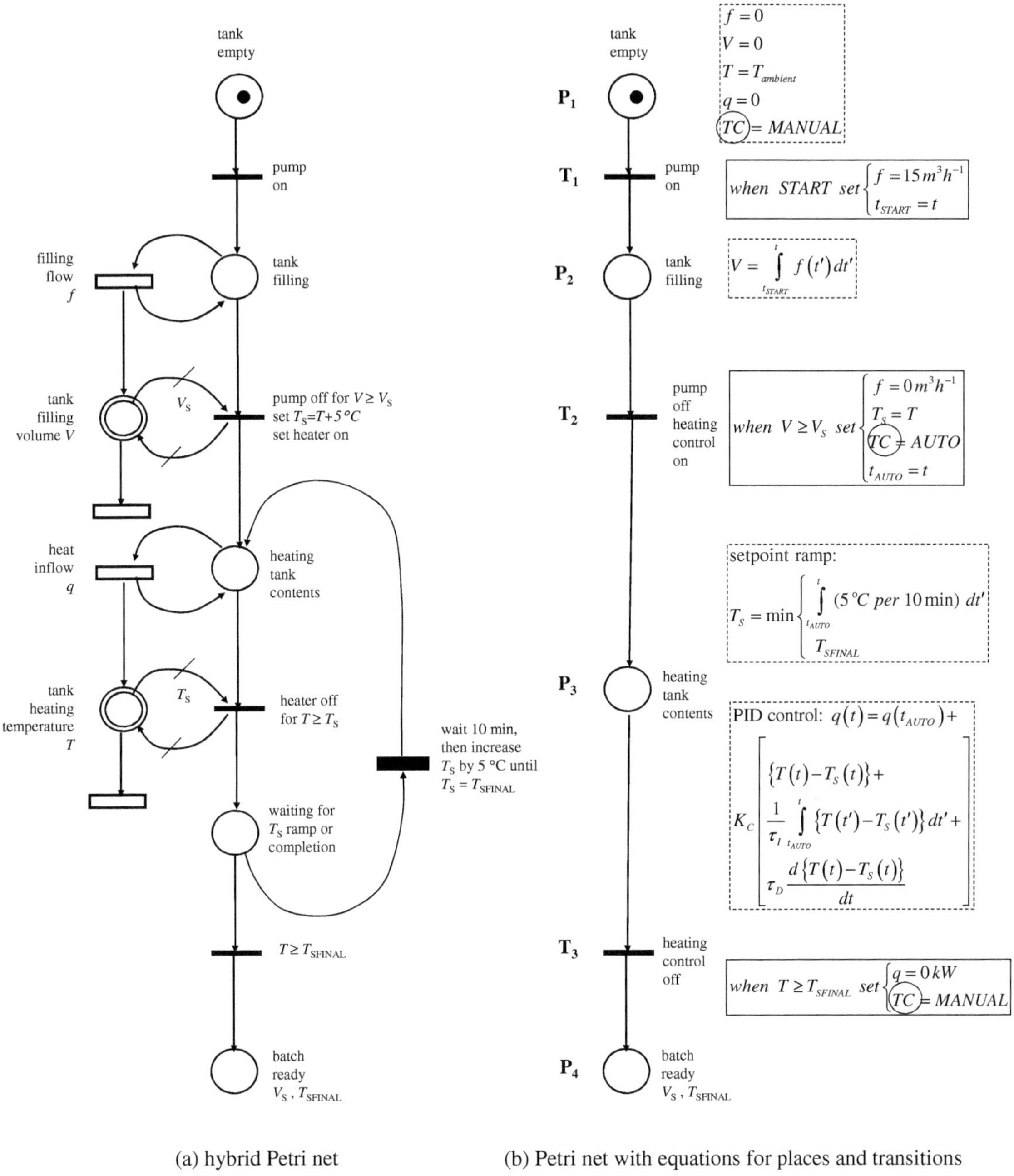

Figure 7.48 Hybrid Petri net and equivalent discrete Petri net assisted by explicit firing conditions for *transitions*, and algebraic and differential equations for *places*: filling a vessel and heating its contents according to a setpoint ramp.

Of course, the gasifier level control in Figure 7.49b focuses on the issues of discrete synchronisation and precedence in the system. Underlying each subsystem will be the continuous dynamics and dynamic control according to the type of separation shown in Figure 7.48b. Yet the important 'control' problem seems to lie on the first-mentioned discrete level. What design philosophy can be brought to bear on this?

Boissel and Kantor (1995) recognised that the Petri net framework was an ideal medium in which to experiment with permutations of the precedence. Each *arc* from *place* to *transition* means exactly the same thing: a token has been delivered to the place and a token can thus be presented at any transition receiving from that place. Likewise, *arcs* proceeding from a *transition* to a *place* can lodge a token at that place once all of the arcs arriving at the transition are presenting a token. Having laid these rules, one can step back from the problem and rather just define what is required:

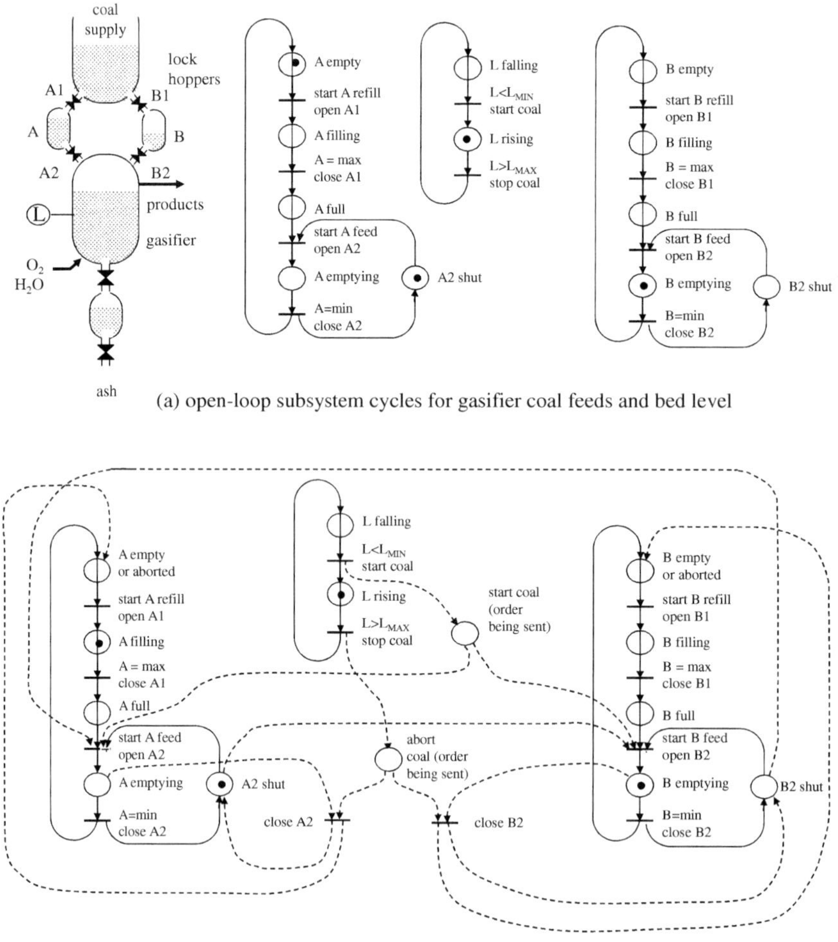

(a) open-loop subsystem cycles for gasifier coal feeds and bed level

(b) gasifier bed level control through dual lockhoppers enforcing exclusive feed

Figure 7.49 Coal feed to a pressurised gasifier through twin lock hoppers which cannot discharge simultaneously for operational reasons: (a) open-loop subsystem cycles; (b) possible closed-loop control.

The subsystem loops have to be able to cycle continuously and must not stop in 'deadlocks', sometimes referred to as 'deadly embraces'. Moreover, any further stipulated rules must be obeyed; for example, in the gasifier problem of Figure 7.49, it is not allowed that both lock hoppers discharge into the gasifier simultaneously.

In theory then, one could experiment with every possible interconnection combination (alternatives to dashed lines in Figure 7.49, for example from *place* 'A full' to *transition* 'start B refill/open B1'). In practice, one is more likely to limit the set of possibilities. Each permutation could be subjected to many random start-up *markings*, and any permutation could be eliminated if it arrives at a deadlock or breaks a rule. The permutations that survive these requirements could be evaluated further in terms of other measures of efficiency. However, for most problems, the number of permutations to be tested becomes too onerous. Thus, Boissel and Kantor (1995) proposed a method of 'simulated annealing' to solve this problem in an approximate (suboptimal) way. A performance measure was given to each trial based on the number of steps it took to reach deadlock or break a rule (good if high) and the number of additional arcs and places required by the scheme (bad if high). The latter criterion recognises the danger of too much complexity in a control scheme. The 'simulated annealing' search method randomly perturbs the lowest cost solution at each step (in this case by altering some connections). The extent of these randomisations gradually decreases on each step according to a slow 'cooling curve' determined by parameter T. The randomisation *away* from the current best solution at each step is of course designed to 'feel out' the neighbourhood to determine whether any better optima lie nearby.

7.14.2 Process Control Representation Using Hybrid Automata

A basic introduction to the use of *automata* in open-loop modelling is given in Section 3.9.6. Like *Petri nets*, the idea of an *automaton* originally focused on discrete systems and grew out of the field of computer science (finite state machines, Turing machines). Indeed, Petri nets are more or less a visualisation of the workings of an automaton. The defining idea is that a number of discrete *states* can exist that can be translated from one to the other by means of *transitions* (Figure 7.50). The attractive idea behind this conceptualisation is that larger systems could be built up out of simpler building blocks which are self-contained in terms of their properties and the rules by which they relate to the rest of the system. There is a strong analogy here to object-oriented programming.

The original definition of an automaton includes the starting state, the possible states and definitions of the causes of transitions between them. This idea has more recently been extended by the inclusion of real-valued state variables. There are several definitions of such *hybrid automata*, but in general they offer the possibility of addressing combined discrete and continuous behaviour (i.e. *hybrid systems*) as well as scope for division into simpler subsystems.

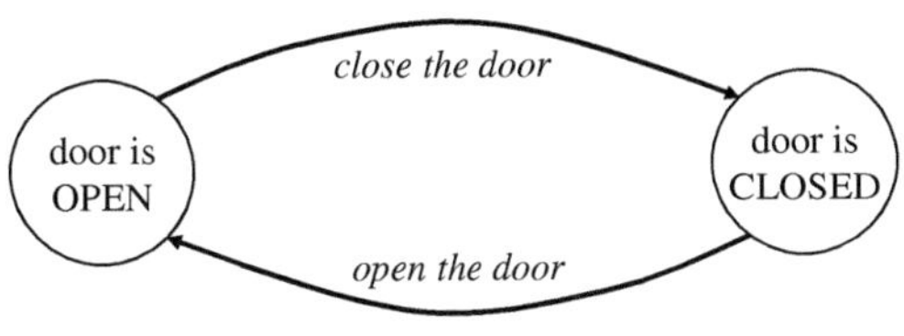

Figure 7.50 Example of a simple automaton.

In hybrid system modelling and control, there is a popular representational paradigm by which the system is divided up into *locations* within which only continuous phenomena occur. Figure 7.51 illustrates how such a concept could be used to handle both discrete inputs and state-dependent behavioural changes (i.e. *mode* changes). Manon, Valentin-Roubinet and Gilles (2002) and Corona, Giua and Seatzu (2004) have used this description of a hybrid system as a basis for optimal control.

Consider a class of hybrid automata based on the structure $H = (\boldsymbol{x}, \boldsymbol{u}, \boldsymbol{L}, \boldsymbol{F}, \boldsymbol{E}, \boldsymbol{M}, \boldsymbol{S})$, defined as follows:

$\boldsymbol{x}$: *vector* of continuous *state variables* x_i, $i = 1, \ldots, n$, common to all locations (but possibly only varying in selected locations);

$\boldsymbol{u}$: *vector* of continuous *control inputs* u_j, $j = 1, \ldots, m$, common to all locations (but possibly only being used in selected locations);

$\boldsymbol{L}$: set of *locations* l_k, $k = 1, \ldots, L$;

$\boldsymbol{F}$: set of *function vectors* $\boldsymbol{f}_k(\boldsymbol{x}, \boldsymbol{u})$, $k = 1, \ldots, L$, such that in location l_k, $\dot{x} = \boldsymbol{f}_k(\boldsymbol{x}, \boldsymbol{u})$;

$\boldsymbol{E}$: matrix of *edge functions* $\boldsymbol{e}_{kp}(\boldsymbol{x})$, $k = 1, \ldots, L$, $p = 1, \ldots, L$, $k \neq p$, such that $\boldsymbol{e}_{l(t)p}(\boldsymbol{x}(t)) \geq 0$, that is for $\boldsymbol{x}$ *inside* location l_k, $\boldsymbol{e}_{kp}(\boldsymbol{x}) \geq 0$ and if this constraint is crossed at time t, $\boldsymbol{x}(t^+) = \boldsymbol{g}_{kp}(\boldsymbol{x}(t))$ and $l(t^+) = l_p$;

$\boldsymbol{G}$: matrix of *function vectors* $\boldsymbol{g}_{kp}(\boldsymbol{x})$, $k = 1, \ldots, L$, $p = 1, \ldots, L$, $k \neq p$, such that $\boldsymbol{x}(t^+) = \boldsymbol{g}_{kp}(\boldsymbol{x}(t^-))$ if the state crosses the constraint $\boldsymbol{e}_{kp}(\boldsymbol{x})$, or the location is *switched* from l_k to l_p at time t;

$\boldsymbol{S}$: matrix of *Boolean parameters* $\boldsymbol{s}_{kp}$, $k = 1, \ldots, L$, $p = 1, \ldots, L$, $k \neq p$, indicating allowed control switching between *locations*, for example if $\boldsymbol{s}_{kp} = \text{TRUE}$, then the location can be *switched* from l_k to l_p.

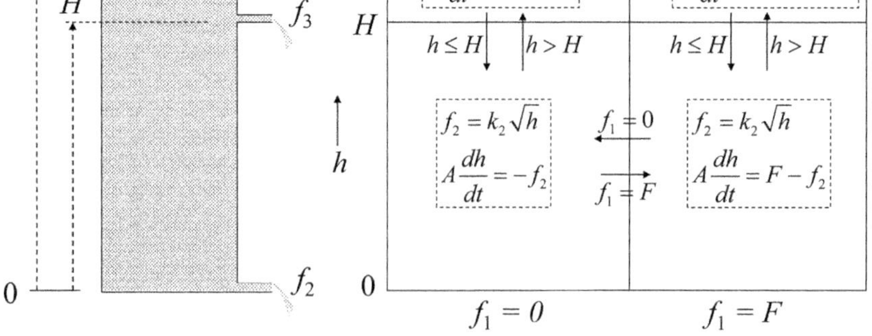

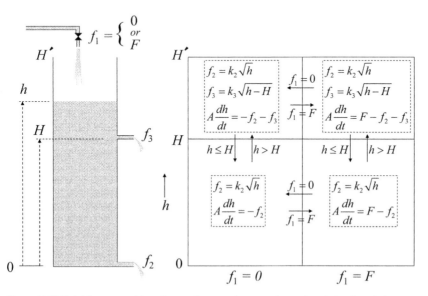

Figure 7.51 Hybrid system with a discrete input and a state-dependent behavioural change represented as continuous *locations* interlinked by guarded *transitions*.

The state of the hybrid automaton at time t is $\{l(t), \boldsymbol{x}(t)\}$ and it starts at $\{l(0), \boldsymbol{x}(0)\}$. Here $l(t)$ is considered to return the index value k that applies at time t, where k lies in the range $1, \ldots, L$. The behaviour thereafter is determined by the input trajectory $\boldsymbol{u}(t)$ together with any *control changes* of the location. One expects that such control changes could be stipulated at any interior point, whilst *autonomous* changes will occur as $\boldsymbol{x}$ arrives at an edge and satisfies a guard condition. The control problem is thus the choice of the continuous input function $\boldsymbol{u}(t)$, together with any discrete location change settings $\boldsymbol{\Sigma}=\{\sigma_1, \sigma_2, \ldots, \sigma_M\}$ and the times at which these are made $\boldsymbol{\tau}=\{\tau_1, \tau_2, \ldots, \tau_M\}$. Here σ_1 refers to location $l_{\sigma 1}$, so $\sigma_1 = l(\tau_1^+)$ and so on, and a receding horizon type of optimal control is being considered with the horizon at time t_F. An objective function for the period 0 to t_F will thus depend only on these input choices and the initial state $\{l(0), \boldsymbol{x}(0)\}$, so the optimal predictive control problem can be stated as

$$\min_{\boldsymbol{u}(t), \boldsymbol{\Sigma}, \boldsymbol{\tau} \text{ for } 0 < t < t_F} J(\boldsymbol{u}(t), \boldsymbol{\Sigma}, \boldsymbol{\tau}, \boldsymbol{x}(0), l(0)) \tag{7.455}$$

such that

$$\dot{\boldsymbol{x}}(t) = \boldsymbol{f}_{l(t)}(\boldsymbol{x}(t), \boldsymbol{u}(t)) \tag{7.456}$$

$$\{l(0), \boldsymbol{x}(0)\} = \{l_0, \boldsymbol{x}_0\} \tag{7.457}$$

$$s_{\sigma_{i-1},\sigma_i} = \text{TRUE} \quad \text{for} \quad i = 1, \ldots, M \quad \text{(location jump allowed)} \tag{7.458}$$

$$l(\tau_i^+) = \sigma_i \quad \text{for} \quad i = 1, \ldots, M \quad \text{(setting event)} \tag{7.459}$$

$$\boldsymbol{x}(\tau_i{}^+) = \boldsymbol{g}_{\sigma_{i-1},\sigma_i} \boldsymbol{x}(\tau_i^-) \quad \text{for} \quad i = 1, \ldots, M \tag{7.460}$$

$$f(\boldsymbol{x}(t) > \boldsymbol{e}_{l(t),p}(\boldsymbol{x}(t))) \quad \text{then} \begin{cases} \boldsymbol{x}(t^+) = \boldsymbol{g}_{l(t),p}(\boldsymbol{x}(t)) \\ l(t^+) = p \end{cases} \quad \text{(event)} \tag{7.461}$$

$$\boldsymbol{u}_{\min} < \boldsymbol{u}(t) < \boldsymbol{u}_{\max} \tag{7.462}$$

where 'box' constraints have been included for the continuous inputs.

To make the task of computing an optimal switching sequence $\boldsymbol{\Sigma}$, $\boldsymbol{\tau}$ and continuous input trajectory $\boldsymbol{u}(t)$ more tractable, one is inclined to respecify the problem in the simplest form required for a specific application. Some specific observations are listed below:

1) If there are *state-dependent mode changes* (*autonomous*), the use of more than one location is necessary, as well as edge transitions with conditions.
2) *Nonlinear behaviour* can be approximated by division of the range of $\boldsymbol{x}$ into several *locations*, and treating as in situation 1. This piecewise linear system is referred to as *piecewise affine*.

$$\dot{\boldsymbol{x}}(t) = \boldsymbol{A}_{l(t)}\boldsymbol{x}(t) + \boldsymbol{B}_{l(t)}\boldsymbol{u}(t) \tag{7.463}$$

3) *If all of the inputs* $\boldsymbol{u}(t)$ *are discrete,* or the use of a limited number of discrete settings is acceptable, the situation could be represented simply as a greater number of *locations* (model modes × input permutations), so for a linear system

$$\dot{\boldsymbol{x}}(t) = \boldsymbol{A}_{l(t)}\boldsymbol{x}(t) + \boldsymbol{v}_{l(t)} \quad \text{where} \quad \boldsymbol{v}_{l(t)} = \boldsymbol{B}_{l(t)}\boldsymbol{u}_{l(t)} \tag{7.464}$$

4) In situation 3, assuming that no transition edges are encountered,

$$\Delta t_{\sigma_i} = \tau_{\sigma_{i+1}} - \tau_{\sigma_i} \tag{7.465}$$

$$\boldsymbol{x}\left(\tau_{\sigma_{i+1}}\right) = \boldsymbol{A}'_{\sigma_i}\boldsymbol{x}\left(\tau_{\sigma_i}\right) + \boldsymbol{v}'_{\sigma_i} \tag{7.466}$$

where from Section 3.9.3.2

$$\boldsymbol{A}'_{\sigma_i} = \mathrm{e}^{\boldsymbol{A}_{\sigma_i}\Delta t_{\sigma_i}} \tag{7.467}$$

$$\boldsymbol{v}'_{\sigma_i} = \left[\mathrm{e}^{\boldsymbol{A}_{\sigma_i}\Delta t_{\sigma_i}} - \boldsymbol{I}\right]\left[\boldsymbol{A}'_{\sigma_i}\right]^{-1}\boldsymbol{v}_{\sigma_i} \tag{7.468}$$

5) In situation 4, fixing also the mode selection times at a fixed interval T,

$$\boldsymbol{x}_{i+1} = \boldsymbol{A}'_{\sigma_i}\boldsymbol{x}_i + \boldsymbol{v}'_{\sigma_i} \tag{7.469}$$

and now only the optimal *location* sequence ($\boldsymbol{\Sigma} = \{\sigma_1, \sigma_2, \ldots, \sigma_M\}$ needs to be found.

6) In most cases, the once-off *transition* of the state as a location is changed (or an edge is crossed), say from location k to location p,

$$\boldsymbol{x}(t^+) = \boldsymbol{g}_{kp}(\boldsymbol{x}(t)) \tag{7.470}$$

will not be necessary. However, it can be used to create a discontinuous jump (e.g. pressure of an air/vapour mixture after ignition), or handle a situation as in Figure 7.52. In this tank flow example,

$$\boldsymbol{x} = \begin{pmatrix} h_1 \\ h_2 \\ h_3 \end{pmatrix}, \quad \boldsymbol{u} = (f) \tag{7.471}$$

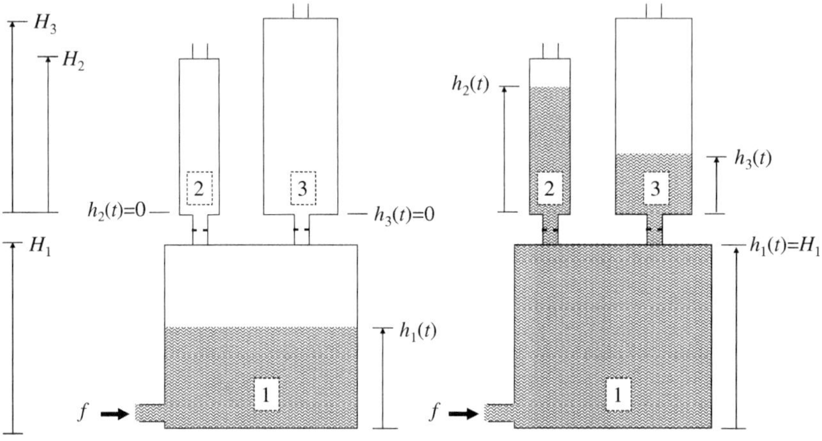

(a) location l_1 : tank 1 not flooded (b) location l_2 : tank 1 flooded

Figure 7.52 System with autonomous mode change requiring translation of state.

$$S = \begin{bmatrix} - & \text{FALSE} \\ \text{FALSE} & - \end{bmatrix} \tag{7.472}$$

$$\boldsymbol{E} = \begin{bmatrix} - & \boldsymbol{e}_{12}(\boldsymbol{x}) \\ \boldsymbol{e}_{21}(\boldsymbol{x}) & - \end{bmatrix} \tag{7.473}$$

$$\boldsymbol{e}_{12}(\boldsymbol{x}) = \begin{pmatrix} H_1 - h_1 \\ -h_2 \\ -h_3 \end{pmatrix} \tag{7.474}$$

$$\boldsymbol{e}_{21}(\boldsymbol{x}) = \begin{pmatrix} h_1 - H_1 \\ h_2 \\ h_3 \end{pmatrix} \tag{7.475}$$

$$\boldsymbol{G} = \begin{bmatrix} - & \boldsymbol{g}_{12}(\boldsymbol{x}) \\ \boldsymbol{g}_{21}(\boldsymbol{x}) & - \end{bmatrix} \tag{7.476}$$

$$\boldsymbol{g}_{12}(\boldsymbol{x}) = \begin{pmatrix} h_1 \\ 0 \\ 0 \end{pmatrix} \tag{7.477}$$

$$\boldsymbol{g}_{21}(\boldsymbol{x}) = \begin{pmatrix} H_1 \\ h_2 \\ h_3 \end{pmatrix} \tag{7.478}$$

7) Corona, Giua and Seatzu (2004) sought an optimal switching of the damping factor between discrete values in a semi-active suspension system (spring and shock absorber). This adjustment instantly changes the model of the system, so the problem is presented as several locations each with its own system model, but with no exogenous input $[\boldsymbol{u}(t) = \boldsymbol{0}]$. In this case, a linear model was adequate, so $\dot{\boldsymbol{x}}(t) = \boldsymbol{A}_{l(t)}\boldsymbol{x}(t)$ and disturbances will decay to zero for a stable $\boldsymbol{A}$. This problem moreover does not have any edge transitions between modes – all mode changes are driven by the location selection sequence in $\boldsymbol{\Sigma}$ – a *switched linear system*. In their approach, they chose

$$S = \begin{bmatrix} 1 & 1 & 0 & 0 & 0 \\ 1 & 1 & 1 & 0 & 0 \\ 0 & 1 & 1 & 1 & \ddots \\ 0 & 0 & 1 & \ddots & 1 \\ 0 & 0 & \ddots & 1 & 1 \end{bmatrix} \tag{7.479}$$

where their locations were arranged in sequence with increasing damping factor models, that is the discrete control choices could only go up or down one damping factor position on each step. The objective of their optimisation, using just a finite M mode switches, was

$$\min_{\boldsymbol{\Sigma}, \boldsymbol{\tau} \text{ for } 0 < t < \infty} J(\boldsymbol{\Sigma}, \boldsymbol{\tau}, \boldsymbol{x}(0), l(0)) = \sum_{i=1}^{M} h_i + \int_0^{\infty} \boldsymbol{x}^{\mathrm{T}}(t')\boldsymbol{Q}_{l(t')}\boldsymbol{x}(t')\mathrm{d}t' \tag{7.480}$$

where h_i is the cost of the ith discrete control change from location $l_{\sigma_{i-1}}$ to location l_{σ_i} at $t = \tau_{\sigma_i}$. For each mode change, the state was varied according to constant matrices $\boldsymbol{M}_{kp}$ using

$$G = \begin{bmatrix} - & g_{12}(\boldsymbol{x}) & - & - \\ g_{21}(\boldsymbol{x}) & - & g_{23}(\boldsymbol{x}) & \ddots \\ - & g_{32}(\boldsymbol{x}) & \ddots & g_{L-1,L}(\boldsymbol{x}) \\ - & \ddots & g_{L,L-1}(\boldsymbol{x}) & - \end{bmatrix} \tag{7.481}$$

$$\boldsymbol{M} = \begin{bmatrix} - & \boldsymbol{M}_{12} & - & - \\ \boldsymbol{M}_{21} & - & \boldsymbol{M}_{23} & \ddots \\ - & \boldsymbol{M}_{32} & \ddots & \boldsymbol{M}_{L-1,L} \\ - & \ddots & \boldsymbol{M}_{L,L-1} & - \end{bmatrix} \tag{7.482}$$

so that for a change from l_k to l_p at $t = \tau_{\sigma_i}$

$$\boldsymbol{x}(\tau^+) = \boldsymbol{g}_{kp}(\boldsymbol{x}(\tau)) = \boldsymbol{M}_{kp}\boldsymbol{x}(\tau) \tag{7.483}$$

The simple structure of this formulation allowed Corona, Giua and Seatzu (2004) to explicitly state the cost of the last of the M switches, and then the cost from the $(M-1)$th step and so on. In this way, they were able to make use of dynamic programming concepts to construct a table of the optimal switch choice and switching time at each of the M switch points, based on the current state feedback $\boldsymbol{x}$.

8) Manon, Valentin-Roubinet and Gilles (2002) also describe a method for optimal switching of a hybrid system. Their objective was to minimise the total amount of time to take a system from an initial state to a final state by switching a given number of times M between available modes. Rather than the objective function in Equation 7.480, they thus used

$$\min_{\boldsymbol{\Sigma}, \boldsymbol{\tau} \text{ for } 0 < t < \infty} J(\boldsymbol{\Sigma}, \boldsymbol{\tau}, \boldsymbol{x}(0), l(0)) = t_{\mathrm{F}} \tag{7.484}$$

with $t_{\mathrm{F}} > \tau_M$, where τ_M is the last time in the sequence of switching times $\boldsymbol{\tau} = \{\tau_1, \tau_2, \ldots, \tau_M\}$. An equality constraint required a given final state $(\boldsymbol{x}(t_{\mathrm{F}}), l(t_{\mathrm{F}}))$. Their application was a continuously fed reactor which switched between two alternative feed streams and associated jacket temperatures. Apart from these two intermediate locations, that is l_1 and l_2, there was just the initial location l_0 (at steady state with a given composition and temperature) and the final location l_3 which was determined only by upper and lower constraints. Apart from the control bounds, the locations were thus determined by different fixed exogenous input values rather than different models, that is

$$\dot{\boldsymbol{x}}(t) = \boldsymbol{A}\boldsymbol{x}(t) + \boldsymbol{B}\boldsymbol{u}_{l(t)} \tag{7.485}$$

7.14.3 Mixed Logical Dynamical Framework in Predictive Control

In a number of publications by Morari and co-workers (e.g. Bemporad and Morari, 1999; Morari, Bemporad and Mignone, 2000; Morari, 2002), a method is described to allow formulation of an optimisation problem for systems involving both continuous dynamics and logical expressions, what the authors called *mixed logical dynamical* (MLD) systems. This of course includes the *hybrid systems* discussed so far, except that Morari and co-workers focused on *linear dynamics*. This is not as limiting as it might appear, because a nonlinear system can be represented by several ranges each with suitable local linear dynamics (i.e. *piecewise affine*), with *autonomous* switching between the models governed by logical expressions. Recall that the term 'autonomous' here refers to logical jumps caused by variation of the system states.

Reconsider the tank with two exits in Figure 7.51. The equations describing that system could, after linearisation, be expressed as

$$\{(h_i > H) \wedge (\text{feed}_i \text{ is on})\} \rightarrow h_{i+1} = h_i + \frac{T}{A}\left[F - k_2' h_i - k_3'(h_i - H)\right] \quad (7.486)$$

$$\{(h_i < H) \wedge (\text{feed}_i \text{ is on})\} \rightarrow h_{i+1} = h_i + \frac{T}{A}\left[F - k_2' h_i\right] \quad (7.487)$$

$$\{(h_i > H) \wedge (\text{feed}_i \text{ is off})\} \rightarrow h_{i+1} = h_i + \frac{T}{A}\left[-k_2' h_i - k_3'(h_i - H)\right] \quad (7.488)$$

$$\{(h_i < H) \wedge (\text{feed}_i \text{ is off})\} \rightarrow h_{i+1} = h_i + \frac{T}{A}\left[-k_2' h_i\right] \quad (7.489)$$

As this type of structure is considered further, use will be made of the following symbols:

$\wedge$: AND (7.490)

$\vee$: OR (7.491)

$\sim$: NOT (7.492)

$\rightarrow$: IMPLIES (if ... then ...) (7.493)

$\leftrightarrow$: IF AND ONLY IF (if and only if ... then ...) (7.494)

$\oplus$: EXCLUSIVE OR (if ... or if ... but not both then ...) (7.495)

Optimising procedures do not accept this type of verbal model description. For example, an LP specification includes *equalities, inequalities* and an *objective function*. Thus, one is next tempted to create *Boolean* variables with values 0 or 1 to represent FALSE or TRUE, δ_1 for $(h > H)$ and δ_2 for (*feed is on*), so Equations 7.486–7.489 can be represented by

$$h_{i+1} = h_i + \frac{T}{A}\left[\delta_2 F - k_2' h_i - \delta_1 k_3'(h_i - H)\right] \quad (7.496)$$

Although F is a constant in this problem, making the first term linear, the term $\delta_1 k_3'(h_i - H)$ is *nonlinear* in two of the variables of interest, δ_1 and h. Though some optimisers might accept this specification, it will complicate the solution. Rather one aims for forms which are linear in all of the variables in order to utilise well-established optimisation algorithms in MILP and MIQP (mixed integer quadratic programming). The term 'mixed integer' refers to the presence of both continuous and discrete variables (such as δ_1 and δ_2, or any integer variable for that matter). The methods developed below seek to replace a nonlinear optimiser constraint such as Equation 7.496 with a combination of linear constraints which achieves the same purpose.

Perhaps it is not yet clear why the 'model' description should have equality and inequality equations in a form which is suitable for an 'optimiser'. What is being envisaged here is that the model will form the basis of a receding horizon predictive controller, which as seen in Section 7.8 will require optimisation of its predicted future output trajectory (within constraints). Typical objective functions penalise deviations from the setpoint trajectory, and the amount of control action effort required. In the case of the tank represented by Equation 7.496, one would be seeking a desirable future sequence of δ_2 (e.g. 0100110 . . .) which gives the minimum deviations of h_i from the setpoint at the future points in time i, where these h_i can be predicted by recursively 'chaining' Equation 7.496.

Morari and co-workers converted systems as in the above example into MLD forms described by linear dynamic equations subject to linear mixed integer inequalities. The propositional logic of equations such as Equations 7.486–7.489 was transformed into the linear inequalities involving integer and continuous variables using methods such as detailed by Cavalier, Pardalos and Soyster (1990) and Raman and Grossmann (1992). Using X_i to represent the ith propositional statement, for example '$h > H'$' or '*the feed is on*', and a Boolean parameter δ_i to represent the truth ($\delta_i = 1$) of falseness ($\delta_i = 0$) of this statement, Morari (2002) provides Table 7.3.

In Table 7.3, ε is a small tolerance, typically the machine precision. The constant m_i is a value below the estimated minimum of $f_i(x)$ and the constant M_i is a value above the estimated maximum of $f_i(x)$. Note that some of the logical conditions require additional inequality or equality conditions, and possibly an additional variable such as z (real). The relations involving the real variable comparisons play a key role in the description of hybrid systems. As an example, consider the case of the tank in Figure 7.51, but limit consideration to only the situation where *feed is on* ($\delta_2 = 1$).

$$\text{EVENT:} \quad [H - h_i \leq 0] \leftrightarrow [\delta_1 = 1] \quad \text{is represented by} \quad \begin{cases} H - h_i \leq H - H\delta_1 \\ H - h_i \geq \varepsilon + (H - H' - \varepsilon)\delta_1 \end{cases} \tag{7.497}$$

Table 7.3 Conversion of logic relations into mixed integer inequalities (Morari, 2002).

Relation	Logic	(In)equalities
AND (∧)	$X_1 \wedge X_2$	$\delta_1 = 1,\ \delta_2 = 1$
OR (∨)	$X_1 \vee X_2$	$\delta_1 + \delta_2 \geq 1$
NOT (~)	$\sim X_1$	$\delta_1 = 0$
XOR (⊕)	$X_1 \oplus X_2$	$\delta_1 + \delta_2 = 1$
IMPLY (→)	$X_1 \rightarrow X_2$	$\delta_1 - \delta_2 \leq 0$
IFF (↔)	$X_1 \leftrightarrow X_2$	$\delta_1 - \delta_2 = 0$
ASSIGN (=, ↔)	$X_3 = X_1 \wedge X_2$ $X_3 \leftrightarrow X_1 \wedge X_2$	$\delta_1 + (1 - \delta_3) \geq 1$ $\delta_2 + (1 - \delta_3) \geq 1$ $(1 - \delta_1) + (1 - \delta_2) + \delta_3 \geq 1$
EVENT	$[f(x) \leq 0] \leftrightarrow [\delta = 1]$	$f(x) \leq M - M\delta$ $f(x) \geq \varepsilon + (m - \varepsilon)\delta$
IF . . . THEN . . . ELSE	IF X THEN $z = f_1(x)$ ELSE $z = f_2(x)\{z = \delta f_1(x) + [1 - \delta]f_2(x)\}$	$(m_2 - M_1)\delta + z \leq f_2(x)$ $(m_1 - M_2)\delta - z \leq -f_2(x)$ $(m_1 - M_2)(1 - \delta) + z \leq f_1(x)$ $(m_2 - M_1)(1 - \delta) - z \leq -f_1(x)$

IF... THEN... ELSE: $h_{i+1} = h_i + \frac{T}{A}\left[F - k_2' h_i - \delta_1 k_3'(h_i - H)\right]$ is represented by

$$\begin{cases} (m_2 - M_1)\delta_1 + h_{i+1} \leq f_2 \\ (m_1 - M_2)\delta_1 - h_{i+1} \leq -f_2 \\ (m_1 - M_2)(1-\delta_1) + h_{i+1} \leq f_1 \\ (m_2 - M_1)(1-\delta_1) - h_{i+1} \leq -f_1 \end{cases}$$

where

$$f_1 = h_i + \frac{T}{A}[F - k'_2 h_i - k'_3(h_i - H)]$$
$$f_2 = h_i + \frac{T}{A}[F - k'_2 h_i] \tag{7.498}$$

and the 'tightest' constant values are

$$m_1 = H + \frac{T}{A}\left[F - k_2' H\right]$$
$$M_1 = H' + \frac{T}{A}\left[F - k_2' H' - k_3'(H' - H)\right]$$
$$m_2 = 0 + \frac{T}{A}[F]$$
$$M_2 = H + \frac{T}{A}\left[F - k_2' H\right]$$

In this example, the 'tightest' values have been used for the extremes m and M. However, any finite values are acceptable as long as they are, respectively, low enough and high enough. So choosing +1000 for both M's and −1000 for both m's, one requires instead

$$\begin{cases} H - h_i \leq 1000 - 1000\delta_1 \\ H - h_i \geq \varepsilon + (-1000 - \varepsilon)\delta_1 \\ -2000\delta_1 + h_{i+1} \leq \left\{h_i + \frac{T}{A}\left[F - k_2' h_i\right]\right\} \\ -2000\delta_1 - h_{i+1} \leq -\left\{h_i + \frac{T}{A}\left[F - k_2' h_i\right]\right\} \\ -2000(1-\delta_1) + h_{i+1} \leq \left\{h_i + \frac{T}{A}\left[F - k_2' h_i - k_3'(h_i - H)\right]\right\} \\ -2000(1-\delta_1) - h_{i+1} \leq -\left\{h_i + \frac{T}{A}\left[F - k_2' h_i - k_3'(h_i - H)\right]\right\} \end{cases} \tag{7.499}$$

in which it is easily seen that $\delta_1 = 1$ forces the equivalence in the bottom two inequalities (yet satisfies the middle two), and $\delta_1 = 0$ forces the equivalence in the middle two inequalities (yet satisfies the bottom two).

In the above example, six inequalities have been required to represent a single IF . . . THEN . . . ELSE . . . case. Whilst one could manage to deal with more compounded logic (e.g. the inclusion of δ_2), it can soon become unwieldy. Since any combinational relation of logic variables can be translated into the *conjunctive normal form*, it is, however, possible to automate the conversion of the hybrid logic into linear inequalities.

Some additional notes on the MLD framework are given below:

1) *Saturation of an output*: $y = \min(cx, y_{MAX})$ that is IF $cx \leq y_{MAX}$ THEN $y = cx$ ELSE $y = y_{MAX}$

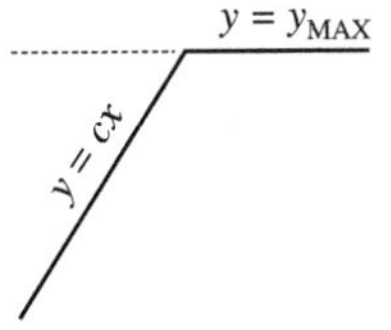

Following the same procedure as above,

$$\begin{cases} y_{\mathrm{MAX}} - cx \le 1000 - 1000\delta \\ y_{\mathrm{MAX}} - cx \ge \varepsilon + (-1000 - \varepsilon)\delta \\ -2000\delta + y \le cx \\ -2000\delta - y \le -cx \\ -2000(1-\delta) + y \le y_{\mathrm{MAX}} \\ -2000(1-\delta) - y \le -y_{\mathrm{MAX}} \end{cases} \tag{7.500}$$

For saturation at both ends of the range (δ_1, δ_2), simplify the result somewhat by making use of the fact that

$$[\delta_1 = 1] \rightarrow [\delta_2 = 0] \tag{7.501}$$

$$[\delta_2 = 1] \rightarrow [\delta_1 = 0] \tag{7.502}$$

2) *Discrete inputs*: For example, $\boldsymbol{x}(t+T) = \boldsymbol{A}\boldsymbol{x}(t) + \boldsymbol{B}\boldsymbol{u}(t)$ where there are only three different constant settings for $\boldsymbol{u}(t)$:

$$[\delta_1(t) = 1] \rightarrow [\boldsymbol{u}(t) = \boldsymbol{u}_1] \tag{7.503}$$

$$[\delta_2(t) = 1] \rightarrow [\boldsymbol{u}(t) = \boldsymbol{u}_2] \tag{7.504}$$

$$[\delta_3(t) = 1] \rightarrow [\boldsymbol{u}(t) = \boldsymbol{u}_3] \tag{7.505}$$

Following Bemporad and Morari (1999), use

$$\boldsymbol{x}(t+T) = \boldsymbol{A}\boldsymbol{x}(t) + \boldsymbol{B}\left[\,\delta_1(t)\boldsymbol{I} \;\vdots\; \delta_2(t)\boldsymbol{I} \;\vdots\; \delta_3(t)\boldsymbol{I}\,\right] \begin{pmatrix} \boldsymbol{u}_1 \\ \hdashline \boldsymbol{u}_2 \\ \hdashline \boldsymbol{u}_3 \end{pmatrix} \tag{7.506}$$

with the inequalities

$$0 \le \delta_1(t) + \delta_2(t) + \delta_3(t) - 1 \tag{7.507}$$

$$0 \le -\delta_1(t) - \delta_2(t) - \delta_3(t) + 1 \tag{7.508}$$

which will ensure that one and only one of δ_1, δ_2, δ_3 will be true at any time, that is $\delta_1(t) + \delta_2(t) + \delta_3(t) = 1$.

3) *Qualitative outputs*: These can be useful when heuristics and rules of thumb are used.
Consider, for example,

$$x(t+T) = ax(t) + bu(t) \tag{7.509}$$

where the output $y(t)$ is to be classified into three qualitative categories:

$$y(t) = \begin{cases} \text{HOT} & \text{for} \quad x(t) > 27\,^{\circ}\text{C} \\ \text{MILD} & \text{for} \quad 15\,^{\circ}\text{C} \le x(t) \le 27\,^{\circ}\text{C} \\ \text{COLD} & \text{for} \quad x(t) < 15\,^{\circ}\text{C} \end{cases} \tag{7.510}$$

Following Bemporad and Morari (1999), represent the following 'staircase' using the EVENT relations in Table 7.3.

$$[x(t) \le 15\,^{\circ}\text{C}] \leftrightarrow [\delta_1(t) = 1] \tag{7.511}$$

$$[x(t) \leq 27\,°\mathrm{C}] \leftrightarrow [\delta_2(t) = 1] \tag{7.512}$$

$$[x(t) > 27\,°\mathrm{C}] \leftrightarrow [\delta_3(t) = 1] \tag{7.513}$$

Then there will also be the requirements

$$[\delta_1(t) = 1] \rightarrow [\delta_2(t) = 1] \tag{7.514}$$

$$[\delta_3(t) = 1] \rightarrow [\delta_2(t) = \delta_1(t) = 0] \tag{7.515}$$

which can be represented using three relations of the IMPLY form in Table 7.3.

4) *General MLD model form*: Consider as an example a discrete linear piecewise affine system (autonomous model switch) which has a continuous input, as well as discrete inputs as in Equation 7.506.

$$\boldsymbol{x}(t+T) = \begin{cases} \boldsymbol{A}_\mathrm{a}\boldsymbol{x}(t) + \boldsymbol{B}_\mathrm{a}\boldsymbol{u}(t) + \delta_1(t)\boldsymbol{C}_\mathrm{a}\boldsymbol{u}_1 + [1-\delta_1(t)]\boldsymbol{C}_\mathrm{a}\boldsymbol{u}_2 & \text{for} \quad \boldsymbol{x}(t) \leq \boldsymbol{e} \\ \boldsymbol{A}_\mathrm{b}\boldsymbol{x}(t) + \boldsymbol{B}_\mathrm{b}\boldsymbol{u}(t) + \delta_1(t)\boldsymbol{C}_\mathrm{b}\boldsymbol{u}_1 + [1-\delta_1(t)]\boldsymbol{C}_\mathrm{b}\boldsymbol{u}_2 & \text{for} \quad \boldsymbol{x}(t) > \boldsymbol{e} \end{cases} \tag{7.516}$$

Choosing

$$[x(t) - e \leq 0] \leftrightarrow [\delta_\mathrm{a}(t) = 1] \tag{7.517}$$

$$\begin{aligned} \boldsymbol{x}(t+T) = {} & \delta_\mathrm{a}(t)\{\boldsymbol{A}_\mathrm{a}\boldsymbol{x}(t) + \boldsymbol{B}_\mathrm{a}\boldsymbol{u}(t) + \delta_1(t)\boldsymbol{C}_\mathrm{a}\boldsymbol{u}_1 + [1-\delta_1(t)]\boldsymbol{C}_\mathrm{a}\boldsymbol{u}_2\} \\ & + [1-\delta_\mathrm{a}(t)]\{\boldsymbol{A}_\mathrm{b}\boldsymbol{x}(t) + \boldsymbol{B}_\mathrm{b}\boldsymbol{u}(t) + \delta_1(t)\boldsymbol{C}_\mathrm{b}\boldsymbol{u}_1 + [1-\delta_1(t)]\boldsymbol{C}_\mathrm{b}\boldsymbol{u}_2\} \end{aligned} \tag{7.518}$$

$$\begin{aligned} \boldsymbol{x}(t+T) = {} & \{\boldsymbol{A}_\mathrm{a} - \boldsymbol{A}_\mathrm{b}\}\delta_\mathrm{a}(t)\boldsymbol{x}(t) + \boldsymbol{A}_\mathrm{b}\boldsymbol{x}(t) + \{\boldsymbol{B}_\mathrm{a} - \boldsymbol{B}_\mathrm{b}\}\delta_\mathrm{a}(t)\boldsymbol{u}(t) \\ & + \delta_\mathrm{a1}(t)\{\boldsymbol{C}_\mathrm{a}\boldsymbol{u}_1 - \boldsymbol{C}_\mathrm{a}\boldsymbol{u}_2 - \boldsymbol{C}_\mathrm{b}\boldsymbol{u}_1 + \boldsymbol{C}_\mathrm{b}\boldsymbol{u}_2\} \\ & + \delta_\mathrm{a}(t)\{\boldsymbol{C}_\mathrm{a}\boldsymbol{u}_2 - \boldsymbol{C}_\mathrm{b}\boldsymbol{u}_2\} + \delta_1(t)\{\boldsymbol{C}_\mathrm{b}\boldsymbol{u}_1 - \boldsymbol{C}_\mathrm{b}\boldsymbol{u}_2\} + \boldsymbol{C}_\mathrm{b}\boldsymbol{u}_2 \end{aligned} \tag{7.519}$$

$$\boldsymbol{x}(t+T) = \boldsymbol{A}_\mathrm{ab}\boldsymbol{z}(t) + \boldsymbol{A}_\mathrm{b}\boldsymbol{x}(t) + \boldsymbol{B}_\mathrm{ab}\boldsymbol{w}(t) + \boldsymbol{B}_\mathrm{b}\boldsymbol{u}(t) + \delta_\mathrm{a1}(t)\boldsymbol{v}_\mathrm{a1} + \delta_\mathrm{a}(t)\boldsymbol{v}_\mathrm{a} + \delta_1(t)\boldsymbol{v}_1 + \boldsymbol{v}_2 \tag{7.520}$$

where

$$\delta_\mathrm{a1}(t) = \delta_\mathrm{a}(t)\delta_1(t) \tag{7.521}$$

$$\boldsymbol{z}(t) = \delta_\mathrm{a}(t)\boldsymbol{x}(t) \tag{7.522}$$

$$\boldsymbol{w}(t) = \delta_\mathrm{a}(t)\boldsymbol{u}(t) \tag{7.523}$$

with constants

$$\boldsymbol{A}_\mathrm{ab} = \boldsymbol{A}_\mathrm{a} - \boldsymbol{A}_\mathrm{b} \tag{7.524}$$

$$\boldsymbol{B}_\mathrm{ab} = \boldsymbol{B}_\mathrm{a} - \boldsymbol{B}_\mathrm{b} \tag{7.525}$$

$$\boldsymbol{v}_\mathrm{a1} = \boldsymbol{B}_\mathrm{a}\boldsymbol{u}_1 - \boldsymbol{B}_\mathrm{a}\boldsymbol{u}_2 - \boldsymbol{B}_\mathrm{b}\boldsymbol{u}_1 + \boldsymbol{B}_\mathrm{b}\boldsymbol{u}_2 \tag{7.526}$$

$$\boldsymbol{v}_\mathrm{a} = \boldsymbol{B}_\mathrm{a}\boldsymbol{u}_2 - \boldsymbol{B}_\mathrm{b}\boldsymbol{u}_2 \tag{7.527}$$

$$\boldsymbol{v}_1 = \boldsymbol{B}_\mathrm{b}\boldsymbol{u}_1 - \boldsymbol{B}_\mathrm{b}\boldsymbol{u}_2 \tag{7.528}$$

$$\boldsymbol{v}_2 = \boldsymbol{B}_\mathrm{b}\boldsymbol{u}_2 \tag{7.529}$$

The model is thus

$$\boldsymbol{x}(t+T) = \boldsymbol{A}_\mathrm{ab}\boldsymbol{z}(t) + \boldsymbol{A}_\mathrm{b}\boldsymbol{x}(t) + \boldsymbol{B}_\mathrm{ab}\boldsymbol{w}(t) + \boldsymbol{B}_\mathrm{b}\boldsymbol{u}(t) + \delta_\mathrm{a1}(t)\boldsymbol{v}_\mathrm{a1} + \delta_\mathrm{a}(t)\boldsymbol{v}_\mathrm{a} + \delta_1(t)\boldsymbol{v}_1 + \boldsymbol{v}_2 \tag{7.530}$$

with the required constraints by Table 7.3:

$$\left.\begin{array}{l} \boldsymbol{x}(t) - \boldsymbol{e} \leq \boldsymbol{M} - \boldsymbol{M}\delta_a(t) \\ \boldsymbol{x}(t) - \boldsymbol{e} \geq \boldsymbol{\varepsilon} + (\boldsymbol{m} - \boldsymbol{\varepsilon})\delta_a(t) \end{array}\right\} \text{ with constants } \boldsymbol{M} > \boldsymbol{x} - \boldsymbol{e},\ \boldsymbol{m} < \boldsymbol{x} - \boldsymbol{e} \quad \text{(for Equation 7.517)} \tag{7.531}$$

$$\left.\begin{array}{l} \delta_a(t) + [1 - \delta_{a1}(t)] \geq 1 \\ \delta_1(t) + [1 - \delta_{a1}(t)] \geq 1 \\ [1 - \delta_a(t)] + [1 - \delta_1(t)] + \delta_{a1}(t) \geq 1 \end{array}\right\} \quad \text{(for Equation 7.521)} \tag{7.532}$$

$$\left.\begin{array}{l} \boldsymbol{z}(t) \leq \delta_a(t)\boldsymbol{M}' \\ \boldsymbol{z}(t) \geq \delta_a(t)\boldsymbol{m}' \\ \boldsymbol{z}(t) \leq \boldsymbol{x}(t) - [1 - \delta_a(t)]\boldsymbol{m}' \\ \boldsymbol{z}(t) \geq \boldsymbol{x}(t) - [1 - \delta_a(t)]\boldsymbol{M}' \end{array}\right\} \text{ with constants } \boldsymbol{M}' > \boldsymbol{x} \text{ and } \boldsymbol{m}' < \boldsymbol{x} \quad \text{(for Equation 7.522)} \tag{7.533}$$

$$\left.\begin{array}{l} \boldsymbol{w}(t) \leq \delta_a(t)\boldsymbol{M}'' \\ \boldsymbol{w}(t) \geq \delta_a(t)\boldsymbol{m}'' \\ \boldsymbol{w}(t) \leq \boldsymbol{u}(t) - [1 - \delta_a(t)]\boldsymbol{m}'' \\ \boldsymbol{w}(t) \geq \boldsymbol{u}(t) - [1 - \delta_a(t)]\boldsymbol{M}'' \end{array}\right\} \text{ with constants } \boldsymbol{M}'' > \boldsymbol{u} \text{ and } \boldsymbol{m}'' < \boldsymbol{u} \quad \text{(for Equation 7.523)} \tag{7.534}$$

Further constraints may be added in a predictive control algorithm, for example upper and lower bounds for $\boldsymbol{x}$ and $\boldsymbol{u}$.

5) *Model predictive control based on the MLD representation of a system*: Putting all of the Boolean variables into a vector $\boldsymbol{\delta}$, the general form of the discrete open-loop model is seen from Equations 7.530–7.534 to be

$$\boldsymbol{x}(t+T) = \boldsymbol{A}_{ab}\boldsymbol{z}(t) + \boldsymbol{A}_b\boldsymbol{x}(t) + \boldsymbol{B}_{ab}\boldsymbol{w}(t) + \boldsymbol{B}_b\boldsymbol{u}(t) + \boldsymbol{D\delta}(t) + \boldsymbol{v}_2 \tag{7.535}$$

with the conditions

$$\boldsymbol{Ex}(t) + \boldsymbol{Fz}(t) \leq \boldsymbol{Gu}(t) + \boldsymbol{Hw}(t) + \boldsymbol{P\delta}(t) + \boldsymbol{q} \tag{7.536}$$

Here $\boldsymbol{x}$ contains the state variables, $\boldsymbol{z}$ auxiliary variables related to the states, $\boldsymbol{w}$ auxiliary variables related to the continuous inputs, $\boldsymbol{u}$ the continuous inputs and $\boldsymbol{\delta}$ Boolean variables for discrete inputs as well as for modes and auxiliary purposes. From a predictive control viewpoint, consider the constant matrices $\boldsymbol{E}$, $\boldsymbol{F}$, $\boldsymbol{G}$, $\boldsymbol{H}$ and $\boldsymbol{P}$, and the constant vector $\boldsymbol{q}$, to have been augmented with extra rows for desired constraints on $\boldsymbol{x}$ and $\boldsymbol{u}$. Then the predictive control problem can be specified as

$$\min_{\substack{\boldsymbol{u}(t+iT),\boldsymbol{\delta}(t+iT) \\ \text{for } i=0,\ldots,N-1}} J(\boldsymbol{u},\boldsymbol{\delta},\boldsymbol{x}(t)) = \sum_{i=0}^{N-1} \left\{ \begin{array}{l} [\boldsymbol{x}(t+iT+T) - \boldsymbol{x}_S]^T \boldsymbol{W}_1 [\boldsymbol{x}(t+iT+T) - \boldsymbol{x}_S] \\ \quad + [\boldsymbol{z}(t+iT) - \boldsymbol{z}_S]^T \boldsymbol{W}_2 [\boldsymbol{z}(t+iT) - \boldsymbol{z}_S] \\ \quad + [\boldsymbol{u}(t+iT) - \boldsymbol{u}_S]^T \boldsymbol{\Lambda}_1 [\boldsymbol{u}(t+iT) - \boldsymbol{u}_S] \\ \quad + [\boldsymbol{w}(t+iT) - \boldsymbol{w}_S]^T \boldsymbol{\Lambda}_2 [\boldsymbol{w}(t+iT) - \boldsymbol{w}_S] \\ \quad + [\boldsymbol{\delta}(t+iT) - \boldsymbol{\delta}_S]^T \boldsymbol{\Lambda}_3 [\boldsymbol{\delta}(t+iT) - \boldsymbol{\delta}_S] \end{array} \right\} \tag{7.537}$$

such that for $i = 0, \ldots, N-1$:

$$x(t+iT+T) = A_{ab}z(t+iT) + A_{b}x(t+iT) + B_{ab}w(t+iT) + B_{b}u(t+iT) + D\delta(t+iT) + v_2 \tag{7.538}$$

and

$$Ex(t+iT) + Fz(t+iT) \le Gu(t+iT) + Hw(t+iT) + P\delta(t+iT) + q \tag{7.539}$$

In practice, only the first $M=1$, 2 or 3 control input steps are optimised. As explained in the MPC discussion of Section 7.8, more than one optimised step increases the controller gain, in the belief that correcting steps will follow. Of course, only the first calculated input is actually used, before the entire calculation is repeated for the next controller time step. The predicted trajectories must be based on the input values (namely $u(t)$ and $\delta_1(t)$ in the above example), remaining constant after the first M optimised input steps. Thus, the remaining values must be replaced in the expressions according to

$$u(t+iT) = u(t+[M-1]T) \quad \text{for} \quad i = M, \ldots, [N-1] \tag{7.540}$$

$$\delta_1(t+iT) = \delta_1(t+[M-1]T) \quad \text{for} \quad i = M, \ldots, [N-1] \tag{7.541}$$

The objective function (Equation 7.537) includes desirable values x_S, z_S, u_S, w_S and δ_S which must be set by the user, taking equilibrium into account, that is the specified combination should aim to achieve a steady state, allowing J to approach zero. An alternative is to suppress input *moves*, rather than deviations from fixed values, using instead terms

$$\begin{aligned}
&\vdots\\
&[u(t+iT) - u(t+iT-T)]^{\mathrm{T}}\Lambda_1[u(t+iT) - u(t+iT-T)]\\
&\quad + \left[w(t+iT) - w(t+iTp-T)\right]^{\mathrm{T}}\Lambda_2\left[w(t+iT) - w(t+iTp-T)\right]\\
&\quad + [\delta(t+iT) - \delta(t+iT-T)]^{\mathrm{T}}\Lambda_3[\delta(t+iT) - \delta(t+iT-T)]\\
&\vdots
\end{aligned} \tag{7.542}$$

Optimal MPC of dynamic systems involving both continuous and discrete variables is sometimes referred to as *mixed integer dynamic optimisation* (MIDO). The problem as specified is in the category that can be solved by *mixed integer quadratic programming.* However, in the worst case, the computer solution time increases exponentially with the number of binary variables in δ (Raman and Grossmann, 1991), since the combinatorial nature of the problem does not lend itself to gradient searches. As noted by Morari (2002), the four major solution approaches are

cutting plane methods;
decomposition methods;
logic-based methods;
branch and bound methods.

For MIQP problems, Fletcher and Leyffer (1998) have found that the major solvers are

generalised Bender's decomposition (GBD);
outer approximation (OA);
LP/QP-based branch and bound;
branch and bound.

There is agreement that the *branch and bound* technique is the most effective for MIQP problems. In this method, each integer choice creates a *branch*. These branches form subproblems which are *fathomed* or left *pending* as the tree is searched. The best subproblem cost so far establishes a *bound* which allows whole sections of the tree to be eliminated as soon as the bound is exceeded by other branches from that node (or if a constraint is violated). Morari, Bemporad and Mignone (2000) note that the MLD formulation tends to result in *sparse* equation sets, so solutions designed for sparse systems are favoured. In Chapter 9, the various optimisation algorithms will be reviewed in a more general context for the solution of processing industry problems.

Since Equations 7.538 and 7.539 are linear, reformulation of the objective function using 1-norms instead of 2-norms, that is

$$\begin{array}{l}\min\\ \boldsymbol{u}(t+iT),\boldsymbol{\delta}(t+iT)\\ \text{for } i=0,\ldots,N-1\end{array} J(\boldsymbol{u},\boldsymbol{\delta},\boldsymbol{x}(t)) = \sum_{i=0}^{N-1}\left\{\begin{array}{l}\boldsymbol{\omega}_1^{\mathrm{T}}|\boldsymbol{x}(t+iT+T)-\boldsymbol{x}_{\mathrm{S}}|\\ +\boldsymbol{\omega}_2^{\mathrm{T}}|\boldsymbol{z}(t+iT)-\boldsymbol{z}_{\mathrm{S}}|\\ +\boldsymbol{\lambda}_1^{\mathrm{T}}|\boldsymbol{u}(t+iT)-\boldsymbol{u}_{\mathrm{S}}|\\ +\boldsymbol{\lambda}_2^{\mathrm{T}}|\boldsymbol{w}(t+iT)-\boldsymbol{w}_{\mathrm{S}}|\\ +\boldsymbol{\lambda}_3^{\mathrm{T}}|\boldsymbol{\delta}(t+iT)-\boldsymbol{\delta}_{\mathrm{S}}|\end{array}\right\} \tag{7.543}$$

allows solution using *mixed integer linear programming*. The requirement for non-negative variables is dealt with by representation of such deviations by two non-negative variables each, for example

$$\Delta\boldsymbol{x}^{+} - \Delta\boldsymbol{x}^{-} = \boldsymbol{x}(t+iT+T) - \boldsymbol{x}_{\mathrm{S}} \tag{7.544}$$

and the solution will ensure that one of these will always be zero in the minimisation of

$$\boldsymbol{\omega}_1^{\mathrm{T}}|\boldsymbol{x}(t+iT+T)-\boldsymbol{x}_{\mathrm{S}}| = \boldsymbol{\omega}_1^{\mathrm{T}}\Delta\boldsymbol{x}^{+} + \boldsymbol{\omega}_1^{\mathrm{T}}\Delta\boldsymbol{x}^{-} \tag{7.545}$$

As described, this MILP solution still requires *real-time* solution on each controller time step. The initial state $\boldsymbol{x}(t)$ in Equations 7.537–7.543 is of course the feedback variable. Bemporad, Borrelli and Morari (2002) took this idea further by *parameterising* the feedback state $\boldsymbol{x}(t)$. A *multiparameter mixed integer linear program* (mp-MILP) was then solved *offline* so that the optimal control setting to be implemented on any step was obtained in terms of the parametric description of the state at that step. This *explicit* control law for real-time implementation was found to give control settings which were *piecewise affine* in terms of the feedback state parameters, just as the open loop was piecewise affine in the reverse direction. The authors remark that the explicit relationship $\boldsymbol{x}(t) \rightarrow$ *required* $\boldsymbol{u}(t)$ *and* $\boldsymbol{\delta}(t)$ provided by the offline solution gives useful insight into the workings of the controller, rather than relying on the usual 'black box' computer optimisation of control actions. For example, the range of state variable values which would demand a certain input switch could be shown as a two-dimensional area on an x_1, x_2 plot. Simple 'control laws' like this are 'cheap' and 'fast', making them feasible for systems in which the full online solution is too expensive or too slow.

7.15 Decentralised Control

Processing plants nowadays are becoming more interactive internally owing to material and energy recycles. These arise from design synthesis techniques such as 'pinch technology' which strive for increased efficiency. As a result, there is a large matrix of *interdependence* of process variables both on each other and on the process inputs. In the process control design, the approach is still largely to consider the plant as a geographical pattern of interconnected *units*, and to focus on coordinated controls solely within these units, such as MIMO control of a distillation column. Indeed, to extend considerations further (e.g. to neighbouring units) renders many of the excellent MIMO techniques too complex and/or slow. The control engineer intuitively separates the various control tasks into a *block diagonal* relationship where a group of MVs is uniquely associated with a group of CVs. This *structural synthesis* is discussed in Section 5.2.1. A number of workers have attempted to develop techniques for the optimal splitting of large dynamic control tasks into smaller ones. These techniques evaluate the stability implications and include proposals for coordination of the subtasks by exchange of information. Bearing in mind the increasing integration of processes, it is important to be aware of some of these formal methods in *decentralised control.*

In this section, discrete linear models will be used for illustrative purposes, but these can be substituted with other forms where appropriate. Figure 7.53 shows a typical division of a large interactive system

$$\boldsymbol{x}(t+T) = \boldsymbol{A}\boldsymbol{x}(t) + \boldsymbol{B}\boldsymbol{u}(t) \tag{7.546}$$

into assumed independent models. The next step is the specification of controllers based on the separate models. The concern in this approach is that the control loops will interact through the off-diagonal blocks in a suboptimal or unstable way. Obviously, the control engineer will attempt to select the subsystems in such a way that the terms in the off-diagonal blocks will be as small as possible. In practice, one controller may be tuned to react slowly, to provide another with sufficient time to 'track' the apparent disturbance.

One of the helpful decompositions is the *lower block triangular* (LBT) form. This is achieved by arranging the system such that the blocks *above* the diagonal blocks in Figure 7.53c are vacant. Then the controller based on $\boldsymbol{u}_1 \rightarrow \boldsymbol{y}_1$ does not need to consider the others. That based on $\boldsymbol{u}_2 \rightarrow \boldsymbol{y}_2$ can make an explicit (feedforward) allowance for the effect of the decision of the preceding controller(s), and so on. Representing the controller by the vector of functions $\boldsymbol{g}$,

$$\boldsymbol{u}_1(t) = \boldsymbol{g}\left(\boldsymbol{A}_{11}, \boldsymbol{B}_{11}, \boldsymbol{y}_1(t), \left\{\boldsymbol{y}_1(t) - \boldsymbol{y}_{1S}\right\}\right) \tag{7.547}$$

$$\boldsymbol{u}_2(t) = \boldsymbol{g}\left(\boldsymbol{A}_{22}, \boldsymbol{B}_{22}, \boldsymbol{y}_2(t), \left\{\boldsymbol{y}_2(t) + \boldsymbol{A}_{21}\boldsymbol{y}_1(t) + \boldsymbol{B}_{21}\boldsymbol{u}_1(t) - \boldsymbol{y}_{2S}\right\}\right) \tag{7.548}$$

$$\boldsymbol{u}_3(t) = \boldsymbol{g}\left(\boldsymbol{A}_{33}, \boldsymbol{B}_{33}, \boldsymbol{y}_3(t), \left\{\boldsymbol{y}_3(t) + \boldsymbol{A}_{31}\boldsymbol{y}_1(t) + \boldsymbol{B}_{31}\boldsymbol{u}_1(t) + \boldsymbol{A}_{32}\boldsymbol{y}_2(t) + \boldsymbol{B}_{32}\boldsymbol{u}_2(t) - \boldsymbol{y}_{3S}\right\}\right) \tag{7.549}$$

$$\vdots$$

Even when the LBT transformation can only be achieved for the $\boldsymbol{A}$ submatrices (not the $\boldsymbol{B}$'s), simple recursive calculations for state feedback controllers such as *pole placement* (Section 7.5) are possible (Levine, 1996).

Some types of decompositions are illustrated in Figure 7.54. Procedures based on the *nested epsilon* structure generally start with a square system $\boldsymbol{M}$ which is transformed into a block diagonal structure by row and column operations according to

$$\boldsymbol{P}^{\mathrm{T}}\boldsymbol{M}\boldsymbol{P} = \boldsymbol{M}_{\mathrm{D}} + \boldsymbol{M}_{\mathrm{C}} \tag{7.550}$$

where $\boldsymbol{M}_{\mathrm{D}}$ is block-diagonal and $\boldsymbol{M}_{\mathrm{C}}$ contains the residual smaller elements. The $\boldsymbol{M}_{\mathrm{C}}$ matrix can be viewed as the superposition of a series of 'nested' regions each containing elements only smaller

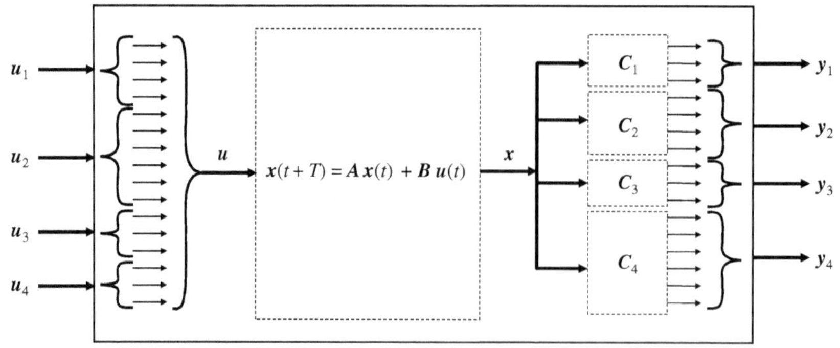

(a) large interactive system viewed as separate subsystems

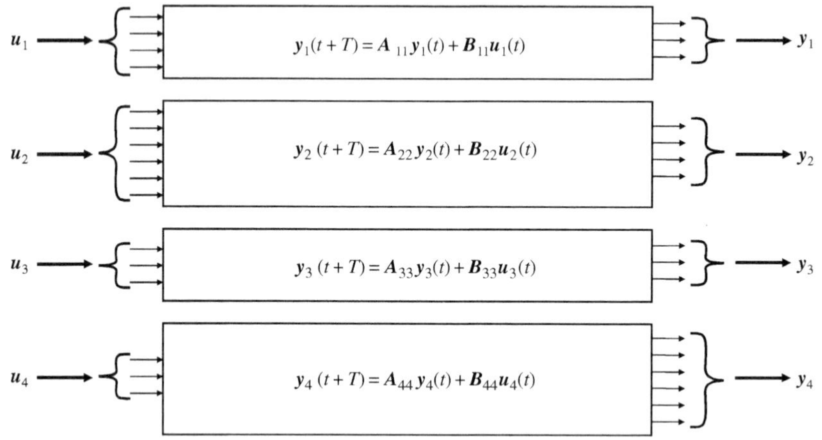

(b) assumed independent subsystems

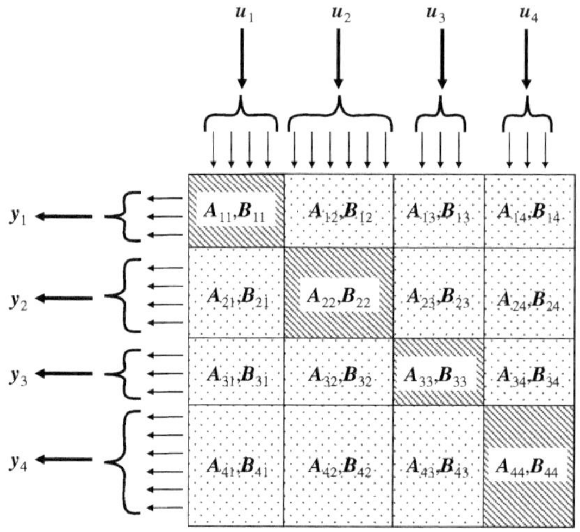

(c) ignored (off-diagonal) dynamics in subsystem view (b)

Figure 7.53 Typical approximation of a large interactive system as independent subsystems showing ignored (off-diagonal) dynamics (for the case of linear discrete models).

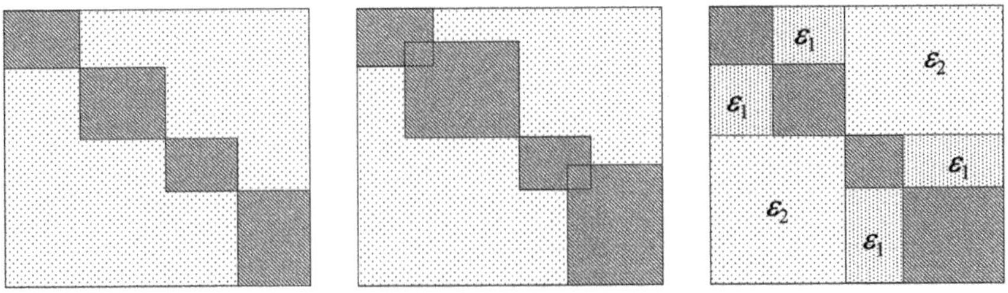

(a) disjoint (b) overlapping (c) nested epsilon (square)

Figure 7.54 Some types of decomposition.

than a succession $\varepsilon_1 > \varepsilon_2 > \varepsilon_3 > \cdots$:

$$\boldsymbol{P}^{\mathrm{T}}\boldsymbol{M}\boldsymbol{P} = \boldsymbol{M}_{\mathrm{D}} + ((((\varepsilon_1\boldsymbol{M}_1) + \varepsilon_2\boldsymbol{M}_2) + \varepsilon_3\boldsymbol{M}_3) + \cdots) \tag{7.551}$$

where each matrix $\boldsymbol{M}_i$ is scaled such that its largest element has size 1. All matrices have the same dimension (i.e. the original full size) but will obviously have occupied regions and blank regions in the indicated patterns. The nested epsilon structure has some advantages for the successive decomposition of a system and in stability analysis (Levine, 1996). A simpler use is in the conceptual representation of a disjoint block diagonal system such as Figure 7.53c, that is

$$\boldsymbol{y}_i(t+T) = \boldsymbol{A}_{ii}\boldsymbol{y}_i(t) + \boldsymbol{B}_{ii}\boldsymbol{u}_i(t) + \varepsilon \sum_{\substack{j=1 \\ j \neq i}}^{N} \left\{\boldsymbol{A}_{ij}\boldsymbol{y}_j(t) + \boldsymbol{B}_{ij}\boldsymbol{u}_j(t)\right\} \tag{7.552}$$

The extremes here are $\varepsilon = 1$ (the full system) and $\varepsilon = 0$ (the disjoint approximation). Kokotovic *et al.* (1969) used this idea to reduce computation in solving the Riccati equation for a linear quadratic regulator (Section 7.6.5).

Keviczky, Borrelli and Balas (2006) describe an approximate method for predictive control of large systems consisting of subsystems which are dynamically decoupled (i.e. no elements in the off-diagonal blocks of Figure 7.54c), but which share common constraints and a common objective function. The optimal trajectory for each block i (up to a receding horizon) is solved together with the optimal trajectories of its 'neighbours'. Only the solution $\boldsymbol{u}_i$ is used, whilst those for block i's neighbours are discarded. Moving on to the next block $i+1$, the process is repeated, depending on that block's closest *neighbour* grouping. In this way, a limited degree of coordination is achieved. Neighbour groupings would be chosen based on the most interactive relationships in the objective function and constraints.

Katebi and Johnson (1997) developed a two-phase approach to the predictive control of large-scale systems by decentralisation. This was to deal with the problem of first estimating unmeasured states, so that they could subsequently be used in the prediction step. A decentralised estimator predicted states in each submodel using separate discrete Kalman filters. On a higher level, iteration of a Lagrange multiplier simultaneously caused convergence of all system states, so that the shared interactive terms attained correct values. This distributed predictor then formed the basis of coordinated optimal predictive controllers in the decentralised MPC scheme.

As in Section 6.4.1, consider a subsystem i (of 1, . . . , N) to have its own states and an observation related by

$$\boldsymbol{x}_i(t+T) = \boldsymbol{A}_{ii}\boldsymbol{x}_i(t) + \boldsymbol{B}_i\boldsymbol{u}_i(t) + \boldsymbol{s}_i(t) + \boldsymbol{\mu}_i(t) \tag{7.553}$$

$$\boldsymbol{y}_i(t+T) = \boldsymbol{C}_i\boldsymbol{x}_i(t+T) + \boldsymbol{\nu}_i(t+T) \tag{7.554}$$

where the *interaction terms* from other subsystems are in

$$\boldsymbol{s}_i(t) = \sum_{\substack{j=1 \\ j \neq i}}^{N} \boldsymbol{A}_{ij}\boldsymbol{x}_j \tag{7.555}$$

and $\boldsymbol{\mu}_i$ and $\boldsymbol{\nu}_i$ are model and measurement Gaussian random error vectors. These are zero-mean vectors and have expected covariances (usually diagonal) $E\{\boldsymbol{\mu}_i^{\mathrm{T}}\boldsymbol{\mu}_i\} = \boldsymbol{Q}_i$ and $E\{\boldsymbol{\nu}_i^{\mathrm{T}}\boldsymbol{\nu}_i\} = \boldsymbol{R}_i$. A second version of the overall system states is held in $\boldsymbol{x}^*$, of which the selection $\boldsymbol{x}_i^*$ correspond to the states in subsystem i. A Hamiltonian for (a) minimisation of the quadratic prediction–observation errors in this subsystem, (b) compliance with the model in Equations 7.553 and 7.554 and (c) simultaneously matching $\boldsymbol{x}_i$ to $\boldsymbol{x}_i^*$ can be written as

$$\begin{aligned} H_i = {} & \frac{1}{2}\left[\boldsymbol{\mu}_i^{\mathrm{T}}(t)\boldsymbol{Q}_i^{-1}\boldsymbol{\mu}_i(t) + \boldsymbol{\nu}_i^{\mathrm{T}}(t+T)\boldsymbol{R}_i^{-1}\boldsymbol{\nu}_i(t+T)\right] \\ & + \boldsymbol{\lambda}_i^{\mathrm{T}}(t+T)\left[\boldsymbol{A}_{ii}\,\boldsymbol{x}_i(t) + \boldsymbol{B}_i\boldsymbol{u}_i(t) + \boldsymbol{s}_i^*(t) + \boldsymbol{\mu}_i(t)\right] \\ & + \boldsymbol{\beta}_i^{\mathrm{T}}\left[\boldsymbol{x}_i(t) - \boldsymbol{x}_i^*(t)\right] \end{aligned} \tag{7.556}$$

where $\boldsymbol{\nu}_i(t+T) = \boldsymbol{y}_i(t+T) - \boldsymbol{C}_i\boldsymbol{x}_i(t+T)$, $\boldsymbol{\lambda}_i$ contains the co-state variables and $\boldsymbol{\beta}_i$ contains the Lagrange multipliers to cause the last term to act as a zero constraint as in Equation 6.34. Note that the interaction terms $\boldsymbol{s}_i^*(t)$ are evaluated using the state values $\boldsymbol{x}^*$. Katebi and Johnson (1997) propose a stepwise solution for estimates of the state $\widehat{x}_i$ based on the Kalman filter:

$$\widehat{x}_i(t+T) = \boldsymbol{A}_{ii}\widehat{x}_i(t) + \boldsymbol{B}_i\boldsymbol{u}_i(t) + \boldsymbol{s}_i^*(t) + \boldsymbol{K}_i(t+T)\boldsymbol{\mu}'(t+T) + \boldsymbol{m}_i\boldsymbol{\beta}_i(t) \tag{7.557}$$

where

$$\boldsymbol{m}_i(t) = \left[\boldsymbol{Q}_i - \boldsymbol{P}_i(t+T)\left\{\boldsymbol{I} + \boldsymbol{C}_i^{\mathrm{T}}\boldsymbol{R}_i^{-1}\boldsymbol{C}_i\boldsymbol{Q}_i\right\}\right]\boldsymbol{A}_{ii}^{\mathrm{T}} \tag{7.558}$$

with $\boldsymbol{P}_i(t+T)$ the updated solution of the corresponding discrete Riccati equation (see Equations 6.72, 6.73 and 6.76) and

$$\boldsymbol{\mu}'(t+T) = \boldsymbol{y}(t+T) - \boldsymbol{C}\widehat{x}(t+T|t) \tag{7.559}$$

in which

$$\widehat{x}(t+T|t) = \boldsymbol{A}\widehat{x}(t) + \boldsymbol{B}\boldsymbol{u}(t) \tag{7.560}$$

A second Hamiltonian for optimal updating of the coordination parameters $\boldsymbol{x}^*$ and $\boldsymbol{\beta}_i$ in an outer iteration gives

$$\boldsymbol{x}^*(t+T) = \widehat{x}(t+T) \tag{7.561}$$

and the $\boldsymbol{\beta}_i$ must be obtained by simultaneous solution of the set of algebraic equations

$$\sum_{\substack{k=1\\k\neq i}}^{N} \boldsymbol{A}_{ki}^{\mathrm{T}}\boldsymbol{A}_{kk}^{-1}\left[\boldsymbol{P}_k\widehat{\boldsymbol{x}}_k(t)-\boldsymbol{\beta}_k(t)\right]=\boldsymbol{\beta}_i(t) \tag{7.562}$$

Equation 7.556 is solved for $i = 1, \ldots, N$, and then Equations 7.560 and 7.561 until convergence. At this point, all of the states are available, so their future values can be predicted recursively using the full system

$$\widehat{\boldsymbol{x}}(t+T) = \boldsymbol{A}\widehat{\boldsymbol{x}}(t) + \boldsymbol{B}\boldsymbol{u}(t) \tag{7.563}$$

$$\widehat{\boldsymbol{y}}(t+T) = \boldsymbol{C}\widehat{\boldsymbol{x}}(t+T) \tag{7.564}$$

Viewed on a subsystem basis, an objective function to be minimised receives contributions according to

$$J=\sum_{i=1}^{N}J_i=\sum_{i=1}^{N}\left\{\sum_{k=1}^{n}\begin{matrix}\left[\widehat{\boldsymbol{y}}_i(t+kT)-\boldsymbol{y}_{i\mathrm{S}}\right]^{\mathrm{T}}\boldsymbol{W}\left[\widehat{\boldsymbol{y}}_i(t+kT)-\boldsymbol{y}_{i\mathrm{S}}\right]\\+\left[\boldsymbol{u}_i(t+kT-T)-\boldsymbol{u}_{i\mathrm{S}}\right]^{\mathrm{T}}\boldsymbol{\Lambda}\left[\boldsymbol{u}_i(t+kT-T)-\boldsymbol{u}_{i\mathrm{S}}\right]\end{matrix}\right\} \tag{7.565}$$

At this point, the predictive control problem is decentralised by individually minimising the J_i as 'lower level problems'. In the same way as for the Kalman filter, a separate set of future control actions is maintained:

$$\boldsymbol{U}^* = \begin{pmatrix}\boldsymbol{u}^*(t+T)\\ \boldsymbol{u}^*(t+2T)\\ \vdots\\ \boldsymbol{u}^*(t+nT)\end{pmatrix} \tag{7.566}$$

where

$$\boldsymbol{u}^*(t+T) = \begin{pmatrix}\boldsymbol{u}_1^*(t+T)\\ \boldsymbol{u}_2^*(t+T)\\ \vdots\\ \boldsymbol{u}_N^*(t+T)\end{pmatrix} \tag{7.567}$$

$$\boldsymbol{u}^*(t+2T) = \begin{pmatrix}\boldsymbol{u}_1^*(t+2T)\\ \boldsymbol{u}_2^*(t+2T)\\ \vdots\\ \boldsymbol{u}_N^*(t+2T)\end{pmatrix} \tag{7.568}$$

$$\vdots$$

and updated on each iteration as an 'upper level problem' until convergence, so that the optimal control trajectories of the N subsystems are coordinated in a similar way to the optimal state estimates.

In a later paper, Vaccarini, Longhi and Katebi (2009) propose methods for dealing with coordination of decentralised controllers in a large-scale setting such as an industrial distributed control system, with local area network communication between the controllers. An *agent* is described as the supervisor and coordinator of one or more controllers. In the discussion above, effectively one agent undertook the task of coordinating the efforts of *all* of the block-diagonal controllers. However, in a fully decentralised system, one would aim to locate agent activity in a decentralised way as in 'peer-to-peer' communication. This clearly has advantages of incremental installation and redundancy. The agent then becomes associated with a specific controller, and its task is to inform other agents of the operation of its own controller, and to obtain and act on information from other agents regarding their controllers, so as to optimise the performance of its own controller.

The environment proposed by these authors is not strongly structured. However, there are some issues of precedence that the designer needs to be aware of. If one imagines a community of predictive controllers (block-diagonal) with known off-diagonal interaction terms, one could structure the liaison as follows.

The controllers run asynchronously, but on each step the respective agents transmit the predicted optimal state trajectories and future moves to the remaining agents. As agents receive this information from their peers, they immediately incorporate it into their 'open-loop' trajectory predictions using the off-diagonal block relationships. Thus, when the next controller step arrives, the next computed control action will automatically account for these 'disturbances'. Depending on the strength of interactions, a scheme like this can of course become unstable, with two controllers affecting one CV alternately raising and lowering their output as they see each other react -180° out of phase. Special damping measures or 'detuning' may be required.

References

Aström, K.J. and Wittenmark, B. (1989) *Adaptive Control*, Addison-Wesley.

Bell, M.L. and Sargent, R.W.H. (2000) Optimal control of inequality constrained DAE systems. *Computers & Chemical Engineering*, **24**, 2385–2404.

Bellman, R. (1957) *Dynamic Programming*, Princeton University Press, (Dover paperback edition).

Bemporad, A., Borrelli, F. and Morari, M. (2002) Model predictive control based on linear programming – the explicit solution. *IEEE Transactions on Automatic Control*, **47** (12), 1974–1985.

Bemporad, A. and Morari, M. (1999) Control of systems integrating logic, dynamics and constraints. *Automatica*, **35**, 407–427.

Beveridge, G.S.G. and Schechter, R.S. (1970) *Optimisation: Theory and Practice*, McGraw-Hill, New York.

Biegler, L.T. (2000) Efficient solution of dynamic optimisations and NMPC problems, in *Nonlinear Predictive Control* (eds F Allgöwer and A. Zheng), Birkhäuser, pp. 219–244.

Biegler, L.T., Cervantes, A.M. and Wächter, A. (2002) Advances in simultaneous strategies for dynamic process optimization. *Chemical Engineering Science*, **57** (4), 575–593.

Bock, H.G., Diehl, M.M., Leineweber, D.B. and Schlöder, J.P. (2000) A direct multiple shooting method for real-time optimization of nonlinear DAE processes, in *Nonlinear Predictive Control* (eds F. Allgöwer and A. Zheng), Birkhäuser, pp. 219–244.

Boissel, O.R. and Kantor, J.C. (1995) Optimal feedback control design for discrete-event systems using simulated annealing. *Computers & Chemical Engineering*, **19** (3), 253–266.

Bojkov, B. and Luus, R. (1994) Application of iterative dynamic programming to time optimal control. *Chemical Engineering Research and Design*, **72** (1), 72–80.

Bojkov, B. and Luus, R. (1995) Time optimal control of high dimensional systems by iterative dynamic programming. *The Canadian Journal of Chemical Engineering*, **73** (3), 380–390.

Cavalier, T.M., Pardalos, P.M. and Soyster, A.L. (1990) Modeling and integer programming techniques applied to propositional calculus. *Computers & Operations Research*, **17** (6), 561–570.

Champagnat, R., Esteban, P., Pingaud, H. and Valette, R. (1998) Petri net based modelling of hybrid systems. *Computers in Industry*, **36**, 139–146.

Chang, T.S. and Seborg, D.E. (1983) A linear programming approach for multivariable feedback control with inequality constraints. *International Journal of Control*, **37**, 583–597.

Chen, H. and Allgöwer, F. (1998) A quasi-infinite horizon nonlinear model predictive control scheme with guaranteed stability. *Automatica*, **34** (10), 1205–1217.

Clarke, D.W., Mohtadi, C. and Tuffs, P.S. (1987) Generalized predictive control. I. The basic algorithm. *Automatica*, **23** (2), 137.

Corona, D., Giua, A. and Seatzu, C. (2004) Optimal control of hybrid automata: an application to the design of a semiactive suspension. *Control Engineering Practice*, **12** (10), 1305–1318.

Corriou, J.-P. (2004) *Process Control – Theory and Applications*, Springer, London.

Cutler, C.R. and Ramaker, B.L. (1979) Dynamic matrix control – a computer control algorithm. AIChE National Meeting, Houston, TX.

Dahlin, E.B. (1968) Designing and tuning digital controllers. *Instrumentation and Control Systems*, **41**, 77–83.

de Larminat, P. (1993) *Automatique: Commande des Systèmes Linéaires*, Hèrmes, Paris.

de Tremblay, M. and Luus, R. (1989) Optimization of non-steady-state operation of reactors. *The Canadian Journal of Chemical Engineering*, **67** (3), 494–502.

Dougherty, D. and Cooper, D. (2003) A practical multiple model adaptive strategy for multivariable model predictive control. *Control Engineering Practice*, **11**, 649–664.

Dumont, G.A., Elnaggar, A. and Elshafei, A. (1993) Adaptive predictive control of systems with time-varying time delay. *International Journal of Adaptive Control*, **7** (2), 91–101.

Fikar, M., Chachuat, B. and Latifi, M.A. (2005) Optimal operation of alternating activated sludge processes. *Control Engineering Practice*, **13**, 853–861.

Findeisen, R., Diehl, M., Uslu, I., Scharzkopf, S., Allgöwer, F., Bock, H.G., Schlöder, J.P. and Gilles, E.D. (2002a) Computation and performance assessment of nonlinear model predictive control. Proceedings of the 42nd IEEE Conference on Decision and Control, Las Vegas, NV, pp. 4613–4618.

Findeisen, R., Imsland, L., Allgöwer, F. and Foss, B.A. (2002b) Output feedback nonlinear predictive control – a separation principle approach. Proceedings of the 15th IFAC World Congress, Barcelona, Spain.

Fletcher, R. and Leyffer, S. (1998) Numerical experience with lower bounds for MIQP branch-and-bound. *SIAM Journal of Optimization*, **8** (2), 604–616.

Garcia, C.E. and Morshedi, A.M. (1984) Quadratic programming solution of dynamic matrix control (QDMC). Proceedings of the American Control Conference, San Diego, CA.

Guiamba, I. and Mulholland, M. (2004) Adaptive linear dynamic matrix control applied to an integrating process. *Computers & Chemical Engineering*, **28**, 2621–2633.

Gupta, Y.P. (1998) Control of integrating processes using dynamic matrix control. *Transactions of the Institution of Chemical Engineers, Part A*, **76**, 465–470.

Higham, J.D. (1968) Single-term control of first- and second-order processes with dead time. *Control*, (February), 2–6.

Hussain, M.A. (1999) Review of the applications of neural networks in chemical process control – simulation and online implementation. *Artificial Intelligence in Engineering*, **13**, 55.

Isermann, R. (1982) Parameter adaptive control algorithms – a tutorial. *Automatica*, **18** (5), 513–528.

Katebi, M.R. and Johnson, M.A. (1997) Predictive control design for large-scale systems. *Automatica*, **33** (3), 421–425.

Keviczky, T., Borrelli, F. and Balas, G.J. (2006) Decentralized receding horizon control for large scale dynamically decoupled systems. *Automatica*, **42**, 2105–2115.

Kokotovic, P., Perkins, W., Cruz, J. and D'Ans, G. (1969) ε-coupling approach for near-optimum design of large scale linear systems. *Proceedings of the IEE, Part D*, **116**, 889–892.

Lee, J.H., Morari, M. and Garcia, C.E. (1994) State space interpretation of model predictive control. *Automatica*, **30** (4), 707–717.

Levine, W.S. (1996) *The Control Handbook*, CRC Press & IEEE Press.

Lin, J.-S. and Hwang, C. (1996) Optimal control of time-delay systems by forward iterative dynamic programming. *Industrial & Engineering Chemical Research*, **35** (8), 2795–2800.

Lin, J.-S. and Hwang, C. (1998) Enhancement of the global convergence of using iterative dynamic programming to solve optimal control problems. *Industrial & Engineering Chemical Research*, **37** (6), 2469–2478.

Lipták, B.G. (2003) *Instrument Engineer's Handbook*, 4th edn (ed. B.G. Lipták), CRC Press.

Luus, R. (1990) Optimal control by dynamic programming using systematic reduction in grid size. *International Journal of Control*, **51**, 995–1013.

Luus, R. (1996) Numerical convergence properties of iterative dynamic programming when applied to high dimensional systems. *Transactions of the Institution of Chemical Engineers*, **74**, 55–62.

Luyben, W.L. (1986) Simple method for tuning SISO controllers in multivariable systems. *Industrial & Engineering Chemical Process Design and Development*, **25**, 654–660.

Maciejowski, J.M. (1989) *Multivariable Feedback Design*, Addison-Wesley, Wokingham, UK.

Mamdani, E.H. (1974) Applications of a fuzzy algorithm for control of a simple dynamic plant. *Proceedings of the IEE*, **121** (12), 1585–1588.

Manon, P., Valentin-Roubinet, C. and Gilles, G. (2002) Optimal control of hybrid dynamical systems: application in process engineering. *Control Engineering Practice*, **10**, 133–149.

Morari, M. (2002) Hybrid system analysis and control via mixed integer optimisation. Chemical Process Control VI, AIChE Symposium Series No. 326, Vol. **98**.

Morari, M., Bemporad, A. and Mignone, D. (2000) A framework for control, state estimation, fault detection and verification of hybrid systems. *Automatisierungstechnik*, **48**, 1–8.

Morshedi, A.M., Cutler, C.R. and Skrovanek, T.A. (1985) Optimal solution of dynamic matrix control with linear programming techniques (LDMC). Proceedings of the American Control Conference, Boston, MA, pp. 199–208.

Mulholland, M. and Prosser, J.A. (1997) Linear dynamic matrix control of a distillation column. South African Institution of Chemical Engineers, 8th National Meeting, Cape Town, April 16–18.

Pontryagin, L.S., Boltyanskii, V.G., Gamkrelidze, R.V. and Mishchenko, E.F. (1961, 1969) *Mathematical Theory of Optimal Control Processes*, Nauka, Moscow (in Russian).

Postlethwaite, B.E. (1994) A model-based fuzzy controller. *Chemical Engineering Research and Design*, **72** (1), 38–46.

Postlethwaite, B.E. and Edgar, C.R. (2000) A MIMO model-based controller. *Transactions of the Institution of Chemical Engineers*, **78** (A), 557–564.

Raman, R. and Grossmann, I.E. (1991) Relation between MILP modelling and logical inference for chemical process synthesis. *Computers & Chemical Engineering*, **15**, 73.

Raman, R. and Grossmann, I.E. (1992) Integration of logic and heuristic knowledge in MINLP optimisation for process synthesis. *Computers & Chemical Engineering*, **16** (3), 155.

Rao, C.V. and Rawlings, J.B. (2000) Linear programming and model predictive control. *Journal of Process Control*, **10** (2–3), 283–289.

Richalet, J., Rault, A., Testud, J.L. and Papon, J. (1978) Model predictive heuristic control: applications to industrial processes. *Automatica*, **14**, 413–428.

Ridley, J.N., Shaw, I.S. and Kruger, J.J. (1988) Probabilistic fuzzy model for dynamic systems. *Electronics Letters*, **24** (14), 890.

Rosenbrock, H.H. (1974) *Computer-Aided Control System Design*, Academic Press, London.

Rumelhart, D.E. and McClelland, J.L. (1986) *Parallel Distributed Processing: Explorations in the Microstructure of Cognition*, vol. **1**, MIT Press, Cambridge, MA, pp. 45–76.

Rusnák, A., Fikar, M., Latifi, M.A. and Mészáros, A. (2001) Receding horizon iterative dynamic programming with discrete time models. *Computers & Chemical Engineering*, **25** (1), 161–167.

Schwefel, H.-P. (1965) Kybernetische Evolution als Strategie der experimentellen Forschung aus der Strömungstechnik. Diplomarbeit, Technische Universität Berlin.

Seborg, D.E., Edgar, T.F., and Shah, S.L. (1986) Adaptive control strategies for process control: a survey. *AIChE Journal*, **32** (6), 881–913.

Shunta, J.P. and Luyben, W.L. (1972) Sampled-data noninteracting control for distillation columns. *Chemical Engineering Science*, **27** (6), 1325–1335.

Sing, C.H. and Postlethwaite, B. (1996) Fuzzy relational model-based control applying stochastic and iterative methods for model identification. *Transactions of the Institution of Chemical Engineers*, **74** (A1), 70–76.

Vaccarini, M., Longhi, S. and Katebi, M.R. (2009) Unconstrained networked decentralised model predictive control. *Journal of Process Control*, **19** (2), 328–339.

Wang, Q.G., Zou, B. and Zhang, Y. (2000) Decoupling Smith predictor design for multivariable systems with multiple time delays. *Chemical Engineering Research and Design Transactions, Part A*, **78** (4), 565–572.

8
Stability and Quality of Control

8.1
Introduction

Control theory has until recently been taught in a way that did not distinguish much between the fields of application, whether those be missiles, aircraft, power systems or chemical processes. The advent of the digital computer has however opened up a vast range of new applications, which have perceptibly moved the emphasis away from the types of things that the classical analysis methods handled. This is particularly the case in the processing industries where the control engineer has a new focus on *advanced process control (APC), model predictive control (MPC) and real time optimisation (RTO)*. Of course, issues of loop stability and controller performance are still important, but in the process industries these are seldom investigated by classical methods such as *frequency response* and *root locus*. The new computer-based work is exclusively in the time domain. In comparison with other fields, the frequencies to be dealt with are very low, often arising from the control loops themselves! 'Good' initial settings for controllers may be estimated theoretically before start-up, but the fine-tuning tends to be based on trial-and-error experience. The signs of instability are early detected as undesirable process fluctuations, and tuned out. In this largely 'regulation' mode, no one is pushing the limits at the edge of instability. One reason for this approach is that the processes are nonlinear and variable, and seldom match a model tightly, as might be the case in mechanical and electrical systems. So there is little point in pushing the control performance close to the limits of a model.

Despite this gloomy view of the relevance of many classical ideas, they are nevertheless important because they underlie the perception and language of the new applications. This has a parallel in the way modern digital control in a DCS or SCADA is represented as a mimic of the old analogue controllers. Nowadays one might say that there is a 'badly damped pole' in the closed-loop control of a distillation column. One is thinking how different terms in a characteristic equation (arising from submodels) can contribute uniquely different behaviours to a system – in this case with a slowly decaying oscillation. One is imagining a position on the complex plane, but there is no way that one is about to embark on a calculation in complex algebra to present to the boss!

In Figure 8.1, a potentially multivariable system is represented in open loop and in closed loop in a classical sense. In terms of modern controllers, this is already a restriction, as the controller often does not work exclusively on the *deviation from setpoint* – more information may be required such as absolute and previous values. The 'real output' is an abstract construct because there is no way of knowing precisely what it is, having to work remotely through sensors, whether they entail human sensing (e.g. vision) or instrument sensing. For example, any

Applied Process Control: Essential Methods, First Edition. Michael Mulholland.

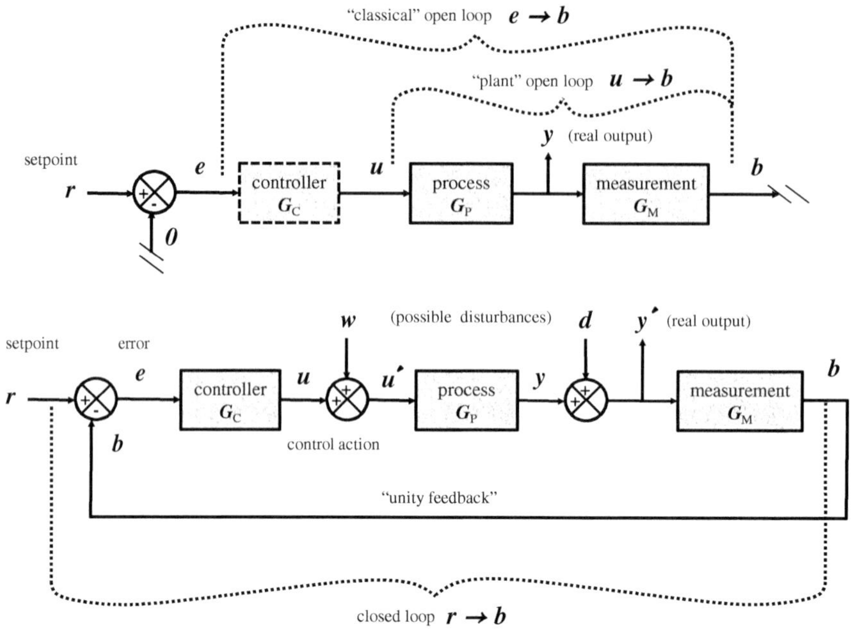

Figure 8.1 Classical views of open-loop and closed-loop systems.

practical sensor will have a response lag. So, the idea of an 'actual output' can be bypassed and loops dealt with as unity feedback. That is, from a 'plant' point of view, the knowledge of what the output is is indistinguishable from the available measurement.

Figure 8.1 also shows the two possible routes used for the entry of disturbances into the system – something that becomes important in the development or testing of algorithms for *disturbance (load) rejection* – the main area of concern in the process industries. The disturbance $\boldsymbol{w}$ is equivalent to a random disturbance of the control action – such as a jittery deviation from the set valve position. The disturbance $\boldsymbol{d}$ is equivalent to deviations in the 'actual' variable, such as liquid surface ripples in level control.

In Section 3.9.2, methods were developed to represent the dynamic alteration of signals using Laplace-domain transfer functions $\boldsymbol{G}(s)$. A series of dynamic elements acting on a signal could be represented as a multiplication of all of the transfer functions.

$$\boldsymbol{B}(s) = \boldsymbol{G}_{M}(s)\boldsymbol{G}_{P}(s)\boldsymbol{G}_{C}(s)\boldsymbol{E}(s) \tag{8.1}$$

The classical open-loop transfer function is

$$\boldsymbol{G}_{O}(s) = \boldsymbol{G}_{M}(s)\boldsymbol{G}_{P}(s)\boldsymbol{G}_{C}(s) \tag{8.2}$$

Ignoring the disturbances in Figure 8.1

$$\boldsymbol{E}(s) = \boldsymbol{R}(s) - \boldsymbol{B}(s) \tag{8.3}$$

$$\boldsymbol{B}(s) = \boldsymbol{G}_{O}(s)\boldsymbol{E}(s) = \boldsymbol{G}_{O}(s)[\boldsymbol{R}(s) - \boldsymbol{B}(s)] \tag{8.4}$$

$$\boldsymbol{B}(s) = [\boldsymbol{I} + \boldsymbol{G}_{O}(s)]^{-1}\boldsymbol{G}_{O}(s)\boldsymbol{R}(s) \tag{8.5}$$

So, the closed-loop transfer function is

$$\boldsymbol{G}_{\mathrm{CL}}(\mathrm{s}) = [\boldsymbol{I} + \boldsymbol{G}_{\mathrm{O}}(\mathrm{s})]^{-1}\boldsymbol{G}_{\mathrm{O}}(\mathrm{s}) \tag{8.6}$$

which is represented in SISO systems as

$$G_{\mathrm{CL}}(\mathrm{s}) = \frac{B(\mathrm{s})}{R(\mathrm{s})} = \frac{G_{\mathrm{O}}(\mathrm{s})}{1 + G_{\mathrm{O}}(\mathrm{s})} \tag{8.7}$$

The view taken above is that of *servo control*, which is to do with the tracking of setpoint variations as in vehicle guidance. In the process industries, *regulator control* is far more important, so one is interested in the behaviour from, say, disturbance $\boldsymbol{d}$ to $\boldsymbol{b}$. Ideally one would like to see this being a zero transfer function! Recalling the linearity assumptions in Section 3.9.2, where the variations were taken as deviations from an operating point, regulator mode can be represented as $\boldsymbol{r}(t) = \boldsymbol{0}$:

$$\boldsymbol{Y}(\mathrm{s}) = \boldsymbol{G}_{\mathrm{P}}(\mathrm{s})\boldsymbol{G}_{\mathrm{C}}(\mathrm{s})\boldsymbol{E}(\mathrm{s}) \tag{8.8}$$

$$\boldsymbol{Y}(\mathrm{s}) = -\boldsymbol{G}_{\mathrm{P}}(\mathrm{s})\boldsymbol{G}_{\mathrm{C}}(\mathrm{s})\boldsymbol{B}(\mathrm{s}) \tag{8.9}$$

$$\boldsymbol{B}(\mathrm{s}) = \boldsymbol{G}_{\mathrm{M}}(\mathrm{s})[\boldsymbol{Y}(\mathrm{s}) + \boldsymbol{D}(\mathrm{s})] \tag{8.10}$$

$$\boldsymbol{B}(\mathrm{s}) = \boldsymbol{G}_{\mathrm{M}}(\mathrm{s})[-\boldsymbol{G}_{\mathrm{P}}(\mathrm{s})\boldsymbol{G}_{\mathrm{C}}(\mathrm{s})\boldsymbol{B}(\mathrm{s}) + \boldsymbol{D}(\mathrm{s})] \tag{8.11}$$

$$\boldsymbol{B}(\mathrm{s}) = [\boldsymbol{I} + \boldsymbol{G}_{\mathrm{O}}(\mathrm{s})]^{-1}\boldsymbol{G}_{\mathrm{M}}(\mathrm{s})\boldsymbol{D}(\mathrm{s}) \tag{8.12}$$

A system may be *open-loop* unstable (e.g. an exothermic reactor) – it will be seen as in Section 3.9.2.3 that this is determined by denominator terms in $\boldsymbol{G}_{\mathrm{O}}(\mathrm{s})$. In the *closed loop* (Equation 8.5 or 8.12), the stability behaviour will be seen to be determined by the 'denominator' term $[\boldsymbol{I} + \boldsymbol{G}_{\mathrm{O}}(\mathrm{s})]$. The interest in stability is not simply whether a system is stable or unstable. Usually as one tries to improve the performance of a controller (e.g. higher gain), one approaches more closely an unstable situation. So, it is useful to have a measure of how close one is to instability as a design criterion.

8.2 View of a Continuous SISO System in the s-Domain

8.2.1 Transfer Functions, the Characteristic Equation and Stability

8.2.1.1 Open-Loop Transfer Functions

In Section 3.9.2.1, it was found that a continuous linear SISO open-loop system could be represented by

$$a_n\frac{\mathrm{d}^n x}{\mathrm{d}t^n} + a_{n-1}\frac{\mathrm{d}^{n-1}x}{\mathrm{d}t^{n-1}} + \cdots + a_1\frac{\mathrm{d}x}{\mathrm{d}t} + a_0 x = b_m\frac{\mathrm{d}^m u}{\mathrm{d}t^m} + b_{m-1}\frac{\mathrm{d}^{m-1}u}{\mathrm{d}t^{m-1}} + \cdots + b_1\frac{\mathrm{d}u}{\mathrm{d}t} + b_0 u \tag{8.13}$$

For *physically realisable* systems, one requires $m \leq n$. Taking the Laplace transform yielded the transfer function

$$G_{\mathrm{O}}(\mathrm{s}) = \frac{X(\mathrm{s})}{U(\mathrm{s})} = \frac{b_m s^m + b_{m-1}s^{m-1} + \cdots + b_1 s + b_0}{a_n s^n + a_{n-1}s^{n-1} + \cdots + a_1 s + a_0} = \frac{N_{\mathrm{O}}(\mathrm{s})}{D_{\mathrm{O}}(\mathrm{s})} \tag{8.14}$$

$$X(\mathrm{s}) = G_{\mathrm{O}}(\mathrm{s})U(\mathrm{s}) \tag{8.15}$$

The numerator $N_O(s)$ and denominator $D_O(s)$ polynomials could be factored to yield

$$G_O(s) = \frac{K(s-z_1)(s-z_2)\cdots(s-z_m)}{(s-p_1)(s-p_2)\cdots(s-p_n)} \tag{8.16}$$

where $K = b_m/a_n$. The z_i are the 'zeros' of the open loop, because $s \to z_i$ causes $G_O(s) \to 0$. The p_i are the 'poles' of the system, because $s \to p_i$ causes $G_O(s) \to \infty$. The *characteristic equation* of the system in Equation 8.14 arises from setting the denominator to zero. Obviously the poles are the roots of the characteristic equation.

Characteristic equation of the open loop:

$$D_O(s) = 0 \tag{8.17}$$

Any pole or zero may be complex. However, it is easily shown that if a complex pole exists, its complex conjugate will also exist as a pole, and likewise for the zeros. The biggest hurdle in getting a linear system to this form would be the presence of *dead time*, which was found in Section 3.9.2.1 to have instead a transfer function $e^{-\tau s}$. In 'frequency-domain' analysis, this can be dealt with explicitly in the form

$$G_O(s) = \frac{K(s-z_1)\cdots(s-z_{m'})}{(s-p_1)\cdots(s-p_{n'})}e^{-\tau s} \tag{8.18}$$

Only where one specifically needs to represent the system entirely in terms of poles and zeros (e.g. root locus), it is necessary to substitute an approximation of $e^{-\tau s}$ in the form of a ratio of two polynomials, for example

first-order Padé approximation:

$$e^{-\tau s} \approx \frac{1-\frac{\tau}{2}s}{1+\frac{\tau}{2}s} \tag{8.19}$$

second-order Padé approximation:

$$e^{-\tau s} \approx \frac{1-\frac{\tau}{2}s+\frac{\tau^2}{12}s^2}{1+\frac{\tau}{2}s+\frac{\tau^2}{12}s^2} \tag{8.20}$$

which will revert the system to the form shown in Equation 8.14 or 8.16.

8.2.1.2 Angles and Magnitutes of s and $G_O(s)$

Consider a general transfer function of the form shown in Equation 8.18. The poles and zeros may be complex, so for a particular choice of s on the complex plane, the vectors represented by the algebraic factors are merely complex numbers with angles and magnitudes as in Figure 8.2.

A complex number $s = a + jb$ has an *angle* and a *magnitude* as follows:

$$\measuredangle(a+jb) = \begin{cases} \tan^{-1}\left(\dfrac{b}{a}\right) & \text{for } a > 0 \quad \text{(first and fourth quadrants)} \\ \pi + \tan^{-1}\left(\dfrac{b}{a}\right) & \text{for } a < 0 \quad \text{(second and third quadrants)} \end{cases} \tag{8.21}$$

$$|a+jb| = \sqrt{a^2+b^2} \tag{8.22}$$

There is a polar notation in which the number is represented as

$$s = |s|, \measuredangle s \quad (\text{e.g. } 1 + j1 = \sqrt{2}, \pi/4) \tag{8.23}$$

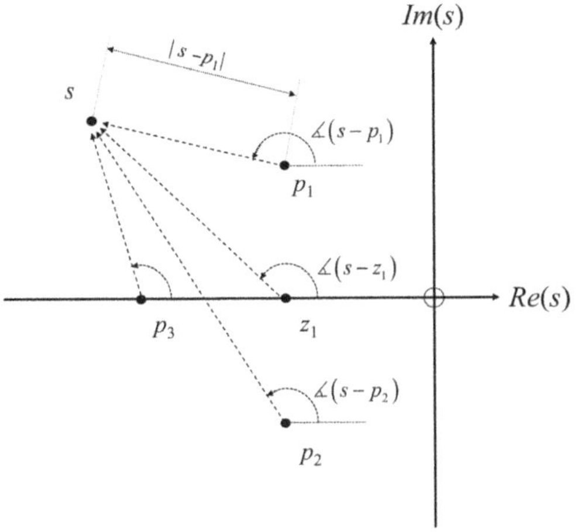

Figure 8.2 Complex vectors representing transfer-function factors.

Recalling that

$$e^{j\theta} = \cos(\theta) + j\sin(\theta) \tag{8.24}$$

and

$$[\cos(\theta)]^2 + [\sin(\theta)]^2 = 1, \tag{8.25}$$

one writes equally

$$s = |s|e^{j\measuredangle s} \quad \text{or} \quad (s - p_1) = |s - p_1|e^{j\measuredangle(s-p_1)} \tag{8.26}$$

Applying this in Equation 8.18 then yields, for example

$$(s - p_1)(s - p_2) = |s - p_1|e^{j\measuredangle(s-p_1)} \cdot |s - p_2|e^{j\measuredangle(s-p_2)} \tag{8.27}$$

$$= |s - p_1| \cdot |s - p_2| \cdot e^{j[\measuredangle(s-p_1)+\measuredangle(s-p_2)]} \tag{8.28}$$

so that

$$|G_O(s)| = \frac{|K| \cdot |s - z_1| \cdot |s - z_2| \cdots |s - z_{m'}|}{|s - p_1| \cdot |s - p_2| \cdots |s - p_{n'}|} e^{-\tau \mathrm{Re}(s)} \tag{8.29}$$

$$\measuredangle G_O(s) = \measuredangle(K) + \measuredangle(s - z_1) + \cdots + \measuredangle(s - z_{m'}) - \measuredangle(s - p_1) - \cdots - \measuredangle(s - p_{n'}) - \tau\,\mathrm{Im}(s) \tag{8.30}$$

This has allowed for the possibility that $K < 0$ (in which case $\measuredangle(K) = \pi$).

8.2.1.3 Open-Loop and Closed-Loop Stability

In Section 3.9.2.1, it was found that the open-loop denominator factors in Equation 8.16 would occur in the partial fraction expansion of the system output regardless of the input:

$$X(s) = \frac{A_1}{(s - p_1)} + \frac{A_2}{(s - p_1)^2} + \cdots + \frac{B}{(s - p_2)} + \cdots + \frac{Z}{(s - p_n)} + \frac{\alpha}{(s - q)} + \frac{\beta}{(s - r)} + \cdots \tag{8.31}$$

Here, the possibility of a repeated pole p_1 is included. The extra α, β, . . . terms arise from the denominator of the input function $U(s)$.

Now consider a particular root p_i of the characteristic equation which may appear n times (i.e. an nth order root). Allow for the possibility that p_i is complex, that is,

$$p_i = a_i + jb_i \tag{8.32}$$

Because the complex conjugate pole must also exist, the following pair of terms will exist in the partial fraction expansion for a root repeated n times

$$\frac{c_i + jd_i}{(s - a_i - jb_i)^n} + \frac{c_i - jd_i}{(s - a_i + jb_i)^n} \tag{8.33}$$

From Table 3.5 the inversion to the time domain is given by

$$L^{-1}\left\{\frac{c_i + jd_i}{(s - a_i - jb_i)^n} + \frac{c_i - jd_i}{(s - a_i + jb_i)^n}\right\} = \frac{2t^{n-1}}{(n-1)!}e^{a_i t}[c_i \cos(b_i t) + d_i \sin(b_i t)] \tag{8.34}$$

Noting Equation 8.31 for repeated poles, similar terms will coexist for n values down to 1, where $0!=1$. Regardless of the order of a pole, or whether it is complex, the process $G_O(s)$ itself will contribute a series of terms in the output which have a factor $e^{a_i t}$, where a_i is the real part of pole i of $G_O(s)$. This factor will start at unity for $t=0$, and with time will increase to infinity if $a_i > 0$, and will decrease to zero if $a_i < 0$. Unbounded growth ($a_i > 0$) for any one pole i determines an *unstable system*. If *all* poles i have $a_i < 0$, (thus causing progressive attenuation of any disturbance), the system is said to be *stable*.

> *Note*: A system is stable if all of its poles lie in the left half of the complex plane, and unstable if any one pole lies in the right half.

These ideas about stability have been developed around a consideration of a nominal 'open-loop' transfer function $G_O(s)$. Any system that can be got into the form of Equation 8.16 can be considered in the same way, whether it be open loop or closed loop. However, the open-loop stability was first considered because classical theories draw many inferences about the closed loop from the open loop. For this to be possible, the open loop must include every element around the loop, as in the $\boldsymbol{e} \rightarrow \boldsymbol{b}$ case at the top of Figure 8.1. From Equation 8.7,

$$G_{CL}(s) = \frac{G_O(s)}{1 + G_O(s)} = \frac{\dfrac{N_O(s)}{D_O(s)}}{1 + \dfrac{N_O(s)}{D_O(s)}} = \frac{N_O(s)}{D_O(s) + N_O(s)} \tag{8.35}$$

This is just a new equation of the form shown in Equation 8.14 having the *characteristic equation*

$$D_O(s) + N_O(s) = a_n s^n + a_{n-1}s^{n-1} + \cdots + a_1 s + a_0 + b_m s^m + b_{m-1}s^{m-1} + \cdots + b_1 s + b_0 = 0 \tag{8.36}$$

which will have n roots because $n \geq m$. More succinctly, the closed-loop characteristic equation is written

$$G_O(s) = -1 \tag{8.37}$$

A review of Equation 8.34 gives useful insight into the contribution of the poles of any process $G(s)$ (open loop or closed loop) to its output behaviour. A time constant is identified with each term according to

$$a_i = -\frac{1}{\tau_i} \tag{8.38}$$

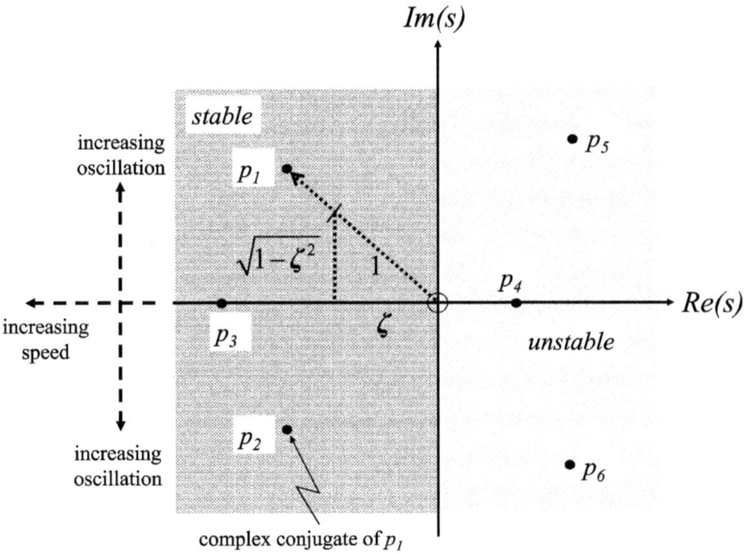

Figure 8.3 Behaviour of terms in the output as dependent on associated poles of the system.

The further the pole lies into the negative real complex half-plane, the shorter the time constant and the faster the response. As the *damping factor* ζ of a pole decreases, it contributes increasing oscillation which takes longer to attenuate (Figure 8.3). In Example 3.9 in Section 3.9.2.2, it was seen how ζ and τ arose in the definition of a standard second-order system – that is $\tau = 1/(-a_i)$ and $\zeta = (-a_i)/\sqrt{a_i^2 + b_i^2}$. (Clearly the a_i and b_i of pole i differ from the a's and b's in Equation 8.36).

8.2.1.4 **Open-Loop and Closed-Loop Steady-State Gain**

Whilst considering the open-loop and closed-loop transfer functions, it is opportune to introduce a useful property based on the *final value theorem*:

$$\lim_{t \to \infty} x(t) = \lim_{s \to 0} s\,X(s) \tag{8.39}$$

For a unit step input

$$\lim_{t \to \infty} x(t) = \lim_{s \to 0} s\,G(s)\frac{1}{s} = \lim_{s \to 0} G(s) \tag{8.40}$$

Thus the *steady-state gain* of a system is got by setting $s \to 0$ in its transfer function. (Later it will be seen that this is just the gain at zero frequency – i.e. steady state). Considering the forms shown in Equations 8.14 and 8.18 of $G_O(s)$, the *open-loop steady-state gain* is:

$$\lim_{s \to 0} G_O(s) = \lim_{s \to 0} \frac{b_m s^m + b_{m-1}s^{m-1} + \cdots + b_1 s + b_0}{a_n s^n + a_{n-1}s^{n-1} + \cdots + a_1 s + a_0} = \frac{b_0}{a_0} \tag{8.41}$$

or

$$\lim_{s \to 0} G_O(s) = \lim_{s \to 0} \frac{K(s - z_1)\cdots(s - z_{m'})}{(s - p_1)\cdots(s - p_{n'})}\,e^{-\tau s} = \frac{K\prod_{i=1}^{m'}(-z_i)}{\prod_{j=1}^{n'}(-p_j)} \tag{8.42}$$

so the *closed-loop steady-state gain* is:

$$\lim_{s \to 0} G_{CL}(s) = \lim_{s \to 0} \frac{N(s)}{N(s) + D(s)}$$

$$= \lim_{s \to 0} \frac{b_m s^m + \cdots + b_0}{b_m s^m + \cdots + b_0 + a_n s^n + \cdots + a_0} = \frac{b_0}{b_0 + a_0} \tag{8.43}$$

or similarly

$$\lim_{s \to 0} G_{CL}(s) = \lim_{s \to 0} \frac{K\prod_{i=1}^{m'}(s - z_i)e^{-\tau s}}{K\prod_{i=1}^{m'}(s - z_i)e^{-\tau s} + \prod_{j=1}^{n'}\left(s - p_j\right)} = \frac{K\prod_{i=1}^{m'}(-z_i)}{K\prod_{i=1}^{m'}(-z_i) + \prod_{j=1}^{n'}\left(-p_j\right)} \tag{8.44}$$

8.2.1.5 Root Locus Analysis of Closed-Loop Stability

In Section 4.2, some basic SISO controllers were considered:

$$\text{P} \quad G_C(s) = K_C \tag{8.45}$$

$$\text{PI} \quad G_C(s) = K_C\left[1 + \frac{1}{\tau_I s}\right] \tag{8.46}$$

$$\text{PID} \quad G_C(s) = K_C\left[1 + \frac{1}{\tau_I s} + \tau_D s\right] \tag{8.47}$$

From Equation 8.2

$$G_O(s) = G_M(s)G_P(s)G_C(s) \tag{8.48}$$

$$= K_C G_O'(s) \tag{8.49}$$

with the controller gain now explicitly factored out of $G_O(s)$. The closed-loop characteristic Equation 8.37 is thus

$$K_C G_O'(s) = -1 \tag{8.50}$$

For the root locus analysis, one must start with the factored form of Equation 8.16, invoking approximations such as Equations 8.19 and 8.20 in the event of dead time. So, the closed-loop characteristic equation is

$$\frac{K_C K'(s - z_1)(s - z_2)\cdots(s - z_m)}{\left(s - p_1\right)\left(s - p_2\right)\cdots\left(s - p_n\right)} = -1 \tag{8.51}$$

which will have n roots since $n \geq m$.

Note: The roots loci of a system are the paths traced out by its closed-loop poles on the complex plane as the system gain increases from 0 to ∞.

The root locus plot is thus a useful aid to understand the effect of increasing controller gain on the closed-loop performance.

The loci of the n roots will start at the open-loop poles (for $K_C = 0$), and as $K_C \to \infty$, m of these will end at the open-loop zeros, with the remaining n-m loci 'heading for infinity' ('large' s). This is what one would expect if Equation 8.51 is to 'balance'. Certainly, where $n > m$, 'large' s would allow the s terms in the denominator to dominate the magnitude, as required to balance the large K_C in the numerator.

Of course, one could generate the root locus plot automatically by stepping K_C in small increments and using a polynomial equation solver on each step to obtain the new pole positions. However, the main features of the plot can be sketched out by inspection of the open-loop transfer function. The procedure for this can provide useful insight to the closed-loop behaviour. For the *standard case* of *negative feedback* ($K_C K' > 0$), one applies the following rules:

1) Factorise the numerator and denominator of the open-loop transfer function, and write down the *closed-loop characteristic equation* in the form shown in Equation 8.51. On the *Im/Re* complex plane, mark the n open-loop poles with '×' and the m open-loop zeros with 'O.' Bear in mind that these may be *multiple-ordered* - that is more than one on the same location.

   The loci will follow paths traced out by positively charged particles in an electrostatic field where the ×'s are positive and the O's are negative. The positively charged particles must be released at the ×'s (but at which 360° position around the 'edge' of an × may need further clarification – see (5) below).

   Because the loci represent the closed-loop poles, if any part of a locus is complex, a mirror-image complex conjugate part must exit – that is the loci must be symmetrical about the real axis (so that associated factors in the characteristic equation leave real coefficients when multiplied).
2) The real axis will be part of the root locus if there is an odd number of ×'s plus O's to the right. Where two real ×'s or two real O's are connected together in this way, there has to be a *breakout point* or a *break-in point*, respectively. The leaving or entering loci do so at right angles to the real axis. The points are determined by solution of

$$\sum_{i=1}^{m} \frac{1}{(s - z_i)} = \sum_{j=1}^{n} \frac{1}{\left(s - p_j\right)} \tag{8.52}$$

3) m of the loci will proceed from ×'s to O's, and the remaining n-m will move outwards along asymptotes which diverge at equally spaced angles θ_k from a *centre of gravity* γ as follows:

$$\gamma = \frac{\sum_{j=1}^{n} p_j - \sum_{i=1}^{m} z_i}{n - m} \tag{8.53}$$

$$\theta_k = \frac{(2k+1)\pi}{n-m} \quad \text{for} \quad k = 1, 2, \ldots, (n-m) \tag{8.54}$$

4) The q *loci departing* from a qth-*ordered* pole p_a do so at the following angles leaving the ×:

$$\theta_k = \frac{1}{q}\left[(2k-1)\pi + \sum_{i=1}^{m} \measuredangle(p_a - z_i) - \sum_{\substack{j=1 \\ j \neq a}}^{n} \measuredangle\left(p_a - p_j\right)\right] \quad \text{for} \quad k = 1, 2, \ldots, q \tag{8.55}$$

   The v *loci approaching* a vth-*ordered* zero z_b do so at the following angles entering the O:

$$\phi_k = \frac{1}{v}\left[(2k-1)\pi + \sum_{j=1}^{n} \measuredangle\left(z_b - p_j\right) - \sum_{\substack{i=1 \\ i \neq b}}^{m} \measuredangle(z_b - z_i)\right] \quad \text{for} \quad k = 1, 2, \ldots, v \tag{8.56}$$

The proofs of these attributes are given by Coughanowr and Koppel (1965). Apart from direct algebraic solution of the characteristic equation, points on the roots loci can also be fixed by applying the *angle criterion* and the *magnitude criterion* to the closed-loop characteristic Equation 8.51, using the results of Section 8.2.1.2:

$$\begin{aligned}\measuredangle G_O(s) &= \measuredangle(K_C K') + \measuredangle(s - z_1) + \cdots + \measuredangle(s - z_m) - \measuredangle(s - p_1) - \cdots - \measuredangle(s - p_n) = \measuredangle(-1)\\ &= (2k+1)\pi\end{aligned} \tag{8.57}$$

where k is an as yet unknown integer, and if $K_C K' > 0$, one obtains:

angle criterion:

$$\measuredangle G_O(s) = \measuredangle(s - z_1) + \cdots + \measuredangle(s - z_m) - \measuredangle(s - p_1) - \cdots - \measuredangle(s - p_n) = \measuredangle(-1) = (2k+1)\pi \tag{8.58}$$

magnitude criterion:

$$|G_O(s)| = \frac{|K_C K'| \cdot |s - z_1| \cdot |s - z_2| \cdots |s - z_m|}{|s - p_1| \cdot |s - p_2| \cdots |s - p_n|} = |-1| = 1 \tag{8.59}$$

The correct odd number of π's in Equation 8.57 is not clear until one is in the vicinity of the solution. Typically s must comply with an additional constraint such as a damping ratio line or magnitude circle which intersects the root locus. This intersection on a rough root locus plot provides a good starting guess of s which is then refined iteratively to comply with the angle criterion shown in Equation 8.58.

Example 8.1

PI Control of a Third-order Process

A process behaves as an under-damped second-order system ($K = 2, \tau = {}^1/_{\sqrt{2}}, \zeta = {}^1/_{\sqrt{2}}$) in series with a first-order system ($K = 1, \tau = {}^1/_2$). It is to be controlled by a PI controller with $\tau_I = 1$.

a. Plot the root locus.

b. Obtain a K_C for the controller to give a damping ratio of $\zeta = {}^1/_{\sqrt{5}}$ (i.e., for the worst-damped poles).

c. At what value of K_C will the closed loop become unstable?
[NB: Use your plot to estimate starting values for the calculations in (b) and (c)]

Solution:

a. Following the steps (1)–(4) of the root locus plotting procedure:

$$G_O(s) = G_C(s) G_{P1}(s) G_{P2}(s) = \left[K_C\left(1 + \frac{1}{\tau_I s}\right)\right]\left[\frac{2}{\tau_1{}^2 s^2 + 2\zeta\tau_1 s + 1}\right]\left[\frac{1}{\tau_2 s + 1}\right] \tag{8.60}$$

$$= \left[K_C\left(1 + \frac{1}{s}\right)\right]\left[\frac{2}{\left({}^1/_{\sqrt{2}}\right)^2 s^2 + 2\left({}^1/_{\sqrt{2}}\right)\left({}^1/_{\sqrt{2}}\right)s + 1}\right]\left[\frac{1}{\left({}^1/_2\right)s + 1}\right] \tag{8.61}$$

$$= K_C\left[\frac{(s+1)}{s}\right]\left[\frac{4}{s^2 + 2s + 2}\right]\left[\frac{2}{s+2}\right] \tag{8.62}$$

Thus, the closed-loop characteristic equation is

$$8K_C \frac{(s+1)}{s(s+1+j)(s+1-j)(s+2)} = -1 \tag{8.63}$$

giving the $n = 4$ pole (X) and $m = 1$ zero (O) locations in Figure 8.4.

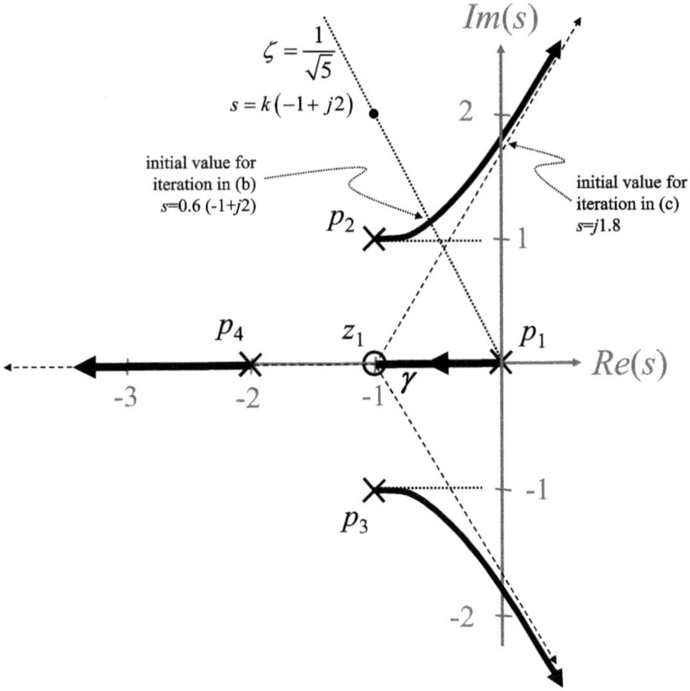

Figure 8.4 Root locus plot for Example 8.1: PI control of a third-order system.

Two parts of real axis are root locus: proceeding from $0 + j0$ to $-1 + j0$, then from $-2 + j0$ to $-\infty + j0$. No break-ins or break-outs.

Centre of gravity of asymptotes:

$$\gamma = \frac{\sum_{j=1}^{n} p_j - \sum_{i=1}^{m} z_i}{n - m} = \frac{[(0 + j0) + (-1 + j1) + (-1 - j1) + (-2 + j0)] - [(-1 + j0)]}{4 - 1} \tag{8.64}$$

$$= (-1 + j0) \tag{8.65}$$

$$\theta_k = \frac{(2k + 1)\pi}{n - m} \quad \text{for} \quad k = 1, 2, \ldots, (n - m) : \theta_1 = \pi; \theta_2 = {}^5/_3\,\pi, \quad \theta_3 = {}^7/_3\,\pi = {}^1/_3\pi \tag{8.66}$$

Angle of departure ($q = 1$):

for $p2 = -1 + j1$:

$$\theta_1 = \frac{1}{1}\left[\pi + \measuredangle(p_2 - z_1) - \measuredangle(p_2 - p_1) - \measuredangle(p_2 - p_3) - \measuredangle(p_2 - p_4)\right] \tag{8.67}$$

$$= \pi + {}^1/_2\,\pi - {}^3/_4\,\pi - {}^1/_2\,\pi - {}^1/_4\pi \tag{8.68}$$

$$= 0 \tag{8.69}$$

for $p3 = -1-j1$: . . . plotted by symmetry

b. In setting the damping ratio of the closed loop, it is implicit that the focus will be on the *worst-damped* loci, *viz.* those starting at p_2, p_3. (One notes that the remaining loci here are critically damped at $\zeta = 1$). To get the required K_C, first find s where the under damped loci intersect the required damping line [$\zeta = {}^1/_{\sqrt{5}} \Rightarrow s = w(-1 + j2)$]. This is done by satisfying the *angle criterion* Equation 8.58 by trial and error. Then K_C results from satisfying the *magnitude criterion* Equation 8.59. The result will be the same for the complex-conjugate locus by symmetry. The angle criterion is

$$\measuredangle(K_C K') + \measuredangle(s - z_1) - \measuredangle(s - p_1) - \measuredangle(s - p_2) - \measuredangle(s - p_3) - \measuredangle(s - p_4) = (2k + 1)\pi \tag{8.70}$$

where k is integer and $K_C K'$ is a positive real for normal negative feedback, so it has an angle of zero. Since $s = w(-1 + j2)$ for the required damping,

$$\begin{aligned}\measuredangle[w(-1 + j2) - (-1 + j0)] - \measuredangle[w(-1 + j2) - (0 + j0)] \\ - \measuredangle[w(-1 + j2) - (-1 + j1)] \\ - \measuredangle[w(-1 + j2) - (-1 - j1)] \\ - \measuredangle[w(-1 + j2) - (-2 + j0)] = (2k + 1)\pi\end{aligned} \tag{8.71}$$

that is one is tempted to write

$$\tan^{-1}\left(\frac{2w}{1 - w}\right) - \tan^{-1}\left(\frac{2w}{-w}\right) - \tan^{-1}\left(\frac{2w - 1}{1 - w}\right) - \tan^{-1}\left(\frac{2w + 1}{1 - w}\right) - \tan^{-1}\left(\frac{2w}{2 - w}\right) = (2k + 1)\pi \tag{8.72}$$

Although the *imaginary* (top) and *real* (bottom) parts of each ratio have been kept intact, and no signs have been cancelled, a problem will arise when the denominator goes negative (i.e., the number lands in the left-hand complex plane). It will be hard to keep track of all of the necessary π shifts. So, it is better to make use of the four-quadrant *ATAN2* function:

$$\begin{aligned}\text{atan2}(1 - w, 2w) - \text{atan2}(-w, 2w) \\ - \text{atan2}(1 - w, 2w - 1) \\ - \text{atan2}(1 - w, 2w + 1) \\ - \text{atan2}(2 - w, 2w) = (2k + 1)\pi\end{aligned} \tag{8.73}$$

Trial and error starting at the apparent plot intercept ($w = 0.6$)

w	0.6	0.65	0.55	0.56	0.562	......
LHS (0)	−191.86	−207.50	−176.43	−179.42	−180.02	......

After the first evaluation it is clear that the aim must be for $k = -1$. Thereafter, it is just a matter of refining w. The point of intersection is given by

$$s = 0.562(-1 + j2) = -0.562 + j1.124 \tag{8.74}$$

Now the *magnitude criterion* Equation 8.59 is required in order to deduce the corresponding K_C value:

$$\frac{|K_C K'| \cdot |s - z_1|}{|s - p_1| \cdot |s - p_2| \cdot |s - p_3| \cdot |s - p_4|} = 1 \tag{8.75}$$

$$\frac{8K_C \cdot |(-0.562 + j1.124) - (-1 + j0)|}{\left\{ \begin{array}{r} |(-0.562 + j1.124) - (0 + j0)| \cdot |(-0.562 + j1.124) - (-1 + j1)| \\ \cdot |(-0.562 + j1.124) - (-1 - j1)| \\ \cdot |(-0.562 + j1.124) - (-2 + j0)| \end{array} \right\}} = 1 \tag{8.76}$$

$$K_C = \frac{1.257 \times 0.455 \times 2.169 \times 1.825 \times 1}{8 \times 1.206} = 0.235 \quad (\text{for a damping ratio of } {}^{1}/_{\sqrt{5}}) \rightarrow \tag{8.77}$$

c. For the value of K_C above which the closed loop will be unstable, one notes that this corresponds to (the first) locus crossing into the right-hand s plane – that is it is the intersection of the locus with the imaginary axis where $s = jw$.

The corresponding angle criterion is

$$\begin{aligned} &\text{atan2}(1, w) - \text{atan2}(0, w) \\ &\qquad - \text{atan2}(1, w - 1) \\ &\qquad - \text{atan2}(1, w - 1) \\ &\qquad - \text{atan2}(2, w) = (2k + 1)\pi \end{aligned} \tag{8.78}$$

which is solved as in **(b)**, starting with $w = 1.8$ from Figure 8.4, to give $w = 1.799$ (for nearest $k = -1$ again). Substituting $s = j1.799$ into the *magnitude criterion* then gives

$$K_C = \frac{1.799 \times 1.280 \times 2.972 \times 2.690 \times 1}{8 \times 2.058} = 1.118 \quad (\text{for } \zeta = 0, \text{ i.e. incipient instability}) \tag{8.79}$$

Example 8.2

Effect of an Extra Pole on the Root Locus Plot of Example 8.1

It is interesting to note the impact of an extra open-loop pole on the above root locus plot of a third-order system under PI control. Adjust Figure 8.4 to reflect an extra first-order process $G_{P3}(s) = 1/(s + 3)$ in the open loop.

1. $m = 1$ and $n = 5$

$$8K_C \frac{(s + 1)}{s(s + 1 + j)(s + 1 - j)(s + 2)(s + 3)} = -1 \tag{8.80}$$

2. Real axis is part of root locus from $(0+j0)$ to $(-1+j0)$ and from $(-2+j0)$ to $(-3+j0)$. Since the latter two are both poles, there must be a *breakout* between them, requiring solution of Equation 8.52:

$$\frac{1}{s}+\frac{1}{(s+1+j)}+\frac{1}{(s+1-j)}+\frac{1}{(s+2)}+\frac{1}{(s+3)}=\frac{1}{(s+1)} \tag{8.81}$$

that is,

$$\frac{1}{s}+\frac{1}{(s^2+2s+2)}+\frac{1}{(s+2)}+\frac{1}{(s+3)}-\frac{1}{(s+1)}=0 \tag{8.82}$$

which is solved by trial and error starting with a good guess of s in the required real interval, *viz.* −2.5. This converges to the solution $s=-2.424$.

Centre of gravity of asymptotes:

$$\gamma=\frac{\sum_{j=1}^{n}p_j-\sum_{i=1}^{m}z_i}{n-m}=\frac{[(0+j0)+(-1+j1)+(-1-j1)+(-2+j0)+(-3+j0)]-[(-1+j0)]}{5-1} \tag{8.83}$$

$$=(-1.5+j0) \tag{8.84}$$

$$\theta_k=\frac{(2k+1)\pi}{n-m}\quad \text{for}\quad k=1,2,\ldots,(n-m) \tag{8.85}$$

giving

$$\theta_1={}^3/_4\,\pi \tag{8.86}$$

$$\theta_2={}^5/_4\pi \tag{8.87}$$

$$\theta_3={}^7/_4\pi \tag{8.88}$$

$$\theta_4={}^9/_4\,\pi={}^1/_4\pi \tag{8.89}$$

Angle of departure ($q=1$):
for p2 = −1 + j1:

$$\theta_1=\frac{1}{1}\left[\pi+\measuredangle(p_2-z_1)-\measuredangle(p_2-p_1)-\measuredangle(p_2-p_3)-\measuredangle(p_2-p_4)-\measuredangle(p_2-p_5)\right] \tag{8.90}$$

$$=\pi+{}^1/_2\,\pi-{}^3/_4\,\pi-{}^1/_2\,\pi-{}^1/_4\pi-0.46\,\textit{radians} \tag{8.91}$$

$$=-26.57^\circ \tag{8.92}$$

for p3 = −1−j1: . . . plotted by symmetry.

The method discussed above for generating the main features of the root locus plot apply to the *standard case* of *negative feedback* – that is, K_CK' in the closed-loop characteristic Equation 8.51 is positive (Figure 8.5). There are situations in which it becomes negative (or a better way of thinking about it is that the '−1' on the right-hand side becomes '+1.' This will obviously change all of the angle

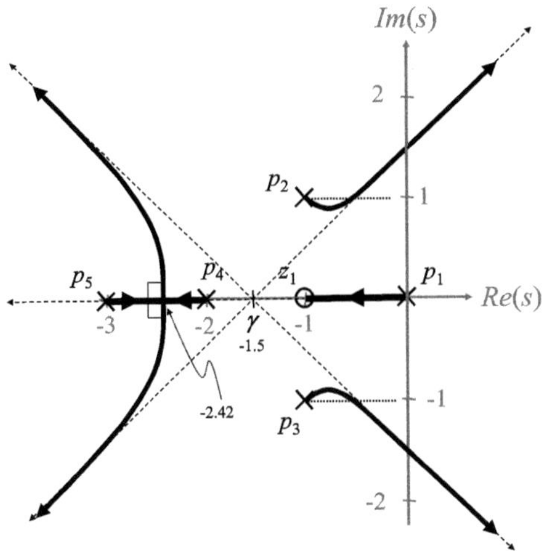

Figure 8.5 Root locus plot for Example 8.2: Example 8.1 extended with an extra pole.

references from '$(2k+1)\pi$' to '$2k\pi$.' There is a similar construction procedure based on this scenario that must be referenced in that case. An example of this occurrence is where a first-order Padé approximation (Equation 8.19) has been used for dead-time lag:

$$e^{-\tau s} \approx \frac{1-\frac{\tau}{2}s}{1+\frac{\tau}{2}s} = -\frac{\left(s-\frac{2}{\tau}\right)}{\left(s+\frac{2}{\tau}\right)} \tag{8.93}$$

which apart from contributing the negative, is also seen to place an open-loop zero in the *right-hand complex plane.* Certainly, as K_C increases, the closed-loop root locus approaching that zero will eventually cause instability.

This case where $K_C K' < 0$ in Equation 8.51 is termed the *complementary case,* and can be thought of as *positive feedback.* It is worthwhile noting how the rules change:

1) No change
2) The real axis will be part of the root locus if there is an *even* number (including zero) of ×'s plus O's to the right. The entry and exit points are determined by Equation 8.52 as before
3) The *centre of gravity* γ is determined by Equation 8.53 as before, but now the asymptotes radiate outwards at angles

$$\theta_k = \frac{2k\pi}{n-m} \quad \text{for} \quad k = 1, 2, \ldots, (n-m) \tag{8.94}$$

4) The q angles of departure from a qth-*ordered* pole p_a become

$$\theta_k = \frac{1}{q}\left[2k\pi + \sum_{i=1}^{m} \measuredangle(p_a - z_i) - \sum_{\substack{j=1 \\ j\neq a}}^{n} \measuredangle\left(p_a - p_j\right)\right] \quad \text{for} \quad k = 1, 2, \ldots, q \tag{8.95}$$

and the ν angles of approach to a νth-*ordered* zero z_b become

$$\phi_k = \frac{1}{\nu}\left[2k\pi + \sum_{j=1}^{n} \measuredangle\left(z_b - p_j\right) - \sum_{\substack{i=1 \\ i\neq b}}^{m} \measuredangle(z_b - z_i)\right] \quad \text{for} \quad k = 1, 2, \ldots, \nu \tag{8.96}$$

With $K_C K'$ negative in Equation 8.51, the *magnitude criterion* shown in Equation 8.59 remains the same, but the *angle criterion* shown in Equation 8.58 obviously becomes

$$\begin{aligned}\measuredangle G_O(s) &= -\pi + \measuredangle(s - z_1) + \cdots + \measuredangle(s - z_m) - \measuredangle\left(s - p_1\right) - \cdots - \measuredangle\left(s - p_n\right) = \measuredangle(-1) \\ &= (2k + 1)\pi\end{aligned} \tag{8.97}$$

that is

$$\measuredangle G_O(s) = \measuredangle(s - z_1) + \cdots + \measuredangle(s - z_m) - \measuredangle\left(s - p_1\right) - \cdots - \measuredangle\left(s - p_n\right) = 2k\pi \tag{8.98}$$

where k is an as yet unknown integer.

8.3 View of a Continuous MIMO System in the *s*-Domain

From Equation 8.6 the closed-loop transfer function is

$$\boldsymbol{G}_{CL}(s) = [\boldsymbol{I} + \boldsymbol{G}_O(s)]^{-1}\boldsymbol{G}_O(s) \tag{8.99}$$

with

$$\boldsymbol{G}_O(s) = \boldsymbol{G}_M(s)\boldsymbol{G}_P(s)\boldsymbol{G}_C(s) \tag{8.100}$$

As an example, consider a process determined by

$$\boldsymbol{G}_P(s) = \begin{bmatrix} \dfrac{3e^{-2s}}{s+1} & \dfrac{1}{s+2} \\ e^{-s} & \dfrac{4}{s^2 + 2s + 2} \end{bmatrix} \tag{8.101}$$

Usually, there is no interaction between *measurement devices* so one might have

$$\boldsymbol{G}_M(s) = \begin{bmatrix} \dfrac{8}{s+8} & 0 \\ 0 & \dfrac{9}{s+9} \end{bmatrix} \tag{8.102}$$

A properly compensating MIMO controller would have off-diagonal terms to handle the interactions, but in a typical multiple-SISO industrial control set-up, interactions are ignored and individual PI controllers might be specified as

$$\boldsymbol{G}_C(s) = \begin{bmatrix} K_{C1}\dfrac{s+2}{s} & 0 \\ 0 & K_{C2}\dfrac{s+3}{s} \end{bmatrix} \tag{8.103}$$

So, in this case

$$\boldsymbol{G}_{\mathrm{O}}(\mathrm{s}) = \begin{bmatrix} K_{C1}\dfrac{e^{-2s}24(s+2)}{s(s+1)(s+8)} & K_{C1}\dfrac{8}{s(s+8)} \\ K_{C2}\dfrac{e^{-s}9(s+3)}{s(s+9)} & K_{C2}\dfrac{36(s+3)}{s(s+9)(s+1+j)(s+1-j)} \end{bmatrix} \tag{8.104}$$

Since $\boldsymbol{A}^{-1} = \frac{adj[A]}{\det[A]}$, from Equation 8.99 the *characteristic equation* of the closed-loop MIMO system is

$$\det[\boldsymbol{I} + \boldsymbol{G}_{\mathrm{O}}(\mathrm{s})] = 0 \tag{8.105}$$

so in this example the poles of the closed loop are obtained by solving

$$\det \begin{bmatrix} 1 + K_{C1}\dfrac{e^{-2s}24(s+2)}{s(s+1)(s+8)} & K_{C1}\dfrac{8}{s(s+8)} \\ K_{C2}\dfrac{e^{-s}9(s+3)}{s(s+9)} & 1 + K_{C2}\dfrac{36(s+3)}{s(s+9)(s+1+j)(s+1-j)} \end{bmatrix} = 0 \tag{8.106}$$

Without the dead-time delays (or if they are approximated as in Equation 8.19 or 8.20), this just requires a polynomial solution, which will give a number of pole positions on the complex s-plane for a particular choice of K_{C1} and K_{C2}. The same principles obviously apply as in the SISO view of Section 8.1, except now the $0 < K_{C1} < \infty$ root loci will vary for each choice of the K_{C2} value, and vice-versa.

8.4
View of Continuous SISO and MIMO Systems in Linear State Space

The approach will be developed for a MIMO system, but the same relationships apply for a SISO system. Referring to the closed-loop system in Figure 8.1, and ignoring disturbances, the continuous linear representation of the process in the state space is

$$\frac{\mathrm{d}\boldsymbol{x}}{\mathrm{d}t} = \boldsymbol{A}\boldsymbol{x} + \boldsymbol{B}\boldsymbol{u} \tag{8.107}$$

$$\boldsymbol{y} = \boldsymbol{C}\boldsymbol{x} \tag{8.108}$$

Ignoring the nicety that $\boldsymbol{b}$ is not $\boldsymbol{y}$, consider that the output of interest $\boldsymbol{y}$ is the same as the feedback signal so the error is

$$\boldsymbol{e} = \boldsymbol{r} - \boldsymbol{y} \tag{8.109}$$

For argument sake, consider every element in the 'controller matrix' to be potentially PID, for example, in a 2-input, 2-output system,

$$u_1 = K_{C11}\left[e_1 + \frac{1}{\tau_{I11}}\int e_1\mathrm{d}t + \tau_{D11}\frac{\mathrm{d}e_1}{dt}\right] + K_{C12}\left[e_2 + \frac{1}{\tau_{I12}}\int e_2\mathrm{d}t + \tau_{D12}\frac{\mathrm{d}e_2}{dt}\right] \tag{8.110}$$

$$u_2 = K_{C21}\left[e_1 + \frac{1}{\tau_{I21}}\int e_1\mathrm{d}t + \tau_{D21}\frac{\mathrm{d}e_1}{dt}\right] + K_{C22}\left[e_2 + \frac{1}{\tau_{I22}}\int e_2\mathrm{d}t + \tau_{D22}\frac{\mathrm{d}e_2}{dt}\right] \tag{8.111}$$

Differentiating

$$\frac{\mathrm{d}\boldsymbol{u}}{\mathrm{d}t} = \boldsymbol{D}\frac{\mathrm{d}^2\boldsymbol{e}}{\mathrm{d}t^2} + \boldsymbol{P}\frac{\mathrm{d}\boldsymbol{e}}{\mathrm{d}t} + \boldsymbol{N}\boldsymbol{e} \tag{8.112}$$

with

$$D = \begin{bmatrix} K_{C11}\tau_{D11} & K_{C12}\tau_{D12} \\ K_{C21}\tau_{D21} & K_{C22}\tau_{D22} \end{bmatrix} \tag{8.113}$$

$$P = \begin{bmatrix} K_{C11} & K_{C12} \\ K_{C21} & K_{C22} \end{bmatrix} \tag{8.114}$$

and

$$N = \begin{bmatrix} K_{C11}/\tau_{I11} & K2/\tau_{I12} \\ K_{C21}/\tau_{I21} & K_{C22}/\tau_{I22} \end{bmatrix} \tag{8.115}$$

Differentiating Equation 8.107 and substituting to close the loop

$$\frac{d^2\boldsymbol{x}}{dt^2} = \boldsymbol{A}\frac{d\boldsymbol{x}}{dt} + \boldsymbol{BD}\frac{d^2\boldsymbol{e}}{dt^2} + \boldsymbol{BP}\frac{d\boldsymbol{e}}{dt} + \boldsymbol{BNe} \tag{8.116}$$

Using Equation 8.109 then one obtains

$$\frac{d}{dt}\begin{pmatrix} \frac{d\boldsymbol{x}}{dt} \\ \boldsymbol{x} \end{pmatrix} = \begin{bmatrix} [\boldsymbol{I}+\boldsymbol{BDC}]^{-1}[\boldsymbol{A}-\boldsymbol{BPC}] & [\boldsymbol{I}+\boldsymbol{BDC}]^{-1}\boldsymbol{BNC} \\ \boldsymbol{I} & \boldsymbol{0} \end{bmatrix}\begin{pmatrix} \frac{d\boldsymbol{x}}{dt} \\ \boldsymbol{x} \end{pmatrix} + \begin{bmatrix} \boldsymbol{BD} & \boldsymbol{BP} & \boldsymbol{BN} \\ \boldsymbol{0} & \boldsymbol{0} & \boldsymbol{0} \end{bmatrix}\begin{pmatrix} \frac{d^2\boldsymbol{r}}{dt^2} \\ \frac{d\boldsymbol{r}}{dt} \\ \boldsymbol{r} \end{pmatrix} \tag{8.117}$$

Representing the forcing input from the setpoint as $\boldsymbol{F}$, the form is

$$\frac{d\boldsymbol{X}}{dt} = \boldsymbol{HX} + \boldsymbol{F} \tag{8.118}$$

where the stability is seen to be determined by the homogeneous part, by taking the Laplace transform

$$s\boldsymbol{X}(s) = \boldsymbol{HX}(s) + \boldsymbol{F}(s) \tag{8.119}$$

$$\boldsymbol{X}(s) = [s\boldsymbol{I} - \boldsymbol{H}]^{-1}\boldsymbol{F}(s) \tag{8.120}$$

As seen in Example 3.9, the *characteristic equation* of this system is

$$\det[s\boldsymbol{I} - \boldsymbol{H}] = 0 \tag{8.121}$$

of which the solutions are clearly the eigenvalues of $\boldsymbol{H}$. As noted in Section 9.2.2.2, these must lie in the left-hand complex plane if the system (in this case the closed loop) is to be stable. If any eigenvalue of

$$\boldsymbol{H} = \begin{bmatrix} [\boldsymbol{I}+\boldsymbol{BDC}]^{-1}[\boldsymbol{A}-\boldsymbol{BPC}] & [\boldsymbol{I}+\boldsymbol{BDC}]^{-1}\boldsymbol{BNC} \\ \boldsymbol{I} & \boldsymbol{0} \end{bmatrix} \tag{8.122}$$

has a positive real part, this closed loop will be unstable. In this example, one can extract the reduced case of *proportional state feedback*, where $\boldsymbol{N}=\boldsymbol{0}$ and $\boldsymbol{D}=\boldsymbol{0}$. Equation 8.117 reduces to

$$\frac{d^2\boldsymbol{x}}{dt^2} = [\boldsymbol{A} - \boldsymbol{BPC}]\frac{d\boldsymbol{x}}{dt} + \boldsymbol{BP}\frac{d\boldsymbol{r}}{dt} \tag{8.123}$$

$$\text{i.e. } \frac{d\boldsymbol{x}}{dt} = [\boldsymbol{A} - \boldsymbol{BPC}]\boldsymbol{x} + \boldsymbol{BP}\,\boldsymbol{r} + \boldsymbol{c}_{\text{const}} \tag{8.124}$$

So, here the K_{Cij} values in the $\boldsymbol{P}$ matrix must be chosen in such a way that the eigenvalues of $[\boldsymbol{A} - \boldsymbol{BPC}]$ lie in suitably damped positions in the left-hand complex plane. Of course, the controller need not have the particular form of Equations 8.110 and 8.111, but the same considerations apply. A number of *pole placement* design techniques are available in the literature – see for example Sections 7.6.1 and 7.6.3.

8.5 View of Discrete Linear SISO and MIMO Systems

Again, the MIMO system is treated as the general case. In Sections 7.5 and 7.8, discrete MIMO models were encountered in the format

$$\boldsymbol{y}(t) = \boldsymbol{f}\{\boldsymbol{y}(t-T), \boldsymbol{y}(t-2T), \ldots, \boldsymbol{u}(t-T), \boldsymbol{u}(t-2T), \ldots\} \tag{8.125}$$

or

$$\boldsymbol{y}(z) = \boldsymbol{G}_P(z)\boldsymbol{u}(z) \tag{8.126}$$

with constant matrices $\boldsymbol{N}_i$ and $\boldsymbol{D}_i$ determining

$$\boldsymbol{G}_P(z) = \boldsymbol{D}^{-1}(z)\boldsymbol{N}(z) \tag{8.127}$$

$$\boldsymbol{D}(z) = \boldsymbol{I} + \boldsymbol{D}_1 z^{-1} + \boldsymbol{D}_2 z^{-2} + \boldsymbol{D}_3 z^{-3} + \cdots \tag{8.128}$$

$$\boldsymbol{N}(z) = \boldsymbol{0} + \boldsymbol{N}_1 z^{-1} + \boldsymbol{N}_2 z^{-2} + \boldsymbol{N}_3 z^{-3} + \cdots \tag{8.129}$$

Where a model arose from *input–output* data (e.g. step responses) $\boldsymbol{D}(z)$ could be made scalar, as in Section 7.8.1. If it arose from state-space modelling, Equation 8.129 would have a state recursive form determined by $\boldsymbol{D}_2 = \boldsymbol{D}_3 = \cdots = \boldsymbol{0}$ and $\boldsymbol{N}_2 = \boldsymbol{N}_3 = \cdots = \boldsymbol{0}$.

A MIMO controller $\boldsymbol{G}_C(z)$ will also have this form, for example, arising from a Tustin approximation as in Section 4.2.3, and likewise the measurement dynamics $\boldsymbol{G}_M(z)$. Taking care with the order in which the signals are processed, the open-loop transfer function is

$$\boldsymbol{G}_O(z) = \boldsymbol{G}_M(z)\boldsymbol{G}_P(z)\boldsymbol{G}_C(z) \tag{8.130}$$

which will again have elements which are ratios of polynomials in z. In Figure 8.1, the closed-loop behaviour is determined by

$$\boldsymbol{b}(z) = [\boldsymbol{I} + \boldsymbol{G}_O(z)]^{-1}\, \boldsymbol{G}_O(z)\boldsymbol{r}(z) \tag{8.131}$$

Since $\boldsymbol{A}^{-1} = \frac{\mathrm{adj}[\boldsymbol{A}]}{\det[\boldsymbol{A}]}$, the *characteristic equation* determining the closed-loop poles is thus

$$\det[\boldsymbol{I} + \boldsymbol{G}_O(z)] = 0 \tag{8.132}$$

Bearing in mind that for sampling interval T,

$$z = \mathrm{e}^{Ts} = \mathrm{e}^{T(a+jb)} = \mathrm{e}^{aT}\mathrm{e}^{jbT} = \mathrm{e}^{aT}\left[\cos(bT) + j\sin(bT)\right] \tag{8.133}$$

one notes that for stability $\mathrm{Re}(s) = a$ must be negative, so the roots of Equation 8.132 must lie within the unit circle on the complex plane ($\mathrm{e}^{aT} < 1$). In Figure 7.11, one notes that b determines the angular deviation from zero of z on the complex plane. The damping ratio is $-a/\sqrt{a^2 + b^2}$, so it will increase as a pole z approaches the origin but will decrease as the angle of z increases.

Example 8.3

Closed-loop Stability for Proportional Feedback Control of a Discrete State Space System

With regard to Equations 8.126–8.129, it was noted that the recursive *state* equations would be

$$\boldsymbol{x}(z) = \boldsymbol{G}_P(z)\boldsymbol{u}(z) \tag{8.134}$$

with

$$\boldsymbol{G}_P(z) = \boldsymbol{D}^{-1}(z)\boldsymbol{N}(z) \tag{8.135}$$

$$\boldsymbol{D}(z) = \boldsymbol{I} + \boldsymbol{D}_1 z^{-1} \tag{8.136}$$

$$\boldsymbol{N}(z) = \boldsymbol{N}_1 z^{-1} \tag{8.137}$$

$$\text{Take } \boldsymbol{G}_C(z) = \boldsymbol{K}_C \tag{8.138}$$

and

$$\boldsymbol{G}_M(z) = \boldsymbol{I} \tag{8.139}$$

$$\text{so } [\boldsymbol{I} + \boldsymbol{G}_O(z)] = \boldsymbol{I} + \left[\boldsymbol{I} + \boldsymbol{D}_1 z^{-1}\right]^{-1} \boldsymbol{N}_1 z^{-1} \boldsymbol{K}_C \tag{8.140}$$

$$= \boldsymbol{I} + [z\boldsymbol{I} + \boldsymbol{D}_1]^{-1} \boldsymbol{N}_1 \boldsymbol{K}_C \tag{8.141}$$

Multiplying by $[z\boldsymbol{I} + \boldsymbol{D}_1]$ and noting that for square matrices $\det[\boldsymbol{AB}] = \det[\boldsymbol{A}]\det[\boldsymbol{B}]$ the closed-loop characteristic Equation 8.132 requires

$$\det[z\boldsymbol{I} + \boldsymbol{D}_1]\det[\boldsymbol{I} + \boldsymbol{G}_O(z)] = \det[z\boldsymbol{I} + \boldsymbol{D}_1 + \boldsymbol{N}_1\boldsymbol{K}_C] = 0 \tag{8.142}$$

so that the closed-loop poles are clearly the *eigenvalues* of $[\boldsymbol{D}_1 + \boldsymbol{N}_1\boldsymbol{K}_C]$. Thus, in the closed loop, as the elements in $\boldsymbol{K}_C$ deviate from zero, the poles are seen to move away increasingly from the open-loop poles (eigenvalues of $\boldsymbol{D}_1$).

For linear discrete state systems with more complex controllers, higher order systems of equations will result in the same way as for the continuous systems in Section 8.4. However, analysis of the effects of the various K_{Cij}, τ_{Iij}, τ_{Dij}, and so on, on the damping and stability of closed-loop poles follows the same pattern as above.

8.6 Frequency Response

So far a view of open-loop and closed-loop stability has been developed around the location of poles on the complex plane. These poles were the solution of the *characteristic equation* (denominator = 0), the importance of which arises from the contribution of the roots to output responses, as becomes evident through a partial fraction expansion. Now an alternative view will be developed based on *frequency*, which offers new measures of relative stability, and has the advantage that it handles dead-time precisely. Noting the similarity of the Laplace transform

$$X(s) = \int_0^t x(t')e^{-st'}dt' \tag{8.143}$$

to the Fourier transform (Section 6.1.2), one expects s to have frequency units (time^{-1}). Indeed, a simple substitution of $j\omega$ for s (where ω is frequency in radians/time) will be seen to provide important information about the *frequency response,* that is the attenuation and phase lag of frequency signals passing through a system.

8.6.1 Frequency Response from G(jω)

Consider a linear SISO system subjected to a steady frequency signal

$$u(t) = A_u \sin(\omega t) \tag{8.144}$$

as in Figure 8.6. According to the condition of stationarity (Section 3.6), the output ultimately becomes a steady oscillation at the same frequency, that is,

$$x(t) = A_x \sin(\omega t + \phi) \tag{8.145}$$

where ϕ is the phase angle. Recalling the identity

$$e^{j\theta} = \cos(\theta) + j\sin(\theta) \tag{8.146}$$

one can write

$$u(t) = A_u \frac{e^{j\omega t} - e^{-j\omega t}}{j2} \tag{8.147}$$

$$x(t) = A_x \frac{e^{j(\omega t+\phi)} - e^{-j(\omega t+\phi)}}{j2} \tag{8.148}$$

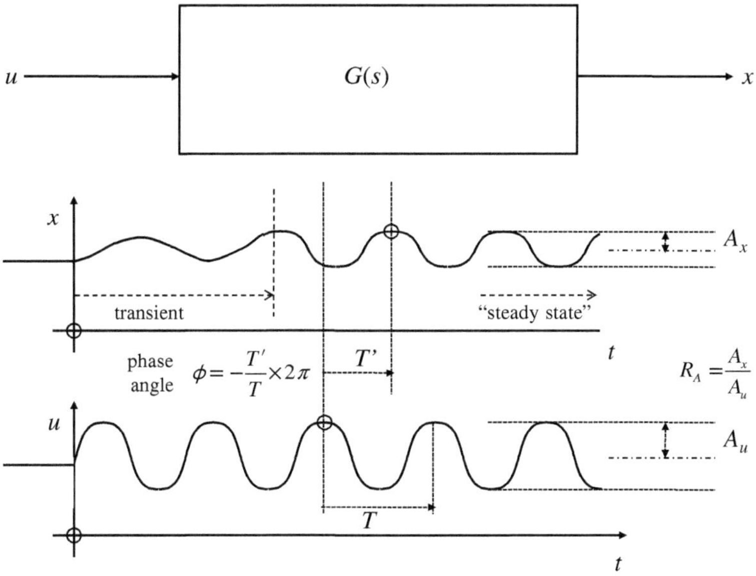

Figure 8.6 Steady frequency signal applied to a linear system G(s).

As in Equation 3.129, consider a linear system with constant coefficients which can be represented by

$$a_n \frac{d^n x}{dt^n} + a_{n-1} \frac{d^{n-1} x}{dt^{n-1}} + \cdots + a_1 \frac{dx}{dt} + a_0 x = b_m \frac{d^m u}{dt^m} + b_{m-1} \frac{d^{m-1} u}{dt^{m-1}} + \cdots + b_1 \frac{du}{dt} + b_0 u \quad (8.149)$$

Substituting u and x using Equations 8.147 and 8.148 obtain

$$\begin{aligned} &\frac{A_x e^{j(\omega t+\phi)}}{j2}\left[a_n(j\omega)^n + a_{n-1}(j\omega)^{n-1} + \cdots + a_0(j\omega)\right] \\ &-\frac{A_x e^{-j(\omega t+\phi)}}{j2}\left[a_n(-j\omega)^n + a_{n-1}(-j\omega)^{n-1} + \cdots + a_0(-j\omega)\right] \\ &\quad = \frac{A_u e^{j\omega t}}{j2}\left[b_m(j\omega)^m + b_{m-1}(j\omega)^{m-1} + \cdots + b_0(j\omega)\right] \\ &\quad - \frac{A_u e^{-j\omega t}}{j2}\left[b_m(-j\omega)^m + b_{m-1}(-j\omega)^{m-1} + \cdots + b_0(-j\omega)\right] \end{aligned} \quad (8.150)$$

Noting that $e^{j(\omega t+\phi)} = e^{j\phi} e^{j\omega t}$, and that multiplication by $e^{j\omega t}$ has the results

$$e^{j\omega t} e^{j(\omega t+\phi)} = e^{j\phi} e^{j2\omega t} \quad (8.151)$$

$$e^{j\omega t} e^{-j(\omega t+\phi)} = e^{-j\phi} \quad (8.152)$$

$$e^{j\omega t} e^{j\omega t} = e^{j2\omega t} \quad (8.153)$$

$$e^{j\omega t} e^{-j\omega t} = 1 \quad (8.154)$$

then Equation 8.150 yields

$$\begin{aligned} &\frac{A_x e^{j\phi} e^{j2\omega t}}{j2}\left[a_n(j\omega)^n + a_{n-1}(j\omega)^{n-1} + \cdots + a_0(j\omega)\right] - f_1(\omega) \\ &\quad = \frac{A_u e^{j2\omega t}}{j2}\left[b_m(j\omega)^m + b_{m-1}(j\omega)^{m-1} + \cdots + b_0(j\omega)\right] - f_2(\omega) \end{aligned} \quad (8.155)$$

The variable t only occurs in $e^{j2\omega t}$ in the first term on each side, and not the f_1 and f_2 terms. Hence, matching the remaining coefficients one has

$$\frac{A_x}{A_u} e^{j\phi} = \frac{b_m(j\omega)^m + b_{m-1}(j\omega)^{m-1} + \cdots + b_0(j\omega)}{a_n(j\omega)^n + a_{n-1}(j\omega)^{n-1} + \cdots + a_0(j\omega)} = G(j\omega) \quad (8.156)$$

Clearly then, at least for this polynomial ratio form of $G(s)$, a steady oscillation passing through the system has a *frequency response* which is related to the input by

$$\textit{Amplitude ratio } R_A(\omega) = |G(j\omega)| \quad (8.157)$$

$$\textit{Phase angle} \quad \phi(\omega) = \angle G(j\omega) \quad (8.158)$$

Actually, a dead-time factor $G_D(s) = e^{-\tau s}$ can be included directly by the substitution $s = j\omega$, bearing in mind that

$$|G_D(j\omega)| = \left|e^{-j\omega\tau}\right| = 1 \quad (8.159)$$

$$\angle G_D(j\omega) = \angle e^{-j\omega\tau} = -\omega\tau \quad (8.160)$$

and

$$|G(j\omega)G_D(j\omega)| = |G(j\omega)| \cdot |G_D(j\omega)| \tag{8.161}$$

$$\angle G(j\omega)G_D(j\omega) = \angle G(j\omega) + \angle G_D(j\omega) \tag{8.162}$$

Factorising the polynomials in Equation 8.156 (see Equation 3.133), an alternative view is

$$G(s) = \left(\frac{b_m}{a_n}\right) \frac{(s - z_1)(s - z_2)\cdots(s - z_m)}{(s - p_1)(s - p_2)\cdots(s - p_n)} \mathrm{e}^{-\tau s} \tag{8.163}$$

so that

$$|G(s)| = \left|\frac{b_m}{a_n}\right| \frac{|s - z_1| \cdot |s - z_2| \cdots |s - z_m|}{|s - p_1| \cdot |s - p_2| \cdots |s - p_n|} 1 \tag{8.164}$$

$$\angle G(s) = \angle\left(\frac{b_m}{a_n}\right) + \angle(s - z_1) + \cdots + \angle(s - z_m) - \angle(s - p_1) - \cdots - \angle(s - p_n) + \angle e^{-\tau s} \tag{8.165}$$

Let

$$B_i \mathrm{e}^{j\theta_i} = (s - z_i)|_{s=j\omega} \quad \text{i.e., } B_i = |j\omega - z_i|,\ \theta_i = \angle(j\omega - z_i),\ i = 1, \ldots, m \tag{8.166}$$

$$A_i \mathrm{e}^{j\phi_i} = (s - p_i)|_{s=j\omega} \quad \text{i.e., } A_i = |j\omega - p_i|,\ \phi_i = \angle(j\omega - p_i),\ i = 1, \ldots, n \tag{8.167}$$

Obviously

$$G(j\omega) = \left|\frac{b_m}{a_n}\right| \frac{B_1 B_2 \ldots B_m}{A_1 A_2 \ldots A_n} \mathrm{e}^{j(\angle[b_m/a_n] + \theta_1 + \theta_2 + \cdots + \theta_m - \phi_1 - \phi_2 - \cdots - \phi_n - \omega\tau)} \tag{8.168}$$

that is

$$R_A(\omega) = |G(j\omega)| = \left|\frac{b_m}{a_n}\right| \frac{B_1 B_2 \cdots B_m}{A_1 A_2 \cdots A_n} \tag{8.169}$$

$$\phi(\omega) = \angle G(j\omega) = \angle[b_m/a_n] + \theta_1 + \theta_2 + \cdots + \theta_m - \phi_1 - \phi_2 - \cdots - \phi_n - \omega\tau \tag{8.170}$$

Example 8.4

Comparison of Full Response and Frequency Response for a First-Order System

Derive the full transient response of a first-order system to a frequency signal starting at $t = 0$ and compare it to the $G(j\omega)$ frequency response result as $t \to \infty$. Make use of the s-domain modelling techniques presented in Section 3.9.2.

$$G(s) = \frac{K}{\tau s + 1} \tag{8.171}$$

$$u(t) = \begin{cases} 0 & \text{for } t < 0 \\ A\sin(\omega t) & \text{for } t \geq 0 \end{cases} \tag{8.172}$$

$$U(s) = \frac{A\omega}{s^2 + \omega^2} \tag{8.173}$$

$$X(s) = G(s)U(s) = \frac{KA\omega}{(\tau s + 1)(s^2 + \omega^2)} = \frac{KA\omega}{(\tau s + 1)(s + j\omega)(s - j\omega)} \tag{8.174}$$

A partial fraction expansion requires

$$\frac{1}{(\tau s + 1)(s + j\omega)(s - j\omega)} = \frac{A'}{\tau s + 1} + \frac{B'}{s + j\omega} + \frac{C'}{s - j\omega} \tag{8.175}$$

"×" by $(\tau s + 1)$ and set $s \to -{}^{1}/_{\tau}$ to get

$$A' = \frac{\tau^2}{1 + \omega^2\tau^2} \tag{8.176}$$

"×" by $(s + j\omega)$ and set $s \to -j\omega$ to get

$$B' = \left(\frac{-1}{2\omega}\right)\frac{(\omega\tau - j)}{(1 + \omega^2\tau^2)} \tag{8.177}$$

so that "by inspection" (complex conjugate) one can write

$$C' = \left(\frac{-1}{2\omega}\right)\frac{(\omega\tau + j)}{(1 + \omega^2\tau^2)} \tag{8.178}$$

It is useful to combine the last two terms before proceeding:

$$\frac{B'}{s + j\omega} + \frac{C'}{s - j\omega} = \frac{-1}{2\omega(1 + \omega^2\tau^2)}\left\{\frac{\omega\tau - j}{s + j\omega} + \frac{\omega\tau + j}{s - j\omega}\right\} \tag{8.179}$$

$$= \frac{1}{(1 + \omega^2\tau^2)}\left\{\frac{1 - \tau s}{s^2 + \omega^2}\right\} \tag{8.180}$$

Using Equations 8.180, 8.176, and 8.175 in Equation 8.174

$$X(s) = \frac{KA\omega}{(1 + \omega^2\tau^2)}\left\{\frac{\tau^2}{\tau s + 1} + \frac{1 - \tau s}{s^2 + \omega^2}\right\} \tag{8.181}$$

$$X(s) = \frac{KA}{(1 + \omega^2\tau^2)}\left\{\frac{\omega\tau}{s + {}^{1}/_{\tau}} + \frac{\omega}{s^2 + \omega^2} - \frac{\omega\tau s}{s^2 + \omega^2}\right\} \tag{8.182}$$

Inverting to the time domain using Table 3.4,

$$x(t) = \frac{KA}{(1 + \omega^2\tau^2)}\left\{\omega\tau e^{-t/\tau} + \sin(\omega t) - \omega\tau\cos(\omega t)\right\} \tag{8.183}$$

The harmonic terms can be combined using the identity

$$[\gamma\cos(\alpha)]\sin(\beta) + [\gamma\sin(\alpha)]\cos(\beta) = \gamma\sin(\alpha + \beta) \tag{8.184}$$

Taking

$$\beta = \omega t \tag{8.185}$$

$$\left.\begin{aligned}\gamma\cos(\alpha) &= 1\\ \gamma\sin(\alpha) &= -\omega\tau\end{aligned}\right\} \Rightarrow \gamma^2 = 1 + \omega^2\tau^2 \tag{8.186}$$

obtain

$$\sin(\omega t) - \omega\tau\cos(\omega t) = \sqrt{1+\omega^2\tau^2}\,\sin(\omega t + \phi) \tag{8.187}$$

$$\text{where } \phi = \angle(1, -\omega\tau) = \tan^{-1}\left(\frac{-\omega\tau}{+1}\right) = -\tan^{-1}(\omega\tau) \tag{8.188}$$

The $\tan^{-1}$ can be used directly in this case because γ is taken as positive and the positive cosine places the angle in the first or fourth quadrant. In general, however, one needs to take careful note of the signs in the expression. Finally, Equation 8.183 can be expressed

$$x(t) = \frac{KA}{\sqrt{1+\omega^2\tau^2}}\left\{\left[\frac{\omega\tau}{\sqrt{1+\omega^2\tau^2}}\right]e^{-t/\tau} + \sin\left(\omega t - \tan^{-1}[\omega\tau]\right)\right\} \tag{8.189}$$

This is the complete transient response, starting at zero according to Equation 8.183. As $t \to \infty$, the transient term disappears and one is left with the *frequency response*

$$x(t) = \frac{KA}{\sqrt{1+\omega^2\tau^2}}\sin\left(\omega t - \tan^{-1}[\omega\tau]\right) \tag{8.190}$$

Comparing with the input Equation 8.172, it is clear that for this first-order process the *amplitude ratio* and *phase angle* are

$$R_A(\omega) = \frac{K}{\sqrt{1+\omega^2\tau^2}} \tag{8.191}$$

$$\phi(\omega) = -\tan^{-1}[\omega\tau] \tag{8.192}$$

Now consider from Equation 8.171

$$G(j\omega) = \frac{K}{j\omega\tau + 1} \tag{8.193}$$

$$R_A(\omega) = |G(j\omega)| = \frac{|K|}{|j\omega\tau + 1|} = \frac{K}{\sqrt{1+\omega^2\tau^2}} \tag{8.194}$$

$$\phi(\omega) = \angle G(j\omega) = \angle K - \angle(1 + j\omega\tau) = 0 - \tan^{-1}(\omega\tau) \tag{8.195}$$

which is in agreement (positive K). Note that the maximum phase lag of this first-order process, at high frequencies, is $\pi/2$, and that a log–log plot of amplitude ratio against frequency will have an asymptotic slope of -1 at high frequencies.

8.6.2
Closed-Loop Stability Criterion in the Frequency Domain

In the root locus analysis of closed-loop stability in Section 8.2.1.5, conclusions were drawn from inspection of the *open-loop* transfer function in the form of the closed-loop characteristic Equation 8.51. The *frequency response* methods are a bit like this because they too focus on the *frequency response of the open loop* in order to infer things about the closed loop. Of course, this open loop has to include every element as the signal passes around the loop (classical open loop in Figure 8.1), including the controller itself.

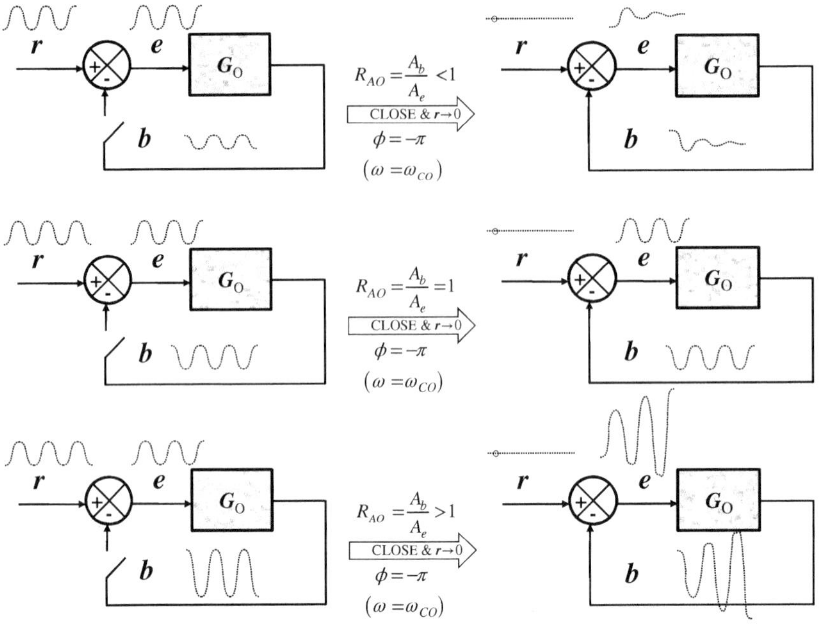

Figure 8.7 Illustration of frequency response stability criterion ($R_{AO} < 1$ when $\phi = -\pi$).

The argument for closed-loop stability in the frequency domain is illustrated in Figure 8.7. Usually (but not always) the *phase lag* ($-\phi$) increases and the amplitude ratio (R_{AO}) decreases as the frequency increases. A few systems (e.g. first order) cannot achieve phase lags as large as π, but for those that do, the particular frequency at which this happens is called the *crossover frequency* (ω_{CO}). If this has been found, and an oscillation of this frequency is fed into the open loop (via the setpoint) as on the left of Figure 8.7, a situation arises where the feedback signal b has the same frequency but is inverted in comparison to e, and has a fixed greater or lesser amplitude, depending on whether $R_{AO}(\omega_{CO})$ is greater than unity or less. Now say that two things are done instantly and simultaneously – the setpoint r is set to zero and the loop is closed. In the case where $R_{AO}(\omega_{CO}) = 1$ there will be no change, because the feedback signal b is the exact inverse of e and it is inverted again at the comparator to replace the setpoint. It is intuitively obvious, though, that if $R_{AO}(\omega_{CO}) < 1$ the oscillation will shrink (stable) and if $R_{AO}(\omega_{CO}) > 1$ the oscillation will expand (unstable). The frequency domain stability criterion could thus be stated:

> "A closedloop will be stable if the openloop, including all elements in the loop, has an amplitude ratio less than unity at its cross-over frequency"[1)]

1) The frequency response definition of stabilty above has an important limitation. It assumes that ϕ and R_{AO} monotonically decrease with increasing frequency. There are cases where they do not, giving rise to the notion of *conditional stability.* It will be seen that the *Nyquist stability criterion* in Section 8.6.4 handles this better.

The situation that has been described might appear rather fortuitous – what if the signals are never in the frequency range that determines $R_{AO} > 1$ and/or $\phi < -\pi$? The problem is that a typical disturbance like an impulse or step is constructed (in Fourier terms) from an infinite range of frequency signals, and the smallest signal with the correct frequency (ω_{CO}) will continue to amplify itself indefinitely in the closed loop.

By measuring or calculating how close the open loop is to having a gain (amplitude ratio) of unity at the crossover frequency, or *vice versa* how close the phase angle is to $-\pi$ when the gain is unity, one is able to obtain a measure of *relative stability* for the closed loop:

$$\text{Gain margin GM} = \frac{1}{R_{AO}(\omega_{CO})} \tag{8.196}$$

$$\text{Phase margin PM} = \pi + \phi|_{R_{AO}=1} \tag{8.197}$$

The condition at which the closed loop is on the verge of instability thus implies that simultaneously GM = 1 and PM = 0. These measures are usually presented in the units of *decibels* and *degrees*

$$\text{GM}[decibel] = 20 \times \log_{10}\{\text{GM}[dimensionless]\} = 0 - R_{AO}(\omega_{CO})[decibel] \tag{8.198}$$

$$\text{PM}\left[degrees\right] = 180 + \phi|_{R_{AO}=0db}\left[degrees\right] = 180 - \text{“phase lag when } R_{AO} = 0\,db\text{”}\left[degrees\right] \tag{8.199}$$

To summarize the stability relationship:

closed-loop stable	GM > 1 (GM > 0 db) PM > 0°
closed loop on the verge of instability	GM = 1 (GM = 0 db) PM = 0°
closed-loop unstable	GM < 1 (GM < 0 db) PM < 0°

Typical choices are GM = 9 db or PM = 40°. It would be coincidental if these were simultaneously met, and indeed, the PM sometimes cannot be calculated, *viz.* in the case where R_{AO} never rises as high as unity. The maximum lags of first- and second-order systems are 90° and 180°, respectively, at very high frequency, so these will not have a crossover frequency, and indeed cannot become unstable if the loop is closed around them.

8.6.3 Bode Plot

The Bode plot consists of two sets of axes, one showing the variation of $R_A(\omega)$ with ω, and the other the phase angle ϕ (ω) with ω. The abscissa ω in both plots is on a logarithmic scale, as well as the R_A ordinate (usually in db). The ϕ ordinate is a linear scale. In Example 8.4 it was shown how the full response to a frequency signal (including the initial transient) could be calculated. In the limit as $t \to \infty$ the same result is obtained for the steady-state response based on $G(j\omega)$. So, there are two ways to arrive at a Bode plot based on the s-domain transfer function of the process. It is, however, more likely that this frequency response will simply be *measured*. Comparison of steady input and output oscillations over a range of frequencies as in Figure 8.6 provides the necessary data. Though the process itself might be measured, the proposed controller (e.g. PI) might only be known in the $G_C(s)$ form, in which case the two separate frequency responses are simply ‘added’ on

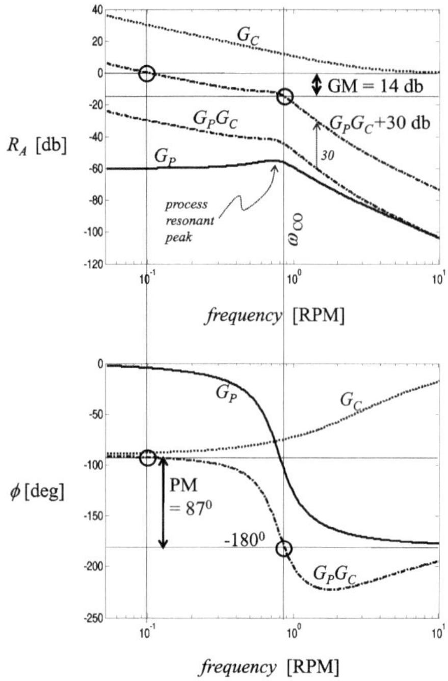

Figure 8.8 Construction of Bode plot to achieve a desired gain margin in Example 8.5. ('Measured' G_P was based on a second-order system with $K = 10^{-3}$, $\tau = 0.2$, $\zeta = 0.3$.)

the Bode plot to predict the open-loop frequency response, and hence the closed-loop stability (Figure 8.8). As in Equations 8.169 and 8.170, the phase angles are added, but the multiplication of magnitudes is of course equivalent to adding their logarithms.

Example 8.5

Controller Design for a Specified Gain Margin Using a Bode Plot

A measured frequency response for a system G_P is given in Figure 8.8. Using an integral time of 0.05 min, specify the gain K_C of a PI controller to achieve a Gain Margin of 14 db for the controlled system.

$$G_C(s) = K_C\left(1 + \frac{1}{\tau_I s}\right) \tag{8.200}$$

$$G_C(j\omega) = K_C\left(1 - \frac{1}{\omega\tau_I}j\right) \tag{8.201}$$

$$|G_C(j\omega)| = K_C\sqrt{1 + \frac{1}{\omega^2\tau_I^2}} \tag{8.202}$$

$$\angle G_C(j\omega) = -\tan^{-1}\left(\frac{1}{\omega\tau_I}\right) \tag{8.203}$$

Now include on Figure 8.8 curves 'G_C' for this PI controller with $\tau_I = 0.05$ min and $K_C = 1$. Then, 'adding' G_P to G_C, get the curves $G_P G_C$. The crossover frequency of this system ω_{CO} is not affected by a change of K_C away from 1, which one sees has the effect of simply shifting the $G_P G_C$ amplitude ratio plot on the log–log plot. In order to obtain the desired Gain Margin of 14 db, one finds that K_C must be increased by 30 db, that is

$$K_C = 1 \times 10^{30/20} = 31.6 \rightarrow \tag{8.204}$$

Figure 8.8 also shows that for this choice of K_C the phase margin will be 87°.

As in the case of root locus (Section 8.2.1.5), the main features of Bode plots can be sketched by inspection. This does not have great practical importance, but the ideas are again an aid to the understanding of system dynamics. Consider a general system in the factored form of Equation 8.163:

$$G(s) = \left(\frac{b_m}{a_n}\right) \frac{(s - z_1)(s - z_2)\cdots(s - z_m)}{(s - p_1)(s - p_2)\cdots(s - p_n)} e^{-\tau s} \tag{8.205}$$

Recall that the frequency response if obtained by the substitution $s \rightarrow j\omega$.

$$G(j\omega) = \left(\frac{b_m}{a_n}\right) \frac{(j\omega - z_1)(j\omega - z_2)\cdots(j\omega - z_m)}{(j\omega - p_1)(j\omega - p_2)\cdots(j\omega - p_n)} e^{-j\omega\tau} \tag{8.206}$$

One can make the following observations:

- $$\lim_{\omega \rightarrow 0} G(j\omega) = \left(\frac{b_m}{a_n}\right) \frac{\prod_{i=1}^{m}(-z_i)}{\prod_{i=1}^{n}(-p_i)} \tag{8.207}$$

  which gives a constant R_A asymptote as $\omega \rightarrow 0$, provided no poles are zero. As $\omega \rightarrow 0$, the phase angle ϕ will be either 0 or π depending on the sign of the expression.
- If q of the poles are zero,

  $$\lim_{\omega \rightarrow 0} G(j\omega) = \text{const} \times \frac{1}{[j\omega]^q} \tag{8.208}$$

  which gives an R_A asymptote of slope $-q$ on a log–log plot, with R_A increasing to ∞ as $\omega \rightarrow 0$. The phase angle in this limit asymptotes to $-q \times \pi/2$ (shifted by π if *const* is negative).
- $$\lim_{\omega \rightarrow \infty} G(j\omega) = \left(\frac{b_m}{a_n}\right) \frac{\prod_{i=1}^{m}(j\omega)}{\prod_{i=1}^{n}(j\omega)}, \tag{8.209}$$

  so $R_A \rightarrow \frac{\omega^m}{\omega^n}$ which is an asymptote of slope m–n on a log–log plot as $\omega \rightarrow \infty$. For physically realizable systems, $m < n$ so the slope is negative and $R_A \rightarrow 0$. As $\omega \rightarrow \infty$, the phase angle ϕ will approach $\angle\left(\frac{b_m}{a_n}\right) - (n - m)\frac{\pi}{2} - \omega\tau$.

8.6.4
Nyquist Plot

Whereas the Bode plot uses two sets of Cartesian axes to express the dependences of R_A and ϕ on the frequency ω, the Nyquist plot, which is similarly plotted for the *open loop*, is rather a single polar plot of the magnitude R_A at the corresponding angle ϕ, as ω varies. Unless the ω values are explicitly marked on this curve, one loses information about the frequency dependence. However, this account jumps ahead. One needs to start by considering the *Nyquist stability criterion* which is based on the *Nyquist contour*. It will be seen that the usual *Nyquist plot* only represents part of the Nyquist contour (where $s = j\omega$, $0 < \omega < +\infty$), but the stability of a loop closed around the plotted transfer function can nevertheless be inferred from such an open-loop plot.

Start by considering the factored form 8.163 of the *open-loop transfer function*

$$G(s) = \left(\frac{b_m}{a_n}\right)\frac{(s-z_1)(s-z_2)\cdots(s-z_m)}{(s-p_1)(s-p_2)\cdots(s-p_n)}e^{-\tau s} \tag{8.210}$$

For an arbitrary choice of point s on the complex plane, a single factor like $(s\text{-}z_i)$ represents a vector pointing from z_i to s as in Figure 8.9. Now let the point s move along a path that encircles z_i in a positive angular direction (anti-clockwise). The angle θ_i of $(s\text{-}z_i)$ will be seen to increase through a net $+2\pi$. If z_i had been outside of the chosen path (or 'contour'), the angle would have oscillated up and down, back to its original value, but completing a *zero* net rotation. One net positive rotation of $(s\text{-}z_i)$ will manifest itself as one net positive rotation of $G(s)$ about its origin, according to Equation 8.170. A positive encirclement of a p_i in the expression 8.210 will conversely contribute one net negative rotation to $G(s)$.

For reasons that will become clearer later, the above phenomenon will now be used to detect the presence of *open-loop zero's* in the right-hand s-plane, without having to factorise the numerator polynomial as in Equation 8.210. Because the *Nyquist contour* has to encircle this entire half-plane,

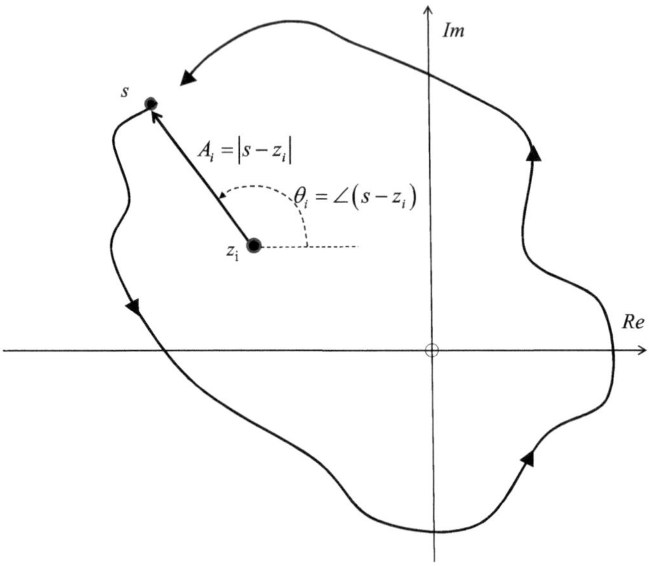

Figure 8.9 Variation of the angle of $(s\text{-}z_i)$ as s encircles z_i.

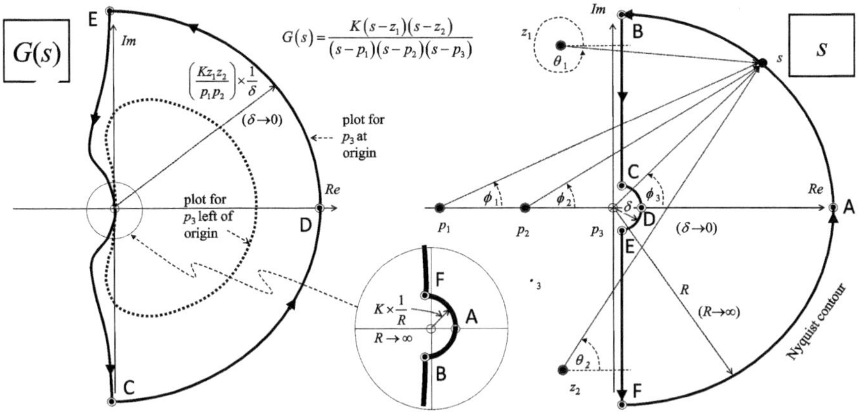

Figure 8.10 Mapping of Nyquist contour [*s*] into Nyquist plot [*G*(*s*)].

it is defined as in Figure 8.10 – that is a semicircle with infinite radius covering the positive real half-plane. Proceeding around the Nyquist contour ABCDEFA one obtains the corresponding trace on the Nyquist plot. The particular plot shown in Figure 8.10 has neither zeros nor poles with positive real parts, so the net rotation about the point $G = (0 + j0)$ is zero.

It is noted that the part of the plot corresponding to DEFA is the complex conjugate of the first half ABCD. Moreover, for physically realizable systems, $m < n$, so the part corresponding to AB disappears at the G-plane origin. Finally, the part corresponding to CD is 'off the G plot.' The upshot of this is that in practice only the BC part is plotted (but one needs to bear in mind what the invisible parts are doing so that rotations can be counted). Thus, as already hinted, the Nyquist plot is simply a polar representation of the frequency response $G(j\omega)$, with the positive angular direction determined by ω proceeding from $+\infty$ down to 0. Some typical Nyquist plots are shown in Figure 8.11.

Now the usefulness of the Nyquist plot, representing an *open-loop* frequency response, for determining *closed-loop* stability, will be explained. Firstly, insist that only *open-loop stable* systems will be considered. That means that not one of the p_i in Equation 8.210 is lying within the Nyquist contour, and therefore these poles cannot contribute any net rotations of $G(s)$ around its origin as s follows the contour. (For poles these would be in a negative angular direction). Consider a new transfer function

$$G'(s) = 1 + G(s) \tag{8.211}$$

This has the same poles as $G(s)$, so the number of positive rotations of $G'(s)$ of its own origin must represent exactly the number of *zeros* of $G'(s)$ which lie in the right-hand s-plane. In practice, one does not draw a new plot for $G'(s)$, but rather just 'imagines' a new coordinate system with its origin at $(-1 + j0)$ on the $G(s)$ plot (Figure 8.12).

Of course, it is very useful to know about these zeros of $1 + G(s)$ because, assuming that $G(s)$ represents the full open loop, these zeros must be poles of the closed loop:

$$G_{CL}(s) = \frac{G(s)}{1 + G(s)} \tag{8.212}$$

In particular, if such closed-loop poles have positive real parts, thus causing one or more rotations of a stable $G(s)$ about $(-1 + j0)$, the closed-loop will be unstable. So the Nyquist stability criterion can be stated

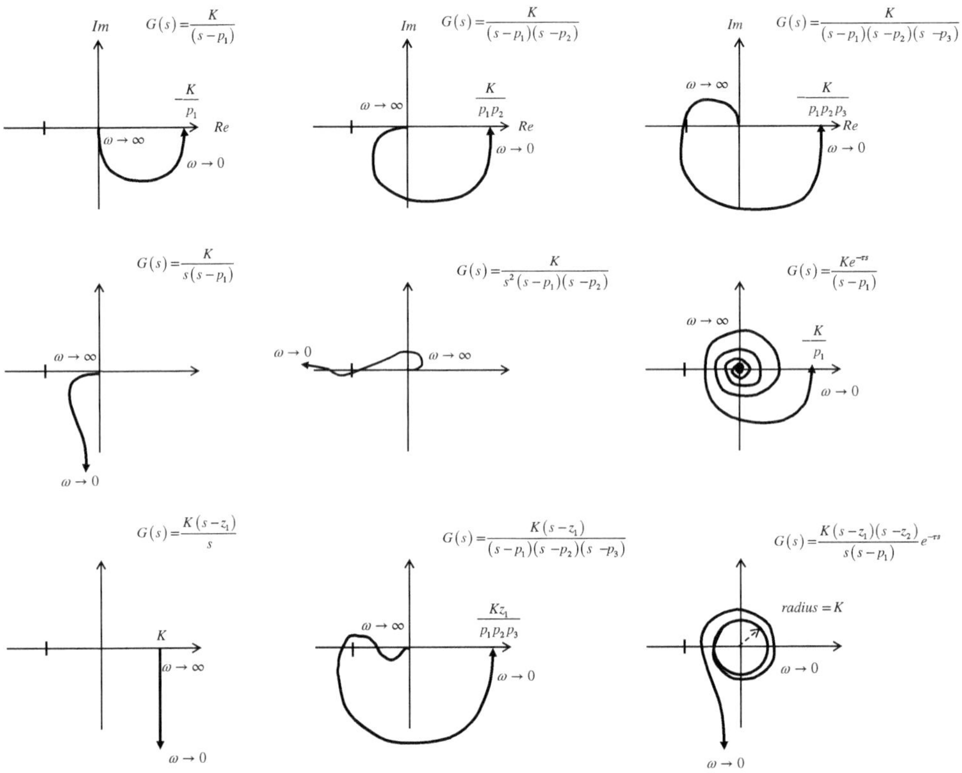

Figure 8.11 Examples of Nyquist plots.

> If $G(s)$ is stable,[2] and $G(s)$ has NO net rotation of the point $(-1+j0)$ whilst s follows the Nyquist contour around the positive real semiplane, then a loop closed around $G(s)$ will be stable. Otherwise it will be unstable.

The Nyquist stability criterion does not have the limitation of the frequency response stability criterion stated in Section 8.6.2. Figure 8.12 is a case in point, because there are two frequencies at which $\phi = -\pi$. As drawn, one of these has $R_A < 1$ and the other has $R_A > 1$. Which would one choose to calculate a gain margin? The Nyquist stability criterion on the other hand has no such confusion. There are clear rotations of $(-1+j0)$ in the drawn plot, so the closed loop will be unstable. However, this is an interesting case of *conditional stability.* If the gain K in the open-loop transfer function is increased (e.g. by increasing K_C), then the whole plot expands proportionally like a balloon. The point $(-1+j0)$ could then occur between the origin and the first crossing of the real axis on the $G(s)$ plot, and this location gives *zero net rotations* – that is the closed-loop system has unusually become stable with an *increase* in open-loop gain! When the definitions of gain margin and phase margin in Equations 8.196 and 8.197

2) There are situations in which unstable systems must be controlled – for example, temperature control of an exothermic reaction. The unstable open-loop poles will cause negative rotation contributions to the observed total rotation of $G(s)$ about $(-1 + j0)$. Provided this negative offset is remembered, the same principles of stability analysis of course apply.

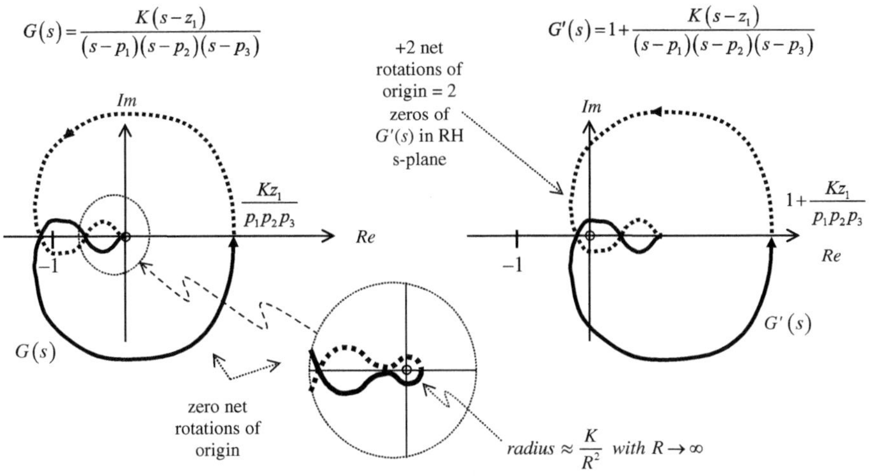

Figure 8.12 Imagined shift of origin when looking for 'encirclements of $-1+j0$.'

are interpreted on a Nyquist plot, one is less likely to err bearing the Nyquist stability criterion in mind. In Figure 8.13, drawn in the positive angular sense for frequency decreasing from infinity to zero, the Nyquist plot closes on the right with an arc of infinite radius as in Figure 8.10. If 'x' were greater than 1, it is clear that there would be an encirclement on the point $(-1+j0)$. Certainly, such an encirclement implies GM<1 (<0 db) and PM < 0° – that is closed-loop instability.

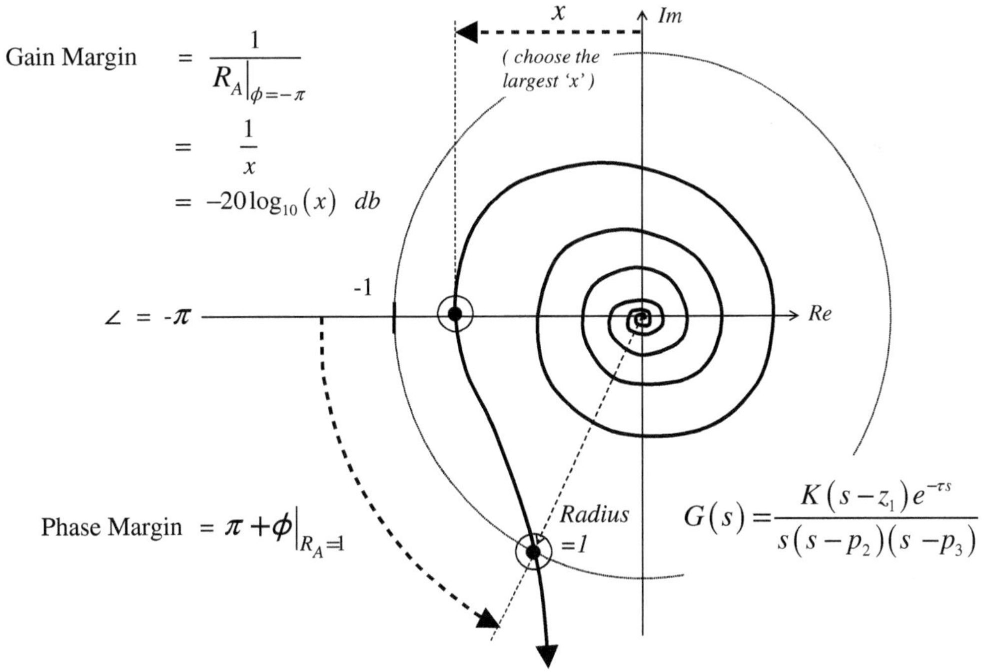

Figure 8.13 Gain margin and phase margin on a Nyquist plot.

Example 8.6

Calculation of Gain Margin and Phase Margin

Given an open-loop transfer function $G_O(s)$, the gain margin and phase margin are more directly *calculated* rather than determined from a Bode or Nyquist plot. For example, consider

$$\begin{aligned} G_O(s) &= \frac{5(s+2)e^{-s}}{s^3+7s^2+12s} \\ &= \frac{5(s+2)e^{-s}}{s(s+3)(s+4)} \end{aligned} \tag{8.213}$$

To avoid losing track of the angular quadrant, if a 4-quadrant tan-1 is not being used, always factorise to first-order factors as above, regardless of complex roots. Now

$$\phi = \angle G_O(j\omega) = 0 + \tan^{-1}\left(\frac{\omega}{2}\right) - \omega - \frac{\pi}{2} - \tan^{-1}\left(\frac{\omega}{3}\right) - \tan^{-1}\left(\frac{\omega}{4}\right) \tag{8.214}$$

Trial and error to find ω_{CO} (care must be taken that the outermost [lowest frequency] crossing is found):

ω[rad time^{-1}]	0	1	2	1.5	1.4	1.41	1.408	1.407
ϕ [deg]	−90	−153.2	−219.85	−186.2	−179.5	−180.19	−180.061	−179.995

Magnitude at ω_{CO}:

$$R_A(\omega) = \frac{5\left(\omega^2+2^2\right)^{1/2}1}{\omega\left(\omega^2+3^2\right)^{1/2}\left(\omega^2+4^2\right)^{1/2}} \tag{8.215}$$

$$R_A(\omega_{CO}) = \frac{5\left(1.407^2+4\right)^{1/2}1}{1.407\left(1.407^2+9\right)^{1/2}\left(1.407^2+16\right)^{1/2}} = 0.618 \tag{8.216}$$

$$\text{Gain margin} = \frac{1}{0.618} = 1.617 = 4.17\ \text{db} \rightarrow \tag{8.217}$$

Trial and error to find $\omega|_{R_A=1}$ (care must be taken that the outermost part of the Nyquist plot is being searched):

ω [rad time^{-1}]	1	0.5	0.9	0.85	0.855	0.852
R_A	0.85	1.68	0.949	1.002	0.997	1.00 026

Phase angle at $\omega|_{R_A=1}$:

$$\phi|_{R_A=1} = 0 + \tan^{-1}\left(\frac{0.852}{2}\right) - 0.852 - \frac{\pi}{2} - \tan^{-1}\left(\frac{0.852}{3}\right) - \tan^{-1}\left(\frac{0.852}{4}\right) \tag{8.218}$$

$$= -2.507\ \text{rad} = -143.6^\circ \tag{8.219}$$

$$\text{Phase margin} = \pi + \phi|_{R_A=1} = 180^\circ - 143.6^\circ = 36.4^\circ \rightarrow \tag{8.220}$$

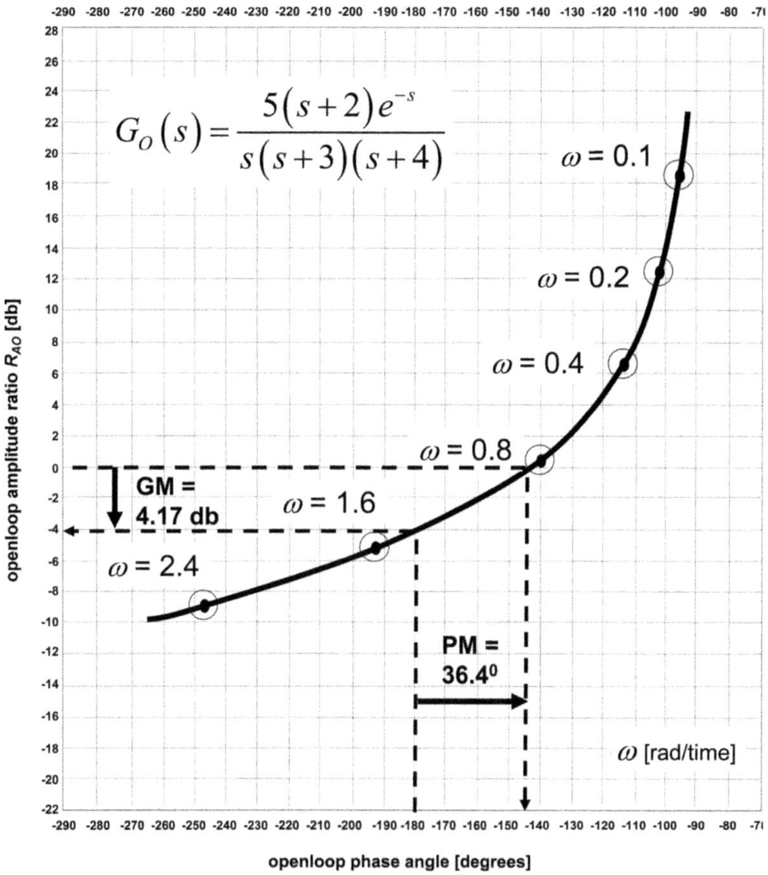

Figure 8.14 Plot of open-loop amplitude ratio versus phase angle at a range of frequencies.

8.6.5
Magnitude versus Phase-Angle Plot and the Nichols Chart

It is sometimes useful to use a 'magnitude versus phase angle' plot of the frequency response, as opposed to the Bode plot (Section 8.6.3) or Nyquist plot (Section 8.6.4). This is a Cartesian representation of the polar Nyquist plot, and it similarly loses information about frequency, unless frequencies are explicitly marked on the curve as in Figure 8.14. The system in Example 8.6 is plotted, and the corresponding closed-loop gain margin and phase margin now determined graphically.

This type of plot is well known in another context, because a particular *open-loop* amplitude ratio and phase angle implies a particular *closed-loop* amplitude ratio and phase angle:

$$G_O(j\omega) = R_{AO}(\omega)e^{j\phi_O(\omega)} \tag{8.221}$$

Consider an arbitrary point (R_{AO},ϕ_O) in this frequency response, where

$$G_O = R_{AO}e^{j\phi_O} \tag{8.222}$$

$$G_O = R_{AO}\left[\cos(\phi_O) + j\sin(\phi_O)\right] \tag{8.223}$$

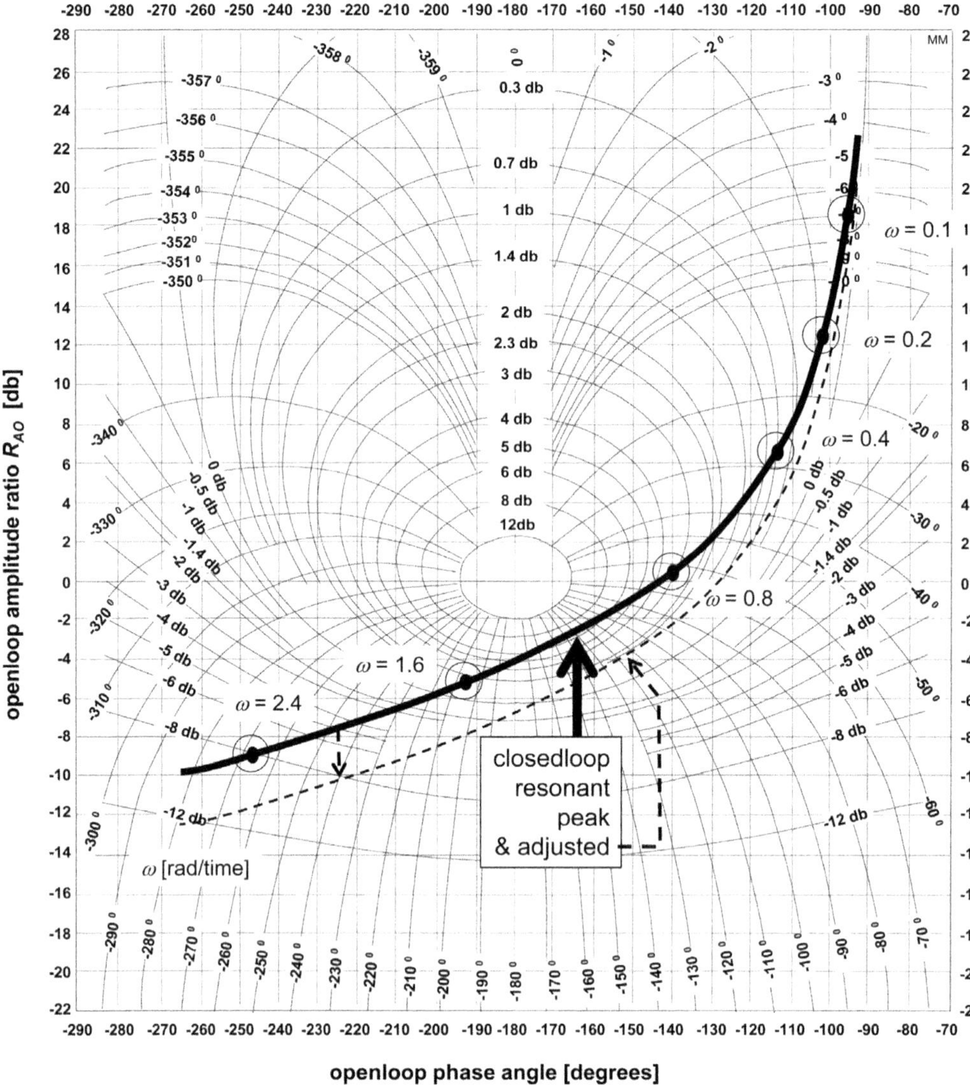

Figure 8.15 Plot of the open-loop system in Example 8.6 on the Nichols chart.

Here the closed loop would map to

$$G_{CL} = \frac{G_O}{1 + G_O} = \frac{\cos(\phi_O) + j\sin(\phi_O)}{{}^1\!/_{R_{AO}} + \cos(\phi_O) + j\sin(\phi_O)} \tag{8.224}$$

So,

$$R_{ACL} = |G_{CL}| = \frac{1}{\sqrt{\left[{}^1\!/_{R_{AO}} + \cos(\phi_O)\right]^2 + \left[\sin(\phi_O)\right]^2}} \tag{8.225}$$

$$\phi_{CL} = \angle G_{CL} = \phi_O - \angle\left[{}^1\!/_{R_{AO}} + \cos(\phi_O), \sin(\phi_O)\right] \tag{8.226}$$

Thus many different (R_{AO},ϕ_O) points in the plane could be used to compute the corresponding (R_{ACL},ϕ_{CL}) values at those points, building two surfaces on which contours of constant R_{ACL} or ϕ_{CL} could be plotted. On the *Nichols chart* (Figure 8.15), both sets of these contours are shown simultaneously. The plot in Figure 8.14 is sometimes called a *Nichols plot,* and it can be superimposed on the *Nichols chart* as shown to give some direct insight into the *closed-loop* behaviour. One notes that the closed loop has a *resonant peak* at frequency $\omega_r \approx 1$ rad time^{-1} (i.e. where the plot is tangential to the constant R_{ACL} contours). A typical design criterion is to set this

$$R_{ACLMAX} \approx 2 \text{ db} \tag{8.227}$$

For the open-loop system of Equations 8.213 and 8.214, which is plotted in Figure 8.15, this requires a 2.5 db reduction in gain, that is K_C must change such that the $K=5$ in this expression becomes

$$K = 5 \times 10^{-2.5/20} = 3.75 \tag{8.228}$$

As seen, this will drop the entire open-loop curve by 2.5 db, giving the desired tangential contact with the 2 db closed-loop contour.

8.7 Control Quality Criteria

Most of the measures used to describe the performance of a controller refer to its setpoint step response as in Figure 8.16. Often this will be the basis for comparison of different settings or different controller algorithms. One appreciates that plant operators will not allow significant stepping in many installations, in which case more general measures such as ISE or IAE (see below) may have to be extracted from normal operating records over longer periods.

1) **Rise time** time to reach setpoint value T_R
2) **Overshoot** expressed as a percentage $\dfrac{A_1}{D} \times 100$

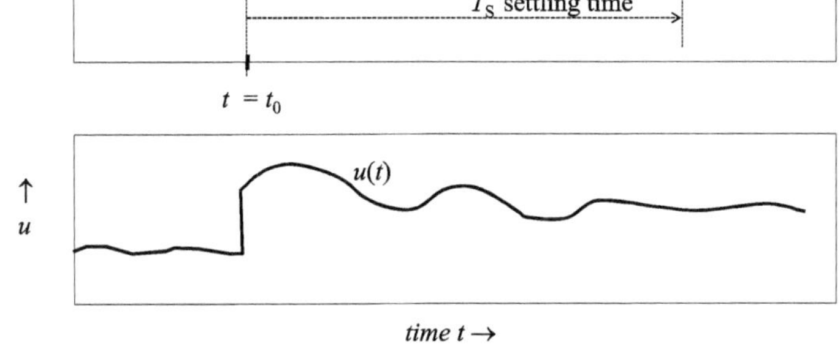

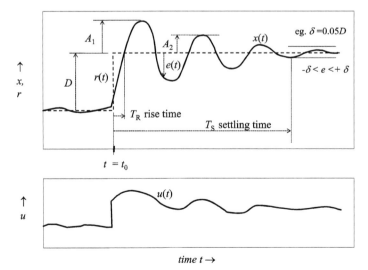

Figure 8.16 Control quality measures based on a setpoint step response.

3) **Decay ratio** ratio of successive amplitudes $\frac{A_2}{A_1}$ (related to closed-loop damping factor)
4) **Settling time** to a defined tolerance – for example 5% as shown: T_S
5) **Offset** not shown in this example, but obviously important in many proportional controllers. Here, one needs to obtain the mean initial output x_{SS1} (before the setpoint step) and the mean settling position x_{SS2}. Then, the percentage offset is $\frac{x_{SS2} - x_{SS1} - D}{D} \times 100$

   Notice that this measure must be relative because the output is likely to *start* and *end* with an offset.
6) **Integrated error** There are several different measures based on integrating the deviation from setpoint over some period of time *following the setpoint step*, but only two will be given here:
   a. ISE (integral of squared error) [also known as QPI (quadratic performance index)]

$$\mathrm{ISE} = \int_{t_0}^{t_0+T} [x(t) - r(t)]^2 \mathrm{d}t \tag{8.229}$$

   One might like to take $t \to \infty$ as a good overall measure of performance, but since there will always be deviations in real situations, one is obliged to fix a finite T for comparison purposes. In the case of systems with offset, eliminate the prejudice of initial offset with

$$\mathrm{ISE} = \int_{t_0}^{t_0+T} [x(t) - x_{SS1} - D]^2 \mathrm{d}t \tag{8.230}$$

   Obviously, if one is forced to compare indices based on different step sizes, the above expressions must be normalized with a factor of $1/D^2$. Bearing in mind the effects of nonlinearity, for comparison purposes a setpoint step should always be made between the same values and in the same direction.

   b. IAE (integral of absolute error)

$$\mathrm{IAE} = \int_{t_0}^{t_0+T} |x(t) - r(t)| \mathrm{d}t \tag{8.231}$$

   Here, one can truly talk of 'the area between the response curve and the setpoint'. The same points arise as under (a) with respect to T, offset and normalisation, except in the case of normalisation the factor will now be of $1/D$.

As mentioned, where actual steps are not allowed on the plant, real-time algorithms such as a 'moving window'-based ISE could continuously monitor the setpoint deviation. Sophisticated methods exist, for example, that relate the setpoint deviation to simultaneous valve (actuator) movements, allowing the identification of valve hysteresis (see Section 2.6.1.5).

8.8 Robust Control

Since the 1980s a lot of work has been done on *robust stability* and *robust performance*. The problem being addressed here is that the design of a controller is based on a *nominal model* of a system – or at least on some fixed nominal behaviour of the equipment, sensors and

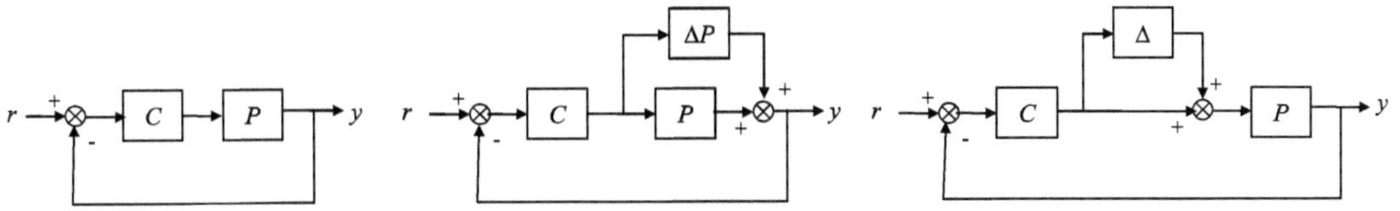

(a) No uncertainty (b) Additive process model uncertainty (c) Multiplicative process model uncertainty

Figure 8.17 Representations of process model uncertainty.

actuators. So what happens when the operation at any time deviates from this behaviour? The aim is to build into the controller certain guarantees of the stability and performance, provided the deviations of the system from the nominal behaviour are within specified bounds. Interestingly, these techniques are developed in the more classical *frequency domain*. Since a good working knowledge of the field requires a significant mathematical background, this section will merely introduce some of the ideas.

In Figure 8.17, two means are proposed for incorporating an unknown model uncertainty, *viz.* *additive* [$\Delta P(s)$], and *multiplicative* [$\Delta(s)$], where it is understood that these deviation functions have a *range* of behaviour, for example as a result of their coefficients varying over some range. The corresponding open-loop transfer functions are

$$\textit{additive}: \quad G_O(s) = [P(s) + \Delta P(s)]C(s) = \delta(s)\tilde{G}_O(s) \tag{8.232}$$

where

$$\tilde{G}_O(s) = P(s)C(s) \tag{8.233}$$

$$\delta(s) = 1 + \frac{\Delta P(s)}{P(s)} \quad \left[\text{or } \boldsymbol{I} + \Delta \boldsymbol{P}(s)\boldsymbol{P}^{-1}(s) \quad \text{for the multivariable case}\right] \tag{8.234}$$

$$\textit{multiplicative}: \quad G_O(s) = P(s)[1 + \Delta(s)]C(s) = \delta(s)\tilde{G}_O(s) \tag{8.235}$$

where

$$\tilde{G}_O(s) = P(s)C(s) \tag{8.236}$$

$$\delta(s) = 1 + \Delta(s) \quad [\text{or } \boldsymbol{I} + \boldsymbol{P}(s)\Delta(s)\boldsymbol{P}^{-1}(s) \quad \text{for the multivariable case}] \tag{8.237}$$

A single-frequency point on the *Nyquist plot* of $\tilde{G}_O(j\omega)$ is thus displaced in both cases, in magnitude and angle, by an uncertain function $\delta(s)$according to

$$|G_O(j\omega)| = |\delta(j\omega)| \cdot |\tilde{G}_O(j\omega)| \tag{8.238}$$

$$\angle G_O(j\omega) = \angle\delta(j\omega) + \angle\tilde{G}_O(j\omega) \tag{8.239}$$

As an example, consider the particular case of a frequency ω_1 at which the magnitude deviation factor and phase deviations are independent, and restricted to ranges $0.8 \rightarrow 1.2$ and $-10° \rightarrow +10°$, respectively (Figure 8.18). A transfer function having this behaviour would be of the form

$$G_O(s) = Ke^{-\tau s}G_2(s) \tag{8.240}$$

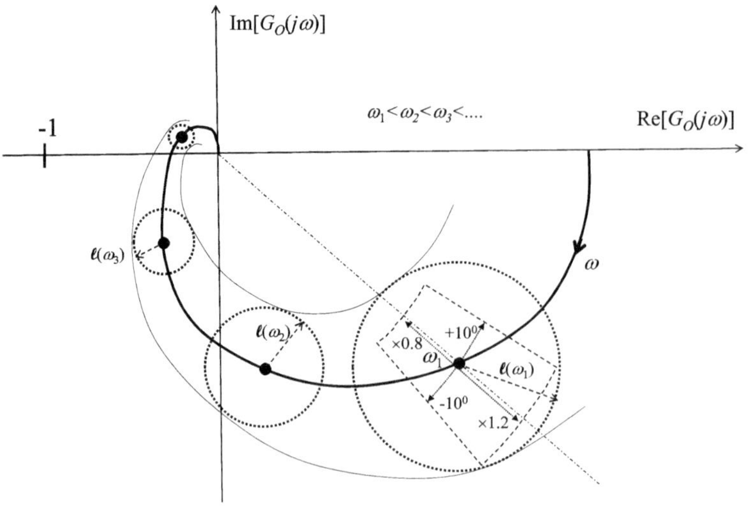

Figure 8.18 Uncertainty region of Nyquist plot point at frequencies ω_i for independent magnitude and phase deviation ranges at these frequencies.

where there is no uncertainty in $G_2(s)$ but K could be from 20% smaller to 20% bigger, and the possible range of τ is such that $\omega_1\tau$ could be from 10° smaller to 10° bigger. Obviously if the uncertain terms lay in other places, such as the coefficients in $G_2(s)$, there would be interdependence of the magnitude and phase deviations, giving a more irregular shape. All these possible shapes representing the uncertainty range cannot be handled, so the practice is to simply approximate the uncertainty with a disc of radius $\ell(\omega)$ centred at $\tilde{G}_O(j\omega)$. Notice that $\ell(\omega)$ is explicitly allowed to vary with ω because one expects a strong dependence of magnitude and phase angle, and thus the *uncertainties* thereof, on the frequency.

In Figure 8.18, it is thus seen that the allowance for uncertain deviations from the nominal $\tilde{G}_O(s)$ defines a region on the complex plane containing all of the possible Nyquist plots of the uncertain system. One is immediately interested in whether this region includes or overlaps the point $(-1+j0)$. If it does not, the system is *robustly stable*. Of course, the same conditions apply as in Section 8.6.4: as ω *decreases*, the net positive encirclements of $(-1+j0)$ represent the number of positive real zeros *minus* positive real poles of $1 + G_O(s)$. First removing any negative rotationary contributions of open-loop unstable poles, if there is still an overlap of $(-1+j0)$ one can conclude that the closed loop lacks robust stability. Bearing in mind that *nominal stability* refers to the nominal transfer function $\tilde{G}_O(s)$, one notes that

- *robust stability* implies *nominal stability*;
- *nominal stability* does not necessarily imply *robust stability*;
- *robust instability* does not necessarily imply *nominal instability*; and
- *nominal instability* implies *robust instability*.

The proposed robust stability criterion requires that for all ω, the distance between the points $\tilde{G}_O(j\omega)$ and $(-1+j0)$ on the complex plane exceed $\ell(\omega)$:

$$\left|1+\tilde{G}_O(j\omega)\right| > \ell(\omega) \quad \text{for } 0 < \omega < \infty \tag{8.241}$$

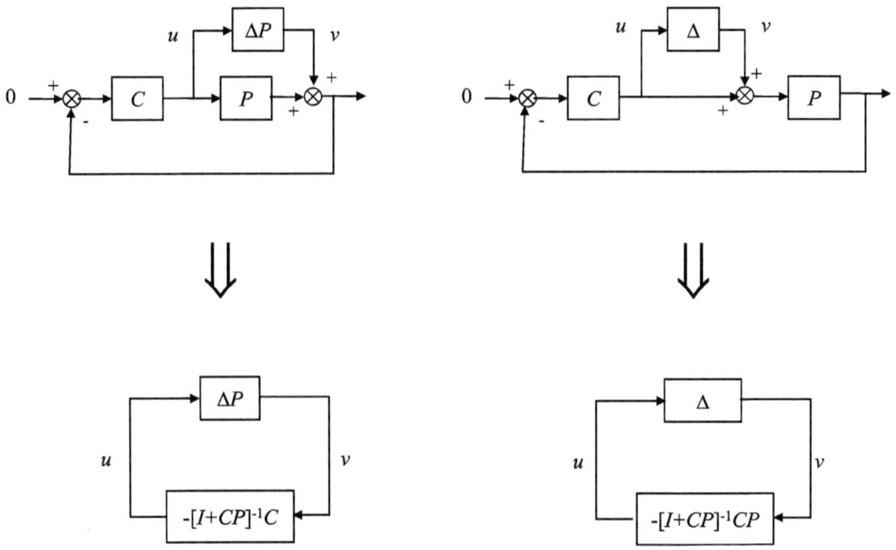

(a) Additive process model uncertainty (b) Multiplicative process model uncertainty

Figure 8.19 'Small gain' representation of the impact of uncertainty in closed loop.

The following discussion follows that of Damen and Weiland (2002). The uncertain control loops in Figure 8.17 can be represented as in Figure 8.19. This easily follows from

$$u = C[-Pu - v] \quad \text{and } v = \Delta Pu \tag{8.242}$$

$$u = CP[-u - v] \quad \text{and } v = \Delta\, u \tag{8.243}$$

The terminology for the various transfer functions is *sensitivity*

$$S = [I + CP]^{-1} \tag{8.244}$$

control sensitivity

$$R = [I + CP]^{-1}C \tag{8.245}$$

complementary sensitivity

$$T = [I + CP]^{-1}CP \tag{8.246}$$

Let $M = -T$. Then the Figure 8.19(b) can be represented as in Figure 8.20.

Small gain theorem: *If* Δ and M are stable, then the loop in Figure 8.20 is asymptotically stable if

$$\|M\Delta\|_\infty < 1$$

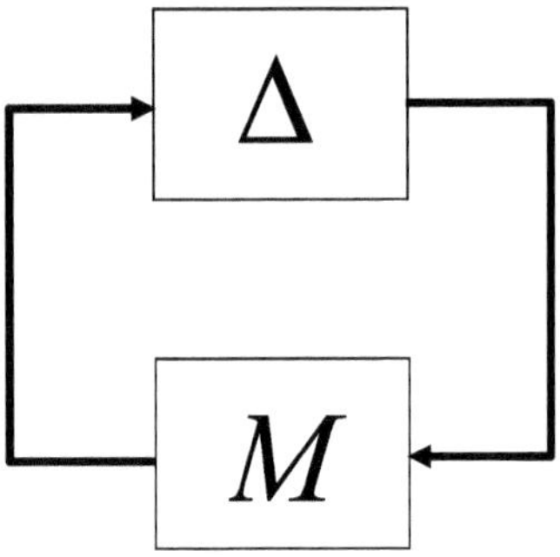

Figure 8.20 General small gain loop.

In this condition, $\|\cdot\|_\infty$ represents the *infinity norm* which for a SISO system $H(s)$ is the maximum amplitude ratio in the frequency range

$$\|H(s)\|_\infty = \max|H(j\omega)| \quad \text{for} \quad 0 < \omega < \infty \quad \text{(SISO)} \tag{8.247}$$

The equivalent expression for a MIMO system is

$$\|\boldsymbol{H}(s)\|_\infty = \sup \overline{\sigma}(\boldsymbol{H}(j\omega)) \quad \text{for} \quad 0 < \omega < \infty \quad \text{(MIMO)} \tag{8.248}$$

The 'sup' stands for *supremum* and effectively means the maximum of $\overline{\sigma}(\boldsymbol{H}(j\omega))$ in the frequency range. The function $\overline{\sigma}(\boldsymbol{H}(j\omega))$ denotes the largest singular value (always real) of the matrix of complex numbers $\boldsymbol{H}(j\omega)$ at any frequency ω.

References

Coughanowr, D.R. and Koppel, L.B. (1965) *Process Systems Analysis and Control*, McGraw-Hill.

Damen, A. and Weiland, S. (2002) Robust Control, Measurement and Control Group, Department of Electrical Engineering, Eindhoven University of Technology, Draft version July 17, 2002.

9 Optimisation

9.1 Introduction

In Chapter 7 a need was identified for the *optimal* choice of control moves in model predictive control (MPC). In the case of the linear quadratic regulator (LQR) (Section 7.6.5), an optimal control policy was obtained explicitly by means of *dynamic programming*. Likewise, an objective function was minimised explicitly to obtain optimal future control moves in the unconstrained version of the *dynamic matrix controller* (DMC) (Section 7.8.2). In the more general context of optimal predictive control of hybrid systems, the focus was on preparation of the *problem specification* for a numerical optimisation solution. Yet other instances of explicit optimisation have been encountered in *plant data reconciliation* (Section 6.3), *Kalman filtering* (Section 6.4) and *recursive least squares model identification* (Section 6.5).

Nowadays, problems of optimisation occur in the above historical ways, *and* in a growing number of new instances in the field of *advanced process control* (APC). At the top of the plant control pyramid in Figure 1.5 are various optimisers that operate either in real-time closed-loop or in an open-loop 'advisory' fashion. These include traditional linear programs to match feedstock/product requirements (e.g. in oil refining), product blenders, event schedulers, supply and product chain optimisers, and such mundane items as how to cut paper machine output into required lengths and widths. In general, these applications are moving away from dynamic control issues, yet they seem to be inextricably linked into the work of the process control engineer. This chapter thus aims to give a basic appreciation of the techniques of optimisation itself, and a few additional applications. Rather than the optimisation of a sequence (as in Chapter 7), the focus will be on the *maximisation or minimisation of a multivariable objective function.*

9.2 Aspects of Optimisation Problems

Some of the features encountered in optimisation problems are illustrated in Figure 9.1.

Objective function: This is a scalar function that can require minimisation or maximisation. It is necessarily linear in *linear programming* (LP) problems, but LP problems *can* nevertheless deal with the minimisation of a modulus, for example $J = \boldsymbol{w}^T|\boldsymbol{x} - \boldsymbol{x}_S|$ by defining $\boldsymbol{r}^+ - \boldsymbol{r}^- = \boldsymbol{x} - \boldsymbol{x}_S$ in the knowledge that LP solutions constrain the search variables as non-negative (or alternatively by just

Applied Process Control: Essential Methods, First Edition. Michael Mulholland.

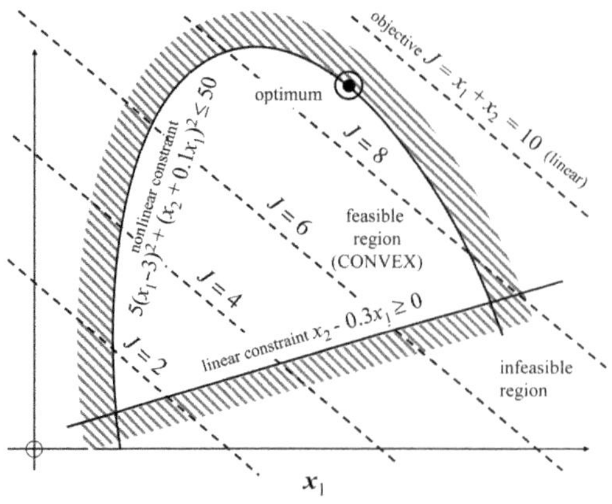

(a) Convex feasible region with linear objective function.

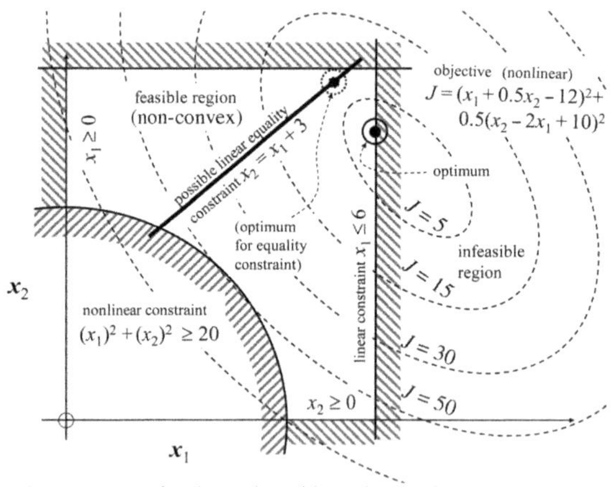

(b) Non-convex feasible region with quadratic objective function.

Figure 9.1 Two-dimensional optimisation problems with a minimisation objective illustrating terms.

setting lower bounds of zero for the elements of $\boldsymbol{r}^+$ and $\boldsymbol{r}^-$). Thus, one in each pair of vector elements will be zero, and $J = \boldsymbol{w}^T[\boldsymbol{r}^+ + \boldsymbol{r}^-]$. Linear constraint problems where the objective function is nonlinear but restricted to a quadratic form can be dealt with using specialised *quadratic programming* (QP) or *mixed integer quadratic programming* (MIQP) solutions.

Constraints: Within the constraints is the *feasible region*. Most solutions require to be *initialised* with values of the optimisation variables $\boldsymbol{x}$ that lie inside the feasible region, and not in the *infeasible region*. This region is bounded (in the multidimensional sense) by *inequality constraints* ($\geq$, $\leq$). There is also the possibility of *equality constraints* that force relationships between some or all of the variables, for example compliance with a process model. A particular problem arises when the feasible region is *non-convex* (in the multidimensional sense), as in Figure 9.1b. Imagine if this were a maximisation problem instead of a minimisation problem. Searches usually follow the steepest gradient, and so could end up trapped on either side of the concave part of the boundary, unable to go back down the gradient to get out. A strict definition of convexity is that all of the points interpolated between any two points on the boundary must lie within the region. This *affine* combination can be viewed in a multipoint sense as $\boldsymbol{x} = \theta_1\boldsymbol{x}_1 + \theta_2\boldsymbol{x}_2 + \theta_3\boldsymbol{x}_3 + \cdots + \theta_N\boldsymbol{x}_N$ where $\theta_1 + \theta_2 + \theta_3 + \cdots + \theta_N = 1$.

Searches: Finding the position within the constraints that minimises or maximises the objective function is performed in three different ways, sometimes in combination with each other:

a) Because of the parallel straight lines defining LP objective function contours, and because the constraints have to be straight lines, the solution always occurs at a *vertex* of the feasible region (in the multidimensional sense, of course). Thus, only the vertices have to be searched. Sometimes two vertices give equal optimum values of the objective function J, in which case the solution is termed *degenerate* (either may be used, but the implication is that values between the two vertices will also be optimal).
b) When nonlinearities are present, a gradient search method is used (*nonlinear programming* (NLP)). The search progresses downhill or uphill in small steps until it cannot get a further improvement. If the objective function surface does not slope monotonically away from the boundary towards a single optimum point, there is a danger that the search could end unwittingly at a *local optimum* instead of the *global optimum*. As seen, this problem could also be caused by a non-convex section of the feasible region boundary.
c) When the search variables have to take on only discrete *integer* or *Boolean* values, a difficult combinatorial problem arises. This is the *integer programming* (IP) problem. No longer can a gradient simply be followed to the optimum result. It is usually prohibitively expensive to test the objective value of every possible combination. Some efficient methods for dealing with this difficulty will be discussed later.

There is a class of problems which occur commonly in the processing industries that involve both continuous and discrete variables. These are the *mixed* (M) problems. When the search variables are linearly arranged, *mixed integer linear programming* (MILP) methods will use features of (a) and (c). When there is nonlinearity, *mixed integer nonlinear programming* (MINLP) methods involve features of (b) and (c).

Relaxed solution: Part of the solution strategy for integer problems and mixed integer problems is to initially *relax* the condition that the integer or Boolean (binary) variables only take on the values of whole numbers. So they are temporarily allowed to find any real values. A search such as LP or NLP then proceeds to find this *relaxed solution*. Sometimes one is lucky, and the best values for the integer variables can land on their integer constraints, so the solution is complete. Otherwise one begins a 'trial and error' process of adding constraints that alternately force the 'integer' variables on to their nearest integer values.

Slack variables: These are introduced in LP solutions to convert inequality constraints into equalities. Thus, if one requires $x \leq 7$, then one demands $x + s = 7$ where the 'simplex' solution for the LP requires that both x and s are non-negative. The *slack variables* have no other use after this than to demonstrate *margins* of compliance with the constraints.

Primal and dual problems: In the development of LP solutions, it was recognised that the *primal* problem of finding N variables that optimally satisfy M constraints was equivalent to the *dual problem* of finding M *dual* variables that optimally satisfy N *dual* constraints.

Dynamic programming and other 'programmings': In Sections 7.6.5 and 7.8.3 *dynamic programming* for optimisation of the future trajectory in MPC was used. A distinction needs to be made here in the sense that dynamic programming is a means of *structuring* an optimisation problem, rather than the optimisation itself. It is applicable in situations where the state is known at each 'step', and therefore the future trajectory is fully defined by the future inputs to the system. Here 'step' can mean anything from a series of points in time to a series of processing stages. In Section 7.6.5 it was seen that Bellman's principle of optimality (of the trajectory) then gave a huge saving in the number of combinations that had to be compared for optimality. In the type of general 'programmings' that will be considered in this chapter, the sense of physical problem structure is lost, and the search variables are considered on an equal basis. The model is a black box, and the sense of precedence and interdependence that can enhance the solution has to manifest itself mathematically.

9.3 Linear Programming

The linear programming problem may be stated:

$$\text{maximise } \boldsymbol{w}^T\boldsymbol{x} \quad \text{such that } \boldsymbol{x} \geq \boldsymbol{0} \tag{9.1}$$

$$\boldsymbol{A}\boldsymbol{x} \leq \boldsymbol{d} \tag{9.2}$$

Minimisation problems, and '≥' constraint functions, are of course accommodated by appropriate sign changes. The *simplex* method for solution of this problem was proposed in 1947 by the American mathematician George Dantzig (1951), and it is considered one of the top 10 algorithms of that century. Bearing in mind that the solution will lie at a vertex, Dantzig's method optimised the choice of neighbouring vertices as the search stepped along the edges of the multidimensional space defined by the constraints (N-dimensional polytope) (Figure 9.2).

The *revised* simplex method is described as follows. Introducing a slack variable vector $\boldsymbol{s}$, one requires to

maximise Z in

$$\begin{bmatrix} 1 & -\boldsymbol{w}^T & \boldsymbol{0}^T \\ \boldsymbol{0} & \boldsymbol{A} & \boldsymbol{I} \end{bmatrix} \begin{pmatrix} Z \\ \boldsymbol{x} \\ \boldsymbol{s} \end{pmatrix} = \begin{pmatrix} 0 \\ \boldsymbol{d} \end{pmatrix} \tag{9.3}$$

$$\text{such that } \boldsymbol{x} \geq \boldsymbol{0} \tag{9.4}$$

$$\boldsymbol{s} \geq \boldsymbol{0} \tag{9.5}$$

A feasible starting point for the solution is

$$\boldsymbol{x} = \boldsymbol{0} \tag{9.6}$$

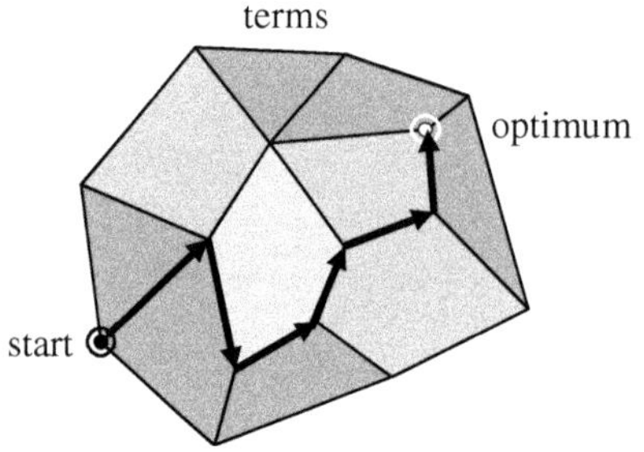

Figure 9.2 Vertex-to-vertex search for optimum along edges of bounded N-dimensional space.

which it is seen requires

$$s = d \tag{9.7}$$

The $\boldsymbol{s}$ variables then start off as the set of *basic variables* $\boldsymbol{b}$ and the $\boldsymbol{x}$ variables start off as the set of *non-basic variables* $\boldsymbol{n}$, that is initially one has the *basic feasible solution:*

$$\boldsymbol{b} = \boldsymbol{s} \quad \text{(will always have the dimension of } \boldsymbol{d}\text{)} \tag{9.8}$$

$$\boldsymbol{n} = \boldsymbol{x} \quad \text{(will always have the dimension of } \boldsymbol{x}\text{)} \tag{9.9}$$

On each iteration, the algorithm swops two members between these sets until there is no further improvement. Thus, $\boldsymbol{b}$ will in due course contain some of the x_i and s_i with the balance in $\boldsymbol{n}$ which is always set to $\boldsymbol{0}$. Thus the bottom partition of Equation 9.3 can be replaced with

$$\left[\boldsymbol{A} \,\vdots\, \boldsymbol{I}\right]\begin{pmatrix} \boldsymbol{x} \\ \hdashline \boldsymbol{s} \end{pmatrix} = \boldsymbol{B}\boldsymbol{b} = \boldsymbol{d} \tag{9.10}$$

where $\boldsymbol{B}$ consists of the columns of $\left[\boldsymbol{A} \,\vdots\, \boldsymbol{I}\right]$ corresponding to the x_i and s_i variables in the *current basic variable vector* $\boldsymbol{b}$, with the columns arranged in the same order as the entries in $\boldsymbol{b}$. So

$$\boldsymbol{b} = \left[\boldsymbol{B}^{-1}\boldsymbol{A} \,\vdots\, \boldsymbol{B}^{-1}\boldsymbol{I}\right]\begin{pmatrix} \boldsymbol{x} \\ \hdashline \boldsymbol{s} \end{pmatrix} = \boldsymbol{B}^{-1}\boldsymbol{d} \tag{9.11}$$

Likewise, defining the vector $\boldsymbol{v}$ (of size $\boldsymbol{b}$) with weights from $\boldsymbol{w}$ corresponding to the current x_i positions in $\boldsymbol{b}$,

$$\boldsymbol{v}^T\boldsymbol{b} = \boldsymbol{w}^T\boldsymbol{x} \tag{9.12}$$

since, as before, the remaining variables are in $\boldsymbol{n} = \boldsymbol{0}$. Thus, multiplying the bottom row of Equation 9.3 by $\boldsymbol{B}^{-1}$, and then further multiplying it by $\boldsymbol{v}^T$ before adding it to the top row, the problem can be viewed as maximising Z in

$$\begin{bmatrix} 1 & \boldsymbol{v}^T\boldsymbol{B}^{-1}\boldsymbol{A} - \boldsymbol{w}^T & \boldsymbol{v}^T\boldsymbol{B}^{-1} \\ \boldsymbol{0} & \boldsymbol{B}^{-1}\boldsymbol{A} & \boldsymbol{B}^{-1} \end{bmatrix}\begin{pmatrix} Z \\ \boldsymbol{x} \\ \boldsymbol{s} \end{pmatrix} = \begin{pmatrix} \boldsymbol{v}^T\boldsymbol{B}^{-1}\boldsymbol{d} \\ \boldsymbol{B}^{-1}\boldsymbol{d} \end{pmatrix} \tag{9.13}$$

$$\text{such that } \boldsymbol{x} \geq \boldsymbol{0} \tag{9.14}$$

$$\boldsymbol{s} \geq \boldsymbol{0} \tag{9.15}$$

Now, noting that

$$Z = \boldsymbol{v}^T\boldsymbol{B}^{-1}\boldsymbol{d} - \left[\boldsymbol{v}^T\boldsymbol{B}^{-1}\boldsymbol{A} - \boldsymbol{w}^T \,\vdots\, \boldsymbol{v}^T\boldsymbol{B}^{-1}\right]\begin{pmatrix} \boldsymbol{x} \\ \hdashline \boldsymbol{s} \end{pmatrix} \tag{9.16}$$

the term $\boldsymbol{v}^T\boldsymbol{B}^{-1}\boldsymbol{A} - \boldsymbol{w}^T$ is a set of gradients $-\left[\mathrm{d}Z/\mathrm{d}\boldsymbol{x}\right]^T$. So to *maximise* Z, one will need to increase any of the $\boldsymbol{x}$ elements corresponding to a negative gradient. On each iteration, a *pivoting rule* is implemented as follows. The $\boldsymbol{x}$ position corresponding to the *largest* negative gradient is added to the *basic* set $\boldsymbol{b}$ – this is called the *entering basic.* At the same time, it is necessary to drop the worst contributor already in the *basic* set into the *non-basic* set.*) This *leaving basic* is determined by the variable that first reaches zero as the $\boldsymbol{x}$ element destined to enter the *basic* set increases. This can

*) The method has been developed in terms of *inequality constraints* only. For *equality constraints*, one would need to ensure that the associated slack variables in $\boldsymbol{s}$ are the first to be made *non-basic*, as in Example 9.1.

be determined from Equation 9.11:

$$\boldsymbol{b} = \left[\boldsymbol{B}^{-1}\boldsymbol{A} \;\vdots\; \boldsymbol{B}^{-1}\boldsymbol{I}\right]\begin{pmatrix}\boldsymbol{x}\\ \boldsymbol{s}\end{pmatrix} \tag{9.17}$$

The $\boldsymbol{b}$ values are first found with the entering x value set to zero (shortcut $\boldsymbol{b} = \boldsymbol{B}^{-1}\boldsymbol{d}$). Then the coefficient for that x in each row divides the b value for its row and the entering x replaces the element of $\boldsymbol{b}$ corresponding to the smallest value.

The $\boldsymbol{B}$ matrix and $\boldsymbol{v}$ vector are updated according to the new membership of $\boldsymbol{b}$ and $\boldsymbol{n}$. Now one is ready for the next iteration, and one repeats until there are no negative gradients left in $\boldsymbol{v}^T\boldsymbol{B}^{-1}\boldsymbol{A} - \boldsymbol{w}^T$. The correspondence between the $\boldsymbol{b}$ and $\boldsymbol{n}$ elements, and the $\boldsymbol{x}$ and $\boldsymbol{s}$ elements, must clearly be tracked at each step, so that the optimised and slack variable values are available at convergence.

Example 9.1

Minimisation of an absolute deviation from setpoint.

The position y of a vehicle after travel time $t = 10$ is determined by $10u$ where u is the chosen speed. Starting from $y = 0$, it is desired to minimise the magnitude of both deviations of y from 50 at $t = 10$ *and* those of u from the best cruising speed 10. Also, it is required that $y < 70$.

$$\text{Let} \quad y = 50 + x_1 - x_2 \tag{9.18}$$

$$u = 10 + x_3 - x_4 \tag{9.19}$$

Then, giving position and speed deviations the same weighting, the problem can be re-stated as follows:

Minimise	$x_1 + x_2 + x_3 + x_4$
such that	$x_i \geq 0$ for $i = 1, \ldots, 4$ (causes one of x_1, x_2 to be 0 and one of x_3, x_4 to be 0)

$$50 + x_1 - x_2 = 100 + 10x_3 - 10x_4 \tag{9.20}$$

$$50 + x_1 - x_2 \leq 70 \tag{9.21}$$

that is

Maximise	$-x_1 - x_2 - x_3 - x_4$
such that	$x_i \geq 0$ for $i = 1, \ldots, 4$

$$x_1 - x_2 - 10x_3 + 10x_4 = 50 \tag{9.22}$$

$$x_1 - x_2 \leq 20 \tag{9.23}$$

In Equations 9.1 and 9.2

$$w = \begin{pmatrix} -1 \\ -1 \\ -1 \\ -1 \end{pmatrix} \tag{9.24}$$

$$A = \begin{bmatrix} 1 & -1 & -10 & 10 \\ 1 & -1 & 0 & 0 \end{bmatrix} \tag{9.25}$$

$$d = \begin{pmatrix} 50 \\ 20 \end{pmatrix} \tag{9.26}$$

From Equation 9.3, maximise Z in

$$\begin{bmatrix} 1 & -1 & -1 & -1 & -1 & 0 & 0 \\ 0 & 1 & -1 & -10 & 10 & 1 & 0 \\ 0 & 1 & -1 & 0 & 0 & 0 & 1 \end{bmatrix} \begin{pmatrix} Z \\ x_1 \\ x_2 \\ x_3 \\ x_4 \\ s_1 \\ s_2 \end{pmatrix} = \begin{pmatrix} 0 \\ 50 \\ 20 \end{pmatrix} \tag{9.27}$$

$$\text{such that } \boldsymbol{x} \geq \boldsymbol{0} \tag{9.28}$$

$$\boldsymbol{s} \geq \boldsymbol{0} \tag{9.29}$$

Initially, $\boldsymbol{n} = \boldsymbol{x}$ (9.30)

$$\text{with } \boldsymbol{x} = \boldsymbol{0} \tag{9.31}$$

$$\text{and } \boldsymbol{b} = \boldsymbol{s} \tag{9.32}$$

whilst

$$\boldsymbol{s} = \boldsymbol{d} \tag{9.33}$$

$$\text{and } \boldsymbol{d} = \begin{pmatrix} 50 \\ 20 \end{pmatrix} \tag{9.34}$$

In this problem there is an equality constraint for which the slack variable ($\boldsymbol{s}_1$) must be swopped to the non-basic set ($\boldsymbol{n}$), where all values are forced to zero, without further ado. Arbitrarily swop with $\boldsymbol{x}_1$. Then:

First iteration:

$$\boldsymbol{n} = \begin{pmatrix} s_1 \\ x_2 \\ x_3 \\ x_4 \end{pmatrix} \tag{9.35}$$

$$\text{and} \quad \boldsymbol{b} = \begin{pmatrix} x_1 \\ s_2 \end{pmatrix} \tag{9.36}$$

For this new system, one requires

$$\boldsymbol{B} = \begin{bmatrix} 1 & 0 \\ 1 & 1 \end{bmatrix} \tag{9.37}$$

$$\text{and } \boldsymbol{v} = \begin{pmatrix} -1 \\ 0 \end{pmatrix} \tag{9.38}$$

$$\text{so } -\left[\frac{\mathrm{d}Z}{\mathrm{d}\boldsymbol{x}}\right]^T = \boldsymbol{v}^T\boldsymbol{B}^{-1}\boldsymbol{A} - \boldsymbol{w}^T \tag{9.39}$$

$$= \begin{pmatrix} -1 & 0 \end{pmatrix} \begin{bmatrix} 1 & 0 \\ -1 & 1 \end{bmatrix} \begin{bmatrix} 1 & -1 & -10 & 10 \\ 1 & -1 & 0 & 0 \end{bmatrix} - \begin{pmatrix} -1 & -1 & -1 & -1 \end{pmatrix} \tag{9.40}$$

$$= \begin{pmatrix} 0 & 2 & 11 & -9 \end{pmatrix} \tag{9.41}$$

So x_4 (most negative gradient) needs to enter the basic set $\boldsymbol{b}$, but what must leave? Before the swop, with $x_4 = 0$, Equation 9.11 gives

$$\boldsymbol{b} = \begin{pmatrix} x_1 \\ s_2 \end{pmatrix} \tag{9.42}$$

$$\boldsymbol{b} = \boldsymbol{B}^{-1}\boldsymbol{d} \tag{9.43}$$

$$\boldsymbol{b} = \begin{bmatrix} 1 & 0 \\ -1 & 1 \end{bmatrix} \begin{pmatrix} 50 \\ 20 \end{pmatrix} \tag{9.44}$$

$$\boldsymbol{b} = \begin{pmatrix} 50 \\ -30 \end{pmatrix} \tag{9.45}$$

and thus consider

$$\boldsymbol{b} = \begin{pmatrix} 50 \\ -30 \end{pmatrix} \tag{9.46}$$

$$\boldsymbol{b} = \begin{bmatrix} \boldsymbol{B}^{-1}\boldsymbol{A} & \vdots & \boldsymbol{B}^{-1}\boldsymbol{I} \end{bmatrix} \begin{pmatrix} \boldsymbol{x} \\ \boldsymbol{s} \end{pmatrix} \tag{9.47}$$

$$\boldsymbol{b} = \begin{bmatrix} 1 & -1 & -10 & 10 & \vdots & 1 & 0 \\ 0 & 0 & 10 & -10 & \vdots & -1 & 1 \end{bmatrix} \begin{pmatrix} 50 \\ 0 \\ 0 \\ x_4 \\ \hdashline 0 \\ -30 \end{pmatrix} \tag{9.48}$$

$$\frac{b_1}{\text{coefficient of } x_4} = 5 \tag{9.49}$$

and

$$\frac{b_2}{\text{coefficient of } x_4} = 3 \tag{9.50}$$

so replace b_2 which corresponds to s_2. Then:

Second iteration:

$$\boldsymbol{n} = \begin{pmatrix} s_1 \\ x_2 \\ x_3 \\ s_2 \end{pmatrix} \tag{9.51}$$

$$\text{and} \quad \boldsymbol{b} = \begin{pmatrix} x_1 \\ x_4 \end{pmatrix} \tag{9.52}$$

For this new system, one requires

$$\boldsymbol{B} = \begin{bmatrix} 1 & 10 \\ 1 & 0 \end{bmatrix} \tag{9.53}$$

$$\text{and } \boldsymbol{v} = \begin{pmatrix} -1 \\ -1 \end{pmatrix} \tag{9.54}$$

so

$$-\left[\frac{\mathrm{d}Z}{\mathrm{d}\boldsymbol{x}}\right]^T = \boldsymbol{v}^T\boldsymbol{B}^{-1}\boldsymbol{A} - \boldsymbol{w}^T \tag{9.55}$$

$$-\left[\frac{\mathrm{d}Z}{\mathrm{d}\boldsymbol{x}}\right]^T = \begin{pmatrix} -1 & -1 \end{pmatrix} \begin{bmatrix} 0 & 1 \\ 0.1 & -0.1 \end{bmatrix} \begin{bmatrix} 1 & -1 & -10 & 10 \\ 1 & -1 & 0 & 0 \end{bmatrix} - \begin{pmatrix} -1 & -1 & -1 & -1 \end{pmatrix} \tag{9.56}$$

$$-\left[\frac{\mathrm{d}Z}{\mathrm{d}\boldsymbol{x}}\right]^T = \begin{pmatrix} 0 & 2 & 2 & 0 \end{pmatrix} \tag{9.57}$$

There are no negative gradients, so the solution has been achieved:

$$\boldsymbol{b} = \begin{pmatrix} x_1 \\ x_4 \end{pmatrix} \tag{9.58}$$

$$\boldsymbol{b} = \boldsymbol{B}^{-1}\boldsymbol{d} \tag{9.59}$$

$$\boldsymbol{b} = \begin{bmatrix} 0 & 1 \\ 0.1 & -0.1 \end{bmatrix} \begin{pmatrix} 50 \\ 20 \end{pmatrix} \tag{9.60}$$

$$\boldsymbol{b} = \begin{pmatrix} 20 \\ 3 \end{pmatrix} \tag{9.61}$$

and obviously

$$\boldsymbol{n} = \begin{pmatrix} s_1 \\ x_2 \\ x_3 \\ s \end{pmatrix} \tag{9.62}$$

$$\boldsymbol{n} = \begin{pmatrix} 0 \\ 0 \\ 0 \\ 0 \end{pmatrix} \tag{9.63}$$

$$\text{Thus, best speed} = u \tag{9.64}$$

$$u = 10 + x_3 - x_4 \tag{9.65}$$

$$u = 10 + 0 - 3 \tag{9.66}$$

$$u = 7 \tag{9.67}$$

$$\text{final distance} = y \tag{9.68}$$

$$y = 50 + x_1 - x_2 \tag{9.69}$$

$$y = 50 + 20 - 0 \tag{9.70}$$

$$y = 70 \rightarrow \tag{9.71}$$

9.4
Integer Programming and Mixed Integer Programming (MIP)

Figure 9.3 illustrates allowed values for two variables in an *integer* and a *mixed integer* optimisation problem. The common approach of *branch and bound* (BB) will initially be considered for these problems. The principle of *branch and bound* could be used in conjunction with any continuous real-variable optimiser, although for linear systems the obvious choice would be a linear programming solver as in Section 9.3. Since the methods start with the *relaxed* solution, in which the integer variables are allowed to assume any real value between their bounds, the *mixed* problem is an obvious extension of the approach. In a *mixed* problem, the real variables simply do not need to be constrained to integer values. A general problem statement (in the linear context) is

$$\begin{aligned}&\text{maximise } \boldsymbol{w}^T\boldsymbol{x} + \boldsymbol{u}^T\boldsymbol{y}\\ &\text{such that } \boldsymbol{x} \geq \boldsymbol{0}\end{aligned} \tag{9.72}$$

$$\boldsymbol{y} \geq \boldsymbol{0} \text{ and integer} \tag{9.73}$$

$$\boldsymbol{Ax} + \boldsymbol{Cy} \leq \boldsymbol{d} \tag{9.74}$$

which as before, using slack variables, translates to

maximise Z in

$$\begin{bmatrix} 1 & -\boldsymbol{w}^T & -\boldsymbol{u}^T & \boldsymbol{0}^T \\ \boldsymbol{0} & \boldsymbol{A} & \boldsymbol{C} & \boldsymbol{I} \end{bmatrix} \begin{pmatrix} Z \\ \boldsymbol{x} \\ \boldsymbol{y} \\ \boldsymbol{s} \end{pmatrix} = \begin{pmatrix} 0 \\ \boldsymbol{d} \end{pmatrix} \tag{9.75}$$

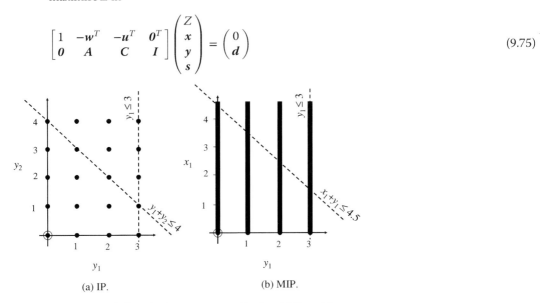

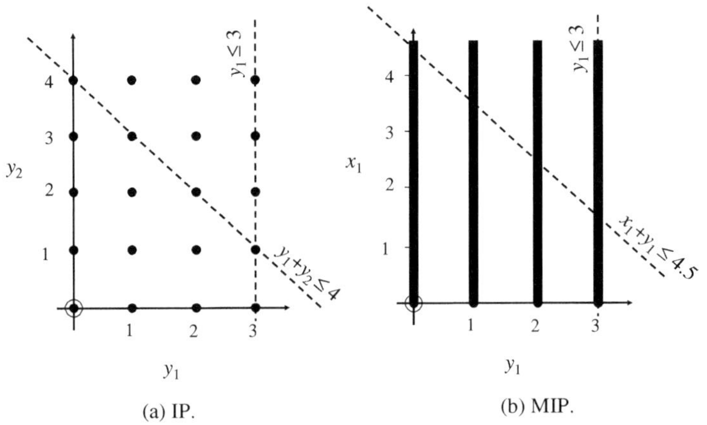

Figure 9.3 Examples of allowed values in two-variable IP and MIP problems.

$$\text{such that } \boldsymbol{x} \geq \boldsymbol{0} \tag{9.76}$$

$$\boldsymbol{y} \geq 0 \text{ and integer} \tag{9.77}$$

$$\boldsymbol{s} \geq \boldsymbol{0} \tag{9.78}$$

The *relaxed* linear solution is performed on each step, meaning that the $\boldsymbol{y}$ values are treated just as the $\boldsymbol{x}$ values, so the approach is the same as in Section 9.3. On the first step, no additional constraints would have been implemented, so the value of the objective function J will be higher than any solution achievable once constraints are inserted to force integer values. If all of the integer variables land on their bounds, then the problem is solved. (The bounds of the $\boldsymbol{y}$ variables would of course be integer, for example if $y_1 \in \{0, 1\}$, then a constrained value would be either 0 or 1.)

For those $\boldsymbol{y}$ elements which turn out to be fractional, one needs to *branch* at each instance:

a) Say the relaxed solution for $y_1 \in \{0, 1\}$ gave $y_1 = 0.7$: then a branch with the constraint $y_1 = 0$ and another with $y_1 = 1$ must be tested.
b) Say the relaxed solution for $y_2 \in \{0, 1, 2, 3, 4\}$ gave $y_2 = 2.3$: then a branch with the constraint $y_2 \leq 2$ and another with $y_2 \geq 3$ must be tested.

Each branch produces a new J value. The decision as to which of these two branches should be *fathomed* further is determined by that which has the highest relaxed J value. If a branch has only integer $\boldsymbol{y}$ values, then work on that branch ends. Likewise, it ends if there is no feasible solution. If the relaxed solution J value of a branch is already less than an integer solution elsewhere, then that branch is abandoned. When no *active* branches are left, that is they all have been either *fathomed*, *ended* or *abandoned*, the search is complete, and the $\boldsymbol{y}$ integer solution with the highest J value determines the optimum.

Example 9.2

A mixed integer linear programming (MILP) problem.

Three different-sized batch reactors are available. Production is restricted to a 9 h working day, so any batch must be started and ended within that period. Only one reactor can be run at a time.

- Reactor 1 takes 5 h and produces 10 tons
- Reactor 2 takes 3 h and produces 5 tons
- Reactor 3 takes 2 h and produces 3 ton

How many times (y_i) should each reactor i be run in order to maximise production?

$$\text{Total production is } Z = 10y_1 + 5y_2 + 3y_3 \tag{9.79}$$

$$\text{Total time is } x_1 = 5y_1 + 3y_2 + 2y_3 \tag{9.80}$$

$$\text{Maximise } Z \text{ such that } x_1 \leq 9 \tag{9.81}$$

The solution calculation sequence is shown in Figure 9.4, giving a production of 16 tons within 9 h.

The *cutting plane* approach, although not saving computation in general, can have advantages in certain cases where the problem structure indicates additional useful bounds. The method again uses the relaxed solution, and adds linear inequality constraints to avoid non-integer results.

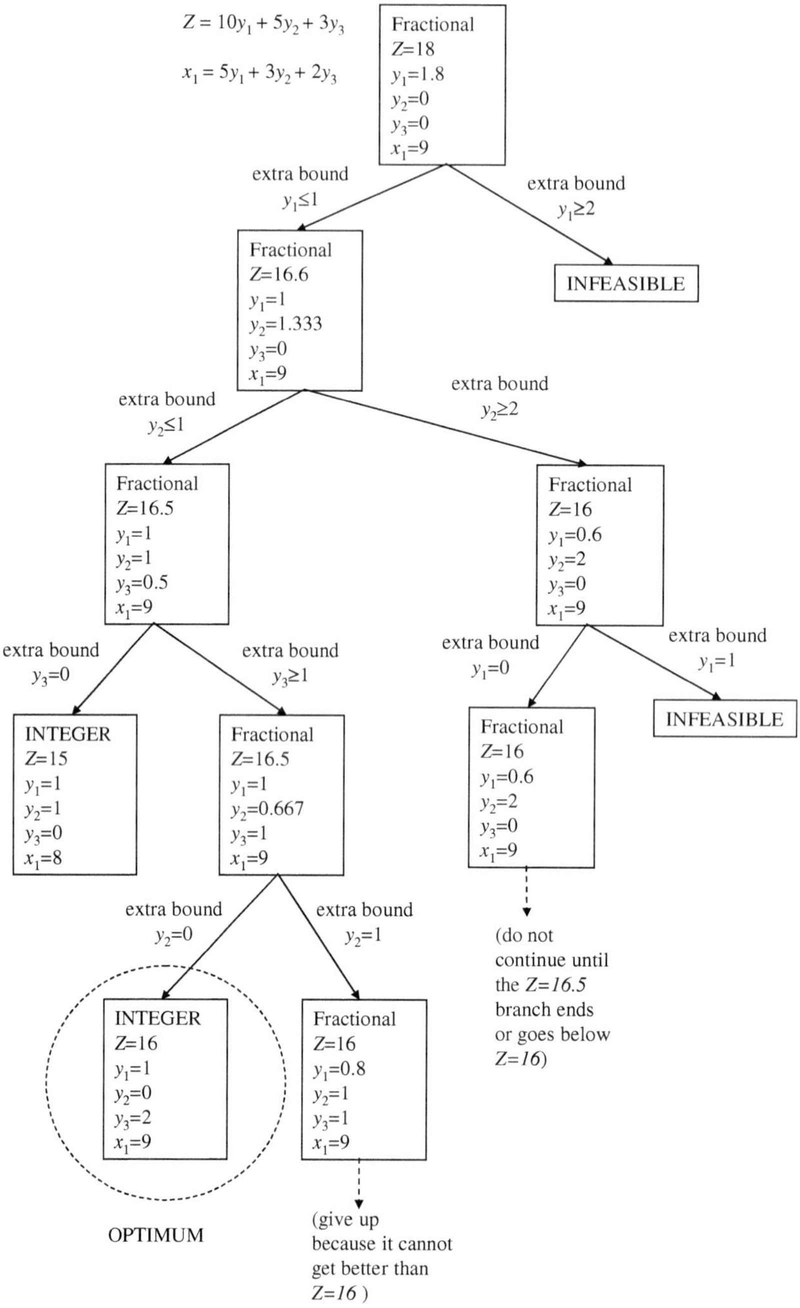

Figure 9.4 Branch and bound solution for MILP problem in Example 9.2.

9.5 Gradient Searches

In general, one requires to maximise or minimise the value of a single objective function, by searching in a multidimensional space determined by the N independent variables which determine the function's value. As seen in Section 9.3, a linear objective function with linear constraints causes the best value of the function to lie at a vertex of the system, that is at an intersection of constraints. However, the general nonlinear problem may have *local extrema*, and it is a drawback of most gradient search techniques that the search will end at one or other such local optimum, depending on the starting point of the search. Some techniques include a degree of randomisation, to allow searches to occasionally move on a contrary gradient, and perhaps escape a local optimum. Alternatively, a number of searches can be started at widely scattered starting points, in order to discover as many optima as possible. Some search techniques require a functional form of the derivative of the objective function, whilst others do not. Another distinguishing factor is the amount of storage required – of order N in some methods and N^2 in others.

9.5.1 Newton Method for Finding a Minimum or a Maximum

Consider a function $f(\boldsymbol{x})$ which is dependent on a number of independent variables represented by the vector $\boldsymbol{x}$. The gradient and Hessian (which is symmetric) are

$$\nabla f(\boldsymbol{x}) = \begin{pmatrix} \dfrac{\partial f}{\partial x_1} \\ \dfrac{\partial f}{\partial x_2} \\ \vdots \\ \dfrac{\partial f}{\partial x_n} \end{pmatrix} \tag{9.82}$$

$$\boldsymbol{H}(\boldsymbol{x}) = \begin{bmatrix} \dfrac{\partial^2 f}{\partial x_1^2} & \dfrac{\partial^2 f}{\partial x_1 \partial x_2} & \cdots & \dfrac{\partial^2 f}{\partial x_1 \partial x_n} \\ \dfrac{\partial^2 f}{\partial x_2 \partial x_1} & \dfrac{\partial^2 f}{\partial x_2^2} & \cdots & \dfrac{\partial^2 f}{\partial x_2 \partial x_n} \\ \vdots & \vdots & \ddots & \vdots \\ \dfrac{\partial^2 f}{\partial x_n \partial x_1} & \dfrac{\partial^2 f}{\partial x_n \partial x_2} & \cdots & \dfrac{\partial^2 f}{\partial x_n^2} \end{bmatrix} \tag{9.83}$$

Assuming that these exist, one can write the approximation

$$f_{i+1} \approx f_i + (\boldsymbol{x}_{i+1} - \boldsymbol{x}_i)^T \nabla f|_i + \frac{1}{2!}(\boldsymbol{x}_{i+1} - \boldsymbol{x}_i)^T \boldsymbol{H}|_i (\boldsymbol{x}_{i+1} - \boldsymbol{x}_i) + \cdots \tag{9.84}$$

$$\nabla f|_{i+1} \approx \boldsymbol{0} + \nabla f|_i + \boldsymbol{H}|_i (\boldsymbol{x}_{i+1} - \boldsymbol{x}_i) + \cdots \tag{9.85}$$

For the extremum, one requires $\nabla f|_{i+1} \to \boldsymbol{0}$, which is used in the iteration

$$\boldsymbol{x}_{i+1} = \boldsymbol{x}_i - \boldsymbol{H}|_i^{-1} \nabla f|_i \tag{9.86}$$

9.5.2 Downhill Simplex Method

Nelder and Mead (1965) devised a simple technique for multidimensional optimisation which required only repeated evaluations of the objective function (i.e. no derivative function). For a search in n-dimensional space, a simplex with $n+1$ vertices is defined by choosing an initial point $\boldsymbol{x}_0$ and creating n neighbouring points by displacing each dimension i in turn by a characteristic length scale λ_i in that direction.

$$\boldsymbol{x}_0 = \begin{pmatrix} x_{01} \\ x_{02} \\ \vdots \\ x_{0n} \end{pmatrix} \tag{9.87}$$

$$\boldsymbol{x}_1 = \begin{pmatrix} x_{01} + \lambda_1 \\ x_{02} \\ \vdots \\ x_{0n} \end{pmatrix} \tag{9.88}$$

$$\boldsymbol{x}_2 = \begin{pmatrix} x_{01} \\ x_{02} + \lambda_2 \\ \vdots \\ x_{0n} \end{pmatrix} \tag{9.89}$$

$$\cdots \boldsymbol{x}_n = \begin{pmatrix} x_{01} \\ x_{02} \\ \vdots \\ x_{0n} + \lambda_n \end{pmatrix} \tag{9.90}$$

The objective function $f(\boldsymbol{x})$ is evaluated at each of these $N+1$ points. For the *minimisation* problem, the point giving the *largest* value of f is *reflected* through the centroid of the n remaining points to an equal distance on the opposite side. For example, in a two-dimensional search, the simplex would be a triangle. The reflection would flip the worst vertex to a point on the other side of the opposite face (Figure 9.5). Thus, A flips to D, B flips to E, D flips to F and so on. One notes that the step size used is too crude to manoeuvre close to the optimum. There are various refinements of the scheme dealing with such things as the approach to the optimum, or crossing a large distance under a steady gradient. Quite clearly the search can result in a mere *local optimum.*

Start by evaluating a possible new apex as follows (use $\alpha = 1$):

$$\boldsymbol{x}_{\text{centroid}} = \frac{1}{n} \sum_{\substack{i=1 \\ i \neq \text{worst}}}^{n+1} \boldsymbol{x}_i \tag{9.91}$$

$$\boldsymbol{x}_{\text{possible}} = \boldsymbol{x}_{\text{centroid}} + \alpha[\boldsymbol{x}_{\text{centroid}} - \boldsymbol{x}_{\text{worst}}] \tag{9.92}$$

If $\boldsymbol{x}_{\text{possible}}$ gives a better result than all present apices, compute also an expanded version of it using $\alpha = 2$. Replace $\boldsymbol{x}_{\text{worst}}$ with the better of these two (*expansion*) and repeat the procedure from the start.

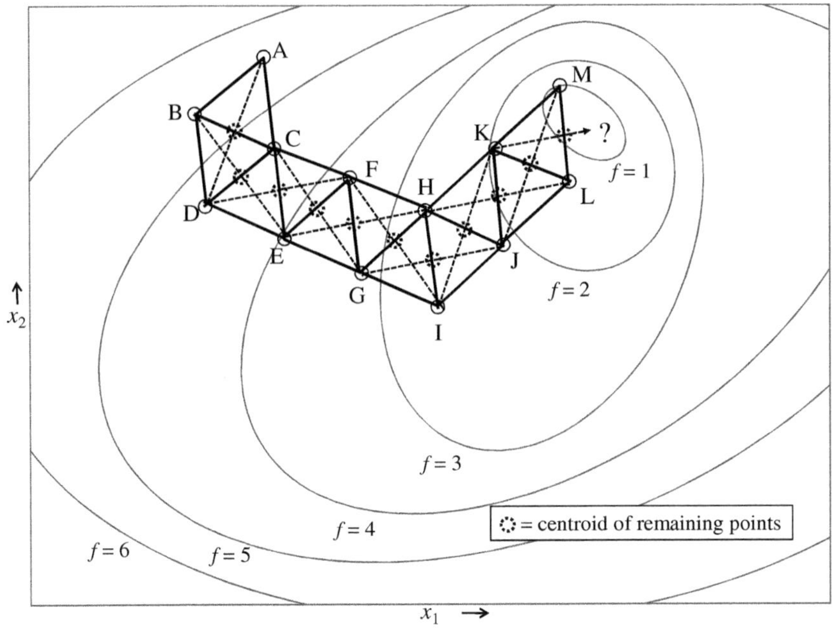

Figure 9.5 Two-dimensional search by downhill simplex starting at ABC and flipping successively to D,E,F,G,H,I,J,K . . .

If $f\,(\boldsymbol{x}_{\text{possible}})$ is better than both the worst and at least the second-worst apices, then replace the worst with $\boldsymbol{x}_{\text{possible}}$ (*reflection*) and repeat the procedure from the start. (If it were only better than the worst, restarting at this new worst apex would just return to the previous worst apex!)

If $\boldsymbol{x}_{\text{possible}}$ gives a worse result than all present apices, compute a contracted version of it using $\alpha = 0.5$ and replace $\boldsymbol{x}_{\text{worst}}$ with it only if it is better (*contraction*), thereafter repeating from the start.

After the above, if no better point has been found, reduce the simplex size whilst retaining the present best point (*reduction*):

$$\boldsymbol{x}_i = \boldsymbol{x}_{\text{best}} + \alpha[\boldsymbol{x}_i - \boldsymbol{x}_{\text{best}}], \quad \text{with } \alpha = 0.5 \text{ and } i = 1, \ldots, n+1, i \neq \text{best} \tag{9.93}$$

Repeat the above sequence from the start until adequately converged.

9.5.3
Methods Based on Chosen Search Directions

The case of a search for maximum or minimum involving one independent variable is easy to understand. Searches in multidimensional space often use a one-dimensional search method in a sequence of directions. As will be seen, the efficiency of such schemes is dependent on the choice of search directions.

Initially consider a one-dimensional search for, say, the value of x which causes the function $f(x)$ to be at a minimum. Figure 9.6a illustrates a crude search in which the range is stepped through until an appropriate change in gradient is detected – after which step sizes could be reduced to refine the location. Conversely, Figure 9.6b illustrates the well-known 'golden section' search in

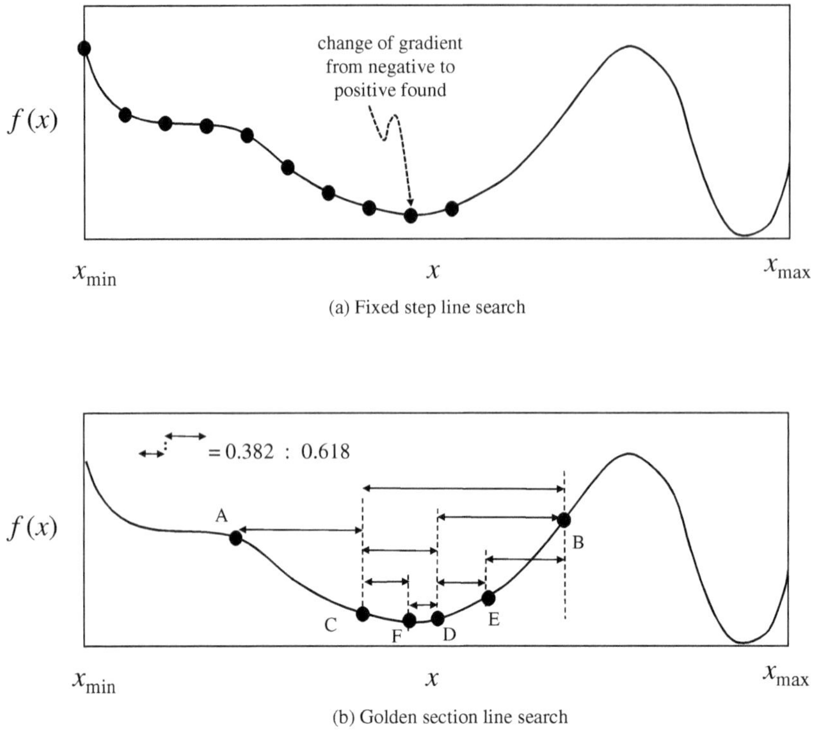

Figure 9.6 Line searches for a minimum.

which the minimum is repeatedly bracketed using the lowest point and its two neighbours. Starting with three points ACB, of which the centre point C is the lowest, a position 0.382 of the distance into the larger interval CB neighbouring the lowest point is used to define a new point D. Again the lowest available point D is selected, and a new point established 0.382 of the way into the larger interval DB neighbouring *that* point. D is still the lowest point, but its larger interval is now DC, resulting in point F and so on. The procedure is akin to the interval-halving method used to find a root of $f(x) = 0$, except that one notes that the interval fractions are 0.382 and 0.618.

The reason for the 'golden section' ratios is illustrated in Figure 9.7. From a previous step, point 1 had been the minimum and 1–2 had been the larger interval (sized D). Thus, point 3 was created at fraction α of D. Then point 3 was found to be minimum, so point 4 was created in 3's larger

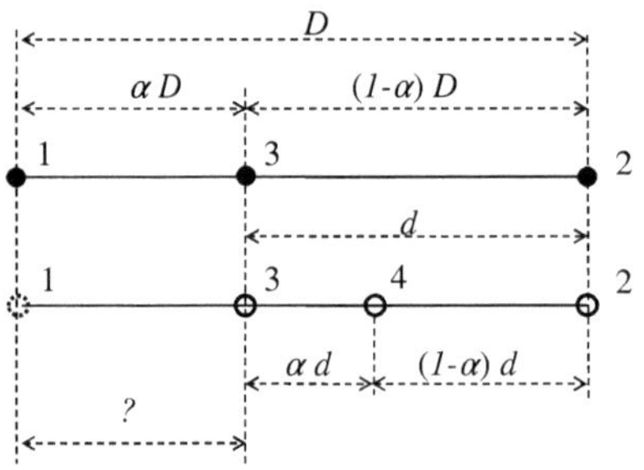

Figure 9.7 Golden section ratios.

interval (sized d) similarly. What now if point 3 is still the minimum? For the two intervals of point 3 to be in the same ratio at this stage, one clearly requires

$$\frac{1 \to 3}{3 \to 4} = \frac{\alpha D}{\alpha d} = \frac{(1-\alpha)d}{\alpha d} \tag{9.94}$$

Since $d = (1 - \alpha)D$, it follows that

$$\alpha = (1-\alpha)^2 \tag{9.95}$$

giving one root between 0 and 1 which is

$$\alpha = \frac{3-\sqrt{5}}{2} \tag{9.96}$$

$$\alpha = 0.382 \tag{9.97}$$

Even if the first three points are not spaced like this, the available intervals will quickly converge to this ratio. The method guarantees that the extremum will successively be bracketed in an interval shrinking by a factor of 0.618.

There are various other methods for one-dimensional line searches. Some, as above, do not need an explicit functional form of the derivative df/dx, whilst others do. In reviewing some of the methods below for optimisation in *multidimensions*, it will be assumed that such a line search algorithm is available.

9.5.3.1 Steepest Descent Method

Here one is searching for an extremum of a scalar function of several independent variables, $f(\boldsymbol{x})$. To illustrate this, consider the two-dimensional case in Figure 9.8.

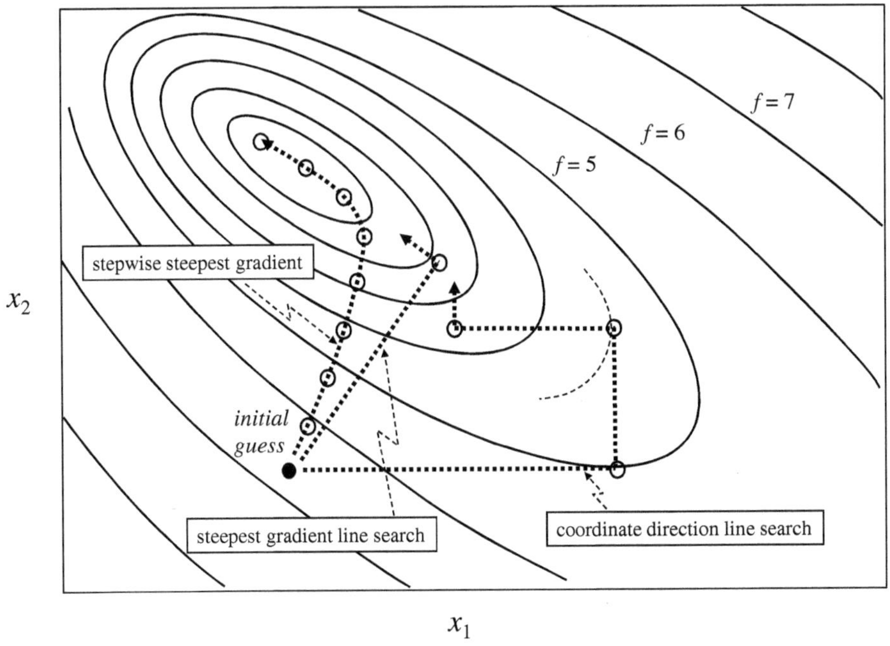

Figure 9.8 Searches based on gradients in two dimensions.

The searches proceed from point-to-point in various ways. Say that one has arrived at point $\boldsymbol{x}_i$. In the simplest case, one might proceed through the coordinate directions j in sequence, either using a line search for the minimum or just taking a fixed step in the downward direction of that coordinate (or even just choosing the coordinate giving the best change in f). In the jth coordinate direction, the one-dimensional function $f_j(x_j)$ is easily established by fixing the values of the other variables according to $\boldsymbol{x}_i$.

$$f_j(x_j) = f(\boldsymbol{x})|_{\boldsymbol{x}=\boldsymbol{x}_i \text{ except for } j\text{th element}} \tag{9.98}$$

Then the search line is determined by

$$\begin{pmatrix} x_1 \\ \vdots \\ x_j \\ \vdots \\ x_n \end{pmatrix}_{i+1} = \begin{pmatrix} x_1 \\ \vdots \\ x_j \\ \vdots \\ x_n \end{pmatrix} + \lambda \begin{pmatrix} 0 \\ \vdots \\ 1 \\ \vdots \\ 0 \end{pmatrix} \tag{9.99}$$

and

$$f = f(\lambda) \text{ on this line} \tag{9.100}$$

This would be the basis of a coordinate line search, as shown in Figure 9.8. If the derivative $\partial f/\partial\lambda$ is required, this might be obtainable as an explicit function. Alternatively, an estimate of the derivative might be obtained using a small perturbation of λ.

The same idea applies in the steepest descent search in a minimisation problem, except that here the elements of $\boldsymbol{x}$ must vary together according to the *initial* gradient:

$$\begin{pmatrix} x_1 \\ \vdots \\ x_j \\ \vdots \\ x_n \end{pmatrix}_{i+1} = \begin{pmatrix} x_1 \\ \vdots \\ x_j \\ \vdots \\ x_n \end{pmatrix} - \lambda \begin{pmatrix} \partial f/\partial x_1 \\ \vdots \\ \partial f/\partial x_j \\ \vdots \\ \partial f/\partial x_n \end{pmatrix}_i \tag{9.101}$$

and again

$$f = f(\lambda) \text{ on this line} \tag{9.102}$$

Actually, this result is not so intuitive, because if λ is viewed to be a common step size, then the increments seem to have the units of f. Nevertheless, if one imagines a sloping plane, for example with a large negative $\partial f/\partial x_1$ and small negative $\partial f/\partial x_2$, the motion will mostly be in the positive x_1 direction, as required.

When the next point is sought as a line search minimum, one notes that both the coordinate direction and the steepest descent methods suffer because they are not recognising local direction changes. At the point where λ determines the minimum value of f, the line must be tangential to the f contours, so the next direction of the search will occur at right angles to the present direction. Thus, a somewhat inefficient orthogonal zigzagging can be expected.

There is an obvious benefit in taking a series of fixed short steps, and re-evaluating the local gradient on each step, as in the stepwise steepest gradient search of Figure 9.8. However, one needs to balance the computational cost of repeatedly re-evaluating the full f function and its gradients. Sophisticated methods have been developed that improve the efficiency of such searches in different situations.

9.5.3.2 Conjugate Gradient Method

As seen above, the steepest descent method results in a series of orthogonal turns which do not in general lead directly to the optimum. The *conjugate gradient method* similarly relies on a series of line minimisations, but it uses information from all previous directions to choose a better direction at each line minimum. The method is based on an approximate quadratic representation of $f(x)$:

$$f(\boldsymbol{x}) = \tfrac{1}{2}\boldsymbol{x}^T \boldsymbol{A}\,\boldsymbol{x} + \boldsymbol{x}^T\boldsymbol{b} + c \tag{9.103}$$

For the extremum, one requires

$$\nabla f(\boldsymbol{x}) = \boldsymbol{A}\,\boldsymbol{x} + \boldsymbol{b} = \boldsymbol{0} \tag{9.104}$$

For $\boldsymbol{A}$ a symmetric positive-definite $n \times n$ matrix, a set of n $\boldsymbol{A}$-orthogonal vectors $\boldsymbol{d}_0$, $\boldsymbol{d}_1$, $\boldsymbol{d}_2$, . . . , $\boldsymbol{d}_{n-1}$ will exist which have the property

$$\boldsymbol{d}_i^T \boldsymbol{A}\,\boldsymbol{d}_k = 0 \quad \text{for} \quad i \neq k \text{ and } i, k = 0, 1, \ldots, n-1 \tag{9.105}$$

This orthogonality across $\boldsymbol{A}$ is described as $\boldsymbol{d}_i$ and $\boldsymbol{d}_k$ being *conjugate* with respect to $\boldsymbol{A}$. In addition, since $\boldsymbol{A}$ is positive-definite, it follows that

$$\boldsymbol{d}_i^T \boldsymbol{A}\,\boldsymbol{d}_i > 0 \quad \text{for } i = 0, 1, \ldots, n-1 \tag{9.106}$$

The basis vectors $\boldsymbol{d}_i$ will also be linearly independent, that is the only way for the following sum,

$$\boldsymbol{x}^* = \alpha_0\boldsymbol{d}_0 + \alpha_1\boldsymbol{d}_1 + \alpha_2\boldsymbol{d}_2 + \cdots + \alpha_{n-1}\boldsymbol{d}_{n-1} \tag{9.107}$$

to equal a zero vector, if all α_i are zero (because multiplying through by $\boldsymbol{d}_i^T\boldsymbol{A}$ would leave only the $\boldsymbol{d}_i^T\boldsymbol{A}\,\boldsymbol{d}_i$ term, which is greater than zero (see Equations 9.105 and 9.106). This then is a means of constructing the solution of Equation 9.104:

$$\boldsymbol{A}\,\boldsymbol{x}^* + \boldsymbol{b} = \boldsymbol{0} \tag{9.108}$$

As in Equation 9.107, if one moves a 'distance' α_i along each conjugate gradient $\boldsymbol{d}_i$, it must be possible to come to the extremum within n moves! What remains to be done is to develop the method for recursively building the $\boldsymbol{d}_i$ vectors. Consider the successive iterations in which $\boldsymbol{x}$ follows each gradient $\boldsymbol{d}$:

$$\boldsymbol{x}_i = \boldsymbol{x}_{i-1} + \alpha_{i-1}\boldsymbol{d}_{i-1} \tag{9.109}$$

From Equation 9.108, the corresponding *residual* on each step is

$$\boldsymbol{r}_i = \boldsymbol{b} - \boldsymbol{A}\,\boldsymbol{x}_i \tag{9.110}$$

Using Equation 9.109, the residual can be updated for movement in this direction by

$$\boldsymbol{r}_i = \boldsymbol{r}_{i-1} - \alpha_{i-1}\boldsymbol{A}\,\boldsymbol{d}_{i-1} \tag{9.111}$$

The α_{i-1} will be chosen to cause the successive residual vectors to be *orthogonal* to each other so that $\boldsymbol{r}_i^T\boldsymbol{r}_{i-1} = 0$ (later it will be seen that this is equivalent to 'minimising' $\boldsymbol{r}_i$). Then, multiplying through by $\boldsymbol{r}_{i-1}^T$,

$$0 = \boldsymbol{r}_{i-1}^T\boldsymbol{r}_{i-1} - \alpha_{i-1}\boldsymbol{r}_{i-1}^T\boldsymbol{A}\boldsymbol{d}_{i-1} \tag{9.112}$$

giving

$$\alpha_{i-1} = -\frac{\boldsymbol{r}_{i-1}^T\boldsymbol{r}_{i-1}}{\boldsymbol{r}_{i-1}^T\boldsymbol{A}\,\boldsymbol{d}_{i-1}} \tag{9.113}$$

Now consider the following scheme for updating the *gradient*:

$$\boldsymbol{d}_i = \boldsymbol{r}_i + \gamma_{i-1}\,\boldsymbol{d}_{i-1} \tag{9.114}$$

For the sequence of gradients $\boldsymbol{d}_i$ to be conjugate with respect to $\boldsymbol{A}$, one requires

$$\boldsymbol{d}_i^T\boldsymbol{A}\,\boldsymbol{d}_{i-1} = 0 \tag{9.115}$$

This can be achieved through the choice of γ_{i-1}: Multiplying through (9.114) from the left by $\boldsymbol{d}_{i-1}^T\boldsymbol{A}$,

$$0 = \boldsymbol{d}_{i-1}^T\boldsymbol{A}\boldsymbol{r}_i + \gamma_{i-1}\boldsymbol{d}_{i-1}^T\boldsymbol{A}\boldsymbol{d}_{i-1} \tag{9.116}$$

and multiplying the transpose of (9.114) from the right by $\boldsymbol{A}\boldsymbol{d}_i$

$$\boldsymbol{d}_i^T\boldsymbol{A}\boldsymbol{d}_i = \boldsymbol{r}_i^T\boldsymbol{A}\boldsymbol{d}_i \tag{9.117}$$

whence

$$\gamma_{i-1} = -\frac{\boldsymbol{r}_i^T\boldsymbol{A}\boldsymbol{d}_{i-1}}{\boldsymbol{r}_{i-1}^T\boldsymbol{A}\boldsymbol{d}_{i-1}} \tag{9.118}$$

Then multiplying (9.111) by $\boldsymbol{r}_i^T$ and by $\boldsymbol{r}_{i-1}^T$ provides substitutions which yield

$$\gamma_{i-1} = \frac{\boldsymbol{r}_i^T\boldsymbol{r}_i}{\boldsymbol{r}_{i-1}^T\boldsymbol{r}_{i-1}} \tag{9.119}$$

Choosing a starting point $\boldsymbol{x}_0$ and an initial gradient $\boldsymbol{d}_0$ (say the steepest descent $-\nabla f|_{x_0}$ at $\boldsymbol{x}_0$), Equations 9.110, 9.111, 9.113, 9.114 and 9.118 could be used in sequence to update the estimate $\boldsymbol{x}_i$. Indeed, quite apart from the present objective of finding an extremum, this turns out to be an efficient method for solution of large sets of linear equations of the form (9.108) (rather than, for example, elimination or inversion), especially when they are sparse.

However, an important aspect of the above algorithm is that it is also an effective method for finding the extrema of more general nonlinear functions that merely *approximate* the quadratic form of Equation 9.103. The matrix $\boldsymbol{A}$ would be the Hermitian used in the truncated Taylor expansion (9.103), but one does not actually have to evaluate it. At $\boldsymbol{x}_0$, $\boldsymbol{r}_0 = -\nabla f|_{x_0}$ which is evaluated directly from $f(\boldsymbol{x})$. Here initialise $\boldsymbol{d}_0 = \boldsymbol{r}_0$ and proceed in this direction using a separate *line minimisation* (Section 9.5.3.1) until the minimum is found. In this way one avoids the use of both (9.113) and (9.111) which require $\boldsymbol{A}$, because α is not required to fix the distance on the line, and the new residual $\boldsymbol{r}_1$ can be evaluated directly as $-\nabla f|_{x_1}$. Then γ_0 is evaluated using (9.119) and the new search direction $\boldsymbol{d}_1$ obtained from (9.114), allowing the above steps to be repeated until convergence.

The procedure described in the above paragraph for general functions $f(\boldsymbol{x})$ can be shown by substitution to exactly parallel the original equations based on the quadratic form (9.103). Additionally, the insistence of orthogonality of the $\boldsymbol{r}_i$ and conjugacy of the $\boldsymbol{d}_i$ is seen as follows to 'minimise' $\boldsymbol{r}_i$ in Equation 9.111. Consider the 'sum of squares' (weighted by the positive definite matrix $\boldsymbol{A}^{-1}$) of the residual elements of $\boldsymbol{r}_i$ whilst moving in the direction $\boldsymbol{d}_{i-1}$:

$$\boldsymbol{r}_i^T\boldsymbol{A}^{-1}\boldsymbol{r}_i = (\boldsymbol{r}_{i-1} - \alpha_{i-1}\boldsymbol{A}\boldsymbol{d}_{i-1})^T\boldsymbol{A}^{-1}(\boldsymbol{r}_{i-1} - \alpha_{i-1}\boldsymbol{A}\boldsymbol{d}_{i-1}) \tag{9.120}$$

$$\frac{d\left(\boldsymbol{r}_i^T\boldsymbol{A}^{-1}\boldsymbol{r}_i\right)}{d\alpha_{i-1}} = \boldsymbol{d}_{i-1}^T(\boldsymbol{r}_{i-1} - \alpha_{i-1}\boldsymbol{A}\,\boldsymbol{d}_{i-1}) = 0 \quad \text{for the minimum} \tag{9.121}$$

$$\alpha_{i-1} = \frac{\boldsymbol{d}_{i-1}^T\boldsymbol{r}_{i-1}}{\boldsymbol{d}_{i-1}^T\boldsymbol{A}\boldsymbol{d}_{i-1}} \tag{9.122}$$

If $\boldsymbol{d}_i$, $\boldsymbol{d}_i$ are conjugate with respect to $\boldsymbol{A}$, substitution in (9.111) and multiplying by $\boldsymbol{d}_{i-1}^T$ shows that $\boldsymbol{d}_{i-1}^T\boldsymbol{r}_i = 0$. Then transposing (9.114) and multiplying from the right by $\boldsymbol{r}_i$ shows $\boldsymbol{d}_i^T\boldsymbol{r}_i = \boldsymbol{r}_i^T\boldsymbol{r}_i$. This allows (9.122) to be expressed in the same form as (9.113), *without* having stipulated that $\boldsymbol{r}_i$, $\boldsymbol{r}_{i-1}$ are orthogonal.

9.6 Nonlinear Programming and Global Optimisation

Apart from the case of a linear system (Section 9.3), the techniques described above suffer a common problem in that they may converge to a local extremum. This may be caused by nonlinear constraints, or the presence of several local extrema within the search region. A number of methods have been developed in an attempt to gain confidence that the solution found is indeed the 'global' optimum. These include deterministic methods (e.g. *branch and bound*), random methods (e.g. *Monte Carlo*) and heuristic methods (e.g. *evolutionary*). There is a large literature on these methods, but in general the best method (and *version* thereof) will be determined by the particular problem structure. To gain some insight into this field, a common deterministic method, *branch and bound*, will be examined more closely. The branch and bound technique has already been discovered in Section 9.4 (IP, MIP). There it played a role in eliminating poor discrete choices (before wasting effort on deeper 'sounding'). The same principle is now employed to discard regions proven *not to contain* the global optimum.

9.6.1 Global Optimisation by Branch and Bound

It has been noted that finding a global optimum will be hindered by the presence of multiple local optima, as well as *concavity* of the search region. In the case of multiple optima, branch

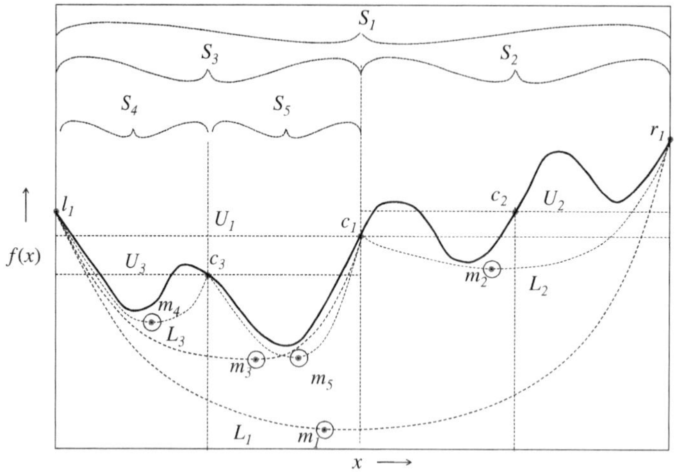

Figure 9.9 Branch and bound sequence to determine the global minimum of a univariate function using a functional fit to determine minima.

and bound techniques allow repeated shrinking of a region surrounding the global optimum. Conversely, for a concave system, branch and bound methods offer a means to compare and discard *convex* subregions achieved by subdivision. The technique will be illustrated using the scalar function $f(x)$ of a single variable x (Figure 9.9), but the same procedure applies to multivariable cases.

The global search for the minimum of $f(x)$ in the interval S_1 proceeds as follows. In each interval one seeks to establish an *upper bound* for the minimum, and a *lower bound* for the minimum. The more closely these can approach the actual minimum in an interval, the more rapid will be the solution. A good *upper bound* could be achieved by an ordinary minimisation in the interval such as in Section 9.5.3. Actually, it is not difficult to set an *upper bound* – any point x in the interval can be chosen and the upper bound set as $f(x)$. Taking this idea further, a small sophistication is shown in Figure 9.9 where the lowest of three points (left l, centre c and right r) is chosen. The *lower bound* is more difficult – how can one be certain that one is lower than every point on the $f(x)$ curve in the interval? The method shown in Figure 9.9 assumes that a *convex* function can be fitted quite closely, such that it is lower than $f(x)$ at all points within the interval, but equal to $f(x)$ at l and r. The *bounding* minimum m can then be found by ordinary minimisation (as in Section 9.5.3) because it is convex.

In the example, the first bounds fitted (for the minimum in the interval S_1) are U_1 and L_1. The interval is then divided into two equal parts to create new intervals S_2 and S_3. (In the multivariable case, it is recommended to divide the largest dimension of S into two parts.) Now first examine the side in which m_1 lies (namely S_3) – again fitting U_3 and L_3. In this case there is a reward that U_3 lies lower than L_2, so that the whole of interval S_2 can be eliminated from further consideration. The process is then repeated leading to U_4, L_4, U_5 and L_5. Although not shown in Figure 9.9, U_5 will be below L_4, leading to another elimination, and it is clear that the solution is rapidly converging to the global minimum.

It remains to discuss the problem of finding the minimum *lower bounding* (or 'underestimating') functions. Here the more general multivariable case will be considered, where the objective value is determined by $f(\boldsymbol{x})$, with $\boldsymbol{x}$ being a vector of independent variables $x_1, x_2, x_3, \ldots, x_n$. Some techniques are presented in the literature, most notably the αBB (alpha branch and bound) method of Adjiman *et al.* (1998a, 1998b). These authors note that $f(\boldsymbol{x})$ can typically be represented as the sum of terms of different properties affecting the convexity as follows:

$$f(\boldsymbol{x}) = \sum_{i=1}^{N_L} L_i(\boldsymbol{x}) + \sum_{i=1}^{N_C} C_i(\boldsymbol{x}) + \sum_{i=1}^{N_B} B_i(\boldsymbol{x}) + \sum_{i=1}^{N_T} T_i(\boldsymbol{x}) + \sum_{i=1}^{N_F} F_i(\boldsymbol{x}) + \cdots \tag{9.123}$$

where L is linear, for example $3x_2 + 5x_4 + \cdots$; C is convex, for example $2x_1^2 + 2x_2^2 + \cdots$; B is bilinear, for example $6x_1x_3 - 2x_2x_4 + \cdots$; T is trilinear, for example $x_4x_6x_7 + 4x_2x_3x_4 + \cdots$; F is fractional, for example $3(x_2/x_3) - 2(x_4/x) + \cdots$.

One method to obtain a *convex lower-bounding function* $\mathcal{L}(\boldsymbol{x})$ is to set

$$\mathcal{L}(\boldsymbol{x}) = \sum_{i=1}^{N_L} L_i(\boldsymbol{x}) + \sum_{i=1}^{N_C} C_i(\boldsymbol{x}) + \sum_{i=1}^{N_B} b_i + \sum_{i=1}^{N_T} t_i + \sum_{i=1}^{N_F} f_i + \cdots \tag{9.124}$$

with additional constraints specified on the unknown variables b_i, t_i, f_i, ... and so on. For example,

$$\text{if } B_1(\boldsymbol{x}) = x_1x_2, \text{ then require} \begin{cases} b_1 \geq x_1^L x_2 + x_2^L x_1 - x_1^L x_2^L \\ b_1 \geq x_1^U x_2 + x_2^U x_1 - x_1^U x_2^U \end{cases} \tag{9.125}$$

where the superscripts L and U refer to the upper and lower limits of the range of each variable in the interval S under consideration. The increasingly complex sets of constraints necessary for t_i, f_i and so on may be obtained from Adjiman *et al.* (1998a).

Adjiman *et al.* (1998a) also present an 'α method' to obtain $\mathcal{L}(\boldsymbol{x})$ as follows:

$$\mathcal{L}(\boldsymbol{x}) = f(\boldsymbol{x}) + \sum_{i=1}^{n} \alpha_i \left(x_i^L - x_i\right)\left(x_i^U - x_i\right) \tag{9.126}$$

where the α_i's are positive scalars. Since the product terms will all be negative, it is clear that $\mathcal{L}(\boldsymbol{x}) \leq f(\boldsymbol{x})$ throughout the interval. It is noted that the quadratic summation term itself is convex. Indeed, all non-convexities in the original function $f(\boldsymbol{x})$ can be 'overpowered' by appropriate choice of the α_i parameters. In this way, one obtains a convex $\mathcal{L}(\boldsymbol{x})$ for which the single minimum can be found, and ensure that this is less than the minimum $f(\boldsymbol{x})$ value in S, since $\mathcal{L}(\boldsymbol{x}) \leq f(\boldsymbol{x})$ here. However, to obtain a tight-fitting lower bound, one has to avoid large α_i's.

Noting Equation 9.83, the Hessian $\boldsymbol{\Lambda}$ of the summation term will be diagonal with the positive α_i's on its axis. Then the Hessian of $\mathcal{L}(\boldsymbol{x})$ is seen to be

$$\boldsymbol{H}_L(\boldsymbol{x}) = \boldsymbol{H}_f(\boldsymbol{x}) + \boldsymbol{\Lambda} \tag{9.127}$$

Now $\mathcal{L}(\boldsymbol{x})$ will be convex if and only if its Hessian is positive semidefinite in the interval. Therefore, the α_i's of the *diagonal shift matrix* $\boldsymbol{\Lambda}$ must be chosen so as to alter the Hessian of $f(\boldsymbol{x})$ to achieve this. Various procedures are given by Adjiman *et al.* (1998a, 1998b) to guide one to a suitable convex underestimating function. These include the possibility of a single uniform α value. Whereas the form (9.126) could be used directly to obtain $\mathcal{L}(\boldsymbol{x})$, they remark that there is a computational advantage in first splitting off the simpler parts according to (9.124) and (9.125). Further procedures are given to deal with nonlinear constraints.

9.7 Combinatorial Optimisation by Simulated Annealing

Here consideration will be given to problems in which combinations of only discrete choices are available to determine the extremum of an objective function. In Section 9.4, IP and MIP were described for linear systems. The technique based on *linear programming* which was outlined there allowed discretisation of the ranges of selected (MIP) or all (IP) independent variables. It was noted that the discretisation introduced the need for *branch and bound* steps in the solution. However, the method was only applied to a *linear system*.

Now consideration will be given to making only discrete choices which may affect an objective function in a *nonlinear* fashion. One immediately anticipates problems with local minima. However, the *simulated annealing* method outlined below will be seen to offer some protection against being trapped at a local extremum, and it often has a good chance of finding the global maximum or minimum.

The analogy with the annealing of metals arises as follows. As a liquid cools, the thermal mobility of molecules decreases. Slow cooling gives them enough time to randomly search out their minimum energy level, which ultimately is a large regular crystal. If the cooling is too rapid, a more amorphous structure results, and the overall energy is left at a higher level. The slow cooling of metals after casting similarly gives atoms an opportunity to align in a stronger structure. Under these random changes, the probability that a system will make an overall *positive* energy jump of ΔE is expressed by the Boltzmann probability:

$$\text{prob}(\Delta E) = e^{-\Delta E/kT} \tag{9.128}$$

Following Metropolis *et al.* (1953), the probability of *negative* overall energy changes ($\Delta E < 0$) is taken as unity, that is it *will* happen. In terms of the present search for the global minimum of an objective function $f(\boldsymbol{x})$, the overall energy E takes the place of the objective function value. Thus, if the present search has reached a point with 'energy' E_i, then 'neighbouring' points are randomly sampled to provide potential E_{i+1} values. For each neighbour tested, the transition is made according to $\text{prob}(E_{i+1} - E_i)$. This probability is compared with a number randomly drawn from a uniform 'top-hat' distribution in the 0–1 range. If the probability exceeds the drawn number, a move is made to the associated neighbouring point. Otherwise, another neighbour is created and tested in the same way.

The idea of a 'neighbouring' point is put forward in the sense that smaller random changes would help convergence, but would also hopefully be enough, in sequence, to lead away from a local minimum. For example, say that the objective was to optimise a sequence of 10 instructions, each of which could be either 0 or 1. Then a 'neighbour' might be considered to arise from changing any one instruction. Such views would depend very much on the particular problem.

One notes that the benefit of the method is that there is always a *chance* to move against the 'gradient' and thus to discover other minima and ultimately the global minimum. The probability distribution used needs to have the general characteristics of Equation 9.128, with an asymptotically decreasing probability of larger positive energy changes. Of course, it is possible to move away from a discovered minimum, and not discover another quite as low. Thus, a record of 'position' and 'value' must be maintained for the best minimum discovered so far. It is then sometimes useful to *restart* at such a position if the search progress later seems poor.

The solution is started with kT considerably larger than the typical ΔE's expected. As the solution proceeds, an *annealing schedule* is implemented in which the 'temperature' T is decreased progressively. In this way, the motion is eventually halted, hopefully in the vicinity of the minimum.

Example 9.3

Reservoir distribution problem: minimisation of an absolute deviation from setpoint by simulated annealing.

Water is to be distributed to two reservoirs by switching a supply flow of 12 $m^3\,h^{-1}$ to either or neither (Figure 9.10). Reservoir 1 has a setpoint of 50 m^3 and reservoir 2 has a setpoint of 100 m^3. The flow can only be switched at hourly intervals. The outflows are steady at 6 and 4 $m^3\,h^{-1}$, respectively. Use simulated annealing to establish an 'optimal' switching sequence over a period of 24 h. Use the average absolute setpoint deviation as the minimisation objective (i.e. the 'energy'). Start with both levels at their setpoints.

Set the objective function to

$$E = \sum_{i=1}^{24} \left|V_{1,i} - V_{SP1,i}\right| + \left|V_{2,i} - V_{SP2,i}\right| \tag{9.129}$$

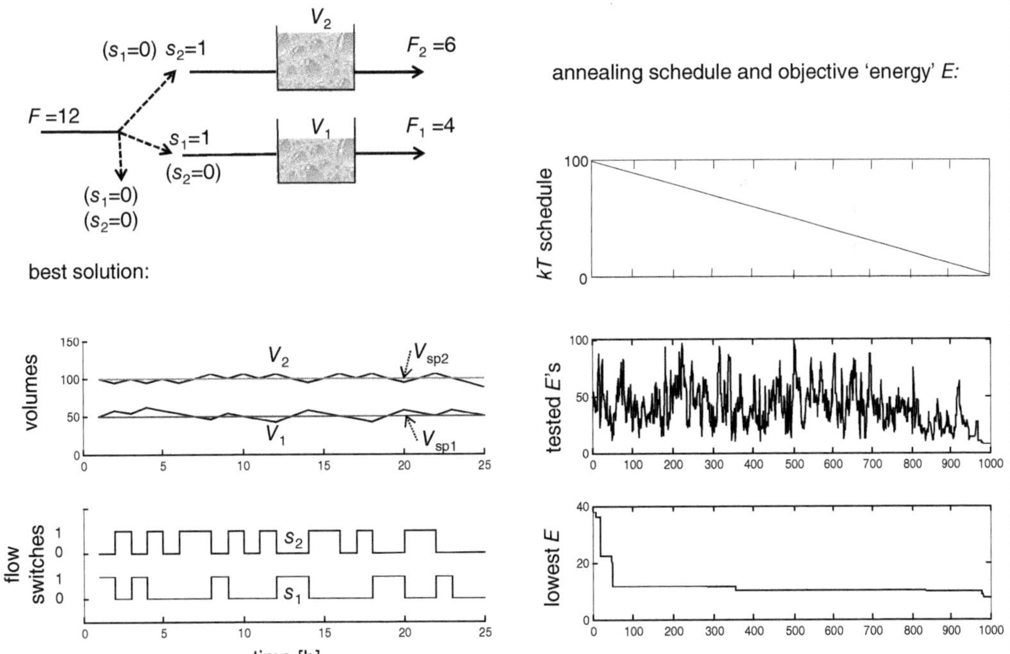

Figure 9.10 Flow switching to maintain reservoir levels – optimised by *simulated annealing*.

The volumes vary as

$$V_{1,i} = V_{1,i-1} + 12s_1 - 4 \quad (9.130)$$

$$V_{2,i} = V_{2,i-1} + 12s_2 - 6 \quad (9.131)$$

Starting with $kT = 100$, the chosen *annealing schedule* requires that kT is linearly reduced to zero over 1000 iterations. On each iteration, a random choice of one of the three switch positions is inserted into five randomly selected positions in the 24 h sequence. The resultant E of this sequence is first screened to see that it does not exceed 10× the minimum E so far. If it does, the algorithm *restarts* from the current *best E* and associated best switching sequence. If a restart is not required, the proposed E is next tested by obtaining its probability:

$$\Delta E = \text{proposed } E - \text{previous } E \quad (9.132)$$

$$\text{prob}(\Delta E) = e^{-\Delta E/kT} \quad (9.133)$$

If prob ≥ 1, this becomes the new E directly. Otherwise, the probability is tested against a number d randomly drawn in the 0–1 range. If prob > d, the proposed E nevertheless becomes the new E, otherwise not.

The effect of this algorithm is a fairly rapid convergence close to a minimum E (see Figure 9.10). The starting point (e.g. random, or $s_1 = s_2 = 0$ for all 24 times) has little effect on this. However, multiple solutions result, and these are all capable of getting close to the minimum, which appears to be just below 10.

9.8 Optimisation by Evolutionary Strategies

The simulated annealing algorithm discussed in Section 9.7 bears some parallels to the genetic algorithm which was applied to predictive control in Section 7.13. The random 'uphill' step, where allowed according to the Boltzmann distribution, replaces the 'mutation' step in genetic algorithms. The 'energy' of the single searching point in simulated annealing becomes the 'fitness' of multiple points in the genetic algorithm population. The genetic algorithm is just one example of an *evolutionary strategy*, that is a strategy based on improving a *population* of points at different positions in multidimensional space. It is this continuously updated *population* that gives evolutionary strategies their special ability to identify the interplay between multiple objectives via the *Pareto front*.

Genetic algorithms mimic the dynamics of an animal or a human population, using principles of natural selection (Darwin, 1859) and the *survival of the fittest*. With each step, the fraction of the population with useful genes increases whilst members with poor genes are eliminated. In this sense the process is an optimisation, so the equivalent mathematical algorithms have found wide application in this area. An advantage of these genetic search techniques is that the trial and error tests can be conducted *in parallel* as individuals move into and out of a common population pool. As mentioned, *several optima* can be monitored simultaneously, based on different functions of the attributes. As in simulated annealing, the list of attributes of interest could have *widely varying data types*, for example in the optimisation of a reactor design, that might be {temperature, lagging, hazardous area classification, number of pipe connections, plant structure} all in one optimisation problem.

The key to genetic algorithms is to keep on perturbing the attributes of individuals left in the population by the creation of children that have some of the attributes from each parent, and by randomly mutating some attributes in the individuals. In this way further points (individuals) on the optimisation surface are randomly created sharing good attributes (genes), and at various 'distances' from the original individuals. The *most fit* individuals in this population are then selected for the next step. As the population improves, the repeated randomisations reduce the danger of finding only a local optimum. As noted by Fleming and Purshouse (2002), evolutionary computing (EC) has its origins in four landmark evolutionary approaches: *evolutionary programming* (EP) (Fogel *et al.*, 1966), *evolution strategies* (ES) (Schwefel, 1965; Rechenberg, 1973), *genetic algorithms* (GA) (Holland, 1975) and *genetic programming* (GP) (Koza, 1992).

9.8.1 Reactor Design Example

To start a solution, a population of individuals is created with randomised attributes, for example one of the individuals in the reactor design example mentioned above might be

{240 C, unlagged, II, 8 connections, serial multiple bed}

The ranges of the attributes are as follows:

Temperature	Lagging	Hazard classification	Pipe connections	Structure
300 °C	Lagged	I	2	Serial multiple bed
↑				
continuous	Unlagged	II	6	Parallel multiple bed
↑				
160 °C		III	8	Single bed
			10	Fluidised bed
				Recirculating bed

Each individual in the population is assigned a *fitness* value on the basis of its attributes. In this case the *profit* (predicted earnings minus capital and operating cost over the lifespan) would be a good measure.

RECOMBINATION (reproduction, crossover):

{240 °C, unlagged, II, 8 conns, serial multiple bed} + {180 °C, unlagged, I, 4 conns, single bed}	→	{240 °C, unlagged, I, 4 conns, single-bed} + {180 °C, unlagged, II, 8 conns, serial multiple bed}

MUTATION:

{220 °C, lagged, III, 2 conns, fluidised bed}	→	{220 °C, lagged, III, 6 conns, fluidised bed}

In the *recombination*, the attributes that are swopped, or chosen in some other way from the parent population, are selected randomly from that population. Likewise, in the *mutation*, one or two attributes are changed by random selection from the attribute list. One notes that the temperature does not fit into this discrete pattern. Providing samples at, say, 20 °C intervals may not give enough resolution. A real number randomisation procedure is considered instead in the *evolutionary strategy* method that is used below. After these changes, the best individuals are selected to become the population on the next step, and the process is repeated. For a fixed population size, the worst individuals would be eliminated, in which case the strategy is termed *elitist*.

Because of the inherent parallelism in the evolutionary approach, it is possible to monitor different objective functions for each member of the population, that is formulate the problem in a *multiobjective* form. So for the reactor above, one could separately demand that both the *profit* and the *production rate* are maximised, retaining a proportion of the fittest individuals under these two separate criteria. Actually, this is not the best practice – Cîrciu and Leon (2010) point out that this has the disadvantage of effectively weighting the objectives according to the selection ratio. Methods such as the *NSGA-II* (fast non-dominated sorting genetic algorithm II) of Deb *et al.* (2002) avoid this problem – see Section 9.8.2 and Example 9.4. Ultimately, these approaches lead to a situation as in Figure 9.11 where the optimisation will move the solution to the edge of the dominated feasible region. The term 'dominated' implies that a better solution is available with respect to at least one of the objectives, for the *same* objective values of all of the other objectives.

A range of *Pareto optimal* (non-dominated) solutions then exist along this boundary for which any change to improve the one objective will detract from another. This is not the same as a *constraint* that one might like to impose on the search variables themselves. A typical method of dealing with such constraints (e.g. temperature <300 °C to protect the catalyst) is to drop the fitness sharply for any search point (attribute combinations) landing outside the constraint. There are also *repair algorithms* to allow re-insertion of such individuals within the feasible region, at some additional cost, that is loss of fitness (a bit like a stay in hospital).

Use of evolutionary optimisation has found many applications in the general area of dynamic process observation and control. These range from real-time searches for optimal PID controller

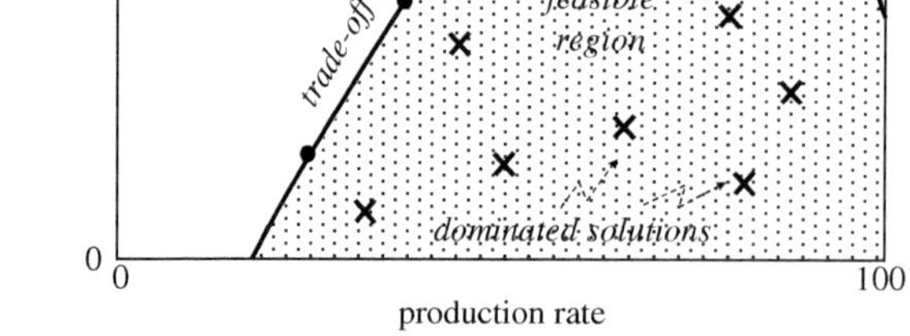

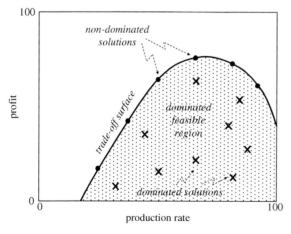

Figure 9.11 Non-dominated (*Pareto optimal*) solutions for a two-objective maximisation optimisation, following Fleming and Purshouse (2002).

parameter settings (Oliveira *et al.*, 1991) to adjusting the $\boldsymbol{Q}$ and $\boldsymbol{R}$ weighting matrix elements in a linear quadratic regulator (Section 7.6.5, Goggos and King, 1996). Kristinsson and Dumont (1992), among others, have used evolutionary algorithms in process model identification.

9.8.2 Non-dominated Sorting Genetic Algorithm (NSGA)

Figure 9.12 illustrates the non-dominated sorting algorithm of Deb *et al.* (2002) for a two-objective minimisation problem. On each step to improve the population, the first non-dominated front is stripped off to reveal a second non-dominated front and so on until all of the individuals in the population are accounted for. The order that these points occurred in the fronts then determines their overall fitness ranking in the population, that is the 'Sorted' list. Algorithmically, this is done as in Figure 9.13. For each individual *I* in the population, a list is made of all of the individuals *(B,E)* which it dominates (i.e. for which it is better in all objective values). Additionally, the total *number* of individuals which dominate *I* is recorded (2). Individuals for which the latter count is zero then form the first non-dominated front. This front is then stripped away, effectively by reducing the number count of each remaining individual, according to the number of dominating individuals removed. The dominating number count of these remaining individuals is then examined for zeros to form the next front and so on.

A selection of the better individuals towards the top of the Sorted list are retained in the population, and also offered the chance to propagate random variations of themselves. As a result, points on a multiobjective value plot gradually move onto a hypersurface defining the best simultaneously attainable objective values, that is the *Pareto front*. The methods used to alter the population are common in GA methods, and include *recombination* (*crossover*) and *mutation*. Problem 9.10 in the accompanying book of problems is a further example of the implementation of these techniques.

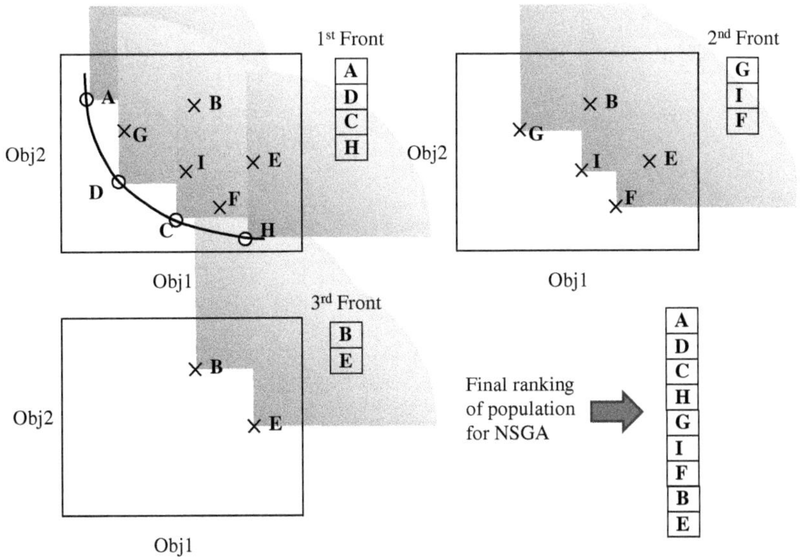

Figure 9.12 Non-dominated sorting sequence for a two-objective minimisation problem.

```
% Now create the non-dominated sorted list for NSGA
% ..supplied with objective values Jmult(i,j)
% for i=1:Nobj (number of objectives) & j=1:Npop (obj values for each)
H=zeros(Npop,Npop);  % H(j,k):List of poimts 'k' that point 'j' dominates
C=zeros(Npop,1);     % C(j):Count of number of points dominating 'j'
c=zeros(Npop,1);     % c(j):Count of number of points which 'j' dominates
F=zeros(Npop,Npop);  % F(j,k):Points 'k' in each front 'j'
cF=zeros(Npop,1);    % cF(jF):Count of number of points in each front jF
for j=1:Npop
    for k=1:Npop
        if k~=j
            if (sum(Jmult(:,k)<Jmult(:,j))==Nobj) % k doms j for all obj?
                C(j)=C(j)+1;
            else
                if (sum(Jmult(:,j)<Jmult(:,k))==Nobj) % j doms k for all obj?
                    c(j)=c(j)+1;
                    H(j,c(j))=k;
                end
            end
        end
    end
end
for jF=1:Npop
    Ctemp=C;
    for k=1:Npop
        if (C(k)==0)
            cF(jF)=cF(jF)+1;  % increment number of points in this front
            F(jF,cF(jF))=k;
            Ctemp(k)=-1;
            for kc=1:c(k)
                Ctemp(H(k,kc))=Ctemp(H(k,kc))-1;
            end
        end
    end
    C=Ctemp;  % ready for the next front
end
% Load the Sorted NSGA vector from F
j=0;
for jF=1:Npop
    if cF(jF)>0
        for kF=1:cF(jF)
            j=j+1;
            mJnsga(j)=F(jF,kF);  % original indices for Sort j=1:Npop
        end
    end
end
```

Figure 9.13 MATLAB® code for the non-dominated sorting algorithm (NSGA) of Deb *et al.* (2002).

Example 9.4

Reservoir distribution problem: genetic algorithm to be applied to the problem in Example 9.3, for both single and double objectives.

Refer to the problem detailed in Example 9.3. Now (a) apply a genetic algorithm to optimise the original objective, and (b) define the two separate volume controls as different objectives (setpoints

$V_{SP1} = 50$, $V_{SP2} = 100$). Using the NSGA (Deb *et al.*, 2002) algorithm, identify the associated Pareto curve. Check the location of the optimal point from (a) on this plot.

Following Example 9.3, set the objective functions to

$$J = \sum_{i=1}^{24} \left|V_{1,i} - V_{SP1,i}\right| + \left|V_{2,i} - V_{SP2,i}\right| \tag{9.134}$$

$$J_1 = \sum_{i=1}^{24} \left|V_{1,i} - V_{SP1,i}\right| \tag{9.135}$$

$$J_2 = \sum_{i=1}^{24} \left|V_{2,i} - V_{SP2,i}\right| \tag{9.136}$$

The volumes vary as

$$V_{1,i} = V_{1,i-1} + 12s_1 - 4 \tag{9.137}$$

$$V_{2,i} = V_{2,i-1} + 12s_2 - 6 \tag{9.138}$$

In the results given below, on each iteration, 20% of the switch settings (S1,S2) were randomly swopped between a parent pair (from the top-ranked half) to create the child pair (inserted into the bottom half). Additionally, 20% of the children were randomly selected for mutation, and of these, each child had 20% of its switch settings randomly changed. For the case (a), with a single objective function, the population rapidly converged to a single optimal point. Figure 9.14 shows an intermediate situation

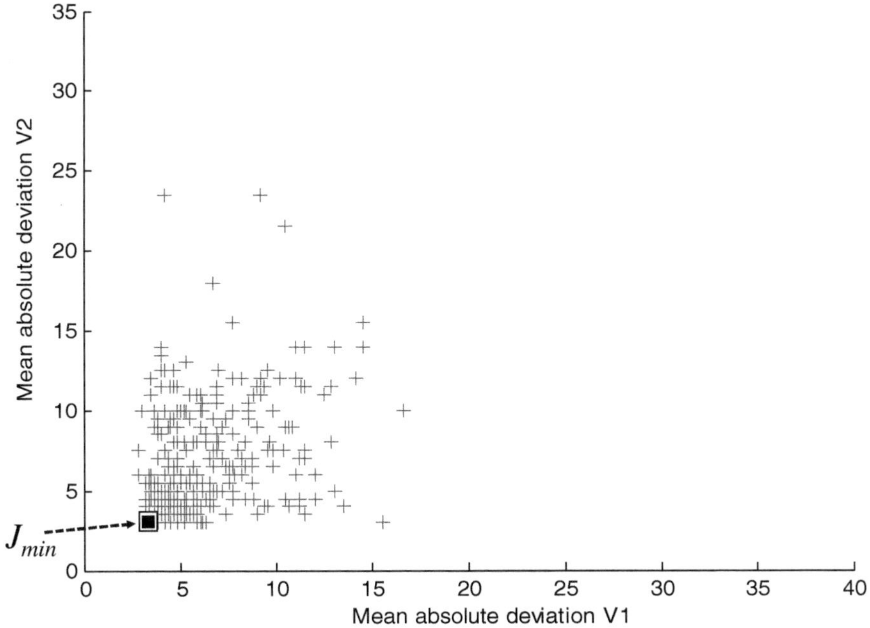

Figure 9.14 Population of 500 for the case (a) in Example 9.4, with a single minimisation objective function $J = J_1 + J_2$ after 50 iterations. The solution rapidly converges to the optimum point, shown as the square at the apex of the distribution.

for a population of 2000 after only 10 iterations. For comparison with the NSGA case, this is shown on axes for the two contributing objective values. The ultimate convergence point is the square symbol at the apex of the distribution.

Conversely, convergence in the case of the NSGA solution is much slower. For the case of (b), a population of just 300 underwent 8000 iterations. Figure 9.15 shows how the non-dominated solutions (circles) have moved onto a distinct Pareto front. In Figure 9.15a, the original adequate water supply of

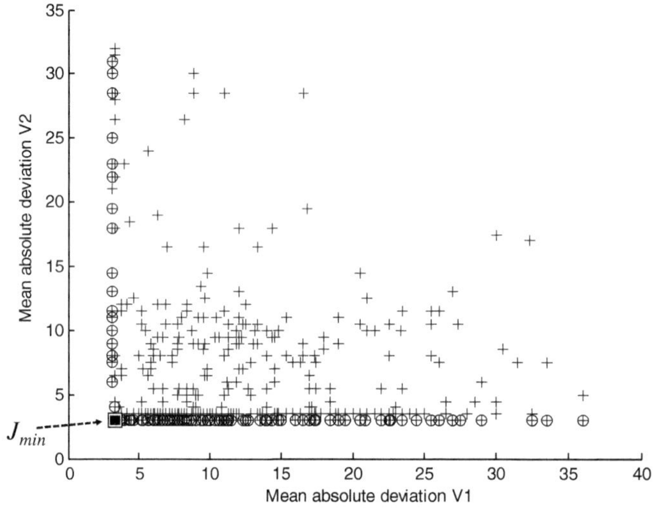

(a) Two-objective case (b) of Example 9.4 with adequate water supply of 12 $m^3 h^{-1}$.

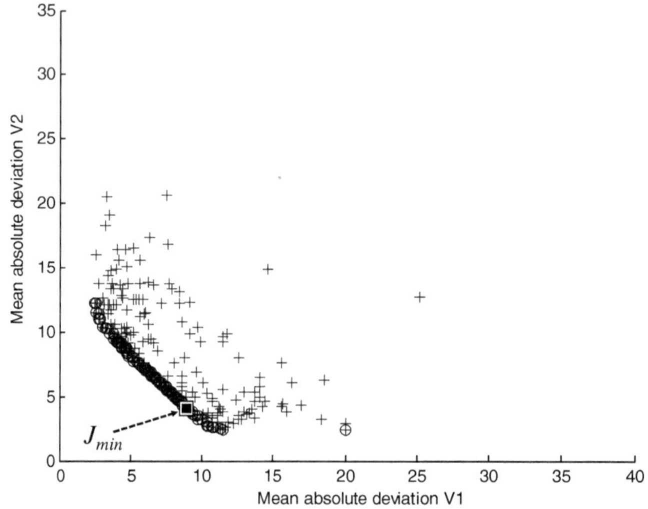

(b) Two-objective case (b) of Example 9.4 with inadequate water supply of 9 $m^3 h^{-1}$.

Figure 9.15 Population of 500 for the two-objective minimization case (b) of Example 9.4 after 50 iterations, using the non-dominated sorting genetic algorithm (NSGA). Non-dominated solutions are shown as circular symbols, and the minimum value of $J = J_1 + J_2$ is represented by a square near the 'ideal point' (apex).

$12\,m^3\,h^{-1}$ is provided, resulting in a negligible region of mutual compromise at the apex of the front. The inability of J_1 and J_2 to get below about 3 anywhere arises from the overshoot/undershoot due to the 1 h switch periods. Conversely, in Figure 9.15b, the supply has been reduced to $9\,m^3\,h^{-1}$, below the required $10\,m^3\,h^{-1}$, giving an extensive region of compromise. The *II* in the *NSGA-II* algorithm (Deb *et al.*, 2002) represents an additional method for spreading out points on the Pareto front in case they are too clustered – not used in this case. Again, a point representing the minimum value of $J = J_1 + J_2$ is shown as a square on this plot – seen to be near the so-called *ideal point* of the Pareto front. The switching sequences and reservoir volume behaviour associated with this point are shown in Figure 9.16.

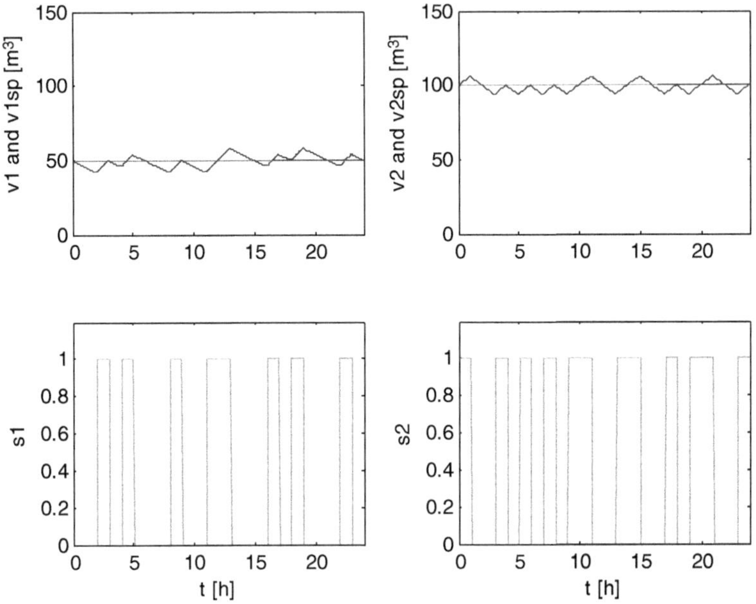

Figure 9.16 System behaviour with an adequate water supply of $12\,m^3\,h^{-1}$ for the point representing the minimum value of $J = J_1 + J_2$ in the case (b) of Example 9.4 – plotted as the square symbol near the 'ideal point' in Figure 9.15a.

9.9 Mixed Integer Nonlinear Programming

In Section 9.4 it was noted that *linear programming* could be used within a branch and bound algorithm to deal with discrete (integer) variables that might be present, *provided the system is linear.* This type of algorithm was referred to as mixed integer linear programming (MILP). The simulated annealing mentioned in Section 9.7 and evolutionary algorithms of Section 9.8 proved capable of handling *discrete* choices in a nonlinear system. In this section, the general problem of a hybrid (continuous-discrete) nonlinear system will be addressed. This type of problem is very common in the processing industries, whether it be in the design phase (e.g. 'pinch'), in production control and optimisation (discrete inputs or behaviour modes – Section 7.14.3) or in product blending and distribution. The solving algorithms are referred to as *mixed integer nonlinear programming.*

The following discussion is based on Grossmann (2002) and Grossmann and Kravanja (1995). The general problem is

$$\text{minimise}\, Z = f(\boldsymbol{x}, \boldsymbol{y}) \tag{9.139}$$

$$\text{such that}\, g_j(\boldsymbol{x}, \boldsymbol{y}) \leq 0, \quad j = 1, 2, \ldots, J \tag{9.140}$$

The $\boldsymbol{x}$ and $\boldsymbol{y}$ are vectors of continuous and discrete variables respectively, and the region is limited by J constraints as shown. The focus will be on the more common problem where the functions f and $\boldsymbol{g}$ are convex and differentiable with respect to $\boldsymbol{x}$. Very often the constraints $\boldsymbol{g}$ are linear, f and $\boldsymbol{g}$ are linear in $\boldsymbol{y}$, and the elements of $\boldsymbol{y}$ are limited to binary (0–1) values.

9.9.1
Branch and Bound Method

As in Section 9.4, a series of 'relaxed' problems is first solved in which some of the integer variables are fixed, and the remaining integer variables are left free as continuous variables between set bounds.

This series of semirelaxed problems is thus based on $k = 0,1,2, \ldots$ different constraint sets acting on the relaxed continuous version $\boldsymbol{y}'$ of $\boldsymbol{y}$. These constraints, where applied, refer to the 'limits' of the ranges of the $\boldsymbol{y}$ elements ($y_{\mathrm{L}i} = 0$, $y_{\mathrm{U}i} = 1$ for binary).

NLP solution:

$$\text{minimise}\, Z_{\mathrm{LB}}^k = f(\boldsymbol{x}, \boldsymbol{y}') \tag{9.141}$$

$$\text{such that}\, g_j(\boldsymbol{x}, \boldsymbol{y}') \leq 0, \; j = 1, 2, \ldots, J \tag{9.142}$$

$$\boldsymbol{y}_{\mathrm{L}} \leq \boldsymbol{y}' \leq \boldsymbol{y}_{\mathrm{U}} \tag{9.143}$$

$$y_i' = y_{\mathrm{L}_i}, \quad \text{for selection } \boldsymbol{I}_{\mathrm{L}}^k \text{ of } i \tag{9.144}$$

$$y_i' = y_{\mathrm{U}_i} \quad \text{for selection } \boldsymbol{I}_{\mathrm{U}}^k \text{ of } i \tag{9.145}$$

The selections $\boldsymbol{I}_{\mathrm{L}}^0$ and $\boldsymbol{I}_{\mathrm{U}}^0$ will be empty sets, so the result Z_{LB}^0 will be the lowest 'lower bound' ever found in the solution. It might happen that this solution causes all y_i' to land on their upper or lower constraints, in which case the problem is immediately solved and no further work is required. In general, however, some y_i' will land at intermediate values, and the sequence of problems $k = 1,2, \ldots$ is then tackled. It starts by constraining one of the intermediates y_i' at y_{L_i} or y_{U_i}. The following NLP solution may leave other intermediates y_i', which are constrained in turn until all y_i' are on their constraints. At this stage that branch is ended, and its Z_{LB}^k value noted. This *fathoming* needs to be continued until all of the possible branches (where y_{L_i} or y_{U_i} was selected) are *sounded*. In the process, immediately the Z_{LB}^k of a branch exceeds that of an already-completed branch, it is abandoned. (The discussion has focussed on binary integer variables – clearly the other branches would need to be tested for integer variables with more than two values.)

9.9.2
Outer Approximation Method (OA)

This method occurs in two phases, repeated for $k=1,2,3,\ldots,K$ different constraint sets which include *all* integer variables. The solution at $k=K$ is based on information obtained for all preceding k values. It starts with a particular choice of the integer settings $\boldsymbol{y}^1$:

a) *NLP solution*

$$\text{minimise } Z^k = f\left(\boldsymbol{x}, \boldsymbol{y}^k\right) \tag{9.146}$$

$$\text{such that } g_j\left(\boldsymbol{x}, \boldsymbol{y}^k\right) \leq 0, \quad j = 1, 2, \ldots, J \tag{9.147}$$

Only the continuous $\boldsymbol{x}$ space is searched for this minimum, yielding a point $\boldsymbol{x}^k$ corresponding to the kth $\boldsymbol{y}$ constraint set. The previous upper bound becomes Z^k_{UB} unless Z^k is less, in which case Z^k becomes Z^k_{UB}.

b) *MIP master problem solution*

This is based on linearisations about point $(\boldsymbol{x}^k, \boldsymbol{y}^k)$ of the objective function f as well as the constraint functions $\boldsymbol{g}$. If these functions are all convex, they will provide a *lower bound* (in the case of f) and an *outer approximation* of the nonlinear feasible region (in the case of $\boldsymbol{g}$), as illustrated in Figure 9.17.

The minimisation problem is now solved about this point, allowing $\boldsymbol{x}$ to vary continuously within constraints, but $\boldsymbol{y}$ to assume only its allowed integer values:

$$\text{minimise } Z^k_{\text{LB}} = \alpha \tag{9.148}$$

$$\text{such that } \alpha \leq Z^k_{\text{UB}} - \varepsilon \tag{9.149}$$

and

$$\alpha \geq f\left(\boldsymbol{x}^i, \boldsymbol{y}^i\right) + \left[\nabla f(\boldsymbol{x}, \boldsymbol{y})|_i\right]^T \begin{pmatrix} \boldsymbol{x} - \boldsymbol{x}^i \\ \boldsymbol{y} - \boldsymbol{y}^i \end{pmatrix} \tag{9.150}$$

$$\boldsymbol{g}\left(\boldsymbol{x}^i, \boldsymbol{y}^i\right) + \left[\nabla \boldsymbol{g}(\boldsymbol{x}, \boldsymbol{y})|_i\right]^T \begin{pmatrix} \boldsymbol{x} - \boldsymbol{x}^i \\ \boldsymbol{y} - \boldsymbol{y}^i \end{pmatrix} \leq \boldsymbol{0} \tag{9.151}$$

for $i = 1, 2, 3 \ldots, k$.

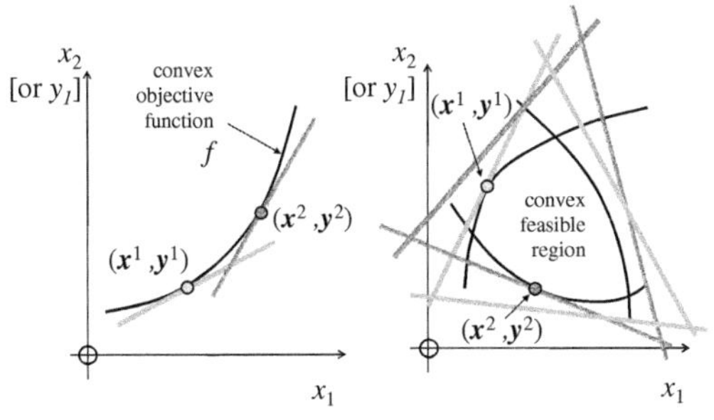

Figure 9.17 Linear approximation of convex objective function and convex constraints following Grossmann (2002) and Grossmann and Kravanja (1995).

Notice that the linearised constraints are accumulated as the solution progresses, giving the type of slightly shifting accumulation of bounds seen in Figure 9.17. These multidimensional bound surfaces are referred to as *cutting planes*.

In practice, one just needs to find a point giving an objective value only *tolerance* ε less than the current upper bound (lowest so far from Equation 9.147). The result of this simple *mixed integer programming* problem is the next integer selection y^{k+1} (the $\boldsymbol{x}$ result is discarded). The solutions (9.146)–(9.147) and (9.148)–(9.151) are repeated until Z_{UB}^k and Z_{LB}^k converge to within a given tolerance.

9.9.3 Comparison of Other Methods

The generalised benders decomposition (GBD) method is similar to the outer approximation method of Section 9.9.2. In the MIP solution, $\boldsymbol{x}$ is fixed at $\boldsymbol{x}^k$, so only the $\boldsymbol{y}$ are searched for their optimal integer values. In addition, constraints in $\boldsymbol{g}$ which did not prove 'active' in the NLP step (in which $\boldsymbol{x}^k$ and $\boldsymbol{y}^k$ are found) are omitted from the MIP. That is to say that only constraints giving $g_j(\boldsymbol{x}^k, \boldsymbol{y}^k) = 0$ are included.

The extended cutting plane (ECP) method does not rely on the solution of NLP subproblems, rather only on an iterative solution of the MIP problem by successively adding the most violated constraint at the predicted point $(\boldsymbol{x}^k, \boldsymbol{y}^k)$.

The choice of algorithm will depend on the particular structure of a problem. The branch and bound method in Section 9.9.1 is attractive only if there are relatively few NLP problems or they are easy to solve. The outer approximation method of Section 9.9.2 generally requires few iterations, but the accumulation of constraints can be problematic. The generalised benders decomposition method requires more cycles than OA, but generally has comparable performance.

9.10 The GAMS® Modelling Environment

A popular means for accessing optimisation programs is through the general algebraic modelling system (GAMS®). The makers of this system have striven to allow a standard problem specification. One then just selects an actual calculation method from a library to which new algorithms and updates are continuously added. Supported problem types include the following:

LP	Linear programming
MIP	Mixed integer programming
NLP	Nonlinear programming
MCP	Mixed complementarity problems
MPEC	Mathematical programs with equilibrium constraints
CNS	Constrained nonlinear systems
DNLP	Nonlinear programming with discontinuous derivatives
MINLP	Mixed integer nonlinear programming
QCP	Quadratically constrained programs
MIQCP	Mixed integer quadratically constrained programs

Table 9.1 Elements of a GAMS® model (Brooke et al., 1988).

Inputs	Outputs
• *Sets* Declaration Assignment of members • *Data* (Parameters, tables, scalars) Declaration Assignment of values • *Variables* Declaration Assignment of type • *Assignment of bounds and/or initial values* (optional) • *Equations* Declaration Definition • *Model* statement • *Solve* statement • *Display* statement (optional)	• Echo print • Reference maps • Equation listings • Status reports • Results

Depending on one's subscription, the methods of several different authors can be available for the solution of any one of these (e.g. ANTIGONE, BARON, CONOPT, CPLEX, DICOPT, GUROBI, and MINOS). These methods are set up in GAMS® with good default parameters, avoiding the need for a deeper knowledge of the principles (however, it is occasionally necessary to access these unique adjustments to get best results).

Example 9.5

Reservoir distribution problem – minimisation of an absolute deviation from setpoint by MIP in GAMS®.

Solve the switched reservoir distribution problem (introduced in Example 9.3) using MIP in the GAMS® environment.

The optimal solution is presented in Figure 9.18. The objective function was constructed as the average absolute deviation from setpoint for reservoir 1, added to that of reservoir 2. The optimal solution by MIP in GAMS® gave an objective value of 5.67 m^3, compared to just under 10.0 m^3 in a typical *simulated annealing* solutions (Figure 9.10), and 6.33 m^3 by the *genetic algorithm* (Figure 9.15). The GAMS® model is in Figure 9.18.

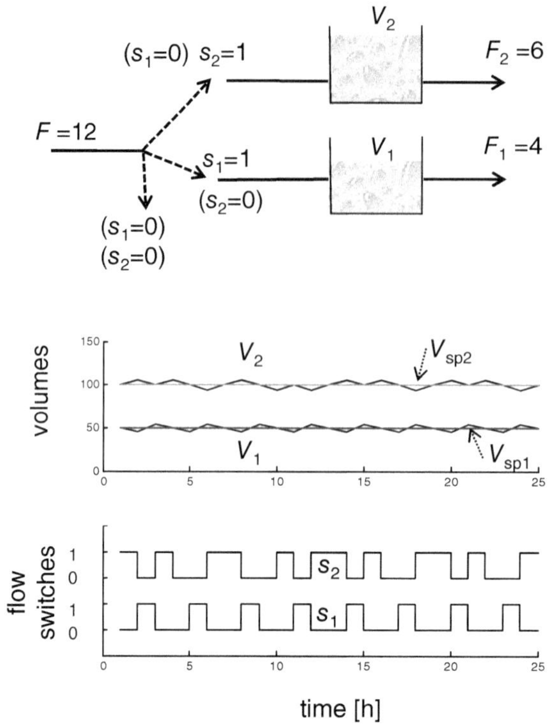

Figure 9.18 Flow switching to maintain reservoir levels – optimised by *MIP* in *GAMS®*.

Following Brooke *et al.* (1988), a GAMS® 'model' consists of the elements listed in Table 9.1. As part of Example 9.5 (Figure 9.18), the associated GAMS® model is provided in Figure 9.19 for MIP solution of the same reservoir distribution problem introduced in Example 9.3 (where it was solved by *simulated annealing*). The modelling environment is based on the structure of a *relational database*. Sets are used to imply relationships between groups of items, thus avoiding the need for 'DO' loops. It is important to distinguish between *parameters* and *variables*, and likewise between *assignments* and *constraints* (called *equations*). There are many useful features in GAMS which cannot be covered in a simple example. However, the GAMS® model in Figure 9.19 will now be discussed in some detail as a basic introduction:

Line 2	on/off upper controls the uppercasing of input lines when echoed to the output (list) file
Line 4	Equivalent to r = {reservoir-A, reservoir-B}. The text names of the set members are here given explicitly in order
Line 5	Implicit creation of a range of text names for set members t = {h1, h2, h3, . . . , h24}
Line 7	Parameter 'vector': both a declaration of a parameter corresponding to each member of r and a setting of values in vertical list format that does not require commas
	As line 7, except horizontal list format (with commas) used to set values
Lines 12–14	Parameter 'vector' values will be set later

```
$title Reservoir Distribution Problem
$offupper

Sets
        r      reservoirs      /reservoir-A, reservoir-B/
        t      time in hours   /h1*h24/;

Parameters
        fout(r)   reservoir outflow
                  /reservoir-A 4
                  reservoir-B 6/
        Vset(r)   reservoir volume setpoint /reservoir-A 50, reservoir-B 100/
        Vmax(r)   reservoir overflow level /reservoir-A 120, reservoir-B 150/
        Vstart(r) starting volume
        t_count(t) index of a given time
        t_first(t) boolean indicator of first time
        Scalar    fin re-directable inflow /12/;
* .......... load indicator of first time
        t_count(t)=ord(t);
        t_first(t)=yes$(t_count(t)=1);
        Vstart(r)=Vset(r);

Variables
        V(r,t)      volumes in the reservoirs at the end of each hour
        s(r,t)      flow switch to reservoir
        posdev(r,t) positive deviations from setpoint
        negdev(r,t) negative deviations from setpoint
        z           average of sum of absolute level deviations from setpoint
        Positive  Variable V
        Positive  Variable posdev
        Positive  Variable negdev
        Binary    Variable s;

Equations
        Vupdate(r,t)   integration of net flow to get volume change
        Max_V(r,t)     overflow limit
        Switch(t)      only one reservoir at a time
        Dev(r,t)       loads the positive or negative deviations from setpoint
        Cost           objective value calculation;

        Vupdate(r,t).. V(r,t) =e= Vstart(r)$t_first(t) + V(r,t-1)$(not t_first(t)) + s(r,t)*fin - fout(r);

        Max_V(r,t)..   V(r,t) =l= Vmax(r);

        Switch(t)..    sum(r,s(r,t)) =l= 1;

        Dev(r,t)..     posdev(r,t)-negdev(r,t)=e= V(r,t)-Vset(r);

        Cost..         z =e= sum((r,t),posdev(r,t)+negdev(r,t))/card(t);

Model   Reservoirs /all/;

Solve   Reservoirs using mip minimizing z;

Display s.l,V.l,z.l;
```

Figure 9.19 *GAMS®* model for *MIP* solution of the switched reservoir distribution problem (Example 9.5).

Line 15	Scalar value setting does not need an element name
Line 17	ord(t) is a set of numbers corresponding to the positions in the t set (needed to set up initial value in the integration)
Line 18	t_first(t) = yes$(t_count(t) = 1) – this loads up a new vector corresponding to the set t, which indicates the first element of t with a Boolean '1'. The yes$(...) returns 1 if the proposition in the bracket is true, else 0 (needed to set up initial value in the integration)
Line 19	Notice that all such assignments apply to all members of the set(s) – that is they are effectively 'DO' loops. Higher dimensional 'arrays', for example Vset(r,t), which might be the desired future setpoint trajectory for the two volumes, could be set as follows:

Table Vset(r,t) volume setpoint trajectories

	h1	h2	h3	. . .	h24
reservoir-A	50	52	54	. . .	96
reservoir-B	100	97	94	. . .	31

Line 21–25	These are the free variables which the solver is able to assign values to in pursuit of the solution. The default is *real* and unbounded, but further type specifications follow which will alter this
Line 23–24	The creation of the 'extra' arrays posdev(r,t) and negdev(r,t) is part of a strategy to avoid use of the abs(...) function in GAMS®, thus preserving linearity (see below)
Line 25	z is a scalar variable required to store the objective function value
Line 26–29	These declarations of type impose the associated constraints on the predeclared variables
Line 31–35	Here the 'names' of the forthcoming *equations* are being declared. Notice that equations implied for a whole set require an associated 'name array' indicating the *domain*
Line 36–40	The *equations* are seen to be the *constraints* which must be complied with. The relational operators used are: =e= equivalence =g= greater than =l= less than
Line 36	Integration to update the volume: A *linear lag operator* ('-') is used to relate a volume to its value for the preceding set position (also available are *linear lead* and *circular* operators). If the item does not exist, it is taken as zero. In this case it is necessary to initialise the starting volume – hence the 'if' conditions $(...). Thus, the Vstart value is used only if the Boolean t_first(t) indicates that the element of t being considered is the first. Conversely, V(r,t-1)$(not t_first(t)) ensures that the normal integration will proceed

Line 38:	Ensures that the flow can only be directed to one reservoir at a time, since s is Boolean. Notice that the first item in the summation bracket sum(r,...) indicates the domain over which the summation is to occur.
Line 39–40:	This is the strategy used to minimise the sum of absolute deviations from setpoint. The arrays posdev(r,t) and negdev(r,t) are constrained to zero and above. Negative deviations are forced to locate in negdev(r,t) (with associated posdev(r,t) element zero), whilst positive deviations are forced to locate in posdev(r,t) (with associated negdev(r,t) element zero). This is attributed to the simultaneous minimisation of posdev(r,t) + negdev(r,t). It is in fact possible to use the abs(...) function available in GAMS® instead. However, this makes the problem nonlinear. That would require a *minlp* solution instead of a *mip* solution. When this was done, using the default settings in CPLEX/DICOPT, the solution terminated at a poor 'local' minimum giving a cost much greater than the 10 achieved by *simulated annealing* (Example 9.3), and the 5.67 achieved in the *mip* solution (Example 9.5).
Line 41	A particular version 'Reservoirs' of the *model* is declared ahead of the solution. Within the associated '/ . . . /', one lists the individual *equation* names which one desires the solution to comply with. Thus, alternative equations could be declared and only selectively used. In this example, one desires to use of all the declared equations
Line 42	This is the instruction for the solution of the particular 'model' Reservoirs to begin. The type of solution procedure to use is given (e.g. mip, minlp), as well as the goal of the optimisation (minimizing or maximizing a particular variable)
Line 43	Request for display output variables which will appear at the end of the output *list*. The '.l' refers to the final 'level' of the variable. Other options include '.m' (marginal value) and '.lo' '.up' (bounds)

9.11 Real-Time Optimisation of Whole Plants

One has now arrived at the top of the process control pyramid represented in Figure 1.5. The kind of strategies that are implemented here will be quite varied and enterprise dependent. Objectives might include energy minimisation, profit maximisation, coordination of production/quality with other plants, or integration into market demands and distribution (Kleinschrodt and Jones, 1996). Figure 9.20 gives an idea of how the scope of 'process control' has expanded with time, spurred on by the advent of fast and reliable computers.

In this section, the focus will be on the type of strategies that are located above the 'advanced process control' layer, and which will pass down setpoints to the lower control layers. These are plant-wide optimisers which generally have a much slower update rate (e.g. hourly) than the dynamic controllers. The field is referred to as *real-time optimisation (RTO)*. The important considerations will be seen to include *steady-state* versus *dynamic* assumptions, *data integrity, model basis, optimisation calculation* and *setpoint implementation*. These aspects will vary a lot from

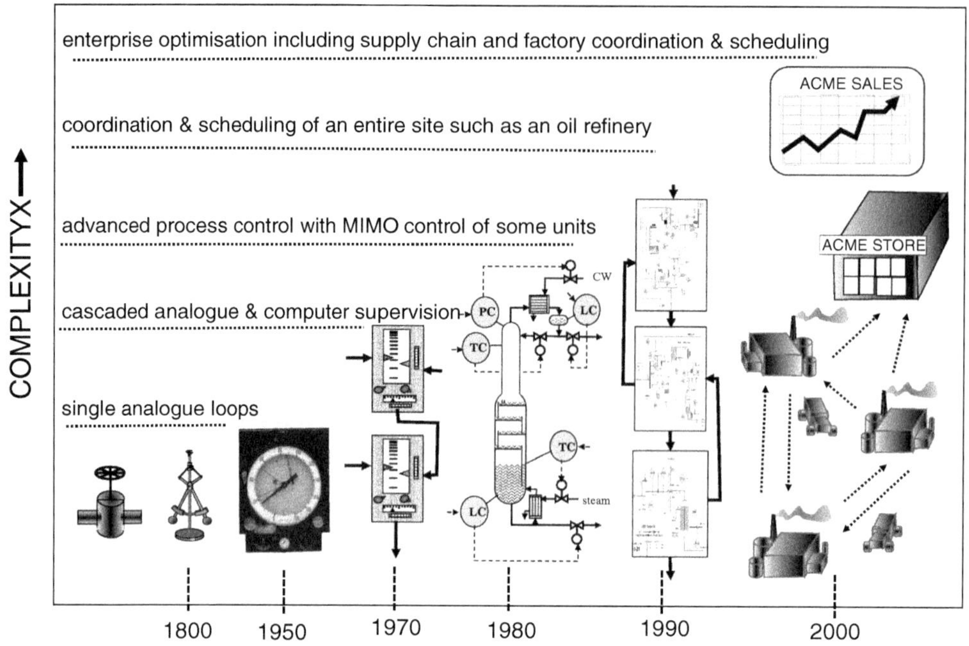

Figure 9.20 Automation trends in the processing industry.

case-to-case, but it is worthwhile noting the general structure of such systems. Figure 9.21 is a typical architecture of the RTO loop adapted from Shokri *et al.* (2009).

Unlike the normal dynamic control functions (regulation) of a plant, where a degree of measurement error usually has little effect, the plant optimiser has to be based on *absolute* representations of the variables in order to truly optimise, say, economically. For example, the difference between the process product value and the process feed value would be sensitive to errors in these flow rates. In most applications a steady-state plant model is used. Often this is an accurate steady-state flowsheet model of the plant which has been fitted to current measurements (e.g. Aspen®, SimSci-Promote®). In the data validation step (gross error detection, data reconciliation – see the aspects already discussed in Section 6.3), it is then necessary to either smooth the plant measurements or to wait for a reasonably steady period. Examples of data reconciliation tools associated with flowsheet models are Honeywell's UNISIM® and SimSci's Datacon® and Sigmafine®. On the other hand, Tosukhowong *et al.* (2004) note that slower responding parts of a plant (e.g. intermediate storages, large inventories) might always be problematic in this regard, so they present a method for a low-frequency dynamic reconciliation that can handle a degree of transient behaviour. Again, the recursive techniques of Kalman filtering (Sections 6.4.1–6.4.3) and model identification (Section 6.5) could be useful if an unsteady-state plant model is available.

Ignoring for the meantime that the plant model might be inaccurate, disagreements of the plant measurements with the model are usually viewed to arise for two distinct reasons (Figure 9.22). Firstly, one can expect *random fluctuations* in measurement signals to result from such things as electronic interference, liquid level disturbances, vibrations and so on. These will be temporary deviations from a mean and can be dealt with by smoothing (longer time constants in the estimators). On the other hand, errors can be long term ('systematic' or 'bias'), arising from instrument

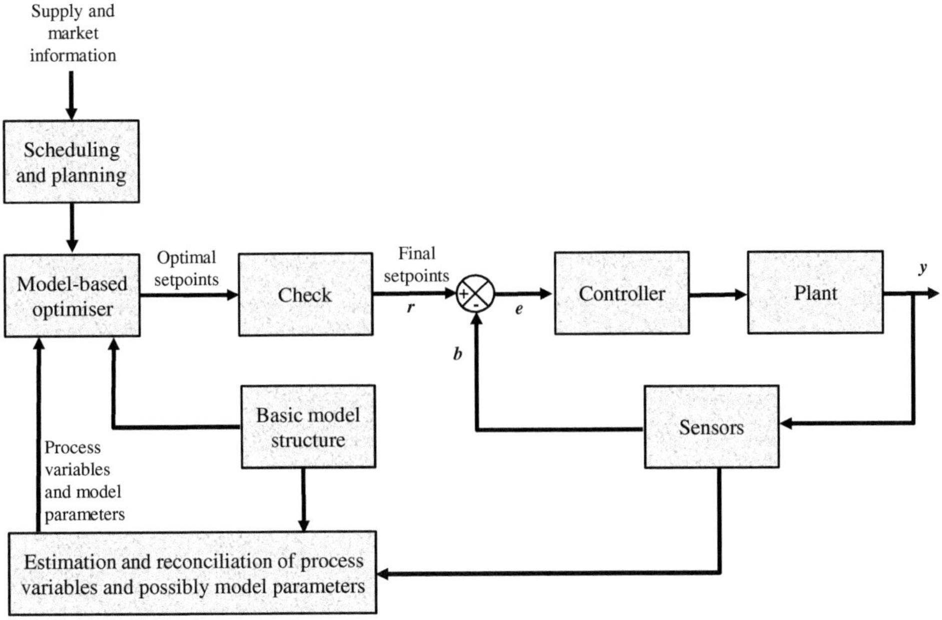

Figure 9.21 Typical architecture of *real-time optimiser* implementation.

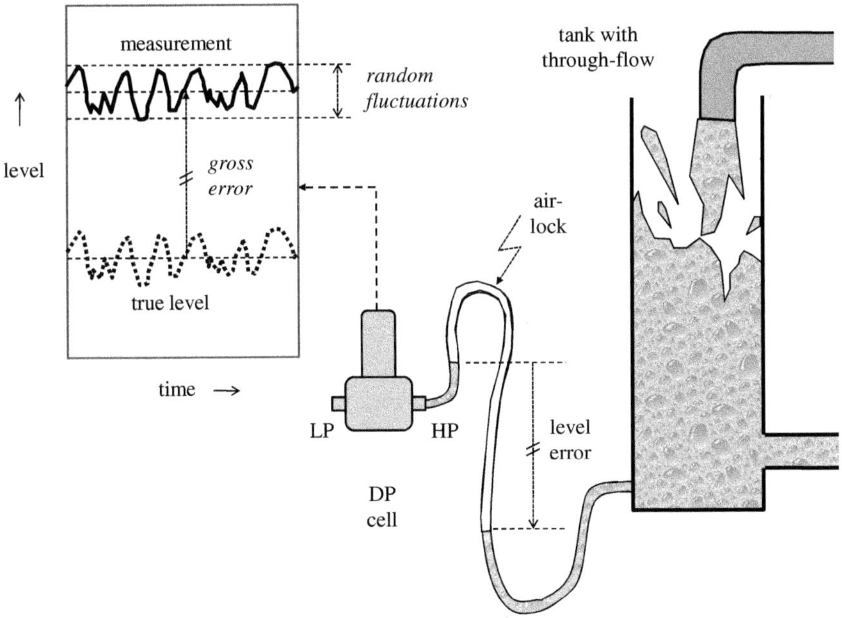

Figure 9.22 Gross versus random error in the measurement of a tank level.

miscalibration/offset, wear, air locks and so on. These are called *gross errors*, and it is important that they are isolated and removed so that they do not cause poor fitting to the remaining measurements in the reconciliation process. A typical reconciliation procedure involves a least-squares fit to find data that are consistent within the model, yet fit in a least-squares sense to measurements at the points where they are available. These individual fittings are weighted according to the confidence one has in each instrument – so these confidence settings would be the first defence to stop a known bad measurement from 'hijacking' the overall fit. Another means of gross error detection would be to locate isolated measurements which, if omitted, would allow a good fit of the remaining measurements. Much work has been done in this general area (e.g. Tjao and Biegler, 1991).

Within the model structure will be unmeasured variables which are merely *implied* by the model in relation to the available measurements, in the same way as unmeasured states can be estimated by a Kalman filter. These can of course be estimated in the fitting process. Indeed, it goes further in the sense that equipment parameters such as heat transfer coefficients or catalyst activities can be adjusted in the same way as unmeasured F,P or T to achieve a better overall fit. Hence, one can view the data validation step to be potentially altering the *model* as well, as in Figure 9.21.

Zhang *et al.* (1995) remark that the general RTO system requires the solution of three *nonlinear programming* (NLP) problems:

1) The *gross error detection/data reconciliation* step is seen as a single problem, based, say, on the mass and heat balance.
2) *Parameter estimation* is seen as a second NLP task, using the results of (1).
3) The results of (1) and (2) then provide a model-consistent view of the operation which is the basis of a third NLP task for *optimisation* of the plant operation.

The computational burden can be substantial, but fortunately, as seen, plant optimisers need to allow time for transients to settle in order to get a clear picture of the process, and thus they only update at large time intervals such as 1 h. The major commercial flowsheet simulators often have associated optimisation tools for (3) above (e.g. SimSci-Promote's SPOT and OPERA suites for Shell). At the crudest level, after combining the economic data with the model, gradient information for the NLP solution could be obtained by perturbation of the variables. A range of approaches is evident in the literature, such as local linearisation at each new operating point in order to find the next optimal point.

Diaz and Bandoni (1996) include binary decision variables in their real-time optimisation of an ethylene plant, requiring *mixed integer nonlinear programming* for solution of (3). The decision variables allowed structural and mode changes to be included in the optimisation, such as switching from steam turbine drives to electrical motor drives. It is worthwhile examining the Diaz and Bandoni (1996) formulation in some detail as a general approach. It will be seen to parallel the MINLP *outer approximation* method introduced in Section 9.9.2.

Recall the general MINLP problem of Equations 9.139–9.140:

$$\text{minimise}\, Z = f(\boldsymbol{x}, \boldsymbol{y}) \tag{9.152}$$

$$\text{such that}\, \boldsymbol{g}(\boldsymbol{x}, \boldsymbol{y}) \leq \mathbf{0} \tag{9.153}$$

where $\boldsymbol{y}$ contains the integer/binary process variables and $\boldsymbol{x}$ contains the continuous process variables. The vector $\boldsymbol{g}$ contains the set of constraint functions relating the elements in $\boldsymbol{x}$ and $\boldsymbol{y}$. (View this as including equivalence constraints also.) Available algorithms usually require that the integer variables only occur as linear terms in f and $\boldsymbol{g}$ – certainly this will simplify the implementation of the OA algorithm in Section 9.9.2. In order to handle the substitution of different functions of the $\boldsymbol{x}$ (e.g. different process items) based on binary (logical) variables, the propositional logic is converted into linear inequalities as in the *mixed logical dynamical* approach in Section 7.14.3 (Williams, 1993). Typically, it is required that

$$g_1(\boldsymbol{x}, y_1) = y_1 h_1(\boldsymbol{x}) + [1 - y_1] h_2(\boldsymbol{x}) - h_3(\boldsymbol{x}) \tag{9.154}$$

$$g_1(\boldsymbol{x}, y_1) = 0 \tag{9.155}$$

However, both y_1 instances occur in nonlinear combination with the h functions. The same result can be achieved by enforcing the four constraints:

$$(m_2 - M_1)y_1 + h_3(\boldsymbol{x}) \leq h_2(\boldsymbol{x}) \tag{9.156}$$

$$(m_1 - M_2)y_1 - h_3(\boldsymbol{x}) \leq -h_2(\boldsymbol{x}) \tag{9.157}$$

$$(m_1 - M_2)(1 - y_1) + h_3(\boldsymbol{x}) \leq h_1(\boldsymbol{x}) \tag{9.158}$$

$$(m_2 - M_1)(1 - y_1) - h_3(\boldsymbol{x}) \leq -h_1(\boldsymbol{x}) \tag{9.159}$$

where m_i is a constant less than the lowest value of $h_i(\boldsymbol{x})$ and M_i is a constant greater than the highest value of $h_i(\boldsymbol{x})$ (these can be guessed). Notice that the terms containing y_1 are now linear. This procedure, taken with a stipulation that the $\boldsymbol{y}$ elements only arise linearly in the objective function, allows the rewriting of Equations 9.152 and 9.153 as

$$\text{minimise } Z = f(\boldsymbol{x}) + \boldsymbol{c}^T \boldsymbol{y} \tag{9.160}$$

$$\text{such that } \boldsymbol{g}(\boldsymbol{x}) + \boldsymbol{B}\boldsymbol{y} \leq \boldsymbol{0} \tag{9.161}$$

As in the OA method discussed in Section 9.8.2, the solution proceeds by first ($k = 1$) guessing a version of $\boldsymbol{y}$, that is $\boldsymbol{y}^1$. Then (9.160)–(9.161) are solved by NLP as

$$\text{minimise } Z^k = f(\boldsymbol{x}) + \boldsymbol{c}^T \boldsymbol{y}^k \tag{9.162}$$

$$\text{such that } \boldsymbol{g}(\boldsymbol{x}) + \boldsymbol{B}\boldsymbol{y}^k \leq \boldsymbol{0} \tag{9.163}$$

The previous upper bound continues as Z^k_{UB} unless the result Z^k is less, in which case Z^k becomes Z^k_{UB}. The solution $\boldsymbol{x}^k$ now becomes the next point about which a linearisation occurs for solution of the MIP problem:

$$\text{minimise } Z^k_{\mathrm{LB}} = \alpha \tag{9.164}$$

$$\text{such that } \alpha \leq Z^k_{\mathrm{UB}} - \varepsilon \tag{9.165}$$

and

$$\alpha \geq f(\boldsymbol{x}^i) + \left[\nabla f(\boldsymbol{x})|_i\right]^T (\boldsymbol{x} - \boldsymbol{x}^i) + \boldsymbol{c}^T \boldsymbol{y} \tag{9.166}$$

$$\boldsymbol{g}(\boldsymbol{x}^i) + \left[\nabla \boldsymbol{g}(\boldsymbol{x})|_i\right]^T (\boldsymbol{x} - \boldsymbol{x}^i) + \boldsymbol{B}\boldsymbol{y} \leq \boldsymbol{0} \tag{9.167}$$

for $i = 1, 2, 3 \ldots, k$.

In practice, one just needs to find a point giving an objective value only *tolerance* ε less than the current upper bound (lowest so far from Equation 9.163). The result of this simple *mixed integer programming* (MIP) problem is the next integer selection $\boldsymbol{y}^{k+1}$ (the $\boldsymbol{x}$ result is discarded). The solutions 9.162–9.163 and 9.164–9.165 are repeated in sequence until Z^k_{UB} and Z^k_{LB} converge to within a given tolerance. It is important to note that the constraints accumulate on each iteration – together forming an 'outer approximation'.

Figure 9.21 shows a final 'check' on the optimal process setpoints computed as above. Here they are subjected to a statistical assessment. Only significantly beneficial changes are forwarded to the process controllers for implementation, in order to avoid frequent upsets. Miletic and Marlin (1996) note that ordinary 'stationary' fluctuations in process measurements will pass through the optimiser and manifest as fluctuations in the calculated setpoints $\boldsymbol{r}$. When the process is otherwise steady, one can use this information to construct a covariance matrix

$$\boldsymbol{Q} = E\{(\boldsymbol{r} - \bar{\boldsymbol{r}})^T(\boldsymbol{r} - \bar{\boldsymbol{r}})\} \tag{9.168}$$

Then in normal operation, the scalar

$$\eta = E\{(\boldsymbol{r}_{\mathrm{n}} - \boldsymbol{r}_{\mathrm{e}})^T \boldsymbol{Q}^{-1}(\boldsymbol{r}_{\mathrm{n}} - \boldsymbol{r}_{\mathrm{e}})\} \tag{9.169}$$

gives an indication of the significance of the difference between a newly calculated setpoint ($\boldsymbol{r}_{\mathrm{n}}$) and one that is in existing use ($\boldsymbol{r}_{\mathrm{e}}$). They propose a statistical significance test using η, which if passed, allows updating of the process setpoints. Beyond this type of 'noise elimination', it would clearly also be useful to make use of the *marginal values* from the optimisation to assess whether a particular change would have adequate economic benefit.

Once an RTO system is installed, several issues need close attention to ensure its ultimate success. All too often the original experts migrate to other jobs leaving the system without backup expertise and without a 'champion'. Karodia *et al.* (1999) achieved excellent benefits and high online times by focusing on these issues. They note that RTO initiatives must be supported from top management down, with all departments (technology, instruments, IT, operations) kept in step. The APC layer which has to implement the RTO setpoints needs to have a high online time. Documentation must be comprehensive and up-to-date. An individual must be appointed to 'champion' the RTO initiative, attending to maintenance and troubleshooting, and moving forward with enhancements.

References

Adjiman, C.S., Androulakis, I.P. and Floudas, C.A. (1998a) A global optimization method, αBB, for general twice-differentiable NLPs – II. Implementation and computational results. *Computers & Chemical Engineering*, **22**, 1159–1178.

Adjiman, C.S., Androulakis, I.P., Floudas, C.A. and Neumaier, A. (1998b) A global optimization method, BB, for general twice-differentiable NLPs – I. Theoretical advances. *Computers & Chemical Engineering*, **22**, 1137–1158.

Brooke, A., Kendrick, D., Meeraus, A. and Raman, R. (1988) *GAMS – A User's Guide*, The Scientific Press, San Francisco, CA.

Cîrciu, M.S. and Leon, F. (2010) Comparative study of multiobjective genetic algorithms. *Buletinul Institutului Politehnic din Iasi*, **LVI (LX)**, 1.

Dantzig, G.B. (1951) Maximization of a linear function of variables subject to linear inequalities, in *Activity Analysis of Production and Allocation* (ed. T.C. Koopmans), John Wiley & Sons, Inc., New York, pp. 339–347.

Darwin, C. (1859) *On the Origin of Species*, John Murray, London.

Deb, K., Pratap, A., Agarwal, S., and Meyarivan, T. (2002) A fast and elitist multiobjective genetic algorithm: NSGA-II. *IEEE Transactions on Evolutionary Computation*, **6** (2), 182–197.

Diaz, M.S. and Bandoni, J.A. (1996) A mixed integer optimization strategy for a large scale chemical plant in operation. *Computers and Chemical Engineering*, **3**, 531–545.

Fleming, P.J. and Purshouse, R.C. (2002) Evolutionary algorithms in control systems engineering: a

survey. *Control Engineering Practice*, **10**, 1223–1241.
Fogel, L., Owens, A., and Walsh, M. (1966) *Artificial Intelligence through Simulated Evolution*, John Wiley & Sons, Inc., New York.
Goggos, V. and King, R.E. (1996) Evolutionary predictive control (EPC). *Computers & Chemical Engineering*, **20** (S), 817–822.
Grossmann, I.E. (2002) Review of nonlinear mixed-integer and disjunctive programming techniques. *Optimization and Engineering*, **3**, 227–252.
Grossmann, I.E. and Kravanja, Z. (1995) Mixed integer nonlinear programming techniques for process systems engineering. *Computers & Chemical Engineering*, **19** (S1), 189–204.
Holland, J.H. (1975) *Adaptation in Natural and Artificial Systems*, University of Michigan Press, Ann Arbor.
Karodia, M.E., Naidoo, S.G., and Appanah, R. (1999) Closed-loop optimization increases refinery margins in South Africa. *World Refining*, **9** (5), 62–64.
Kleinschrodt, F.J. and Jones, J.D. (1996) Industrial vision for process optimization. *Computers & Chemical Engineering*, **20** (S), 473–483.
Koza, J.R. (1992) *Genetic Programming: On the Programming of Computers by Means of Natural Selection*, MIT Press.
Kristinsson, K. and Dumont, G.A. (1992) System identification and control using genetic algorithms. *IEEE Transactions on Systems, Man and Cybernetics*, **22**, 1033–1046.
Metropolis, N., Rosenbluth, A.W., Rosenbluth, M.N., Teller, A.H. and Teller, E. (1953) Equation of state calculation by fast computing machines. *The Journal of Chemical Physics*, **21** (6), 1087–1092.
Miletic, I.P. and Marlin, T.E. (1996) Results analysis for real-time optimization (RTO): deciding when to change the plant operation. *Computers and Chemical Engineering*, **20**, S1077–S1082.
Nelder, J.A. and Mead, R. (1965) A simplex method for function minimization. *The Computer Journal*, **7**, 308–313.
Oliveira, P., Sequeira, J., and Sentieiro, J. (1991) Selection of controller parameters using genetic algorithms, in S.G. Tzafestas (ed.), *Engineering Systems with Intelligence: Concepts, Tools and Applications*, Kluwer Academic Publishers, Dordrecht, pp. 431–438.
Rechenberg, I. (1973) Evolutionsstrategie: Optimierung technischer Systeme nach Prinzipien der biologischen evolution. Dissertation, Fromman-Holzboog.
Schwefel, H.-P. (1965) *Kybernetische Evolution als Strategie der experimentellen Forschung aus der Strömungstechnik*, Technische Universität Berlin, Diplomarbeit.
Shokri, S., Hayati, R., Marvast, M. Ahmadi, Ayazi, M. and Ganji, H. (2009) Real-time optimization as a tool for increasing petroleum refineries profits. *Petroleum & Coal*, **51** (2), 110–114.
Tjao, I.B. and Biegler, L.T. (1991) Simultaneous strategies for data reconciliation and gross error detection of nonlinear systems. *Computers & Chemical Engineering*, **15** (10), 679–690.
Tosukhowong, T., Lee, J.M., Lee, J.H. and Lu, J. (2004) An introduction to a dynamic plant-wide optimization strategy for an integrated plant. *Computers & Chemical Engineering*, **29** (1), 199–208.
Williams, H.P. (1993) *Model Building in Mathematical Programming*, 3rd edn, John Wiley & Sons, Inc., New York.
Zhang, Z., Pike, R.W. and Hertwig, T.A. (1995) An approach to on-line optimization of chemical plants. *Computers & Chemical Engineering*, **19** (S1), 305–310.

Index